SECOND EDITION

Molecular Genetics
of Bacteria

SECOND EDITION

Molecular Genetics of Bacteria

Larry Snyder and
Wendy Champness

Department of Microbiology and Molecular Genetics
Michigan State University
East Lansing, Michigan

ASM
PRESS

WASHINGTON, D.C.

Cover photograph (courtesy of Richard Losick and Masaya Fujita, Department of Molecular and Cellular Biology, Harvard University) illustrates the cellular localization of a Bacillus subtilis *sporulation-specific transcription factor,* σ^E, *as visualized with protein fusions to the "green fluorescent protein." Pro-*σ^E *first localizes to the septal and cytoplasmic membranes of sporulating cells at the stage of polar septation as shown in the cells on the right side of the insert. To the left, the mature* σ^E *is present in the cytoplasm in the large chamber of the sporangium, where it directs mother-cell-specific transcription.*

Address editorial correspondence to ASM Press, 1752 N St. NW, Washington, DC 20036-2904, USA

Send orders to ASM Press, P.O. Box 605, Herndon, VA 20172, USA
Phone: 800-546-2416; 703-661-1593
Fax: 703-661-1501
E-mail: books@asmusa.org
Online: www.asmpress.org

Copyright © 2003   ASM Press
American Society for Microbiology
1752 N St. NW
Washington, DC 20036-2904

Library of Congress Cataloging-in-Publication Data

Snyder, Larry.
Molecular genetics of bacteria / Larry Snyder and Wendy Champness.—2nd ed.
p. cm.
Includes bibliographical references and index.
ISBN 1-55581-204-X (hardcover)
1. Bacterial genetics. 2. Bacteriophage—Genetics. 3. Molecular genetics. I. Champness, Wendy. II. Title.

QH434.S59 2002
572.8'293—dc21

2002018697

10 9 8 7 6 5 4 3 2 1

Cover and interior design: Susan Brown Schmidler
Cover illustration: Terese Winslow

The Genetic Code

First position	Second position				Third position
	U	C	A	G	
U	Phe	Ser	Tyr	Cys	U
	Phe	Ser	Tyr	Cys	C
	Leu	Ser	Stop	Stop	A
	Leu	Ser	Stop	Trp	G
C	Leu	Pro	His	Arg	U
	Leu	Pro	His	Arg	C
	Leu	Pro	Gln	Arg	A
	Leu	Pro	Gln	Arg	G
A	Ile	Thr	Asn	Ser	U
	Ile	Thr	Asn	Ser	C
	Ile	Thr	Lys	Arg	A
	Met	Thr	Lys	Arg	G
G	Val	Ala	Asp	Gly	U
	Val	Ala	Asp	Gly	C
	Val	Ala	Glu	Gly	A
	Val	Ala	Glu	Gly	G

Nucleotide Structure

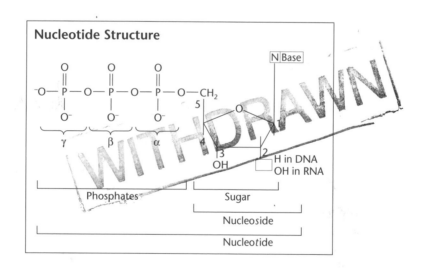

Names of Nucleic Acid Subunits

Base	Nucleoside	Nucleotide	Abbreviation	
			RNA	DNA
Adenine	Adenosine	Adenosine triphosphate	ATP	dATP
Guanine	Guanosine	Guanosine triphosphate	GTP	dGTP
Cytosine	Cytidine	Cytidine triphosphate	CTP	dCTP
Thymine	Thymidine	Thymidine triphosphate		dTTP
Uracil	Uridine	Uridine triphosphate	UTP	

Contents

Boxes

Preface

We were motivated to prepare a second edition of our textbook *Molecular Genetics of Bacteria* because of the favorable response to the first edition and because of the important advances made in this field since the original edition was published. The second edition retains much of the organization and style of the first edition. The order of topics is unchanged as is the emphasis on experimental approaches: features popular with most instructors using the first edition. Each chapter retains a chapter summary, problems (with answers), and suggested readings, all of which are updated. We continue to use "boxes" to present material of related interest to each topic without breaking the continuity of the narrative. The material in these boxes can serve as a starting point for special reports or to link with material from other courses that the students might be taking. One change in the organization of the second edition is that much of the material in the last four chapters of the first edition, which presented some more sophisticated experiments and some current applications, has been updated and incorporated into the earlier chapters. As a result, the chapters usually begin with a descriptive treatment of each topic and end with some relatively technical molecular genetic experiments that led to the knowledge. This organization allows instructors to assign only early sections of each chapter if appropriate but retains the usefulness of the textbook for more advanced undergraduate and beginning graduate classes. This allows the book to be more concise since each subject need be introduced only once. We added a final chapter that illustrates comprehensively how the techniques and concepts of bacterial molecular genetics discussed in earlier chapters have been used to study biological phenomena such as protein translocation and sporulation.

The material in each chapter is substantially updated and reflects the exciting developments in the field of bacterial molecular genetics and emphasizes the relationships of prokaryotic and eukaryotic cell biology and development. In particular, important progress has been made since the first edition in understanding chromosome segregation and cell division in bacteria; the close relationship of protein secretion and conjugation both between bacteria and from bacteria into eukaryotes; the intimate relationship between the "three Rs"—replication, recombination, and repair; the universality of repair mechanisms and mutagenic DNA polymerases and their roles in cancer; the techniques of genomics, microarrays, and bioinformatics and their applications in bacterial molecular genetics; as well as many other important updates. With these changes, the textbook will continue to provide an appropriate up-to-date treatment of bacterial molecular genetics for undergraduate and beginning graduate courses and will also continue to provide a good foundation and reference guide for scientists in the many fields of biology and engineering who depend on the concepts and techniques developed with prokaryotes.

An extraordinary number of researchers have made major contributions to the field of bacterial molecular genetics. We could not reasonably expect students to learn even a fraction of their names, and we could not possibly do justice to all of their important contributions. Therefore, we include only those names that have become icons in the field because they are associated with certain seminal experiments (e.g., Meselson and Stahl or Luria and Delbrück), models (e.g., Jacob and Monod), or a structure (e.g., Watson and Crick). We redress our omissions somewhat in the suggested readings, where we give some of the original references to the developments under discussion.

In writing the second edition, we have benefited from the help and advice of a large number of colleagues around the world. Many instructors using the first edition pointed out errors and communicated advice. Many additional colleagues generously read or discussed material in their areas of specialty. We would particularly like to thank our colleague at Michigan State University, Lee Kroos, for his careful reading of the new *Bacillus subtilis* sporulation section. We also benefited from the input of many of our own undergraduates, who used the first edition in our classes. We especially thank Andrea Hartlerode and Heather Hall, who pointed out areas they found confusing or in need of additional explanation. However, we take full responsibility for any mistakes, misconceptions, or omissions.

As before, it was a great pleasure to work with the professionals at ASM Press, who repeated their magic act of transforming our primitive text and illustrations into a printed book. For the first edition, as neophyte authors, we depended on the expert advice of the former director of ASM Press, Patrick Fitzgerald. In preparing the second edition, we have been indebted to the current director, Jeff Holtmeier, for his encouragement, enthusiasm, and patience. We have also had the good fortune to work again with a number of the same professionals who did a masterful job with the first edition, including Susan Birch, Production Manager, who oversaw the entire process; Yvonne Strong, who copyedited the manuscript and illustrations; Susan Brown Schmidler, who created the book and cover design; and Terese Winslow, who created the cover illustration. Finally, we especially thank Berta Steiner, president of Bermedica Production, Ltd., who managed this publication project and supervised the work of Precision Graphics artists who rendered our hand-drawn sketches into clear attractive figures.

Introduction

THE GOAL OF THIS TEXTBOOK is to introduce the student to the field of bacterial molecular genetics. Bacteria are relatively simple organisms, and some are quite easy to manipulate in the laboratory. For these reasons, many methods in molecular biology and recombinant DNA technology have been developed around bacteria, and these organisms often serve as model systems for understanding cellular functions and developmental processes in more complex organisms. Much of what we know about the basic molecular mechanisms in cells, such as translation and replication, has originated with studies in bacteria. This is because such central cellular functions have remained largely unchanged throughout evolution. Ribosomes have a similar structure in all organisms, and many of the translation factors are highly conserved. The DNA replication apparatus of all organisms contains features in common such as sliding clamps and editing functions, which were first described in bacteria and their phages. Chaperones that help other proteins fold and topoisomerases that change the topology of DNA were first discovered in bacteria and phages. Studies of repair of DNA damage and mutagenesis in bacteria have also led the way to an understanding of such pathways in eukaryotes. Excision repair systems, mutagenic polymerases, and mismatch repair systems are similar in all organisms and have recently been implicated in some types of human cancers.

Also, recent evidence indicates that the cell biology of bacteria might be much more complex and more like that of eukaryotes than previously believed. For a long time it has been possible to observe the seemingly purposeful movement of constituents on the cytoskeleton within eukaryotic cells. However, bacterial cells, being much smaller, were thought to be merely "bags of enzymes" and to rely on mere diffusion to move their cellular constituents around. Now new technologies make it possible to observe movement within bacterial cells, revealing, for example, that some proteins involved in cell division and partitioning oscillate from one end of the cell to the other during the cell cycle (see chapter 1). Bacteria even have proteins related to the proteins of the cytoskeleton, including a cell division protein called FtsZ, which is similar structurally to the tubulins that make up microtubules, and a protein called MreB, which helps give bacterial cells their structure and forms filaments like actin (see Jones et al., and van den Ent et al., Suggested Reading). It has been speculated that we might be entering another stage in biology similar to the early days of molecular genetics, when studies with bacteria led the way to the discovery of new principles of cell biology that are common to all organisms.

However, bacteria are not just important as laboratory tools to understand higher orgamisms; they are important and interesting in their own right. For instance, they play an essential role in the ecology of the Earth. They are the only organisms that can "fix" atmospheric nitrogen, that is, convert N_2 to ammonia, which can be used to make nitrogen-containing cellular constituents such as proteins and nucleic acids. Without bacteria, the natural nitrogen cycle would be broken. Bacteria are also central to the carbon cycle of the Earth because of their ability to degrade recalcitrant natural polymers such as cellulose and lignin. Bacteria and some types of fungi thus prevent the Earth from being buried in plant debris and other carbon-containing material. Toxic compounds including petroleum, many of the chlorinated hydrocarbons, and other products of the chemical industry can also be degraded by bacteria. For this reason, these organisms are essential in water purification and toxic waste cleanup. Moreover, bacteria produce most of the naturally occurring so-called greenhouse gases, such as methane and carbon dioxide, which are in turn degraded by other types of bacteria. This cycle helps maintain climate equilibrium. Bacteria have even had a profound effect on the geology of the Earth, being responsible for some of the major iron ore and other types of deposits in the Earth's crust.

Another unusual feature of bacteria is their ability to live in extremely inhospitable environments, many of which are devoid of life except for bacteria. These organisms are the only ones living in the Dead Sea, where the water's salt concentration is very high. Some types of bacteria live in hot springs at temperatures close to the boiling point of water, and others survive in atmospheres devoid of oxygen, such as eutrophic lakes and swamps.

Bacteria that live in inhospitable environments sometimes enable other organisms to survive in those environments through symbiotic relationships. For example, symbiotic bacteria make life possible for tubular worms next to hydrothermal vents on the ocean floor, where the atmosphere is hydrogen sulfide rather than oxygen. In this symbiosis, the bacteria fix carbon dioxide by using the reducing power of the hydrogen sulfide given off by the hydrothermal vents, thereby furnishing food in the form of high-energy carbon compounds for the worms. Symbiotic cyanobacteria allow fungi to live in the Arctic tundra in the form of lichens. The bacterial partners in the lichens fix atmospheric nitrogen and make carbon-containing molecules through photosynthesis to allow their fungal partners to grow on the tundra in the absence of nutrient-containing soil. Symbiotic nitrogen-fixing *Rhizobium* and *Azorhizobium* spp. in the nodules on the roots of legumes and some other types of higher plants allow plants to grow in nitrogen-deficient soils. Other types of symbiotic bacteria digest cellulose to allow cows and other ruminant animals to live on a diet of grass. Chemiluminescent bacteria even generate light for squid and other marine animals, allowing individuals to find each other in the darkness of the deep ocean.

Bacteria are also worth studying because of their role in disease. They cause many human, plant, and animal diseases, and new ones are continuously appearing. Knowledge gained from the molecular genetics of bacteria will help in the development of new ways to treat or otherwise control these diseases.

Bacteria and their phages (i.e., viruses that infect bacteria) are also the source of many useful substances such as many of the enzymes used in biotechnology and other industries. Moreover, bacteria make antibiotics and chemicals such as benzene and citric acid.

In spite of substantial progress, we have only begun to understand the bacterial world around us. Bacteria are the most physiologically diverse organisms on Earth, and the importance of bacteria to life on Earth and the potential uses to which bacteria can be put can only be guessed at. Thousands of different types of bacteria are known, and new insights into their cellular mechanisms and their applications constantly emerge from research with bacteria. Moreover, it is estimated that less than 1% of the types of bacteria living in the soil and other environments have ever been isolated. Who knows what interesting and useful functions the undiscovered bacteria might have. Clearly, studies with bacteria will continue to be essential to our future efforts to understand, control, and benefit from the biological world around us, and bacterial molecular genetics will be an essential tool in these efforts. But before discussing this field, we must first briefly discuss the evolutionary relationship of the bacteria to other organisms.

The Biological Universe

The Eubacteria

According to a present view, all organisms on Earth belong to three major divisions: the eubacteria, the archaea (formerly archaebacteria), and the eukaryotes (Figure 1). Most of the familiar bacteria such as *Escherichia coli*, *Streptococcus pneumoniae*, and *Staphylococcus aureus* are **eubacteria**. These organisms can differ greatly in

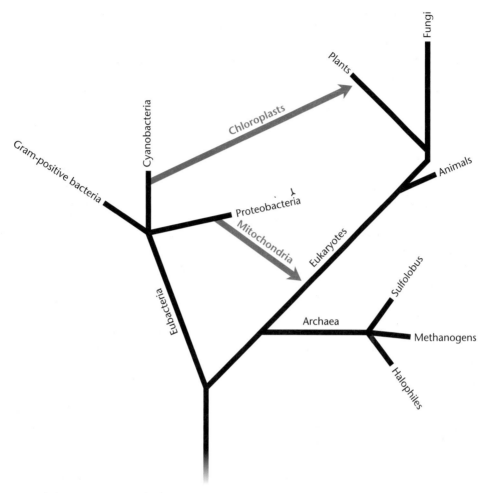

Figure 1 The evolutionary tree showing the points of divergence of eubacteria, archaea, and eukaryotes. The points of transfer of the eubacterial symbiotic organelles—mitochondria and chloroplasts—to eukaryotes are shown in gold.

their physical appearance. Although most are single celled and rod shaped or spherical, some are multicellular and undergo complicated developmental cycles. The cyanobacteria (formerly called blue-green algae) are eubacteria, but they have chlorophyll and can be filamentous, which is why they were originally mistaken for algae. The antibiotic-producing actinomycetes, which include *Streptomyces* spp., are also eubacteria, but they form hyphae and stalks of spores, making them resemble fungae. Another eubacterial group, the *Caulobacter* spp., have both free-swimming and sessile forms that attach to surfaces through a holdfast structure. One of the most dramatic-appearing eubacteria of all is the genus *Myxococcus*, which can exist as free-living single-celled organisms but can also aggregate to form fruiting bodies much like slime molds. Eubacterial cells are usually much smaller than the cells of higher organisms, but recently a eubacterium was found that is 1 mm long, longer than even most eukaryotic cells. Because eubacteria come in so many shapes and sizes, they cannot be dis-

tinguished by their physical appearance but only by biochemical criteria and by the absence of organelles.

GRAM-NEGATIVE AND GRAM-POSITIVE EUBACTERIA

The eubacteria can be further divided into two major subgroups, the **gram-negative** and **gram-positive** eubacteria. This division is based on the response to a test called the Gram stain. Gram-negative eubacteria will be pink after this staining procedure, whereas gram-positive bacteria will turn deep blue. The difference in staining reflects the fact that only gram-negative eubacteria have a well-developed outer membrane. However, the difference between these groups seems to be more fundamental than the possession of an outer membrane. Individual types of gram-negative bacteria are in general more closely related to other gram-negative bacteria than they are to gram-positive bacteria, suggesting that the eubacteria separated into these two groups long before modern bacterial species arose.

The Archaea

The **archaea** (formerly called archaebacteria) are single-celled organisms that resemble eubacteria but are very different biochemically. The archaea are mostly represented by extremophiles (or "extreme-condition-loving" organisms) that, as their name implies, live under extreme conditions where other types of organisms cannot survive, such as at the very high temperatures in sulfur springs, at high pressures on the ocean floor, and at very high osmolality such as in the Dead Sea. Some of the archaea also perform relatively unusual biochemical functions such as making methane.

The separate classification of archaea from the true bacteria or eubacteria is fairly recent and is mostly based on the sequence of their ribosomal RNAs (rRNAs) and the structures of their RNA polymerase and lipids (see Olsen et al., Suggested Reading). In fact, some evidence obtained by comparing the sequences of translation factors and membrane ATPases suggests that the archaea may be more closely related to eukaryotes than they are to eubacteria (see Figure 1 and Iwabe et al., Suggested Reading). The archaea themselves form a very diverse group of organisms and are sometimes divided into two different kingdoms.

Although substantial progress is being made, much less is known about the archaea than about the eubacteria. The examples in this book come mostly from the eubacteria, which will be referred to as just "bacteria" throughout.

The Eukaryotes

The **eukaryotes** are members of the third kingdom of organisms on Earth. They include organisms as seemingly diverse as plants, animals, and fungi. The name "eukaryotes" is derived from their nuclear membrane. They usually have a nucleus (the genus *Giardia* is a known exception), and the word "karyon" in Greek means "nut"—which is what the nucleus must have resembled to early cytologists. The eukaryotes can be unicellular like yeasts and protozoans and some types of algae, or they can be multicellular like plants and animals. In spite of their widely diverse appearances, lifestyles, and relative complexity, however, all eukaryotes are remarkably similar at the biochemical level, particularly in their pathways for macromolecular synthesis.

The Prokaryotes and the Eukaryotes

Organisms on Earth are also sometimes divided into prokaryotes and eukaryotes. This classification is based on whether the organism has a nucleus and some other organelles. In contrast to eukaryotes, both archaea and eubacteria lack a nuclear membrane, which caused them to be lumped together as **prokaryotes**, which means "before the nucleus." They were given this name because they were thought to be the most primitive organisms, existing before the development of "higher organisms," or eukaryotes that have a nucleus.

The presence or absence of a nuclear membrane greatly influences the mechanisms available to make proteins in the cell. Messenger RNA (mRNA) synthesis and translation can occur simultaneously in prokaryotes, since no nuclear membrane separates the ribosomes (which synthesize proteins) from the DNA. However, in most eukaryotes, the DNA is physically separated from the ribosomes. Therefore, mRNA made in the nucleus must be transported through the nuclear membrane before it can be translated into protein in the cytoplasm, and transcription and translation cannot occur simultaneously.

Besides lacking a nucleus, prokaryotic cells also lack many other cellular constituents common to eukaryotes, including mitochondria and chloroplasts, as well as such organelles as the Golgi apparatus and the endoplasmic reticulum. The absence of mitochondria, chloroplasts, and most organelles generally gives prokaryotic cells a much simpler appearance under the microscope.

MITOCHONDRIA AND CHLOROPLASTS

All eukaryotic cells contain mitochondria. In addition, plant cells and some unicellular eukaryotic cells contain chloroplasts. The mitochondria of eukaryotic cells are the sites of efficient ATP generation through respiration, and the chloroplasts are the sites of photosynthesis.

Recent evidence suggests that the mitochondria and chloroplasts of eukaryotes are descended from free-living eubacteria that formed a symbiosis with eukaryotes. In fact, these organelles resemble bacteria in many ways. For instance, they contain DNA that encodes ribosomes, transfer RNAs (tRNAs), translation factors, and other components of the transcription-translation apparatus. Even more striking, the mitochondrially encoded and chloroplast-encoded RNAs and proteins, as well as the membranes of the organelles, more closely resemble those of the eubacteria than they do those of eukaryotes.

Not only is there little doubt of the eubacterial origin of mitochondria and chloroplasts, but also it is possible to guess to which eubacterial families they are most closely related. Comparisons of the sequences of highly conserved organelle genes, such as those for the rRNAs, with those of eubacteria suggest that mitochondria are descended from the proteobacteria and that chloroplasts are descended from the cyanobacteria (Figure 1).

Mitochondria and chloroplasts may have come to be associated with early eukaryotic cells when these cells engulfed eubacteria to take advantage of their superior energy-generating systems or the ability to obtain energy from light through photosynthesis. The engulfed bacteria eventually lost many of their own genes and thereby their autonomy and became permanent symbionts of the eukaryotic cells. In fact, modern-day eukaryotes called dinoflagellates sometimes engulf cyanobacteria, allowing the dinoflagellates to photosynthesize.

What Is Genetics?

Genetics can be simply defined as the manipulation of DNA to study cellular and organismal functions. Since DNA encodes all of the information needed to make the cell and the complete organism, the effects of changing this molecule can give clues to the normal functions of the cell and organism.

Before the advent of methods for manipulating DNA in the test tube, the only genetic approaches available for studying cellular and organismal functions were those of **classical genetics**. In this type of analysis, mutants (i.e., individuals that differ from the normal, or wild-type, members of the species by a certain observable attribute, or phenotype) that are altered in the function being studied are isolated. The changes in the DNA, or mutations, responsible for the altered function are then localized in the chromosome by genetic crosses. The mutations are then grouped into genes by allelism tests to determine how many different genes are involved. The functions of the genes can then sometimes be deduced from the specific effects of the mutations on the organism. The ways in which mutations in genes involved in a biological system can alter the biological system provide clues to the normal functioning of the system.

Classical genetic analyses continue to contribute greatly to our understanding of developmental and cellular biology. A major advantage of the classical genetic approach is that mutants altered in a function can be isolated and characterized without any a priori understanding of the molecular basis of the function. Classical genetic analysis is also often the only way to determine how many gene products are involved in a function and, through suppressor analysis, to find other genes whose products may interact either physically or functionally with the products of these genes.

The development of **molecular genetic techniques** has greatly expanded the range of methods available for studying genes and their functions. These techniques include methods for isolating DNA and identifying the regions of DNA that encode particular functions, as well as methods for altering or mutating DNA in the test tube and then returning the mutated DNA to cells to determine the effect of the mutation on the organism.

The approach of first cloning a gene and then altering it in the test tube before reintroducing it into the cells to determine the effect of the alterations is sometimes called **reverse genetics** and is essentially the reverse of a classical genetic analysis. In classical genetics, a gene is known to exist only because a mutation in it has caused an observable change in the organism. With the molecular genetic approach, a gene can be isolated and mutated in the test tube without any knowledge of its function. Only after the mutated gene has been returned to the organism does its function become apparent.

Rather than one approach supplanting the other, molecular genetics and classical genetics can be used to answer different types of questions, and the two approaches often complement each other. In fact, the most remarkable insights into biological functions have sometimes come from a combination of classical and molecular genetic approaches.

Bacterial Genetics

In bacterial genetics, genetic techniques are used to study bacteria. Applying genetic analysis to bacteria is no different in principle from applying it to other organisms. However, the methods that are available differ greatly. Some types of bacteria are relatively easy to manipulate genetically. As a consequence, more is known about some bacteria than is known about any other organism on Earth. Some of the properties of bacteria that facilitate genetic experiments are listed below.

Bacteria Are Haploid

One of the major advantages of bacteria for genetic studies is that they are **haploid**. This means that they have only one copy or **allele** of each gene. This property makes it much easier to identify cells with a particular type of mutation.

In contrast, most higher organisms are diploid, with two alleles of each gene, one on each homologous chromosome. Most mutations are recessive, which means that they will not cause a phenotype in the presence of a normal copy of the gene. Therefore, in diploid organisms, most mutations will have no effect unless both copies of the gene in the two homologous chromosomes have the mutation. Backcrosses between different organisms with the mutation are usually required to produce offspring with the mutant phenotype, and even then only some of the progeny of the backcross will have the mutated gene in both homologous chromosomes. With a haploid organism like a bacterium, however, most mutations have an immediate effect, and there is no need for backcrosses.

Short Generation Times

Another advantage of some bacteria for genetic studies is that they have very short generation times. The **generation time** is the length of time the organism takes to reach maturity and give rise to offspring. If the generation time of an organism is too long, it can limit the number of possible experiments. Some strains of the bacterium *E. coli* can reproduce themselves every 20 min under ideal conditions. With such rapid multiplication, cultures of the bacteria can be started in the morning and the progeny can be examined later in the day.

Asexual Reproduction

Another advantage of bacteria is that they multiply asexually, by cell division. Sexual reproduction, in which individuals of the same species must mate with each other to give rise to progeny, can complicate genetic experiments because the progeny are never identical to their parents. To achieve purebred lines of a sexually reproducing organism, a researcher must repeatedly cross the individuals with their relatives. However, if the organism multiplies asexually by cell division, all the progeny will be genetically identical to their parent and to each other. Genetically identical organisms are called **clones**. Some lower eukaryotes and some types of plants can also multiply asexually to form clones.

Colony Growth on Agar Plates

Genetic experiments often require that numerous individuals be screened for a particular property. Therefore, it helps if large numbers of individuals of the species being studied can be propagated in a small space.

With some types of bacteria, thousands, millions, or even billions of individuals can be screened on a single agar-containing petri plate. Once on an agar plate, these bacteria will divide over and over again, with all the progeny remaining together on the plate until a visible lump or **colony** has formed. Each colony will be composed of millions of bacteria, all derived from the original bacterium and hence all clones of the original bacterium.

Colony Purification

The ability of some types of bacteria to form colonies through the multiplication of individual bacteria on plates allows **colony purification** of bacterial strains and mutants. If a mixture of bacteria containing different mutants or strains is placed on an agar plate, individual mutant bacteria or strains in the population will each multiply to form colonies. However, these colonies may be too close together to be separable or may still contain a mixture of different strains of the bacterium. However, if the colonies are picked and the bacteria are diluted before replating, discrete colonies that result from the multiplication of individual bacteria may appear. No matter how crowded the bacteria were on the original plate, a pure strain of the bacterium can be isolated in one or a few steps of colony purification.

Serial Dilutions

To count the number of bacteria in a culture or to isolate a pure culture, it is often necessary to obtain discrete colonies of the bacteria. However, because bacteria are so small, a concentrated culture contains billions of bacteria per milliliter. If such a culture is plated directly on a petri plate, the bacteria will all grow together and discrete colonies will not form. **Serial dilutions** offer a practical method for diluting solutions of bacteria before plating to obtain a measurable number of discrete colonies. The principle is that if smaller dilutions are repeated in succession, they can be multiplied to produce the total dilution. For example, if a solution is diluted in three steps by adding 1 ml of the solution to 99 ml of water, followed by adding 1 ml of this dilution to another 99 ml of water and finally by adding 1 ml of the second dilution to another 99 ml of water, the final dilution is $10^{-2} \times 10^{-2} \times 10^{-2} = 10^{-6}$, or one in a million. To achieve the same dilution in a single step, 1 ml of the original solution would have to be added to 1,000 liters (about 250 gallons) of water. Obviously, it is more convenient to handle three solutions of 100 ml each than to handle a solution of 250 gallons, which weighs about 1,000 lb!

Selections

Probably the major advantage of bacterial genetics is the opportunity to do **selections**, by which very rare mutants and other types of strains can be isolated. To select a rare strain, billions of the bacteria are plated under conditions where only the desired strain, not the bulk of the bacteria, can grow. In general, these conditions are called the **selective conditions**. For example, a nutrient may be required by most of the bacteria but not by the strain being selected. Agar plates lacking the nutrient then present selective conditions for the strain, since only the strain being selected will multiply to form a colony in the absence of the nutrient. In another example, the desired strain may be able to multiply at a temperature that would kill most of the bacteria. Incubating agar plates at that temperature would provide the selective condition. After the strain has been selected, a colony of the strain can be picked and colony purified away from other contaminating bacteria under the same selective conditions.

The power of selections with bacterial populations is awesome. Using a properly designed selection, a single bacterium can be selected from among billions placed on an agar plate. If we could apply such selections to humans, we could find one individual in the entire human population of the Earth.

Storing Stocks of Bacterial Strains

Most types of organisms must be continuously propagated; otherwise they will age and die off. Propagating organisms requires continuous transfers and replenishing of the food supply, which can be very time-consuming. However, many types of bacteria can be stored in a dormant state and therefore do not need to be continuously propagated. The conditions used for storage depend upon the type of bacteria. Some bacteria sporulate and so can be stored as dormant spores. Others can be stored by being frozen in glycerol or being dried. Storing organisms in a dormant state is particularly convenient for genetic experiments, which often require the accumulation of large numbers of mutants and other strains. The strains will remain dormant until the cells are needed, at which time they can be revived.

Genetic Exchange

Genetic experiments with an organism usually require some form of exchange of DNA or genes between members of the species. Most types of organisms on Earth are known to have some means of genetic exchange, which presumably accelerates evolution and increases the adaptability of a species.

Exchange of DNA from one bacterium to another can occur in one of three ways. In **transformation,** DNA released from one cell enters another cell of the same species. In **conjugation,** plasmids, which are small autonomously replicating DNA molecules in bacterial cells, transfer DNA from one cell to another. Finally, in **transduction,** a bacterial virus accidentally picks up DNA from a cell it has infected and injects this DNA into another cell. The ability to exchange DNA between strains of a bacterium makes possible genetic crosses and complementation tests as well as the tests essential to genetic analysis.

Phage Genetics

Some of the most important discoveries in genetics have come from studies with the viruses that infect bacteria, called **bacteriophages,** or **phages** for short. Phages are not alive; instead, they are just genes wrapped in a protective coat of protein and/or membrane, as are all viruses. Because phages are not alive, they cannot multiply outside a bacterial cell. However, if a phage encounters a type of bacterial cell that is sensitive to phages, the phage will enter the cell and direct it to make more phage.

Phages are usually identified by the holes, or **plaques,** they form in layers of sensitive bacteria. In fact, the name "phage" (Greek for "eat") derives from these plaques, which look like eaten-out places. A plaque can form when a phage is mixed with large numbers of susceptible bacteria and the mixture is placed on an agar plate. As the bacteria multiply, one may be infected by the phage, which will multiply and eventually break open or **lyse** the bacterium, releasing more phage. As the surrounding bacteria are infected, the phage will spread, even as the bacteria multiply to form an opaque layer called a **bacterial lawn.** Wherever the original phage infected the first bacterium, the plaque will disrupt the lawn, forming a

clear spot on the agar. Despite its appearance, this location will contain millions of the phage.

Phages offer many of the same advantages for genetics as bacteria. Thousands or even millions of phages can be put on a single plate. Also, like bacterial colonies, each plaque contains millions of genetically identical phage. By analogy to the colony purification of bacterial strains, individual phage mutants or strains can be isolated from other phages through **plaque purification.**

Phages Are Haploid

Phages are, in a sense, haploid, since they have only one copy of each gene. As with bacteria, this property makes isolation of phage mutants relatively easy, since all mutants will immediately exhibit their phenotypes without the need for backcrosses.

Selections with Phages

Selection of rare strains of a phage is possible; as with bacteria, it requires conditions under which only the desired phage strain can multiply to form a plaque. For phage, these selective conditions may be a bacterial host in which only the desired strain can multiply or a temperature at which only the phage strain being selected can multiply. Note that the bacterial host must be able to multiply under the same selective conditions; otherwise, a plaque cannot form.

As with bacteria, selections allow the isolation of very rare strains or mutants. If selective conditions can be found for the strain, millions of phages can be mixed with the bacterial host and only the desired strain will multiply to form a plaque. A pure strain can then be obtained by picking the phage from the plaque and plaque purifying the strain under the same selective conditions.

Crosses with Phages

Phage strains can be crossed very easily. The same cells are infected with different mutants or strains of the phage. The DNA of the two phages will then be in the same cell, where the molecules can interact genetically with each other, allowing genetic manipulations such as gene mapping and allelism tests.

A Brief History of Bacterial Molecular Genetics

Because of the ease with which they can be handled, bacteria and their phages have long been the organisms of choice for understanding basic cellular phenomena, and their contributions to this study are almost countless. The following chronological list should give a feeling for the breadth of these contributions and the central position that bacteria have occupied in the development of modern molecular genetics. Some original references are given at the end of the chapter under Suggested Reading.

Inheritance in Bacteria

In the early part of this century, biologists agreed that inheritance in higher organisms follows Darwinian principles. According to Darwin, changes in the hereditary properties of organisms occur randomly and are passed on to the progeny. In general, the changes that happen to be beneficial to the organism are more apt to be passed on to subsequent generations.

With the discovery of the molecular basis for heredity, Darwinian evolution now has a strong theoretical foundation. The properties of organisms are determined by the sequence of their DNA, and as the organisms multiply, changes in this sequence sometimes occur randomly and without regard to the organism's environment. However, if a random change in the DNA happens to be beneficial in the situation in which the organism finds itself, the organism will have an improved chance of surviving and reproducing.

As late as the 1940s, many bacteriologists believed that inheritance in bacteria was different from inheritance in other organisms. It was thought that rather than enduring random changes, bacteria could adapt to their environment by some sort of "directed" change and that the adapted organisms could then somehow pass on the change to their offspring. Such beliefs were encouraged by the observations of bacteria growing under selective conditions. For example, in the presence of an antibiotic, all the bacteria in the culture soon become resistant to the antibiotic. It seemed as though the resistant bacterial mutants appeared in response to the antibiotic.

One of the first convincing demonstrations that inheritance in bacteria follows Darwinian principles was made in 1943 by Salvador Luria and Max Delbrück (see Suggested Reading). Their work demonstrated that particular phenotypes, in their case resistance to a virus, occur randomly in a growing population, even in the absence of the virus. By the directed-change or adaptive-mutation hypothesis, the resistant mutants should have appeared only in the presence of the virus.

The demonstration that inheritance in bacteria follows the same principles as inheritance in higher organisms set the stage for the use of bacteria in studies of basic genetic principles common to all organisms.

Transformation

As discussed at the beginning of the Introduction, most organisms exhibit some mechanism for exchanging genes. The first demonstration of genetic exchange in bacteria was made by Fred Griffith in 1928. He was studying two variants of pneumococci, now called *Streptococcus pneumoniae*. One variant formed smooth-appearing colonies on plates and was pathogenic in mice. The other variant formed rough-appearing colonies on plates and did not kill mice. Only live, and not dead, smooth-colony-forming bacteria could cause disease, since the disease requires that the bacteria multiply in the infected mice. However, when Griffith mixed dead smooth-colony formers with live rough-colony formers and injected the mixture into mice, the mice became sick and died. Moreover, he isolated live smooth-colony formers from the dead mice. Apparently, the dead smooth-colony formers were "transforming" some of the live rough-colony formers into the pathogenic, smooth-colony-forming type. The "transforming principle" given off by the dead smooth-colony formers was later shown to be DNA, since addition of purified DNA from the dead smooth-colony formers to the live rough-colony formers in a test tube transformed some of the rough type to the smooth type (see Avery et al., Suggested Reading). This method of exchange is called **transformation**, and this experiment provided the first direct evidence that genes are made of DNA. Later experiments by Alfred Hershey and Martha Chase in 1952 (see Suggested Reading) showed that phage DNA alone is sufficient to direct the synthesis of more phages.

Conjugation

In 1946, Joshua Lederberg and Edward Tatum (see Suggested Reading) discovered a different type of gene exchange in bacteria. When they mixed some strains of *E. coli* with other strains, they observed the appearance of recombinant types that were unlike either parent. Unlike transformation, which requires only that DNA from one bacterium be added to the other bacterium, this means of gene exchange requires direct contact between two bacteria. It was later shown to be mediated by plasmids and is named **conjugation**.

Transduction

In 1953, Norton Zinder and Joshua Lederberg discovered yet a third mechanism of gene transfer between bacteria. They showed that a phage of *Salmonella enterica serovar typhimurium* could carry DNA from one bacterium to another. This means of gene exchange is named **transduction** and is now known to be quite widespread.

Recombination within Genes

At the same time, experiments with bacteria and phages were also contributing to the view that genes were linear arrays of nucleotides in the DNA. By the early 1950s, recombination had been well demonstrated in higher organisms, including fruit flies. However, recombination was believed to occur only between mutations in different genes and not between mutations in the same gene.

This led to the idea that genes were like "beads on a string" and that recombination is possible between the "beads," or genes, but not within a gene. In 1955, Seymour Benzer disproved this hypothesis by using the power of phage genetics to show that recombination is possible within the *r*II genes of phage T4. He mapped numerous mutations in the *r*II genes, thereby demonstrating that genes are linear arrays of mutable sites in the DNA. Later experiments with other phage and bacterial genes showed that the sequence of nucleotides in the DNA directly determines the sequence of amino acids in the protein product of the gene.

Semiconservative DNA Replication

In 1953, James Watson and Francis Crick published their structure of DNA. One of the predictions of this model is that DNA replicates by a semiconservative mechanism, in which specific pairing occurs between the bases in the old and the new DNA strands. In 1958, Matthew Meselson and Frank Stahl used bacteria to confirm that DNA replicates by this semiconservative mechanism.

mRNA

The existence of mRNA was also first indicated by experiments with bacteria and phage. In 1961, Sydney Brenner, François Jacob, and Matthew Meselson used phage-infected bacteria to show that ribosomes are the site of protein synthesis and confirmed the existence of a "messenger" RNA that carries information from the DNA to the ribosome.

Genetic Code

Also in 1961, phages and bacteria were used by Francis Crick and his collaborators to show that the genetic code is unpunctuated, three lettered, and redundant. These researchers also showed that not all possible codons designated an amino acid and that some were nonsense. These experiments laid the groundwork for Marshall Nirenberg and his collaborators to decipher the genetic code, in which a specific three-nucleotide set encodes one of 20 amino acids. The code was later verified by the examination of specific amino acid changes due to mutations in the lysozyme gene of T4 phage.

The Operon Model

François Jacob and Jacques Monod published their operon model for the regulation of the lactose utilization genes of *E. coli* in 1961 as well. They proposed that a repressor blocks RNA synthesis on the *lac* genes unless the inducer, lactose, is bound to the repressor. Their model has served to explain gene regulation in other systems, and the *lac* genes and regulatory system continue to be used in molecular genetic experiments, even in systems as far removed from bacteria as animal cells and viruses.

Enzymes for Molecular Biology

The early 1960s saw the start of the discovery of many interesting and useful bacterial and phage enzymes involved in DNA and RNA metabolism. In 1960, Arthur Kornberg demonstrated the synthesis of DNA in the test tube by an enzyme from *E. coli*. The next year, a number of groups independently demonstrated the synthesis of RNA in the test tube by RNA polymerases from bacteria. From this time on, other interesting and useful enzymes for molecular biology were isolated from bacteria and their phages, including polynucleotide kinase, DNA ligases, topoisomerases, and many phosphatases.

From these early observations, the knowledge and techniques of molecular genetics exploded. For example, in the early 1960s, techniques were developed for detecting the hybridization of RNA to DNA and DNA to DNA on nitrocellulose filters. These techniques were used to show that RNA is made on only one strand in specific regions of DNA, which later led to the discovery of promoters and other regulatory sequences. By the late 1960s, restriction endonucleases had been discovered in bacteria and shown to cut DNA in specific places (see Linn and Arber, Suggested Reading). By the early 1970s, these restriction endonucleases were being exploited to introduce foreign genes into *E. coli* (see Cohen et al., Suggested Reading), and by the late 1970s, the first human gene had been expressed in a bacterium. Also in the late 1970s, methods were developed to sequence DNA by using enzymes from phages and bacteria.

In 1988, a thermally stable DNA polymerase from a thermophilic bacterium was used to invent the technique called the polymerase chain reaction (PCR). This extremely sensitive technique allows the amplification of genes and other regions of DNA, facilitating their cloning and study.

These examples illustrate that bacteria and their phages have been central to the development of molecular genetics and recombinant DNA technology. Contrast the timing of these developments with the timing of comparable major developments in physics (early 1900s) and chemistry (1920s and 1930s), and you can see that molecular genetics is arguably the most recent major conceptual breakthrough in the history of science.

What's Ahead

This textbook emphasizes how molecular genetic approaches can be used to solve biological problems. As an educational experience, the methods used and the interpretation of experiments are at least as important as the conclusions drawn. Therefore, whenever possible, the experiments that led to the conclusions will be presented. The first two chapters of the textbook mostly review the concepts of macromolecular synthesis that are essential to understanding bacterial molecular genetics. The next seven chapters are more descriptive accounts of the mechanisms of mutations and gene exchange in bacteria and phages, giving overviews of the experiments that contributed to this knowledge where feasible. Chapters 10 through 13 describe the current state of knowledge about the molecular basis of recombination, repair, mutagenesis, and gene regulation in bacteria, again presenting experimental evidence where feasible. Finally, Chapter 14 gives more detailed examples of how classical and molecular genetic analyses have been used to study gene structure and regulation in bacteria.

SUGGESTED READING

Avery, O. T., C. M. MacCleod, and M. McCarty. 1944. Studies on the chemical nature of the substance inducing transformation of pneumococcal types. I. Induction of transformation by a desoxyribonucleic acid fraction isolated from pneumococcus type III. *J. Exp. Med.* **79**:137.

Brenner, S., F. Jacob, and M. Meselson. 1961. An unstable intermediate carrying information from genes to ribosomes for protein synthesis. *Nature* (London) **190**:576.

Cairns, J., G. S. Stent, and J. D. Watson. 1966. *Phage and the Origins of Molecular Biology.* Cold Spring Harbor Laboratory Press, Cold Spring Harbor, N.Y.

Cohen, S. N., A. C. Y. Chang, H. W. Boyer, and R. B. Helling. 1973. Construction of biologically functional bacterial plasmids *in vitro. Proc. Natl. Acad. Sci. USA* **70**:3240–3244.

Crick, F. H. C., L. Barnett, S. Brenner, and R. J. Watts-Tobin. 1961. General nature of the genetic code for proteins. *Nature* (London) **192**:1227–1232.

Hershey, A. D., and M. Chase. 1952. Independent functions of viral protein and nucleic acids in growth of bacteriophage. *J. Gen. Physiol.* **36**:393.

Iwabe, N., K. Kuma, M. Hasegawa, S. Osawa, and T. Miyata. 1989. Evolutionary relationship of archaebacteria, eubacteria and eukaryotes inferred from phylogenetic trees of duplicated genes. *Proc. Natl. Acad. Sci. USA* **86**:9355–9359.

Jacob, F., and J. Monod. 1961. Genetic regulatory mechanisms in the synthesis of proteins. *J. Mol. Biol.* **3**:3183.

Jones, L. J. F., R. Carballido-López, and J. Errington. 2001. Control of cell shape in bacteria: Helical, actin-like filaments in *Bacillus subtilis. Cell* **104**:913–922.

Lederberg, J., and E. L. Tatum. 1946. Gene recombination in *E. coli. Nature* (London) **158**:558.

Linn, S., and W. Arber. 1968. Host specificity of DNA produced by *Escherichia coli.* X. *In vitro* restriction of phage fd replicative form. *Proc. Natl. Acad. Sci. USA* **59**:1300–1306.

Luria, S., and M. Delbrück. 1943. Mutations of bacteria from virus sensitivity to virus resistance. *Genetics* **28**:491–511.

Meselson, M., and F. W. Stahl. 1958. The replication of DNA in *Escherichia coli. Proc. Natl. Acad. Sci. USA* **44**:671.

Nirenberg, M. W., and J. H. Matthei, 1961. The dependence of cell-free protein synthesis in *E. coli* upon naturally occurring or synthetic polynucleotides. *Proc. Natl. Acad. Sci. USA* **47**:1588–1602.

Olby, R. 1974. *The Path to the Double Helix.* Macmillan Press, London, United Kingdom.

Olsen, G. J., C. R. Woese, and R. Overbeek. 1994. The winds of (evolutionary) change: breathing new life into microbiology. *J. Bacteriol.* **176**:1–6. (Minireview.)

Schrodinger, E. 1944. *What is Life? The Physical Aspect of the Living Cell.* Cambridge University Press, Cambridge, United Kingdom.

van den Ent, F., L. A. Amos, and J. Löwe. 2001. Prokaryotic origin of the actin cytoskeleton. *Nature* (London) **413**:39–44.

Watson, J. D. 1968. *The Double Helix.* Atheneum, New York, N.Y.

Zinder, N. D., and J. Lederberg. 1952. Genetic exchange in *Salmonella. J. Bacteriol.* **64**:679–699.

PART **I**

Genes: Replication and Expression

1 Macromolecular Synthesis: Chromosome Structure, Replication, and Segregation

2 Macromolecular Synthesis: Gene Expression

THE MODERN SCIENCE OF GENETICS requires an understanding of the structure and synthesis of DNA, RNA, and proteins, called **macromolecules** because of their large size. Much has been learned in recent years about how macromolecules are made and how their structures influence their properties. Chapters 1 and 2 give the necessary background information about the structure and synthesis of macromolecules in the eubacteria, with occasional references to the eukaryotes and archaea in those instances where important differences are known to exist. There are many useful biochemistry and molecular biology textbooks which deal with the structure and synthesis of macromolecules in more detail; we present an overview of these topics with a special emphasis on topics which are more relevant to bacterial molecular genetics.

Macromolecular Synthesis: Chromosome Structure, Replication, and Segregation

DNA Structure

THE SCIENCE OF MOLECULAR GENETICS began with the determination of the structure of DNA. Experiments with bacteria and phages (i.e., viruses that infect bacteria) in the late 1940s and early 1950s, as well as the presence of DNA in chromosomes of higher organisms, had implicated this macromolecule as the hereditary material (see Introduction). In the 1930s, biochemical studies of the base composition of DNA by Erwin Chargaff established that the amount of guanine always equals the amount of cytosine and the amount of adenine always equals the amount of thymine, independent of the total base composition of the DNA. In the early 1950s, X-ray diffraction studies by Rosalind Franklin and Maurice Wilkins showed that DNA is a double helix. Finally, in 1953, Francis Crick and James Watson put together the chemical and X-ray diffraction information in their famous model of the structure of DNA. This story is one of the most dramatic in the history of science and has been the subject of many historical treatments, some of which are listed at the end of this chapter.

Figure 1.1 illustrates the Watson-Crick structure of DNA, in which two strands wrap around each other to form a double helix. These strands can be extremely long, even in a simple bacterium, extending up to 1 mm—a thousand times longer than the bacterium itself. In a human cell, the strands that make up a single chromosome (which is one DNA molecule) are hundreds of millimeters, or many inches, long.

The Deoxyribonucleotides

If we think of DNA strands as chains, deoxyribonucleotides form the links. Figure 1.2 shows the basic structure of deoxyribonucleotides, called deoxynucleotides for short. Each is composed of a **base**, a **sugar**, and a

13

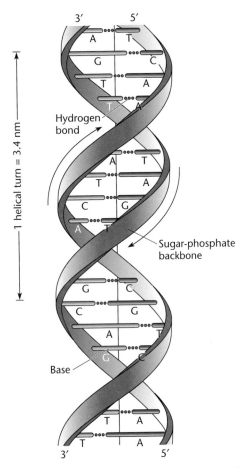

Figure 1.1 A schematic drawing of the Watson-Crick structure of DNA, showing the helical sugar-phosphate backbones of the two strands held together by hydrogen bonding between the bases.

phosphate group. DNA bases are **adenine (A), cytosine (C), thymine (T),** and **guanine (G),** which have either one or two rings, as shown in Figure 1.2. The bases with two rings (A and G) are the **purines,** and those with only one ring (T and C) are **pyrimidines.** A third pyrimidine, uracil (U), replaces thymine in RNA. The carbons and nitrogens making up the rings of the bases are numbered sequentially, as shown in the figure. All four DNA bases are attached to the five-carbon sugar **deoxyribose.** This sugar is identical to ribose, which is found in RNA, except that it does not have an oxygen attached to the second carbon, hence the name *deoxyri-bose.* The carbons in the sugar of a nucleotide are also numbered 1, 2, 3, and so on, but they are primed to distinguish them from the carbons in the bases (Figure 1.2). The nucleotides also have one or more phosphate groups attached to a carbon of the deoxyribose sugar, as shown. The carbon to which the phosphate group is attached is indicated, although if it is attached to the 5'

carbon (the usual situation), the carbon to which it is attached is often not stipulated.

The components of the deoxynucleotides have special names. A **deoxynucleoside** (rather than *-tide*) is a base attached to a sugar but lacking a phosphate. Without phosphates, the four deoxynucleosides are called **deoxyadenosine, deoxycytidine, deoxythymidine,** and **deoxyguanosine.** As shown in Figure 1.2, the deoxynucleotides have one, two, or three phosphates attached to the sugar and are known as deoxynucleoside *mono*phosphates, *di*phosphates, or *tri*phosphates, respectively. The individual deoxynucleoside monophosphates, called deoxyguanosine monophosphate, etc., are often abbreviated dGMP, dAMP, dCMP, and dTMP, where the d stands for *deoxy;* the G, A, C, or T stands for the base; and the MP stands for *mono*phosphate. In turn, the *di*phosphates and *tri*phosphates are abbreviated dGDP, dADP, dCDP, dTDP, and dGTP, dATP, dCTP, dTTP, respectively. Collectively, the four deoxynucleoside triphosphates are often referred to as dNTPs.

The DNA Chain

Phosphodiester bonds join each deoxynucleotide link in the DNA chain. As shown in Figure 1.3, the phosphate attached to the last (5') carbon of the deoxyribose sugar of one nucleotide is attached to the third (3') carbon of the sugar of the next nucleotide, thus forming one strand of nucleotides connected 5' to 3', 5' to 3', etc.

The 5' and 3' Ends

Clearly, at the ends of DNA will be nucleotides that are linked to only one other nucleotide. At one end of the DNA chain, a nucleotide will have a phosphate attached to its 5' carbon that does not connect it to another nucleotide. This end of the strand is called the **5' end** or the **5' phosphate end** (Figure 1.3B). On the other end, the last nucleotide lacks a phosphate at its 3' carbon. Because it only has a hydroxyl group (the OH in Figure 1.3B), this end is called the **3' end** or the **3' hydroxyl end.**

Base Pairing

The sugar and phosphate groups of DNA form what is often called a **backbone** to support the bases, which jut out from the chain. This structure allows the bases to form hydrogen bonds with each other, thereby holding together two separate nucleotide chains (Figure 1.3B). This role of the four bases was first suggested by their ratios in DNA.

First, Erwin Chargaff found that no matter the source of the DNA, the concentration of guanine (G) always equals the concentration of cytosine (C) and the concentration of adenine (A) always equals the concentration of thymine (T). These ratios, named Chargaff's Rules, gave

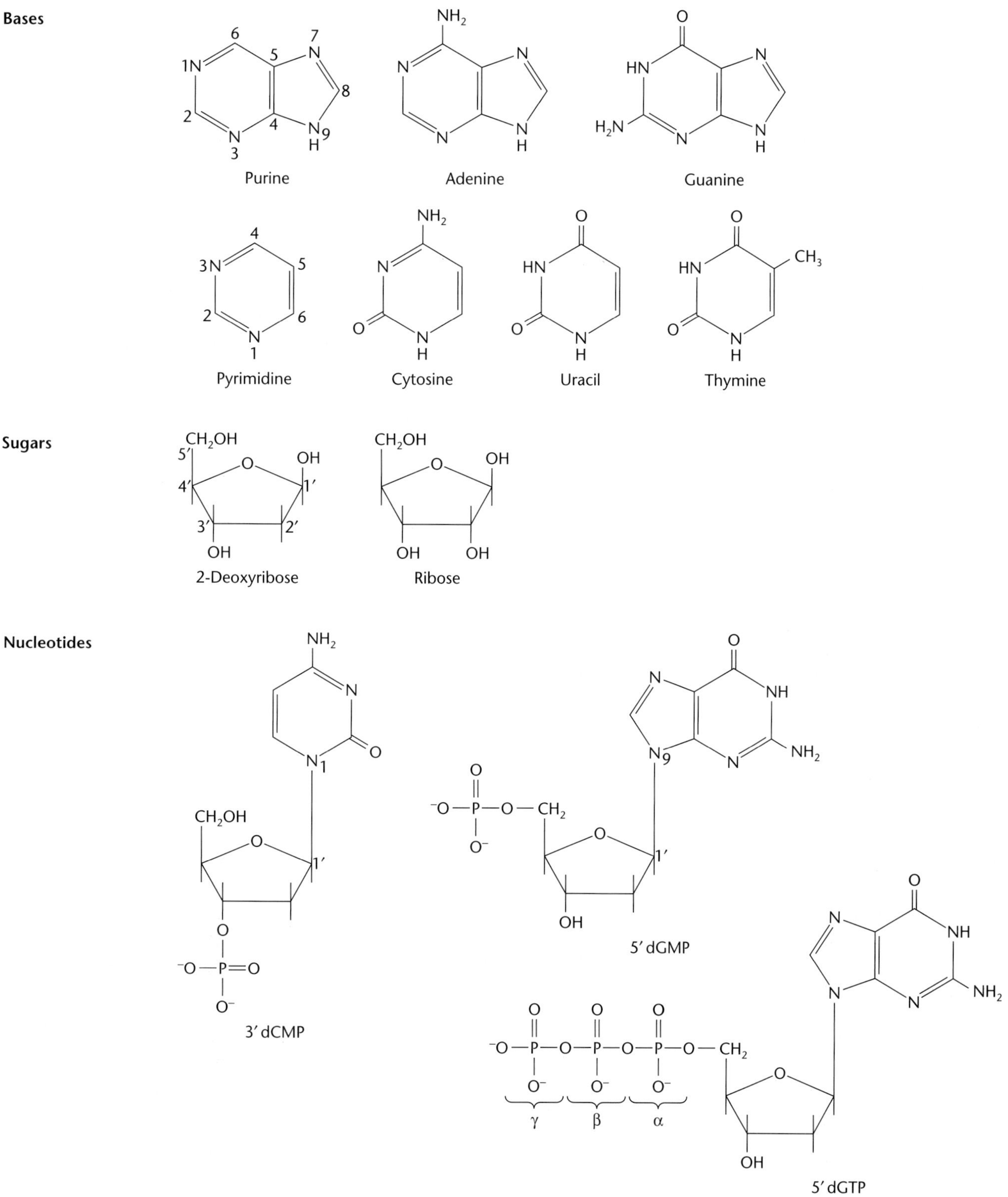

Bases

Purine

Adenine

Guanine

Pyrimidine

Cytosine

Uracil

Thymine

Sugars

2-Deoxyribose

Ribose

Nucleotides

3′ dCMP

5′ dGMP

5′ dGTP

Figure 1.2 The chemical structures of deoxynucleotides, showing the bases and sugars and how they are assembled into a deoxynucleotide.

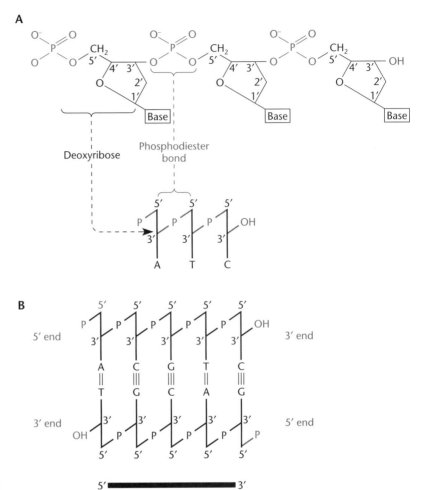

Figure 1.3 (A) Schematic drawing of a DNA chain showing the 3'-to-5' attachment of the phosphates to the sugars, forming phosphodiester bonds. (B) Two strands of DNA bind at the bases in an antiparallel arrangement of the phosphate-sugar backbones.

Watson and Crick one of the essential clues to the structure of DNA. They proposed that the two strands of the DNA are held together by specific hydrogen bonding between the bases in opposite strands, as shown in Figure 1.4. Thus, the amounts of A and T and of C and G are always the same because A's will pair only with T's and G's will pair only with C's to hold the DNA strands together. Each A-and-T pair and each G-and-C pair in DNA is called a **complementary base pair,** and the sequences of two strands of DNA are said to be **complementary** if one strand always has a T where there is an A in the other strand and a G where there is a C in the other strand.

It did not escape the attention of Watson and Crick that the complementary base-pairing rules essentially explain heredity. If A pairs only with T and G pairs only with C, then each strand of DNA can replicate to make a complementary copy of itself, so that the two replicated DNAs will be exact copies of each other. Offspring containing these DNAs would have the same sequence of nucleotides in their DNAs as their parents and so would be exact copies of their parents.

Antiparallel Construction

As mentioned at the beginning of this section, the complete DNA molecule consists of two long chains wrapped around each other in a double helix (Figure 1.1). The double-stranded molecule can be thought of as being like a circular staircase, with the alternating phosphates and deoxyribose sugars forming the railings and the bases connected to each other forming the steps. However, the two chains run in opposite directions, with the phosphates on one strand attached 5' to 3', 5' to 3', etc., to the sugars and those on the other strand attached 3' to 5', 3' to 5', etc. This arrangement is called **antiparallel**. In addition to phosphodiester bonds running in opposite directions, the antiparallel construction causes the 5' phosphate end of one strand and the 3' hydroxyl end of the other to be on the same end of the double-stranded DNA molecule (Figure 1.3B).

The Major and Minor Grooves

Because the two strands of the DNA are wrapped around each other to form a double helix, the helix will

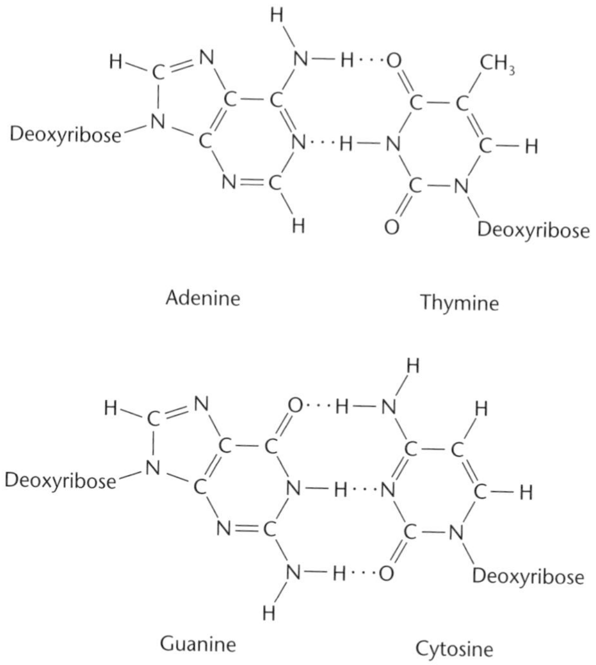

Adenine   Thymine

Guanine   Cytosine

Figure 1.4 The two complementary base pairs found in DNA. Two hydrogen bonds form in adenine-thymine base pairs. Three hydrogen bonds form in guanine-cytosine base pairs.

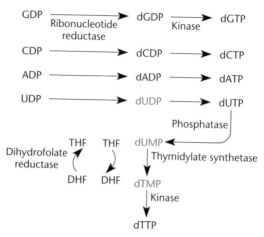

Figure 1.5 The pathways for synthesis of the deoxynucleotides from the ribonucleotides. Some of the enzymes referred to in the text are identified. THF, tetrahydrofolate; DHF, dihydrofolate.

have two grooves between the two strands (Figure 1.1). One of these grooves is wider than the other, so it is called the **major groove**. The other, narrower groove is called the **minor groove**. Most of the modifications to DNA that are discussed in this and later chapters occur in the major groove of the helix.

Mechanism of DNA Replication

The molecular details of DNA replication are probably similar in all organisms on Earth. The basic process of replication involves **polymerizing**, or linking, the nucleotides of DNA into long chains or strands, using the sequence on the other strand as a guide. Because the nucleotides must be made before they can be put together into DNA, the nucleotides are called the **precursors** of DNA synthesis.

Deoxyribonucleotide Precursor Synthesis

The precursors of DNA synthesis are the four deoxyribonucleoside triphosphates, dATP, dGTP, dCTP, and dTTP. The triphosphates are synthesized from the corresponding ribose nucleoside diphosphates by the pathway shown in Figure 1.5. In the first step, the enzyme **ribonucleotide reductase** reduces (i.e., removes an oxygen from) the ribose sugar to produce the deoxyribose sugar by changing the hydroxyl group at the 2′ position (the

second carbon) of the sugar to a hydrogen. Then an enzyme known as a **kinase** adds a phosphate to the deoxynucleoside diphosphate to make the deoxynucleoside triphosphate precursor.

The deoxynucleoside triphosphate dTTP is synthesized by a somewhat different pathway from the other three. The first step is the same. Ribonucleotide reductase synthesizes the nucleotide dUDP (deoxyuridine diphosphate) from the ribose UDP. However, from then on, the pathway differs. A phosphate is added to make dUTP, and the dUTP is converted to dUMP by a phosphatase that removes the phosphates. This molecule is then converted to dTMP by the enzyme thymidylate synthetase, using tetrahydrofolate to donate a methyl group. Kinases then add two phosphates to the dTMP to make the precursor dTTP.

Deoxynucleotide Polymerization

The complex process of DNA replication involves many enzymes and other cellular components. In the end, complementary copies of each of the extremely long strands of a double-stranded DNA must be made. In this section, we discuss the obstacles that DNA replication must overcome and how the various functions involved overcome these obstacles.

DNA POLYMERASES

The properties of the DNA polymerases, the enzymes that actually join the deoxynucleotides together to make the long chains, are the best guides to an understanding of the replication of DNA. These enzymes make DNA by linking one deoxynucleotide to another to make a long chain of DNA. This process is called DNA **polymerization**, hence the name DNA *polymer*ases.

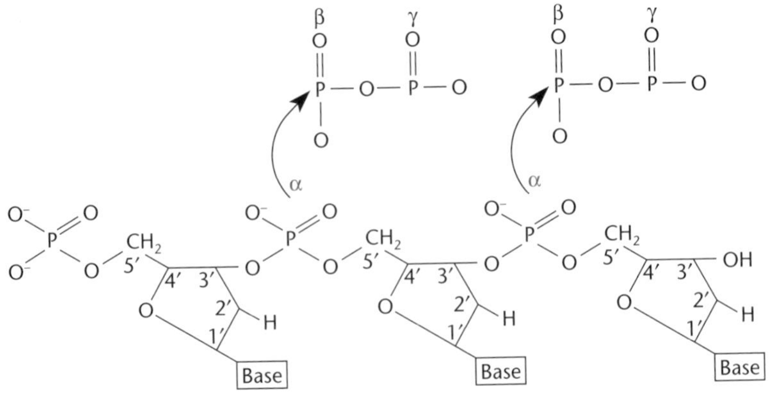

Figure 1.6 Polymerization of the deoxynucleotides during DNA synthesis. The β and γ phosphates of each deoxynucleoside triphosphate are cleaved off to give energy for the polymerization reaction.

Figure 1.6 shows the basic process of DNA polymerization by DNA polymerase. The DNA polymerase attaches the first phosphate (called α) of one deoxynucleoside triphosphate to the 3′ carbon of the sugar of the next deoxynucleoside triphosphate, in the process releasing the last two phosphates (called the β and γ phosphates) of the first deoxynucleoside triphosphate to produce energy for the reaction. Then the α phosphate of another deoxynucleoside triphosphate is attached to the 3′ carbon of this deoxynucleotide, and the process continues until a long chain is synthesized.

DNA polymerases also need a **template strand** to direct the synthesis of the new strand. As mentioned in the base-pairing section, complementary base pairing dictates that wherever there is a T in the template strand, an A will be inserted in the strand being synthesized, and so forth according to the base-pairing rules. The DNA polymerase can move only in the 3′-to-5′ direction on the template strand, linking deoxynucleotides in the new strand in the 5′-to-3′ direction. When replication is completed, the product is a new double-stranded DNA with antiparallel strands, one strand of which is the old template strand and one strand of which is the newly synthesized strand.

There are two DNA polymerases which participate in normal DNA replication in *E. coli*; they are called DNA polymerase III and DNA polymerase I (see Table 1.1). DNA polymerase III is a large complex composed of the enzyme which polymerizes the nucleotides in a complex with accessory proteins, which make new DNA at the replication fork. DNA polymerase I is responsible for removing RNA primers in the lagging strand and filling the gaps between Okazaki fragments (see below). It also plays a role in DNA repair, as we discuss in chapter 11.

NUCLEASES

Enzymes which degrade DNA strands by breaking the phosphodiester bonds are just as important in replication as the enzymes which polymerize DNA by forming phosphodiester bonds between the nucleotides. These bond-breaking enzymes, called **nucleases** can be grouped into two major categories. One type can initiate breaks in the middle of a DNA strand and so are called **endonucleases,** from a Greek word meaning "within," and the other type can remove nucleotides only from the ends of DNA strands and so are called **exonucleases,** from a Greek word meaning "outside." Exonucleases can in turn be divided into two groups. Some exonucleases can only degrade from the 3′ end of a DNA strand, degrading DNA in the 3′-to-5′ direction. These are called 3′ exonucleases; one example of their activity is their role in the editing function associated with DNA polymerase I and III, which is discussed below. Other exonucleases, called 5′ exonucleases, degrade DNA strands only from the 5′ end, an example being the 5′ exonuclease activity of DNA polymerase I which removes RNA primers during replication. Nucleases can also leave either 3′ or 5′ phosphate deoxynucleotides, depending upon which side of the phosphodiester bond they cut.

DNA LIGASES

DNA ligases are enzymes which form phosphodiester bonds between the ends of chains of DNA, another reaction that DNA polymerases cannot perform for themselves. During replication, these enzymes join the 5′ phosphate at the end of one DNA chain to the 3′ hydroxyl at the end of another chain to make a longer continuous chain.

PRIMASES

Another type of enzyme which performs a reaction that DNA polymerases cannot perform for themselves is the group of **primases**. These are enzymes that make RNA **primers** to initiate the synthesis of new strands of DNA. DNA polymerases cannot start the synthesis of a new strand of DNA; they can only attach deoxynucleotides to a preexisting 3′ OH group. The 3′ OH group to which

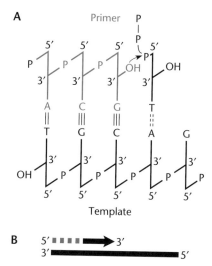

Figure 1.7 Functions of the primer and template in DNA replication. (A) The DNA polymerase adds deoxynucleotides to the 3′ end of the primer by using the template strand to direct the selection of each base. (B) Simple illustration of 5′ → 3′ DNA synthesis. Dotted line, primer.

TABLE 1.1	Proteins involved in *E. coli* DNA replication	
Protein	**Gene**	**Function**
DnaA	*dnaA*	Initiator protein; primosome (priming complex) formation
DnaB	*dnaB*	DNA helicase
DnaC	*dnaC*	Delivers DnaB to replication complex
SSB	*ssb*	Binding to single-stranded DNA
Primase	*dnaG*	RNA primer synthesis
DNA Pol I	*polA*	Primer removal; gap filling
DNA Pol III (holoenzyme)		
α	*dnaE*	Polymerization
ε	*dnaQ*	3′ → 5′ editing
RNase H	*rnhA*	Removes RNA primers
θ	*holE*	Present in core (αεθ)
β	*dnaN*	Sliding clamp
τ[a]	*dnaX*	Organizes complex; joins leading and lagging DNA PolIII
γ[b]	*dnaX*	Binds clamp loaders and SSB protein
δ	*holA*	Clamp loading
δ′	*holB*	Clamp loading
χ	*holC*	Binds SSB
ψ	*holD*	Binds SSB
DNA ligase	*lig*	Sealing DNA nicks
DNA gyrase		Supercoiling
α	*gyrA*	Nick closing
β	*gyrB*	ATPase

[a]Full-length product of the *dnaX* gene.
[b]Shorter product of the *dnaX* gene produced by translational frameshifting (see Box 2.4).

DNA polymerase adds a deoxynucleotide is called the primer (Figure 1.7).

The requirement of a primer for DNA polymerase creates an apparent dilemma in DNA replication. When a new strand of DNA is synthesized, there is no DNA upstream (i.e., on the 5′ side) to act as a primer. The cell usually solves this problem by using RNA as the primer to initiate the synthesis of new strands. Unlike DNA polymerase, RNA polymerase does not require a primer to initiate the synthesis of new strands. The RNA primers that are used to initiate the synthesis of new strands of DNA are either made by the RNA polymerase, which makes all the other RNAs including mRNA, tRNA, and rRNA, or by the special enzymes called primases. During DNA replication, special enzymes recognize and remove the RNA primer (see below).

ACCESSORY PROTEINS

Ten different proteins travel with the DNA polymerase as part of a **DNA replication complex** as the polymerase moves along the template strand. These proteins are called **DNA polymerase accessory proteins;** together with the polymerizing activity, they form the **DNA polymerase III holoenzyme.** One of these accessory proteins forms a clamp that helps keep the DNA polymerase from falling off the template strand. Others help the clamp load on the DNA or are exonucleases that serve an **editing function** to correct mistakes the DNA polymerase may make (see the section on editing, below). Table 1.1 lists many of the DNA replication proteins.

Semiconservative Replication

The process of replication described above is called **semiconservative replication** because each time a DNA molecule replicates, the two old strands are *conserved* to become part of two new DNA molecules. Each of the new molecules will consist of one old conserved strand and one newly synthesized strand, so that each new molecule is only partly conserved (or *semi*conserved). The next time the DNA molecule replicates, each strand will serve as a template, becoming the conserved strand in a new double-stranded molecule.

THE MESELSON-STAHL EXPERIMENT

The semiconservative mechanism was suggested by the structure of DNA. However, other mechanisms of DNA replication are also possible. Soon after Watson and Crick published their structure for DNA, Matthew Meselson and Frank Stahl performed an experiment that showed that DNA does replicate by the proposed semiconservative mechanism.

One prediction of the semiconservative replication hypothesis is that after replication, each DNA molecule should have one newly synthesized strand and one old strand from the original molecule. If it could be demonstrated that newly synthesized DNA had one new strand and one old strand, this prediction would have been fulfilled. Figure 1.8 illustrates the details of their experiment. First, they chose the heavy isotopes of nitrogen, carbon, and hydrogen, three of the types of atoms contained in DNA, as markers for new strands. DNA synthesized from these heavy isotopes is more dense than DNA synthesized from the normal atoms.

To incorporate the heavy isotopes into newly synthesized DNA, they grew the bacterium *Escherichia coli* on a medium containing the heavy isotopes instead of one containing the normal isotopes for about the time the cells took to divide. Then they extracted the DNA from the cells and analyzed its density by means of density gradient equilibrium centrifugation. As Figure 1.8 shows, the density of the DNA will be reflected by the position of the DNA in the gradient; heavier DNA, composed of atoms of the heavy isotopes, bands farther down. If the DNA replicates by a semiconservative mechanism, then after one cycle of replication in the presence of the heavy

Figure 1.8 The Meselson-Stahl experiment. Synthesis of a "light" complementary strand gives a double-stranded DNA of hybrid light and heavy density. The light-light, heavy-light, and heavy-heavy DNAs can be separated by equilibrium density centrifugation.

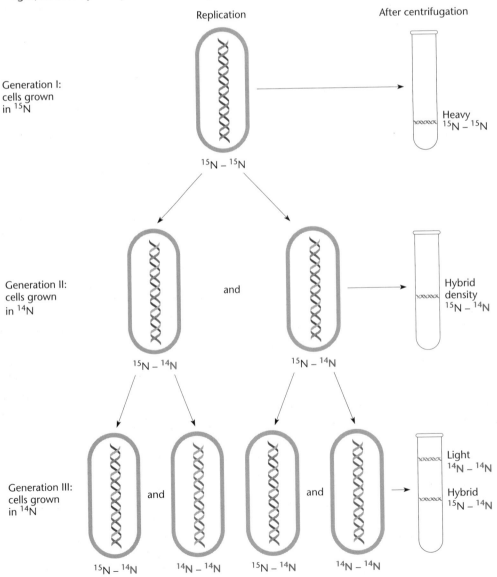

isotopes, one strand should be made from precursors with the heavy isotopes and the newly replicated molecule should then consist of one heavy strand and one light strand. Consequently, its density will be intermediate between those of light DNA, with no heavy isotopes, and heavy DNA, in which both new strands contain heavy isotopes. After a short time of replication, Meselson and Stahl did observe DNA of intermediate density, composed of one light strand and one heavy strand; thus, their results supported the semiconservative mechanism of replication.

Replication of Double-Stranded DNA

The replication of most long DNA molecules such as bacterial DNA begins at one point and moves in both directions from there. In the process, the two old strands of DNA are separated and used as templates to synthesize new strands. The structure where the two strands are separated and the new synthesis is occurring is referred to as the **replication fork**. The DNA polymerase enzyme cannot separate the two strands of a bacterial chromosome, replicate them, and separate the daughter DNAs by itself. Many other proteins are required for replication, as we discuss in this section.

HELICASES AND HELIX-DESTABILIZING PROTEINS

One task the DNA polymerase cannot perform in the replication of double-stranded DNA is the separation of the strands of DNA at the replication fork. The strands of DNA must be separated for the two strands to serve as templates. The bases of the DNA are inside of the double helix, where they are not available to pair with the incoming deoxynucleotides, to direct which nucleotide will be inserted at each step. The strands of DNA are separated by proteins called **DNA helicases** (see Singleton and Wigley, Suggested Reading). Some of these proteins form a ring around one strand of the DNA and suck the strand through the ring; others "snowplow" along the DNA, separating the strands as they go. It takes a lot of energy to separate the strands of DNA, and helicases cleave a lot of ATP to form ADP in the process. There are about 20 different helicases in *E. coli*, and each helicase works in only one direction, either the 3′-to-5′ or the 5′-to-3′ direction. The hexameric DnaB helicase that normally separates the strands of DNA ahead of the replication fork sucks one strand through itself in the 5′-to-3′ direction as it opens the strands of DNA ahead of the replication fork (see Figure 1.12). We discuss some other helicases in later chapters in connection with recombination and repair.

Once the strands of DNA have been separated, they also must be prevented from coming back together (or from annealing to themselves if they happen to be complementary over short regions). The separation of the strands is maintained by proteins called **helix-destabilizing proteins** or single-strand-binding proteins. These are proteins that bind preferentially to single-stranded DNA and prevent double-stranded helical DNA from re-forming prematurely.

OKAZAKI FRAGMENTS AND THE REPLICATION FORK

Another problem in replicating double-stranded DNA is created because the two strands are antiparallel. As already discussed, in one strand the phosphates connect the sugars 3′ to 5′ and in the other strand they connect the sugars 5′ to 3′. However, DNA polymerase can move only in the 3′-to-5′ direction on the template strand, synthesizing the new DNA in the 5′-to-3′ direction. How can the replication fork move in one direction on a double-stranded DNA and make complementary copies of both strands at the same time? Because of the antiparallel structure, the DNA polymerase on one of the two strands would have to be moving in the wrong direction overall.

This problem is overcome by replicating the two strands differently (Figure 1.9). On one template strand, DNA polymerase III initiates synthesis from an RNA primer and moves along the template DNA in the 3′-to-5′ direction. This newly synthesized strand is referred to as the **leading strand** (Figure 1.9). To replicate the other strand, the DNA polymerase must wait until the DNA strands have been separated by the DnaB helicase before it can load on the DNA, which is why this is called the **lagging-strand** synthesis (Figure 1.9). This DNA polymerase makes short pieces called **Okazaki fragments** in the opposite direction to that in which the fork is moving. Synthesis of each Okazaki fragment requires a new RNA primer, about 10 to 12 nucleotides long. In *E. coli*, these primers are synthesized by DnaG primase, which will produce a new RNA primer about once every 2 kilobases, recognizing the sequence 3′-GTC-5′ and beginning synthesis opposite the T. These RNA primers are then used to prime DNA synthesis by the DNA polymerase III, which continues until it encounters another piece being synthesized further along the DNA (Figure 1.10). However, before these short pieces can be joined to make a long continuous strand of DNA, the short RNA primers must be removed. Most of the RNA primer is removed by an enzyme called RNase H, which removes the RNA strand of an DNA:RNA double helix (Table 1.1). Then DNA polymerase I comes into play. Using its concerted 5′ exonuclease and DNA polymerase activities (Figure 1.11), DNA polymerase I removes what remains of each RNA primer and replaces it with

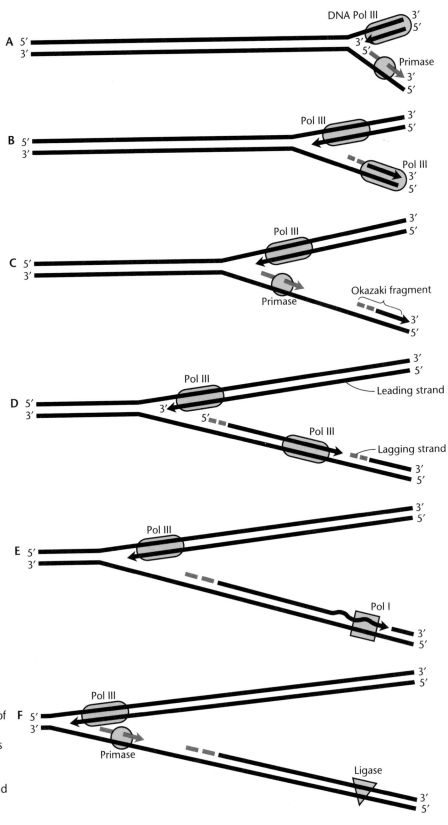

Figure 1.9 Discontinuous synthesis of one of the two strands of DNA during chromosome replication. The functions of the primase and ligase are shown. The short RNA primers in the lagging strand are removed after replication and resynthesized as DNA by using upstream DNA as primer.

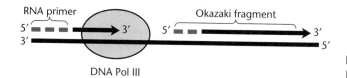

Figure 1.10   Synthesis of short Okazaki fragments from an RNA primer.

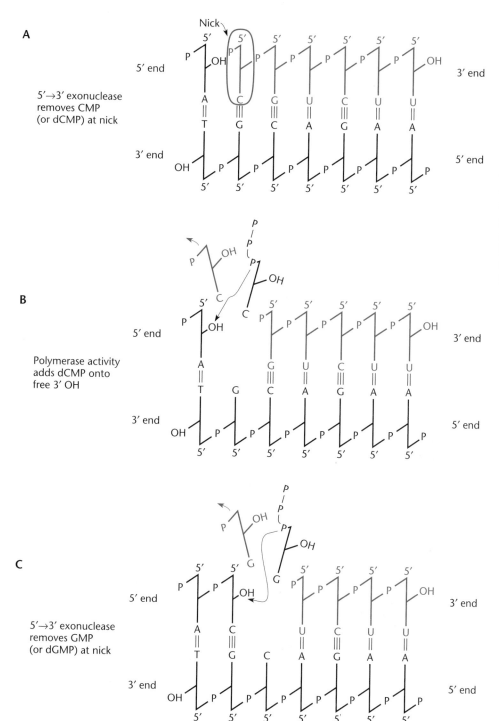

A

Nick

5′→3′ exonuclease removes CMP (or dCMP) at nick

B

Polymerase activity adds dCMP onto free 3′ OH

C

5′→3′ exonuclease removes GMP (or dGMP) at nick

Nick translation

Figure 1.11   DNA polymerase I can remove the nucleotides of an RNA primer by using its "nick translation" activity. (A) A break, or nick, in the DNA strand occurs when the DNA polymerase III holoenzyme incorporates the last deoxynucleotide before it encounters a previously synthesized RNA primer. (B) In the example, the 5′-to-3′ exonuclease activity of DNA polymerase I removes the CMP at the nick and its DNA polymerase activity incorporates a dCMP onto the free 3′ hydroxyl. (C) This process continues, moving the nick in the 5′-to-3′ direction.

DNA, using the upstream Okazaki fragment as a primer. The Okazaki fragments are then joined together by DNA ligase as the replication fork moves on, as shown in Figure 1.9. By using RNA rather than DNA to prime the synthesis of Okazaki fragments, the cell lowers the mistake rate (see below).

What actually happens at the replication fork is more complicated than suggested by this simple picture. For one thing, this picture ignores the overall topological restraints on the DNA that is replicating. The **topology** of a molecule refers to its position in space. Because the circular DNA is very long and its strands are wrapped around each other, pulling the two strands apart introduces stress into other regions of the DNA in the form of **supercoiling.** Unless the two strands of DNA were free to rotate around each other, supercoiling would cause the chromosome to look like a telephone cord wound up on itself. To relieve this stress, enzymes called **topoisomerases** undo the supercoiling ahead of the replication fork. We shall discuss DNA supercoiling and topoisomerases later in the chapter.

Recent evidence indicates that the picture of the two strands of DNA replicating independently is too simple. Rather than replicating independently, the two DNA polymerase III holoenzymes replicating the leading and lagging strands are joined to each other through their τ subunits (Table 1.1). To accommodate the fact that the two DNA polymerases must move in opposite directions and still remain joined, the lagging-strand template loops out as the Okazaki fragment is synthesized. The loop is then relaxed as the sliding clamp releases the lagging-strand polymerase, allowing the DNA polymerase to hop ahead to the next RNA primer to begin synthesizing the next Okazaki fragment (Figure 1.12). This has been referred to as the "trombone" model of replication because the loops forming and contracting at the replication fork resemble the "oom-paa-oom-paa" of the musical instrument. The situation is probably similar in all bacteria and even higher organisms, although in some other bacteria including *Bacillus subtilis* and in eukaryotes, different DNA polymerases are used to replicate the leading and lagging strands (see Dervyn et al., Suggested Reading).

THE GENES FOR REPLICATION PROTEINS
Most of the genes for replication proteins have been found by isolating mutants defective in DNA replication but not RNA or protein synthesis. Since a mutant cell that cannot replicate its DNA will die, any mutation (for definitions of mutants and mutations, see the section on replication errors, below, and chapter 3) that inactivates a gene whose product is required for DNA replication will kill the cell. Therefore for experimental purposes, only a type of mutant called a **temperature-sensitive mutant** can

be usefully isolated with mutations in DNA replication genes. These are mutants in which the product of the gene is inactive at one temperature but active at another. The mutant cells can be propagated at the temperature at which the protein is active. Then, the effects of inactivating the protein can be tested by shifting to the other temperature. We shall discuss the molecular basis for temperature-sensitive mutants in more detail in chapter 3.

The immediate effect of a temperature shift on a mutant with a mutation in a DNA replication gene depends on whether the product of the gene is continuously required for replication at the replication forks or is involved only in the initiation of rounds of replication. For example, if the mutation is in a gene for DNA polymerase III or in the gene for the DnaG primase, replication will immediately cease. However, if the temperature-sensitive mutation is in a gene whose product is required only for initiation of DNA replication, for example, the gene for DnaA or DnaC (see section on initiation of chromosome replication, below), the replication rate for the population will slowly decline. Unless the cells have been somehow synchronized in their cell cycle, each cell will be at a different stage of replication, with some cells having just finished a round of replication and other cells having just begun a new round. Cells in which rounds of chromosome replication were under way at the time of the temperature shift will complete their replication but not start a new round. Therefore, the rate of replication will decrease until the rounds of replication in all the cells are completed.

Replication Errors

To maintain the stability of a species, replication of the DNA must be almost free of error. Otherwise, changes in the DNA sequence called **mutations** will occur, and these changes will be passed on to subsequent generations. Depending on where these changes occur, they can severely alter the protein products of genes or other cellular functions. To avoid such instability, the cell has mechanisms that reduce the error rate.

As DNA replicates, the wrong base is sometimes inserted into the growing DNA chain. For example, Figure 1.13 shows the incorrect incorporation of a T opposite a G. A base pair in which the bases are paired wrongly is called a **mismatch.** Mismatches can occur when the bases take on forms called **tautomers,** which pair differently from the normal form of the base (see chapter 3). After the first replication in Figure 1.13, the mispaired T will usually be in its normal form and pair correctly with an A, causing a GC-to-AT change in the sequence of one of the two progeny DNAs and thus changing the base pair at that position on all subsequent copies of the mutated DNA molecule.

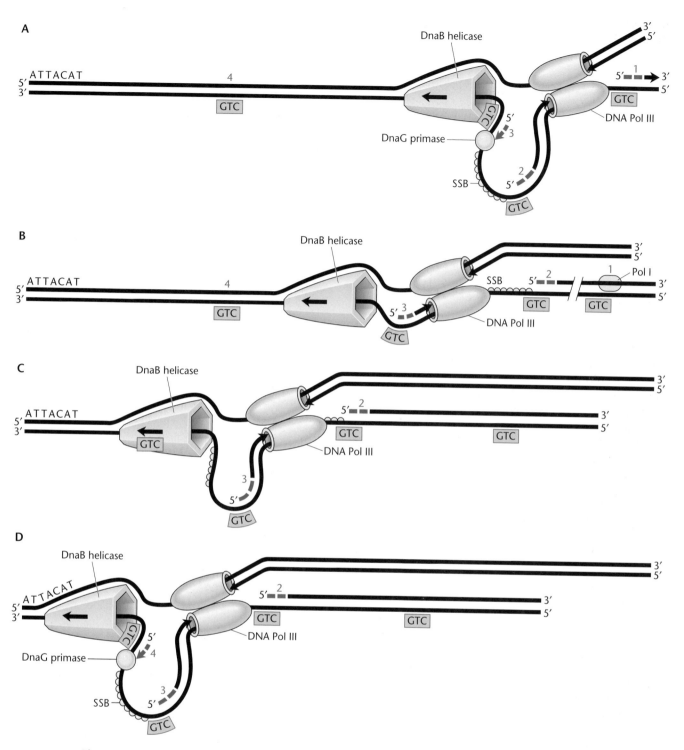

Figure 1.12 A "trombone" model for how both the leading and lagging strands of the DNA helix might be simultaneously replicated at the replication fork. SSB, single-strand-binding protein. RNA primers are shown in gold. (A) PolIII holoenzyme is synthesizing lagging-strand DNA from primer 2 and has just run into primer 1. (B) Lagging-strand PolIII has been released from the template at primer 1 and hopped ahead, reassembling on the DNA at primer 3 to synthesize an Okazaki fragment. Both the leading- and lagging-strand PolIII enzymes have remained bound to each other during the release and reassembly process. Meanwhile, PolI is removing primer 1 and replacing it with DNA. (C) PolIII is continuing lagging-strand synthesis from primer 3. (D) PolIII has completed the Okazaki fragment and has run into primer 2. PolIII will now dissociate from the DNA and reassemble at primer 4. In all panels, the arbitrary sequence ATTACAT shows the progress of the replication fork. The sequence 3'GTC5' boxed in gold shows where the primer synthesis initiates. The length of primers and Okazaki fragments is not drawn to scale. Part A adapted with permission from T. A. Baker and S. H. Wickner, *Annu. Rev. Genet.* **26**:447–478, 1992.

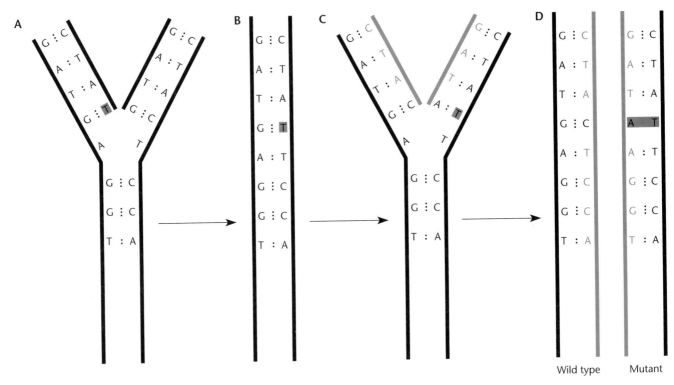

Figure 1.13 Mistakes in base pairing can lead to changes in the DNA sequence called mutations. A T mistakenly put opposite a G during replication (A) can lead to an AT base pair replacing a GC in the progeny DNA (B to D).

Editing

One way the cell reduces mistakes during replication is through **editing** functions. Sometimes these functions are performed by separate proteins, and sometimes they are part of the DNA polymerase itself! Editing proteins are aptly named because they go back over the newly replicated DNA looking for mistakes, recognizing and removing incorrectly inserted bases (Figure 1.14). If the last nucleotide inserted in the growing DNA chain creates a mismatch, the editing function will stop the replication until the offending nucleotide is removed. The replication then continues, inserting the correct nucleotide. Because the DNA chain grows in the 5′-to-3′ direction, the last nucleotide added is at the 3′ end. The enzyme activity that removes this nucleotide is therefore called a **3′ exonuclease.** The editing proteins probably recognize a mismatch because the mispairing (between T and G in the example) causes a minor distortion in the structure of the double-stranded helix of the DNA.

In many cellular organisms, some viruses, and some repair DNA polymerases, for example, DNA polymerase I, the 3′ exonuclease editing activity is part of the DNA polymerase itself. In bacterial chromosome replication, the editing functions are accessory proteins encoded by separate genes whose products travel along the DNA with the DNA polymerase during replication. In *E. coli,* the 3′ exonuclease editing function is encoded by the *dnaQ* gene (Table 1.1), and *dnaQ* mutants, also called *mutD* mutants (i.e., cells with a mutation in this gene that inactivates the 3′ exonuclease function), show much higher rates of spontaneous mutagenesis than do the **wild-type,** or normally functioning, cells. Because of their high spontaneous mutation rates, *mutD* mutants of *E. coli* are often used to introduce random mutations into plasmids and bacteriophages.

RNA PRIMERS AND EDITING

The importance of the editing functions in lowering the number of mistakes during replication may explain why DNA replication is primed by RNA rather than by DNA. When the replication of a DNA chain has just initiated, the helix may be too short for distortions in its structure to be easily recognized by the editing proteins. The mistakes may then go uncorrected. However, if the first nucleotides inserted in a growing chain are ribonucleotides rather than deoxynucleotides, an RNA primer will be synthesized rather than a DNA primer. The RNA primer can be removed and resynthesized as DNA by

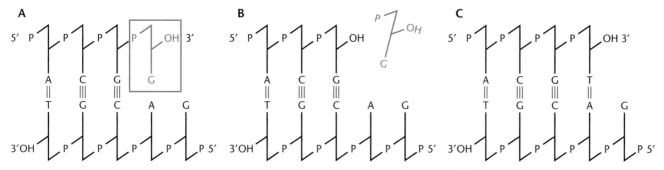

Figure 1.14   The editing function of DNA polymerase. (A) A G is mistakenly put opposite an A while the DNA is replicating. (B and C) The DNA polymerase stops while the G is removed and replaced by a T before the replication continues.

using preexisting upstream DNA as primer. Under these conditions, the editing functions will be active and mistakes will be avoided.

Methyl-Directed Mismatch Repair

Sometimes, the wrong base pair will be inserted into DNA in spite of the vigilance of the editing functions. However, the cell still has another chance to prevent a permanent mistake or mutation: the wrong base can be recognized by another repair system called the **mismatch repair system**. This system recognizes the mismatch and removes it as well as DNA in the same strand around the mismatch, leaving a gap in the DNA that is refilled by the action of DNA polymerase, which inserts the correct nucleotide.

The mismatch repair system is very effective at removing mismatches from DNA. However, by itself, it would not lower the rate of spontaneous mutagenesis unless it repaired the correct strand of DNA at the mismatch. In the example shown in Figure 1.13, a T was mistakenly incorporated opposite an G. If the mismatch repair system changes the T to the correct C in the *newly* replicated DNA, a GC base pair will be restored at this position and no change in the sequence or mutation will have occurred. However, if it repairs the G in the *old* DNA in the mismatch to an A, the mismatch will have been removed and replaced by a AT base pair with correct pairing, but a GC-to-AT change would have occurred in the DNA sequence at the site of the mismatch, creating a mutation. To prevent mutations, the mismatch repair system must have some way of distinguishing the newly synthesized strand from the old strand, so that it can repair the correct strand.

The state of methylation of the DNA strands allows the mismatch repair system of *E. coli* to distinguish the new from the old strands after replication. In *E. coli*, the A's in the symmetric sequence GATC/CTAG are methylated at the 6′ position of the larger of the two rings of the

adenine base. These methyl groups are added to the bases by the enzyme **deoxyadenosine methylase** (Dam methylase), but this occurs only *after* the nucleotide has been incorporated into the DNA. Since DNA replicates by a semiconservative mechanism, the newly synthesized strand of the GATC/CTAG will remain temporarily unmethylated after replication of a region containing this sequence. The DNA at this site is said to be **hemimethylated** if the bases on only one strand are methylated. Figure 1.15 shows that a hemimethylated GATC/CTAG tells the mismatch repair system which strand is newly synthesized and should be repaired. For this reason, the repair system is called the **methyl-directed mismatch repair system.**

The mismatch repair system also repairs many types of chemical damage to DNA, which we discuss in more detail in chapter 11. Methylation of DNA in *E. coli* plays other roles as well, including protecting against cutting by restriction endonucleases and helping to time the initiation of chromosomal DNA replication (see below).

Role of Editing and Mismatch Repair in Maintaining Replication Fidelity

It is possible to estimate how much each of the repair systems lowers the mistake rate of replication. On the basis of normal mutation rates, the *E. coli* replication mechanism probably makes a mistake about once every 10^{10} times it incorporates a nucleotide. Since there are about 4.7×10^6 nucleotides in each strand of *E. coli* DNA, the cell makes $1/10^{10} \times 4.7 \times 10^6 = 4.7 \times 10^{-4}$ mistake every time it replicates a chromosome. In other words approximately 1 in every 2,000 progeny bacteria will have a mistake in its DNA, since the entire bacterial DNA must replicate once every time the cell divides. Mistake levels like these are apparently low enough to be tolerated, and some mistakes may even be desirable because they increase diversity in the population and speed up evolution.

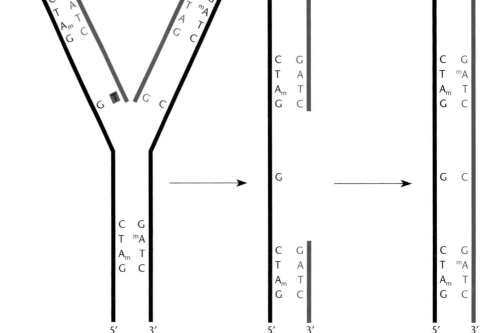

Figure 1.15 The methyl-directed mismatch repair system. (A) The newly replicated DNA contains a GT mismatch, and the GATC sequences are not methylated. (B) The repair system excises the mismatched T and adjacent bases of the unmethylated strand. (C) The gap has been filled in, and the GATC sequences have been methylated.

In contrast, mutant bacteria (i.e., those that differ from normal, or wild-type, bacteria by a definable mutation; see chapter 3) that lack either the editing functions or the mismatch repair system have unacceptably high mistake rates. An *E. coli* bacterium that lacks the editing function will make a mistake on average once every 10^6 times a nucleotide is added during replication, or 10,000 times more frequently than the wild-type bacteria. This means that every such cell will suffer an average of about five mutations each time its DNA replicates. If the mismatch repair system is also inactivated, the mistake rate will be increased another 10-fold or more, for an average of 50 or more mistakes every time the DNA replicates and the cell divides. It would not be very long before these bacteria were severely compromised by mutations and bore little resemblance to their ancestors. By lowering the spontaneous mutation rate, DNA correction systems are important for maintaining replication fidelity and the stability of the species.

Replication of the Bacterial Chromosome and Cell Division

So far we have discussed the details of DNA replication, but we have not discussed how the bacterial DNA as a whole replicates, nor have we discussed how the replication process is coordinated with division of the bacterial cell. To simplify the discussion, we shall consider only bacteria that grow as individual cells and divide by bi-nary fission to form two cells of equal size, even though this is far from the only type of multiplication observed among bacteria.

The replication of the bacterial DNA occurs during the cell division cycle. The **cell division cycle** is the time during which a cell is born, grows larger, and divides into two progeny cells. **Cell division** is the process by which the larger cell splits into the two new cells. The **division time,** or **generation time,** is the time that elapses from the point when a cell is born until it divides. This time is usually approximately the same for all the individuals in the population under certain growth conditions. Before cell division the original cell is called the **mother cell** and after division the two progeny cells are the **daughter cells.**

Structure of the Bacterial Chromosome

The DNA molecule of a bacterium that carries most of its normal genes is commonly referred to as its **chromosome,** by analogy to the chromosomes of higher organisms. This name distinguishes the molecule from plasmid DNA, which can be almost as large as chromosomal DNA but usually carries genes that are not always required for growth of the bacterium (see chapter 4).

Most bacteria only have one chromosome; in other words, there is only one *unique* DNA molecule per cell that carries most of the normal genes. This does not mean that there is necessarily only one copy of the chromosomal DNA in bacterial cells. Bacterial cells whose chromosomes replicate but for some reason do not divide have more than one copy of the chromosomal DNA per cell. However, these DNAs are not unique, since they are derived from the same original molecule.

The structure of bacterial DNA differs significantly from that of the chromosomes of higher organisms. For example, DNA in the chromosomes of most bacteria is **circular** (for exceptions, see Box 1.1), with a circumference of approximately 1 mm. In contrast, eukaryotic chromosomes are linear with free ends. As we shall discuss later, the circularity of bacterial chromosomal DNA allows it to replicate in its entirety without using telomeres, as eukaryotic chromosomes do, or terminally redundant ends, as some phages do. Another difference between the DNA of bacteria and eukaryotes is that the DNA in eukaryotes is wrapped around proteins called histones to form nucleosomes. Bacteria contain histone-like proteins including HU, HN-S, Fis, and IHF around which DNA is often wrapped. However, in general, DNA is much less structured in bacteria than in eukaryotes.

Replication of the Bacterial Chromosome

The replication of the circular bacterial chromosome initiates at a unique site in the DNA called the **origin** of chromosomal replication, or *oriC,* and proceeds in both directions around the circle. On the *E. coli* chromosome, *oriC* is located at 84.3 min. At the positions where polymerases add the nucleotides, the double-stranded DNA splits and forms two new double-stranded DNAs. As mentioned earlier, the place in DNA at which replication is occurring is known as the replication fork. The two replication forks proceed around the circle until they meet and **terminate** chromosomal replication. As discussed below in the section on termination, many bacteria do not terminate replication at a unique site in the DNA but, rather, where the two replication forks meet. Each time the two replication forks proceed around the circle and meet, a **round of replication** has been completed and two new DNAs, called the **daughter DNAs,** are created.

Initiation of Chromosome Replication

Much has been learned about the molecular events occurring during the initiation of replication. Some of this information has a bearing on how the initiation of chromosome replication is regulated and serves as a model for the interaction of proteins and DNA.

Two types of functions are involved in the initiation of chromosome replication. One consists of the sites or sequences on DNA at which proteins act to initiate replication. These are called *cis*-acting sites. The prefix *cis* means "on this side of," and these sites act only on the same DNA. The proteins involved in initiation of replication are called *trans*-acting functions. The prefix *trans*

means "on the other side of," and these functions can act on any DNA in the same cell, not just the DNA from which they were made. These concepts will be used again as the book progresses.

ORIGIN OF CHROMOSOMAL REPLICATION
One *cis*-acting site is the *oriC* site, at which replication initiates. The sequence of *oriC* is well defined and is similar in most bacteria. Figure 1.16 shows the structure of the origin of replication of *E. coli.* Less than 250

Figure 1.16 Structure of the *oriC* region of *E. coli.* Shown are the position of the AT-rich region and the position of some of the DnaA-binding sequences (DnaA boxes) referred to in the text.

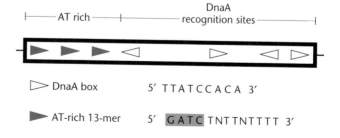

bp of DNA is required for initiation at this site. Within *oriC*, similar sequences of 9 bases, the so-called **DnaA boxes,** are repeated four times. In addition, regions 13 bp long with a higher than average AT base pair frequency are repeated three times. These two types of repeated sequences are thought to be very important for the initiation of chromosome replication.

INITIATION PROTEINS
Many *trans*-acting proteins are also required for the initiation of DNA replication, including the DnaA, DnaB, and DnaC proteins. DnaA is required only for initiation, but both DnaB and DnaC are also required for primer synthesis once DNA replication is under way. Many proteins used in other cellular functions are also involved, such as the primase (DnaG) and the normal RNA polymerase that makes most of the RNA in the cell.

Figure 1.17 outlines how DnaA, DnaB, DnaC, and other proteins participate in the initiation of chromosome replication. In the first step, 10 to 12 molecules of the DnaA protein bind to the DnaA boxes in the *oriC* region. This has the effect of wrapping the DNA around the aggregated DnaA proteins, as shown in the figure. The bending helps separate the strands of the DNA in the region of the bend as shown.

Once the strands are partially opened, the DnaB protein binds to the *oriC* region with the help of the DnaC protein. This binding is also aided by supercoiling at the origin (see the section on supercoiling, below) and by the helix-destabilizing protein, or single-strand-binding protein (SSB in Figure 1.17), which helps keep the helix from re-forming. The DnaB protein is a helicase that opens the strands further for priming and replication. The DnaC protein may then leave. This structure, with many copies of DnaA as well as DnaB and other proteins, is called a **primosome.** The primosome may help the DnaG primase or another RNA polymerase to synthesize an RNA primer to start replication.

RNA PRIMING OF INITIATION
Initiation of DNA replication requires RNA primers, but which RNA polymerase makes the primer for initiating leading-strand synthesis is not completely clear. The RNA polymerase that synthesizes most of the RNA molecules, including mRNA, in the cell (see chapter 2) is needed to initiate rounds of replication. However, its role may be to help separate the strands of DNA in the *oriC* region by transcribing through this region, because the strands may have to be separated before the DnaA protein can bind. In this case, the RNA primers themselves may be synthesized by the DnaG protein, the same RNA polymerase that makes RNA primers for lagging-strand synthesis at the replication fork. Alternatively, the normal

RNA polymerase may make the primer that initiates leading-strand synthesis, while the DnaG protein makes only the primers that initiate lagging-strand synthesis. More experiments are needed to answer this question.

Termination of Chromosome Replication

After the replication of the chromosome initiates in the *oriC* region and proceeds around the circular chromosome in both directions, the two replication forks must meet somewhere on the other side of the chromosome and the two daughter chromosomes must separate. Do they meet and terminate replication at a certain well-defined site in the DNA, or do they terminate replication wherever they happen to meet? Also, are specific proteins required for termination of chromosome replication?

As with most cellular processes, the process of termination of chromosome replication is especially well understood in *E. coli*. In this bacterium, chromosome replication usually terminates in a certain region but not at a well-defined unique site. This termination region, called *ter*, contains clusters of sites called *ter* sequences, which are only 22 bp long. These sites act somewhat like the one-way gates in an automobile parking lot, allowing the replication forks to pass through in one direction but not in the other.

Figure 1.18 shows how the one-way nature of *ter* sequences causes replication to terminate in the *ter* region. In the illustration, two *ter* site clusters called *terA* and *terB* bracket the termination region. Replication forks can pass site *terA* in the clockwise direction but not the counterclockwise direction. The opposite is true for *terB*. Thus, the clockwise-moving replication fork can pass through *terA*, but if it gets to *terB* before it meets the counterclockwise-moving fork, it will stall because it cannot move clockwise through *terB*. Similarly, the replication fork moving in the counterclockwise direction will stall at site *terA* and wait for the clockwise-moving replication fork. When the counterclockwise and clockwise replication forks meet, at *terA*, *terB*, or somewhere in between them, the two forks will terminate replication, releasing the two daughter DNAs. This picture is somewhat oversimplified because each replication fork passes through a gauntlet of many *ter* sites, each of which only slows it down, so that it eventually stops. While one replication fork is slowing down at a succession of *ter* sites, the other replication fork has time to make it around the chromosome to meet it at a position more or less opposite the replication origin.

Encountering a *ter* DNA sequence, by itself, is not sufficient to stop the replication fork. Proteins are also required to terminate replication at *ter* sites. These proteins, called terminus utilization substance (Tus) in *E. coli* and replication terminator protein (RTP) in

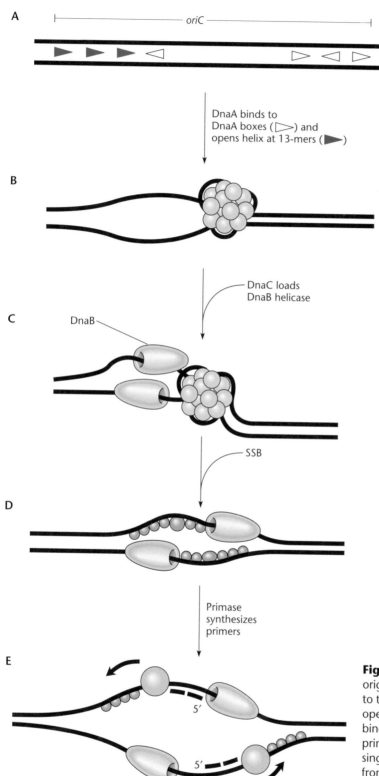

A oriC

DnaA binds to
DnaA boxes (▷) and
opens helix at 13-mers (▶)

B

DnaC loads
DnaB helicase

C DnaB

SSB

D

Primase
synthesizes
primers

E

5′

5′

DnaG primase

Figure 1.17   (A) Initiation of replication at the *E. coli*
origin (*oriC*) region. (B) A number of DnaA proteins bind
to the origin, wrapping the DNA around themselves and
opening the helix. (C) DnaC helps the DnaB helicase to
bind. (D and E) The DnaG primase synthesizes RNA
primers, initiating replication in both directions. SSB,
single-strand-binding protein. Adapted with permission
from T. A. Baker and S. H. Wickner, *Annu. Rev. Genet.*
26:447–478, 1992.

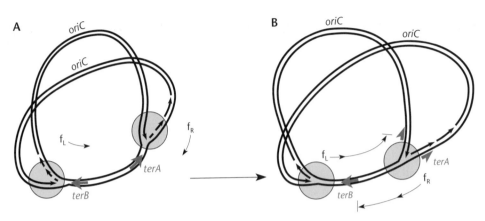

Figure 1.18 Termination of chromosome replication in *E. coli*. (A) The replication forks that start at *oriC* can traverse *terA* and *terB* in only one direction, opposite to that indicated by the arrows. (B) When they meet, between or at one of the two clusters, chromosome replication terminates. f$_L$ is the fork that initiated to the left and moved in the counterclockwise direction. f$_R$ is the fork that initiated to the right and moved in the clockwise direction.

B. subtilis, are thought to bind to the *ter* sites and stop the replicating helicase (DnaB in *E. coli*) that is separating the strands of the DNA ahead of the replication fork. These proteins may also assist in the orderly separation of the two newly synthesized daughter DNAs to prevent any free 3' ends from being used as primers to continue replication (see Bussiere and Bastia, Suggested Reading). However, in spite of these seeming advantages, at least some types of bacteria do not seem to absolutely need their *ter* sites. As evidence, the entire *ter* region of the *B. subtilis* chromosome can be deleted and the bacteria can still multiply, albeit more slowly. Apparently, the chromosome can terminate replication fairly reliably wherever the forks happen to meet. There are probably subtle advantages to terminating chromosome replication in the *ter* region opposite the *oriC* origin of replication, but, as is often the case, these advantages are not apparent in a laboratory situation.

Segregation of the Two Daughter DNAs after Replication

Once the DNA has replicated, the two daughter DNAs must be separated or **segregated**, each into one of the daughter cells, when the cell divides. Segregation could be difficult to achieve if the two daughter DNAs have become joined by recombination during the cell cycle or if they are spread throughout the cell and have become tangled during replication. We shall discuss each of these problems separately.

RESOLUTION OF DIMERIZED CHROMOSOMES
Sometimes the two circular daughter DNAs have become joined in a chromosome **dimer**, in which the two daughter DNAs are joined end to end to form a double-

length circle. Such dimerized chromosomes are fairly common and may arise by recombination between the two daughter DNAs or by replication restarts at DNA damage (see Box 1.2). Such dimers obviously cause problems for chromosome segregation because the two daughter chromosomes cannot be separated if they are part of the same larger molecule!

If dimerized chromosomes can be created by recombination, they can also be resolved into the individual chromosomes by a second recombination. The general recombination system will resolve the dimers by recombination anywhere within the repeated chromosomes. However, the general recombination system can both create and resolve dimers depending on how many crossovers occur between the daughter DNAs. An odd number of crossovers occurring between any two sequences on the two daughter DNAs in the dimer will resolve the dimer, but an even number of crossovers will recreate a dimer.

In *E. coli*, where it is best understood, and probably in other bacteria as well, a very clever mechanism is used to resolve chromosome dimers. Rather than using the general recombination system, the cell uses a **site-specific recombination system** to resolve chromosome dimers. The advantages of this system are that it will resolve dimers into the individual chromosomes but will not create new dimers and that its action is coordinated with division of the cells. The Xer recombination system consists of two proteins called XerC and XerD and a specific site in the chromosome called *dif*. If two copies of the *dif* site occur on the same DNA, such as occurs when the chromosome is dimerized, the Xer proteins will promote a recombination between the two *dif* sites, resolving the dimer into the individual chromosomes. To ensure that there is

BOX 1.2

Restarting Replication Forks

Once replication forks have been created, they proceed around the chromosome, making the two daughter DNAs as they go. In an ideal world, a replication fork, once formed, should be able to replicate the entire chromosome, making two complete copies of the chromosome. However, DNA is not static in the cell and is constantly being altered, either by chemical damage from inside and outside influences or by participation in other cellular functions such as recombination. When a replication fork encounters such an alteration in the DNA, it is blocked and falls off, in much the same way as damage to a railroad track or repairs to the track can block the passage of a train. This is not a trivial problem for the cell because if it cannot finish the round of replication it will die. It has been estimated that *E. coli* replication forks encounter some obstacle in almost every round of chromosome replication.

What happens when the replication fork encounters an alteration in the DNA depends upon the nature of the alteration. If the alteration is chemical damage to a base that prevents proper pairing with the complementary base, the replication fork might stall, leaving a gap opposite the damage (see Figure). Such a gap, in which the DNA is single stranded for a stretch, could be repaired by recombination with the other daughter DNA, as will be described in chapter 11. Alternatively, if the alteration is a break in one of the two strands, the replication fork may proceed past the break, causing both strands in one of the two daughter DNAs to be broken. This type of damage can also be repaired, using the other daughter DNA, by a different recombination pathway called "double strand break repair," which is also outlined in chapters 10 and 11. The problem comes when the replication machinery has to reassemble and reinitiate synthesis after the damage has been repaired. Normally, the replication apparatus can initiate synthesis of DNA only at the *oriC* region, which is somewhere else in the chromosome. Recent evidence suggests that proteins called the primosome proteins PriA, PriB, and PriC, as well as another protein named DnaT, may cooperate with the recombination functions in reinitiating replication at the block. These proteins were first found because they are required to initiate the replication of some phages (see chapter 7). The PriA, PriB, PriC, and DnaT proteins are thought to help reassemble the replication apparatus, using the single-stranded 3′ OH end of DNA that invades the double-stranded DNA to form the "D-loop" recombination intermediate as the primer to initiate new DNA synthesis (see figure).

Until these discoveries, it was generally believed that recombination and replication were separate functions involving DNA. Replication allowed copies of the DNA to be made

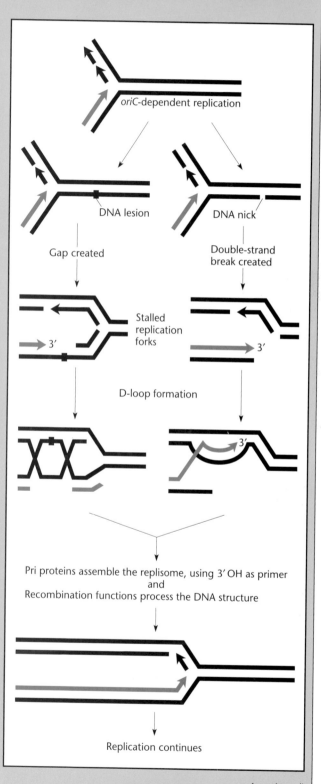

oriC-dependent replication

DNA lesion   DNA nick

Gap created   Double-strand
                              break created

Stalled replication forks

3′   3′

D-loop formation

3′

Pri proteins assemble the replisome, using 3′ OH as primer
and
Recombination functions process the DNA structure

Replication continues

(continued)

BOX 1.2 (continued)

Restarting Replication Forks

to pass on to progeny, while recombination allowed the species to try new combinations of genes, thereby accelerating evolution. Some bacterial viruses (phages) were known to use recombination to initiate replication, but this was thought to be the exception, not the rule. However, it is now apparent that recombination and replication are intimately related. In fact, the major role of recombination, in

bacteria at least, may be to replicate the entire chromosome, by allowing replication to continue past damage and other alterations to the DNA.

Reference
Cox, M. M., M. F. Goodman, K. N. Kreuzer, D. J. Sherratt, S. J. Sandler, and K. J. Marians. 2000. The importance of repairing stalled replication forks. *Nature* **404**:37–41.

normally only one *dif* site in the cell until just before cell division, the *dif* site is located close to the *ter* region, so it will not be replicated until just before the chromosome has completed replication and just before the cell divides (see below). As added insurance, the activity of the Xer site-specific recombination system is also made dependent on the formation of the division septum. Only when a division septum begins to form and a protein called FtsK binds to the division septum is this recombination system activated (see Steiner et al., Suggested Reading). There are many other site-specific recombination systems that play specialized roles in bacterial cells, and we discuss many of them in subsequent chapters.

DECATENATION

Replicating DNAs can also become joined to each other through the formation of **catenenes**, in which the daughter DNAs become interlinked like the links on a chain. These interlinks could form as the natural result of terminating a round of chromosome replication or could be caused by topoisomerases passing the strands of the two DNAs through each other (see "Topoisomerases" below). Once such interlinks are formed, the only way to unlink them is to break both strands of one of the two DNAs and pass the two strands of the other DNA through the break. The break must then be resealed. This double-strand passage, called **decatenation**, is one of the reactions performed by type II topoisomerases (see Figure 1.25). In *E. coli*, a type II topoisomerase called **topo IV** is thought to be responsible for removing most of the interlinks between the daughter DNAs after replication (see Zechiedrich et al., Suggested Reading). Recent evidence suggests that topo IV also removes positive supercoils ahead of the replication fork (see Khodursky et al., Suggested Reading).

CONDENSATION

By itself, just passing the two daughter DNAs through each other would not necessarily have the effect of sepa-

rating the two interlinked DNAs. As many new interlinks would be created as would be removed, since the topo IV enzyme has no way of knowing whether it is removing or creating interlinks when it passes two DNAs through each other. To completely decatenate the two DNAs, it is necessary to simultaneously separate them as their strands are being passed through each other, much like fishing line should be reeled in as it is untangled, to separate the untangled part from the part which is still tangled.

One way to separate the two daughter DNAs from each other is to **condense** them in different parts of the cell as they are untangled. If they are more condensed, they will not overlap as much in the cell and so they will be less apt to become interlinked. Condensation of chromosomes prior to mitosis has been known for a long time in eukaryotic cells, where the chromosomes are clearly visible. Now we know that bacteria also condense their daughter DNAs to make them easier to separate prior to division, even though it is more difficult to visualize the condensation of bacterial chromosomes because of their small size.

One way bacteria condense DNAs is through **supercoiling** (see Figure 1.24). In bacteria, all DNAs are negatively supercoiled, which means that DNA is twisted in the opposite direction to the Watson-Crick helix, creating underwinds. As we discuss in more detail below, the underwinds introduce stress into the DNA, causing it to wrap up on itself, much like a rope will wrap up on itself if the two ends are rotated in opposite directions. This twisting occurs in loops in the DNA, causing the DNA to be condensed into a smaller space.

Proteins called **condensins** also help condense the DNA in the cell, making the daughter DNA molecules easier to separate. Condensins are long molecules which bind DNA on each of their ends, thus forming clamps to hold the DNA together in large loops and condensing the DNA into a smaller space. Because they help condense DNA into chromosomes, condensins are sometimes called structural maintenance of chromosome (SMC) proteins. Condensins have been known for some time to occur in eukaryotes, where the condensation of the large chromosomes is clearly evident, but were only recently discovered in bacteria, where the chromosomes are more difficult to visualize. The condensin of *E. coli*, called MukB, was found because mutations in its gene interfere with chromosome segregation. MukB was suspected of condensing the DNA because this protein and supercoiling of the DNA can compensate for each other in allowing proper segregation of the daughter chromosomes into daughter cells. It was reasoned that they could compensate for each other only if they both did the same thing, i.e., condense the daughter DNAs, making them easier to separate (see Holmes and Cozzarelli, Suggested Reading). *B. subtilis* also has a condensin, which is more similar in amino acid sequence to the eukaryotic condensins and so was also named SMC (see Britton et al., Suggested Reading). Some types of bacteria are now known to have condensins related to the MukB of *E. coli*, and others have condensins related to the SMC of *B. subtilis*. It is not clear why the proteins which perform such an essential role in cells as condensins would be so different in different types of bacteria. Usually, proteins which perform such central cellular functions are closely related to each other, even if they are in distantly related species of bacteria. Perhaps the various functions of chromosome segregation and condensation are divided up differently in different types of bacteria, with MukB and SMC playing different but overlapping roles (see the discussion of Muk proteins in chromosome partitioning, below).

Partitioning of the Chromosome after Replication

Not only must the two daughter chromosomes be segregated after replication, they must also be segregated in such a way that each daughter cell gets one of the two copies of the chromosome. Otherwise, one daughter cell would get two chromosomes and the other would be left with no chromosome and eventually would die. Since chromosomeless cells appear so infrequently, there must be some process that directs one copy of each of the daughter chromosomes to one daughter cell. This apportionment of one daughter chromosome to each of the two daughter cells is called **partitioning**. In spite of extensive study, the process of chromosome partitioning is still not understood very well. However, some of the genes whose products play roles in this process are now being identified, and a lot of relevant information is being accumulated. This is one of the most exciting areas of current research in bacterial cell biology, and a number of surprises have arisen from this work.

THE Par PROTEINS

Early work has concentrated on the functions of the so-called partitioning proteins, the products of the *par* genes. The Par functions were first discovered in plasmids, which are small DNA molecules in bacterial cells that replicate independently of the chromosome (see chapter 4). Because they exist independently of the chromosome, plasmids must also have a system for partitioning, otherwise they would often be lost from cells when the cells divide. The Par systems of plasmids consist of two proteins, usually called ParA and ParB, and the *parS* site on the DNA at which they act. The exact role of the Par proteins is unknown, but the ParB protein is known to bind to the *parS* site on the DNA of the plasmid, which presumably helps direct the plasmids to the daughter cells. Even less is known about the role of ParA, but it has an ATPase activity, which cleaves ATP to ADP. Cleavage of ATP often provides energy for reactions, so this ATPse activity may provide energy for some reaction related to plasmid partitioning.

When genes similar to the *par* genes of plasmids were found in the chromosomes of many types of bacteria, they were presumed to perform a similar role in partitioning the chromosomes during cell division. In some types of bacteria at least, mutants that lack these Par proteins are defective in partitioning their chromosomes. The situation is clearest in *B. subtilis* and *Caulobacter crescentus*, two species of bacteria in which this system has been studied most extensively.

In *B. subtilis*, the proteins analogous to the ParA and ParB proteins of plasmids are called Soj and Spo0J, respectively. There are also a number of *parS*-like sites close to the origin of chromosome replication, and Spo0J has been shown to bind to these sites, as expected if it is analogous to ParB. In addition, plasmids lacking their own Par system but containing the *parS* site of the *B. subtilis* chromosome are more faithfully partitioned into daughter cells, but only if both Spo0J and Soj are present in the cell, suggesting that both proteins help in the partitioning of plasmids. However, while Spo0J is required for normal partitioning of the chromosome, Soj is not. This suggests either that the *soj* gene product does not perform the same role in chromosome partitioning that ParA performs in plasmid partitioning or that some other protein can substitute for Soj in chromosome partitioning.

Attempts have also been made to determine where Soj and Spo0J are in the cell. If these functions are involved in pulling the chromosomes apart, they should be at the ends, or "poles," of the cell prior to cell division. Amazingly, rather than being localized at one pole or the other, the Soj proteins oscillate in unison, first to one end of the cell and then to the other, during the cell cycle (see Box 1.3).

BOX 1.3

The Oscillation of Proteins Involved in Partitioning and Cell Division

Using standard phase-contrast microscopy, it has long been possible to observe the movement of proteins and other cellular components within living eukaryotic cells. Such directed movement of proteins could not be observed within bacterial cells, which are generally much smaller than eukaryotic cells. It was even theorized that directed movement is not necessary within bacterial cells since random diffusion should be rapid enough to allow proteins and other cellular constituents to reach their final destinations in these smaller cells in a timely fashion. Accordingly, bacteria were generally considered to be merely "bags of enzymes" with little cellular organization. With the development of new technology involving the use of genetic fusions to green fluorescent protein (GFP), it is now possible to monitor the movement of proteins within individual bacterial cells in real time during the cell cycle. The position of the protein in the cell can be monitored under the microscope by observing the fluorescence given off by the GFP part of a fusion protein. Using this technology to study proteins involved in cell division in bacteria, it was discovered, to everyone's surprise, that some proteins including the MinC, MinD, and MinE proteins of *E. coli*, involved in selecting the site of formation

of the division septum, oscillate from one end of the cell to the other during the cell cycle, with a periodicity of 10 to 20 s. *B. subtilis* also contains Min proteins analogous to those in *E. coli*, which also play a role in division site selection, but these proteins do not seem to oscillate. However, the Soj protein involved in chromosome partitioning in *B. subtilis* does relocate. The significance of the oscillation of proteins involved in cell division and chromosome partitioning is unknown, but observations like these emphasize the extent of our naivete about even these relatively simple cells.

References

Hu, Z., and J. Lutkenhaus. 1999. Topological regulation of cell division in *E. coli* involves rapid pole to pole oscillation of the division inhibitor MinC under the control of MinD and MinE. *Mol. Microbiol.* **34:**82–90.

Marston, A. L., and J. Errington. 1999. Dynamic movement of the ParA-like Soj protein of *B. subtilis* and its dual role in nucleoid organization and developmental regulation. *Mol. Cell* **4:**673–682.

Quisel, J. D., D. C.-H Lin, and A. D. Grossman. 1999. Control of development by altered localization of a transcription factor in *B. subtilis*. *Mol. Cell* **4:**665–672.

Raskin, D. M., and P. A. deBoer. 1999. Rapid pole-to-pole oscillation of a protein required for directing division to the middle of *Escherichia coli*. *Proc. Natl. Acad. Sci. USA* **96:**4971–4976.

As an aside, the *par* genes of *B. subtilis* were given the names *spo0J* and *soj* because they were first discovered as a result of their role in sporulation (see chapter 14). Mutations in the *spo0J* gene block sporulation at an early state and mutations in the *soj* gene overcome this block and allow sporulation, so the gene was named *soj* for "suppressor of J" (see chapter 3 for the genetic definition of a suppressor). It makes sense that Par functions like Soj and Spo0J should be required for sporulation. The chromosomes must also be partitioned during sporulation since the spore must also get one copy of the chromosome. However, Spo0J and Soj also play another, apparently unrelated, essential role in sporulation, which complicates the issue.

Work with *C. crescentus* has shed yet more light on the role of Par functions in partitioning. There are advantages to using *Caulobacter* species for studying the bacterial cell cycle because they are easier to **synchronize** in the cell cycle. To "synchronize" cells means to establish conditions where all of the cells in a culture are the same "age" or at the same stage in the cell cycle at the same time. This is a big advantage because, generally, cells growing in a culture are all of different ages or

stages in the cell cycle. If the cells in a culture are all going through the cell cycle in lockstep, their ages are identical; therefore, if at any time during this period the cells are fixed for microscopy, the age of any cell we look at will be known. Another advantage is that all the cells can be harvested at any particular time and their contents can be examined biochemically to determine what cellular constituents are made at each time of the cell cycle.

To determine when ParA and ParB are synthesized during the *Caulobacter* cell cycle, synchronized cultures of *Caulobacter* with all the cells at the same stage of the cell cycle were harvested and the amounts of ParA and ParB in the cells at this stage were measured (see Mohl and Gober, Suggested Reading). They found that ParA and ParB are always present in the cells, independent of the stage of the cell cycle. They also determined the position of ParA and ParB in cells at different stages by immunofluorescence analysis of fixed cells. In this technique, specific antibodies fused to a fluorescent dye are used to detect proteins under the microscope. During most of the cell cycle, ParA and ParB are spread out over the entire cell. However, immediately before the cell divides, they concentrate at the cell poles, where they could either help pull the chromosomes apart or hold them during cell division.

THE Muk PROTEINS

It is possible that the Muk proteins of *E. coli* play a more direct role in chromosome partitioning, in addition to their role in helping condense the daughter chromosomes prior to their segregation (see above). Recent evidence indicates that the Muk proteins concentrate at the replication forks rather than in the condensed chromosomes, as would be expected of condensins. Also, the Muk proteins are similar in sequence to motor proteins such as myosin, which make sliding filaments to allow movement within cells. If the Muk proteins do form some sort of sliding filaments, these might help pull the chromosomes apart during cell division. Close relatives of the Muk proteins have not been found in very many species of bacteria other than *E. coli*, but they might exist in other forms. Obviously, much more work on partitioning of chromosomes in bacteria remains to be done, and this is an exciting field.

Where Are the Replication Forks?

The location where chromosome replication occurs in the cell might also give clues to how the chromosomes segregate. Chromosome DNA replication in *E. coli* cells is performed by DNA polymerase III with the help of all of its accessory proteins that make up the replication fork. Therefore, if we knew where DNA polymerase III was in the cell, we would know where replication was occurring. The position of the replicative DNA polymerase in *B. subtilis* (called PolC) during the cell cycle was determined by genetically fusing the DNA polymerase to green fluorescent protein (GFP) and then seeing where the fluorescence was located (see Lemon and Grossman, Suggested Reading, and Box 1.3). Two possibilities suggest themselves. Either the chromosome replicates at the midpoint of the cell and then the two daughter DNAs are quickly pulled apart as soon as a round of replication is completed and just before the cell divides, or the two replication forks move apart continuously as the cell grows and the chromosomes replicate. The above experiments revealed that the first model is correct. The two replication forks stay together at the midpoint of the cell until replication is complete. Meanwhile, the two origins of replication move toward the poles of the cell as soon as they emerge from the replication apparatus.

Cell Division

Much has also been learned about how the bacterial division septum forms. The most important protein in this process is a protein called FtsZ, which forms a ring around the midpoint of the cell. The FtsZ protein then attracts other proteins, which help form the **division septum,** which eventually squeezes the cell apart into the two daughter cells.

THE Min PROTEINS

Less clear is why FtsZ forms division septa only at the midpoint of the cell. In *E. coli*, three proteins called MinC, MinD, and MinE are known to be involved in this site selection. The *min* genes of *E. coli* were found because mutations in these genes can cause the division septa to form in the wrong places, sometimes pinching off smaller cells called minicells. Apparently, in the absence of the Min proteins, division septa can form in places other than in the middle of the cell where they are supposed to. This suggested that the Min proteins would be localized in the ends of the cell, where they could prevent FtsZ from forming a division septum anywhere but the middle of the cell. However, when the localization of the Min proteins in the cell was studied, using GFP fusions to the Min proteins, a very surprising result was in store: the Min proteins oscillate from one pole of the cell to the other during the cell cycle (see Box 1.3). The MinC and MinD proteins oscillate the most, collecting at one end of the cell and then all moving to the other end. MinD may drive the oscillation of MinC, which may in turn inhibit the formation of the FtsZ ring. It has been hypothesized that the purpose of the oscillation may be to ensure that the concentration of the MinC division inhibitor is highest at the poles, where it needs to inhibit FtsZ, and lowest in the middle of the cell, where it is just passing through. The MinE protein forms a ring which oscillates back and forth in the middle of the cell, apparently pulled somehow by the oscillation of MinC and MinD. The significance of these MinE rings is unclear since they do not seem to have anything directly to do with FtsZ ring formation. It is almost as though the Min proteins play the role of "division site policemen," constantly scanning the cell to make sure that FtsZ doesn't loiter and form a septum somewhere it is not supposed to. In fact, they are more like "Keystone Cops" in this role, chasing each other from one pole to the other as the cell goes through its cycle.

The situation becomes even more confusing when one considers the Min proteins of other bacteria. Proteins similar to the MinC and MinD proteins of *E. coli* have also been found in *B. subtilis*, although MinE seems to be lacking in this bacterium. *B. subtilis* also has another protein, named DivIVA, which plays a role in this process. Mutations in the *min* genes of *B. subtilis* also allow division septa to form at the ends, where they are not supposed to. However, the Min proteins of *B. subtilis* do not oscillate but, rather, just gather at the poles.

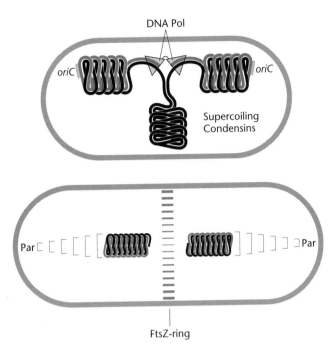

Figure 1.19 Model for replication and partitioning of chromosomes during the cell cycle. (Top) Replication occurs at a factory in the middle of the cell, and the newly replicated DNA moves out toward the poles, where it is condensed by supercoiling and condensins. (Bottom) When replication is complete, the Par functions have moved the newly replicated daughter DNAs to the poles of the cell before a FtsZ ring forms at the middle of the cell and the cell divides. DNA Pol, DNA polymerase III; *oriC*, the origin of replication; Par, Par functions. For details, see the text.

Instead, it is the Par functions of *B. subtilis* that seem to relocate, although with a slower periodicity than the Min proteins of *E. coli* (see above and Box 1.3). *C. crescentus* seems to lack Min proteins altogether, although it is possible that this role could be played by other, unrelated proteins. Figure 1.19 shows an overview of how *E. coli* chromosomes replicate and are segregated in bacteria prior to division. The specific proteins involved may differ in other bacteria.

Coordination of Cell Division with Replication of the Chromosome

It is not sufficient to know how chromosomes replicate and are segregated into the daughter cells prior to division. Something must coordinate the replication of the chromosome with division of the cells. If the cells divided before the replication of the chromosome was completed, there would not be two complete chromosomes to segregate into the daughter cells and one cell would end up without a complete chromosome. The mechanism by which cell division is coordinated with

replication of the DNA is still not understood, but there is a lot of relevant information.

TIMING OF REPLICATION IN THE CELL CYCLE

It is important to know when replication occurs during the cell cycle. Experiments were designed to determine the relationship between the time of chromosome replication and the cell cycle in *E. coli* (see Helmstetter and Cooper, Suggested Reading). The conclusions are still generally accepted, so it is worth going over them in some detail.

These scientists recognized that if the DNA content of cells at different stages in the cell cycle could be measured, it would be possible to determine how far chromosome replication had proceeded at that time in the cell cycle. Since bacterial cells are too small to allow observation of DNA replication in a single cell, it was necessary to measure DNA replication in a large number of cells. However, cells growing in culture are all at different stages in their cell cycles. Therefore, to know how far replication had proceeded at a certain stage in the cell cycle, it was necessary to **synchronize** cells in the population so that all were the same age or point in their life cycle at the same time.

Helmstetter and Cooper accomplished this by using what they called a bacterial "baby machine." Their idea was to first label the DNA of a growing culture of bacterial cells by adding radioactively labeled nucleosides and then fix the bacterial cells on a membrane. When the cells on the filter divided, one of the two daughter cells would no longer be attached and would be released into the medium. All of the daughter cells released at a given time would be newborns and so would be the same age. This means that cells that divided to release the daughter cells at a given time would also be the same age and would have DNA in the same replication state. The amount of radioactivity in the released cells is then a measure of how much of the chromosome had replicated in cells of this age. This experiment was done under different growth conditions to show how the timing of replication and the timing of cell division are coordinated under different growth conditions.

Figure 1.20 shows the results of these experiments. For convenience, the following letters were assigned to each of the intervals during the cell cycle. The letter I denotes the time from when the last round of chromosome replication initiated until a new round begins. The letter C is the time it takes to replicate the entire chromosome, and the letter D is the time from when a round of chromosome replication is completed until cell division occurs. The top of the figure shows the relationship of I, C, and D when the cells are growing very slowly with a generation time of 70 min. Under these conditions, I is 70 min, C is 40 min, and D is 20 min. However, when the cells are growing in

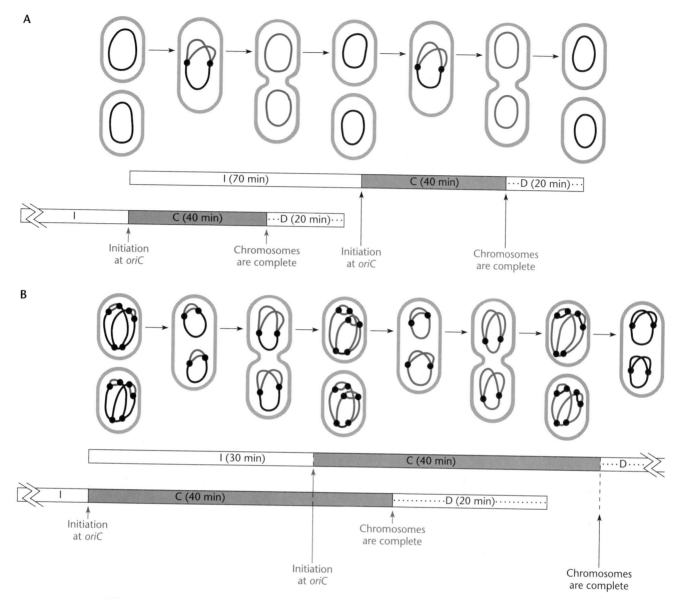

Figure 1.20   The timing of DNA replication during the cell cycle, with two different generation times. Only the time between initiations (I) changes. See the text for definitions of I, C, and D.

a richer medium and they are dividing more rapidly with a generation time of only 30 min, the pattern changes. The C and D intervals remain about the same, but the I interval is much shorter, only about 30 min.

Some conclusions could be drawn from these data. One conclusion is that the C and D intervals remain about the same independent of the growth rate. At 37°C, the time it takes the chromosome to replicate is always about 40 min, and it takes about 20 min from the time a round of replication terminates until the cell divides. However, the I interval gets shorter when the cells are growing faster and have shorter generation times. In fact, the I interval is approximately equal to the genera-

tion time—the time it takes a newborn cell to grow and divide. This makes sense because, as we shall discuss later, initiation of chromosome replication occurs every time the cells reach a certain size. They will reach this size once every generation time, independent of how fast they are growing.

Another point apparent from the data is that in cells growing rapidly with a short generation time, the I interval can be shorter than the C interval. If I is shorter than C, a new round of chromosomal DNA replication will begin before the old one is completed. This explains the higher DNA content of fast-growing cells than of slow-growing cells. It also explains the observation that genes closer to the origin of replication are present in more copies than are genes closer to the replication terminus.

Despite providing these important results, this elegant analysis does not allow us to tell whether division is coupled to initiation or termination of chromosomal DNA replication. The fact that the I interval always equals the generation time suggests that the events leading up to division are set in motion at the time a round of chromosome replication is initiated and are completed 60 min later independent of how fast the cells are growing. However, it is also possible that the act of termination of a round of chromosome replication sets in motion a cell division 20 min later. More experiments are needed to resolve these issues.

Timing of Initiation of Replication

A new round of replication must be initiated each time the cell divides, or the amount of DNA in the cell would increase until the cells were stuffed full of it or decrease until no cell had a complete copy of the chromosome. Clearly, initiation of replication is exquisitely timed. In cells growing very rapidly, in which the next rounds of replication initiate before the last ones are completed, so that they contain a number of origins of replication, all of the origins in the cell "fire" simultaneously, indicating tight control.

A number of attempts have been made to correlate the timing of initiation of chromosome replication with other cellular parameters during the cell cycle. Most such evidence points to initiation of replication being tied to cell mass. After cells divide, their mass or weight continuously increases until they divide again. The initiation of chromosome replication occurs each time the cell achieves a certain mass, the **initiation mass**. If cells are growing faster in richer medium, they are larger and achieve the initiation mass sooner than do smaller, slower-growing cells, explaining why new rounds of chromosome replication occur before the termination of previous rounds in faster-growing cells but not in slower-growing cells. However, these experiments by themselves do not explain what it is about the cell mass that triggers initiation.

ROLE OF DnaA PROTEIN

Most evidence indicates that the timing of initiation of chromosome replication is tied to the intracellular concentration of DnaA protein. This makes sense. The DnaA protein is the first protein to bind to each origin during initiation. Many copies of DnaA bind to each origin, thereby opening the DNA strands and allowing the DnaB helicase protein to bind and the replication fork to form (Figure 1.17). Therefore, initiation of chromosome replication can occur only when there are sufficient copies of DnaA in the cell to allow many copies of DnaA to bind to each origin. However, what matters is not the absolute amount of DnaA in the cell. What matters is the *ratio* of the amount of DnaA protein to the number of origins of replication in the cell. According to this model, as the cell grows, the amount of the DnaA protein in the cell progressively increases but the number of origins of replication stays the same since no new initiations are occurring. Finally, the ratio of DnaA protein to origins reaches the critical number and initiation occurs.

We can add more detail to this model. There are three DnaA binding sequences, "DnaA boxes," in each *oriC* region (Figure 1.17). These sites can be recognized because they have the nucleotide sequence TT(A/T)TNCACA reading in the 5'-to-3' direction, where (A/T) means either A or T and N means any of the four bases. Because their nucleotide sequences are somewhat different, these three sites bind DnaA protein with different affinities (i.e., tightness). Initiation will occur only if all three boxes are occupied by DnaA. Then and only then will additional copies of DnaA "pile on," as shown in Figure 1.17, and initiation will occur.

So let's see what this model predicts will happen as the cells go through the cell cycle. Initially, after a round of chromosome replication has just initiated, the ratio of DnaA proteins to origins of replication is low and at most only one or two of the DnaA boxes is occupied. As DnaA protein accumulates but the number of origins of replication remains the same, the ratio of DnaA protein to origins steadily increases. Finally, there is enough DnaA that all three DnaA boxes are occupied and initiation occurs. Once initiation occurs, there are twice as many origins as previously, so the ratio of DnaA protein to origins drops to half of what it was and the process repeats itself. This model is consistent with evidence that either artificially decreasing the amount of DnaA in the cell or increasing the number of copies of the DnaA boxes will delay initiation of rounds of chromosome replication. There also may be other factors at play. DnaA will bind either ATP or ADP, but only the ATP-bound form will initiate replication. The ATP/ADP ratio is higher in faster-growing cells, which may help ensure that initiation occurs only in growing cells. Also, there is some evidence that the sliding clamp which forms on the DNA once the replication fork forms can inhibit the binding of more DnaA, preventing another initiation immediately.

HEMIMETHYLATION AND SEQUESTRATION

At least some types of bacteria, including *E. coli* and *Caulobacter*, have yet another means of delaying initiation of new rounds of chromosome replication until the cell divides. In these bacteria at least, methylation of the DNA plays a role in delaying initiation. In *E. coli*, in which this is best understood, the methylation is due to the Dam methylase, the same enzyme involved in mismatch repair (see "Methyl-Directed Mismatch Repair" above). The Dam methylase methylates the two A's in the DNA sequence GATC/CTAG. The methylation occurs only after the DNA is synthesized, so that the A in a newly synthesized strand of this sequence will not be immediately methylated. Intriguingly, the sequence GATC/CTAG appears 11 times in only 245 bp in the *oriC* region of the chromosome, much more often than would be expected by chance alone. Furthermore, the promoter region of the *dnaA* gene, the region in which mRNA synthesis initiates for the DnaA protein, also has GATC/CTAG sequences, and no DnaA protein will be synthesized unless these sequences are fully methylated.

Figure 1.21 depicts a model of how the Dam methylase may sequester *oriC* regions after initiation. Immedi-ately after an *oriC* region has been used to initiate replication, the GATC/CTAG sequences in the *oriC* region will be hemimethylated; only the A in the old strand of the sequence will be fully methylated. According to the model, the hemimethylated *oriC* region is sequestered by binding to the membrane, a process that renders it nonfunctional for the initiation of new rounds of replication and prevents it from being further methylated by the Dam methylase. In addition, a protein called SeqA (sequestration protein A) may be involved in binding hemimethylated *oriC* regions to the membrane.

There is some evidence to support this role of methylation in *oriC* sequestration and regulation of DnaA protein synthesis after initiation (see Crooke, Suggested Reading). For instance, the GATC/CTAG sequences in the *oriC* region and in the promoter region for the *dnaA* gene remain hemimethylated much longer after replication than GATC/CTAG sequences elsewhere in the chromosome. Also, increasing the amount of Dam methylase in the cell causes premature initiation of replication, as might be expected if higher than normal levels of Dam methylase fully methylate the GATC/CTAG sequences, which would unsequester *oriC* sooner than the hemimethylated sequences would allow. Finally, hemimethylated *oriC* regions bind more readily to membranes than do fully methylated *oriC* regions in the presence of SeqA (see Slater et al., Suggested Reading). Many more details of the regulation of chromosome replication initiation remain to be uncovered.

Figure 1.21 Model of the sequestration of the *oriC* region of the *E. coli* chromosome after initiation of chromosome replication. Before initiation, the *oriC* region is methylated in both strands. After initiation, only one of the two strands is methylated; hence, the region is hemimethylated. A protein called SeqA helps bind the hemimethylated *oriC* to the membrane, thereby sequestering it and preventing further initiation and methylation. The newly synthesized strand is shown in gold, and the methylated bases are starred. Only 1 of the 11 GATC/CTAG sequences in the *oriC* region is shown.

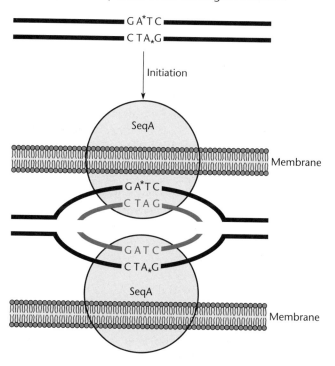

Supercoiling

The Bacterial Nucleoid

As mentioned at the beginning of this chapter, the DNA of even a simple bacterium is approximately 1 mm long, while bacteria themselves measure only micrometers in length. Therefore, the DNA is about 1,000 times longer than the bacterium itself and must be condensed to fit in the cell, but it also must be folded in such a way that it is available for transcription, recombination, and other functions.

Figure 1.22 shows a picture of a thin section of an *E. coli* cell. The chromosome is not spread all over the cell but is condensed in only one part. As the chromosome replicates, this condensed mass becomes larger until it finally separates into two masses of DNA, just before cell division.

Condensed bacterial DNA isolated from bacteria is shown in Figure 1.23. This condensed structure is called the bacterial **nucleoid**. The nucleoid is composed of 30 to 50 loops of DNA emerging from a more condensed region, or **core.** Seeing this tangle of loops, it is difficult

a

0,5 μm

b

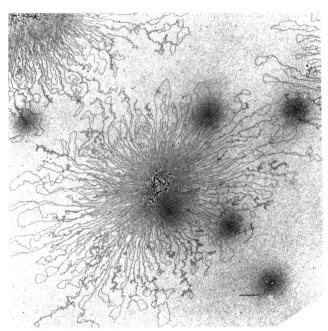

Figure 1.22 Thin section of *E. coli* showing condensed DNA. From F. C. Neidhardt, R. Curtiss III, J. L. Ingraham, E. C. C. Lin, K. B. Low, B. Magasanik, W. S. Reznikoff, M. Riley, M. Schaechter, and H. E. Umbarger (ed.), Escherichia coli *and* Salmonella: *Cellular and Molecular Biology*, 2nd ed., ASM Press, Washington, D.C., 1996.

to imagine that the DNA in this complicated structure is actually one long, continuous circular molecule.

Supercoiling in the Nucleoid

One of the most noticeable features of the nucleoid is that most of the DNA loops are twisted up on themselves. This twisting is the result of **supercoiling** of the DNA, as discussed previously.

Figure 1.24 illustrates supercoiling. In this example, the ends of a DNA molecule have been rotated in opposite directions and the DNA has become twisted up on itself to relieve the stress. The DNA will remain supercoiled as long as its ends are constrained and so cannot rotate, and a circular DNA has no free ends that can rotate. Therefore, a supercoiled circular DNA will remain supercoiled, but a linear DNA will immediately lose its supercoiling unless the ends are somehow constrained.

Even a circular DNA will lose its supercoiling if we cut one of the strands of the DNA, thereby allowing the strands to rotate around each other. The phosphodiester bond connecting the two deoxyribose sugars on the other strand will serve as a swivel and rotate, resulting in **relaxed** (i.e., not supercoiled) DNA. A DNA with a phosphodiester bond broken in one of the two strands is said to be **nicked**.

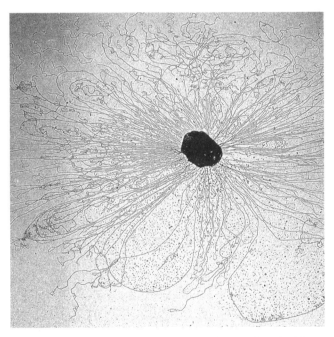

Figure 1.23 Electron micrographs of bacterial nucleoids. From F. C. Neidhardt, R. Curtiss III, J. L. Ingraham, E. C. C. Lin, K. B. Low, B. Magasanik, W. S. Reznikoff, M. Riley, M. Schaechter, and H. E. Umbarger (ed.), Escherichia coli *and* Salmonella: *Cellular and Molecular Biology*, 2nd ed., ASM Press, Washington, D.C., 1996.

When nucleoids are prepared, some of the loops are usually relaxed, probably by nicks introduced during the extraction process. The fact that only some, and not all, of the loops of DNA in the nucleoid are relaxed tells us

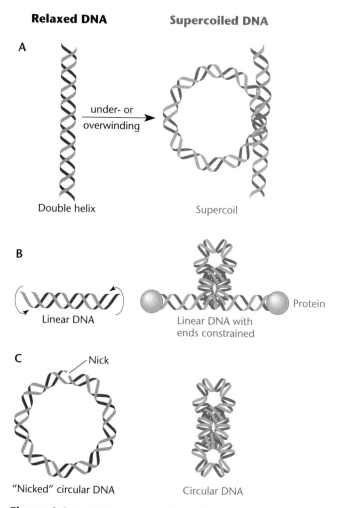

Figure 1.24 (A) Supercoiled DNA. (B) Twisting of the ends in opposite directions causes linear DNA to wrap up on itself. The supercoiling will be lost if the ends of the DNA are not somehow constrained. (C) A break, or nick, in one of the two strands of a circular DNA will relax the supercoils.

something about the structure of the nucleoid. A break or nick in a circular DNA should relax the whole DNA unless portions of the molecule are periodically attached to something that prevents rotation of the strands, such as condensins.

SUPERCOILING OF NATURAL DNAs
It is possible to estimate the supercoiling of natural DNAs. According to the Watson-Crick structure, the two strands are wrapped around each other about once every 10.5 bp to form the double helix. Therefore, in a DNA of 2,100 bp, the two strands should be wrapped around each other about 2,100/10.5 = 200 times. In a supercoiled DNA of this size, however, the two strands will be wrapped around each other either more or less than

200 times. If they are wrapped around each other more than once every 10.5 bp, the DNA is said to be **positively supercoiled;** if less than once every 10.5 bp, **negatively supercoiled.**

Most DNA in bacteria is negatively supercoiled, with an average of one negative supercoil for every 300 bp, although there are localized regions of higher or lower negative supercoiling. Also, in some regions, such as ahead of a transcribing RNA polymerase, the DNA may be positively supercoiled.

SUPERCOILING STRESS
Some of the stress due to supercoiling of the DNA, which causes it to twist up on itself, can be relieved if the DNA is wrapped around something else such as proteins at the same time. Sailors know about this effect. If you twist a rope around something as you roll it up, it will not try to unroll itself again when you are finished. Wrapping DNA around proteins in the cell is called constraining the supercoils. Unconstrained supercoils will cause stress in the DNA, which can be relieved by twisting the DNA up on itself, as shown in Figure 1.24, and making the DNA more compact. The stress due to unconstrained supercoils can have other effects as well, for example, helping to separate the strands of DNA during reactions such as replication, recombination, and initiation of RNA synthesis at promoters.

Topoisomerases

The supercoiling of DNA in the cell is modulated by enzymes called **topoisomerases** (see Wang, Suggested Reading) All organisms have these proteins, which manage to remove the supercoils from a circular DNA without permanently breaking either of the two strands. They perform this feat by binding to DNA, breaking one or both of the strands, and passing the DNA strands through the break before resealing it. As long as the enzyme holds the cut ends of the DNA so that they do not rotate, this process, known as **strand passage,** will either introduce or remove supercoils from DNA.

The topoisomerases are classified into two groups, type I and type II (Figure 1.25). These two types differ in how many strands are cut and how many strands pass through the cut. The type I topoisomerases cut one strand and pass the other strand through the break before resealing the cut. The type II topoisomerases cut both strands and pass two other strands from somewhere else in the DNA or even another DNA through the break before resealing it. Hence, type I topoisomerases change DNA one supercoil at a time, whereas type II topoisomerases change DNA two supercoils at a time, as shown in Figure 1.25.

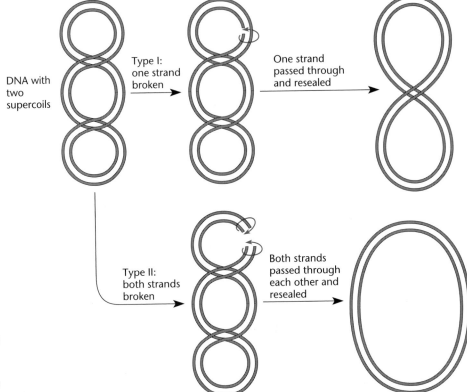

Figure 1.25 Action of the two types of topoisomerases. The type I topoisomerases break one strand of DNA and pass the other strand through the break, removing one supercoil at a time. The type II topoisomerases break both strands and pass another part of the same DNA through the breaks, introducing or removing two supercoils at a time.

TYPE I TOPOISOMERASES
Bacteria have several type I topoisomerases. The major bacterial type I topoisomerase removes negative supercoils from DNA. In *E. coli* and *Salmonella enterica serovar Typhimurium*, the *topA* gene encodes this type I topoisomerase. As expected, DNA isolated from *E. coli* with a *topA* mutation is more highly negatively supercoiled than normal.

TYPE II TOPOISOMERASES
Bacteria also have more than one type II topoisomerase. Because type II topoisomerases can break both strands and pass two other DNA strands through the break, they can either separate two linked circular DNA molecules or link them up. Linkage sometimes happens after replica-

tion or recombination. The major type II topoisomerase in bacteria is called **gyrase** instead of topoisomerase II because rather than removing negative supercoils like most type II topoisomerases, this enzyme adds them. Gyrase acts by first wrapping the DNA around itself and then cutting the two strands before passing another part of the DNA through the cuts, thereby introducing two negative supercoils. Adding negative supercoils increases the stress in the DNA and so requires energy; hence, gyrase needs ATP for this reaction.

The gyrase of *E. coli* is made up of four polypeptides, two of which are encoded by the *gyrA* gene, whereas the other two are encoded by *gyrB*. These genes were first identified by mutations that make the cell resistant to antibiotics that affect gyrase (see Table 1.2). The GyrA

TABLE 1.2	Antibiotics that block replication	
Antibiotic	**Source**	**Target**
Trimethoprim	Chemically synthesized	Dihydrofolate reductase
Hydroxyurea	Chemically synthesized	Ribonucleotide reductase
5-Fluorodeoxyuridine	Chemically synthesized	Thymidylate synthetase
Nalidixic acid	Chemically synthesized	*gyrA* subunit of gyrase
Novobiocin	*Streptomyces sphaeroides*	*gyrB* subunit of gyrase
Mitomycin C	*Streptomyces caespitosus*	Cross-links DNA

subunits seem to be responsible for breaking the DNA and holding it as the strands pass through the cuts. The GyrB subunits have the ATP site that furnishes the energy for the supercoiling.

As mentioned above, the other major type II topoisomerase in *E. coli*, Topo IV, decatenates daughter chromosomes after infection, allowing them to be segregated into the daughter cells.

Antibiotics That Affect Replication and DNA Structure

Antibiotics are substances that block the growth of cells. Many antibiotics are naturally synthesized chemical compounds made by soil microorganisms, especially actinomycetes, to help them compete with other soil microorganisms. Consequently, this group of compounds has a broad spectrum of activity and target specificity. Antibiotics have proven useful in understanding cellular functions as well as in disease therapy. Many stop the growth of bacteria by specifically blocking DNA replication or by changing the molecule's structure. Table 1.2 lists a few representative antibiotics that affect DNA, along with their targets in the cell and their sources. Because some parts of the replication machinery have remained relatively unchanged throughout evolution, many of these antibiotics work against essentially all types of bacteria. Some even work against eukaryotic cells and so are used as antifungal agents and in tumor chemotherapy.

Antibiotics That Block Precursor Synthesis

As discussed above, DNA is polymerized from the deoxynucleoside triphosphates. Any antibiotic that blocks the synthesis of these nucleotide precursors will block DNA replication.

INHIBITION OF DIHYDROFOLATE REDUCTASE

Some of the most important precursor synthesis blockers are antibiotics that inhibit the enzyme dihydrofolate reductase. One such compound, trimethoprim, works very effectively in bacteria, and the antitumor drug methotrexate (amethopterin) inhibits the dihydrofolate reductase of eukaryotes. Methotrexate is used as an antitumor agent.

Antibiotics, like trimethoprim, that inhibit dihydrofolate reductase kill the cell by depleting it of tetrahydrofolate, which is needed for many biosynthetic reactions. This inhibition is overcome, however, if the cell lacks the enzyme thymidylate synthetase, which synthesizes dTMP; therefore, most mutants that are resistant to trimethoprim have mutations that inactivate the *thyA* thymidylate synthesase gene. The reason is apparent from the pathway for dTMP synthesis shown in Figure 1.5. Thymidylate synthetase is solely responsible for converting tetrahydrofolate to dihydrofolate when it transfers a methyl group from tetrahydrofolate to dUMP to make dTMP. The dihydrofolate reductase is the only enzyme in the cell that can restore the terahydrofolate needed for other biosynthetic reactions. However, if the cell lacks thymidylate synthetase, there is no need for a dihydrofolate reductase to restore tetrahydrofolate. Therefore, inhibition of the dihydrofolate reductase by trimethoprim will have no effect, thus making *thyA* mutant cells resistant to the antibiotic. Of course, if the cell lacks a thymidylate synthetase, it cannot make its own dTMP and must be provided with thymidine in the medium so that it can replicate its DNA.

There is more than one mechanism by which cells can achieve trimethoprim resistance. They can have an altered dihydrofolate reductase to which trimethoprim cannot bind or can have more copies of the gene so that they make more enzyme. Some plasmids and transposons carry genes for resistance to trimethoprim. These genes encode dihydrofolate reductases that are much less sensitive to trimethoprim and so can act even in the presence of large concentrations of the antibiotic.

INHIBITION OF RIBONUCLEOTIDE REDUCTASE

The antibiotic hydroxyurea inhibits the enzyme ribonucleotide reductase, which is required for the synthesis of all four precursors of DNA synthesis (Figure 1.5). The ribonucleotide reductase catalyzes the synthesis of the deoxynucleoside diphosphates dCDP, dGDP, dADP, and dUDP from the ribonucleoside diphosphates, an essential step in deoxynucleoside triphosphate synthesis. Mutants resistant to hydroxyurea have an altered ribonucleotide reductase.

COMPETITION WITH dUMP

5-Fluorodeoxyuridine and the related 5-fluorouracil have monophosphate forms resembling dUMP, the substrate for the thymidylate synthetase. By competing with the natural substrate for this enzyme, they inhibit the synthesis of deoxythymidine monophosphate. Mutants resistant to these compounds have an altered thymidylate synthetase. These are useful antibiotics for the treatment of fungal as well as bacterial infections.

Antibiotics That Block Polymerization of Nucleotides

The polymerization of deoxynucleotide precursors into DNA would also seem to be a tempting target for antibiotics. However, there seem to be surprisingly few antibiotics that directly block this process. Most antibiotics that block polymerization do so indirectly, by binding to DNA or by mimicking the deoxynucleotides and causing chain termination, rather than by inhibiting the DNA polymerase itself.

DEOXYNUCLEOTIDE PRECURSOR MIMICS

Dideoxynucleotides are similar to the normal deoxynucleotide precursors except that they lack a hydroxyl group on the 3′ carbon of the deoxynucleotide. Consequently, they can be incorporated into DNA, but then replication stops because they cannot link up with the next deoxynucleotide. These compounds are not useful antibacterial agents, probably because they are not phosphorylated well in bacterial cells. However, this property of prematurely terminating replication has made them useful for DNA sequencing by the so-called "dideoxy" method (see below).

CROSS-LINKING

Mitomycin C blocks DNA synthesis by cross-linking the guanine bases in DNA to each other. Sometimes the cross-linked bases are in opposing strands. If the two strands are attached to each other, they cannot be separated during replication. Even one cross-link in DNA that is not repaired will prevent replication of the chromosome. This antibiotic is also a useful antitumor drug, probably for the same reason.

Antibiotics That Affect DNA Structure

ACRIDINE DYES

The acridine dyes include proflavin, ethidium, and chloroquine. These compounds insert between the bases of DNA and thereby cause frameshift mutations and inhibit DNA synthesis, particularly in the kinetoplasts of trypanosomes and the mitochondria of eukaryotic cells.

Their ability to insert themselves between the bases in DNA has made acridine dyes very useful in genetics and molecular biology. We discuss some of these applications in later chapters. Members of this large family of antibiotics have long been used as antimalarial drugs.

THYMIDINE MIMIC

5-Bromodeoxyuridine (BUdR) is similar to thymidine and is efficiently phosphorylated and incorporated in its place. However, BUdR incorporated into DNA often mispairs and increases replication errors. DNA containing BUdR is also more sensitive to some wavelengths of ultraviolet (UV) light (which makes BUdR useful in enrichment schemes for isolating mutants; see chapter 3). Moreover, DNA containing BUdR has a different density from DNA containing exclusively thymidine (another feature of BUdR that is useful in experiments).

Antibiotics That Affect Gyrase

The gyrase in bacteria is a target for many different antibiotics. These antibiotics will kill the bacterial cell because gyrase is required for bacterial growth. Because this enzyme is similar among all bacteria, these antibiotics have a broad spectrum of activity and will kill many types of bacteria.

GyrA INHIBITION

Nalidixic acid specifically binds to the GyrA subunit, which is involved in cutting the DNA and in strand passage. This activity makes nalidixic acid and its many derivatives, including oxolinic acid and chloromycetin, very useful antibiotics. Another antibiotic that binds to the GyrA subunit, ciprofloxacin (Cipro) is used for treating gonorrhea and anthrax.

The mechanism of killing by these antibiotics is not completely understood. They are known to cause degradation of the DNA and, after mild-detergent treatment, can cause the DNA to become covalently linked to gyrase, presumably trapping it in an intermediate state in the process of strand passage. Bacteria resistant to nalidixic acid have an altered *gyrA* gene.

GyrB INHIBITION

Novobiocin and its more potent relative coumermycin bind to the GyrB subunit, which is involved in ATP binding. These antibiotics do not resemble ATP, but by binding to the gyrase, they somehow prevent ATP cleavage, perhaps by changing the conformation of the enzyme. Mutants resistant to novobiocin have an altered *gyrB* gene.

Molecular Biology Manipulations with DNA

In addition to their basic science interest, the meticulous studies of the mechanism of DNA replication in bacteria and the enzymes involved in DNA replication discussed above have led to many practical applications in molecular biology. These applications have had profound effects on many aspects of our everyday lives including medicine, agriculture, and even law enforcement. We review some of these applications in this section.

Restriction Endonucleases

Among the most useful enzymes that alter DNA are the restriction endonucleases. These are enzymes which recognize specific sequences in DNA and cut the DNA in or close to the recognition sequence. They are usually accompanied by methylating activities that modify DNA by methylating the DNA in the recognition sequence, making it immune to cutting by the endonuclease activity. These enzymes are made exclusively by bacteria, and their role seems to be to defend against phages by cutting incoming unmodified phage DNA. Some of them may also play a role in preventing the loss of plasmids by killing the cell if it is cured of the plasmid (see "Plasmid Addiction" in Box 4.3).

The restriction endonucleases have been divided into three groups, types I, II, and III. These enzymes differ mostly in the relationship between their methylating and cutting activities. Of these three types, the type II enzymes have proven to be most useful because the methylating activity can be separated from the cutting activity. Hundreds of type II enzymes are known, and many of them can be purchased from biochemical supply companies. What makes them so useful is that they each recognize their own specific sequence in DNA and then cut the DNA at or close to the recognition sequence. These recognition sequences are often either 4, 6, or 8 base pairs long. The sequences recognized by some restriction endonucleases are shown in Table 1.3.

USING RESTRICTION ENDONUCLEASES TO CREATE RECOMBINANT DNA

One of the properties of some type II restriction endonucleases that make them so useful is that the sequences they recognize read the same in the 5′-to-3′ direction on both strands. Such a sequence is said to have a twofold rotational symmetry because it reads the same if you rotate it through 180° and read the other strand. Because the sequence reads the same on both strands, the restriction endonuclease binds to the identical sequence on both strands and then cuts the two strands at the same place in the sequence. For example, the restriction endonuclease *Hin*dIII (so called because it was the third restriction endonuclease found in *Haemophilus influenzae*) recognizes the 6-base-pair sequence 5′AAGCTT3′/3′TTCCAA5′ and cuts between the two A's on each strand (Table 1.3), which you can see is in the same place in the sequences of the two strands read in the 5′-to-3′ direction. Such a break is called a "staggered break" because the breaks in the two strands are not exactly opposite each other in the DNA, which has the effect of leaving short single-stranded ends on both ends of the broken DNA. Because the original sequence which was cut had a twofold rotational symmetry, both of these single-stranded ends will have the same sequence read from 5′ to 3′, in this case 5′AGCT3′. The two single-stranded ends are complementary to each other and so can pair with each other. More importantly, each can pair with the single-stranded end of any other DNA that had been cut with the same restriction endonuclease, because they will all have the single-stranded ends with the same sequence. These single-stranded ends are called **sticky ends** because they can pair with (or "stick" to) any other single-stranded ends with the complementary sequence. Other restriction nucleases might leave the same sticky ends even if they recognize a somewhat different sequence. Such restriction endoucnucleases are said to be **compatible**. When two sticky ends pair with each other, they leave a double-stranded DNA with staggered nicks in the two strands, which can then be sealed by DNA ligase, as illustrated in Figure 1.26. The new DNA which has been created this way is called **recombinant DNA** because two DNAs have been recombined into new sequence combinations. While other methods can be used to make recombinant DNA, this method involving restriction endonucleases has been and continues to be the most generally applicable.

CLONING AND CLONING VECTORS

There is only one molecule of a recombinant DNA when it first forms. In order for it to be useful, we need lots of copies of the recombinant DNA molecule. This is the function of **cloning vectors**. A cloning vector is a DNA which has its own origin of replication and is capable of independent replication in the cell. DNAs which have an *ori* sequence that makes them capable of independent replication in the cell are called **replicons**. The process of cloning a piece of DNA into a circular plasmid with its own *ori* sequence is illustrated in Figure 1.27. Once it has been joined to another piece of DNA and introduced into a cell, the cloning vector will replicate itself, along with the piece of DNA to which it is joined, making many exact copies of the original DNA molecule. These exact replicas of the piece of DNA are called **DNA clones** in analogy to the genetic replicas of an organism that are made when an organism replicates itself asexually. Plasmids and phages, in particular, are capable of independent replication in their bacterial hosts, and some of these have been modified to serve as convenient cloning vectors. We discuss some examples of cloning vectors and their relative advantages in subsequent chapters.

DNA Libraries

A **DNA library** is a collection of DNA clones that includes all, or at least almost all, the DNA sequences of the organism. One way to make a DNA library is with restriction endonucleases. The entire DNA of the organism is cut with a restriction endonuclease, and the pieces

TABLE 1.3	Recognition sequences of restriction endonucleases
Enzyme	**Recognition sequence**[a]
*Sau*3A	*GATC/CTAG*
*Bam*HI	G*GATCC/CCTAG*G
*Eco*RI	G*AATTC/CTTAA*G
*Pst*I	CTGCA*G/G*ACGTC
*Hin*dIII	A*AGCTT/TTCGA*A
*Sma*I	CCC*GGG/GGG*CCC
*Not*I	GC*GGCCGC/CGCCGG*CG

[a]Asterisks indicate where the endonucleases cut.

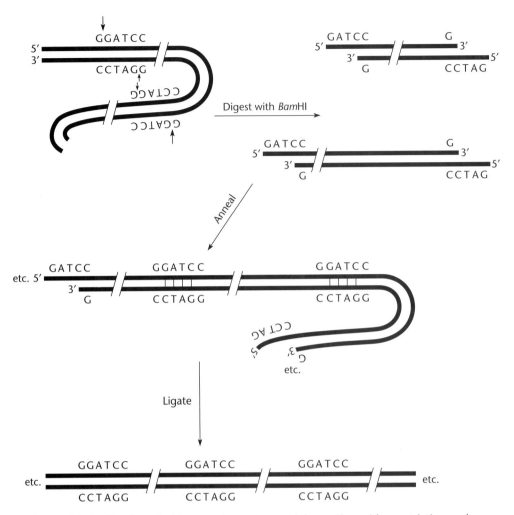

Figure 1.26 Creation of sticky complementary ends by cutting with a restriction endonuclease. The two single-stranded ends can pair with each other, and the nicks can be sealed by DNA ligase.

are ligated into a cloning vector cut with a compatible enzyme. The mixture is then transformed or transfected into cells, and the transformants or plaques are pooled. If the collection is large enough, every DNA sequence of the organism will be represented somewhere in the pooled clones, and the library is complete. The trick then is to find the clone you want out of all of the clones in the library, and we shall touch on some methods in this and subsequent chapters.

The number of clones required to make a complete DNA library of an organism depends on the complexity of the DNA of that organism. For example, if we make a library of *E. coli* DNA cut with an enzyme that recognizes six base pairs (a six-hitter) like *Eco*RI, we will need about $4.5 \times 10^6 / 4 \times 10^3 \approx 1,100$ different clones, since *E. coli* DNA contains approximately 4.5×10^6 bp and a six-hitter like *Eco*RI cuts the DNA about once every

4,000 bp. In contrast, a library of λ DNA should require only about $5 \times 10^4 / 4 \times 10^3 \approx 13$ clones, since λ DNA has only about 50,000 bp. An important point is that these are minimum estimates of the number of clones required to make a library of the DNA of the organism; not all clones will be equally represented in the library because of random statistical fluctuation. Also, some pieces may be easier to clone, for example, because they are smaller or contain no genes whose products are toxic to the cell.

PHYSICAL MAPPING

Another important use of restriction endonucleases is in **physical mapping**. In analogy to genetic mapping, where the approximate position of mutations is determined by genetic crosses (see the discussion of genetic analysis in chapter 3), physical mapping is the process of determining the exact position of particular sequences in the nu-

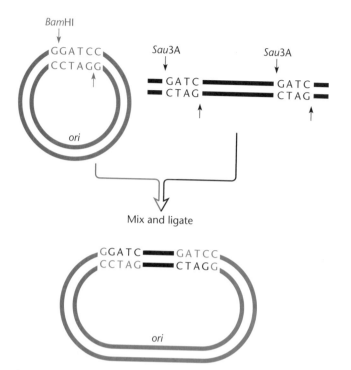

Figure 1.27 DNA cloning. The compatible restriction endonucleases *Sau*3A and *Bam*HI were used to clone a piece of DNA into a cloning vector. The DNA to be cloned was cut with *Sau*3A and ligated into a cloning vector cut with *Bam*HI. The piece of DNA inserted into the cloning vector cannot replicate by itself, since it lacks an *ori* region; however, once it is inserted into the cloning vector, it will replicate each time the cloning vector replicates.

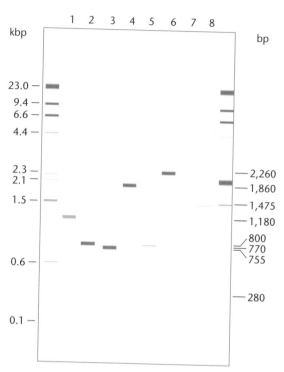

Figure 1.28 Agarose gel electrophoresis of fragments of DNA. Smaller fragments will migrate faster on the gel so will move farther in the same amount of time. The outside lanes are marker DNAs of known size for comparison. Adapted from E. L. Rosey, M. J. Kennedy, D. K. Petrella, R. G. Ulrich, and R. J. Yancey, Jr., *J. Bacteriol.* **177**:5959–5970, 1995.

cleotide sequence of the DNA. Because restriction endonucleases cut the DNA only at specific sequences, they create unique-sized pieces for each DNA molecule. From the size of these pieces, it is possible to determine where the recognition sequences for the restriction endonuclease must have been on the original DNA. By comparing the sizes of the pieces left by a number of different restriction endonucleases, it is possible to order the restriction sites with respect to each other and construct a physical map of the DNA. Figures 1.28 and 1.29 illustrate the reasoning behind the physical mapping of restriction sites in a DNA.

RESTRICTION SITE POLYMORPHISMS

While all the members of a particular species are very similar genetically, there are minor differences called polymorphisms between individuals. These minor genetic differences are reflected in differences in the sequence of DNA from each organism, in particular in differences in the location of restriction sites. These differences are called **restriction fragment length polymorphisms**

(RFLPs) and can be due to deletions or insertions between the positions of the sites or inversions of DNA containing the sites. RFLPs in the DNA can be useful for determining the ancestry of a particular individual, mapping genetic diseases, and, in forensic science, identifying the person who committed a crime and left behind some blood or other material containing their DNA.

Hybridizations

Many applications in molecular genetics have come from our knowledge of the structure of DNA and how it is synthesized and held together. The two strands of DNA are held together in the double helix by hydrogen bonds between the complementary bases. Heating double-stranded DNA or treating it at high pH will disrupt these hydrogen bonds and cause the two strands to separate. If the temperature is then lowered or the pH is returned to neutral, the complementary sequences will eventually find each other and a new double helix will form. Two strands of RNA or a strand of DNA and a strand of RNA can also be held together by such a double-stranded helix, provided that their sequences are complementary. This process is called **hybridization**. Under

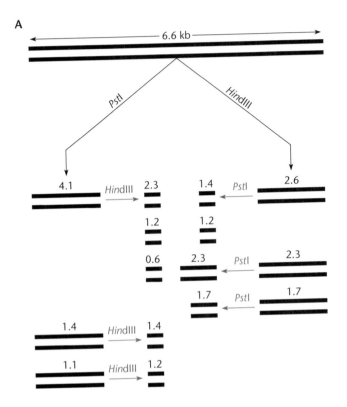

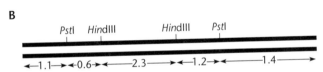

Figure 1.29 A method for mapping the restriction sites on a DNA fragment. The original fragment of 6.6 kb (6.6×10^3 bp) contains two recognition sites for the restriction endonuclease HindIII and two sites for PstI. (A) The fragment is digested with HindIII and PstI separately, and the isolated fragments are digested with the other restriction endonuclease (in gold). (B) From the size of the fragments, the order of the sites must have been as shown.

tion can be detected provided that the second DNA or RNA has been labeled somehow, for example with radioactivity or fluorescent chemicals. Similar techniques can be used to detect the binding of specific antibodies to proteins fixed on the filter or even the binding of proteins to DNA fixed on a filter and vice versa.

Some of the most useful techniques in molecular biology involve filter hybridization (Figure 1.30), in which the membrane filter receives a replica of the molecules on a gel or of the colonies or plaques on a petri plate: the filter

Figure 1.30 Hybridization methods. (A) Method of Southern blot hybridization. In step 1, DNA is isolated and digested with a restriction endonuclease. In steps 2 and 3, after electrophoresis (step 2), the DNA is transferred and fixed to a filter (step 3). In step 4, the filter is hybridized with a probe. Only bands complementary to the probe appear as dark bands because the signal detection procedure reveals the radioactivity or reactive chemical in the probe.

A Step 1   DNA isolation and digestion

Step 2   DNA electrophoresis

Step 3   DNA transfer

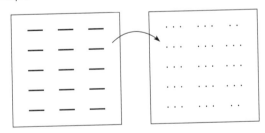

Step 4   DNA hybridrization

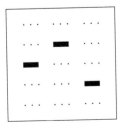

optimal conditions, two strands of DNA or RNA will form a double-stranded helix only if their sequences are almost perfectly complementary, making hybridization very specific and sensitive and allowing the detection of RNA or DNA of a particular sequence among thousands of other sequences.

BLOTS AND PLATE HYBRIDIZATIONS

Most methods of hybridization utilize a membrane filter made of nitrocellulose or some other related compound. First a single-stranded DNA or RNA is fixed to the membrane. Then a solution containing another DNA or RNA is added to the membrane. If the second RNA or DNA hybridizes to the RNA or DNA fixed to the membrane, it too will become fixed to the membrane. The hybridiza-

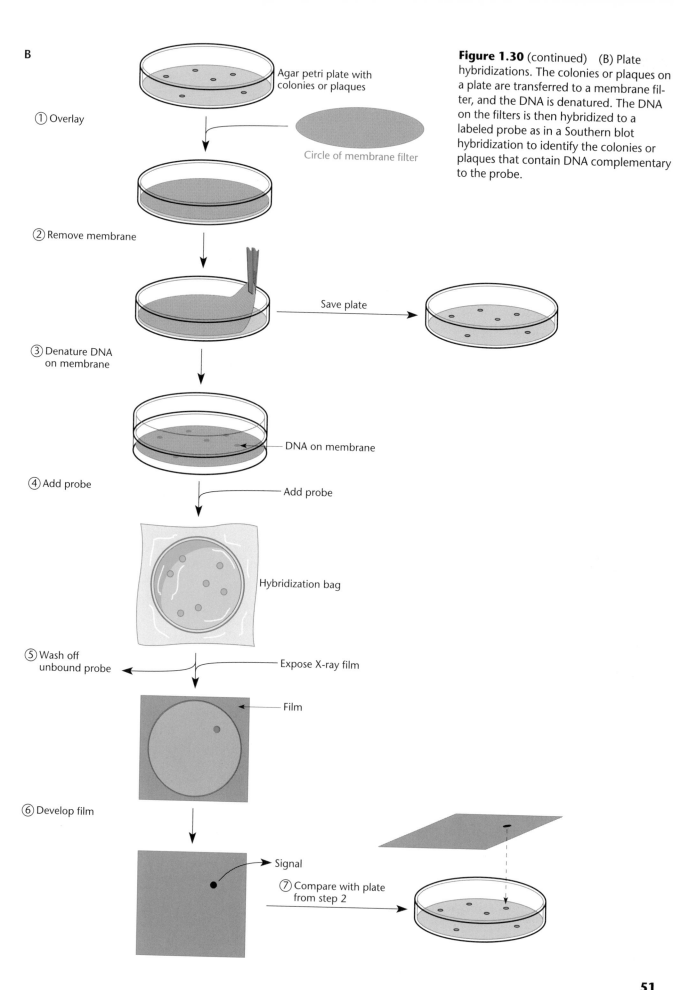

B

① Overlay

Agar petri plate with colonies or plaques

Circle of membrane filter

② Remove membrane

③ Denature DNA on membrane

Save plate

④ Add probe

DNA on membrane

Add probe

Hybridization bag

⑤ Wash off unbound probe

Expose X-ray film

⑥ Develop film

Film

Signal

⑦ Compare with plate from step 2

Figure 1.30 (continued) (B) Plate hybridizations. The colonies or plaques on a plate are transferred to a membrane filter, and the DNA is denatured. The DNA on the filters is then hybridized to a labeled probe as in a Southern blot hybridization to identify the colonies or plaques that contain DNA complementary to the probe.

is layered on the gel or plate, and the macromolecules (DNA, RNA, or proteins) are transferred to their same position on the filter as they were on the gel or petri plate. The transfer can be by diffusion, capillary action, or application of an electric field, depending upon the application. Transfer of DNA, RNA, or proteins from a gel to a filter is called **blotting,** and the filter containing the replica is a **blot.** The blot can then be hybridized to a labeled probe to determine the location of particular DNA or RNA sequences or proteins on the original gel or plate.

The first such procedure for transferring DNA bands from a gel to a filter was named a **Southern blot** because it was developed by Ed Southern. When similar procedures were developed for blotting RNA and proteins, they were whimsically given the names of other directions on the compass: **Northern blots** for RNA blots, **Western blots** for protein blots, and so on.

The principle behind a Southern blot and an example of such a blot are shown in Figures 1.30 and 1.31. In the left-hand panel of Figure 1.31, a mixture of DNA fragments has been applied to an agarose gel and subjected to an electric field. The fragments separate on the basis of their size, with the smaller fragments moving faster. After electrophoresis, the fragments can be stained with ethidium bromide and the gel photographed to show the positions of all of the bands (Figure 1.31, lanes A to G). If a blot is made of this gel and the blot is hybridized to DNA of a particular sequence, the only bands that will show are the bands which contain DNA complementary

to the probe (Figure 1.31, lanes a to g). For detailed protocols of this and other types of blotting, consult a cloning manual (see Suggested Reading).

Applications of the Enzymes Used in DNA Replication

As mentioned earlier, meticulous work on the properties of the enzymes involved in DNA replication has not only increased our understanding of DNA replication but also led directly to many important practical applications. As discussed in the introductory chapter, many of these enzymes were first detected and purified from bacteria and phage-infected bacteria, but their applications extend to the molecular genetics of all organisms. We discuss a few of the more prominent applications here.

DNA POLYMERASES

The properties of DNA polymerases have been exploited in many applications in molecular genetics. As discussed earlier in this chapter, these enzymes all extend a primer polynucleotide chain by attaching the 5′ phosphate of an incoming deoxynucleotide to the 3′ end of the growing primer chain. They can only extend primers which are hybridized to a template DNA, and the choice of which deoxynucleotide to add at each step is determined by complementary base pairing between the incoming deoxynucleotide and the template DNA, leading to synthesis of DNA which is a complementary copy of the template.

Figure 1.31 Results of a Southern blot hybridization. Lanes A to G show the total DNA in each lane. Lanes a to g show the bands that hybridize to a specific probe. Markers are shown in lane C. Reprinted from C. Kao and L. Snyder, *J. Bacteriol.* **170:**2056–2062, 1988.

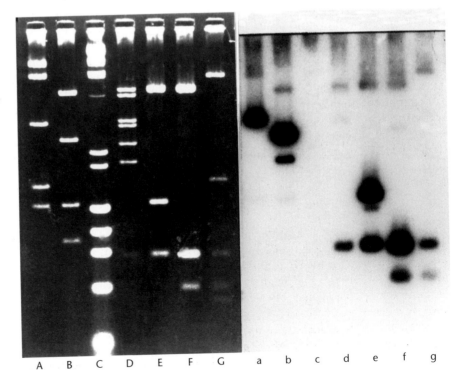

A B C D E F G a b c d e f g

DNA Sequencing

One important application of DNA polymerases is in DNA sequencing, the process of determining the sequence of deoxynucleotides in DNA. This technology has received much publicity recently with the sequencing of the entire almost 1-meter-long human DNA, containing many billions of deoxynucleotides, the so-called Human Genome Project (see Box 1.4). The entire DNA sequence has also been determined for one plant and a large number of bacteria and viruses, as well as some lower eukaryotes that are important model systems in cell and developmental biology. Sequencing of entire genomes requires a tremendous amount of technical ingenuity but the principle behind it is quite simple, involving knowledge of the properties of DNA polymerases.

The methods for DNA sequencing used today are based on a method developed by Fred Sanger, who received a Nobel Prize for this work (his second—the first was for sequencing a protein). This method is called the dideoxy method and is based on the chain-terminating property of the dideoxynucleotides. The dideoxynucleotides are like the normal deoxynucleotides, except that they have hydrogens instead of hydroxyl groups at the 3′ positions of the deoxyribose sugar. The dideoxynucleotides can be phosphorylated to give the dideoxynucleoside triphosphates, which will then be incorporated into DNA by DNA polymerases which are unable to tell the difference between them and the normal deoxynucleotides. However, because it lacks a hydroxyl group at the 3′ position, an incorporated dideoxynucleotide cannot be joined to the 5′ phosphate of the next deoxynucleotide and so the growing DNA chain will terminate. From the length of the DNA chain which is made, we can deduce which base must have been next in the template DNA, because the base in the chain-termininating dideoxynucleotide which was last added was the complement of this base. In the original method on which all the later variations are based, four separate polymerizing reactions were run, each containing a small amount of one of the four dideoxynucleoside triphosphates (ddTTP, ddGTP, ddATP, or ddCTP) mixed with the normal deoxynucleoside triphosphates. A short DNA primer complementary to a known sequence adjacent to the unknown DNA sequence is hybridized to the DNA, and DNA polymerase is added. The DNA polymerase will extend the primer, making a DNA which is complementary to the template. Each time the DNA polymerase encounters a base in the template DNA that is complementary to the dideoxynucleotide used in that reaction, there is a chance that the dideoxynucleotide instead of the normal deoxynucleotide will be incorporated into the chain and the growth of that chain will be terminated. Since each reaction mixture contains a different dideoxynucleoside triphosphate, each of the four reactions will produce a set

of shortened DNA chains of different lengths determined by the positions of the complementary nucleotides in the template DNA. Therefore, the sequence of the template DNA can be determined by measuring the lengths of the shortened chains in each of the reactions.

A genome must be broken up into smaller regions before it can be sequenced in its entirety. The sequences of the smaller regions can then be assembled into the sequence of the larger region, using various techniques (see Box 1.4).

Site-Specific Mutagenesis

Another exploitation of the properties of DNA polymerases is in site-specific mutagenesis. These methods allow the investigator to make a desired change at a particular site in the sequence of DNA rather than relying on more traditional methods of mutagenesis which more or less randomly cause mutations and do not target them to a particular site. Most methods for site-specific mutagenesis rely on synthetic DNA primers which are mostly complementary to the sequence of the DNA being mutagenized except for the desired mutational change. When this primer is hybridized to the DNA and used to prime the synthesis of new DNA by the DNA polymerase, the DNA synthesized will have the same sequence as the template except for the change in the attached primer. This method of site-specific mutagenesis can be used only to make minor changes, such as single-base-pair changes, in the DNA sequence, because if the sequence of the primer is altered too much, it will no longer hybridize to the template DNA being mutagenized.

The difficult part of site-specific mutagenesis is in replicating the mutated DNA and selecting it among the myriad of DNA molecules that have not been mutated. Many methods have been developed to accomplish this, one of which is the "two-primer" method illustrated in Figure 1.32. The DNA to be mutagenized is cloned into a cloning vector with a unique restriction site as shown. The entire cloning vector containing the clone is then replicated using two primers. One primer is complementary to the sequence being mutagenized, except that it contains the desired mutational change. The other primer is complementary to the region of the cloning vector containing the unique restriction site, except that it has a single-base change in the recognition sequence for the restriction endonuclease. Both these primers are present at high concentrations, so that usually the replication of the DNA will use both primers rather than only one or the other. After the DNA polymerase has made a copy of the DNA and the DNA has been introduced into cells and allowed to replicate, the DNA is isolated and cut with the restriction nuclease. The DNAs not descended from a DNA that had replicated from the primers will be cut and eliminated. The DNAs descended from a DNA

BOX 1.4

Bacterial Genomics

Several efforts, known as genome projects, are under way to sequence the DNA of entire organisms. The most famous of these is the **Human Genome Project**, as a result of which the sequence of the billions of deoxynucleotides making up the almost 1-meter (or 1-yard)-long DNA of a human is now known. The genomic DNA of the lambda phage was the first to be sequenced, and now many other virus DNAs have been sequenced. At the time of this writing, more than 50 microbial genomes have been sequenced and published and several hundred more will be finished in the near future (http://www.ornl.gov/microbialgenomes/). Microbes of particular medical or environmental importance or ones which serve as model systems for cell and molecular biology have tended to be chosen for sequencing.

The method used for most genomic sequencing projects is random shotgun sequencing. This approach is illustrated in the flowchart. First, random pieces of the genome are cloned to produce libraries of the total DNA. These pieces are then sequenced typically until the same sequence is obtained eight times, to provide an eightfold coverage of the genome. The sequencing is now automated. Computerized analysis of the sequences allows their assembly into so-called "contigs," which are overlapping sequences. The sequences can then be put in order. Finally, sequencing of unrepresented regions which have no contigs on one side or the other is usually necessary to "close the gaps." Once the genome is sequenced, it can be "annotated," which means that it is analyzed for features such as open reading frames, RNA-encoding genes, repeat sequences, consensus sequences, etc. (see Annotation Resources, below). Ongoing advances in bioinformatics are allowing more and more information to be derived from such sequence information.

There are several reasons to sequence the entire DNA of a living organism. One is to make its identification as unambiguous as possible, for example in epidemiology, so as to trace the source of a disease. The source of the *B. anthracis* strain used in the bioterrorism attack in the United States could be unambiguously determined this way. Another is to compare the sequences of different types of organisms in order to derive their common features. The sequences of more and more genes have been entered into databases, and proteins with the same function often have sequences, called motifs, in common (see chapter 2). Therefore, from the sequence of the gene alone, it is often possible to search the databases to find similar proteins of known function. This may make it possible to guess the functions of many of the gene products of the organism. A number of insights into evolution may emerge. For example, there may be interesting patterns in how genes of similar function are organized in different species. Also, the minimum number of genes required to make an organism might be determined by comparing the total genomes of many different organisms. Only the genes shared by all types of organisms would be absolutely required to make a living organism. The sizes of the genomes of some common bacteria are shown in the table.

Flowchart for Genomic Sequencing
1 **Isolate genomic DNA**
• DNA should be in pieces of >20 kb
2 **Shear DNA**
• DNA fragments will be of random lengths
3 **Size fractionate DNA**
• Collect fragments in size range from 1.5 to 2 kb
4 **Construct plasmid library**
• Inserts are the 1.5- to 2-kb genomic DNA
5 **Randomly sequence inserts**
• Sequencing process is highly automated
• Need ~15,000 sequence runs of ~500 to 600 bases each per megabase of genome DNA
6 **Assemble randomly generated sequence information in contiguous segments ("contigs")**
7 **Close gaps with directed sequencing**
• Several hundreds of reactions needed
8 **Analyze sequence**
• Bioinformatics allows "annotation" of ORFs, etc.

Size Range of Genomes		
Organism	Size (Mbp)	Comments
Mycoplasma genitalium	0.58	Smallest cell genome
Treponema pallidum	1.14	Causes syphilis
Helicobacter pylori	1.67	Causes duodenal ulcers
Sulfolobus solfataricus	2.25	Found in Yellowstone hot springs
Bacillus subtilis	4.20	Soil bacterium; "model" for development
Escherichia coli	4.64	Intestinal bacterium; "model" for genetics and physiology
Pseudomonas aeruginosa	6.26	Causes respiratory infections
Streptomyces coelicolor	8.4	Antibiotic-producing soil bacterium

(continued)

BOX 1.4 (continued)

Bacterial Genomics

References

1. Fleischmann, R. D., M. D. Adams, O. White, R. A. Clayton, E. F. Kirkness, A. R. Kerlavage, C. J. Bult, J.-F. Tomb, B. A. Dougherty, J. M. Merrick, K. McKenney, G. Sutton, W. Fitzhugh, C. Fields, J. D. Gocayne, J. Scott, R. Shirley, L.-I. Liu, A. Glodek, J. M. Kelley, J. F. Weidman, C. A. Phillips, T. Spriggs, E. Hedblom, M. D. Cotton, T. R. Utterback, M. C. Hanna, D. T. Nguyen, D. M. Saudek, R. C. Brandon, L. D. Fine, J. L. Fritchman, J. L. Fuhrmann, N. S. M. Geoghagen, C. L. Gnehm, L. A. McDonald, K. V. Small, C. M. Fraser, H. O. Smith, and J. C. Venter. 1995.

Whole-genome random sequencing and assembly of *Haemophilus influenzae* Rd. *Science* **269**:496–512.

2. Franguel, L., K. E. Nelson, C. Buchrueserm, A. Danchin, P. Glaser, and F. Kunst. 1999. Cloning and assembly strategies in bacterial genome projects. *Microbiology* **145**:2625–2634.

Annotation Resources

Comprehensive Microbial Resource (http://www.tigr.org/tigr-scripts/CMR2/CMRHomePage.spl).

EcoCyc and MetaCyc (http://ecocyc.org).

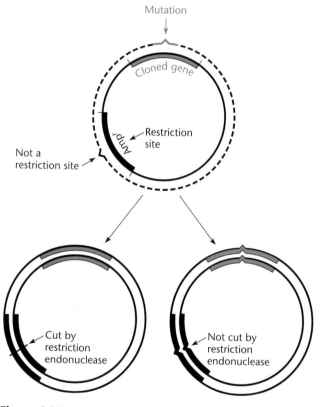

Figure 1.32   Use of two primers to eliminate the wild-type sequence after site-specific mutagenesis. See the text for details.

replicated from the two primers will not be cut and will survive. No such selection is 100% effective, and a few of the surviving DNA clones might still have to be sequenced to find one with the desired mutation.

Polymerase Chain Reaction

One of the most useful technical applications involving DNA polymerases is the **polymerase chain reaction** (**PCR**). This technology makes it possible to selectively amplify regions of DNA out of much longer DNAs. It is called the polymerase chain reaction because each newly synthesized DNA serves as the template for more DNA synthesis in a sort of chain reaction until large amounts of DNA have been amplified from a single DNA molecule. The power of this method is that it can be used to detect and amplify sequences from just a few molecules of DNA from any biological specimen, for example a drop of blood or a single hair; this has made it very useful in criminal investigations to identify the perpetrators of crimes on the basis of DNA typing. However, for our purposes here, it also has many other applications including the physical mapping of DNA, gene cloning, mutagenesis, and DNA sequencing.

The principles behind the use of PCR to amplify a region of DNA are outlined in Figure 1.33. PCR takes advantage of the same properties of DNA polymerases that are important in other applications, i.e., their ability to make a complementary copy of a DNA template starting from the 3′ hydroxyl of a primer DNA. PCR uses two primers complementary to sequences on either side of the region to be amplified. One primer has the same sequence in the 5′-to-3′ direction as one of the strands on one side of the region to be amplified, and the other has the complementary sequence to this strand on the other side of the region to be amplified, but written in the opposite direction. Thus, the two primers will prime the synthesis of DNA in opposite directions over the region to be amplified. The DNA is denatured to separate the strands and hybridized to the primers. One primer will prime the synthesis of DNA over the region to be amplified and will continue polymerizing past the region. If the two strands are again separated by heating and the temperature is again lowered, the other primer can then hybridize to this newly synthesized strand and prime replication back over the region. Now, however, the DNA polymerase will run off the end of the template DNA when it reaches the

Figure 1.33 The steps in a PCR. In the first cycle, the template is denatured by heating. Primers are added; they hybridize to the separated strands for the synthesis of the complementary strand. The strands of the DNA are separated by the next heating cycle, and the process is repeated. The DNA polymerase will survive the heating steps because it is from a thermophilic bacterium. The DNA sequence will be amplified approximately 10^9-fold.

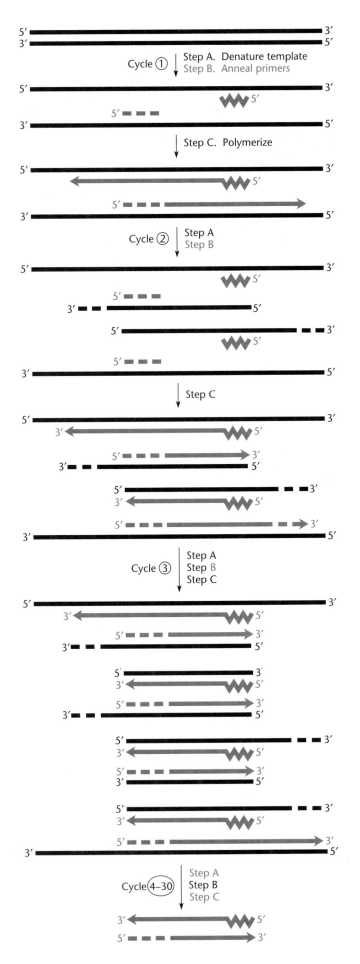

end of the first primer sequence, leading to the synthesis of a short piece of DNA with the primer sequences on both ends. If this DNA is then heated to separate the strands, this shorter piece can bind another primer DNA, which can then prime the synthesis of the complementary strand of this shorter DNA. This process of heating and cooling can be repeated 30 or 40 times until large numbers of copies of the particular DNA region have accumulated, beginning from one or very few longer DNA molecules containing the region.

In principle, any DNA polymerase could be used to perform such an amplification. However, most DNA polymerases would be inactivated by the high temperatures required to separate the strands of double-stranded DNA, making it necessary to add fresh DNA polymerase after each heating step. This is where the DNA polymerases from thermophilic bacteria such as *Thermus aquaticus* come to the rescue. These bacteria normally live at very high temperatures, and so their DNA polymerase, called the *Taq* polymerase, can survive the high temperatures needed to separate the strands of DNA, obviating the need to add new DNA polymerase at each step. We can just mix the primers, a tiny amount of biological material containing DNA with the region to be amplified, and the *Taq* polymerase; set a thermocycler to heat and cool over and over again; and come back a few hours later. Voilà, we should have large amounts of the region of the amplified DNA, which we can detect on a gel.

PCR mutagenesis. PCR can also be used either to make specific changes in a DNA sequence or to randomly mutagenize a region of DNA. Making specific changes by PCR is similar to the other means of site-specific mutagenesis. A complementary primer is made but with the desired change in the sequence. When the polymerase uses the primer to amplify the region, the specific change will be made in the sequence.

PCR can be used to make random changes in a sequence because the *Taq* polymerase makes many mistakes, particularly in the presence of manganese ions, because it lacks an editing function. In fact, the mistake level during normal amplification by *Taq* polymerase is so high that clones made from PCR fragments should usually be sequenced to be certain that no unwanted mutations have been introduced.

Introduction of restriction sites. PCR is also useful for adding sequences, such as restriction sites for cloning, to the ends of the amplified fragment. Although the primers used for PCR amplification must be complementary to the sequence being amplified at the 3′ end, they need not be complementary at their 5′ end. Therefore, the 5′ end of the primer sequence can include, for example, the recognition sequence for a specific restriction endonuclease, making it easier to clone the PCR amplified fragment (see above).

Figure 1.34 illustrates how we can use PCR amplification to introduce restriction sites at the ends of an amplified fragment.

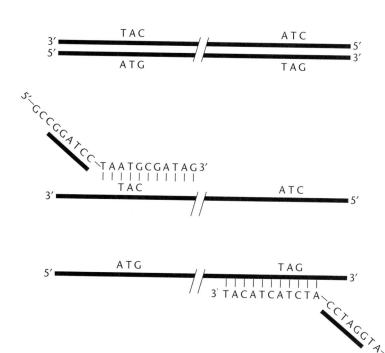

Figure 1.34 Use of PCR to add convenient restriction sites to the ends of an amplified fragment. The primers contain sequences at their 5′ ends that are not complementary to the gene to be cloned but, rather, contain the sequence of the cleavage site for *Bam*HI (underlined). The amplified fragment will contain a *Bam*HI cleavage site at both ends. See the text for details.

SUMMARY

1. DNA consists of two strands wrapped around each other in a double helix. Each strand consists of a chain of nucleotides held together by phosphates joining their deoxyribose sugars. Because the phosphate joins the third carbon of one sugar to the fifth carbon of the next sugar, the DNA strands have a directionality, or polarity, and have distinct 5′ phosphate and 3′ hydroxyl ends. The two strands of DNA are antiparallel, so that the 5′ end of one is on the same end as the 3′ end of the other.

2. DNA is synthesized from the precursor deoxynucleoside triphosphates by DNA polymerase. The first phosphate of each nucleotide is attached to the 3′ hydroxyl of the next deoxynucleotide, giving off the terminal two phosphates to provide energy for the reaction.

3. DNA polymerases require both a primer and a template strand. The pairing of the bases between the incoming deoxynucleotide and the base on the template strand dictates which deoxynucleotide will be added at each step, with A always pairing with T and G always pairing with C. The DNA polymerase synthesizes DNA in the 5′-to-3′ direction, moving in the 3′-to-5′ direction on the template.

4. DNA polymerases cannot put down the first deoxynucleotide, so RNA is usually used to prime the synthesis of a new strand. Afterward, the RNA primer is removed and replaced by DNA using upstream DNA as a primer. The use of RNA primers helps reduce errors by allowing editing.

5. DNA polymerase does not synthesize DNA by itself but needs other proteins to help it replicate DNA. These other proteins are helicases that separate the strands of the DNA, ligases to join two DNA pieces together, primases to synthesize RNA primers, and other accessory proteins to keep it on the DNA and reduce errors.

6. Both strands of double-stranded DNA are usually replicated from the same end, so that the overall direction of DNA replication is from 5′ to 3′ on one strand and from 3′ to 5′ on the other strand. Because DNA polymerase can polymerize only in the 5′-to-3′ direction, it must replicate one strand in short pieces and ligate these afterward to form a continuous strand. The short pieces are called Okazaki fragments. The two DNA polymerases replicating the leading and lagging strands remain bound to each other in a process called the trombone model of replication.

7. The DNA in a bacterium that carries most of the genes is called the bacterial chromosome. The chromosome of most bacteria is a long, circular molecule that replicates in both directions from a unique origin of replication *oriC*. Replication of the chromosomes initiates each time the cells reach a certain size. For fast-growing cells, new rounds of replication initiate before old ones are completed. This accounts for the fact that fast-growing cells have a higher DNA content than slower-growing cells.

8. Chromosome replication terminates, and the two daughter DNAs separate, when the replication forks meet. Multiple *ter* sites that act as "one-way gates" delay movement of the replication forks on the chromosome. Proteins which are inhibitors of the DnaB helicase stop replication at these sites.

9. To separate the daughter DNAs after replication, dimerized chromosomes, created by recombination between the daughter DNAs, are resolved by XerC and XerD, a site-specific recombination system which promotes recombination between the *dif* sites on the daughter chromosomes. Topo IV decatenates the intertwined daughter DNAs by passing the double-stranded DNAs through each other.

10. The daughter chromosomes are segregated by condensing the DNAs through supercoiling by DNA gyrase and by condensins which hold the DNA in large supercoiled loops.

11. Chromosome replication occurs at the midpoint of the cell. After replication is complete, the two daughter DNAs quickly partition to the poles of the cell through the action of partitioning (Par) proteins.

12. The FtsZ protein forms a ring at the midpoint of the cell, attracting other proteins, which form the division septum. The Min proteins prevent the formation of FtsZ rings anywhere in the cell other than in the middle.

13. Initiation of a round of chromosome replication occurs once every time the cell divides. Initiation occurs when the ratio of DnaA protein to origins of replication reaches a critical number. In some bacteria including *E. coli*, related enterics, and *Caulobacter crescentus*, new initiations are prevented by hemimethylation of the newly replicated DNA at the origin and by sequestration until the cell divides again.

14. The chromosomal DNA of bacteria is usually one long, continuous circular molecule about 1,000 times as long as the cell itself. This long DNA is condensed in a small part of the cell called the nucleoid. In this structure, the DNA loops out of a central condensed core region. Some of these loops of DNA are negatively supercoiled. In *E. coli*, most DNAs have one supercoil about every 300 bases.

15. The enzymes that modulate DNA supercoiling in the cell are called topoisomerases. There are two types of topoisomerases in cells. Type I topoisomerases remove supercoils one at a time by breaking only one strand and passing the other strand through the break. Type II topoisomerases remove or add supercoils two at a time by breaking both strands and passing another region of the DNA through the break. The enzyme responsible for adding the negative supercoils to DNA in bacteria is a type II topoisomerase called gyrase. Topo IV decatenates daughter DNAs after replication. It may also remove positive supercoils ahead of the replication fork.

(continued)

SUMMARY (continued)

16. Type II restriction endonucleases recognize defined sequences in DNA and cut at a defined position in or near the recognition sites, which has made them very useful in physical mapping of DNA and in DNA cloning. Most type II restriction endonucleases cut at the same position in both strands of symmetric sequences. If the cuts are not immediately opposite each other, single-stranded sticky ends will form that can hybridize to any other sticky end cut with the same or a compatible enzyme. This property has made these enzymes very useful for DNA cloning and for DNA manipulations in vitro.

17. The physical map of a DNA shows the actual location of sequences in DNA including the position of recognition sequences for restriction endonucleases.

18. Cloned DNA consists of multiple copies of a DNA sequence descended from a single molecule. If a piece of DNA is inserted into a cloning vector, the DNA will replicate along with the vector, making millions of clones of the original DNA.

19. A DNA library is a collection of clones that, among themselves, contain all the DNA sequences of the organism.

20. Knowledge of the properties of the enzymes involved in DNA replication has led to many applications including DNA sequencing, site-specific mutagenesis, and PCR.

21. Some antibiotics block DNA replication or affect the structure of DNA. The most useful of these inhibit the synthesis of deoxynucleotides or inhibit the gyrase enzyme. Antibiotics that inhibit deoxynucleotide synthesis include inhibitors of dihydrofolate reductase (trimethoprim, methotrexate) and inhibitors of ribonucleotide reductase (hydroxyurea). Inhibitors of gyrase include novobiocin and nalidixic acid. Acridine dyes also affect the structure of DNA by intercalating between the bases and are used as antimalarial drugs as well as in molecular biology.

QUESTIONS FOR THOUGHT

1. Some viruses avoid the problem of lagging-strand synthesis by replicating the individual strands of the DNA in the 3′-to-5′ direction simultaneously from both ends so that eventually the entire molecule will be replicated. Why don't bacterial chromosomes replicate this way?

2. Why are DNA molecules so long? Wouldn't it be easier to have lots of shorter pieces of DNA? What problems would this present for the cell?

3. Why do cells have DNA as their hereditary material instead of RNA?

4. What effect would shifting a temperature-sensitive mutant with a mutation in the *dnaA* gene for initiator protein DnaA have on the rate of DNA synthesis? Would the rate drop linearly or exponentially? Would the slope of the curve be affected by the growth rate of the cells at the time of the shift? Explain.

5. The gyrase inhibitor novobiocin inhibits the growth of almost all types of bacteria. What would you predict about the gyrase of the bacterium, *Streptomyces sphaeroides*, that makes this antibiotic? How would you test your hypothesis?

6. How do you think chromosome replication and cell division are coordinated in bacteria like *E. coli*? How would you go about testing your hypothesis?

7. Why is termination of chromosome replication so sloppy that the *ter* region is nonessential for growth and there has to be more than one *ter* site in each direction to completely stop the replication fork?

PROBLEMS

1. You are synthesizing DNA on the template 5′ACCTTAC-CGTAATCC3′ from an upstream primer. You add three of the deoxynucleotides but leave out the fourth deoxynucleotide, deoxycytosine triphosphate, from the reaction. What DNA would you make? Draw a picture.

2. You are synthesizing DNA from the same template and with the same upstream primer, but instead of just deoxythymidine triphosphate you add an equal mixture of the inhibitor of replication, dideoxythymidine triphosphate, and the normal nucleotide deoxythymidine triphosphate in addition to the other three deoxynucleoside triphosphates. What DNAs would you make? Draw a picture.

3. You are growing *E. coli* with a generation time of only 25 min. How long will the I periods, C periods, and D periods be?

Draw a picture showing when the various events occur during the cell cycle.

4. You are growing *E. coli* with a generation time of 90 min. Now how long will the I, C, and D periods be? Draw a picture.

5. You are measuring the supercoiling of the nucleoid from a *topA* mutant of *E. coli* which lacks the major type I topoisomerase and comparing it with the normal *E. coli* (without the *topA* mutation). Would you expect there to be more or fewer negative supercoils in the nucleoid of the mutant? Why?

6. Outline how you would use PCR to introduce a desired base pair change in a particular DNA sequence.

SUGGESTED READING

Blakely, G., G. May, R. McCulloch, L. K. Arciszewska, M. Burke, S. T. Lovett, and D. J. Sherratt. 1993. Two related recombinases are required for site-specific recombination at *dif* and *cer* in *E. coli. Cell* **75**:351–361.

Britton, R. A., D. C. Lin, and A. D. Grossman. 1998. Characterization of a prokaryotic SMC protein involved in chromosome partitioning. *Genes Dev.* **12**:1254–1259.

Bussiere, D. E., and D. Bastia. 1999. Termination of DNA replication of bacterial and plasmid chromosomes. *Mol. Microbiol.* **31**:1611–1618. (Microreview.)

Cohen, S. N., A. C. Y. Chang, H. W. Boyer, and R. B. Helling. 1973. Construction of biologically functional bacterial plasmids in vitro. *Proc. Natl. Acad. Sci. USA* **70**:3240–3244.

Christensen, B. B., T. Atlung, and F. G. Hansen. 1999. DnaA boxes are important elements in setting the initiation mass of *Escherichia coli. J. Bacteriol.* **181**:2683–2688.

Crooke, E. 1995. Regulation of chromosome replication in *E. coli*: sequestration and beyond. *Cell* **82**:877–880. (Minireview.)

Danna, K. J., and D. Nathans. 1971. Specific cleavage of simian virus 40 DNA by restriction endonuclease of *Haemophilus influenzae. Proc. Natl. Acad. Sci. USA* **68**:2913–2917.

Dervyn, E., C. Suski, R. Daniel, C. Bruand, J. Chapuis, J. Errington, L. Janniere, and S. D. Ehrich. 2001. Two essential DNA polymerases at the bacterial replication fork. *Science* **294**:1716–1719.

Harry, E. J. 2001. Bacterial cell division: regulating Z-ring formation. *Mol. Microbiol.* **40**:795–803. (Microreview.)

Helmstetter, C. E., and S. Cooper. 1968. DNA synthesis during the division cycle of rapidly growing *Escherichia coli* B/r. *J. Mol. Biol.* **31**:507–518.

Holmes, V. F., and N. R. Cozzarelli. 2000. Closing the ring: links between SMC proteins and chromosome partitioning, condensation, and supercoiling. *Proc. Natl. Acad. Sci. USA* **97**:1322–1324. (Commentary.)

Khodursky, A. B., B. J. Peter, M. B. Schmid, J. DeRisi, D. Botstein, P. O. Brown, and N. R. Cozzarelli. 2000. Analysis of topoisomerase function in bacterial replication fork movement: Use of DNA microarrays. *Proc. Natl. Acad. Sci. USA* **97**:9419–9424.

Lemon, K. P., and A. D. Grossman. 1998. Localization of bacterial DNA polymerase: evidence for a factory model of replication. *Science* **282**:1516–1519.

Linn, S. 1996. The DNases, topoisomerases, and helicases of *Escherichia coli* and *Salmonella*, p. 764–772. *In* F. C. Neidhardt, R. Curtiss III, J. L. Ingraham, E. C. C. Lin, K. B. Low, B. Magasanik, W. S. Reznikoff, M. Riley, M. Schaechter, and H. E. Umbarger (ed.), Escherichia coli *and* Salmonella: *Cellular and Molecular Biology*, 2nd ed. ASM Press, Washington, D.C.

Linn, S., and W. Arber. 1968. Host specificity of DNA produced by *Escherichia coli* X: *in vitro* restriction of phage fd replicative form. *Proc. Natl. Acad. Sci. USA* **59**:1300–1306.

Mohl, D. A., and J. W. Gober. 1997. Cell cycle-dependent polar localization of chromosomal partitioning proteins in *Caulobacter crescentus. Cell* **88**:675–684.

Olby, R. 1974. *The Path to the Double Helix.* Macmillan Press, London, United Kingdom.

Sawitzke, J., and S. Austin. 2001. An analysis of the factory model from chromosome replication and segregation in bacteria. *Mol. Microbiol.* **40**:786–794. (Microreview.)

Singleton, M. R., and D. B. Wigley. 2002. Modularity and specialization in Superfamily 1 and 2 helicases. *J. Bacteriol.* **184**:1819–1826. (Minireview.)

Slater, S., S. Wold, M. Lu, E. Boye, K. Skarsted, and N. Kleckner. 1995. The *E. coli* SeqA protein binds *oriC* in two different methyl-modulated reactions appropriate to its roles in DNA replication initiation and origin sequestration. *Cell* **82**:927–936.

Steiner, W. W., G. Liu, W. D. Donachie, and P. L. Kuempel. 1999. The cytoplasmic domain of the FtsK protein is required for resolution of chromosome dimers. *Mol. Microbiol.* **31**:579–583.

Wang, J. C. 1996. DNA topoisomerases. *Annu. Rev. Biochem.* **65**:635–692.

Watson, J. D. 1968. *The Double Helix.* Atheneum, New York, N.Y.

Watson, J. D., and F. H. C. Crick. 1953. Molecular structure of nucleic acids. *Nature* (London) **171**:737–738.

Zechiedrich, E. L., A. B. Khodursky, and N. R. Cozzarelli. 1997. Topoisomerase IV, not gyrase, decatenates products of site-specific recombination in *Escherichia coli. Genes Dev.* **11**:2580–2592.

Cloning Manual

Sambrook, J., and D. Russell. 2001. *Molecular Cloning: A Laboratory Manual*, 3rd ed. Cold Spring Harbor Laboratory Press, Cold Spring Harbor, N.Y.

2

Macromolecular Synthesis: Gene Expression

UNCOVERING THE MECHANISM of protein synthesis, and therefore of gene expression, was one of the most dramatic accomplishments in the history of science. The process of protein synthesis is sometimes called the **Central Dogma** of molecular biology, which states that information in DNA is copied into RNA to be translated into protein. We now know of many exceptions to the Central Dogma. For example, information does not always flow from DNA to RNA but sometimes in the reverse direction, from RNA to DNA. The information in RNA is often changed after it has been copied from the DNA (see Box 2.1). Moreover, the information in DNA may be translated differently depending on where it is in a gene. Despite these exceptions, however, the basic principles of the Central Dogma remain sound.

This chapter outlines the process of protein synthesis and gene expression. The discussion is meant to be only a broad overview but one that is sufficient to understand the chapters that follow. For more detailed treatments, consult any modern biochemistry textbook.

Overview

DNA carries the information for the synthesis of RNA and proteins in regions called **genes**. The first step in expressing a gene is to **transcribe**, or copy, an RNA from one strand in that region. The word *transcription* is descriptive because the RNA is copied in the same language as DNA, a language written in a sequence of nucleotides. If the gene carries information for a protein, this RNA transcript is called **messenger RNA (mRNA)**. An mRNA is a messenger because it carries the gene's message to a **ribosome**. Once on the ribosome, the information in the mRNA can be **translated** into the protein. *Translation* is another descriptive word because one language—the sequence of nucleotides in DNA and RNA—is interpreted into a different language—a sequence of amino acids in a protein. The mRNA is translated as it moves along the ribosome, 3 nucleotides at a time. Each 3-nucleotide sequence, called a **codon,** carries information for a specific amino acid. The assignment of each of the possible codons to amino acids is called the genetic code and is shown in Table 2.1 (also see Box 2.4 below).

The actual translation from the language of nucleotide to amino acid sequences is performed by small RNAs called transfer RNAs (tRNAs) and enzymes called aminoacyl-tRNA synthetases. The enzymes attach specific amino acids to each tRNA. Then a tRNA specifically pairs with a codon in the mRNA as it moves through the ribosome, and the amino acid it carries is inserted into the growing protein. The tRNA pairs with the codon in the mRNA through a complementary three-nucleotide sequence called the **anticodon.** The base-pairing rules for codons and anticodons are basically the same as the base-pairing rules for DNA replication, except that RNA has uracil (U) rather than thymine (T) and the pairing between the last of the three bases in the codon and the anticodon is less stringent.

This basic outline of gene expression leaves many important questions unanswered. How does mRNA synthesis begin and end at the correct places and on the correct strand in the DNA? Similarly, how does translation start and stop at the correct places on the mRNA? What actually happens to the tRNA and ribosomes during translation? The answers to these questions and many others are important for the interpretation of genetic experiments, so we will discuss the structure of RNA and proteins and the processes by which they are synthesized in much more detail.

The Structure and Function of RNA

In this section, we review the basic components of RNA and how it is synthesized. We also review how structure varies among different types of cellular RNAs and the role each type plays in cellular processes.

Types of RNA

There are many different types of RNA in cells. Some of these, including mRNA, ribosomal RNA (rRNA), and tRNA, are involved in protein synthesis. Each of these types of RNAs has special properties that we discuss later in the section. Others are involved in regulation and replication.

RNA Precursors

RNA is similar to DNA in that it is composed of a chain of nucleotides. However, RNA nucleotides contain the sugar ribose instead of deoxyribose. These five-carbon sugars differ in the second carbon, which is attached to a hydroxyl group in ribose rather than the hydrogen found in deoxyribose (see chapter 1). Figure 2.1A shows the structure of a **ribonucleoside triphosphate,** so named because of the different sugar.

TABLE 2.1	The genetic code				
First position (5′ end)	**Second position**				**Third position (3′ end)**
	U	**C**	**A**	**G**	
U	Phe	Ser	Tyr	Cys	U
	Phe	Ser	Tyr	Cys	C
	Leu	Ser	Stop	Stop	A
	Leu	Ser	Stop	Trp	G
C	Leu	Pro	His	Arg	U
	Leu	Pro	His	Arg	C
	Leu	Pro	Gln	Arg	A
	Leu	Pro	Gln	Arg	G
A	Ile	Thr	Asn	Ser	U
	Ile	Thr	Asn	Ser	C
	Ile	Thr	Lys	Arg	A
	Met	Thr	Lys	Arg	G
G	Val	Ala	Asp	Gly	U
	Val	Ala	Asp	Gly	C
	Val	Ala	Glu	Gly	A
	Val	Ala	Glu	Gly	G

A

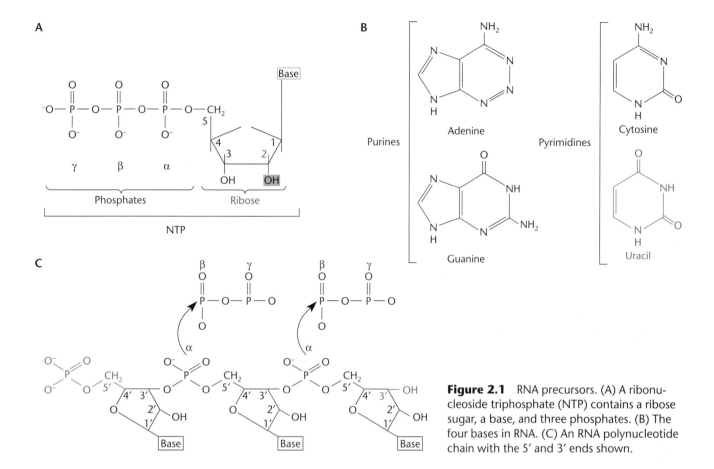

Figure 2.1 RNA precursors. (A) A ribonucleoside triphosphate (NTP) contains a ribose sugar, a base, and three phosphates. (B) The four bases in RNA. (C) An RNA polynucleotide chain with the 5′ and 3′ ends shown.

The only other difference between RNA and DNA chains when they are first synthesized is in the bases. Three of the bases—adenine, guanine, and cytosine—are the same, but RNA has uracil instead of the thymine found in DNA (Figure 2.1B). The RNA bases can also be modified later, as we discuss below.

Figure 2.1C shows the basic structure of an RNA polynucleotide chain. As in DNA, RNA nucleotides are held together by phosphates that join the 5′ carbon of one ribose sugar to the 3′ carbon of the next. This arrangement ensures that, as with DNA chains, the two ends of an RNA polynucleotide chain will be different from each other, with the 5′ end terminating in a phosphate and the 3′ end terminating in a hydroxyl. When it is first synthesized, the 5′ end of an RNA chain has three phosphates attached to it, although two of the phosphates are usually removed soon after.

According to convention, the sequence of bases in RNA is given from the 5′ to the 3′ end, which is actually the direction in which the phosphates are attached 3′ to 5′, 3′ to 5′, etc. Also, by convention, regions in RNA that are closer to the 5′ end in a given sequence are referred to as **upstream** and regions that are 3′ as **downstream** because RNA is both made and translated in the 5′-to-3′ direction.

RNA Structure

Except for the sequence of bases and minor differences in the pitch of the helix, little distinguishes one DNA molecule from another. However, RNA chains generally have more structural properties than DNA, tending to be folded into complex structures, and can be extensively modified.

PRIMARY STRUCTURE

All RNA is created equal. No matter what their function, all RNA transcripts are made the same way, from a DNA template. Only the sequences of their nucleotides and their lengths are different. The sequence of nucleotides in RNA is the **primary structure** of the RNA. In some cases, the primary structure of an RNA is changed after it is transcribed from the DNA (see the section on processing and modification, below).

SECONDARY STRUCTURE

Unlike DNA, RNA is usually single stranded. However, pairing between the bases in some regions of the molecule may cause it to fold up on itself to form a double-stranded region. Such double-stranded regions are called

BOX 2.1

RNA Editing: If You Don't Get It Right the First Time

One of the major tenets of the Central Dogma is that RNA is a complementary copy of DNA. The RNA can be processed and modified after it is made; large regions can even be removed (see Box 2.3) but the sequence of what remains is largely unchanged. However, sometimes this tenet does not hold and the sequence of the RNA is changed (edited) after the RNA is made. Probably the best definition of RNA editing is that the final RNA sequence could have been encoded in the DNA but wasn't, to distinguish it from modification. So far, such RNA editing has been observed only in lower eukaryotes, including the mitochondria of trypanosomes, plants, and cellular slime molds. One of the most dramatic and best studied examples of RNA editing occurs in the kinetoplasts (mitochondria) of trypanosomes. The mRNA, as it is made as a complementary transcript of the DNA, lacks many of the uridine nucleotides of the final mRNA and so other uridines have to be added after it is synthesized. Therefore, as a result of this editing, the sequence of the mature mRNA which is translated into protein is not complementary to the DNA template. The cell contains RNAs called guide RNAs, which direct the addition and removal of uridines to make the final mRNA. These guide RNAs are complementary to the final edited mRNA, and, essentially, the RNAs are repaired to match the sequence of the guide RNA, although the process is not completely understood. The purpose of RNA editing remains a mystery. Why not just encode the RNA correctly in the DNA to begin with? One idea is that RNA editing may no longer serve a useful purpose and may be a fossil remnant of early protein synthesis, much like our appendix is a remnant of the cloaca left over from the evolution of our digestive tract. Perhaps it dates from early in life on earth before the ozone layer formed and organisms were exposed to higher levels of UV irradiation than they are now. Adjacent thymines in the DNA are particularly sensitive to UV irradiation (see chapter 11), and so to avoid having to encode adjacent uridines in the DNA, they were edited in afterwards, although this would not explain why uridines are edited out of the RNA as well as edited in. It also does not explain why the genes for the guide RNAs can contain adjacent thymines, but the guide RNAs are made from another type of DNA called maxicircles, which exist in many copies in the cell, so they may be less sensitive to DNA damage.

Reference

Gott, J. M., and R. B. Emeson. 2000. Functions and mechanisms of RNA editing. *Annu. Rev. Genet.* **34**:499–531.

the **secondary structure** of the RNA. All RNAs, including mRNAs, probably have extensive secondary structure.

Figure 2.2 shows an example of RNA secondary structure, in which the sequence 5'AUCGGCA3' has paired with the complementary sequence 5'UGCUGAU3' somewhere else in the molecule. As in double-stranded DNA, the paired strands of RNA are antiparallel; i.e., pairing occurs only when the two sequences are complementary when read in opposite directions (5' to 3' and 3' to 5') and the double-stranded RNA forms a helix, although in the slightly different A form rather than the B form of a DNA:DNA helix. However, the pairing rules for double-stranded RNA are slightly different from the pairing rules for DNA. In RNA, guanine can pair with uracil as well as with cytosine. Because these GU pairs do not share hydrogen bonds, they do not contribute to the stability of the double-stranded RNA. Thus, the GU pair shown in Figure 2.2 does not disrupt the helix, although it does not help hold the structure together.

Each base pair that forms in the RNA makes the secondary structure of the RNA more stable. Consequently, the RNA will generally fold so that the greatest number of continuous base pairs can form. The stability of a structure can be predicted by adding up the energy of all its hydrogen bonds that contribute to the structure. By eye, it is very difficult to predict what regions of a long RNA will pair to give the most stable structure. However, computer software is available that, given the sequence of bases (primary structure) of the RNA, will predict the most stable secondary structure.

TERTIARY STRUCTURE

Double-stranded regions of RNAs created by base pairing are stiffer than single-stranded regions. As a result, an RNA that has secondary structure will have a more rigid shape than one without double strands. Also, the intermingled paired regions cause the RNA to fold back on itself extensively. One type of tertiary structure occurs when the unpaired region in a hairpin such as that shown in Figure 2.2 pairs with another region of the same RNA molecule to form a knot. Such a structure is called a pseudoknot, rather than a real knot, because it is held together only by hydrogen bonds, which are more easily broken than covalent chemical bonds. Together, these effects give many RNAs a well-defined three-dimensional shape, called its **tertiary structure.** Proteins or other cellular constituents

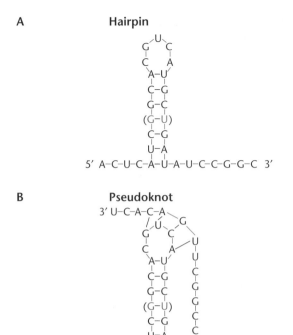

Figure 2.2 Secondary structure in an RNA. (A) The RNA folds back on itself to form a hairpin loop. The presence of a G:U pair (in parentheses) does not disrupt the structure. (B) Different regions of the RNA can also pair with each other to form a pseudoknot. In the example, the loop of the hairpin is pairing with another region of the RNA. The gold dashes show the phosphate-ribose backbone; black dashes the hydrogen bonds. Details are given in the text.

recognize RNA forms by their tertiary structure, which also gives ribozymes their enzymatic activity (see Box 2.2).

RNA Processing and Modification

The alterations to the RNA molecule caused by secondary and tertiary structure are examples of **noncovalent changes**, because only hydrogen bonds, not chemical (**covalent**) bonds, are formed or broken. However, once the RNA is synthesized, covalent changes can occur during **RNA processing** and **RNA modification**.

RNA processing involves forming or breaking phosphate bonds in the RNA after it is made. For example, the terminal phosphates at the 5′ end may be removed, or the RNA may be cut into smaller pieces and even religated into new combinations, requiring the breaking and making of many phosphate bonds (see Box 2.2).

RNA modification, by contrast, involves altering the bases or sugars of RNA. Examples include methylation of the bases or sugars of rRNA and enzymatic alteration of the bases of tRNA. In eukaryotes, "caps" of inverted methylated nucleotides are added to the 5′ ends of some types of mRNA. In bacteria, mRNAs are not

capped, and only the stable rRNAs and tRNAs are extensively modified.

PROCESSING AND MODIFICATION OF tRNA

The tRNAs are probably the most highly processed and modified RNAs in cells. Figure 2.3 shows a "mature" tRNA that was originally cut out of a much longer molecule that may also have included the rRNAs. Then, some of the bases were modified by specific enzymes, creating altered bases such as pseudouracil and thiouracil. Finally, an enzyme called CCA transferase added the sequence CCA to the 3′ end. Clearly, much had to be done to this molecule after it was synthesized so that it could become a functional, mature tRNA.

Transcription

Transcription is the synthesis of RNA on a DNA template. The process of transcription is probably fairly similar in all organisms, but it is best understood in bacteria.

Bacterial RNA Polymerase

The transcription of DNA into RNA is the work of **RNA polymerase**. In bacteria, the same RNA polymerase makes all the cellular RNAs, including rRNA, tRNA, and mRNA. There are approximately 2,000 molecules of this RNA polymerase in each bacterial cell. Only the primer RNAs of Okazaki fragments are made by a different RNA polymerase. In contrast, eukaryotes have three nuclear RNA polymerases, as well as a mitochondrial RNA polymerase, which make their cellular RNAs.

Figure 2.4 shows a schematic structure of a typical bacterial RNA polymerase, which has five subunits and a molecular weight of more than 400,000, making it one of the largest bacterial enzymes. The enzyme consists of five subunits: two identical α subunits, two very large subunits called β and β′, and the σ factor. The α, β, and β′ subunits are permanent parts of the RNA polymerase, but the σ factor is required only for initiation and cycles off the enzyme after initiation of transcription. Without the σ factor, the RNA polymerase is called the core enzyme. With it, it is called the holoenzyme. Eukaryotic and archaeal RNA polymerases have more subunits and seem more complex, although their basic structure is very similar to that of bacterial RNA polymerase.

Transcription Initiation

Much like DNA polymerase (see chapter 1), the RNA polymerase makes a complementary copy of a DNA template, building a chain of RNA by attaching the 5′ phosphate of a ribonucleotide to the 3′ hydroxyl of the one preceding it (Figure 2.5). However, in contrast to DNA polymerase, RNA polymerases do not need a preexisting primer to initiate the synthesis of a new chain of RNA.

BOX 2.2

Parasitic DNAs: RNA Introns and Protein Inteins

The chromosomal DNA of all organisms abounds with parasitic DNA elements, so named because they cannot replicate themselves but can replicate only when the host DNA replicates. These parasitic DNAs often have few functions except for the ability to move from one DNA to another and thereby parasitize new hosts (see Box 9.1). When such a parasitic DNA integrates into a region of the DNA encoding a protein or RNA, it will disrupt the coding sequence, which is why these elements are sometimes called **intervening sequences.** Sometimes an intervening sequence will insert into the coding sequence for an essential RNA or protein. Like all good parasites, these DNA elements do as little harm to their host as possible, which makes sense since the parasite is dependent upon the host for its own survival. Many of them minimize damage to their host by splicing their sequences out of RNAs and proteins made from the DNA so that the RNAs and protein will be functional. Intervening sequences that splice themselves out of the RNA are called **introns,** while those that splice themselves out of the protein product of a gene are called **inteins.** The sequences upstream and downstream of the intron or intein in a gene that are rejoined following splicing are called **exons** and **exteins**, respectively.

The two types of RNA introns in bacteria are called group 1 and group 2, based on their mechanism of splicing. (see the figure panel A). In group I introns, a free guanosine nucleoside or nucleotide residue initiates the splicing by breaking the RNA at the 5′ end of the intron, called the 5′ splice site, initiating a series of phosphodiester bond transfers that complete the splicing process. Group II intron splicing is similar except that the initiating nucleotide is a specific adenine base internal to the intron, creating a characteristic "lariat" structure of the intron. This type of splicing is more analogous to mRNA splcing in eukaryotes, as shown. The group I introns are typically found in bacteriophage protein-coding regions and in the tRNA genes of bacteria. While common in lower eukaryotes, group II introns are much rarer in bacteria and are typically found only in other movable elements such as conjugative plasmids and transposons (see chapters 5 and 9). Both groups of introns are typically self-splicing, meaning that they are capable of splicing themselves out of the RNA without the help of proteins. The RNA of an intron is therefore an enzyme. RNA enzymes are called **ribozymes** to distinguish them from the more common protein enzymes. Other known ribozymes are some RNases and the 23S rRNA peptidyltransferase (see the text).

Even though many of these introns are themselves enzymes and can splice themselves out of the RNA, many en-

code maturase proteins that help them fold into the structure required for splicing. In the group II introns, these maturase proteins are also the reverse transcriptases and, in combination with the "lariat" intron RNA, form the DNA endonuclease that cuts the target DNA during **retrohoming** (the process of the intron moving from the RNA to the same site in another DNA that lacks it). Retrohoming is essentially the reverse of splicing in that the intron splices itself *into* a DNA rather than *out* of an RNA as in splicing. In a process somewhat analogous to splicing, some lower eukaryotes such as trypanosomes and nematodes also attach short RNA sequences to the 5′ ends of their mRNA after synthesis, which may help their translation and stability.

Protein **inteins** are parasitic DNAs like self-splicing introns, except that they splice themselves out of the protein product of the gene rather than out of the mRNA. Inteins probably also exist in all organisms from bacteria to humans. These inteins self-splice themselves out of a protein by the mechanism shown in the figure (panel B). The first amino acid in the intein is always cysteine or serine. This amino acid can be rearranged so that it is attached to the amino acid upstream through its side chain rather than by a normal peptide bond. This new bond is then attacked by the last amino acid in the intein, which is always an asparagine, to form a branched protein as shown. Then the cysteine or serine just downstream of the intein in the extein then attacks this bond, removing the branch. Because the first amino acid in the downstream extein must always be a serine or cysteine, inteins are always located just upstream of a codon for one of these two amino acids, otherwise the intein could not be spliced out of the protein precisely.

One particularly interesting example of protein intein splicing is in the split *dnaE* gene for the replicative DNA polymerase of a strain of the cyanobacterium *Synechocystis,* where intein splicing brings two widely separated parts of a gene product together to form an active enzyme. This is called *trans* splicing because it occurs between two different gene products rather than within the same gene product. The *dnaE* gene for the DNA polymerase in this strain of bacteria is split into two parts separated by 745,226 bp of DNA. Apparently, an intein once integrated into the gene for the DNA polymerase and could splice itself out of the protein. Then another large DNA was inserted into the intein, splitting it into two widely separated parts. However, the two parts of the intein can still find each other and perform the splicing reaction as shown in the figure, joining the two parts of the DNA polymerase together to make the active

(continued)

BOX 2.2 (continued)

Parasitic DNAs: RNA Introns and Protein Inteins

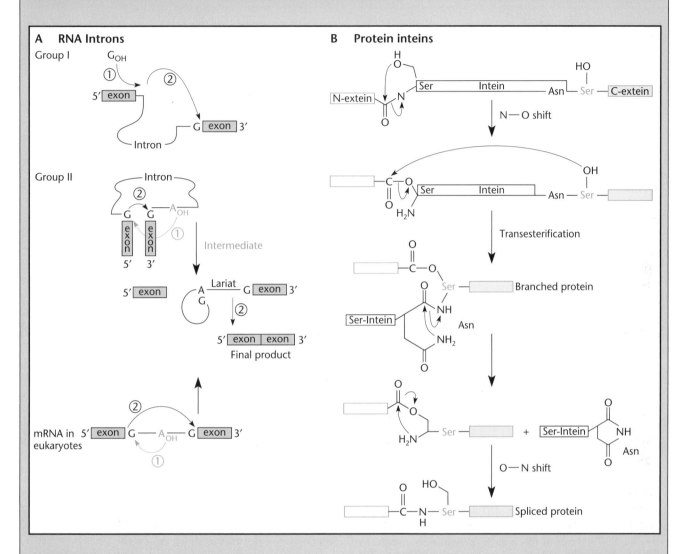

A   RNA Introns

B   Protein inteins

enzyme. RNA introns are known to perform similar *trans*-splicing reactions, and *trans* splicing should find many applications in molecular genetics.

References

Martinez-Abarca, F., and N. Toro. 2000. Group II introns in the bacterial world. *Mol. Microbiol.* **38:**917–926. (Microreview.)

Vepritskiy, A. A., I. A. Vitol, and S. A. Nierzwicki-Bauer. 2002. Novel group I intron in the tRNA^Leu (UAA) gene of a proteobacterium isolated from a deep subsurface environment. *J. Bacteriol.* **184:**1481–1487.

Wu, H., Z. Hu, and X. Q. Liu. 1998. Protein trans-splicing by a split intein encoded in a split DnaE gene of *Synechocystis* sp. PCC6803. *Proc. Natl. Acad. Sci. USA* **95:**9226–9231.

Xu, M.-Q., and F. B. Perler. 1996. The mechanism of protein splicing and its modulation by mutation. *EMBO J.* **15:**5146–5153.

PROMOTERS

RNA transcripts are copied from only selected regions of the DNA, rather than from the whole molecule, so the RNA polymerase must start making an RNA chain from a double-stranded DNA only at certain sites. These DNA regions are called **promoters,** and the RNA polymerase will recognize a particular T or C in the promoter region as a **start point** of transcription. Thus, usually the first base in the chain is an A or a G laid down opposite to a T or C, respectively (Figure 2.5).

The RNA polymerase recognizes different types of promoters on the basis of which type of σ factor is attached.

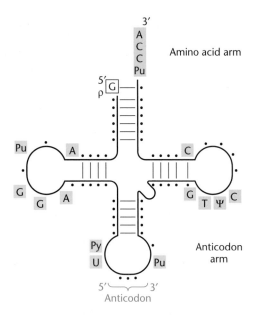

Figure 2.3 The structure of mature tRNAs, showing secondary structure and some modifications. Shown is the base pairing that holds the molecule together and some of the standard modifications. Ψ is the modified base pseudouracil. tRNAs also contain thymine (T) and thiouracil among other modifications. Boxed letters represent the bases found in the same places in all tRNAs. If a pyrimidine or purine is usually at a position, it is designated Py or Pu, respectively. The CCA end, where an amino acid is attached, is boxed.

The most common promoters are those recognized by the RNA polymerase with σ^{70}. The σ factors are named for their size, and this one has a molecular weight of 70,000 (or a molecular mass of 70 kDa).

Even promoters of the same type are not identical to each other, but they do share certain sequences known as **consensus sequences** by which they can be distinguished. Figure 2.6 shows the consensus sequence of the σ^{70} promoter in *E. coli*. The promoter sequence has two important regions: a short AT-rich region about 10 base pairs upstream of the transcription start site, known as the **−10 sequence,** and a region about 35 base pairs upstream of the start site, called the **−35 sequence.** The σ^{70} factor must bind to both sequences to start transcription.

The Polymerization Reaction

To begin transcription, the RNA polymerase binds to the promoter sequence and separates the strands of the DNA, exposing the bases. Unlike DNA polymerase, which requires helicases to separate the strands, the RNA polymerase can complete this step by itself. The complex formed when the RNA polymerase first binds to the DNA is called the **closed complex,** because the DNA strands are still wrapped around each other "closed" in a tight helix. Once the strands of DNA have been pulled apart, i.e., "opened," by RNA polymerase, the complex is called the **open complex.** Then a ribonucleoside triphosphate complementary to the nucleotide at the transcription start site bonds to the template. The second ribonucleoside triphosphate comes in, pairing with the next complementary base in the DNA template.

Figure 2.4 The structure of bacterial RNA polymerase. The core enzyme is composed of two α subunits, β and β' subunits, and an ω subunit. Subunit boundaries not shown. The holoenzyme containing the σ subunit, shown in gold, interacts with the promoter. After initiation, the σ subunit cycles off the RNA polymerase core. The regions where the DNA moves through the polymerase and where the newly synthesized RNA emerges are shown. The channel through which the ribonucleotides gain access to the polymerizing activity of the enzyme is the secondary channel shown by an arrow. Adapted with permission from R. A. Mooney and R. Landick, RNA Polymerase Unveiled, *Cell* **98:**687–690, 1999. Copyright 1999, Elsevier Science.

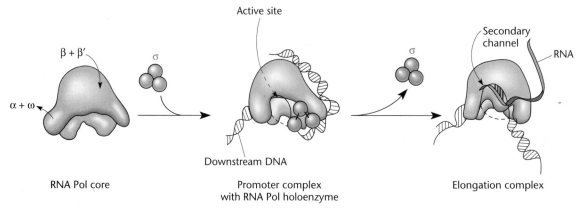

A

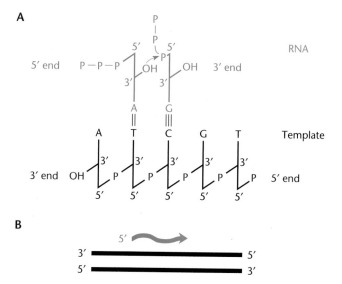

B

Figure 2.5 RNA transcription. (A) The polymerization reaction in which the newly synthesized RNA pairs with the template strand of DNA during transcription. (B) RNA polymerase synthesizes RNA in the 5'-to-3' direction, moving 3' to 5' on the template. RNA is shown as a wavy line, and both strands of DNA are shown as straight lines.

RNA polymerase catalyzes the reaction in which the α phosphate of the second nucleotide joins with the 3' hydroxyl of the first nucleotide. Then the third nucleotide comes in and bonds to the second, and so forth.

The strands of DNA in a region that is transcribed are named to reflect the sequence of the RNA made from that region. The template strand of DNA that is copied is called the **transcribed strand**. The other strand, which has the same sequence as the RNA copy, is called the **coding strand**. As shown in Figure 2.7, the sequence of a gene is usually written as the sequence of the coding strand.

As the RNA chain begins to grow, the RNA polymerase holoenzyme releases its σ factor, and the four-

subunit enzyme continues moving along the DNA strand in the 3'-to-5' direction, polymerizing RNA in the 5'-to-3' direction. Inside an opening approximately 18 bases long, called a **transcription bubble,** the elongating RNA and the complementary strand of DNA pair with each other to form a DNA-RNA hybrid of 8 base pairs, which has a double-helix structure similar to that of a double-stranded DNA molecule. At the end of the polymerization process, an antiparallel complementary copy, or **transcript,** has been made on one strand of the DNA in a particular region. Figure 2.8 shows an overview of the transcription process.

It was once assumed that after initiation occurs the RNA polymerase moves along the DNA at a uniform rate, polymerizing nucleotides into RNA. However, it is now known that the RNA polymerase often pauses as it moves along the DNA and sometimes even backs up (backtracks) before continuing. RNA polymerase pausing can be caused by at least two mechanisms (see Artsimovitch and Landick, Suggested Reading). One way is when a hairpin forms in the RNA (Figure 2.2) that has just emerged from the RNA polymerase. The hairpin can interact retroactively with the RNA polymerase by binding to a region of the β subunit of RNA polymerase (Figure 2.4) called the "flap." This somehow causes the 3' OH end of the RNA to be displaced from the active center, causing transcription to pause. Hairpins that have this effect on the RNA polymerase are called "pause hairpins." They often occur close to the 5' end of mRNAs and help regulate the transcription of operons by a process called attenuation, in which translation of the mRNA by a ribosome disrupts the hairpin and allows the RNA polymerase to continue (see chapter 12).

RNA hairpins also play a role in transcription termination where the RNA polymerase and newly synthesized RNA are released from the DNA rather than merely pausing before continuing. However, in transcription termination, the RNA hairpin may not bind to

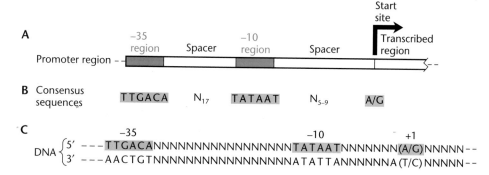

Figure 2.6 (A) Typical structure of a σ^{70} bacterial promoter with −35 and −10 regions. (B) The consensus sequences of such promoters. (C) The relationship of these sequences with respect to the start site of transcription on the two strands. RNA synthesis typically starts with an A or a G, and no primer is required. N, any nucleotide.

Figure 2.7 The coding strand has the same sequence as the mRNA. The template strand is the strand to which the mRNA is complementary if read in the 3′-to-5′ direction.

the RNA polymerase but the hairpin could form instead of the DNA:RNA hybrid required to stabilize the RNA polymerase on the DNA (see under "Transcription Termination" below).

Another type of pausing occurs when the RNA polymerase backtracks. This often happens at places where the RNA:DNA duplex that forms in the RNA polymerase is particularly weak, for example where there is a string of U's in the RNA, since a U:A pair is weaker than a C:G pair. When the RNA polymerase backtracks, the extra RNA has nowhere to go and may back up into the pore through which the rNTPs (ribonucleoside triphosphates) enter, stopping polymerization. Proteins called GreA and GreB can degrade this extra RNA, allowing the polymerase to retrace its steps and continue. We discuss GreA and GreB in more detail in chapter 8. Pausing due to backtracking is also used in a type of regulation in which the backtracked RNA polymerase remains paused until a transcription factor binds to it and allows it to continue.

All of this starting, stopping, and backing up slows the RNA polymerase, reducing the amount of RNA that is made from a gene. Therefore, genes whose products must be made in large amounts, such as the rRNA genes (see below), sometimes have special mechanisms to reduce pausing. The rRNAs have sequences called antitermination sites, which bind to the RNA polymerase and preempt the binding of hairpins that appear further along in the emerging rRNA. The rRNA genes may have an additional motivation for reducing pausing; to avoid ρ-dependent termination of transcription. The ρ factor can terminate transcription if the RNA polymerase pauses and the RNA is not being translated (Figure 2.10) and the rRNAs are not translated (see below).

Transcription Termination

Once the RNA polymerase has initiated transcription at a promoter, it will continue along the DNA, polymerizing ribonucleotides, until it encounters a transcription termination site in the DNA. These sites are not necessarily at the end of individual genes. In bacteria, more than one

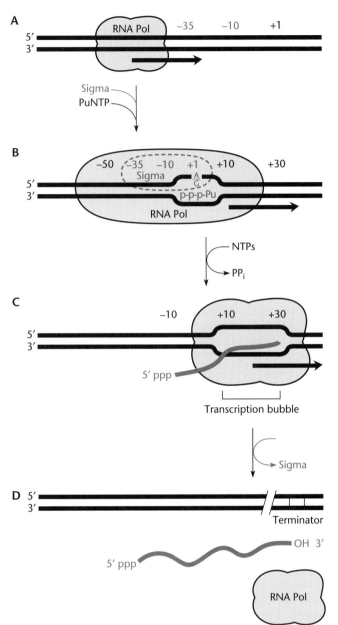

Figure 2.8 Transcription begins at a promoter and ends at a transcription terminator. (A and B) RNA polymerase (Pol) moves along the DNA until it recognizes a promoter. Transcription begins when the strands of DNA are opened at the promoter. (C) As the RNA polymerase moves along the DNA, polymerizing ribonucleotides, it forms a transcription bubble containing an RNA-DNA double-stranded hybrid, which helps hold the RNA polymerase on the DNA. The sigma factor is released. (D) The RNA polymerase encounters a transcription terminator and comes off the DNA, releasing the newly synthesized RNA.

gene is often transcribed into a single RNA, so that a transcription termination site will not occur until the end of the cluster of genes being transcribed. Even if only a single gene is being transcribed, the transcription termination site may occur far downstream of the gene.

Bacterial DNA has two basic types of transcription termination sites: **factor independent** and **factor dependent**. As their names imply, these types are distinguished by whether they work with just RNA polymerase and DNA alone or need other factors before they can terminate transcription.

FACTOR-INDEPENDENT TERMINATION

The factor-independent transcription terminators are easy to recognize because they share similar properties. As shown in Figure 2.9A, a typical factor-independent terminator sequence consists of two sequences. The first is an inverted repeat, which is one sequence followed by another that is similar when read in the opposite direction on the other strand, as though the sequence were turned around, or "inverted." The inverted repeat is followed by a short string of A's. Transcription then terminates somewhere in the string of A's in the DNA, leaving a string of U's at the end of the RNA.

Figure 2.9 also shows how a factor-independent transcription terminator might work. When an inverted repeat is transcribed into RNA, the RNA forms a **hairpin** (Figure 2.9B). According to this model, the RNA hairpin destroys the RNA-DNA hybrid in the transcription bubble that helps hold the RNA polymerase on the DNA because the hairpin is more stable than the DNA-RNA hybrid in the transcription bubble and therefore may pull the RNA from the RNA-DNA hybrid. Then, when this destabilized RNA polymerase hits the AT-rich region, the AU base pairs that form are less stable than the AT base pairs in the original DNA, so the RNA polymerase and RNA spontaneously fall off the DNA, terminating transcription.

FACTOR-DEPENDENT TERMINATION

The factor-dependent transcription terminators have very little sequence in common with each other and so are not readily apparent. *E. coli,* in which factor-dependent termination is best understood, has three transcription termination factors, Rho (ρ), Tau (τ), and NusA. Since τ and NusA are not as specific and their role is less well understood, we shall concentrate on the ρ factor. The ρ factor probably exists in all types of bacteria, so this type of termination is probably universal.

Any model for how the ρ factor terminates transcription at ρ-dependent termination sites has to incorporate the following facts about ρ-dependent termination. First, ρ usually causes the termination of RNA synthesis only

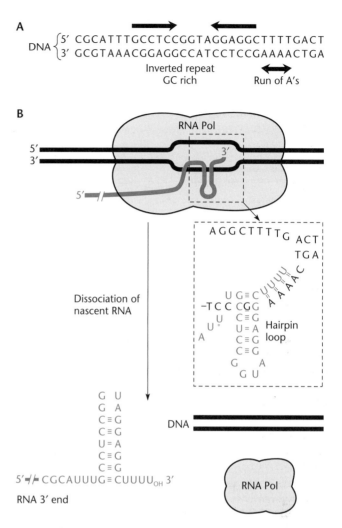

Figure 2.9 Transcription termination at a factor-independent termination site (A). (B) The RNA hairpin loop forms because the inverted repeat in the RNA is more stable than the DNA-RNA hybrid in the transcription bubble. Loss of the transcription bubble destabilizes the polymerase on the DNA, causing it to come off at a string of A's in the template.

if the RNA is not being translated. In bacteria, which lack a nuclear membrane, transcription and translation can occur at the same time, so this is an important criterion (see Introduction). Second, ρ is an RNA-dependent ATPase, so it will cleave ATP to get energy, but this will occur only if RNA is present. Finally, ρ is also an RNA-DNA helicase. It is similar to the DNA helicases that separate the strands of DNA during replication, but it will unwind only a double helix with RNA in one strand and DNA in the other.

Figure 2.10 illustrates a current model for how ρ terminates transcription: First, ρ binds to certain sequences in the RNA before the RNA polymerase reaches a ρ-dependent termination site. These sequences,

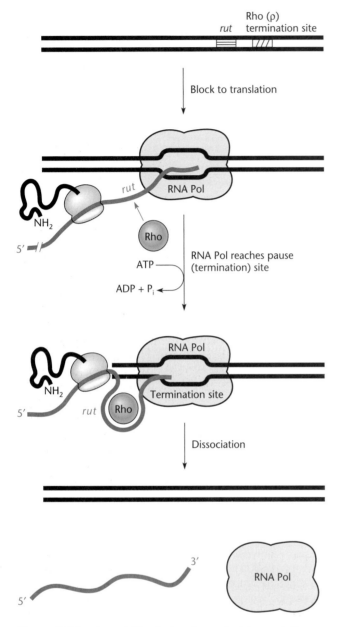

Figure 2.10 A model for factor-dependent transcription termination at a ρ termination site. The ρ-factor attaches to the mRNA at a *rut* site if the mRNA is not being translated and then chases the RNA polymerase along the mRNA until it catches up when the RNA polymerase pauses at a ρ-termination site. The ρ factor then dissociates the RNA-DNA hybrid in the transcription bubble, causing the RNA polymerase to come off the DNA.

called *rut* (for *r*ho *ut*ilization sites), are not well characterized but have a lot of C's and not much secondary structure. If that portion of the RNA is being translated, the ρ protein cannot bind to the *rut* sequences because the ribosomes get in the way. However, if a ribosome is not translating, the ρ protein can bind to the RNA; it

then moves along it in the 5′-to-3′ direction, chasing the RNA polymerase. This movement requires the energy that ρ obtains from cleaving ATP. If ρ catches up with the RNA polymerase because the enzyme pauses, or stalls, at the termination site, the RNA-DNA helicase activity of ρ will unwind the RNA-DNA hybrid, causing the RNA polymerase to be released from the DNA and transcription to terminate. The coupling of transcription termination to translation blockage ensures that after translation terminates at the end of an RNA transcript, transcription of the gene will stop as soon as the next ρ-dependent transcription signal is encountered. This type of ρ-dependent termination occurs at the end of transcribed regions and also accounts for polarity (see the section on polarity below).

According to these models, the basic difference between a factor-independent termination site and a ρ-dependent one is the dissociation of the RNA-DNA hybrid in the transcription bubble. The factor-independent site has a symmetric sequence that results in an RNA-RNA loop that is more stable than the alternative RNA-DNA hybrid and so prevents it from forming, whereas at the ρ-dependent site, the ρ protein dissociates the RNA-DNA hybrid.

rRNAs and tRNAs and Their Synthesis

Transcription of the genes for all the RNAs of the cell is basically the same. However, rRNAs and tRNAs have special roles in protein synthesis, so their fate after transcription differs from that of mRNAs.

The ribosomes are some of the largest organelles in bacterial cells and are composed of both proteins and RNA. Bacterial ribosomes contain three types of rRNA: 16S, 23S, and 5S. S (from "Svedberg," the name of the person who pioneered this way of measuring the sizes of molecules) is a measure of how fast a molecule sediments in an ultracentrifuge. In general, the higher the S value, the larger the RNA. This designation has persisted even though this method of measuring molecular size is used less often than previously.

In some strains of bacteria, the 23S rRNA is broken into two pieces. Nevertheless, the ribosomes containing the broken rRNA still function because the overall structure of the ribosome holds the pieces of rRNA together. The breaks occur because the rRNA genes contain parasitic DNAs whose sequences are cut out of the rRNA (see Box 2.2). However, unlike better behaved RNA introns, these introns are not spliced back together again after they leave the RNA.

In addition to their structural role in the ribosome, the rRNAs play a direct role in translation. The 23S rRNA is the peptidyltransferase enzyme, which joins amino acids into protein on the ribosome, making it a ribozyme (see Box 2.2). The 16S RNA is directly involved in both initiation and termination of translation.

The rRNAs and tRNAs are the most common RNAs in cells for two reasons. In a rapidly growing bacterial cell, about half of the total RNA synthesis is devoted to making these RNAs. Also, the rRNAs and tRNAs are far more stable than mRNA. They are not usually degraded until a long time after they are synthesized. With this combination of high synthesis rate and high stability, the rRNAs and tRNAs together amount to more than 95% of the total RNA in a bacterial cell.

Not only do the rRNAs physically associate in the ribosome, but also they are synthesized together as long precursor RNAs containing all three species, or forms, of rRNAs. The precursors often contain one or more tRNAs as well (Figure 2.11). After the precursor RNA is synthesized, the individual rRNAs and tRNAs are cut from it. At some point during the processing, the RNAs are modified to make the mature rRNAs and tRNAs.

Ribosomes are the site of protein synthesis; therefore, cells can increase their growth rate by increasing their number of ribosomes, because the faster a cell makes proteins, the faster it grows. In many bacteria, the coding sequences for the rRNAs are repeated in 7 to 10 different places around the genome. Duplication of these genes leads to higher rates of rRNA synthesis in these bacteria. However, although the precursor RNAs encoded by these different regions have identical rRNAs, they each encode different tRNAs and so-called spacer regions.

Proteins

It is the proteins that do most of the work of the cell. Most of the enzymes that make and degrade energy sources and make cell constituents are proteins. Also, proteins make up much of the structure of the cell. Be-cause of these diverse roles, cells have many more types of proteins than other types of constituents. Even in a relatively simple bacterium, there are thousands of different types of proteins, and most of the DNA sequences in bacteria are dedicated to genes encoding proteins.

Protein Structure

Unlike DNA and RNA, which consist of a chain of nucleotides held together by phosphodiester bonds between the sugars and phosphates, proteins consist of chains of 20 different amino acids held together by **peptide bonds.** Figure 2.12 shows an example of a peptide bond between two amino acids. The peptide bond is formed by joining the amino group (NH_2) of one amino acid to the carboxyl group (COOH) of another. These amino acids will in turn be joined to other amino acids by the same type of bond, making a chain. A short chain of amino acids is called an **oligopeptide,** and a long one is called a **polypeptide.**

Like RNA and DNA, polypeptide chains have direction. However, in polypeptides, the direction is defined by their carboxyl and amino groups. One end of the chain, the **amino terminus** or **N terminus,** will have an unattached amino group. The amino acid at this end is called the **N-terminal amino acid.** On the other end of the polypeptide, the unattached carboxyl group is called the **carboxyl terminus** or **C terminus,** and the amino acid is called the **C-terminal amino acid.** As we shall see, proteins are synthesized from the N terminus to the C terminus.

Protein structure terminology is the same as that for RNA structures. Proteins have primary, secondary, and tertiary structures, as well as quatenary structures. All of these are shown in Figure 2.13.

Figure 2.11 The precursor of rRNA. The long molecule contains the 16S, 23S, and 5S rRNAs, as well as one or more tRNAs. Nucleases cut the individual RNAs out of the long precursor after it is synthesized.

Figure 2.12 Two amino acids joined by a peptide bond. The bond connects the amino group on one amino acid to the carboxylic acid (carboxyl) group on the one preceding it. R is the side group of the amino acid that differs in each type of amino acid.

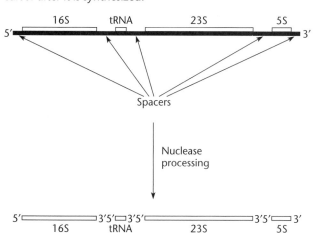

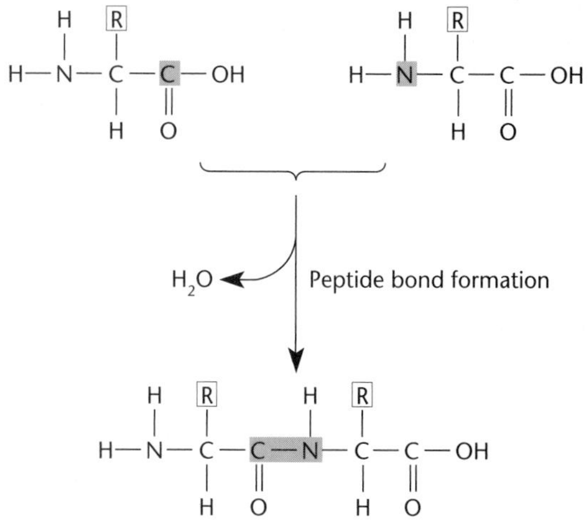

Primary structure

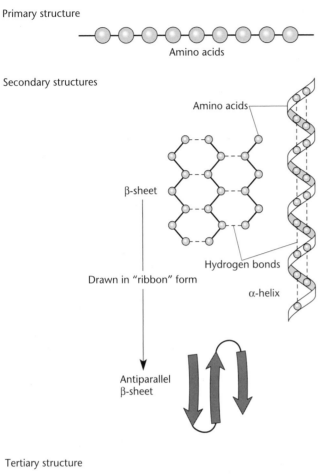

Amino acids

Secondary structures

Amino acids

β-sheet

Hydrogen bonds

Drawn in "ribbon" form

α-helix

Antiparallel
β-sheet

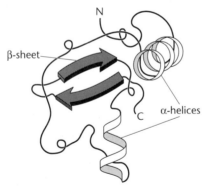

Tertiary structure

N

β-sheet

C

α-helices

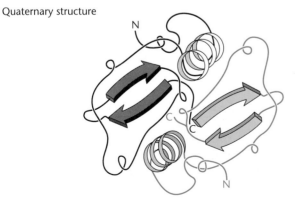

Quaternary structure

N

C
C

N

PRIMARY STRUCTURE

Primary structure refers to the sequence of amino acids and the length of a polypeptide. Because polypeptides are made up of 20 amino acids instead of just four nucleotides, as in RNA, many more primary structures are possible for polypeptides than for RNA chains.

SECONDARY STRUCTURE

Also like RNA, polypeptides can have a **secondary structure**, in which parts of the chain are held together by hydrogen bonds. However, because many more types of pairings are possible between amino acids than between nucleotides, the secondary structure of a polypeptide is more difficult to predict. The two basic forms of secondary structures in polypeptides are α-helices, where a short region of the polypeptide chain forms a helix owing to the pairing of each amino acid with the one before and the one after it, and β-sheets, in which stretches of amino acids pair with other stretches to form loops or sheetlike structures (Figure 2.13). Computer software is available to help predict which secondary structures of a polypeptide are possible on the basis of its primary structure. However, none of these programs are entirely reliable, and a technique called X-ray crystallography is the only way to be certain of the secondary structure of a polypeptide.

TERTIARY STRUCTURE

Polypeptides usually also have a well-defined **tertiary structure,** in which they fold up on themselves, with hydrophobic amino acids such as leucine and isoleucine, which are not very soluble in water, on the inside and charged amino acids such as glutamate and lysine, which are more water soluble, on the outside.

QUATERNARY STRUCTURE

Proteins made up of more than one polypeptide chain also have **quaternary structure.** Such proteins are called **multimeric proteins.** When the polypeptides are the same, the protein is a **homomultimer.** When they are different, the protein is a **heteromultimer.** Other names reflect the number of polypeptides composing the protein. For example, the term **homodimer** describes a protein composed of two identical polypeptides whereas **heterodimer** describes a protein made of two different chains encoded by different genes. The names trimer, tetramer, and so on refer to increasing numbers of polypeptides.

Figure 2.13 Primary, secondary, tertiary, and quatenary structures of proteins.

The polypeptide chains in a protein are usually held together by hydrogen bonds. The only covalent chemical bonds in most proteins are the peptide bonds holding adjacent amino acids together to form the polypeptide chains. As a result, if the protein is heated, it will fall apart into its individual polypeptide chains. However, some proteins must be unusually stable; these include extracellular enzymes, which must be able to function in the harsh environment outside the cell. Such proteins are often also held together by **disulfide bonds** between cysteine amino acids in the protein. We shall discuss how disulfide bonds are formed and broken later in the chapter.

Even though the secondary, tertiary, and quaternary structures are usually specific to a protein, determining the structure of even simple proteins is laborious and generally requires analysis of X-ray diffraction patterns of the crystallized protein.

Translation

The translation of the sequence of nucleotides in mRNA to the sequence of amino acids in a protein occurs on the ribosome.

As mentioned in the section on RNA, the ribosome is one of the largest and most complicated structures in cells, consisting of both three different RNAs and over 50 different proteins. It is also one of the major constituents of the cell, and much of the cell's capacity goes to making ribosomes. Each cell contains thousands of ribosomes, with the actual number depending upon growth conditions. It is also one of the most evolutionarily highly conserved structures in cells, having remained largely unchanged in shape and structure from bacteria to humans. For this reason the sequence of the rRNAs is often used in molecular phylogeny, to classify species (see below).

The ribosome is actually an enormous enzyme which performs the complicated role of polymerizing specific amino acids into polypeptide chains, using the information in mRNA as a guide. As such, a better name for it might have been amino acid polymerase, in analogy to DNA and RNA polymerases. It was given the historical name "ribosome" because it is large enough to have been visualized under the electron microscope, and so it was called a "some" (for body), and because it contains ribonucleotides. The recent determination of the structure of ribosomes has led to insights into how it performs its function of polymerizing amino acids (see Figure 2.17 below).

RIBOSOMAL SUBUNITS
Figure 2.14 shows the components of a ribosome. The complete ribosome, called the 70S ribosome, consists of two subunits, the 30S subunit, which contains 16S rRNA, and the 50S subunit, which contains both 23S and 5S rRNA. Each subunit also contains **ribosomal proteins;** the 30S subunit contains 21 different proteins, while the 50S subunit contains 31 different proteins. Like the different terms for rRNA, the names of ribosomal subunits are derived from their sedimentation rates. The 30S and 50S subunits normally exist separately in the cell; only when they are translating an mRNA do they come together to form the complete 70S ribosome.

The two ribosomal subunits play very different roles in translation. To initiate translation, the 30S subunit binds to the mRNA. Then the 30S ribosome binds to the 50S subunit, to make the 70S ribosome. From this point on, the role of the 30S subunit is mostly to help select the correct tRNA for each codon while the 50S subunit does most of the work of forming the peptide bonds and translocating the tRNAs from one site on the ribosome to another (see below). The 70S ribosome moves along the mRNA, allowing tRNA anticodons to pair with the mRNA codons and translate the nucleic acid chain into a polypeptide. After the polypeptide chain is completed, the ribosome separates again into the 30S and 50S subunits. We discuss the role of the subunits in more detail below in the section on initiation of translation.

Details of Protein Synthesis

In this section, we return to the process of translation in more detail. First, we discuss reading frames. Then we discuss translation elongation, or what happens as the 70S ribosome moves along the mRNA, translating its nucleotides into amino acids. Finally, we discuss how translation is initiated and terminated.

READING FRAME
As discussed in the overview of translation, each three-nucleotide sequence, or codon, in the mRNA encodes a specific amino acid, and the assignment of the codons is known as the genetic code. Because there are three nucleotides in each codon, an mRNA can be translated in three different frames in each region. Usually, initiation of translation at a initiator codon establishes the **reading frame of translation**. Once translation has begun, the ribosome moves three nucleotides at a time through the coding part of the mRNA. If the translation is occurring in the proper frame for protein synthesis, we say the translation is in the **zero frame** for that protein. If the translation is occurring in the wrong reading frame, it can be displaced either back by one nucleotide in each codon (the −1 frame) or forward by one nucleotide (+1 frame). Sometimes, translational frameshifts will occur that change the reading frame even after translation has initiated.

BOX 2.3

Traffic Jams on mRNA: Removing Stalled Ribosomes with tmRNA

When a ribosome reaches a nonsense codon in frame, the release factors release it, along with the finished polypeptide, from the mRNA. But what happens if the ribosome gets to the end of an mRNA before it encounters a nonsense codon? This might happen fairly often, because mRNA is constantly being degraded and transcription often terminates prematurely The release factors can function only at a nonsense codon, and so the ribosome should stall on the mRNA. Not only would this cause a traffic jam and use up ribosomes, but also the protein that is being made will be defective because it is shorter than normal, and accumulation of defective proteins may cause problems for the cell. This is where a small RNA called tmRNA comes to the rescue. As the name implies, tmRNA is both a tRNA and an mRNA (see the figure). It can be loaded with alanine like a tRNA but also contains a short ORF terminating in a nonsense codon like an mRNA. If the ribosome reaches the end of an mRNA without encountering a nonsense codon, the tmRNA enters the A site of the stalled ribosome, and alanine is inserted as the next amino acid of the polypeptide. Then, by a process not well understood, the ribosome shifts from translating the ORF on the mRNA to translating the ORF on the tmRNA where it soon encounters the nonsense codon. The release factors then release the ribosome and the truncated polypeptide containing a short "tag" sequence of only about 10 amino acids encoded by the tmRNA. The tag sequence which has been attached to the carboxy end of the truncated polypeptide is recognized by an enzyme called Clp protease which degrades the entire defective polypeptide so that it cannot cause problems for the cell. In some cases, tmRNA-mediated degradation may play a regulatory role, allowing the degradation of proteins until they are needed (see Abo et al. and Withey and Friedman, below).

References

Abo, T., T. Inada, K. Ogawa, and H. Aiba. 2000. SsrA-mediated tagging and proteolysis of LacI and its role in the regulation of the *lac* operon. *EMBO J.* **19:**3762–3769.

Gillet, R., and B. Felden. 2001. Emerging views on the tmRNA-mediated protein tagging and ribosome rescue. *Mol. Microbiol.* **42:**879–885. (Microreview.)

Keller, K. C., P. R. Waller, and R. T. Sauer. 1996. Role of a peptide tagging system in degradation of proteins synthesized from damaged messenger RNA. *Science* **271:**990–993.

Withey, J. and D. Friedman. 1999. Analysis of the role of *trans*-translation in the requirement of tmRNA for λ*imm*^P22 growth in *Escherichia coli. J. Bacteriol.* **181:**2148–2157.

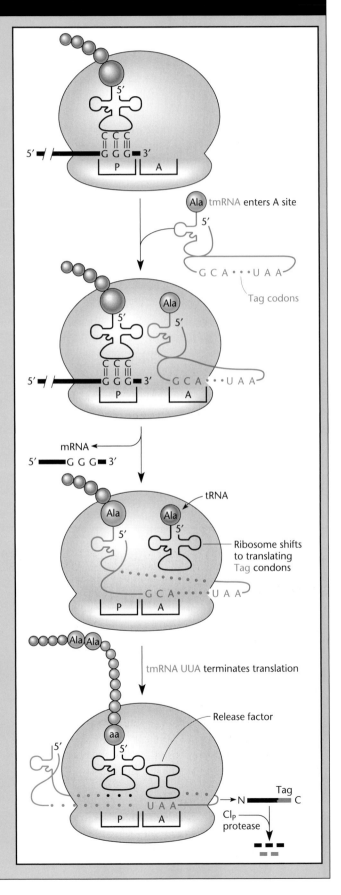

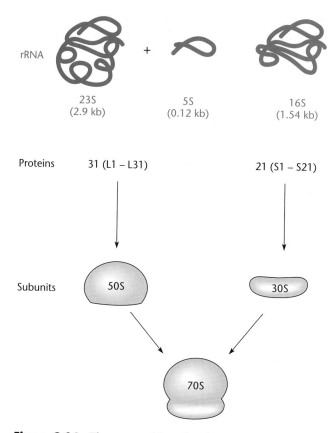

rRNA

23S
(2.9 kb)

5S
(0.12 kb)

16S
(1.54 kb)

Proteins   31 (L1 – L31)   21 (S1 – S21)

Subunits   50S   30S

70S

Figure 2.14 The composition of a ribosome containing one copy each of the 16S, 23S, and 5S rRNAs as well as many proteins. The proteins of the large 50S subunit are designated L1 to L31. The proteins of the small 30S subunit are designated S1 to S21. The simple subunit shapes shown here will be used as icons to represent ribosomes in illustrations throughout the textbook.

TRANSLATION ELONGATION

Before translation can begin, a specific amino acid is attached to each tRNA by its **cognate aminoacyl-tRNA synthetase** (Figure 2.15). Each of these enzymes specifically recognizes only one type of tRNA, hence the name *cognate*. How each cognate tRNA-synthetase recognizes its own tRNA varies, but the anticodon (i.e., the three tRNA nucleotides that base pair with the complementary mRNA sequence [see the overview section]) is not the only determinant. Often, if the anticodon changes in a given tRNA, the cognate synthetase will still attach the amino acid for the original tRNA, and that amino acid will be inserted for a different codon in the mRNA. This is the reason for nonsense suppression, which is discussed in chapter 3. Finally, the tRNA with its amino acid becomes bound to a protein called **translation elongation factor Tu (EF-Tu)**.

During translation, the ribosome moves 3 nucleotides at a time along the mRNA in the 5′-to-3′ direction, allowing tRNAs carrying amino acids (aminoacylated tRNAs) to pair with the larger molecule. Which tRNA can enter the ribosome depends on the sequence of the mRNA codon occupying the ribosome at that time. At a particular place in one of its loops, the tRNA must have 3 nucleotides that are complementary to the mRNA bases, so that the tRNA and mRNA can pair with each other. As mentioned above, this tRNA sequence is called the anticodon (Figure 2.16). To pair, the two RNA sequences must be complementary when read in opposite directions. In other words, the 3′-to-5′ sequence of the anticodon must be complementary to the 5′-to-3′ sequence of the codon.

BOX 2.4

Exceptions to the Code

Normally, translation begins at an initiation codon and then proceeds three bases at a time through the mRNA in the 5′-to-3′ direction, inserting the amino acid at a codon according to the genetic code. However, there are exceptions to this general rule. Although the code is mostly universal, in some situations a codon can mean something else. We gave the example of initiation codons that encode different amino acids when internal to a gene (see the text). Also, some organelles and primitive microorganisms use a different code. Mitochondria are notorious for using different code words for some amino acids and for termination. Mitochondria from different species sometimes even use different codons from each other! Also, some protozoans use the nonsense codons UAA and UAG for glutamine. In these organisms, UGA is the only nonsense codon. Some

yeasts of the genus *Candida*, the causative agent of thrush, ringworm, and vaginal yeast infections, recognize the codon CUG as serine instead of the standard leucine. In bacteria, the only known exceptions to the Universal Code involve the codon UGA. Usually, UGA is a nonsense codon that causes termination of translation at the end of an ORF or coding sequence. However, it encodes the amino acid tryptophan in bacteria of the genus *Mycoplasma*, which are responsible for some plant and animal diseases.

Some exceptions to the code occur only at specific sites in the mRNA. For example, UGA can encode the rare amino acid selenocysteine in some contexts. This amino acid exists at one or a very few positions in certain bacterial and eukaryotic

(continued)

BOX 2.4   (continued)

Exceptions to the Code

proteins. It has its own unique aminoacyl-synthetase, translation elongation factor Tu (EF-Tu), and tRNA, to which the amino acid serine is added and converted into selenocysteine. Then the selenocysteine is inserted for the codon UGA, but only at a very few, unique positions in proteins. But how does the tRNA insert selenocysteine at the few correct UGA codons instead of at the numerous other UGAs, which usually signify the end of a polypeptide? The answer seems to be that the specific selenocystyl EF-Tu recognizes the sequence around the selenocystyl codon and that only if the sequence around the UGA codon is right will this EF-Tu allow its tRNA to enter the ribosome. It is a mystery why the cell goes to so much trouble to insert selenocysteine in a specific site in only a very few proteins. In some instances where selenocysteine was replaced by cysteine, the mutated protein still functioned. Nevertheless, this amino acid has persisted throughout evolution, existing in a bacterial as well as human proteins.

Other examples of specific deviations from the code are high-level frameshifting and readthrough of nonsense codons. In high-level frameshifting, the ribosome can back up one base or go forward one base before continuing translation. High-level frameshifting usually occurs where there are two cognate codons next to each other in the RNA, for example, in the sequence UUUUC, where both UUU and UUC are phenylalanine codons that are presumably recognized by the same tRNA through wobble. Then the ribosome with the tRNA bound can slip back one codon before it continues translating, creating a frameshift. Sites at which high-level frameshifting occurs, "shifty sequences," usually have common features. They have a secondary structure such as a pseudoknot in the RNA just downstream of the frameshifted region, which causes the ribosome to pause (Figure 2.2). They also have a Shine-Dalgarno sequence just upstream of the frameshifted site, to which the ribosome then binds through its 16S rRNA, shifting the ribosome 1 nucleotide on the mRNA and causing the frameshift. Sometimes both the normal protein and the frameshifted protein, which will have a different carboxy end, can function in the cell. Frameshifting can also allow the readthrough of nonsense codons to make "polyproteins," as occurs in many retroviruses such as the AIDs virus. Moreover, high-level frameshifting can play a regulatory role, for example, in the regulation of the RF2 gene in *E. coli*. The RF2 protein causes release of the ribosome at the nonsense codons UGA and UAA (see "Translation Ter-

mination" below). The gene for RF2 in *E. coli* is arranged so that its function in translation termination can be used to regulate its own synthesis through frameshifting. How long the ribosome pauses at a UGA codon depends upon the amount of RF2 in the cell. If there is a lot of RF2 in the cell, pausing will be brief and the polypeptide will be quickly released by RF2. If there is less RF2, the ribosome will pause for longer, allowing time for a –1 frameshift. The RF2 protein is translated in the –1 frame, so this is the correct frame for translation of RF2 and more RF2 will be made.

In the most extreme cases of frameshifting, the ribosome can skip large sequences in the mRNA. This is known to occur in gene *60* of bacteriophage T4 and the *trpR* gene of *E. coli*. Somehow, the ribosome quits translating the mRNA at a certain codon and "hops" to the same codon further along. Presumably, the secondary and tertiary structures of the mRNA between the two codons cause the ribosome to hop. In the case of gene *60* of T4, the hopping occurs almost 100% of the time and the protein that results is the normal product of the gene. In the *E. coli trpR* gene, the hopping is less efficient and the physiological significance of the hopped form is unknown.

High-level readthrough of nonsense codons can also give rise to more than one protein from the same ORF. Instead of stopping at a particular nonsense codon, the ribosome often continues making a longer protein. Examples are the synthesis of the head proteins in the RNA phage Qβ and the synthesis of Gag and Pol proteins in some retroviruses. Many plant viruses also make readthrough proteins. Again, it seems to be the sequence around the nonsense codon that destines it for high-level readthrough. However, it is important to emphasize that these are all exceptions and normally the codons on an mRNA are translated one after the other from the translational initiation region.

References

Alam, S. L., J. F. Atkins, and R. F. Gesteland. 1999. Programmed ribosomal frameshifting: much ado about knotting! *Proc. Natl. Acad. Sci. USA* **96:**14177–14179. (Commentary).

Bock, A., K. Forchhammer, J. Heider, W. Leinfelder, G. Sawers, B. Veprek, and F. Zinoni. 1991. Selenocysteine: the 21st amino acid. *Mol. Microbiol.* **5:**515–520. (Microreview.)

Gesteland, R. F., and J. F. Atkins. 1996. Recoding: dynamic reprocessing of translation. *Annu. Rev. Biochem.* **65:**741–768.

Maldonada, R., and A. J. Herr. 1998. Efficiency of T4 Gene 60 translational bypassing. *J. Bacteriol.* **180:**1822–1830.

If the anticodon is complementary to the mRNA codon passing through the ribosome, the complex of tRNA plus EF-Tu can enter, binding to a site called the **A site** (for acceptor site) on the ribosome (Figure 2.17). The pairing of only three bases is enough to direct the right tRNA to the A site on the ribosome; in fact, sometimes the pairing

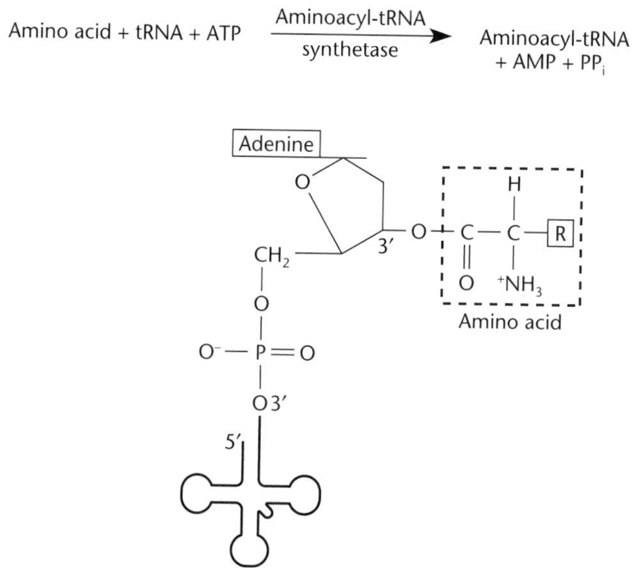

$$\text{Amino acid + tRNA + ATP} \xrightarrow[\text{synthetase}]{\text{Aminoacyl-tRNA}} \text{Aminoacyl-tRNA} + \text{AMP} + PP_i$$

Figure 2.15 Aminoacylation of a tRNA by its cognate aminoacyl-tRNA synthetase.

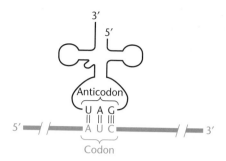

Figure 2.16 Complementary pairing between a tRNA anticodon and an mRNA codon.

of only two bases is sufficient to direct the anticodon-codon interaction (see the section on wobble, below). However, the hydrogen bonding in base pairs is not strong enough to stably hold the tRNA at this site. Apparently, the presence of the bound EF-Tu and the structure of the ribosome at the A site help to stabilize the binding. The tRNA is bound between the 30S and 50S subunits of the ribosome such that the anticodon loop is in communication with the mRNA in the 30S subunit and the acceptor end of the aminoacylated tRNA containing the bound amino acid is in communication with the 23S rRNA in the 50S subunit. After the aminoacylated tRNA is bound to the A site, the GTP on EF-Tu is cleaved to GDP and the EF-Tu is released from the ribosome.

A number of enzymes and sites on the ribosome then participate in the elongation of the polypeptide chain. A simple schematic view of this process is as follows. After the aminacylated tRNA is bound at the **A site** of the ribosome, the **peptidyltransferase**, which is actually the ribozyme 23S rRNA (28S in eukaryotes) (see Box 2.2) catalyzes the formation of a peptide bond between the incoming amino acid at the A site and the growing polypeptide at an adjacent site called the **P site**, thereby transferring the polypeptide to the tRNA at the A site of the ribosome. Another enzyme, **translation elongation factor G (EF-G)** or the **translocase**, then enters the ribosome and moves or translocates the polypeptide-containing tRNA from the A site to the P site, displacing the tRNA at the P site and making room for another aminoacylated tRNA to enter the A site. The tRNA which has been displaced then moves to yet another site, the **E site**

before it exits the ribosome. In the meantime, the mRNA is moving through the ribosome 3 nucleotides at a time and each tRNA can remain in contact with its own codon through its anticodon sequence as it marches through the ribosome, thereby discouraging frameshifting.

According to one attractive model, there are distinct A and P sites on *both* the 30S and 50S subunits of the ribosome. The anticodon end of the tRNA binds to the sites on the 30S subunit, while the acceptor CCA end, to which the amino acid or polypeptide is attached, binds to the sites on the 50S subunit. A tRNA bound to the A site on the 30S subunit and the corresponding A site on the 50S subunit is said to be bound to the A/A site, while one bound to the A site on the 30S subunit but the P site on the 50S subunit is bound to the A/P site, etc. The incoming aminoacylated tRNA first binds to the A site on the 30S subunit through its anticodon end. The CCA end of tRNA is still bound to EF-Tu and is therefore "masked." Once EF-Tu cycles off, the CCA end is free to bind to the A site on the 50S ribosome, so the aminoacylated tRNA is now bound to the A/A sites. The peptide bond then forms, concomitant with the movement of the CCA end of the tRNA to the P site on the 50S ribosome, and so it is now bound to the A/P sites. EF-G then moves the anticodon end of the tRNA to the P site on the 30S, so the tRNA is now bound to the P/P site. There may be other dual sites, called E/E sites, to which the tRNA temporarily binds before it exits the ribosome. One attractive feature of this model is that it allows the growing polypeptide chain to stay fixed at the P site on the 50S subunit and exit through the channel in the 50S subunit as it grows while a progression of tRNAs "sashay" through the ribosome, making contacts with the different sites, like folkdancers on a promenade.

Interestingly, recent structural studies have indicated that the translation factors EF-Tu and EF-G and even the release factors (see below) may mimic each other in their various states, allowing them to bind to the same limited A and P sites on the ribosome to perform their very different roles in translation (see Nyborg et al., Suggested Reading).

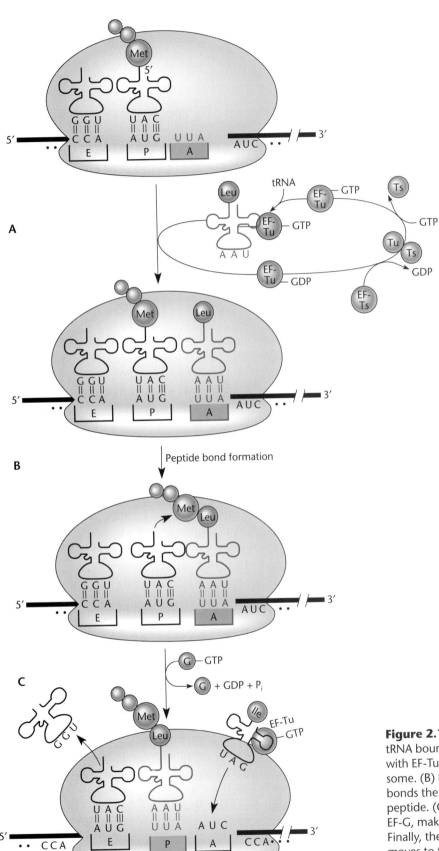

Figure 2.17 Overview of translation. (A) The tRNA bound to its amino acid and complexed with EF-Tu comes into the A site on the 30S ribosome. (B) Petidyltransferase on the 50S ribosome bonds the next amino acid to the growing polypeptide. (C) The tRNA is moved to the P site by EF-G, making room at the A site for another tRNA. Finally, the tRNA, now stripped of its amino acid, moves to the E site before exiting the ribosome. The incoming leucyl-tRNA is shown in gold.

The translation of even a single codon in an mRNA requires a lot of energy. First, ATP must be cleaved for an aminoacyl-tRNA synthetase to attach an amino acid to a tRNA (Figure 2.15). Also, EF-Tu requires that a GTP be cleaved to GDP before it can be released from the ribosome after the tRNA is bound (Figure 2.17A). Yet another GTP must be cleaved to GDP for the EF-G to move the tRNA with the attached polypeptide to the P site (Figure 2.17C). In all, the energy of three or possibly four nucleoside triphosphates is required for each step of translation.

STRUCTURE OF THE RIBOSOME

A variety of physical techniques, combined with much indirect information accumulated over the years from genetics and biochemistry, have revealed many details of the overall structure of the ribosome. Recently, the crystal structures of the individual subunits and the entire 70S ribosome have been determined and correlated with the earlier indirect information. A number of laboratories participated in this project, and this awesome achievement will go down in history as one of the major milestones in molecular biology. We can only review a few of the most salient features here.

The two subunits of the ribosome are rather round, with a flat side which binds to the other subunit, leaving a gap between them. It is through this gap that the aminoacylated rRNAs enter and pass through the ribosome, contributing their amino acid to the growing polypeptide chain. The polypeptide chain being synthesized passes out through a channel running through the 50S subunit. This channel is long enough to hold a chain of about 70 amino acids, and so a polypeptide of this length must be synthesized before the N-terminal end of a protein first emerges from the ribosome. The 50S ribosomal subunit is rather rigid with no moveable parts, but the 30S subunit has three domains or regions that can move relative to each other during translation.

The rRNAs play many of the most important roles in the ribosome, and the ribosomal proteins seem to be present mostly to give rigidity to the structure, helping cement the rRNAs in place. This has contributed to speculation that RNAs were the primoridial enzymes and that proteins only came along later in the earliest stages of life on Earth. The 23S rRNA rather than a ribosomal protein also performs the enzymatic function which forms the peptide bonds. A region of the 23S rRNA is the peptidyltransferase enzyme, which forms the peptide bonds between the carboxyl end of the growing polypeptide and the amino group of the incoming amino acid. Thus, 23S rRNA is an enzyme or ribozyme (see Box 2.2). The 23S rRNA also forms most of the channel in the 50S subunit through which the growing polypeptide passes. The 16S RNA has a region close to

its 3′ end that base pairs with the Shine-Dalgarno region in the translation initiation region in the mRNA to initiate translation (see below). A structure of the ribosome illustrating some of the features we have discussed is shown in Figure 2.18.

The Genetic Code

As mentioned in the overview, the **genetic code** determines which amino acid will be inserted into a protein for each three-nucleotide set, or **codon**, in the mRNA. More precisely, the genetic code is the assignment of each possible combination of three nucleotides to one of the 20 amino acids. The code is universal (with a few minor exceptions, see below), meaning that it is the same in all organisms, from bacteria to humans. The assignment of each codon to its amino acid appears in Table 2.1.

REDUNDANCY

In the genetic code, more than one codon often encodes the same amino acid. This feature of the code is called **redundancy**. There are $4 \times 4 \times 4 = 64$ possible codons that can be made of four different nucleotides taken three at a time. Thus, without redundancy, there would be far too many codons for only 20 amino acids.

WOBBLE

Codons that encode the same amino acid often differ only by their third base, which is why they tend to be together in the same column when the code is presented as in Table 2.1. This pattern of **redundancy** in the code is due to less stringent pairing or **wobble** between the first base in the anticodon on the tRNA and the last base in the codon on the mRNA (remember that RNA sequences are always given 5′ to 3′ and the pairing of strands of RNA, like DNA, is antiparallel). As a consequence of wobble, the same tRNA can pair with more than one of the codons for a particular amino acid, so there can be fewer types of tRNA than there are codons. For example, even though there are two codons for lysine, AAA and AAG, *E. coli* makes do with only one tRNA for lysine, which, because of wobble, can respond to both lysine codons.

The binding at the third position is not totally random, however, and certain rules apply (Figure 2.19). For example, a G in the first position of the anticodon might pair with either a C or a U in the third position of the codon but not with an A or a G, explaining why UAU and UAC, but not UAA or UAG, are codons for tyrosine. Looking at the figure, we might predict the existence of a tyrosine tRNA with the anticodon GUA. The rules for wobble are difficult to predict, however, because the bases in tRNA are sometimes modified, and a modified base in the first position of an anticodon can have altered pairing properties.

A tRNA

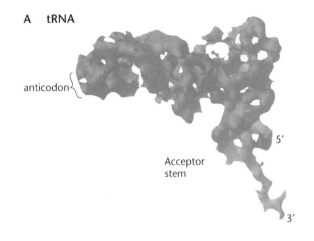

B 30S subunit   50S subunit

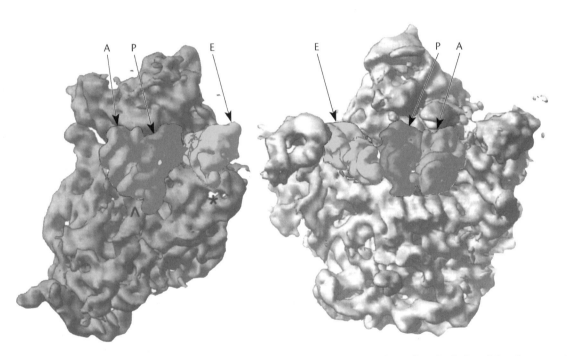

Figure 2.18 The structures of a tRNA and the ribosome. (A) The structure of a tRNA bound to the P site of the ribosome. The anticodon loop is on the left, and the 3' acceptor end where the polypeptide is attached is at the bottom. (B) The two subunits of the ribosome separated to show the channel between them through which the tRNAs move. The 30S subunit is on the left, and the 50S subunit is on the right. The tRNAs bound at the A, P, and E sites are indicated in gold. Panel A adapted with permission from M. M. Yusupov, G. Z. Yusupova, A. Baucom, K. Lieberman, T. N. Earnest, J. H. Cate, and H. F. Noller, *Science* **292**:883–896, 2001. Panel B adapted with permission from J. H. Cate, M. M. Yusupov, G. Yusupova, T. N. Earnest, and H. F. Noller, *Science* **285**:2100, 1999. Copyright, 1999, 2001, American Association for the Advancement of Science.

NONSENSE CODONS

Not all codons stipulate an amino acid; of the 64 possible, only 61 nucleotide combinations actually encode an amino acid. The other three, UAA, UAG, and UGA, are **nonsense codons** in most organisms. The nonsense codons are usually used to terminate translation at the end of genes (see the section on termination of transla-tion below). If an mRNA lacks a nonsense codon, a ribosome will stall (see Box 2.3).

AMBIGUITY

In general, each codon specifies a single amino acid, but some can specify a different amino acid depending upon where they are in the mRNA. For example, the codons

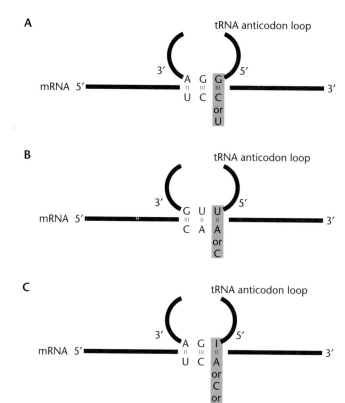

Figure 2.19 Wobble pairing between the anticodon on the tRNA and the codon in the mRNA. Many pairing interactions are possible in the third position of the codon. Alternative pairings for the anticodon base are shown: guanine (A), uracil (B), and inosine (a purine base found only in tRNAs) (C).

AUG and GUG encode formylmethionine if they are at the beginning of the coding region but they encode methionine or valine, respectively, if they are internal to the coding region. The codons CUG, UUG, or even AUU also sometimes encode formylmethionine if they are at the beginning of a coding sequence.

UGA is another exception. This codon is usually used for termination but encodes the amino acid selenocysteine in a few positions in genes (see Box 2.4) and encodes tryptophan in some types of bacteria.

CODON USAGE

Just because more than one codon can encode an amino acid does not mean that all the codons will be used equally in all organisms. The same amino acid may be preferentially encoded by a different codon in different organisms. This codon preference may reflect higher concentrations of certain tRNAs or may be related to the base composition of the DNA of the organism. While mammals and other higher eukaryotes have an average GC content of about 50% (so that there are about as

many AT base pairs in the DNA as there are GC base pairs), some bacteria and their viruses have very high or very low GC contents. How the GC content can influence codon preference is illustrated by some members of the genera *Pseudomonas* and *Streptomyces*. These organisms have GC contents of almost 75%. To maintain such high GC contents, the codon usage of these bacteria favors the codons that have the most G's and C's for each amino acid.

Translation Initiation

The process of initiating the synthesis of a new polypeptide chain is very different from the process of translation once it is under way. For example, the 30S, but not the 50S, ribosomal subunit works with other factors unique to initiation. Initiation of translation in bacteria is somewhat different from initiation in eukaryotic organisms, and we shall point out some of these differences as we go along.

TRANSLATIONAL INITIATION REGIONS

In the chain of thousands of nucleotides that make up an mRNA, the ribosome must bind and initiate translation at the correct site. If the ribosome starts working at the wrong initiation codon, the protein will have the wrong N-terminal amino acid or the mRNA will be translated out of frame and all of the amino acids will be wrong. Hence, mRNA have sequences called **translational initiation regions (TIRs)** that flag the correct first codon for the ribosome. In spite of extensive research, it is still not possible to predict with 100% accuracy whether a sequence is a TIR. However, some general features of TIRs are known.

Initiation Codon

All TIRs have an **initiation codon,** which codes for a definable amino acid. The three bases in these codons are usually AUG or GUG but in rare cases are CUG, UUG, or other sequences. There is even one known case in *E. coli* of a gene with the initiation codon AUU.

The initiation codon does not have to be the first sequence in the mRNA chain. In fact, the 5′ end of the mRNA may be some distance from the TIR and the initiation codon; this region is called the 5′ **untranslated region,** or **leader sequence.**

Regardless of which amino acid these sequences call for in the genetic code (Table 2.1), if they are serving as initiation codons, they encode methionine (actually formylmethionine [see below]) as the N-terminal amino acid. After translation, this methionine is usually cut off (see the section on removal of the methionine, below). Notice that for the initiation codons, there seems to be "wobble in reverse," with the first position being the one that can wobble instead of the third position. The significance of this is unknown.

Shine-Dalgarno Sequence

Given that the initiation codons code for amino acids other than methionine when internal to a coding region, the presence of one codon is clearly not enough to define a translational initiation region. These sequences may also occur out of frame, in which case they would not read as an amino acid. They could even appear in an mRNA sequence that is not translated at all. Obviously, other regions around these three bases must help define them as a place to begin translation.

Many bacterial genes have 5 to 10 nucleotides on the 5′ side (upstream) of the initiation codon that define a TIR. These sequences, named the **Shine-Dalgarno (S-D) sequence** after the two scientists who first noticed them, are complementary to short sequences within certain regions of the 16S RNA. Figure 2.20 shows an example of a typical bacterial TIR with a characteristic Shine-Dalgarno sequence. In all likelihood, by pairing with their complementary sequences on the 16S rRNA, S-D sequences define TIRs by properly aligning the mRNA on the ribosome. However, these sequences are not always so easy to identify because they can be very short and do not have a distinct sequence, being complementary to different regions of the 16S rRNA. Moreover, not all bacterial genes have S-D sequences. The initiation codon sometimes resides at the extreme 5′ end of the mRNA, leaving no room for an S-D sequence. In such cases, the sequence that interacts with the 16S rRNA of the ribosome may be downstream of the initiation codon.

Because of this lack of universality, often the only way to be certain that translation is initiated at a particular initiation codon is to sequence the N terminus of the polypeptide to see if the N-terminal amino acids correspond to the codons immediately adjacent to the putative initiation codon.

The steps in the initiation of translation at a TIR are outlined in Figure 2.21. It can be seen that the process of initiation of translation at a TIR is very different from the process of elongation and requires different factors. In bacteria, initiation of translation also requires a unique

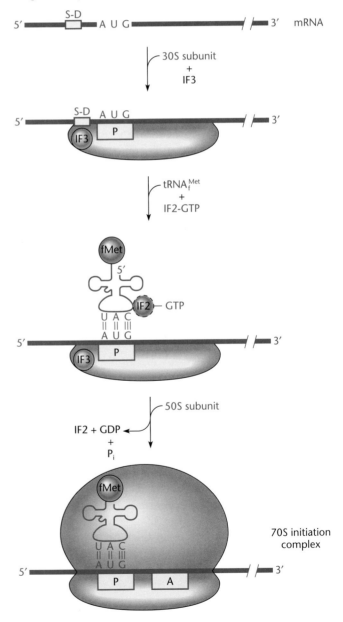

Figure 2.21 Initiation of translation. The 30S subunit and the mRNA associate. IF2 brings fMet-tRNA$_f^{Met}$ into the P site of the 30S ribosome. The 50S subunit then binds. Another aminoacylated-tRNA can enter the A site. S-D, Shine-Delgarno sequence.

Figure 2.20 Structure of a typical bacterial translation initiation region (TIR) showing the pairing between the S-D sequence in the mRNA and a short sequence close to the 3′ end of the 16S rRNA. The initiator codon, typically AUG or GUG, is 5 to 10 bases downstream of the S-D sequence. N designates any base.

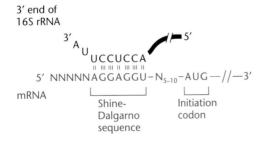

aminoacyl-tRNA, the formylmethionine tRNA (fMet-tRNA$_f^{Met}$). This unique aminoacyl-tRNA has a formyl group attached to the amino group of the methionine (see Figure 2.22), making it resemble a peptidyl-tRNA rather than a normal aminoacyl-tRNA. This causes it to bind to the P site rather than the A site of the ribosome, which is an important step in initiation, as we discuss below.

For initiation to occur, the 70S ribosome must first be separated or dissociated into its smaller 30S and 50S subunits. This dissociation occurs after the termination step of translation (see below), so that the only 70S ribosomes in the cell are those actually engaged in the act of translating an mRNA. Therefore, ribosomes are continuously cycling between the 70S ribosome and the 30S and 50S subunits depending upon whether they have initiated translation. This is called the **ribosome cycle.**

To initiate translation, first the 30S subunit binds to the mRNA at a TIR and then an fMet-tRNA$_f^{Met}$ binds. Two initiation factors, called IF1 and IF2, help the fMet-tRNA$_f^{Met}$ bind to the 30S subunit, perhaps to the 30S part of the P site discussed earlier. One of the initiation factors, IF1, recognizes the anticodon on the tRNA$_f^{Met}$, so that only a tRNA with this anticodon can bind to the 30S subunit. The other initiation factor, IF2, recognizes the formyl group on the methionine, so that this tRNA can bind only if it is formylated. A third initiation factor, IF3, is already bound to the 30S subunit and helps keep the ribosome subunits apart until after the fMet-tRNA$_f^{Met}$ and the mRNA have already bound. This initiation factor also plays a role in recognizing the initiator codon. Only if the anticodon on the tRNA$_f^{Met}$ pairs properly with the initiator codon (give or take a little wobble at the 5′ base) will the IF3 allow the 50S ribosomal subunit to bind to the complex to form the 70S ribosome (see Meinnel et al., Suggested Reading). The 70S ribosome has now been assembled, but the A site is left unoccupied to bind one of the aminoacyl-tRNAs depending upon which codon is just downstream of the initiator codon. The peptide bond then forms between the carboxyl group of the formylmethionine and the amino group of the second amino acid by the peptidyltransferase reaction, as shown in Figure 2.23. The translocase, EF-G, then moves the tRNA with the dipeptide attached to the P site and the tRNA$_f^{Met}$, now stripped of its fMet, to the E site, and translation is underway.

The Initiator tRNA

The initiator fMet-tRNA$_f^{Met}$ is synthesized somewhat differently from the other aminoacyl-tRNAs. Like other tRNAs, this special tRNA has its own aminoacyltransferase, which attaches methionine to the tRNA$_f^{Met}$. Then an enzyme called **transformylase** adds a formyl group to the amino group of the methionine on the tRNA to form fMet-tRNA$_f^{Met}$.

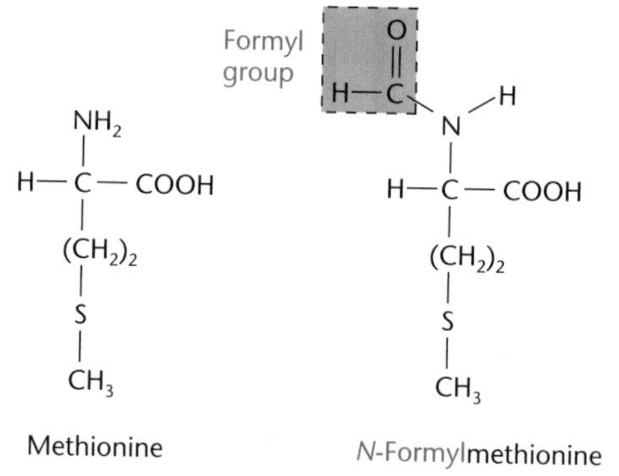

Figure 2.23 The peptidyltransferase reaction.

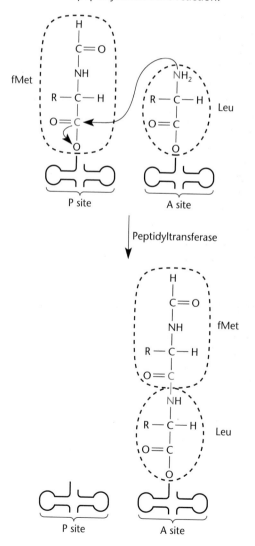

Figure 2.22 Comparison between methionine (Met) and N-formylmethionine (fMet).

Removal of the Formyl Group
and the *N*-Terminal Methionine

Normally, polypeptides do not have a formyl group attached to their N terminus. In fact, they usually do not even have methionine as their N-terminal amino acid. The formyl group is removed from the polypeptide after it is synthesized by a special enzyme called **peptide deformylase** (Figure 2.24). The N-terminal methionine is also usually removed by an enzyme called **methionine aminopeptidase**.

TRANSLATION INITIATION IN ARCHAEA AND EUKARYOTES

Translation initiation in the archaea is similar to that in the eubacteria. Like bacteria, archaea use well-defined ribosome-binding sites and formylmethionine for initiation of translation. In contrast, eukaryotes do not seem to use special ribosome-binding sites but usually use the first AUG from the 5′ end of the mRNA as the initiation codon. Also, although eukaryotes have a special methionine tRNA that responds to the first AUG codon, the methionine attached to the eukaryotic initiator tRNA is never formylated. The first methionine is, however, usually removed by an aminopeptidase after the protein is synthesized. Eukaryotes also seem to need many more initiation factors than do bacteria, although the exact role of most of these initiation factors is unknown. It is interesting that the mechanism of translation initiation in the archaea operates more like its counterpart in eubacteria while transcription initiation in the archaea seems more like that in eukaryotes. Thus, in some ways the archaea seem more like eukaryotes and in others they seem more like eubacteria (see Box 2.5).

Translation Termination

Once initiated, translation proceeds along the mRNA, one codon at a time, until the ribosome encounters the codon UAA, UAG, or UGA. These codons do not encode an amino acid, so they have no corresponding tRNA. When a ribosome comes to a nonsense codon, translation stops. Similar to the positioning of translation initiators, the nonsense codon that terminates translation may not be at the end of the mRNA molecule. The region between this last codon and the 3′ end of the mRNA is called the **3′ untranslated region**.

RELEASE FACTORS

In addition to a codon for which there is no tRNA, termination of translation requires **release factors**. These proteins recognize the specific nonsense codons and promote the release of the ribosome from the mRNA. In *E. coli*, there are four translation release factors, called RF1, RF2, RF3, and RF4. Of these, RF1 and RF2 respond to specific nonsense codons. RF1 responds to UAA and UAG, whereas RF2 responds to UAA and UGA.

Figure 2.25 outlines the process of translation termination. After translation stops at the nonsense codon, the release factors free the polypeptide chain. Termination is more efficient when it occurs in the proper **context**, that is, when a nonsense codon is surrounded by certain sequences.

RELEASE OF THE POLYPEPTIDE

How the last tRNA is removed from the polypeptide after termination remains a mystery. An enzyme called **peptidyl tRNA hydrolase** can remove tRNA from a polypeptide, but this enzyme is not associated with the ribosome, where normal release is thought to occur. The peptidyl tRNA hydrolase is also not essential for growth, as you might expect of a protein playing such an important role in translation. Instead, tRNA removal might normally be caused by EF-G in combination with another ribosomal protein called **ribosome release factor**.

Polycistronic mRNA

In bacteria and archaea, the same mRNA can encode more than one polypeptide. Such mRNAs, called **polycistronic mRNAs**, must have more than one TIR to

Figure 2.24 Removal of the N-terminal formyl group by peptide deformylase (A) and the N-terminal methionine by methionine aminopeptidase (B).

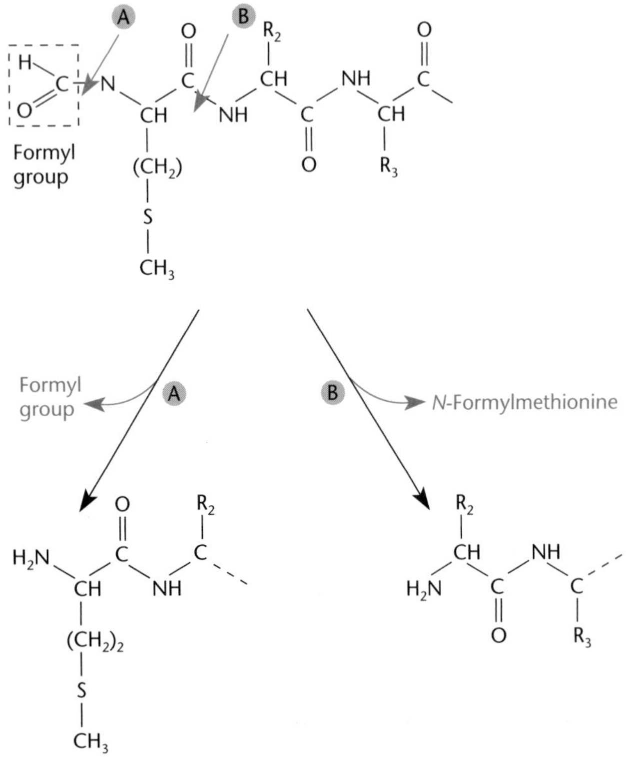

BOX 2.5

Molecular Phylogeny

The advent of DNA sequencing has made possible the sequencing of genes from a number of different organisms. A comparison of these sequences has given rise to the science of molecular phylogeny, which is based on the relatedness of sequences from different organisms rather than just on similarities in physiological and morphological characteristics. The idea is that the more recently two types of organisms separated during evolution, the more similar will be the sequences of their genes. Two properties of the translational machinery have made these components particularly useful for such studies. First, some components of the translational apparatus are very highly conserved. Second, particularly in bacteria, these components are probably not often exchanged horizontally during evolution. In other words, these are not genes that are likely to be carried by plasmids or transposons.

The conservation of components of the translation apparatus is so high that evolutionary trees can be made that include eukaryotes and archaebacteria. Such trees are usually not too different from what has been obtained from physiological and other comparisons, but there are sometimes surprises. For example a 1-mm-long organism found in sea clams around thermal vents in the sea floor was shown to be a bacterium on the basis of its 16S rRNA sequence. Also, the sequence of the translation elongation factors led to the suggestion that the archaebacteria are more closely related to eukaryotes than they are to other bacteria, prompting the change of their name to archaea.

References

Ibawe, N., K.-I. Kuma, M. Hasegawa, S. Osawa, and T. Migata. 1989. Evolutionary relationships of archaebacteria, eubacteria and eukaryotes inferred from phylogenetic trees of duplicated genes. *Proc. Natl. Acad. Sci. USA* **86:**9355–9359.

Pace, N. R. 1997. A molecular view of microbial diversity and the biosphere. *Science.* **276:**734–740.

Woese, C. R. 1987. Bacterial evolution. *Microbiol. Rev.* **51:**221–271.

allow simultaneous translation of more than one sequence of the mRNA.

The name "polycistronic" is derived from *cistron*, which is the genetic definition of the coding region for each polypeptide, and *poly*, which means many. Figure 2.26 shows a typical polycistronic mRNA, in which the coding sequence for one polypeptide is followed by the coding sequence for another. The space between two coding regions can be very short, and the coding sequences may even overlap. For example, the coding region for one polypeptide may end with the nonsense codon UA*A*, but the last A may be the first nucleotide of the initiator codon *A*UG for the next coding region. Even if the two coding regions overlap, the two polypeptides on an mRNA can be translated independently by different ribosomes.

Polycistronic mRNAs do not exist in eukaryotes, in which TIRs are much less well defined and translation usually initiates at the AUG codon closest to the 5′ end of the RNA. In eukaryotes, the synthesis of more than one polypeptide from the same mRNA usually results from differential splicing of the mRNA or to high-level frameshifting during the translation of one of the coding sequences (see Box 2.4 and the section on reading frames, above). Polycistronic RNA leads to phenomena unique to bacteria, i.e., polarity and translational coupling, which are described in the following sections.

Figure 2.25 Termination of translation at a nonsense codon. A specific release factor interacts with the ribosome stalled at the nonsense codon, causing dissociation of the ribosome from the mRNA.

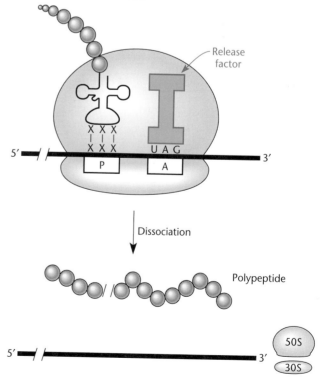

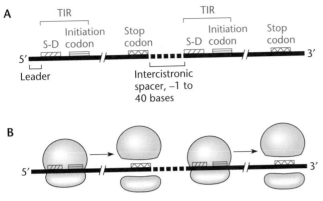

Figure 2.26   Structure of a polycistronic mRNA. The coding sequence for each polypeptide is between the TIR and the stop codon. The region 5′ of the first initiation codon is called the leader sequence, and the untranslated region between a stop codon for one gene and the next translational initiation region is known as the intercistronic spacer. S-D, Shine-Dalgarno sequence.

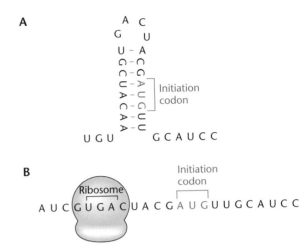

Figure 2.27   Model for translational coupling in a polycistronic mRNA. The secondary structure of the RNA blocks translation of the second polypeptide (A) unless it is disrupted by a ribosome translating the first coding sequence (B).

TRANSLATIONAL COUPLING

Two or more polypeptides encoded by the same polycistronic mRNA can be translationally coupled. Two genes are **translationally coupled** if translation of the upstream gene is required for translation of the gene immediately downstream.

Figure 2.27 shows an example of how two genes could be translationally coupled. The TIR including the AUG initiation codon of the second gene is inside a hairpin on the mRNA, and so it cannot be recognized by a ribosome. However, a ribosome arriving at the UGA stop codon for the first gene can open up this secondary structure, allowing another ribosome to bind and initiate translation on the second gene. Thus, translation of the second gene depends on the translation of the first.

POLAR EFFECTS ON GENE EXPRESSION

Some mutations that affect the expression of a gene in a polycistronic mRNA can have secondary effects on the transcription of downstream genes. Such mutations are said to exert a polar effect on gene expression. There are several types of mutations that can result in polar effects. One type of mutation that can cause a polar effect is an insertion mutation that carries a transcriptional terminator. For example, if a transposon "hops" into a polycistronic trancription unit, the trancriptional terminators on the transposon will prevent the transcription of genes downstream in the same polycistronic transcription unit. Likewise, a "knockout" of a gene by insertion of an antibiotic resistance gene with a transcriptional terminator will cause a polar effect on the genes downstream in the same transcription unit.

A second type of mutation that can cause a polar effect is a mutation that disrupts translation so that ribosomes dissociate. Within an ORF, a change of an amino acid codon to a nonsense codon can cause ribosome dissociation. Then a downstream gene expressed on the same mRNA that is translationally coupled to the upstream gene will not be translated as described above. Additional mutations that can cause this type of effect are frameshift mutations and deletion mutations that shift the reading frame. Ribosomes that are translating out of frame are likely to encounter a nonsense triplet and so dissociate.

ρ-DEPENDENT POLARITY

Recall that translation of mRNAs in bacteria normally occurs simultaneously with transcription and that the mRNA is translated in the same 5′ to 3′ direction as it is transcribed. Moreover, ribosomes often load onto a TIR as soon as it is vacated by the preceding ribosome, so the mRNA is coated with translating ribosomes. If a nonsense mutation causes dissociation of ribosomes, the abnormally naked mRNA downstream may be targeted by the transcription termination factor ρ, which may find an exposed *rut* sequence in the mRNA and cause transcription termination, as shown in Figure 2.28. The nonsense mutation will then have prevented the downstream gene by preventing its transcription as shown. Such ρ-dependent polarity effects are relatively rare because the effect occurs only if a *rut* sequence recognizable by ρ and a ρ-dependent terminator lie between the point of the mutation and the next downstream TIR.

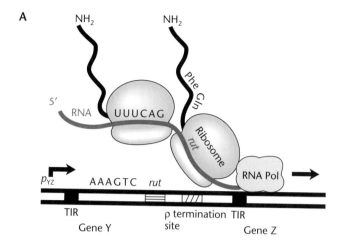

A

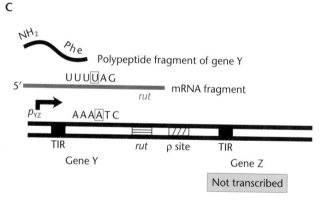

B

C

Figure 2.28 Polarity in transcription of a polycistronic mRNA transcribed from p_{YZ}. (A) Normally the *rut* site is masked by ribosomes translating the mRNA of gene Y. (B) If translation is blocked in gene Y by a mutation that changes the codon CAG to UAG (boxed in gold) the ρ factor can cause transcription termination before the RNA polymerase reaches gene Z. (C) Only fragments of the normal gene Y protein and mRNA are produced, and gene Z is not even transcribed into mRNA.

Superficially, translational coupling and polarity due to transcription termination have similar effects; in both cases, blocking the translation of one polypeptide affects the synthesis of another polypeptide normally encoded on the same mRNA. However, as we have seen, the molecular bases of the two phenomena are completely different.

Protein Folding

Translating the mRNA into a polypeptide chain is only the first step in making an active protein. To be an active protein, the polypeptide must fold into its final conformation. This is the most stable state of the protein and is determined by the primary structure of its polypeptides. In principle, every protein would eventually fold into its most stable structure. However, without the help of other factors, folding might take too long for the protein to be useful.

Chaperones

Some proteins, called **chaperones,** help other proteins fold into their final conformation. Some chaperones are dedicated to the folding of only one other protein, while others are general chaperones, which help many different proteins fold. We only discuss general chaperones here.

THE DnaK PROTEIN AND OTHER Hsp70 CHAPERONES

The **Hsp70** family of chaperones is the most prevalent and ubiquitous type of general chaperone, existing in all types of cells with the possible exception of some Archaea (see Bakau and Horwich, Suggested Reading). These chaperones are also highly conserved evolutionarily, being almost the same size and having almost the same amino acid sequence whether they come from a human cell or a bacterial cell. These chaperones are called the Hsp70 proteins because they are about 70 kDa in size and because their synthesis is induced, along with many other proteins, if cells are subjected to a sudden increase in temperature or "heat shock" (Hsp70: heat shock protein 70 kDa). They are made after a heat shock to help refold proteins that have been denatured by the sudden increase in temperature. This type of chaperone was first discovered in *E. coli,* where it was given the name **DnaK** because it is required to assemble the DNA replication apparatus of λ phage and so is required for λ DNA replication. This name for the Hsp70 chaperone in *E. coli* is still widely used in spite of being a misnomer because the chaperone has nothing directly to do with DNA. In its role as a heat shock protein, the DnaK protein of *E. coli* also functions as a cellular thermometer, regulating the synthesis of other proteins in response to a heat shock (see chapter 13).

The Hsp70-type chaperones help proteins fold by binding to certain hydrophobic regions in a nascent protein as it emerges from the ribosome and keeping these regions from binding to each other prematurely as the protein folds. The Hsp70 proteins have an ATPase activity which, by cleaving bound ATP to ADP, helps the chaperone periodically bind to, and dissociate from, the hydrophobic regions of the protein they are helping fold. The hydrophobic regions of proteins are composed mostly of amino acids that are nonpolar and so are not soluble in water (hydrophobic means "water-hating" (see the inside cover of the book for a list of the nonpolar amino acids). Stretches of nonpolar amino acids tend to be in the inside of the final folded protein, where they are less exposed to the water surrounding the protein. If these hydrophobic regions become bound to each other prematurely while the protein is being made or if they become exposed and bind to each other if the protein is unfolded by heating, the protein subsequently will not be able to fold properly and might precipitate, This is what happens when you cook an egg and the proteins precipitate into a hard white mass.

The Hsp70-type chaperones are helped in their protein-folding role by smaller proteins called **cochaperones.** These cochaperones were named DnaJ and GrpE in *E. coli,* again for historical reasons. The DnaJ cochaperone helps DnaK recognize some substrates and to cycle on and off proteins by regulating its ATPase activity. It can also sometimes function as a chaperone by itself. The GrpE protein is a nucleotide exchange protein (NEP), which helps regenerate the ATP-bound form of DnaK from the ADP-bound form, allowing the cycle to continue.

TRIGGER FACTOR

Given the prevalence and central role of DnaK in the cell, it came as a surprise that *E. coli* can live without its major Hsp70 chaperone DnaK. *E. coli* mutants that lack DnaK still multiply, albeit slowly. One reason why cells lacking DnaK are not dead seems to be that another chaperone, called **trigger factor,** can sometimes substitute for DnaK in protein folding. Much less is known about trigger factor, but it seems to be more closely associated with the ribosome and helps proteins fold as they emerge from the ribosome (see Teter et al., Suggested Reading).

THE Hsp60 CHAPERONINS

Another type of protein that helps other proteins fold is the chaperonin family. These proteins are also induced by heat shock and are about 60 kDa in size so they are called the **Hsp60 chaperonins** (see Bakau and Horwich, Suggested Reading). The first Hsp60 chaperonin was also discovered in *E. coli,* where it was named GroEL because of its role in folding the head protein of lambda phage, called the E protein. These proteins also have a smaller cochaperonin called Hsp10, which in *E. coli* is named GroES.

These proteins are called chaperonins rather than chaperones because they use a very different mechanism from the true chaperones to help other proteins fold. Rather than binding to the hydrophobic regions of other proteins as they emerge from the ribosome and preventing their premature folding, the chaperonins take up the entire improperly folded protein and then release it after helping the protein fold properly. To perform this feat, the Hsp60 chaperonin (GroEL) forms very large structures with two cylinders made up of many copies of the Hsp60 protein (see Figure 2.29). A denatured protein is taken into the chamber of one of the two cylinders, while the other unoccupied cylinder plays some sort of regulatory role. Once the protein is in the chamber, a cap made up of the cochaperonin Hsp10 (GroES) forms on the cylinder. The hydrophobic regions of the unfolded protein can then rearrange in the more favorable hydrophobic environment of the chamber to form the properly folded protein. Cleavage of ATP then removes the cap, evicting the properly folded protein.

Unlike the Hsp70-type chaperones, which exist in all cells, the Hsp60 type chaperonins seem to be confined to the eubacteria. They have not been found in the cytoplasm of eukaryotes but have been found only in their mitochondria and chloroplasts, which makes sense since these organelles are descended from eubacteria (see the introductory chapter). It is not clear why eukaryotes do not use cochaperonins to help fold their proteins in the cytoplasm, but this may reflect the generally larger size

Figure 2.29 Chaperonins. The GroEL (Hsp60)-type chaperonin multimers form two connected cylinders (shown separately). A denatured protein enters the chamber in one of the cylinders, and the chamber is then capped by the cochaperonin GroES (Hsp10). The denatured protein can then be helped to fold in the chamber. The other chamber plays a regulatory role. Details are given in the text.

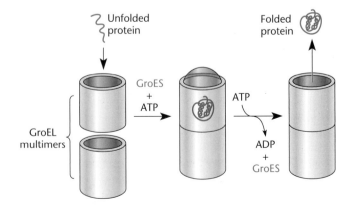

of eukaryotic proteins, which might make this type of folding impractical.

Membrane Proteins and Protein Export

Almost one-fifth of all the proteins that are made in a bacterium do not remain inside the cell; they leave the cytoplasm and enter the membranes that surround the cell or leave the cell altogether in the form of extracellular proteins. Specific words are used to designate each of these processes. Proteins that pass through the inner membrane into the periplasm or outer membrane of gram negative bacteria are **exported proteins**. Proteins that pass completely out of the cell into the external environment are **secreted proteins**. The process of passing a protein through a membrane is **translocation.**

Proteins which reside in the inner membrane of gram negative bacteria are called **inner membrane proteins.** These proteins often extend through the membrane a number of times and have stretches that are in the periplasm and other stretches that are in the cytoplasm. The stretches that traverse the membrane have mostly uncharged, nonpolar (hydrophobic) amino acids, which make them more soluble in the membranes, which are made up of lipids and so are very hydrophobic. A stretch of about 20 amino acids is long enough to extend from one side to the other of the bipolar lipid membrane, and such stretches in proteins are called the **transmembrane domains.** The other stretches between them are called the **cytoplasmic domains** and the **periplasmic domains,** depending upon whether they extend into the cytoplasm on one side of the membrane or into the periplasm on the other side. Transmembrane proteins are very important because they allow communication from outside of the cell to the cytoplasm and we discuss some of them in chapter 13 in the section on two-component global regulatory systems.

Secreted proteins and even transmembrane proteins contain many amino acids which are either polar or charged (basic or acidic), which makes it difficult for them to pass through the membranes. They must be helped in their translocation through the membrane by other specialized proteins that are dedicated to this purpose. Some of these proteins form a channel in the membrane with a hydrophilic core through which hydrophilic regions of proteins can pass. Some transport systems make their own channel, while others use the more general channel called the **translocase.** We discuss some of these systems below.

THE Sec SYSTEM

Many of the proteins that help other proteins through the membranes are part of the **Sec system.** A current picture of how the Sec system helps a protein pass through the inner membrane is outlined in Figure 2.30. The genes for this system were first found in a search for *E. coli* mutants defective in protein transport into the periplasm. This was an elegantly designed selection, and we discuss it in detail in chapter 14.

Some of the proteins of the Sec system form the channel in the membrane called the translocase, through which proteins can pass (Figure 2.30A). The channel is made up of three proteins, SecY, SecE, and SecG. Other proteins called SecD and SecF, are essential only at lower temperatures but help in the process of translocation. They lie on the periplasmic side of the channel and help release translocated proteins from the channel into the periplasmic space. They may also help close the channel after the protein has been transported, to prevent other cellular components from passing through the channel.

The **SecYEG channel** is highly evolutionarily conserved and exists in all organisms. While eukaryotes have many such channels, the translocase which helps transported proteins into the endoplasmic reticulum of eukaryotic cells is the one most similar to the SecYEG channel of bacteria. This makes sense, since the endoplasmic reticulum plays a role in protein translocation similar to the role played by the inner membrane of bacteria.

THE SIGNAL SEQUENCE

A defining feature of proteins that are to be exported through the membrane by the general secretory or Sec system is the presence at their N terminus of a short hydrophobic sequence of amino acids called the **signal sequence** (Figure 2.30A). While the amino acids in this short sequence differ from one type of protein to another, all the proteins are made up of mostly hydrophobic amino acids. The signal sequence is removed by a protease as the protein passes through the SecYEG channel. The most prevalent of the proteases which clip off signal sequences in *E. coli* is the **Lep protease** (for "leader peptide protease"), but there is at least one other more specialized protease called LspA, which removes the leader sequence from lipoproteins destined for the outer membrane. Proteins that are destined to be exported but have just been synthesized and so still retain their signal sequence are called **presecretory proteins.** A presecretory protein will become somewhat shorter before it reaches its final destination. The shortening of a protein after it is synthesized is easy to detect and is often taken as evidence that the protein is exported by the Sec system.

The Targeting Factors

The targeting factors are factors that recognize proteins to be transported into or through the inner membrane and help target them to the membrane. Enteric gram-negative

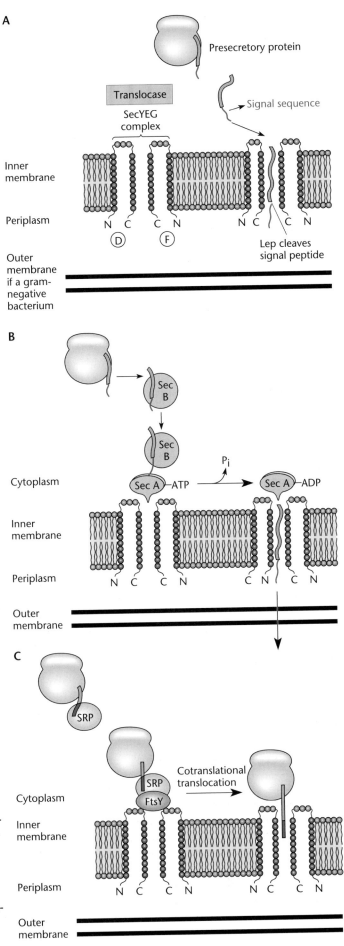

Figure 2.30 Protein translocation in gram-negative bacteria. (A) The translocase contains the SecY, SecE, and SecG proteins, shown as a simplified channel. The hydrophobic membrane-spanning regions of proteins are shown in gold. The hydrophobic signal sequence of a presecretory protein is also shown in gold.(B) The SecB-SecA pathway. The presecretory protein binds to SecB. SecA helps SecB pass the protein into the SecYEG channel as described in the text. (C) The SRP pathway. The SRP binds to the N-terminal hydrophobic region (shown in gold) of an inner membrane protein, immediately after it emerges from the ribosome. Further translation is coupled to translocation into the membrane. Details are given in the text.

bacteria like *E. coli* have at least two separate systems that target proteins to the membranes. One of these is the SecB-SecA system, which seems to be limited to bacteria. In fact, the SecB protein many be limited to gram-negative bacteria while the SecA protein is more general. The other system is the **SRP (signal recognition particle)** system, which may exist in all organisms including humans.

THE SecB-SecA PATHWAY

Proteins that have a true signal sequence are most often targeted by the SecB-SecA system in bacteria. The SecB protein is a specialized chaperone which can bind to presecretory proteins even after they are synthesized and can prevent them from folding prematurely; this may be required to leave the signal sequence exposed. The SecB chaperone then passes the unfolded protein to SecA, which functions somehow to facilitate the association of the protein with the SecYEG channel, perhaps by binding to the signal sequence and to the channel in the membrane simultaneously (Figure 2.30B). After SecA binds to the channel, the ATP on SecA is cleaved to ADP, which causes SecA to cycle off the translocase channel and drives the protein into the channel. SecB is not an essential protein, and the cell can use DnaK or other general chaperones as substitutes for SecB to transport some proteins.

THE SRP PATHWAY

The SRP pathway for protein targeting generally targets proteins that are to remain in the inner membrane rather than presecretory proteins that are transported through the membrane and beyond. It consists of a particle made up of both a small **4.5S RNA** and at least one protein, Ffh, as well as a specific receptor on the membrane, called FtsY in *E. coli*, to which it binds. This membrane protein is sometimes referred to as the docking protein because it "docks" proteins targeted by the SRP pathway on the membrane.

Figure 2.30C illustrates how the SRP system works. Rather than specifically recognizing a signal sequence, the SRP recognizes a longer hydrophobic region close to the N terminus, which will often become the first transmembrane segment (TMS) of the protein (see above). Binding of the SRP to the hydrophobic N terminus of the protein stops translation of the protein until it binds to the FtsY receptor in the membrane. Translation of the protein then continues, feeding the protein directly into the membrane as it is translated. This is called **cotranslational translocation** because the protein is translated as it is inserted into the membrane. The cotranslation of inner membrane proteins may be essential. The long hydrophobic transmembrane domains of the inner membrane would tend to fold too tightly and precipitate if they were synthesized completely in the more polar hydrophilic environment of the cytoplasm before they are transported. Presecretory proteins are generally less hydrophobic, so they can be held in a partially unfolded state by chaperones like SecB until they can be transported (see Lee and Bernstein, Suggested Reading).

What happens after the docking protein binds proteins directed by the SRP system to the membrane is less clear. Some longer inner membrane proteins may be directed to the YEG translocase channel by a process which may involve another membrane protein called YidC while others may be directed to other membrane proteins or directly to the membrane itself.

It is interesting to compare the targeting systems of eukaryotes to *E. coli*, the bacterium in which the targeting systems are best understood. The SRP was first described in eukaryotes where it is much larger, consisting of a 300-nucleotide RNA and eight proteins. Also, the SRP of eukaryotes targets both membrane and presecretory proteins to the endoplasmic reticulum (which plays an analogous role in protein transport to the inner membrane of bacteria) and seems to recognize both proteins with a true signal sequence and membrane proteins, which merely have a long hydrophobic N-terminal segment (TMS). Furthermore, the SRP of eukaryotes also stops translation of both secreted and inner membrane proteins enforcing cotranslational transport of both. As mentioned, not all types of bacteria have analogs of SecA and SecB. It will be interesting to compare other types of bacteria to see whether they all have two separate targeting systems, like *E. coli*, or whether they are more like eukaryotes and have only one.

Protein Secretion

Some translocated proteins do not stop when they reach the inner membrane, the periplasmic space, or the outer membrane but keep going until they are outside the cell. As mentioned above, these are called **secreted proteins.** They include extracellular enzymes that degrade large molecules such as polysaccharides so that the smaller degradation products can be transported back into the cell to be used as food. Other examples of secreted proteins include proteins that are secreted directly from the bacterial cell into a eukaryotic host cell to help pathogenic bacteria establish an infection and the relaxosomes that are attached to plasmids which are secreted from one bacterium to another or from a bacterium to a plant cell (see below and chapter 5).

There are basically four known types of secretion systems in gram-negative bacteria: type I, type II, type III, and type IV. The type I systems are dedicated to secreting only one protein, such as the hemolysin of *E. coli*. The type II systems are represented by the cholera toxin secretion system of *Vibrio cholerae*. The toxin is first secreted through the inner membrane by the Sec pathway discussed above and then uses its own complex structure to pass through the outer membrane. Once the toxin is outside the cell, its B subunit can help its A subunit enter a eukaryotic cell, where it acts as a toxin. Type IV secretion systems are ancestrally related to plasmid conjugation systems and will be discussed in chapter 5. The most dramatic secretion systems are the type III secretion systems. These systems form large, syringe-like multiprotein complexes that operate independently of the normal *sec* system and inject the protein directly through both membranes into a eukaryotic cell. They have attracted much attention recently because of their role in bacterial pathogenesis and their similarity in different types of pathogenic bacteria including both plant and animal pathogens (see Box. 2.6).

Disulfide Bonds

Another characteristic of proteins that are exported to the periplasm or secreted outside the cell is that many of them have **disulfide bonds** between cysteines. The sulfur atom of a cysteine in a disulfide bond is in its oxidized form, while the sulfur atom of an unbound cysteine is in its reduced form. These disulfide bonds can be either between two cysteines in the same polypeptide chain or between cysteines in different polypeptide chains. Secreted proteins need the covalent disulfide bonds to hold them together in the harsh environments of the periplasm and outside the cell. The disulfide bonds are formed as the proteins pass through the oxidizing environment in the periplasmic space between the inner and outer membranes of gram-negative bacteria by proteins called **disulfide oxidoreductases.** In *E. coli,* the major periplasmic enzymes that form disulfide bonds are **DsbA** and **DsbC** (for "disulfide bond A and C"). The Dsb proteins contain the motif Cys-X-X-Cys (two cysteines separated by two other amino acids, where X can be any amino acid), and the two cysteines normally form a disulfide bond with each other. These proteins can then form disulfide bonds in other proteins in the periplasm by an exchange reaction in which they exchange their disulfide bond for a disulfide bond in the protein being exported, as shown in Figure 2.31. In other words, the disulfide bond in DsbA or DsbC is reduced to individual cysteines when the proteins oxidize two cysteines in the protein being exported to form a new disulfide bond. This is why they are called oxidoreductases. The two cysteines in DbsA and DbsC are in turn reoxidized by other oxidoreductases, DsbB and DbsD, to form a disulfide bond. Parts of these latter two oxidoreductases extend through the inner membrane to the cytoplasm, which allows them to be rereduced in turn by the thioredoxin in the cytoplasm to continue the cycle.

Not only do proteins which are found inside the cell in the cytoplasm lack disulfide bonds, but also this type of bond cannot normally form in the cytoplasm. This is because of the "reducing atmosphere" inside the cytoplasm due to the presence of high concentrations of the small reducing molecules, mostly glutathione and thioredoxin. In fact, the appearance of disulfide bonds in some cytoplasmic regulatory proteins is taken as a signal by the cell that oxidizing chemicals are accumulating in the cell and that proteins should be made to combat the potentially lethal oxidative chemical stress.

Regulation of Gene Expression

The previous sections have reviewed how a gene is expressed in the cell, from the time mRNA is transcribed from the gene until the protein product of the gene reaches its final destination in or outside the cell and has its effect (i.e., is expressed). The products of different genes are made in vastly different amounts, depending upon how much of the product of the gene is required by the cell. This is sometimes referred to as the copy number of the protein and is determined by many factors including the strength of the promoter and the strength of the TIR on the mRNA. The amount of any particular gene product made by the cell also often varies depending upon the state in which the cell finds itself. In general, genes are expressed in the cell only when their products are needed by the cell and then only as much as is required to make the amount of product needed by the cell. This saves energy and prevents the products of different genes from interfering with each other. The process by which the output of genes is changed depending upon the state of the cell is called the **regulation of gene expression** and can occur at any stage in the expression of the gene. Genes whose products regulate the expression of other genes are called **regulatory genes.** The product of a regulatory gene can either inhibit or stimulate the expression of a gene. If it inhibits the expression, the regulation is negative; if it stimulates the expression, the regulation is positive. A regulatory protein need not necessarily be one or the other, however; some regulatory gene products are both positive and negative regulators depending upon the situation. Sometimes the product of a regulatory gene can regulate the expression of only one other gene and sometimes it can regulate the expression of many genes. The set of genes regulated by the same regulatory gene product is called a **regulon.** Sometimes a gene can also regulate its own expression as well as the expression of one or more other genes. If a gene product regulates its own expression, it is said to be **autoregulated.** We discuss the molecular mechanisms of regulation of gene expression in much more detail in chapter 12, but in this chapter we

BOX 2.6

Type III Secretion Systems and Bacterial Pathogenesis

As mentioned in the text, gram-negative bacteria contain four different types of secretion systems dedicated to secreting proteins out of the cell. Three of these, I to III, are illustrated in the figure. Type IV secretion systems are addressed in chapter 5 in the discussion of conjugation. Probably the most striking of the secretion systems are the type III systems, dedicated to secreting pathogenicity proteins directly into the cytoplasm of eukaryotic cells. Composed of about 20 proteins, these secretion systems take up certain proteins in the cytoplasm of the bacterium and secrete them directly across both bacterial membranes and through the eukaryotic cell membrane into the cytoplasm of the eukaryotic host cell. The identifying mark of proteins to be secreted by at least some of these systems is not a signal sequence on the N terminus of the protein, as it is for the *sec* system, but, rather, a sequence at the 5′ end of the mRNA encoding the protein. Apparently, the ribosome stops immediately following the translation of the 5′ end of the mRNA and continues only when the 5′ end of the mRNA binds to the channel of the type III secretion system, secreting the protein as it is synthesized.

One striking feature of type III secretion systems is how similar they are in both animal and plant pathogens. They exist in many gram-negative animal pathogens, including *Salmonella* and *Yersinia,* but are also found in many plant pathogens including *Erwinia* and *Xanthomonas.* In all these bacteria, the parts of the secretion systems involved in getting the secreted protein through the bacterial membranes are very similar. Where they differ is in the part that penetrates the eukaryotic cell wall to allow injection through the wall into the host cell cytoplasm. Animal and plant cells are surrounded by very different cell walls, and so a syringe that can penetrate the membrane of a mammalian cell would be expected to be very different from a syringe that can penetrate a plant cell wall.

Many of the proteins secreted into eukaryotic cells by type III secretion systems are involved in subverting the host defenses against infection by bacteria. In animals, one of the first lines of defense against infecting bacteria are the macrophages, phagocytic white blood cells that engulf invading bacteria and destroy them by emitting a burst of oxidizing compounds. When an invading bacterium such as *Yersinia pestis,* the infamous causative agent of bubonic plague, is approached by a macrophage, its type III secretion system will secrete a number of proteins called Yop proteins directly into the macrophage, diverting the macrophage from its purpose of engulfing the bacterium. The Yop proteins have a remarkably intimate understanding of how a macrophage works. For example, one of the Yop proteins is a tyrosine phosphatase which removes phosphates from proteins in a signal transduction system in the macrophage, blocking the signal to take up the bacterium and preventing the burst of oxidizing compounds that kill the bacterium.

Plants use very different defense mechanisms against bacteria, and so plant pathogens have to adapt their strategy accordingly. Plants defend themselves against infection by inducing necrosis or destruction of the infected tissue and inducing phenolic compounds which destroy the bacterium. This is called a hypersensitive response and is induced by proteins called Avr (for "avirulence") proteins that are injected into the plant cell by the type III secretion system. In a susceptible plant, these Avr proteins do not elicit the hypersensitive response.

References

Anderson, D. M., and O. Schneewind. 1997. A mRNA signal for the type III secretion of Yop proteins by *Yersinia enterocolitica. Science* **278:**1140–1143. (Also see the review **Silhavy, T. J.,** Death by lethal injection, **278:**1085–1086, in the same volume.)

Hueck, C. J. 1998. Type III protein secretion systems in bacterial pathogens of animals and plants. *Microbiol. Mol. Biol. Rev.* **62:**379–433.

Galan, J. E., and A. Collmer. 1999. Type III secretion machines: bacterial devices for protein delivery into host cells. *Science* **284:**1322–1328.

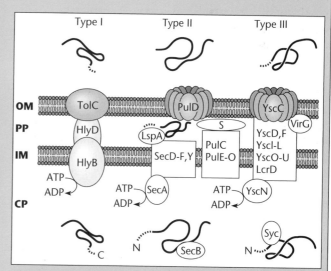

The three types of secretion systems in gram-negative bacteria. The example shows the type I secretion system for α-hemolysin, the type II secretion system for pullulanase, and the type III secretion system for Yop secretion by *Yersinia.* The specific gene products involved are shown, but their functions are not identified here. Reprinted from C. J. Hueck, *Microbiol. Mol. Biol. Rev.* **62:**379–433, 1998, with permission from the American Society for Microbiology.

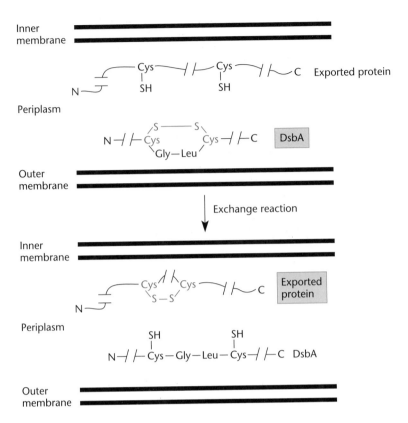

Figure 2.31 Disulfide bond formation in the periplasm. Oxidoreducatases in the periplasm exchange disulfide bonds (in gold) with the protein as it enters the periplasm. Only DsbA is shown. Details are given in the text.

briefly review some basic concepts needed to understand the next chapters.

TRANSCRIPTIONAL REGULATION

Usually the expression of a gene is regulated by controlling the amount of mRNA that is made on the gene. This is called **transcriptional regulation.** It makes sense to regulate gene expression at this level since it is wasteful to make mRNA on a gene if the expression of the gene is going to be blocked at a later stage anyway. Also, bacterial genes are often arranged in an **operon** (see above), so the same mRNA can be made on a number of genes, whose products perform related functions, simultaneously. The expanded definition of an operon is all of the genes whose products are translated from the same mRNA plus the promoter and other *cis*-acting sequences required for expression and regulation of the genes of the operon. The regulatory gene is normally not considered part of the operon unless it is contranscribed with the other genes of the operon. Sometimes a regulatory gene is part of an operon it regulates, in which case the regulatory gene is autoregulated.

The regulation of transcription of an operon usually occurs at the start point of transcription, at the promoter. Whether a gene is expressed depends upon whether the promoter for the gene is used to make mRNA. Transcriptional regulation at the promoter for

a gene can be either negative or positive, depending upon whether the regulatory gene product is a transcriptional **repressor** or a transcriptional **activator,** respectively. The difference between regulation of transcription by repressors and activators is illustrated in Figure 2.32. A repressor binds to the DNA at an **operator** sequence close to, or even overlapping, the promoter and somehow prevents RNA polymerase from using the promoter, usually either by physically obstructing access to the promoter by the RNA polymerase or by bending the promoter so that the RNA polymerase is unable to bind to it. An activator, in contrast, binds upstream of the promoter at an **upstream activator site (UAS)**, where it can help the RNA polymerase bind to the promoter or help open the promoter after the RNA polymerase binds. Whether a repressor represses transcription or an activator activates transcription depends upon the state of the repressor or activator. Some regulatory proteins bind small molecules called **effectors,** which affects their activity. Effectors are often molecules that can be used by the cell if it turns on the operon or an essential metabolite that does not have to be made by the cell if it is present in the medium. If the small molecule effector causes the transcription of the operon to be turned on, for example by binding to a repressor and changing it so that the repressor can no longer bind to the DNA, the small mole-

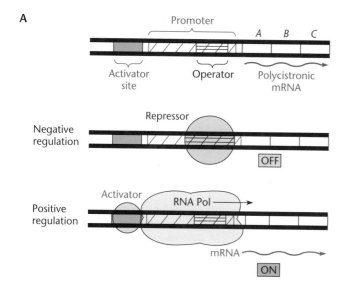

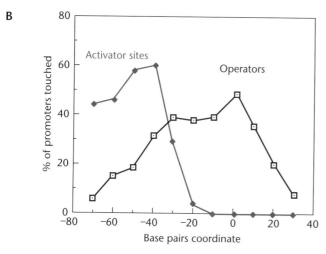

Figure 2.32 (A) The two general types of transcriptional regulation. In negative regulation, a repressor binds to an operator and turns the operon off. In positive regulation, an activator protein binds upstream of the promoter and turns the operon on. (B) Graph showing the usual locations of activator sites relative to operators. Activator sites are usually farther upstream. Each datum point indicates the middle of the known region on the DNA where a regulatory protein binds. Zero (0) on the *x* axis marks the start point of transcription. Adapted from J. Collado-Vides, B. Magasanik, and J. D. Gralla, *Microbiol. Rev.* **55**:371–394, 1991.

cule is called an **inducer**. If, by binding to a repressor, the effector causes the operon to be turned off, it is called a **corepressor**. The regulatory molecule can also have its activity changed by being phosphorylated by another protein in the cell in response to a certain set of conditions (see chapter 13, "Two-Component Regulatory Systems"). In this type of regulation, a phosphate (PO₄) group is transferred from an amino acid in an-

other protein called a phosphotransferase to an amino acid in the regulatory protein in response to some environmental condition, changing the activity of the regulatory protein.

POSTTRANSCRIPTIONAL REGULATION

More rarely, expression of a gene can be regulated in other ways besides by repressors and activators acting at the promoter for the gene. Sometimes transcription begins at the promoter but soon terminates unless certain conditions are met. This type of regulation is called **attenuation**. Alternatively, the translation of the gene may be inhibited even after the mRNA has been made. This is called **translational regulation**. The protein product of the gene may even have its activity regulated after it is made. It may be degraded by other proteins called proteases if it is not needed; it may have its activity altered by being phosphorylated, methylated, or adenoribosyated, depending upon the conditions in which the cell finds itself; or the product of a pathway may inhibit the activity of an enzyme in the pathway, by a process called **feedback inhibition**. In general, a type of regulation of gene expression which operates after the mRNA for a gene has been made is called **posttranscriptional regulation**. We discuss the many ways of regulating gene expression in subsequent chapters.

Useful Concepts

We have introduced a lot of detail in this chapter, so it is worth reviewing some of the most important concepts and words. As with any field, molecular genetics has its own jargon, and in order to follow a paper or seminar in molecular genetics, this jargon must be very familiar.

Figure 2.33 shows a typical gene with a promoter and transcription terminator. The mRNA is transcribed beginning at the promoter and ending at the transcription terminator. The direction on the DNA or RNA is indicated by the direction of the phosphate bonds between the carbons on the ribose or deoxyribose sugars in the backbone of the polynucleotide. These carbons are labeled with a prime to distinguish them from the carbons in the bases of the nucleotides. On one end of the RNA, the 5' carbon of the terminal nucleotide is not joined to another nucleotide by a phosphate bond. Therefore, this is called the **5' end**. Similarly, the other end is called the **3' end** because the 3' carbon of the last nucleotide on this end is not joined to another nucleotide by a phosphate bond. The direction on DNA or RNA from the 5' end to the 3' end is called the **5'-to-3' direction**. An RNA polymerase molecule synthesizes mRNA in the 5'-to-3' direction, moving 3' to 5' on the **transcribed strand** of DNA. The opposite strand of DNA from the transcribed strand

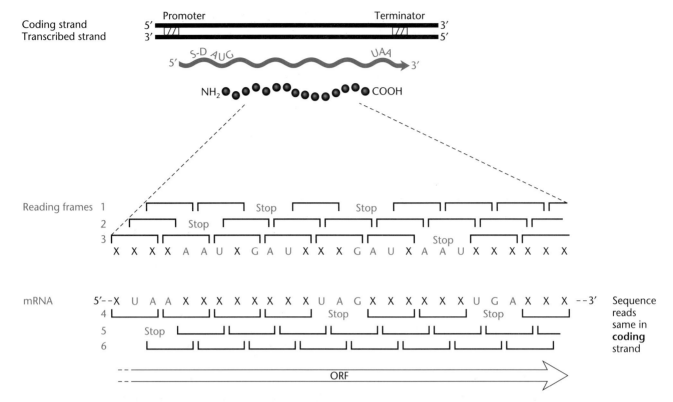

Figure 2.33 Relationship between a gene in DNA and the coding sequence in mRNA. There are a total of six different sequences in the two strands of DNA that may contain ORFs, but generally only one ORF encodes a polypeptide in each region.

has the same sequence and 5′ to 3′ polarity as the RNA so it is called the **coding strand.** Sequences of DNA in the region of a gene are usually shown as the sequence of the coding strand if this is known. Also, if a gene product is made from the region of the DNA and if the coding strand is known, the relative positions of sequences are given as though they were in a river flowing in the 5′ to 3′ direction. A sequence in the 5′ direction of another sequence on the coding strand is **upstream** of that sequence, while a sequence in the 3′ direction is **downstream.** Therefore, the promoter for a gene and the S-D sequences are both upstream of the initiation codon, while the termination codon and the transcription termination sites are both downstream.

The positions of nucleotides around a promoter are numbered as shown in Figure 2.32. The position of the first nucleotide in the RNA is called the start point and is given the number +1; the distance in nucleotides from this point to another point is numbered negatively or positively, depending upon whether the second site is upstream or downstream of the start point, respectively. We have already used this numbering system in Figure 2.6, which shows a σ^{70} promoter with consensus sequences at −10 and −35 relative to the start point of transcription. Note that these definitions can be used to

describe only a region of DNA which is known to encode an RNA or protein, where we know which is the sense and which is the transcribed strand. Otherwise, what is upstream on one strand of DNA is downstream on the other strand.

Because mRNAs are both made and translated in the 5′-to-3′ direction, an mRNA can be translated while it is still being made. We have discussed how this can lead to ρ-dependent polarity and is used to regulate the synthesis of RNA on some genes in bacteria by a process called attenuation (see chapter 12). Regulation through coupling of transcription and translation is probably not possible in eukaryotes, in which the mRNA must pass out of the nucleus into the cytoplasm, where the ribosomes reside, before the mRNA can be translated.

It is important to distinguish promoters from **translational initiation regions (TIR)** and to distinguish **transcription termination** sites from **translation termination** sites. Figure 2.33 also illustrates this difference. Transcription begins at the 5′ end of the mRNA at the promoter, but the place where translation begins, the TIR, can be some distance from the 5′ end of the mRNA. The untranslated region on the 5′ end of an mRNA upstream of the TIR is called the 5′ **untranslated region** or **leader**

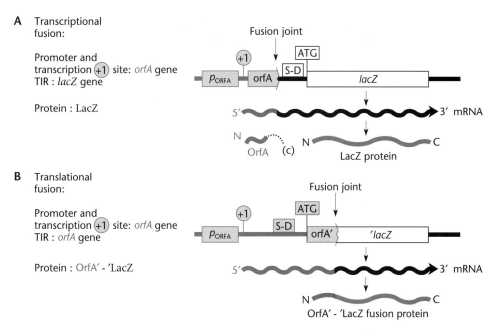

Figure 2.34 Transcriptional and translational fusions. In both types of fusions, transcription begins at the +1 site at the p_{ORFA} promoter upstream of OrfA. (A) In a transcriptional fusion, both the upstream OrfA coding region and the downstream *lacZ* reporter gene are translated from their own TIRs. Only the TIR for *lacZ* is shown as S-D and ATG (boxed). The translation of the upstream OrfA continues until it encounters a nonsense codon in frame, as indicated by the dashed line. (B) In a translational fusion, a fusion protein is translated from the TIR upstream of OrfA to make a fusion protein containing the LacZ reporter protein fused to the product of the upstream OrfA.

region and can be quite long. Similarly, a nonsense codon in the reading frame for the protein is not a transcription terminator, only a translation terminator. The transcription terminator and therefore the 3′ end of the mRNA may be some distance from the nonsense codon which terminates translation of the mRNA. The distance from the last nonsense codon to the 3′ end of the mRNA is the **3′ untranslated region.** These distinctions are dramatically illustrated in the case of polycistronic mRNAs, which encode more than one polypeptide. These mRNAs have a separate TIR and nonsense codon for each gene and can have noncoding or untranslated sequences upstream of, downstream of, and between the genes. Eukaryotes do not seem to have polycistronic mRNAs, possibly because their TIRs are less well defined, so that they cannot be recognized unless they are at the 5′ end of the mRNA.

Open Reading Frame

The concept of an **open reading frame (ORF)** is very important, particularly in this age of genomics. As discussed above, a **reading frame** in DNA is a succession of nucleotides in the DNA taken three at a time, the same way the genetic code is translated. Each DNA sequence will have six reading frames, three on each strand, as illustrated in Figure 2.33. An ORF is a string of potential codons for amino acids in DNA unbroken by nonsense codons in one of these reading frames. Computer software will show where all the ORFs are in a sequence, and most DNA sequences will have many ORFs on both strands, although most of these will be short. The region shown in Figure 2.33 contains many ORFs, but only the longest, in frame 6, is likely to encode a polypeptide. However, the presence of even a long ORF in a DNA sequence does not necessarily indicate that the sequence encodes a protein, and fairly long ORFs will often occur by chance. More information is usually required to establish which, if any, of the ORFs in a sequence encode(s) a protein.

If an ORF does encode a polypeptide, it will begin with a TIR, but, as we have discussed above, TIRs are sometimes difficult to identify. Clues to whether an ORF is likely to encode a protein may come from the choice of the third base in the codon for each amino acid in the ORF. Because of the redundancy of the code, an organism has many choices of codons for each amino acid, but each organism prefers to use some codons over others (see the section on codon usage, above, and Table 2.1).

One more direct way to determine if an ORF actually encodes a protein is to ask which polypeptides are made from the DNA in an in vitro transcription-translation system. These systems use extracts of cells, typically of *E. coli*, from which the DNA has been removed but the RNA polymerase, ribosomes, and other components of the translation apparatus remain. When DNA with the ORFs under investigation is added to these extracts, polypeptides can be synthesized from the added DNA. If the size of one of these polypeptides corresponds to the size of an ORF on the DNA, this ORF probably encodes a protein. Another way is to make a translation fusion of a reporter gene to the ORF and see if the reporter gene is expressed (see the following section).

Transcriptional and Translational Fusions

Probably the most convenient way to determine which of the possible ORFs on the two strands of DNA in a given region are translated into proteins is to make **transcriptional** and **translational fusions** to the ORFs. These methods make use of **reporter genes** such as *lacZ* (β-galactosidase), *lux* (luciferase), *cat* (chloramphenicol acetyltransferase), or other genes whose products are easy to detect. Figure 2.34 illustrates the concepts of translational and transcription fusions.

An ORF can be translated only if it is transcribed into RNA. Transcriptional fusions can be used to determine whether this has occurred. To make a transcriptional fusion, a reporter gene with the sequence for a TIR but no promoter of its own is fused immediately downstream of the promoter. If the gene is transcribed into mRNA, the reporter gene will also be transcribed and its product will be detectable in the cell. Transcriptional fusions also offer a convenient way of determining how much mRNA is made on a coding sequence. In general, the more reporter gene product that is made in the transcriptional fusion, the more mRNA was made on the coding sequence. We discuss the use of transcriptional fusions in studying the regulation of operons in subsequent chapters.

In a translational fusion, the two coding sequences are cloned in such a way that they are translated in the same frame and there are no nonsense codons between them. Translation beginning at a TIR upstream of one of the coding sequences will proceed through the other coding sequence, making a **fusion protein** that contains both polypeptide sequences. The coding sequence can be fused either to the amino terminus of the reporter gene product or to its carboxy terminus. The reporter gene product can then be assayed as before to determine how much of the fusion protein has been made. The reporter gene product must retain its activity even when fused to the potential polypeptide encoded by the ORF; otherwise it will not be detectable. Many reporter genes have been chosen because their products remain active even

when fused to other polypeptides. Translation fusions are also often used to attach affinity tags to proteins to use in their purification (see the following section).

EXPRESSION VECTORS

One important application of transcriptional and translational fusions is in **expression vectors.** It is often desirable to make large amounts of a protein for biochemical or structural studies. We may have cloned the gene for a protein which we want to purify, but the protein is normally made in small amounts and/or in an organism from which it is difficult to purify proteins. Or we may not have a convenient activity for the protein to help in its purification. To synthesize large amounts of the protein, we can clone the gene into an expression vector, which is a cloning vector that has been designed so that a gene cloned into it will be transcribed and/or translated from the cloning vector. We can then express the protein product of the gene in larger amounts from the expression vector or fuse it to other proteins that are much easier to purify.

Most expression vectors are designed to work in *E. coli*. These expression vectors can be divided into two groups: **transcription vectors** and **translation vectors**. Transcription vectors transcribe the cloned gene into mRNA from a promoter on the expression vector, but the cloned gene must be translated from its own TIR; therefore, genes from bacteria distantly related to *E. coli* or genes from eukaryotes, in general, cannot be expressed from a transcription vector. In a translation vector, the ORF to be expressed is translationally fused to the TIR on the vector so that it will be translated in *E. coli*, independently of the source of the gene. Translation vectors will translate essentially any gene, but the gene must be cloned into the cloning vector in such a way that its ORF is translationally fused in frame to the TIR on the cloning vector. To reproduce the protein exactly, it is necessary to know the DNA sequence of the gene and the position where translation of the gene normally begins so that it can be fused to the TIR in such a way that the N terminal amino acids will remain unchanged. PCR offers a convenient way to make such fusions (see chapter 1). Other problems may also be encountered when expressing foreign proteins in *E. coli*. For example, only self-splicing introns will be removed from the mRNA since *E. coli* does not have RNA-splicing systems. If the gene contains introns, it might be necessary to perform a PCR amplification of the gene for cloning from the mRNA from which the introns have already been removed, using a technique called reverse transcription PCR (RT-PCR). Also, most types of eukaryotic modifications to proteins such as glycosylation will not occur in *E. coli*.

pET-15b sequence landmarks

T7 promoter	453–469
T7 transcription start	452
His·Tag coding sequence	362–380
Multiple cloning sites	
(*Nde*I - *Bam*HI)	319–335
T7 terminator	213–259
pBR322 origin	3882
bla coding sequence	4643–5500

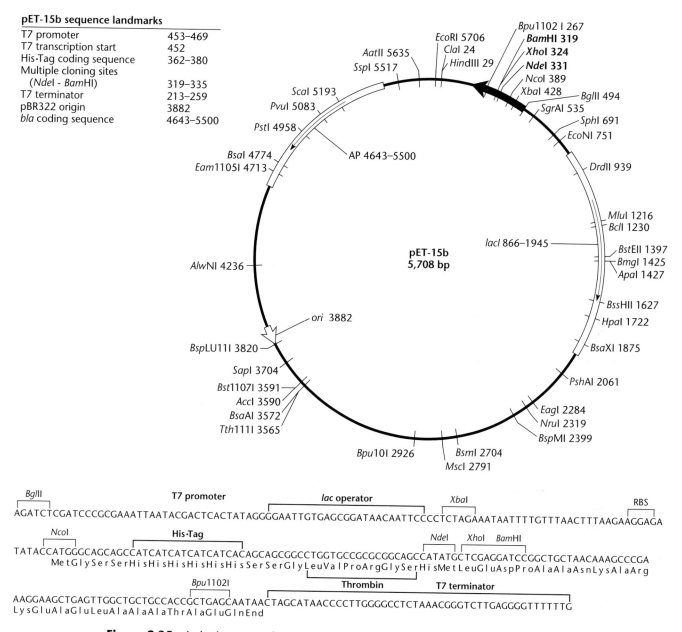

Figure 2.35 A cloning vector for attaching a His tag to a protein. If the ORF for a gene is cloned into the *Bam*HI site on the vector in such a way that it will be translated in the right frame from the upstream TIR (RBS in the figure), a chain of six histidines will be attached to the N terminus of the protein. The protein can then be easily purified on nickel-containing columns. The His tag can subsequently be cut off the purified protein by adding the specific protease thrombin. This type of vector uses the T7 phage promoter and a strong TIR to synthesize large amounts of the fusion protein. We shall discuss the use of such phage-derived cloning vectors in chapter 7. Adapted with permission from the *Novagen Catalog,* p. 108, Novagen, Madison, Wis., 1997.

Affinity Tags

Another current application of gene fusions involves attachment of **affinity tags** to proteins. This is a powerful technology because it allows the purification of unfamiliar protein products of genes that have been cloned. Affinity tags have some property that makes them very easy to purify. If the coding sequence for a protein we want to purify has been translationally fused to the coding sequence of an affinity tag, we can purify the tag and the protein will come along for the ride. One commonly used affinity vector attaches a string of six or eight histidines called a His tag to either the amino or carboxy terminus of the protein whose coding sequence has been cloned into the vector. Such a cloning vector is shown in Figure 2.35. Histidine binds strongly to nickel, and so the string of histidines will bind to a column that contains nickel. The procedure is to break open cells containing the fusion protein and pass the extracts through a column to which nickel is bound. Only the fusion protein containing the attached His tag will remain on the column; the other thousands of types of proteins will all pass through. The fusion protein can then be eluted by washing with a solution containing high concentrations of imidazole, which also binds to nickel, so will displace the His tag and the protein to which it is fused from the column. In this way, the fusion protein can be separated from most of the other proteins in the extract in a single step. Some of the affinity vectors have been designed so that they also include a sequence of a few amino acids that is the site of cleavage for a specific protease like thrombin. This sequence will be introduced in the fusion between the affinity tag and the protein being purified, so that cleavage of the purified fusion protein with the protease will remove the affinity tag from the protein. The purified protein, or almost an exact replica of the native protein, can then be used to make specific antibodies or in any other application requiring a purified protein.

Inducible Vectors

Many expression vectors express the cloned gene from a regulated promoter such as the *lac* promoter. The protein product of the cloned gene will then be synthesized only when the promoter is turned on. These inducible vectors are particularly useful if the protein product of a cloned gene is toxic to the cell. Even relatively nontoxic proteins can kill the cell if they are made in very large amounts from an expression vector. If the promoter from which the cloned gene is transcribed is regulated, the cells can be grown before the gene is induced. Even if the cells die after the induction, they will still contain large amounts of the cloned gene product. The *lac* and *L-ara* promoters are often used as inducible promoters, and we discuss the regulation of these promoters in chapter 12.

Antibiotics That Block Transcription and Translation

As in chapter 1, we devote the remainder of this chapter to a discussion of antibiotics because these compounds not only are useful therapeutic agents but also allow mutants to be isolated for genetic studies. Studies on how antibiotics affect transcription and translation have greatly contributed to our understanding of these processes.

Antibiotic Inhibitors of Transcription

Some of the components of the transcription apparatus are the targets of many antibiotics used in treatment of bacterial infections and in tumor therapy. Some of these antibiotics are made by soil bacteria and fungi and some have been synthesized chemically. Table 2.2 lists examples, along with their sources and their targets.

INHIBITORS OF RIBONUCLEOSIDE TRIPHOSPHATE SYNTHESIS
Some antibiotics that inhibit transcription do so by inhibiting the synthesis of the ribonucleoside triphosphates. An example is azaserine, which inhibits purine biosynthesis.

Uses
Azaserine and other antibiotics that block the synthesis of the ribonucleotides are usually not specific to transcription, since the ribonucleotides, including ATP and GTP, have many other uses in the cell. This lack of

TABLE 2.2	Antibiotics that block RNA synthesis	
Antibiotic	**Source**	**Target or action**
Streptolydigin	*Streptomyces lydicus*	β subunit of RNA polymerase
Actinomycin D	*Streptomyces antibioticus*	Binds DNA
Rifampin	*Nocardia mediterranei*	β subunit of RNA polymerase
Bleomycin	*Streptomyces verticulus*	Cuts DNA

specificity limits the usefulness of these antibiotics for studying transcription, although some of them have other uses.

INHIBITORS OF RNA SYNTHESIS INITIATION

Rifamycin and its more commonly used derivative, rifampin, block transcription by binding to the β subunit of RNA polymerase and specifically blocking the initiation of RNA synthesis.

Uses

The property of blocking only initiation of transcription has made these antibiotics very useful in the study of transcription. For example, they have been used to analyze the steps in initiation of RNA synthesis and to study the stability of RNA and proteins in the cell. These antibiotics are useful therapeutic agents in the treatment of tuberculosis and other bacterial infections because they inhibit the RNA polymerases of essentially all types of bacteria but not the RNA polymerases of eukaryotes, so that they are not toxic to humans and animals. Accordingly, many derivatives have been made from them.

Resistance

In rifampin-resistant mutants, one or more amino acids in the β subunit of RNA polymerase have been changed so that rifampin can no longer bind. Such chromosomal mutations conferring resistance to rifampin and other streptovarcin-type antibiotics are fairly common and have limited the usefulness of these antibiotics somewhat.

INHIBITORS OF RNA ELONGATION

Streptolydigin also binds to the β subunit of the RNA polymerase of bacteria but can block RNA synthesis after it is under way. It has a weaker affinity for RNA polymerase than does rifampin, and so it blocks transcription only when added at higher concentrations, which limits its usefulness.

Bicyclomycin targets the transcription terminator protein ρ and prevents transcription termination.

INHIBITORS THAT AFFECT THE DNA TEMPLATE

Actinomycin D and bleomycin block transcription by binding to the DNA. After bleomycin binds, it also nicks the DNA. While such drugs have been useful for studying transcription in bacteria, they are not very useful in antibacterial therapy because they are not specific to bacteria and are very toxic to humans and animals. They are, however, used in antitumor therapy.

Antibiotic Inhibitors of Translation

Because it is somewhat different from the eukaryotic translation apparatus, the translation apparatus of bacteria is a particularly tempting target for antibacterial drugs. In fact, antibiotics that inhibit translation are among the most useful of all the antibiotics, and some of them are household words. Commonly used antibiotics directed against the translation apparatus of bacteria are listed in Table 2.3, which also lists their target and source. Some of these antibiotics are also very useful in combating fungal diseases and in cancer chemotherapy. We discuss some of these antibiotics in more detail below.

INHIBITORS THAT MIMIC tRNA

Puromycin mimics tRNA with an amino acid attached (aminoacylated tRNA). It enters the ribosome as an aminoacylated tRNA, and the peptidyltransferase attaches it to the growing polypeptide. However, it does not translocate properly from the A site to the P site, and the peptide with puromycin attached to its carboxy terminus is released from the ribosome, terminating translation.

TABLE 2.3	Antibiotics that block translation	
Antibiotic	**Source**	**Target**
Puromycin	*Streptomyces alboniger*	A site of ribosome
Kanamycin	*Streptomyces kanamyceticus*	16S rRNA
Neomycin	*Streptomyces fradiae*	16S rRNA
Streptomycin	*Streptomyces griseus*	30S ribosome
Thiostrepton	*Streptomyces azureus*	23S rRNA
Gentamicin	*Micromonospora purpurea*	16S rRNA
Tetracycline	*Streptomyces rimosus*	A site of ribosome
Chloramphenicol	*Streptomyces venezuelae*	Peptidyltransferase
Erythromycin	*Saccharopolyspora erythraea*	23S rRNA
Fusidic acid	*Fusidium coccineum*	Translation elongation factor G
Kirromycin	*Streptomyces collinus*	Translation elongation factor Tu

Uses

Studies with puromycin have contributed greatly to our understanding of translation. The model of the A and P sites in the ribosome and the concept that the 50S ribosome contains the enzyme for peptidyl bond formation, which was recently shown to be the 23S RNA itself, came from studies with this antibiotic. Puromycin is not a very useful antibiotic for treating bacterial diseases, however, because it also inhibits translation in eukaryotes, making it toxic in humans and animals.

INHIBITORS OF PEPTIDE BOND FORMATION

Chloramphenicol inhibits translation by binding to ribosomes and inhibiting the peptidyltransferase reaction, preventing the formation of peptide bonds. Chloramphenicol probably binds to the 23SrRNA, which contains the peptidyltransferase activity, although ribosomal proteins are also part of the binding site.

Uses

Chloramphenicol is effective at low concentrations and therefore has been one of the most useful antibiotics for studying cellular functions. For example, it has been used to determine the time in the cell cycle when proteins required for cell division and for initiation of chromosomal replication are synthesized. It is also quite useful in treating bacterial diseases since it is not very toxic for humans and animals, being fairly specific for the translation apparatus of bacteria. Whatever toxicity it does have may result from inhibition of the translational apparatus of mitochondria, which is similar to the translation apparatus of bacteria (see the introduction to this book). Chloramphenicol is bacteriostatic, which means that it stops the growth of bacteria without actually killing them. Such antibiotics should not be used in combination with antibiotics that depend upon cell growth for their killing activity, such as penicillin, since they neutralize the effect of these other antibiotics.

Resistance

It takes many mutations in ribosomal proteins to make bacteria resistant to chloramphenicol, so that resistant mutants are very rare. Some bacteria have enzymes that inactivate chloramphenicol. The genes for these enzymes are often carried on plasmids and transposons, interchangeable DNA elements that will be discussed in subsequent chapters. The best-characterized chloramphenicol resistance gene is the *cat* gene of the transposon Tn9, whose product is an enzyme that specifically acetylates (adds an acetyl group to) chloramphenicol, thus inactivating it. The *cat* gene has been used extensively as a reporter gene to study gene expression in both bacteria and eukaryotes and has been introduced into many plasmid cloning vectors.

Erythromycin is a member of a large group of antibiotics called the macrolide antibiotics. These antibiotics may also inhibit translation by binding to the 23S rRNA and blocking either the peptidyltransferase reaction or the translocation step, causing the peptidyl-tRNA to dissociate from the ribosome.

Uses

Erythromycin and other macrolide antibiotics have been among the most useful antibiotics in treating bacterial diseases including *Legionella* and *Mycoplasma* infections.

Resistance

Some bacterial mutants resistant to macrolides have an altered L4 ribosomal protein. Some plasmids and transposons confer resistance to the macrolide antibiotics by methylating the 23S rRNA to prevent binding of the antibiotic or by acetylating or cleaving the antibiotics to inactivate them. The increasingly high levels of natural resistance to these antibiotics due to overuse has begun to limit their usefulness.

INHIBITORS OF BINDING OF AMINOACYLATED tRNA TO THE A SITE

Tetracycline inhibits translation by inhibiting the binding of the tRNA to the A site of the ribosome.

Uses

Tetracycline is another useful antibiotic for treating bacterial diseases, although it is somewhat toxic to humans because it also inhibits the eukaryotic translation apparatus.

Resistance

In some types of bacteria, ribosomal mutations confer low levels of resistance to tetracycline by changing protein S10 of the ribosome. Also, transposons and plasmids sometimes carry genes that confer resistance to tetracycline. One of these genes, *tetM*, carried by the conjugative transposon Tn916 and its relatives (see chapter 5), encodes an enzyme that confers resistance by methylating certain bases in the 16S rRNA. The *tetM* gene is ubiquitous; related genes occur in both gram-positive and gram-negative bacteria. Other tetracycline resistance genes, such as the *tet* genes carried by transposon Tn10 and plasmid pSC101 of *E. coli*, encode membrane proteins that confer resistance by pumping tetracycline out of the cell. These tetracycline resistance genes are extensively used as reporter genes and as markers for genetic analysis in *E. coli*, and the *tetA* gene from pSC101 has been introduced into many plasmid cloning vectors (see chapter 4). However, the *tet* genes of Tn10 and pSC101 are specific for *E. coli* and do not confer tetracycline resistance in many other types of bacteria, which limits their usefulness in bacteria other than *E. coli.*

INHIBITORS OF TRANSLOCATION

Fusidic acid specifically inhibits translation elongation factor G (EF-G, called EF-2 in eukaryotes), probably by preventing its dissociation from the ribosome after GTP cleavage. It has been very useful in studies of the function of ribosomes. In *E. coli,* mutations that confer resistance to fusidic acid are in the *fusA* gene, which encodes EF-G.

Kanamycin and its close relatives neomycin and gentamicin are members of a larger group of antibiotics, the aminoglycoside antibiotics, which also includes streptomycin. Their mechanism of action is somewhat obscure, but they seem to affect some aspect of translocation. They also cause misreading of mRNA by the ribosome, and the high level of translation errors they cause could be lethal.

Uses

Aminoglycosides have a very broad spectrum of action, and some of them inhibit translation in plants and animal cells as well as in bacteria. For example, the ability of neomycin to block translation in plants and animals has made it very useful in biotechnology, where it is used to select transgenic plants containing bacterial genes conferring resistance to these antibiotics (see the discussion of resistance below).

Resistance

Bacterial mutants resistant to aminoglycosides are quite rare, and multiple mutations are required to confer high levels of resistance. The fact that resistant mutants are rare has contributed to the usefulness of kanamycin and its relatives in biotechnology. Some genes confer resistance to these antibiotics by encoding enzymes that phosphorylate or acetylate them. For example, the *neo* gene for kanamycin and neomycin resistance, from transposon Tn5, phosphorylates these antibiotics. The *neo* gene has been very important in genetics and biotechnology because it expresses kanamycin resistance in almost all gram-negative bacteria and even makes plant and animal cells resistant to kanamycin (more accurately, to a derivative, G418), provided that it is transcribed and translated in the plant or animal cells.

INHIBITORS WHICH BIND TO 23S RNA

Thiostrepton and other thiopeptide antibiotics block translation by binding to 23S rRNA in the region of the ribosome involved in the peptidyltransferase reaction. Thiostrepton is specific to gram-positive bacteria; it does not enter gram-negative bacterial cells.

Uses

Thiostrepton has limited usefulness because it is is not very soluble. It is used mostly in veterinary medicine and agriculture.

Resistance

Most thiostrepton-resistant mutants are missing the L11 ribosomal protein from the 50S ribosomal subunit. This protein seems not to be required for protein synthesis but plays a role in guanosine tetraphosphate (ppGpp) synthesis (see chapter 13). Other mutations confer resistance by changing nucleotides 1067 and 1095 in the 23S rRNA; these nucleotides presumably are close to where the antibiotic binds. Plasmids and transposon genes can confer thiostrepton resistance by methylating ribose sugars of the 23S rRNA in certain positions. Eukaryotes may be insensitive to this antibiotic because the ribose sugars of the analogous eukaryotic 28S rRNAs are normally extensively methylated.

INHIBITORS WHICH BIND TO THE 30S RIBOSOME

Of all the antibiotics that block translation, the mechanism of action of streptomycin has been studied most extensively. Streptomycin binds very tightly to the 30S ribosome, probably directly to the 16S RNA, and distorts the A site, thereby preventing the correct positioning of fmet-tRNA$_f^{Met}$ for initiation. Other effects are evident at lower concentrations; these include an increased misreading of mRNA, causing the cell to accumulate defective proteins. However, most cell killing probably results from the alteration of the A site.

Uses

Streptomycin is one of the most useful antibiotics because it is specific for bacteria and works at low concentrations. It is used for treating bacterial diseases of humans and animals as well as of plants. For example, it is sometimes sprayed on apple orchards to control fire blight, a disease of apples.

Resistance

The molecular basis for chromosomal streptomycin resistance in bacteria has also been well characterized. *E. coli* mutants that are streptomycin resistant have an altered 30S ribosomal protein, S12, which is the product of the *rpsL (strA)* gene. Somehow, the altered S12 protein prevents streptomycin binding to the 16S RNA. Mutations that change the *rpsL* gene can make cells almost totally resistant to streptomycin.

Some plasmids and transposons can also carry genes that confer resistance to streptomycin by encoding enzyme products that specifically inactivate the antibiotic. Some of these enzymes adenylate (transfer an AMP to) the antibiotic, while others phosphorylate it.

SUMMARY

1. RNA is a polymer made up of a chain of ribonucleotides. The bases of the nucleotides—adenine, cytosine, uracil, and guanine—are attached to the five-carbon sugar ribose. Phosphate bonds connect the sugars to make the RNA chain, attaching the third (3′) carbon of one sugar to the fifth (5′) carbon of the next sugar. The 5′ end of the RNA is the nucleotide that has a free phosphate attached to the 5′ carbon of its sugar. The 3′ end has a free hydroxyl group at the 3′ carbon, with no phosphate attached. RNA is both made and translated from the 5′ end to the 3′ end.

2. After they are synthesized, RNAs can undergo extensive processing and modification. Processing is when phosphate bonds are broken or new phosphate bonds are formed. Modification is when the bases or the sugars of the RNA are chemically altered, for example, by methylation. The rRNAs and tRNAs, but not the mRNAs, of bacteria are extensively modified.

3. The primary structure of an RNA is its sequence of nucleotides. The secondary structure is formed by hydrogen bonding between bases in the same RNA to give localized double-stranded regions. The tertiary structure is the three-dimensional shape of the RNA due to the stiffness of the double-stranded regions of secondary structure. All RNAs including mRNA, rRNA, and tRNA probably have secondary and tertiary structure.

4. The enzyme responsible for making RNA is called RNA polymerase. One of the largest enzymes in the cell, the bacterial RNA polymerase, has four subunits plus another detachable subunit, the σ factor, which comes off after the initiation of transcription.

5. Transcription begins at well-defined sites on DNA called promoters. What type of promoter is used depends upon the type of σ factor bound to the RNA polymerase.

6. Transcription stops at sequences in the DNA called transcription terminators, which can be either factor dependent or factor independent. The factor-independent terminators have a string of A's that follows a symmetric sequence. This sequence allows the RNA transcribed from that region to fold back on itself to form a loop or hairpin, which causes the RNA molecule to fall off the DNA template. The factor-dependent terminators do not have such a well-defined sequence. The ρ protein is the best-characterized termination factor in *E. coli*.

7. Most of the RNA in the cell falls into three groups: messenger (mRNA), ribosomal (rRNA), and transfer (tRNA). Of these, mRNA is very unstable, existing for only a few minutes before being degraded. rRNAs in bacteria are further divided into three types: 16S, 23S, and 5S. Both rRNA and tRNA are very stable and account for about 95% of the total RNA. Other RNAs include the primers for DNA replication and small RNAs involved in regulation or RNA processing.

8. Ribosomes, the site of protein synthesis, are made up of two subunits, the 30S subunit and the 50S subunit, as well as many proteins. The 16S rRNA is in the 30S subunit, while the 23S and 5S rRNAs are in the 50S subunit.

9. Polypeptides are chains of the 20 amino acids, which are held together by peptide bonds between the amino group of one amino acid and the carboxyl group of another. The amino terminus or N terminus of the polypeptide has the amino acid with an unattached amino group. The carboxyl terminus or C terminus of a polypeptide has the amino acid with a free carboxyl group.

10. Translation is the synthesis of polypeptides from mRNA. During translation, the mRNA moves in the 5′-to-3′ direction along the ribosome three nucleotides at a time. Three reading frames are possible depending on how the ribosome is positioned at each triplet.

11. The genetic code is the assignment of each possible three-nucleotide codon sequence in mRNA to 1 of 20 amino acids. The code is redundant, with more than one codon sometimes encoding the same amino acid. Because of wobble, the first position of the tRNA anticodon does not have to be complementary to the third position of the antiparallel codon sequence.

12. Initiation of translation occurs at translation initiation regions (TIRs) on the mRNA that consist of an initiation codon, usually AUG or GUG, and often a Shine-Dalgarno (S-D) sequence, a short sequence that is complementary to part of the 16S rRNA and precedes the initiation codon.

13. The first tRNA to enter the ribosome is a special methionyl-tRNA called fMet-tRNA$_f^{Met}$, which carries the amino acid formylmethionine. After the polypeptide has been synthesized, the formyl group and often the first methionine are removed.

14. Translation termination occurs when one of the terminator or nonsense codons UAA, UAG, or UGA is encountered as the ribosome moves down the mRNA. Proteins called ribosome release factors (RFs) are also required for release of the polypeptide.

15. The primary structure of a polypeptide is the sequence of amino acids in the polypeptide. Proteins can be made up of more than one polypeptide chain, which can be the same as or different from each other. The secondary structure results from hydrogen bonding of the amino acids to form α-helical regions and β-sheets. Tertiary structure refers to how the chains fold up on themselves, and quatenary structure refers to one or more different polypeptide chains folding up on each other. Proteins can also be held together by disulfide linkages between cysteines in the protein. Generally only proteins that are exported into the periplasm or out of the cell have disulfide bonds.

(continued)

SUMMARY (continued)

16. Proteins that help other proteins fold are called chaperones. The most ubiquitous chaperones are the Hsp70 chaperones, called DnaK in *E. coli*, which are almost the same in all types of cells from bacteria to humans. These chaperones bind to the hydrophobic regions of proteins and prevent them from associating prematurely. They are aided by their smaller cochaperones, DnaJ and GrpE, which help in binding to proteins and cycling ADP off the chaperone, respectively. Other proteins, called Hsp60 chaperonins, also help proteins fold, but by a very different mechanism. One Hsp60 chaperonin, called GroEL in *E. coli*, forms large cylindrical structures with internal chambers that take up unfolded proteins and help them refold properly. A cochaperonin called GroES forms a cap on the cylinder after the unfolded protein is taken up. The chaperonins are only found in bacteria and in the organelles of eukaryotes.

17. The process of passing proteins through membranes is called transport. Proteins which pass through the inner membrane into the periplasm are said to be exported. Proteins which pass out of the cell are secreted. The *sec* system, responsible for transporting many proteins into and through the inner membrane, consists of the SecYEG channel in the inner membrane through which proteins pass. Proteins to be transported through the inner membrane and whose final destination is the inner membrane are recognized by different targeting factors. The SecB-SecA system recognizes proteins that are to be transported through the membrane by the *sec* system after they are synthesized. These proteins characteristically have a short hydrophobic signal sequence at their N terminus which is cleaved off by a peptidase as the protein passes through the SecYEG channel in the inner membrane. These targeting factors are unique to bacteria. Another targeting system, the signal recognition particle (SRP), specifically recognizes proteins destined to remain in the inner membrane. The SRP binds to a hydrophobic region close to the N terminus of the protein as it emerges from the ribosome. This binding somehow stops translation of the protein until the SRP binds to a receptor on the membrane. Translation can then continue, feeding the protein into the membrane as it is synthesized. The SRP targeting system is more universal, being found in a modified form also in eukaryotes, where it plays a more general role in protein transport, also transporting proteins with a signal sequence.

18. In gram-negative bacteria, proteins which are secreted out of the cell often have specialized structures to help them pass through the outer membrane. The most dramatic of these are the type III secretion systems of pathogenic bacteria, which act like syringes to inject the protein through both bacterial membranes and directly into the eukaryotic host cell, either plant or animal.

19. An open reading frame (ORF) is a string of amino acid codons in DNA unbroken by a nonsense codon. In vitro transcription-translation systems or transcriptional and translation fusions are often required to prove that an ORF in DNA actually encodes a protein.

20. The strand of DNA from which the mRNA is made is the transcribed strand. The opposite strand, which has the same sequence as the mRNA, is the coding strand.

21. A sequence 5′ on the coding strand of DNA is said to be upstream, whereas a sequence 3′ is downstream.

22. The TIR sequence of a gene does not necessarily occur at the beginning of the mRNA. The 5′ end of the mRNA is called the 5′ untranslated region. Similarly, the sequence downstream of the nonsense codon is the 3′ untranslated region.

23. Because mRNA is both transcribed and translated in the 5′-to-3′ direction, it can be translated as it is transcribed in bacteria, which have no nuclear membrane.

24. Bacteria often make polycistronic mRNAs with more than one polypeptide coding sequence on an mRNA. This makes possible polarity of transcription and translation coupling, phenomena unique to bacteria.

25. The expression of genes is regulated, depending upon the conditions the cell finds itself in. This regulation can be either transcriptional or posttranscriptional. Transcriptional regulation can be either negative or positive depending upon whether the regulatory protein is a repressor or activator, respectively. A repressor binds to an operator or operators close to the promoter and prevents transcription from the promoter. An activator binds to an upstream activator sequence (UAS), upstream of the promoter, and allows transcription from the promoter. Transcriptional regulation can also occur after the RNA polymerase leaves the promoter, as attenuation or antitermination of transcription. Posttranscriptional regulation can occur at the level of translation of the mRNA, stability of the mRNA, or processing and modification of the gene product.

26. Gene fusions have many uses in modern molecular genetics. They can be either transcriptional or translational fusions. In a transcriptional fusion, the two coding regions are transcribed into the same mRNA but each is translated from its own TIR. In a translational fusion, the two coding regions are fused to each other so that they are translated in the same frame and with no nonsense codons between them. A translational fusion makes a fusion protein with the two polypeptides fused to each other.

27. Expression vectors are designed to allow the synthesis of the product of a cloned gene in a convenient host such as *E. coli*. They can be either transcription or translation vectors. In transcription vectors, the cloned gene is transcribed from a promoter on the vector but translated from its own

(continued)

SUMMARY (continued)

TIR. Translation vectors also contain a TIR from which the gene can be translated. Affinity vectors are translation vectors that fuse a polypeptide which is easily purified to the protein product of the cloned gene. Some expression vectors have inducible promoters, so that the cloned gene will be expressed only when the inducer is added.

28. Many naturally occurring antibiotics attack components of the transcription and translation apparatuses. Some of the more useful are rifampin, streptomycin, tetra-cycline, thiostrepton, chloramphenicol, and kanamycin. Besides treating bacterial infections and being used in tumor chemotherapy, such antibiotics have also helped us understand the workings of the transcription and translation apparatuses. In addition, the genes that confer resistance to these antibiotics have served as selectable genetic markers and reporter genes in molecular genetics studies of both bacteria and eukaryotes.

QUESTIONS FOR THOUGHT

1. Which do you think came first, DNA, RNA, or protein? Why?

2. Why is the genetic code universal?

3. Why do you suppose prokaryotes have polycistronic messages but eukaryotes do not?

4. Why do you suppose mitochondrial genes differ in their genetic code from chromosomal genes?

5. Why is selenocysteine inserted into proteins of almost all organisms but only into a few sites in a few proteins in these organisms?

6. Why do so many antibiotics inhibit the translation process as opposed to, say, amino acid biosynthesis?

7. Why do you think bacteria and organelles derived from them have Hsp60-type chaperonins to help them fold their proteins but this type of chaperonin does not exist in the eukaryotic cytoplasm?

8. Why do some proteins have specialized systems of their own for membrane transport instead of using the general secretory *sec* system? Why do all proteins not use the *sec* system?

9. List all the reasons you can think of why bacteria would regulate the expression of their genes.

PROBLEMS

1. What is the longest open reading frame (ORF) in the mRNA sequence 5′AGCUAACUGAUGUGAUGUCAACGUCCUACUCUAGCGUAGUCUAAAG3′? Remember to look in all three frames.

2. Where do you think translation is most likely to start in the mRNA sequence 5′UAAGUGAAAGAUGUGAAUGAAGUAGCCACCAAAGUCACUAAUGCUUCCAACA3′? Why?

3. Which of the following is more likely to be a factor-independent transcription termination site? Note that in each case, only the transcribed strand of the DNA is shown in the 3′-to-5′ direction.

a. 3′AACGACTAGTACGACATACTAGTCGTTGGCAAAAAAAAATGCA5′

b. 3′ACTAGCCTAAGCATCTTGCATCAGGCACAGAAAAAAAAAATCGCA5′

4. Design a 20-nucleotide PCR primer that could be used as the upstream primer to introduce a *Bam*HI restriction site to clone a protein-coding sequence that begins with ATGUUGCGATTU to fuse it downstream of a His tag in which translation begins with ATGCCGCATCATCATCATCATCATGGATCCT. The six histidine codons are underlined, and a *Bam*HI restriction endonuclease recognition site is shown in bold.

5. What would be the effect of a mutation that inactivates the regulatory protein of an operon on the expression of the operon if

a. The regulation is negative?

b. The regulation is positive?

SUGGESTED READING

Artsimovitch, I., and R. Landick. 2000. Pausing by bacterial RNA polymerase is mediated by mechanistically distinct classes of signals. *Proc. Natl. Acad. Sci. USA* **97:**7090–7095.

Baku, B., and A. L. Horwich. 1998. The Hsp70 and Hsp60 chaperone machines. Review. *Cell* **92:**351–366.

Ban, N., P. Nissen, J. C. Jansen, M. Capel, P. B. Moore, and T. A. Steitz. 1999. Placement of protein and RNA structures into a 5A resolution map of the 50S ribosomal subunit. *Nature* **400:**841–847.

Cate, J. H., M. M. Yusupov, G. Z. Yusupova, T. N. Earnest, and H. F. Noller. 1999. X-ray crystal structures of 70S ribosome functional complexes. *Science* **285:**2095–2104.

Draper, D. 1996. Translation initiation, p. 902–908. *In* F. C. Neihardt, R. Curtiss III, J. L. Ingraham, E. C. C. Lin, K. B. Low, B. Magasanik, W. S. Reznikoff, M. Riley, M. Schaechter, and H. E. Umbarger (ed.), Escherichia coli *and* Salmonella: *Cellular and Molecular Biology,* 2nd ed. ASM Press, Washington, D.C.

Gabashvili, I. S., R. K. Agrwal, C. M. T. Spahn, R. A. Grassucci, D. I. Svergun, and J. Frank. 2000. Solution structure of the *E. coli* 70S ribosome at 11.5A resolution. *Cell* **100:**537–549.

Green, R., and H. Noller. 1997. Ribosomes and translation. *Annu. Rev. Biochem.* **66:**679–716.

Herskovitz, A., E. S. Bochkareva, and E. Bibi. 2000. New prospects in studying the bacterial signal recognition pathway. *Mol. Microbiol.* **38:**927–939. (Microreview.)

Lee, H. C., and H. D. Bernstein. 2001. The targeting pathway of *Escherichia coli* presecretory and intergral membrane proteins is specficied by the hydrophobicity of the targeting signal. *Proc. Natl. Acad. Sci. USA* **98:**3471–3476.

Manting, E. H., and A. J. M. Driessen. 2000. *Escherichia coli* translocase: the unraveling of a molecular machine. *Mol. Microbiol.* **37:**226–238. (Microreview.)

Maguire, B. A., and R. A. Zimmermann. 2001. The ribosome in focus. Minireview. *Cell* **104:**813–816.

Meinnel, T., C. Sacerdot, M. Graffe, S. Blanquet, and M. Springer. 1999. Discrimination by *Escherichia coli* initiation factor IF3 against initiation on non-canonical codons relies on complementarity rules. *J. Mol. Biol.* **290:**825–837.

Mooney, R. A., and R. Landick. 1999. RNA polymerase unveiled. Minireview. *Cell* **98:**687–690.

Nyborg, J., P. Nissen, M. Kjeldgaard, S. Thirup, G. Polekhina, B. F. C. Clark, and L. Reshetnikova. 1996. Structure of the ternary complex of Ef-Tu: macromolecular mimicry in translation. *TIBS* **21:**81–82.

Teter, S. A., W. A. Houry, D. Ange, T. Tradler, D. Rockaband, G. Fischer, P. Blum, C. Georgopoulos, and F. U. Hartl. 1999. Polypeptide flux through bacterial Hsp70: DnaK cooperates with trigger factor in chaperoning nascent chains. *Cell* **97:**755–765.

Yusupov, M. M., G. Z. Yusupova, A. Baucom, K. Lieberman, T. N. Earnest, J. H. Cate, and H. F. Noller. 2001. Crystal structure of the ribosome at 5.5A resolution. *Science* **292:**883–896.

PART II

Genes and Genetic Elements

ALL OF THE FEATURES of a living organism are determined either directly or indirectly by the products of the genes of the organism. The geneticist manipulates the genes of an organism to study how the organism normally performs its various functions and perhaps to alter these functions in potentially useful ways.

Not all genes are a normal part of the DNA of the organism. Some genes are carried on DNA elements. A DNA element is a DNA sequence which is not common to all of the members of a species of organism and can sometimes be exchanged from one organism to another. Plasmids, transposons, and bacteriophages are DNA elements found in bacteria. These DNA elements can exist either as an integral part of the normal chromosomal DNA or autonomously in the cell as a self-replicating entity. While these DNA elements can be considered parasites and depend upon the host cell for their replication and maintenance, they often carry genes that change the properties of the bacterium and sometimes confer advantages on the bacterium. Bacteria are also unusual among organisms on Earth in that they depend on their DNA elements to exchange genetic information from one bacterium to another. Detailed knowledge about bacterial DNA elements has been exploited to study gene structure and function in both bacteria and eukaryotic cells.

In part II, we show how mutations, which are changes in the DNA sequence, manifest themselves in bacteria. We also discuss the properties of the different types of mutations and how they can be distinguished. We give the properties of the different types of DNA elements that exist in bacteria and how these DNA elements affect the bacteria that carry them and how they can transfer themselves as well as normal bacterial genes between members of the population. These chapters give the necessary background for later chapters where we give more specific examples to illustrate how this knowledge can be used to study many of the functions of bacteria.

3

Mutations and Genetic Analysis

As mentioned in the introductory chapter, the relative ease with which bacteria can be handled genetically has made them very useful model systems for understanding many life processes, and much of the information on basic macromolecular synthesis that we discussed in the first two chapters came from genetic experiments with bacteria. In this chapter, we introduce the genetic concepts and definitions that will be used in later chapters.

Definitions

In genetics, as in any field of knowledge, we need definitions. However, words do not mean much when taken out of context, and so here we define only the most basic terms. We will define other important terms as we go along; these appear in boldface.

Terms Used in Genetics

These words are common to all types of genetic experiments, whether with prokaryotic or eukaryotic systems.

MUTANT

The word **mutant** refers to an organism that is the direct offspring of a normal member of the species (the **wild type**) but is different. Organisms of the same species isolated from nature that have different properties are usually not called mutants but, rather, variants or **strains,** because, even if one of the strains has recently arisen from the other in nature, we have no way of knowing which one is the mutant and which is the wild type.

PHENOTYPE

The phenotypes of an organism are all the observable properties of that organism. Usually in genetics, the term **phenotype** means **mutant phenotype**, or the characteristics of the mutant organism that differ from those of the wild type. The corresponding normal property is sometimes referred to as the **wild-type phenotype**.

GENOTYPE

The **genotype** of an organism refers to the actual sequence of its DNA. If two organisms have the same genotype, they are genetically identical to each other. Identical twins have almost the same genotype. If two organisms differ by only one mutation, they are said to be **isogenic** except for that mutation.

MUTATION

A **mutation** refers to any heritable change in the DNA sequence. Practically every imaginable type of change is possible, and all changes are called mutations. However, the word "heritable" must be emphasized. Changes or damage that are repaired, so that the original sequence is restored, are not inherited. Hence, only a permanent change in the sequence of deoxynucleotides constitutes a mutation.

ALLELE

Different forms of the same gene are called **alleles**. For example, if one form of a gene has a mutation and the other has the wild-type sequence, the two forms of the gene are different alleles of the same gene. In this case, one gene is a mutant allele and the other gene is the wild-type allele. Diploid organisms can have two different alleles of the same gene, one on each homologous chromosome. The term "allele" can also refer to genes with the same or similar sequences that appear at the same chromosomal location in closely related species. However, similar gene sequences occurring in different chromosomal locations are not alleles; rather, they are **copies** of the gene.

USE OF GENETIC DEFINITIONS

The following example illustrates the use of the definitions in the previous section. For an explanation of the methods, see the introductory chapter.

A culture of *Pseudomonas fluorescens* normally grows as bright green colonies on agar plates. However, suppose that one of the colonies is colorless. It probably arose through multiplication of a **mutant** organism. The **mutant phenotype** is "colorless colony," and the corresponding **wild-type phenotype** is "green colony." The mutant bacterium that formed the colorless colony probably had a **mutation** in a gene for an enzyme required to

make the green pigment. Perhaps the mutation consists of a base pair change in the gene, causing the insertion of a wrong amino acid in the polypeptide, and the resulting enzyme cannot function. Thus, the mutant and wild-type bacteria have different **alleles** of this gene, and we can refer to the gene in the colorless colony-forming bacteria as the **mutant allele** and the gene in the green-colony-forming bacteria as the **wild-type allele**.

In the example above, we only know that a mutation has occurred because of the lack of color. However, recall that any heritable change in the DNA sequence is a mutation, and so mutations can occur without changing the organism's phenotype. Many such changes, called **silent mutations**, have been found by sequencing the DNA directly.

Genetic Names

There are some commonly accepted rules for naming mutants, phenotypes, and mutations in bacteria, although different publications sometimes use different notations. We use the terms recommended by the American Society for Microbiology.

NAMING MUTANT ORGANISMS

The mutant organism can be given any name as long as the designation does not refer specifically to the phenotype or the gene thought to have been mutated. This rule helps to avoid confusion if the gene with the mutation is introduced into another strain or if other mutations occur or are transferred into the original organism. Quite often, someone who has isolated a mutant names it after himself or herself, giving it his or her initials and a number (e.g., *Escherichia coli* AB2497). This notation is not intended to gratify the person's ego but to inform others where they can obtain the mutant strain and get advice about its properties. If another mutation alters the mutant strain, this new strain is usually given another name, such as *E. coli* AB2498.

NAMING GENES

Bacterial genes are designated by three lowercase italicized letters that usually refer to the function of the gene's product, when this is known. For example, the name *his* refers to a gene whose product is an enzyme required to synthesize the amino acid histidine. Sometimes more than one gene encodes a product with the same function, or an enzymatic pathway requires more than one different polypeptide. In these cases, a capital letter designating each individual gene follows the three lowercase letters. For example, the *hisA* and *hisB* genes both encode polypeptides required to synthesize histidine. A mutation that inactivates either gene will make the cell unable to synthesize histidine.

NAMING MUTATIONS

Hundreds of different types of mutations can occur in a single gene, and so all alleles of a particular gene have a specific allele number. For example, *hisA4* refers to the *hisA* gene with mutation number 4, and the *hisA* gene with mutation number 4 is referred to as the *hisA4* allele.

If a mutation is known to inactivate the product of a gene, a superscript minus sign, or simply the word "mutation," may be added to the gene or allele name. For example, *hisA⁻* or a *hisA* mutation inactivates the product of the *hisA* gene. Alternatively, the designation *hisA⁺* refers specifically to the wild-type form of the *hisA* gene, which encodes a functional gene product.

Different nomenclatural rules apply if a mutation is a deletion or insertion. We shall defer a discussion of these rules until we discuss these types of mutations (see below).

NAMING PHENOTYPES

Phenotypes are also denoted by three-letter names, but the letters are not italicized and the first letter is capitalized. As with genotypes, superscripts are often used to distinguish mutant from wild-type phenotypes. For example, His⁻ describes the phenotype of an organism with a mutated *his* gene that will not grow without histidine in its environment. The corresponding wild-type organisms will grow without histidine, so they are phenotypically His⁺. Another example, Rif^r, describes resistance to the antibiotic rifampin, which blocks RNA synthesis (see chapter 2). A mutation in the *rpoB* gene, which encodes a subunit of the RNA polymerase, makes the cell resistant to this antibiotic. The corresponding wild-type phenotype is rifampin sensitivity, or Rif^s.

Useful Phenotypes in Bacterial Genetics

What phenotypes are useful for genetic experiments depends on the organism being studied. For bacterial genetics, the properties of the colonies formed on agar plates are the most useful phenotypes (see the introductory chapter).

The visual appearance of colonies sometimes provides useful mutant phenotypes, such as the colorless colony discussed earlier. Colonies formed by mutant bacteria might also be smaller than normal or smooth instead of wrinkled. The mutant bacterium may not multiply to form a colony at all under some conditions, or, conversely, it may multiply when the wild type cannot.

Many mutant phenotypes have been used to study cellular processes such as DNA recombination and repair, mutagenesis, and development. The following sections describe a few of the more commonly used phenotypes. In later chapters, we shall discuss many more types of mutants and demonstrate how mutations can be used to study life processes.

Auxotrophic Mutants

Some of the most useful bacterial mutants are **auxotrophic mutants**, or auxotrophs. Of the two types of these mutants, one cannot multiply without a particular growth supplement that is not required by the original, wild-type isolate. For example a His⁻ auxotrophic mutant cannot grow unless the medium is supplemented with the amino acid histidine, while the wild type could grow without added histidine. Similarly, a Bio⁻ auxotrophic mutant cannot grow without the vitamin biotin, which is not needed by the wild type.

The other type of auxotrophic mutant cannot use a particular substance for growth that can be used by the wild type. For example, the wild-type bacteria may be able to use the sugar maltose as a sole carbon and energy source, but a Mal⁻ auxotrophic mutant must be given another carbon and energy source.

Even though these two types of auxotrophs seem opposite, their molecular basis is similar. In both types, a mutation has altered a gene encoding an enzyme of a metabolic pathway, thereby inactivating the enzyme. The only difference is that in the first case, the inactivated enzyme was in a **biosynthetic** pathway, which is required to synthesize a substance, while in the second case, the inactivated enzyme was in a **catabolic** pathway, which is required to degrade a substance to use it as a carbon and energy source.

ISOLATING AUXOTROPHIC MUTANTS

Figure 3.1 shows a simple method for isolating mutants auxotrophic for histidine and biotin. In this experiment, eight colonies from plate 1, which contains all the nutrients the bacteria need, including histidine and biotin, were picked up with a loop and transferred onto two other plates. These plates are the same as plate 1 except that plate 2 lacks biotin but has histidine and plate 3 lacks histidine but has biotin. The bacteria from most of the colonies can multiply on all three types of plates. However, the bacteria in colony 2 grow only on plate 2; they are mutants that require added histidine, that is, they are His⁻. However, these mutants do not require biotin, and so are Bio⁺. Similarly, the bacteria in colony 6 are Bio⁻ but His⁺, since they can grow on plate 3 but not on plate 2. Under real conditions, mutants that require histidine or biotin would not be this frequent, and thousands of colonies would have to be tested to find one mutant that required histidine, biotin, or indeed any growth supplement not required by the wild type.

In principle, it should be possible to find auxotrophic mutants unable to synthesize any compound required

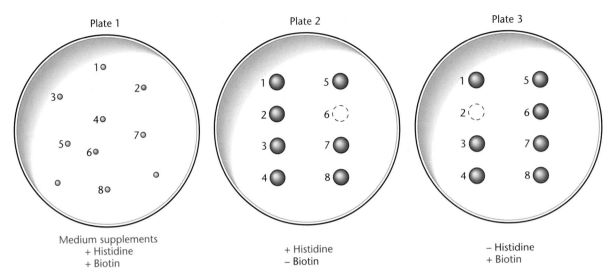

Figure 3.1 Detection of auxotrophic mutants. Colonies were scraped with a loop from plate 1 and transferred onto plates 2 and 3. Colony 6 contains bacteria that cannot multiply without biotin and so are Bio⁻. The bacteria in colony 2 are His⁻.

for growth or unable to use any carbon and energy source. However, auxotrophic mutants must be supplied with the compound they cannot synthesize, and these compounds must enter the cell. Yet many bacteria cannot take in some compounds that have a high electrical charge, such as nucleotides, so that some types of auxotrophs are very difficult to isolate.

Conditional Lethal Mutants

Auxotrophic mutants can be isolated because they have mutations in genes whose products are required under only certain conditions. The cells can be grown under conditions where the product of the mutated gene is not required and tested under conditions where it is required. However, many gene products of the cell are essential for growth no matter what conditions the bacteria find themselves in. The genes that encode such functions are called **essential genes**. Examples of essential genes include those for RNA polymerase, ribosomal proteins, DNA ligase, and helicases. Cells with mutations that inactivate essential genes cannot be isolated unless the mutations inactivate the gene under only some conditions. Hence, any mutants that are isolated will have **conditional lethal mutations**, because these DNA changes are lethal only under some conditions.

TEMPERATURE-SENSITIVE MUTANTS

The most generally useful conditional lethal mutations in bacteria are mutations that make the mutant **temperature sensitive** for growth. Usually, such mutations change an amino acid of a protein so that the protein no longer functions at higher temperatures but still func-

tions at lower temperatures. The higher temperatures are called the **nonpermissive temperatures** for the mutant, whereas the temperatures at which the protein still functions are the **permissive temperatures** for the mutant.

Mutations can affect the temperature stability of proteins in various ways. Often, an amino acid required for the protein's stability at the nonpermissive temperature is changed, causing the protein to unfold, or denature, partially or completely. The protein could then remain in the inactive state or be destroyed by cellular proteases that remove abnormal proteins. If the protein remains, it sometimes spontaneously renatures (refolds) when the temperature is lowered; then growth can resume immediately. With other mutations, the protein is irreversibly denatured and must be resynthesized before growth can resume.

The temperature ranges used to isolate temperature-sensitive mutations depend on the organism. Bacteria can be considered poikilothermic, or cold-blooded, organisms, with a cell temperature that varies with the outside temperature. Therefore, their proteins are designed to function over a wide range of temperatures. However, different species of bacteria differ greatly in their preferred temperature range. For example, a "mesophilic" bacterium such as *E. coli* may grow well in a range of temperatures from 20 to 42°C. In contrast, a "thermophilic" bacterium such as *Bacillus stearothermophilus* may grow well only between 42 and 60°C. For *E. coli*, a temperature-sensitive mutation may make a protein nonfunctional at 42°C, the nonpermissive temperature, but not at 33°C, the permissive temperature. For *B. stearothermophilus*, the temperature-sensitive mutation may make a protein inactive at the nonpermissive temperature of 55°C but leave it functional at the permissive temperature of 47°C.

Isolating Temperature-Sensitive Mutants

In principle, temperature-sensitive mutants are as easy to isolate as auxotrophic mutants. If a mutation that makes the cell temperature sensitive occurs in a gene whose protein product is required for growth, the cells will stop multiplying at the nonpermissive temperature. To isolate such mutants, the bacteria are incubated on a plate at the permissive temperature until colonies appear and then the colonies are transferred to a plate incubated at the nonpermissive temperature. Bacteria that can form colonies at the permissive temperature but not at the nonpermissive temperature are temperature-sensitive mutants. However, temperature-sensitive mutants are usually much rarer than auxotrophic mutants. Many changes in a protein will inactivate it, but very few will make a protein functional at one temperature and nonfunctional at another. We shall discuss the frequency of occurrence of different types of mutations later in the chapter.

COLD-SENSITIVE MUTANTS

Cells with proteins that fail to function at lower temperatures are called specifically **cold-sensitive mutants**. Mutations that make a bacterium cold sensitive for growth are often in genes whose products must form a larger complex such as the ribosome. The increased movement at the higher temperature may allow the mutated protein, despite its altered shape, to enter the complex, but it will not be able to do so at lower temperatures.

NONSENSE MUTATIONS

Mutations that change a codon in a gene to one of the three nonsense codons—UAA, UAG, or UGA—can also be conditional lethal mutations. A nonsense mutation will cause translation to stop within the gene unless the cell has a "nonsense suppressor" tRNA, as we shall explain later in the chapter. Because nonsense mutations are more generally useful in viral genetics than bacterial genetics, we discuss them in more detail in chapter 7.

Resistant Mutants

Among the most useful types of bacterial mutants to isolate are resistant mutants. If a substance kills or inhibits the growth of a bacterium, mutants resistant to the substance can often be isolated merely by plating the bacteria in the presence of the substance.

The numerous mechanisms of resistance depend on the basis for toxicity and on the options available to prevent the toxicity (examples are given in Table 3.1). For example, the mutation may destroy a cell surface receptor to which the toxic substance must bind to enter the cell. If the substance cannot enter the cell, the mutant will not be killed by it. Alternatively, a mutation might change the "target" affected by the substance inside the cell. For example, an antibiotic might normally bind to a ribosomal protein and affect protein translation. However, if the antibiotic cannot bind to a mutant (but still functional) protein, it cannot kill the cell. An example of such a resistance mutation is a mutation to streptomycin resistance in *E. coli* (Table 3.1) The antibiotic streptomycin binds to the 16S rRNA in the 30S subunit of the ribosome and blocks translation. However, some mutations in the gene for the S12 protein, *rpsL* (for "ribosomal protein small subunit L") prevent streptomycin from binding to the ribosome but do not inactivate the S12 protein. These mutations therefore confer streptomycin resistance on *E. coli*. In some cases, the substance added to the cells is not toxic until one of the cell's own enzymes changes it. A mutation inactivating the enzyme that converts the nontoxic substance into the toxic one could make the cell resistant to that substance.

Inheritance in Bacteria

Salvador Luria and Max Delbrück were among the first people to attempt to study inheritance quantitatively in bacteria. They published a now-classic paper in the journal

TABLE 3.1	Some resistance mutations	
Substance	**Toxicity**	**Resistance mutation**
Bacteriophage T1	Infects and kills	Inactivates *tonB* outer membrane protein; phage cannot absorb
Streptomycin	Binds to ribosomes; inhibits translation	Changes ribosomal protein S12 so that it no longer binds
Chlorate	Converted to chlorite, which is toxic	Inactivates nitrate reductase, which converts chlorate to chlorite
High concentrations of valine, no isoleucine	Feedback inhibits acetolactate synthetase; starves for isoleucine	Activates a valine-insensitive acetolactate synthetase

Genetics in 1943. This paper is still very much worth reading and is listed in the Suggested Reading section at the end of the chapter. As discussed in the introductory chapter, the experiments and reasoning of Luria and Delbrück helped debunk what was then a popular misconception among bacteriologists. At the time of the Luria and Delbrück studies, it was generally believed that bacteria were different from other organisms on Earth in their inheritance. It was generally accepted that heredity in higher organisms followed "neo-Darwinian" principles. According to Darwin, random mutations occur, and if one happened to confer a desirable phenotype, organisms with this mutation would be selected by the environment and become the predominant members of the population. Undesirable as well as desirable mutations would continuously occur, but only the desirable mutations would be passed on to future generations.

However, many bacteriologists believed that heredity in bacteria followed different principles. They thought that bacteria, rather than changing as the result of random mutations, somehow "adapt" to the environment by a process of directed change, after which the adapted organism would pass the adaptation on to its offspring. This process is called Lamarckian Inheritance, and belief in it was encouraged by the observation that all the bacteria in a culture exposed to a toxic substance seem to become resistant to that substance in response to it (Figure 3.2).

The Luria and Delbrück Experiment

The Luria and Delbrück experiment was designed to test two hypotheses for how mutants arise in bacterial cultures: the **random-mutation** hypothesis and the **directed-change** hypothesis. The random-mutation hypothesis predicts that the mutants appear prior to the addition of the selective agent, whereas the directed-change hypothesis predicts that mutants appear only in response to a selective agent.

Figure 3.3 illustrates that mutations that occur early in the culture will have a disproportionate effect on the

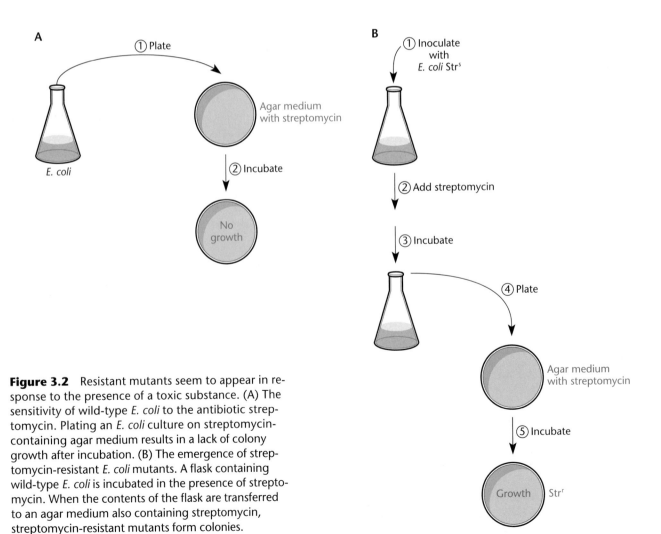

Figure 3.2 Resistant mutants seem to appear in response to the presence of a toxic substance. (A) The sensitivity of wild-type *E. coli* to the antibiotic streptomycin. Plating an *E. coli* culture on streptomycin-containing agar medium results in a lack of colony growth after incubation. (B) The emergence of streptomycin-resistant *E. coli* mutants. A flask containing wild-type *E. coli* is incubated in the presence of streptomycin. When the contents of the flask are transferred to an agar medium also containing streptomycin, streptomycin-resistant mutants form colonies.

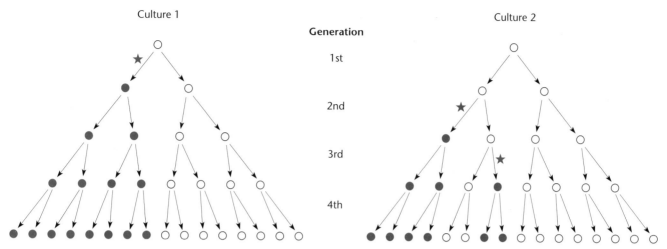

Culture 1   **Generation**   Culture 2

1st

2nd

3rd

4th

Figure 3.3 Earlier mutations give rise to more mutant progeny in a growing culture. Only one mutation occurred in culture 1, but it gave rise to eight mutant progeny because it occurred in the first generation. In culture 2, two mutations occurred, one in the second generation and one in the third. However, because these mutations occurred later, they gave rise to only six mutant progeny. The mutant cells are shaded.

number of mutants in the culture. In culture 1, only one mutation occurred, but this mutation gave rise to eight resistant mutants because it occurred early. In culture 2, two mutations arose, but they gave rise to only six resistant mutants because they occurred later.

To determine if mutants appear before or after addition of a selective agent, one can grow several cultures in the absence of the selective agent, add the agent to all the cultures at the same time, and then measure the fraction of bacteria resistant to the selective agent in each culture. If the random-mutation hypothesis is correct, the number of mutant colonies will vary among all the cultures, depending on when the mutations occurred, as was illustrated in Figure 3.3. In contrast, if the directed-change hypothesis is correct, each bacterium will have the same chance of becoming a mutant, but only after the selective agent is added, and so the same percentage of the bacteria should become resistant in all the cultures. Therefore, a result in which the number of mutants per culture varies greatly will favor the random-mutation hypothesis but a result in which the number of mutants in a series of cultures is about the same will favor the directed-change hypothesis.

In their experiments, Luria and Delbrück used *E. coli* as the bacterium and bacteriophage T1 as the selective agent. As shown in Table 3.1 and Figure 3.4, phage T1 will kill wild-type *E. coli*, but a mutation in the gene for an outer membrane protein called TonB can make these cells resistant to killing by the phage. If bacteria are spread on an agar plate with the phage, only those resistant to the phage will multiply to form a colony. All the

others will be killed. The number of colonies on the plate is therefore a measure of the number of bacteria resistant to the bacteriophage in the culture.

Figure 3.5 shows the two experiments that Luria and Delbrück performed, which seem superficially similar

Figure 3.4 When bacteriophage T1 infects wild-type *E. coli*, it binds to a receptor in the outer membrane, protein TonB (Table 3.1). After phage replication, the *E. coli* cell is lysed and new phage are released. A mutation in the *tonB* gene results in an altered (mutant) receptor to which T1 can no longer bind or eliminates the receptor, and so the cells survive.

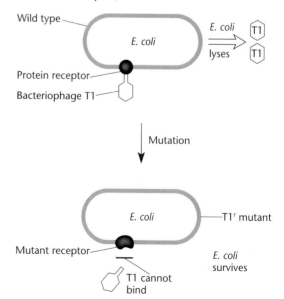

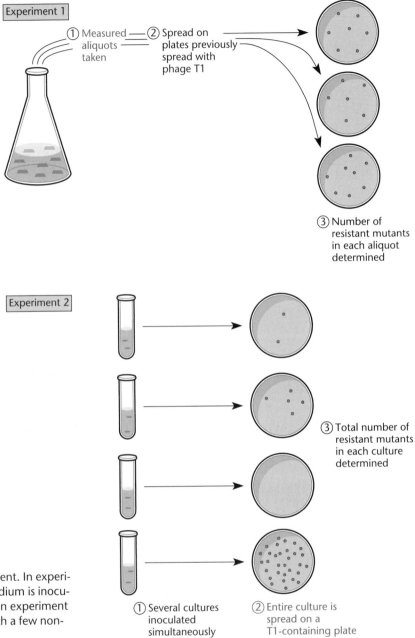

Figure 3.5 The Luria and Delbrück experiment. In experiment 1, a single flask containing standard medium is inoculated with bacteria and incubated overnight. In experiment 2, a number of smaller cultures are started with a few non-mutant bacteria. See the text for details.

but gave very different results. In experiment 1, they started one culture of bacteria. After incubating it, they took out small aliquots and plated them with and without phage T1 to measure the number of resistant mutants as well as the total number of bacteria in the culture. They then calculated the fraction of resistant mutants. In experiment 2, they started a large number of relatively smaller cultures. After incubating these cultures, they measured the number of resistant mutants and the total number of bacteria in each culture.

Table 3.2 displays some representative results. In experiment 1, the number of resistant mutants in each aliquot is almost the same, subject only to sampling errors and statistical fluctuations. However, in experiment 2, a very large variation in the number of resistant bacteria per culture was found. Some cultures had no resistant mutants, while some had many. One culture even had 107 resistant mutants! Luria and Delbrück referred to this and the other mutant-rich cultures as "jackpot" cultures. Apparently, these are cultures in which a mutation to resistance occurred very early. Hence, these results fulfill the predictions of the random-mutation hypothesis. In contrast, the directed-change hypothesis predicts that the results of the two experiments should be the same,

TABLE 3.2	The Luria and Delbrück experiment		
Experiment 1		**Experiment 2**	
Aliquot no.	No. of resistant bacteria	Culture no.	No. of resistant bacteria
1	14	1	1
2	15	2	0
3	13	3	3
4	21	4	0
5	15	5	0
6	14	6	5
7	26	7	0
8	16	8	5
9	20	9	0
10	13	10	6
		11	107
		12	0
		13	0
		14	0
		15	1
		16	0
		17	0
		18	64
		19	0
		20	35

and certainly no jackpot cultures should appear in the second experiment. Box 3.1 presents these predictions in statistical terms.

The Newcombe Experiment

The analysis of Luria and Delbrück was fairly sophisticated mathematically and so was not generally understood. Some people still held to the belief that bacteria were somehow different from other organisms in their inheritance. Consequently, Howard Newcombe in 1951 devised an experiment that was conceptually simpler and so convinced many skeptics. In his experiment, he also used *E. coli* mutants resistant to phage T1.

The principle behind Newcombe's experiment is that if the random-mutation hypothesis is correct, mutants should be **clonal**; that is, one mutant bacterium will give rise to more, even in the absence of the selective agent (see the introductory chapter). However, if the directed-change hypothesis is correct, mutants should not be clonal, because they will not be multiplying to form a colony before the selective agent is added. Instead, all the mutants will first appear at the time the selective agent is added.

To detect clones of resistant bacteria, Newcombe analyzed cultures grown on agar plates. According to the random-mutation hypothesis, the number of colonies due to resistant mutants on an agar plate will vary de-

---

BOX 3.1

Statistical Analysis of the Number of Mutants per Culture

A simple statistical analysis shows us that the number of mutants in experiment 2 of Luria and Delbrück does not follow a normal distribution. If the number of mutants per culture follows a normal distribution, the variance would be approximately equal to the mean:

$$\text{Variance} = \sum_{i=1}^{n} \frac{(M_i - \overline{M})^2}{n-1} = \text{mean} = \overline{M} = \sum_{i=1}^{n} \frac{M_i}{n}$$

where M_i is the number of mutants in each culture and n is the number of cultures.

In experiment 1 of Luria and Delbrück, the variance was 18.23 and the mean was 16.7, so they are approximately equal and vary owing to statistical fluctuations and pipetting errors. In experiment 2, however, the variance was 752.38 and the mean was 11.35—very different values. Therefore, the number of mutants per culture does *not* follow a normal distribution, and the result is not consistent with the directed-change hypothesis; however, it is consistent with the random-mutation or neo-Darwinian hypothesis.

pending on whether the colonies are left alone or disturbed by having been spread out on the plate. When the colonies on a plate are not disturbed, all the descendants of a particular bacterium will remain together in the same colony. However, if the colonies on a plate are disturbed, each resistant bacterium should give rise to a separate colony of resistant bacteria. Consequently, a spread plate will have many more resistant colonies than an unspread plate. However, the directed-change hypothesis predicts that the mutants need not arise from each other (i.e., are not clonal), so that the number of resistant colonies on the undisturbed and disturbed plates should be about the same, because the resistant bacteria will have appeared only at the time the phage were added.

How Newcombe did his experiment is illustrated in Figure 3.6, and some of his actual data are given in Table 3.3. He first spread the same number of bacteria (an average of 5.1×10^4) on plates and incubated them. (He actually used many more plates than the six pictured in Figure 3.6, but this number serves as an illustration.) After 5 h,

he removed three of the plates and sprayed one with the virus without disturbing the bacteria. This is the unsp (*unsp*read) plate in Table 3.3. He treated the second plate in the same way, except that he spread the bacteria around before spraying them with the virus. This is the sp (*sp*read) plate in Table 3.3. He washed the bacteria off the third plate, diluted them, and plated them without the virus to determine the total number of bacteria on the plates at this time. After a 6-h incubation, he took the remaining three plates out of the incubator and subjected them to the same treatment. He incubated all the plates overnight, and the next day he counted the colonies produced by phage-resistant mutants. The data in Table 3.3 show that the spread plates have many more resistant colonies than the corresponding unspread plates, a result supporting the random-mutation hypothesis. The difference became greater the longer the plates were incubated. After 5 h, only a few resistant colonies had appeared, and there was not much difference between the unspread and spread plates (8 and 13 colonies, respectively), because the mutants had not had much time to multiply, and so

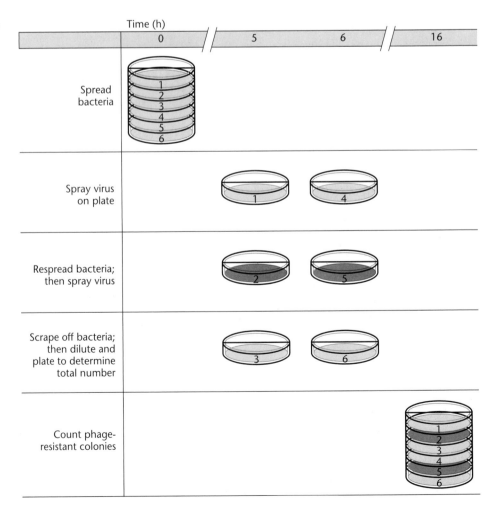

Figure 3.6 The Newcombe experiment. See the text for details.

TABLE 3.3	The Newcombe experiment		No. of resistant colonies[a]	
Incubation time (h)	No. of bacteria plated	Ending no. of bacteria	unsp	sp
5	5.1×10^4	2.6×10^8 (plate 3)	8 (plate 1)	13 (plate 2)
6	5.1×10^4	2.8×10^9 (plate 6)	49 (plate 4)	3,719 (plate 5)

[a]unsp, unspread; sp, spread.

each colony contained few resistant bacteria. However, after 6 h, the numbers of unspread and spread colonies are very different (49 and 3,719, respectively). Note also that the number of resistant bacteria, as measured by the resistant colonies on the spread plates, increased faster than the total population. At 5 h, there were 13 resistant bacteria in 2.6×10^8 total bacteria, so the fraction of resistant bacteria was $13/(2.6 \times 10^8)$, or 1 mutant for every 2×10^7 bacteria. By 6 h, there were 3,719 resistant mutants in 2.8×10^9 total bacteria, so that the fraction was $3,719/(2.8 \times 10^9)$, or 1 mutant for every 7×10^5 bacteria. Therefore, the fraction of resistant bacteria rose about 30-fold in just 1 h. In other words, the number of resistant mutants increased about 30 times faster than the total number of bacteria during this hour. This fulfills a prediction of the random-mutation hypothesis, as we explain below in the section on mutation rates.

The Lederbergs' Experiment

The experiments that really buried the directed-change hypothesis, at least as the sole explanation for some types of resistant mutants, were the replica-plating experiments of the Lederbergs (see Lederberg and Lederberg, Suggested Reading). They spread millions of bacteria on a plate without an antibiotic and allowed the bacteria to form a lawn during overnight incubation. This plate was then replicated onto another plate containing the antibiotic. After incubating the antibiotic-containing plate, the Lederbergs could determine where antibiotic-resistant mutants had arisen on the original plate by aligning the two plates and marking the regions on the first plate where antibiotic-resistant mutants had grown on the second. They cut these regions out of the original plate, diluted the bacteria, and repeated the experiment. This time, there were many more resistant mutants than previously. Eventually, by repeating this process, they obtained a pure culture of bacteria, all of which were resistant to the antibiotic even though they had never been exposed to it! Therefore, the bacteria must have acquired the resistance independently of exposure to the antibiotic and passed the resistance on to their offspring.

The experiments of Luria and Delbrück, Newcombe, and the Lederbergs proved that at least some types of mutants of *E. coli* arise through Darwinian inheritance.

Mutation Rates

As defined above, a mutation is any heritable change in the DNA sequence of an organism, and we usually know a mutation has occurred because of a phenotypic change in the organism. The **mutation rate** can be loosely defined as the chance of mutation to a particular phenotype. Mutation rates can differ because mutations to some phenotypes occur much more often than mutations to other phenotypes. When many possible different mutations in the DNA can give rise to a particular phenotype, the chance that a mutation to that phenotype will occur is relatively high. However, if only a very few types of mutations can cause a particular phenotype, the mutation rate for that phenotype will be relatively low.

For example, the spontaneous mutation rate for the His⁻ phenotype is hundreds of thousands of times higher than the mutation rate for Str^r. Approximately 11 gene products are required for histidine biosynthesis, and each has hundreds of amino acids, many of which are essential for activity. Changing any of these amino acids will inactivate the enzyme. By contrast, streptomycin resistance results from a change in one of very few amino acids in a single ribosomal protein, S12, so that the mutation rate for streptomycin resistance is very low. Hence, a mutation to Str^r will occur spontaneously in about 1 in 10^{10} to 10^{11} cells, whereas a mutation to His⁻ will occur in about 1 in 10^6 to 10^7 cells.

Generally, if the mutation rate for a phenotype is high, the phenotype probably results from inactivation of the product of a gene or genes. If the mutation rate is low, the phenotype probably is due to a subtle change in the properties of a gene product. An extremely high mutation rate to a particular phenotype may indicate not a mutation but rather the loss of a plasmid or prophage or the occurrence of some programmed recombination event such as inversion of an invertible sequence. We shall discuss plasmids and prophages and other gene rearrangements in subsequent chapters.

Calculating Mutation Rates

To calculate mutation rates, we shall define them as the chance of a mutation each time a cell grows and divides. This is a reasonable definition because, as discussed in chapter 1, DNA replicates once each time the cell divides,

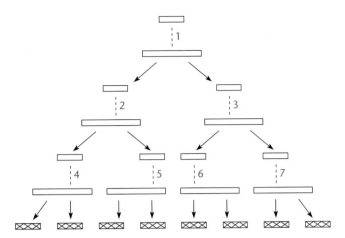

Figure 3.7 The number of cell generations (7) equals the total number of cells in an exponentially growing culture (8) minus the number at the beginning (1).

and most mutations occur during this process. The number of times a cell grows and divides in a culture is called the number of **cell generations**. Therefore, the mutation rate is the number of mutations to a particular phenotype that have occurred in a growing culture divided by the total number of cell generations that have occurred in the culture during the same time.

DETERMINING THE NUMBER OF CELL GENERATIONS

The total number of cell generations that have occurred in an exponentially growing culture is easy to calculate and is simply the the total number of cells in the culture minus the number of cells in the starting innoculum. To understand this, look at Figure 3.7. In this illustration, a culture that was started from one cell multiplies to form eight cells in seven cell divisions, or cell generations. This number equals the final number of cells (8) minus the number of cells at the beginning (1). In general, the number of cell generations that have occurred in the culture equals $N_2 - N_1$ if N_2 equals the number of cells at time 2 and N_1 equals the number of cells at time 1.

Therefore, from the definition, the mutation rate (a) is given by

$$a = \frac{m_2 - m_1}{N_2 - N_1}$$

where m_2 and m_1 are the number of mutations in the culture at time 2 and time 1, respectively.

Usually, a culture is started with a few cells and ends with many, so we can often ignore the initial cells and just call the number of cell generations N, where N is the total number of cells in the culture. Then the mutation

rate equation can be simplified to

$$a = m/N$$

where m is the number of mutations that have occurred in the culture and N is the number of bacteria. This equation assumes that there were no or an insignificant number of mutants in the culture when it was started, which is likely if the culture was started with only a few cells.

DETERMINING THE NUMBER OF MUTATIONS THAT HAVE OCCURRED IN A CULTURE

From the equations above, it looks as though it might be easy to calculate the mutation rate. The total number of mutations must simply be divided by the total number of cells. The problem comes in determining the number of mutational events that have occurred in a culture, because mutant cells, not mutational events, are usually what are detected. Moreover, recall from Figure 3.3 that one mutant cell resulting from a single mutational event can give rise to many mutant cells, depending upon when the mutation occurred during the growth of the culture. Therefore, one cannot determine the number of mutations in a culture merely by counting the number of mutant cells. However, in some cases the number of mutant cells can form the basis of a calculation of the number of mutational events and, by extension, the mutation rate.

Using the Data of Luria and Delbrück To Calculate the Mutation Rate

Luria and Delbrück used the data in Table 3.2 to calculate the mutation rate to T1 phage resistance. They assumed that even though the number of mutants per culture does not follow a normal distribution, the number of mutations per culture should, because each cell has the same chance of acquiring a mutation to T1 phage resistance each time it grows and divides. For convenience, the Poisson distribution can be used to approximate the normal distribution. According to the Poisson distribution, if P_i is the probability of having i mutations in a culture, then

$$P_i = \frac{m^i e^{-m}}{i!}$$

where m is the average number of mutations per culture, the number they wanted to know. Therefore, if they knew how many cultures had a certain number of mutations, they could calculate the average number of mutations per culture.

By this method and with the data from experiment 2 in Table 3.2, Luria and Delbrück calculated the average number of mutations per culture. The data give the number of T1-phage-resistant mutants per culture but do not indicate how many of the cultures had one, two, three, or more mutations. For example, cultures with one mutant probably had one mutation, but others, even the one with 107 mutants, might also have had only one mutation. However, how many cultures had zero mutations is clear—those with zero mutants, or 11 of 20. Therefore, the probability of having zero mutations equals 11/20. Applying the formula for the Poisson distribution, the probability of having zero mutations is given by

$$\frac{11}{20} = \frac{m^0 e^{-m}}{0!} = \frac{1 \cdot e^{-m}}{1} = e^{-m}$$

and $m = -\ln 11/20 = 0.59$. Therefore, in this experiment, an average of 0.59 mutation occurred in each culture. From the equation for mutation rate,

$$a = m/N = 0.59/5.6 \times 10^8 = 1.06 \times 10^{-9}$$

Therefore, there is 1.06×10^{-9} mutation per cell generation if there were a total of 5.6×10^8 total bacteria per culture. In other words, a mutation for phage T1 resistance will occur about once every 10^9, or every billion, times a cell divides.

Calculating the Mutation Rate from Newcombe's Data

Newcombe's data can be used more directly to calculate the number of mutations per culture (refer to Figure 3.6). On the unspread plate, each mutation will give rise to only one resistant colony. Therefore, the number of resistant colonies on his unspread plates equals the number of mutational events that have occurred at the time of incubation.

According to Newcombe's data in Table 3.3, from 5 to 6 h there were $49 - 8 = 41$ new resistant colonies on the unspread plates. Therefore, 41 mutations to phage resistance must have occurred during that time interval. During this time, the total number of bacteria went from 2.6×10^8 to 2.8×10^9 based on the total number of bacteria on the plates (Table 3.3). From the equation for mutation rate,

$$a = \frac{m_2 - m_1}{N_2 - N_1} = \frac{49 - 8}{2.8 \times 10^9 - 2.6 \times 10^8}$$

the mutation rate to T1 resistance is $41/(2.54 \times 10^9) = 1.6 \times 10^{-8}$ mutation per cell generation. In other words, Newcombe's data indicate that a mutation to resistance to phage T1 will occur a little more than once every hundred million times a cell divides. Notice that New-

combe's data give a mutation rate about 10 times higher than that derived from the data of Luria and Delbrück. This discrepancy can be explained by the phenotypic lag, as we explain later.

Using the Increase in the Fraction of Mutants To Measure Mutation Rates

As mentioned, Newcombe's data fulfilled one prediction of the random-mutation hypothesis, i.e., that the number of mutants should increase faster than the total population. In other words, the fraction of mutants in the population should increase as the population grows.

It seems surprising that the total number of mutants increases faster than the total population until one thinks about where mutants come from. If the multiplication of old mutants were the only source of mutants, the fraction of mutants would remain constant, or even drop, if the mutants do not multiply as rapidly as the normal type (which is often the case). However, new mutations occur constantly, and their progeny are also multiplying. Therefore, new mutations are continuously adding to the total number of mutants.

This fact can also be used to measure mutation rates. The higher the mutation rate, the faster the proportion of mutants will increase (Figure 3.8). In fact, if we plot the fraction of mutants (M/N) against time (in doubling times), as in Figure 3.8, the slope of this curve will be the mutation rate. In theory, this fact could be used to calculate mutation rates. In practice, however, mutation rates are usually low and the bacteria we can conveniently work with are few, so each new mutation makes too large a contribution to the number of mutants and we do

Figure 3.8 The fraction of mutants increases as a culture multiplies, and the slope is the mutation rate. M is the number of mutants, N is the total number of cells, and time is the total time elapsed divided by the time it takes the culture to double in mass (i.e., the doubling time).

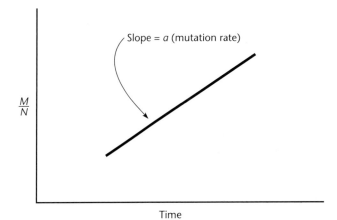

not get a straight line. To make this method practicable, we would have to work with trillions of bacteria in a large chemostat.

The fact that other mutations are causing some mutants to become wild type again (in a process known as reversion [see later sections]) will affect the results shown in Figure 3.8. However, reversion of mutants becomes significant only when the number of mutants is very large and the number of mutants times the mutation rate back to the wild type (called the reversion rate) begins to approximate the forward mutation rate times the number of nonmutant bacteria. At earlier stages of culture growth, the vast majority of bacteria will be nonmutant. Also, for reasons that we discuss later in the chapter, reversion rates are generally much lower than forward mutation rates. Therefore, the reversion of mutants to the wild type can generally be ignored.

PHENOTYPIC LAG

Some of the difficulty in accurately determining mutation rates results from **phenotypic lag**. Most phenotypes are not immediately evident after a mutation but appear some time later. The length of the lag depends on the molecular basis for the phenotype.

Mutations to phage T1 resistance would be expected to show a phenotypic lag. Recall that resistance to phage T1 derives from the alteration or loss of the protein product of a gene, *tonB*. This is an outer membrane protein to which the phage binds to start the infection. The mutant bacteria survive because they lack the wild-type protein in their outer membrane, so they cannot be infected by the virus. However, when the mutation first occurs, the mutant bacteria still have wild-type TonB in their outer membrane and so are T1 sensitive. Only after a few generations is all the wild-type TonB diluted out, so that the progeny cells can no longer absorb virus and are resistant.

It may seem that phenotypic lag would not be a serious problem in measuring mutation rates, because bacteria go through so many generations in a culture. However, in an exponentially growing culture, one-half of all mutations occur in the last generation time. Therefore, when the bacteria are plated with the selective agent, in our example phage T1, more than half of the mutations will not be counted because they have not yet been expressed and the bacteria are still sensitive. Obviously, ignoring more than half of all mutations will introduce a significant error in the mutation rate.

Some methods for measuring mutation rates are also influenced by differences in the growth rate of mutants relative to the original or wild type. Quite often, mutants grow measurably more slowly than the wild type even under nonselective conditions. Note that the two methods that have been described for determining mutation rates will not be affected by such differences.

PRACTICAL IMPLICATIONS OF POPULATION GENETICS

The fact that the proportion of mutants increases as the culture grows presents both opportunities and problems in genetics. This fact can be advantageous in the isolation of a rare mutant such as one resistant to streptomycin. If we grow a culture from a few bacteria and plate 10^9 bacteria on agar containing streptomycin, we might not find any resistant mutants, since they occur at a frequency of only about 1 in 10^{11} cell generations. However, if we add a large number of bacteria to fresh broth, grow the broth culture to saturation, and then repeat this process a few times, the fraction of streptomycin-resistant mutants will increase. Then when we plate 10^9 bacteria, we may find many streptomycin-resistant mutants.

Because the fraction of all types of mutants increases as the culture multiplies, if we allow a culture to go through enough generations, it will become a veritable "zoo" of different kinds of mutants—virus resistant, antibiotic resistant, auxotrophic, and so on. To deal with this problem, most researchers store cultures under nongrowth conditions (e.g., as spores or lyophilized cells or in a freezer) that still maintain cell viability. An alternative is to periodically colony purify bacteria in the culture to continuously isolate the progeny of a single cell (see the introductory chapter). The progeny of a single cell are not likely to be mutated in a way that could confound our experiments.

Summary

Two very important points emerge from this discussion of mutations and mutation rates. First, measuring mutation rates is not as simple as one might think. The mutation rate is not simply the number of mutants with a particular phenotype divided by the total number of organisms in the culture. To calculate the mutation rate, we must use special methods to measure the number of mutations or must apply statistical methods to the data. Second, mutants of all kinds accumulate in cultures as we grow them. Consequently, it is best to store bacteria without growing them or to periodically isolate a single cell before mutants have had a chance to become a significant proportion of the total population.

Types of Mutations

As defined above, any heritable change in the sequence of nucleotides in DNA is a mutation. A single base pair may be changed, deleted, or inserted; a large number of base pairs may be deleted or inserted; or a large region of the DNA may be duplicated or inverted. Regardless of how many base pairs are affected, a mutation is considered to be a **single mutation** if only one error in replication, recombination, or repair has altered the DNA sequence.

As discussed earlier in this chapter, to be considered a mutation, the change in the DNA sequence must be permanent. Damage to DNA, by itself, is not a mutation, but a mutation can occur when the cell attempts to repair damage or replicate over it and a strand of DNA is synthesized that is not completely complementary to the original sequence. The wrong sequence will then be faithfully replicated through subsequent generations and thus be a mutation.

Lethal changes in the DNA sequence (as also mentioned earlier) do occur but cannot be scored as mutations since the cells do not survive. Ordinarily, to be scored as a mutation, the change must be heritable and so cannot be lethal. For example, deletion of a gene required for growth is usually lethal because bacteria are haploid and usually have only one gene of each type. If the gene is deleted, the organism cannot multiply and will die without leaving progeny. Therefore, such a deletion will not be scored as a mutation.

The properties and causes of the different types of mutations are probably not very different in all organisms, but they are more easily studied with bacteria. A geneticist can often make an educated guess about what type of mutation is causing a mutant phenotype merely by observing some of its properties.

One property that distinguishes mutations is whether or not they are **leaky**. The term "leaky" means something very specific in genetics. It means that in spite of the mutation, the gene product still retains some activity.

Another property of mutations is whether or not they **revert**. If the sequence has been changed to a different sequence, it can often be changed back to the original sequence by a subsequent mutation. The organism in which a mutation has reverted is called a **revertant**, and the **reversion rate** is the rate at which the mutated sequence in DNA returns to the original wild-type sequence.

Usually, the reversion rate is much lower than the mutation rate that gave rise to the mutant phenotype. As an illustration, consider the previously discussed example, histidine auxotrophy (His⁻). Any mutation that inactivates any of the approximately 11 genes whose products are required to make histidine will cause a His⁻ phenotype. Since thousands of changes can result in this phenotype, the mutation rate for His⁻ is relatively quite high. However, once a *his* mutation has occurred, the mutation can revert only through a change in the mutated sequence that restores the original sequence. Everything else being equal, the reversion rate to His⁺ re-

vertants would be expected to be thousands of times lower than the forward mutation rate to His⁻.

Some types of revertants are very easy to detect. For example, His⁺ revertants can be obtained by plating large numbers of His⁻ mutants on a plate with all the growth requirements except histidine. Most of the bacteria cannot multiply to form a colony. However, any His⁺ revertants in the population will multiply to form a colony. The appearance of His⁺ colonies when large numbers of a His⁻ mutant are plated would be evidence that the *his* mutation can revert.

Base Pair Changes

A **base pair change** is when one base pair in DNA, for example a GC pair, is changed into another base pair, for example an AT pair. Base pair changes can be classified as **transitions** or **transversions** (Figure 3.9). In a transition, the purine (A and G) in a base pair is replaced by the other purine and the pyrimidine (C and T) is replaced by the other pyrimidine. Thus, an AT pair would become a GC pair or a CG would become a TA. In a transversion, by contrast, the purines change into pyrimidines and vice versa. For example, a GC could become a TA, or a CG could become an AT.

BASE PAIR CHANGES RESULTING FROM MISPAIRING

Base pair changes can be the result of mistakes in replication, recombination, or repair. Figure 3.10A shows an example of mispairing during replication. In this example, a T instead of the usual C is mistakenly placed opposite a G as the DNA replicates. In the next replication, this T will usually pair correctly with an A, causing a GC-to-AT transition in one of the two daughter DNAs. Mistakes in pairing may occur because the bases are sometimes in a different form called the enol form, which causes them to pair differently (Figure 3.10B).

Mispairing between a purine and a pyrimidine will cause a transition, whereas mispairing between two purines or two pyrimidines will cause a transversion. Because a pyrimidine in the enol form still pairs with a purine and a purine in the enol form still pairs with a pyrimidine, mispairing during replication usually leads to transition mutations. Furthermore, all four bases can undergo the shift to the enol form, and either the base in the DNA template or the incoming base can be in the enol form and cause mispairing. Thus, the thymine in the enol form pictured in Figure 3.10B might be in the template, in which case the transition would be AT to GC, or it could be the incoming base, resulting in a GC-to-AT transition.

Mistakes during replication leading to mutations are not random, however, and some sites are much more

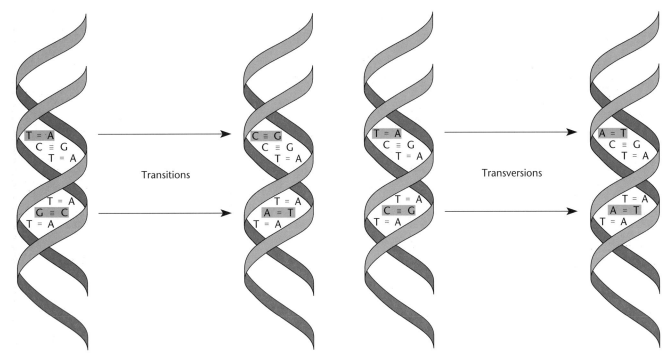

Figure 3.9 Transitions versus transversions. The mutations are shown in gold.

prone to base pair changes than are others. Mutation-prone sites are called **hot spots.**

Mispairing occurs fairly often during replication, and it is of obvious advantage for the cell to reduce the number of base pair change mutations that occur during replication. In chapter 1, we discussed some of the mechanisms that cells use to reduce these base pair changes, including editing and methyl-directed mismatch repair.

DEAMINATION OF BASES IN DNA

Deamination, or the removal of an animo group, can also cause base pair changes. This is particularly true of cytosine, which becomes uracil when deaminated, since the only difference between cytosine and uracil is the amino group at the 6 position of the cytosine ring (see chapters 1 and 2 for structures). However, uracil pairs with adenine instead of guanine. Therefore, unless the uracils due to deamination of cytosines are removed from DNA, they will cause CG-to-TA transitions the next time the DNA replicates.

Because of the special problems caused by deamination of cytosine, cells have evolved a special mechanism for removing uracil from DNA whenever it appears (Figure 3.11). An enzyme called uracil-N-glycosylase, the product of the *ung* gene in *E. coli*, recognizes the uracil as unusual in DNA and removes the uracil base. The DNA strand in the region where the uracil was removed is then degraded and resynthesized, and the cor-

rect cytosine is inserted opposite the guanine. As expected, *ung* mutants of *E. coli* show high rates of spontaneous mutagenesis, and most of the mutations are GC-to-AT transitions.

OXIDATION OF BASES

Reactive forms of oxygen such as peroxides and free radicals are given off as by-products of oxidative metabolism, and these forms can react with and alter the bases in DNA. A common example is the altered guanine base, 8-oxoG, which sometimes mistakenly pairs with adenine instead of cytosine, causing GC-to-TA transversion mutations. We discuss repair systems specific to damage such as deamination and oxidation in more detail in chapter 11.

CONSEQUENCES OF BASE PAIR CHANGES

Whether or not a base pair change causes a detectable phenotype depends, of course, on where the mutation occurs and what the actual change is. Even a change in an open reading frame (ORF) that encodes a polypeptide may not result in an altered protein. If the mutated base is the third in a codon, the amino acid inserted into the protein may not be different because of degeneracy of the code (see the section on the genetic code in chapter 2). Mutations in the coding region of a gene that do not change the amino acid sequence of the polypeptide product are called **silent mutations.**

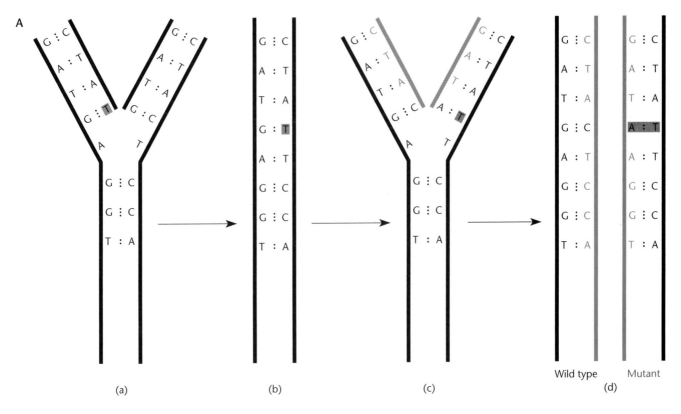

Figure 3.10 (A) A mispairing during replication can lead to a base pair change in the DNA (shown in gold). The gold line shows the second generation of newly replicated DNA. The mispaired base is boxed. (B) Pairing between guanine and an alternate enol form of thymine can cause G-T mismatches.

The change may also occur in a region that does not encode a polypeptide but, rather, is a regulatory sequence such as an operator or promoter. Alternatively, the mutation may occur in a region that has no detectable function. We shall first discuss mutations that change the coding region of a polypeptide.

MISSENSE MUTATIONS
Most base pair changes in bacterial DNA will cause one amino acid in a polypeptide to be replaced by another. These mutations are called **missense mutations** (Figure 3.12). Not all of them inactivate proteins. If the original and new amino acids have similar properties, the change may have little or no effect on the activity of a protein. For example, a missense mutation changing an acidic amino acid, such as glutamate, into another acidic amino acid, such as aspartate, may have less effect on the functioning of the protein than does a mutation that substitutes a basic amino acid, such as arginine, for an acidic one.

The consequences also depend on which amino acid is changed. Certain amino acids in any given protein sequence will be more essential to activity than others, and a change at one position can have much more effect than a change elsewhere. Investigators often use this fact to determine which amino acids are essential for activity in different proteins. Such methods, called site-specific mutagenesis, were discussed in Chapter 1.

NONSENSE MUTATIONS
Instead of changing a codon into one coding for a different amino acid, base pair changes sometimes produce

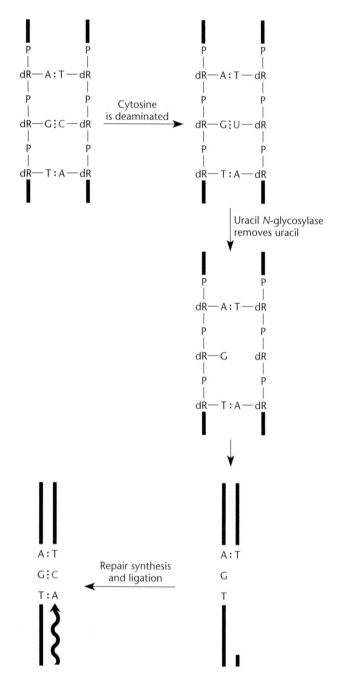

Figure 3.11 Removal of deaminated cytosine (uracil) from DNA by uracil-*N*-glycosylase. The uracil base is cleaved off, and the DNA strand is degraded and resynthesized with cytosine.

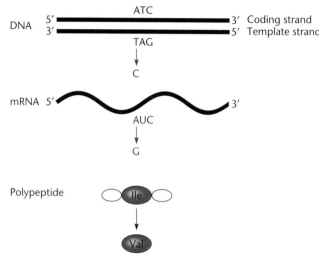

Figure 3.12 Missense mutation. A mutation that changes T to C in the DNA template strand will result in an A-to-G change in the mRNA. The mutant codon GUC will be translated as valine instead of isoleucine.

these codons are normally recognized by release factors (see chapter 2), which cause release of the translating ribosome and the polypeptide chain. Therefore, if a mutation to one of the nonsense codons occurs in an ORF for a protein, the protein translation will terminate prematurely at the site of the nonsense codon and the shortened or truncated polypeptide will be released from the ribosome (Figure 3.13). For this reason, nonsense mutations are sometimes called "chain-terminating mutations." These mutations almost always inactivate the protein product of the gene in which they occur. If, however, they occur in a noncoding region of the DNA or in a region that encodes an RNA rather than a protein, such as a gene for a tRNA, they are indistinguishable from other base pair changes.

The three nonsense codons—and their corresponding mutations—are sometimes referred to by color designations: **amber** for UAG, **ochre** for UAA, and **umber** or **opal** for UGA. These names have nothing to do with the effects of the mutation. Rather, when nonsense mutations were first discovered at Cal Tech, their molecular basis was unknown. The investigators thought that descriptive names might be confusing later on if their interpretations were wrong, so they followed the lead of physicists with their "quarks" and "barns." The first nonsense mutations to be discovered, to UAG, were called amber mutations. Following suit, UAA and UGA mutations were also named after colors—ochre and umber, respectively.

one of the nonsense codons, UAA, UAG, or UGA. These changes are called **nonsense mutations**.

While nonsense mutations are base pair changes and have the same causes as other base pair changes, the consequences are very different. Because the nonsense codons are normally used to signify the end of a gene,

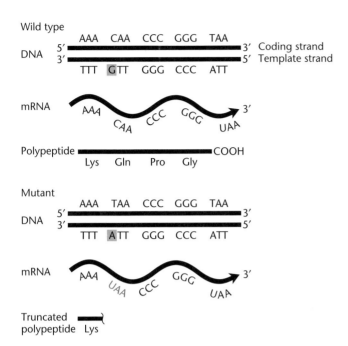

Figure 3.13 Nonsense mutation. Changing codon CAA, encoding glutamine (Gln), to UAA, a nonsense codon, causes truncation of the polypeptide gene product.

Wild type

5'--- C A A U C C C G G ——etc.——→

       Gln   Ser   Arg

Mutant

5'--- C A A A U C C C G ——→

       Gln   Ile   Pro   All amino acids
                                      downstream
                                      are changed

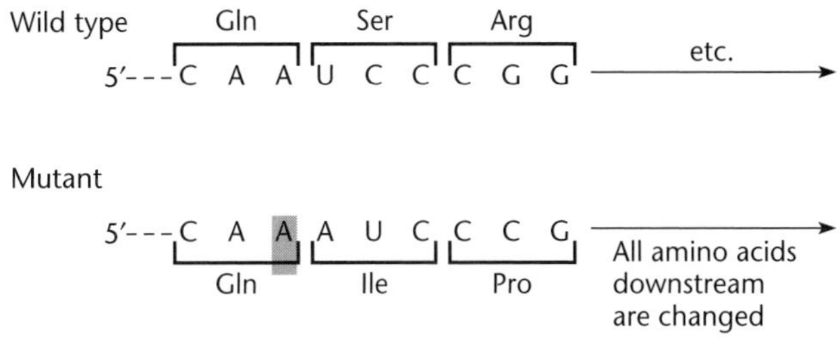

Figure 3.14 Frameshift mutation. The wild-type mRNA is translated glutamine (Gln)-serine (Ser)-arginine (Arg)-, etc. Addition of an A (boxed) would shift the reading frame, so that the codons would be translated glutamine (Gln)-isoleucine (Ile)-proline (Pro)-, etc., with all downstream amino acids being changed.

PROPERTIES OF BASE PAIR CHANGE MUTATIONS

Base pair changes are often leaky. A substituted amino acid may not work nearly as well as the original at that position in the chain, but the protein can retain some activity. Even nonsense mutations are usually somewhat leaky because sometimes an amino acid will be inserted for a nonsense codon, albeit at a low frequency. In wild-type *E. coli*, UGA tends to be most leaky, followed by UAG; the nonsense codon UAA tends to be the least leaky.

Base pair mutations also revert. If the base pair has been changed to a different base pair, it can also be changed back to the original base pair by a subsequent mutation. Moreover, base pair changes are a type of point mutation because they map to a particular "point" on the DNA, as we discuss later.

Frameshift Mutations

A high percentage of all spontaneous mutations are **frameshift mutations** (Figure 3.14). This type of mutation occurs when a base pair or a few base pairs are removed from or added to the DNA, causing a shift in the reading frame if they occur in an ORF encoding a polypeptide. Because the code is three lettered, any addition or subtraction that is not a multiple of 3 will cause a frameshift in the translation of the remainder of the gene. For example, adding or subtracting 1, 2, or 4

base pairs will cause a frameshift, but adding or subtracting 3 or 6 base pairs will not. Mutations that remove or add base pairs are usually called frameshift mutations even if they do not occur in an ORF and do not actually cause a frameshift in the translation of a polypeptide.

CAUSES OF FRAMESHIFT MUTATIONS

Spontaneous frameshift mutations often occur where there is a short repeated sequence that can slip. As an example, Figure 3.15 shows a string of AT base pairs in the DNA. Since any one of the A's in one strand can pair with any T in the other strand, the two strands could slip with respect to each other, as in the illustration. Slippage during replication could leave one T unpaired, and an AT base pair would be left out on the other strand when it replicates. Alternatively, the slippage could occur before the base was added, and an extra AT base pair could appear in one strand as shown.

PROPERTIES OF FRAMESHIFT MUTATIONS

Frameshift mutations are usually not leaky and almost always inactivate the protein, because every amino acid in the protein past the point of the mutation will be wrong. The protein will also most often be truncated, because a nonsense codon will usually be encountered while the gene is being translated in the wrong frame. Because, in general, 3 of the 64 codons are the nonsense codons, one of these should be encountered by chance about every 20 codons when the region is being translated in the wrong frame.

Another property of frameshift mutations is that they revert. If a base pair has been subtracted, one can be added to restore the correct reading frame and vice versa. More often, frameshift mutations do not revert but are suppressed by the addition or subtraction of a base pair close to the site of the original mutation that

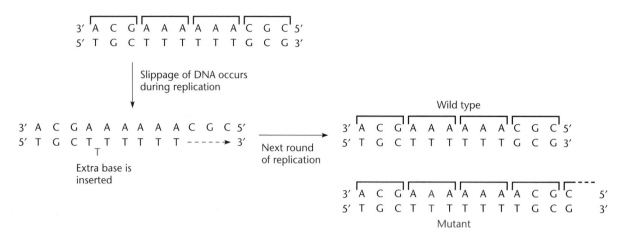

Figure 3.15 Slippage of DNA at a repeated sequence can cause a frameshift mutation.

restores the original reading frame. We shall discuss this means of frameshift suppression later in this chapter. Finally, frameshift mutations are a type of point mutation.

Some types of pathogenic bacteria apparently take advantage of the frequency and high reversion rate of frameshift mutations to avoid host immune systems. In such bacteria, genes required for the synthesis of cell surface components that are recognized by host immune systems often have repeated sequences. Consequently, these genes can be turned off and on by frameshift mutations and subsequent reversion. We discuss how one such frameshift mutation aids in the synthesis of virulence gene products by *Bordetella pertussis*, the causative agent of whooping cough, in chapter 13. Frameshift mutations are also used to reversibly inactivate genes of *Haemophilus influenzae* and *Neisseria gonorrhoeae*, which cause spinal meningitis and gonorrhea, repectively.

Deletion Mutations

Deletion mutations may be very long, removing thousands of base pairs and possibly many genes. Often, the only limitation on these mutations in bacterial DNA is that they cannot delete any essential genes, since haploid bacteria generally only have one copy of each gene. Some deletions in bacteria can be quite long, however, since bacterial genomes often possess long stretches of genes that can be deleted without losing cell viability. Deletion mutations often constitute a high percentage of spontaneous mutations; for example, in *E. coli*, almost 5% of all spontaneous mutations are deletions.

CAUSES OF DELETIONS
Deletions can be caused by recombination between different regions of the DNA. Recombination usually occurs between the same regions in two DNAs because the two DNAs will be identical in the same region. However, recombination can sometimes mistakenly occur between two different regions if they are similar enough in sequence. The strands of two DNA molecules are broken and rejoined in new combinations, hence the name **recombination**. This process is discussed later in this chapter and in detail in chapter 10.

As shown in Figure 3.16, deletions will result from recombination between two sequences that are **direct repeats**, that is, two sequences that are similar or identical when read in the 5′-to-3′ direction on the same strand of DNA. Bacterial DNA contains several types of repeats, the longest of which include insertion sequence (IS) elements and the rRNA genes, which are often repeated in many places in the DNA. IS elements are transposons and are discussed in chapter 9.

As shown in Figure 3.16, a deletion can occur when different copies of a direct repeated sequence in daughter DNAs mistakenly pair with each other. The two regions are then broken and rejoined, removing the sequence between the two direct repeats as shown. This is sometimes called "unequal crossing over," because the two DNAs are not equally aligned during the recombination. Alternatively, the two direct repeats on the same DNA molecule could pair, "looping out" the intervening sequences. Breaking and rejoining the DNA would remove the looped-out sequences as shown. For purposes of illustration, the directly repeated sequences shown in the figure are much shorter than would normally be required for recombination. Usually, direct repeats that promote mistaken recombination are hundreds or even thousands of base pairs long.

PROPERTIES OF DELETION MUTATIONS
Deletions have very distinctive properties. They are usually not leaky; deleting part or all of a gene will usually totally inactivate the gene product. Mutations that inactivate more than one gene simultaneously are most often

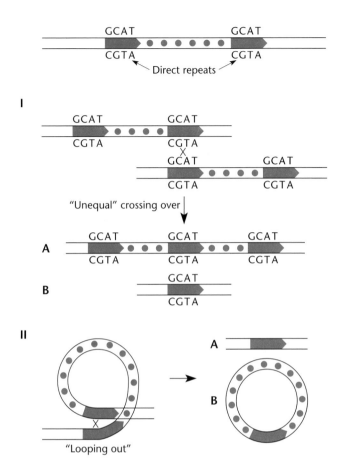

Figure 3.16 Recombination between directly repeated sequences can cause deletion mutations. (I) The recombination can occur between repeated sequences in different DNAs, resulting in a duplication (A) or a deletion (B). (II) Alternatively, it can occur between repeated sequences in the same DNA, resulting in a deletion (A) and the looped-out deleted segment (B).

deletions. Moreover, deletion mutations sometimes fuse one gene to another, sometimes putting one gene under the control of another.

The most distinctive property of long deletion mutations is that they never revert. Every other type of mutation will revert at some frequency, but for a deletion to revert, the missing sequence would somehow have to be found and reinserted. Deletions also behave differently from point mutations in genetic crosses, not mapping to a single point. This property can be very useful in some types of genetic mapping, as we discuss in later chapters.

NAMING DELETION MUTATIONS

Deletion mutations are named differently from other mutations. The Greek letter for "d," Δ, for *del*etion, is written in front of the gene designation and allele number, e.g., Δ*his8*. Often, deletions remove more than one gene, and so if known, the deleted regions are shown,

followed by a number to indicate the particular deletion. For example, Δ(*lac-proAB*)*195* is deletion number 195 extending through the *lac* and *proAB* genes on the *E. coli* chromosome. Often a deletion removes one or more known genes but extends into a region of unknown genes, so that the endpoints of the deletion are not known. In this case, the deletion is often named after the known gene. For example, the Δ*his8* deletion may delete the entire *his* operon but also extend an unknown distance into neighboring genes.

Inversion Mutations

Sometimes a DNA sequence is not removed, as in a deletion, but, rather, is flipped over, or **inverted**. After such an **inversion**, all the genes in the inverted region face in the opposite orientation.

CAUSES OF INVERSIONS

Inversions are caused in the same way as deletions, by recombination between repeats. However, recombination between inverted sequences rather than directly repeated sequences produces inversions. Inverted repeats read almost the same in the 5′-to-3′ direction on opposite strands (see chapter 1). Also unlike deletions, the recombination that produces inversions must occur between two regions on the same DNA (Figure 3.17).

Figure 3.17 Recombination between inverted repeats can cause inversion mutations. The order of genes within the inverison is reversed after the recombination.

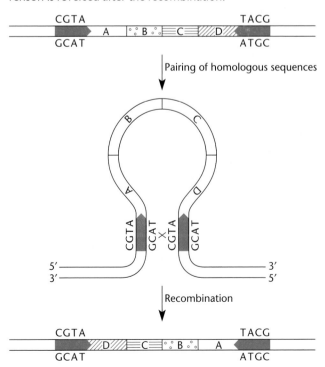

PROPERTIES OF INVERSION MUTATIONS

Unlike deletions, inversion mutations can generally revert. Recombination between the inverted repeats that caused the mutation will "reinvert" the affected sequence, recreating the original order. However, an inversion might occur between very short inverted repeats or repeats that are not exactly the same. Then the recombination event would have to occur between the exact bases involved in the first recombination to restore the correct sequence. Such a recombination could be a very rare event, and reversions of such a mutation would be very rare.

Inversion mutations often cause no phenotype. If the inversion involves a longer sequence, including many genes, generally the only affected regions will be those in the **inversion junctions**, where the recombination occurred. Most of the genes in the inverted region will still be intact, although they will be present in the reverse order. Consequently, even very long inversion mutations often cause no obvious phenotypes. Like deletions, inversion mutations sometimes fuse one gene to another gene. This property provides a mechanism for detecting them. The occurrence of inversions in evolution is discussed in Box 3.2.

BOX 3.2

Inversions and the Genetic Map

Even a single large inversion mutation will cause a dramatic change in the genetic map, or order of genes in the DNA, of an organism. The order of all the genes will be reversed between the sites of the recombination that led to the inversion. We would also expect inversions to be fairly frequent because repeated sequences often exist in inverted orientation with respect to each other. In spite of this, inversions seem to have occurred very infrequently in evolution. As evidence, consider the genetic maps of *S. enterica* serovar Typhimurium and *E. coli*. These bacteria presumably diverged billions of generations ago. Nevertheless, the maps are very similar except for one short inverted sequence between about 25 and 27 min on the *E. coli* map. At present, we can only speculate on why the genetic maps are so highly conserved. Perhaps organisms with this gene order have some selective advantage, or perhaps other sequences in the DNA cannot be inverted without disadvantaging the organism.

Termination of chromosome replication after sites like *terA* and *terB* (see chapter 1) may help explain why so few large inversions seem to have occurred in the evolution of bacteria. There may be sequences that resemble *terA* and *terB* distributed around the chromosome, but because they are on the wrong strand, they do not cause termination. However, an inversion mutation would reverse their orientation, so that if the *terA* site were preceded by a *terB*-like site, the DNA between the two sites would not be replicated. This situation would be lethal. However, there are probably other explanations for the rarity of large inversions.

Reference
Mahan, M. J., and J. R. Roth. 1991. Ability of a bacterial chromosome to invert is dictated by included material rather than flanking sequences. *Genetics* **129**:1021–1032.

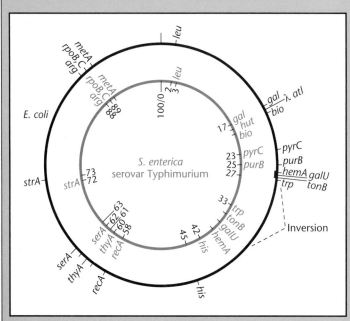

Genetic maps of *S. enterica* serovar Typhimurium and *E. coli*, showing a high degree of conservation. The region from *hemA* to the 40-min position is inverted in *E. coli* relative to that in serovar Typhimurium.

NAMING INVERSIONS

A mutation known to be an inversion is given the letters IN followed by the genes in which the inversion junctions occur, provided that these are known, followed by the number of the mutation. For example, IN(*purB-trpA*)3 is inversion number 3 in which the inverted region extends from somewhere within the gene *purB* to somewhere within the gene *trpA*.

Tandem Duplication Mutations

In a duplication mutation, a sequence is copied from one region of the DNA to another. The most common, a **tandem duplication**, consists of a sequence immediately followed by its duplicate. Tandem duplications occur frequently and can be very long.

CAUSES OF TANDEM DUPLICATIONS

Like deletions, tandem duplications can result from recombination between directly repeated sequences in DNA. In fact, as shown in Figure 3.18, they are probably often created at the same time as a deletion. Pairing between two directly repeated sequences in different DNAs, followed by recombination, can give rise to a tandem duplication and a deletion as the two products.

PROPERTIES OF TANDEM DUPLICATION MUTATIONS

Although the mechanism by which tandem duplications arise is probably similar to the mechanism that creates

Figure 3.18 Formation of tandem duplication mutations by recombination between directly repeated sequences on different DNA molecules. A deletion, the reciprocal recombinant, is created at the same time as the duplication. The symbol ⚡ designates the duplication junction.

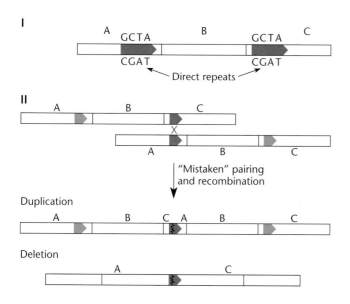

deletions and inversions, except that the recombination that creates a tandem duplication occurs between direct repeats that must be on different DNAs, the properties of tandem duplications are very different. Tandem duplication mutations that occur within a single gene will usually inactivate the gene and not be leaky. However, if the duplicated region is long enough to include one or more genes, no genes will be inactivated, including those in which the recombination occurred—the **duplication junctions**. This conclusion may seem surprising, but consider the example shown in Figure 3.18. Direct repeats in genes A and C on different DNAs pair with each other. The repeats in the two DNAs are then broken and rejoined to each other, creating a duplication in one DNA and a deletion in the other. Only part of gene A exists in the duplicate, but an entire gene A exists upstream. Conversely, only part of gene C exists upstream, but the entire gene exists in the duplicate. There are now two copies of gene B, both of which are unaltered. Therefore, intact genes A, B, and C still exist after the duplication, and there would be no indication that a mutation had even occurred unless there happened to be a phenotype associated with the presence of two copies of gene B.

Like deletions and inversions, duplications can sometimes fuse two genes to put expression of one gene under the control of a different gene. In the example in Figure 3.18, part of gene A has been fused to gene C, which might put genes A and B under the control of gene C.

The most characteristic property of tandem duplications is that they are very unstable and revert at a high frequency. Even though the recombinations that lead to a duplication are usually rare, recombination anywhere within the duplicated segments can delete them, restoring the original sequence. The instability of tandem duplications is often the salient feature that allows their identification.

ROLE OF TANDEM DUPLICATION MUTATIONS IN EVOLUTION

Tandem duplication mutations may play an important role in evolution. Ordinarily, a gene cannot change without loss of its original function, and if the lost function was a necessary one, the organism will not survive. However, when a duplication has occurred, there will be two copies of the genes in the duplicated region, and now one of these is free to evolve to a different function. This mechanism would allow organisms to acquire more genes and become more complex. However, how tandem duplications could persist long enough for some of the duplicated genes to evolve is not clear.

Insertion Mutations

Insertion mutations are caused by the insertion of a large piece of DNA into a region, usually by transposons "hopping" into the DNA. **Transposons** are DNA elements that can promote their own movement from one place in the DNA to another. In doing so, they create insertion mutations. Although these elements are usually thousands of base pairs long, sometimes only part of a transposon moves, or hops, producing a shorter insertion. Indeed, the movement of relatively short transposons, known as **insertion elements**, produces the majority of insertion mutations. These elements, which are only about 1,000 bp long, carry no easily identifiable genes. Most bacteria carry several insertion elements in their chromosome (see chapter 9).

PROPERTIES OF INSERTION MUTATIONS

Insertion of DNA into a gene almost always inactivates the gene; therefore, insertion mutations are usually not leaky. Transposons also contain many transcription termination sites, and so their insertion results in polarity (see chapter 2), which prevents the transcription of genes normally copied onto the same mRNA as the gene with the insertion. Finally, insertion mutations seldom revert, because the inserted DNA must be precisely removed, with no DNA sequences remaining. These last two unusual properties of insertion mutations led to their discovery.

SELECTING INSERTION MUTATIONS

A significant percentage of all spontaneous mutations are insertion mutations, but their phenotypes are difficult to distinguish from those of other types of mutations. However, transposons can serve as useful tools in genetics experiments, because many carry a selectable gene, such as one for antibiotic resistance. The insertion of such a transposon into a cell's DNA will then make the cell antibiotic resistant and easy to isolate. Moreover, transposon insertions are relatively easy to map, both genetically and physically. The methods of transposon mutagenesis are central to bacterial molecular genetics and biotechnology and so will be discussed in some detail in later chapters.

NAMING INSERTION MUTATIONS

An insertion mutation in a particular gene is represented by the gene name, two colons, and the name of the insertion. For example, *galK*::Tn5 denotes the insertion of the transposon Tn5 into the *galK* gene. If more than one Tn5 insertion exists in *galK*, the mutations can be numbered to distinguish them (e.g., *galK35*::Tn5). When insertion mutations are constructed for use in genetic experiments, they are denoted with the capital Greek letter Ω (omega) followed by the name of the insertion. For example, pBR322Ω::*kan* is a kanamycin resistance gene inserted into plasmid pBR322.

Reversion versus Suppression

Reversion mutations are often detected through the restoration of a mutated function. As discussed above, a reversion actually restores the original sequence of a gene. However, sometimes the function that was lost because of the original mutation can be restored by a second mutation elsewhere in the DNA. Whenever one mutation in the DNA relieves the effect of another mutation, that mutation has been **suppressed** and the second mutation is called a **suppressor mutation**. The following sections present some of the mechanisms of suppression.

Intragenic Suppressors

Suppressor mutations in the same gene as the original mutation are called **intragenic suppressors**, from the Latin prefix "intra" meaning "within." These mutations can restore the activity of a mutant protein by many means. For example, the original mutation may have made an unacceptable amino acid change that inactivated the protein, but changing another amino acid somewhere else in the polypeptide could restore the protein's activity. This form of suppression is not uncommon and is often interpreted to indicate an interaction between the two amino acids in the protein.

The suppression of one frameshift mutation by another frameshift mutation in the same gene is another example of intragenic suppression. If the original frameshift resulted from the removal of a base pair, the addition of another base pair close by could return translation to the correct frame. The second frameshift can restore the activity of the protein product provided that ribosomes, while translating out of frame, do not encounter any nonsense codons or insert any amino acids that alter the activity of the protein.

Intergenic Suppressors

Intergenic (or **extragenic**) **suppressors** do not occur in the same gene as the original mutation. The prefix "inter" comes from the Latin for "between." There are many ways in which intergenic suppression can occur. The suppressing mutation may restore the activity of the mutated gene product or provide another gene product to take its place. Alternatively, it may alter another gene product with which the original gene product must interact in a complementary way so that now the two mutated gene products can again interact properly.

One way an intergenic suppressing mutation may restore the viability of the cell is by preventing the accumulation of a toxic intermediate. If the gene for a step in a biochemical pathway is mutated, a toxic intermediate in that pathway can accumulate, causing cell death. However, a suppressing mutation in another gene of the pathway may prevent the accumulation, allowing the cell to survive even though it still has the original mutation. The suppression of *galE* mutations by *galK* mutations provides an illustration of such an intergenic suppressor. Cells with *galE* mutations are galactose sensitive (galactosemic), and their growth is inhibited by galactose in the medium. The reason is apparent from the pathway for galactose utilization shown in Figure 3.19. Many types of cells use the sugar galactose by first converting it to glucose. In the first step, galactose is phosphorylated by the product of the *galK* gene, a galactose kinase. The second step is the transfer of galactose 1-phosphate to uridine diphosphoglucose (UDPglucose) by the product of the *galT* gene, a transferase. The glucose produced is used as a carbon and energy source. The third step is the isomerization of the galactose on UDPgalactose to UDPglucose by the product of the *galE* gene, an isomerase. The newly synthesized UDPglucose can then cycle back into the pathway to convert more galactose to glucose.

Cells with *galE* mutations are galactose sensitive (galactosemic), because the absence of the GalE epimerase will permit the accumulation of both phosphorylated galactose and UDPgalactose, which are toxic to cells in high concentrations. Consequently, if we plate large numbers of a *galE* mutant strain on plates containing galactose, most of the cells will be inhibited. However, a few mutants will multiply to make colonies. Most of these mutants will not have undergone reversion of the *galE* mutation but will be double mutants with the original *galE* mutation and a suppressing *galK* mutation. The *galK* mutation blocks the first step of the pathway, so that no toxic intermediates accumulate. Revertants with reversion of the original *galE* mutation could also grow. However, we would expect *galE* reversion mutations to be much rarer than *galK* suppression mutations

because many changes will inactivate the *galK* gene but only one base pair change will cause the *galE* mutation to revert. Also, the *galE*⁺ revertants can be distinguished from *galE galK* double mutants because *galE*⁺ revertants will be Gal⁺ and will grow on galactose as the sole carbon and energy source. In contrast, the *galE galK* double mutants are still Gal⁻, and so another carbon source must be provided in the medium.

Nonsense Suppressors

Nonsense suppressors are another type of intergenic suppressor. A **nonsense suppressor** is usually a mutation in a tRNA gene that changes the anticodon of the tRNA product of the gene, so that it now recognizes a nonsense codon. In Figure 3.20, for example, the gene for a tRNA with the anticodon 3′GUC5′ (so that it normally recognizes the glutamine codon 5′CAG3′) mutates, causing the anticodon to become 3′AUC5′. This altered anticodon can pair with the nonsense codon UAG instead of CAG. However, the anticodon mutation does not significantly change the tertiary shape of the tRNA, which means that the cognate aminoacyl-tRNA synthetase will still load it with glutamine. Therefore, this mutated tRNA will bind with the amber codon UAG, allowing insertion of glutamine into the growing polypeptide instead of translation termination. This can lead to synthesis of the active polypeptide and suppression of the amber mutation.

The mutated tRNA is called a **nonsense suppressor tRNA**, and nonsense suppressors themselves are referred to as amber suppressors, ochre suppressors, or umber suppressors depending on whether they suppress UAG, UAA, or UGA mutations, respectively. Table 3.4 lists several *E. coli* nonsense suppressor tRNAs.

Nonsense suppressors can also be classified as **allele-specific suppressors** because they suppress only one type of allele of a gene, that is, one with a particular type of nonsense mutation. In contrast, the *galK* suppressing mutations discussed above suppress any *galE* mutation and so are not allele specific.

SUPPRESSED POLYPEPTIDES ARE NOT NORMAL

The polypeptide synthesized as the result of a nonsense suppressor is not always fully active. Usually, the amino acid inserted at the site of a nonsense mutation is not the same amino acid that was coded for by the original gene. This changed amino acid will sometimes cause the polypeptide to be almost inactive or temperature sensitive.

TYPES OF NONSENSE SUPPRESSORS

Not all tRNA genes can be mutated to form a nonsense suppressor. Generally, if there is only one type of tRNA to respond to a particular codon, the gene for that tRNA

Figure 3.19 The pathway to galactose utilization in *E. coli* and most other organisms. *galK* mutations suppress *galE* mutations because they prevent the accumulation of the toxic intermediates galactose 1-phosphate and UDP galactose.

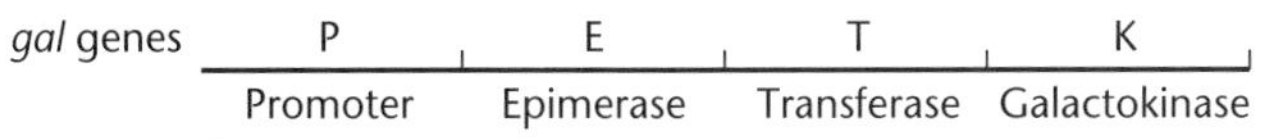

gal genes

P	E	T	K
Promoter	Epimerase	Transferase	Galactokinase

Pathway

1 Galactose + ATP $\xrightarrow{\text{GalK}}$ Galactose-1-PO₄ + ADP

2 Galactose-1-PO₄ + UDPglucose $\xrightarrow{\text{GalT}}$ UDPgalactose + glucose

3 UDPgalactose $\xrightarrow{\text{GalE}}$ UDPglucose

A Wild type

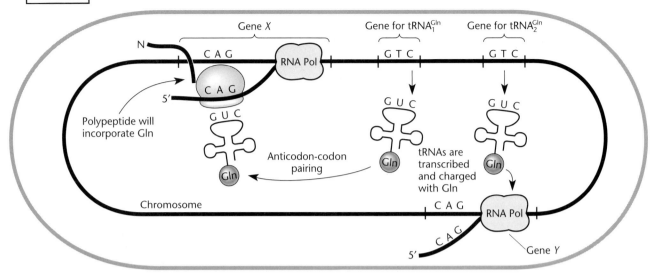

B Mutant_A with nonsense mutation in gene X

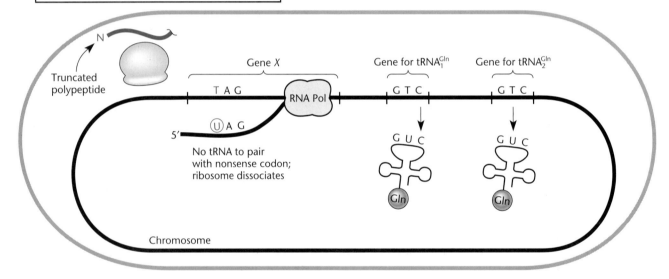

C Mutant_B with nonsense mutation in gene X and nonsense suppressor mutation in gene for tRNA₁^Gln

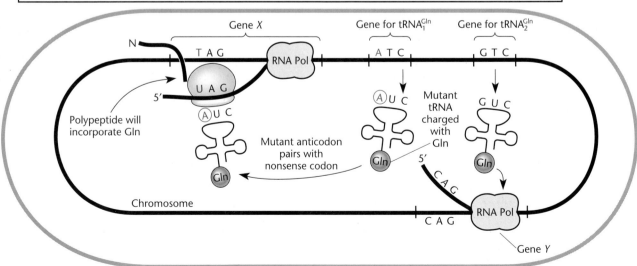

TABLE 3.4	Some *E. coli* nonsense suppressor tRNAs		
Suppressor name	**tRNA**	**Anticodon change**	**Suppressor type**
supE	tRNA^Gln	CU<u>G</u>-CUA	Amber
supF	tRNA^Tyr	<u>G</u>UA-CUA	Amber
supB	tRNA^Gln	U<u>U</u>G-UUA	Ochre/amber
supL	tRNA^Lys	UU<u>U</u>-UUA	Ochre/amber

cannot be mutated to make a suppressor tRNA. The original codon to which the tRNA responded would be "orphaned," and no tRNA would respond to it in an mRNA. In the example, two different tRNAs encoded by different genes recognize the codon CAG, one of which continues to recognize CAG after the other has been mutated to recognize UAG.

Wobble (see chapter 2) offers the only exception to the rule that a tRNA can be mutated to a nonsense suppressor only if there is another tRNA to respond to the original codon. Because of wobble, the same tRNA can sometimes respond both to its original codon and to one of the nonsense codons. For example, in a particular organism, there may be only one tRNA that recognizes the codon for tryptophan, 5'UGG3'. If the anticodon, 3'ACC5', is mutated to ACU, by wobble the tRNA might be able to recognize both the tryptophan codon UGG and the nonsense codon UGA, so that the suppressor strain could be viable. Wobble also allows the same suppressor tRNA to recognize more than one nonsense codon. In *E. coli,* all naturally occurring ochre suppressors also suppress amber mutations. From the wobble rules (see chapter 2), we know that a suppressor tRNA with the anticodon AUU could recognize both the UAG and UAA nonsense codons in mRNA (Table 3.4). Note that Table 3.4 anticodons are written 5'-3'.

EFFICIENCY OF SUPPRESSION

Nonsense suppression is never complete, because the nonsense codons are also recognized by release factors, which free the polypeptide from the ribosome (see chapter 2). Therefore, translation of the complete protein depends on the outcome of a race between the re-

lease factors and the suppressor tRNA. If the tRNA can base pair with the nonsense codon before the release factors terminate translation at that point, translation will continue. Sequences around the nonsense codon influence the outcome of this race and determine the efficiency of suppression of nonsense mutations at particular sites.

NONSENSE SUPPRESSOR STRAINS ARE USUALLY SICK

It would seem that nonsense suppressors would tend to translate through the proper nonsense codons at the ends of genes, resulting in proteins longer than normal. However, because nonsense suppressors are never 100% efficient, some of the correct proteins will always be synthesized. Moreover, since the efficiency of suppression depends upon the sequence of nucleotides in the gene around the nonsense codon, the nonsense codons at the ends of genes presumably have a "context" that favors termination rather than suppression. Also, often more than one type of nonsense codon lies in frame at the end of genes, presumably to avoid suppression by any particular tRNA suppressor.

Nevertheless, cells do pay a price for nonsense suppression. Cells with nonsense suppressors usually grow more slowly. Only lower organisms such as bacteria, fungi, and roundworms seem to tolerate nonsense suppressors, which are known to be lethal in higher organisms, including fruit flies and humans.

Genetic Analysis

One of the cornerstones of modern biological research is genetic analysis. Gregor Mendel performed the first

Figure 3.20 Formation of a nonsense suppressor tRNA. (A) Gene *X* and gene *Y* contain CAG codons, encoding glutamine. Other codons are not shown. The bacterium also has two different tRNA genes inserting glutamine for the CAG codon. Only the anticodons of the tRNAs are shown. (B) A mutation occurs in gene *X* (shown in gold), changing the CAG codon to UAG and causing synthesis of a truncated polypeptide (also shown in gold). (C) A suppressor mutation in the gene for one of the two tRNAs changes its anticodon so that it now pairs with the nonsense codon UAG. The translational machinery will now sometimes insert glutamine for the UAG nonsense codon in gene *X,* allowing synthesis of the complete polypeptide. The anticodon of the other tRNA still pairs with the CAG codon, allowing synthesis of the gene *Y* protein and the products of other genes carrying the CAG codon.

genetic analysis of a cellular function almost 150 years ago, when he crossed wrinkled peas with smooth peas and counted the number of progeny of each type. The methods of genetic analysis have become considerably more sophisticated since then and are still central to research in cell and developmental biology. The first information about many basic cellular and developmental processes often comes from a genetic analysis of the process. Advantages of the genetic approach are that it requires few assumptions and can be applied to any type of organism, even ones about which little to nothing is known. Genetic analysis is still the only way to determine how many gene products are involved in a function and to obtain a preliminary idea of the role of each gene product in the function. Through suppressor analysis, it also offers one of the best ways to ascertain which gene products interact with each other in performing the function. Genetic analysis is covered in general genetics textbooks, and we shall review the basic principles here only as they apply to bacteria and phages. Bacteria and their phages are ideal for demonstrating basic genetic principles, which is why so many discoveries in basic genetics have been made using these organisms. However, it is important to keep in mind that the basic principles discussed here are universal, applying equally to all organisms including humans, and that only the details of the performance of a genetic analysis differ from one type of organism to another.

Isolating Mutants

As discussed in the Introduction, a classical genetic analysis begins with finding mutants in which the function is altered. This process is called the **isolation of mutants** because the mutant organisms are somehow found and separated or "isolated" from the myriad of normal or nonmutant organisms with which they are associated. A major reason why bacteria and phages are such excellent genetic subjects is the relative ease with which mutants can be isolated. One reason is that we do not need to clone bacteria and phages like sheep; they clone themselves. They multiply asexually, which means that when an individual multiplies, its progeny are genetically identical clones of itself. Bacteria and viruses are also generally haploid, meaning they have only one allele of each gene. This makes the effects of even recessive mutations immediately apparent, obviating the need for backcrosses with immediate relatives to obtain homozygous individuals which show the effects of the mutation. Bacteria and phages also multiply on petri plates, and numbers equivalent to the entire human population on Earth can be placed on a single petri plate, facilitating the isolation of even very rare mutants.

TO MUTAGENIZE OR NOT TO MUTAGENIZE?

The first step in obtaining a collection of mutants for a genetic analysis is to decide whether to allow the mutations to occur spontaneously or to mutagenize the organism. Spontaneous mutations occur normally as mistakes in DNA replication, but the frequency of mutations can be greatly increased by treating the cells with some chemicals or with some types of irradiation. Treatments such as chemicals or UV irradiation, which cause mutations, are said to be **mutagenic,** and agents which cause mutations are **mutagens.** In general, treatments which damage DNA are mutagenic, but it is important to keep in mind that damage to DNA is not a mutation. Mutations are heritable changes in the sequence of normal deoxynucleotides in the DNA (see the definition of a mutation, above). Damaged DNA may mispair more frequently during replication, causing mutations, or mutations may arise during misguided attempts by the cell to repair the damage. We discuss mutagenesis in much more detail in chapter 11.

Both spontaneous and induced mutations have advantages in a genetic analysis. To decide whether to mutagenize the cells and, if so, which mutagen to use, we must first ask how frequent the mutations are likely to be. Spontaneous mutations are usually much rarer than induced mutations and so are more difficult to isolate. Therefore, to isolate very rare types of mutants or ones for which there is no good selection, we might have to use a mutagen. On the other hand, mutants containing spontaneously arising mutations are less likely to contain more than one different mutation, and the presence of multiple mutations can confuse the analysis later.

One major advantage of inducing mutations by using mutagens is that mutagens often induce only a particular type of mutation. Spontaneous mutations can be base pair changes, frameshifts, duplications, insertions or deletions. However, the acridine dye mutagens, such as acriflavin, cause only frameshift mutations, and base analogs, such as 2-aminopurine, cause only base pair changes. Therefore, the use of a particular mutagen may make it possible to restrict the mutations to the type desired.

Isolating Independent Mutations

For an effective genetic analysis, mutants defective in a function should have mutations that are as representative as possible of all the mutations that can cause the phenotype. If the strains in a collection of mutants carry many different mutations, we can get a better idea of how many genes can be mutated to give the phenotype and how many types of mutations can cause the phenotype. A general rule is that if some genes are represented by only one mutation, then, by the Poisson distribution discussed above, many other genes in which mutations could give the same phenotype have no mutations and have probably been missed.

There are two ways to ensure that the maximum number of different mutations are represented in a collection of mutants. One way is to avoid picking **siblings,** which are organisms that are descendents of the same original mutant. Two sibling mutants will always have the same mutation. The best way to avoid picking siblings is to isolate only one mutant from each of a number of different cultures, all started from nonmutant bacteria. If two mutants arose in different cultures, their mutations must have arisen independently and they could not be siblings. Another way is to use more than one mutagen. All mutagens have preferred "hot spots" and tend to mutagenize some sites more than others (see the discussion of mutational spectra in the *r*II genes of T4 phage in chapter 7). If all the mutants are obtained with the same mutagen, many of them will have mutations in the same hot spot, but mutants obtained with different mutagens will tend to have different mutations.

Selecting Mutants

Even after mutagenesis, mutants are rare and still must be found among the myriad of individuals that remain normal for the function. The process of finding mutants is called **screening.** Screening for mutants is usually the most creative part of a genetic analysis. One must anticipate the phenotypes that mutations in the genes for a particular function might cause. This is where the geneticist earns her or his pay, because predicting what types of mutations are possible and how to select them often requires intuition as well as rational thinking, but it is one of the more enjoyable aspects of genetics. For example, what do you imagine would be the phenotype of mutants defective in protein transport through the membrane? We discuss specific examples of screening for this and other types of mutants in later chapters.

Screening for mutant bacteria and phages usually involves finding **selective conditions** to distinguish the mutants from the original type. These are usually conditions under which either the mutant or the wild type will not be able to multiply to form a colony or a plaque. Agar plates or media with selective conditions are called selective plates and selective media, respectively. Selections can be either positive or negative. In a **positive selection,** selective conditions are chosen under which the mutant but not the original type can multiply. Figure 3.21 shows an example of a positive selection for His⁺ revertants of a *his* mutation. A *galK* mutation is another example of a type of mutation that can be selected by a positive selection, by plating a *galE* mutant on media containing galactose (see above).

In a **negative selection,** selective conditions are used under which the wild type but not the mutant can grow. Screening for mutants is much easier with positive selections. In a sense, negative selections are not really selections at all, because the selective conditions are being used to screen for the mutants rather than to eliminate all other organisms that are not mutated in the same way. Nevertheless, we shall use the common terms in this discussion.

ISOLATING MUTANTS BY NEGATIVE SELECTIONS

Most mutants, such as those that are auxotrophic or temperature sensitive, can be isolated only by using negative selections. Most of an organism's gene products help it to multiply. Therefore, mutations that inactivate a gene product are more likely to make the organism unable to multiply under a given set of conditions rather than able to multiply when the wild type cannot.

To isolate mutants by negative selection, the bacteria are first plated on a nonselective plate, on which both the mutant and the wild type can multiply. When the

Figure 3.21 Positive selection of a His⁺ revertant. A His⁻ mutant bacterium is plated on minimal media with all the growth requirements except histidine. Any colonies that form after the plate is incubated are due to His⁺ revertants that can multiply without histidine in the medium.

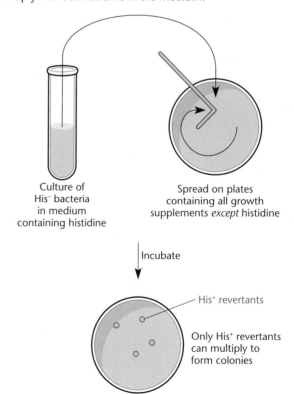

Culture of His⁻ bacteria in medium containing histidine

Spread on plates containing all growth supplements *except* histidine

Incubate

His⁺ revertants

Only His⁺ revertants can multiply to form colonies

colonies have developed, some of the bacteria in each colony are transferred to a selective plate to determine which colonies contain mutant bacteria that cannot multiply to form a colony under those conditions. Once such a colony has been identified, the mutant bacterial strain can be retrieved from the corresponding colony on the original nonselective plate. Figure 3.1 showed the detection of two types of auxotrophic mutants, His⁻ (unable to make histidine) and Bio⁻ (unable to make biotin), by negative selections.

Replica Plating

Because of the general rarity of mutants, many colonies usually have to be screened to find a mutant when using negative selection. **Replica plating** can be used to streamline this process and is illustrated in Figure 3.22. A few hundred bacteria are spread on a nonselective plate, and the plate is incubated to allow colonies to form. A replica is then made of this plate by inverting the plate and pressing it down over a piece of fuzzy cloth, such as velveteen. Then a selective plate is inverted and pressed down over the same cloth so that the colonies are transferred from the cloth to the selective plate. After the selective plate has been incubated, it can be held in front of the original nonselective plate to identify colonies that did not reappear on the selective plate. The missing colonies presumably contain descendants of a mutant bacterium that are unable to multiply on the selective plate. The mutant bacteria can then be taken from the colony on the original, nonselective plate. Replica plating was used by the Lederbergs to demonstrate that bacteria behave by the principles of Darwinian inheritance, as discussed earlier in the chapter.

Enrichment

If a type of bacterial mutant being sought is rare, finding it by negative selection can be very laborious, even with replica plating. No more than about 500 bacteria can be spread on a plate and still give discrete colonies. So, for example, if the mutant occurs at a frequency of 1 in 1 million, more than 2,000 plates might have to be replicated to find a mutant!

Many fewer colonies need to be screened if the frequency of mutants is first increased through mutant **enrichment**. This method depends on the use of antibiotics such as ampicillin and 5-bromouracil (5-BU) that kill growing but not nongrowing cells. Ampicillin inhibits cell wall synthesis and causes a growing bacterial cell literally to grow out of its skin and lyse. A mutant cell that was not growing while the ampicillin was present will not grow out of its skin and so will not be killed. 5-BU also kills only growing cells but by a very different mechanism. DNA containing 5-BU (an analog of

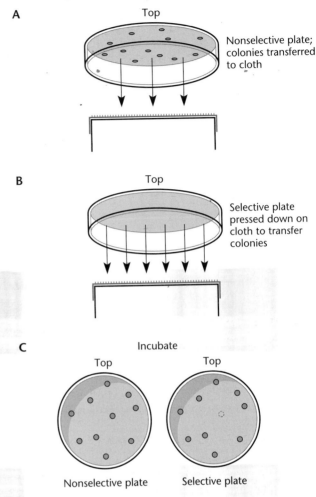

Figure 3.22 Replica plating. (A) A few hundred bacteria are spread on a nonselective plate, and the plates are incubated to allow colonies to form. The plate is then inverted over velveteen cloth to transfer the colonies to the cloth. (B) A second plate is then inverted and pressed down over the same cloth and incubated. (C) Both plates after incubation. The dotted circle indicates the position of a colony missing from the selective plate. See the text for details.

thymine) is much more sensitive to UV light than is normal DNA containing only thymine. Cells replicate their DNA only while they are growing, so they will take 5-BU into their DNA and become more UV sensitive only if they were growing while 5-BU was present in the medium.

To enrich for mutants that cannot grow under a particular set of selective conditions, the population of mutagenized cells is placed under the selective conditions in which the desired mutants stop growing. Meanwhile, the nonmutant wild-type cells will continue to multiply. The antibiotic—either ampicillin or 5-BU—is then added to kill any multiplying cells. The cells are then filtered or centrifuged to remove the antibiotic and transferred to nonselective conditions. The mutant cells will have survived preferentially because they were not growing in the presence of the antibiotic; therefore, they will have become a higher percentage of the population. No enrichment is 100% effective; however, even if the enrichment makes the mutant only 100 times more frequent, only 1/100 as many colonies and therefore 1/100 as many plates must be replicated to find a mutant after an enrichment. In the example given above, after an enrichment we would need to replicate only 20 plates instead of 2,000 to find a mutant.

Unfortunately, enrichments cannot be applied to all types of mutants. Some mutants are killed by the selective conditions and so cannot be enriched by these procedures. To be enriched, the mutant must still be alive and resume multiplying after it is removed from the selective conditions.

Complementation

Once we have a collection of mutants, we can further analyze them by genetic tests to characterize the mutations they have. One type of analysis we can perform is the complementation test. To perform a complementation test, we must put two copies of the regions of DNA containing two different mutations into cells and see what effect this has on the phenotypes of the mutations. With a diploid organism, which contains two homologous chromosomes of each type, this is no problem. With phages and other viruses, it is also no problem because we can infect cells with two different mutants simultaneously. However with bacteria, which are naturally haploid, complementation tests are more difficult. We must find a way to put only that part of the chromosome containing the mutations into the cells to make a partial diploid organism that has two copies of that region while remaining haploid for the rest of its genes. This can be done with plasmids or prophages which coexist with the chromosome. A gene cloned into a plasmid can exist in two copies when it is introduced into a cell already carrying a copy of the gene in its chromosome (see chapter 1). We discuss genetic ways of making partial diploids with bacteria in the chapters on plasmid conjugation (chapter 5) and lysogenic phages (chapter 8).

ALLELISM TESTS

One application of complementation is its use in determining how many different genes (or regions encoding a particular gene product) can be mutated to give a particular phenotype. Another name for this is an **allelism test,** because we are asking whether any two mutations are allelic, i.e., whether they affect the same gene (see above).

Returning to our example of histidine biosynthesis, assume that we have isolated a collection of mutants, all of which exhibit the His⁻ phenotype, and want to know how many genes they represent. This should tell us how many enzymes (or, more accurately, separate polypeptides, since some enzymes are composed of more than one different polypeptide [see chapter 2]) are required to make the amino acid histidine and allow the cell to multiply in the absence of histidine in the medium. Each of these polypeptides should be encoded by a different gene, and if our collection of mutations is large and varied enough, each of these genes should be inactivated by at least one of our mutations. The allelism test is performed on the mutations two at a time, as illustrated in Figure 3.23. If the two mutations are in different genes, each DNA can furnish the polypeptide that cannot be

Figure 3.23 Complementation tests for allelism. Three mutations, hisA1, hisA2, and hisB3, are being tested to determine which of them are allelic. (A) The hisA1 and hisB3 mutations are in different DNAs in the same cell, and the two mutations are in different genes whose products are required for synthesis of histidine. The DNA with the hisA1 mutation can make HisB, and the DNA with the hisB3 mutation can make HisA; hence, the cell is His⁺. (B) The hisA1 and hisA2 mutations are in the same gene. Neither DNA can make HisA; hence, the cell is His⁻.

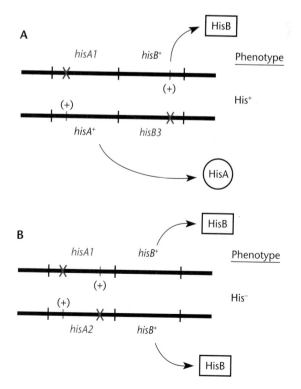

furnished by the other, so that all the polypeptides will be present and the diploid cell will be phenotypically His⁺. If, however, the two mutations are allelic (see above), neither DNA can make that gene product, the two mutations cannot complement each other, and the cells remain phenotypically His⁻. We can then extend this analysis to include the other mutations in our collection, two at a time, to place them in complementation groups and determine how many total genes or complementation groups are represented in the collection of mutations.

Usually these rules apply, and complementation between two mutations indicates that the mutations are in different genes while lack of complementation indicates they are in the same gene. However, complementation occasionally occurs between two mutations even though they are in the same gene. Complementation between two mutations in the same gene is called **intragenic complementation** and usually occurs only if the protein product of the gene is a homodimer or homomultimer composed of more than one polypeptide product of the gene (see chapter 2 for a definition of multimer). Also, the polypeptide may have more than one functional domain, with one domain of the polypeptide having one activity and the other domain being responsible for the other activity. Then a mutation in either domain will inactivate the protein if the organism is haploid for the gene and the protein can be assembled only from the one type of mutant polypeptide. However, if the organism is diploid for the gene and one copy has a mutation that inactivates one domain and the other copy has a mutation that inactivates the other domain, some proteins will be assembled from the different mutant polypeptides rather than just from the polypeptides with one mutation or the other. Such a mixed protein will have both activities and may be active. However, intragenic complementation is rare and occurs only between certain mutations in the gene; it is usually interpreted to mean that the product of the gene is a homomultimer.

Sometimes complementation does not occur between two mutations even though they are in different genes. This can happen if the mutation in one of the two genes is polar on the other gene or if the two genes are translationally coupled. Then a mutation which terminates translation in one gene can prevent the transcription or translation of another gene downstream of it and transcribed into the same mRNA (see chapter 2 for explanations of polarity and translational coupling). Table 3.5 outlines the interpretation of complementation experiments and their possible complications.

RECESSIVE OR DOMINANT

Complementation can also be used to tell if a mutation is **recessive** or **dominant** to the wild-type allele. A recessive

TABLE 3.5	Interpretation of complementation tests
Test result	**Possible explanations**
x and y complement	Mutations are in different genes Intragenic complementation has occurred[a]
x and y do not complement	Mutations are in the same gene One of the mutations is dominant One of the mutations affects a regulatory site or is polar

[a]See the text for an explanation of intragenic complementation. This is a less likely explanation than the mutations being in different genes.

mutation is subordinate to the wild type, so it does not exert its phenotype if the wild-type allele of the gene is present. A dominant mutation will exert its phenotype even if the wild type allele is present. Recessive mutations have generally inactivated the gene product, while dominant mutations often have subtly changed the gene product so that it can function in a situation where the wild type cannot function. Recessive mutations are much more common than dominant mutations because many more types of changes in the DNA will inactivate the gene product than change it in some subtle way. Whether a particular type of mutation in a gene that gives a particular phenotype is dominant or recessive can tell us something about the normal functioning of the gene product.

To determine whether a mutation is recessive or dominant, we make a diploid cell which has both the wild-type allele and the mutant allele and ask whether the wild-type phenotype or the mutant phenotype prevails. Let's return again to the example of the *his* pathway to illustrate the difference between recessive and dominant mutations. Most mutations which make the cell His⁻ have inactivated one of the enzymes required to make histidine. These mutations will all be recessive to the wild type because, in the presence of the wild-type allele for each of the genes, all the enzymes required to make histidine will be made and the cell will be His⁺, the phenotype of the wild-type alleles. However, assume there is an inhibitor of the pathway, say an analogue of histidine which binds to the first enzyme of the pathway and inactivates it. In the presence of this inhibitor, the cell will also be unable to make histidine and will be His-, so the phenotype caused by the wild-type allele in the presence of the inhibitor is His-. However, a mutation in the gene for the enzyme can make the enzyme insensitive to the inhibitor, so that the mutant cell can make histidine even in the presence of the inhibitor. The phenotype of the cell containing this mutation is His⁺ in the presence of the inhibitor. The mutant enzyme might continue to function to make histidine in the presence of the inhibitor even if the sensitive wild-type enzyme is present. If so, in a diploid containing both the mutant and wild-type alleles, the phenotype would be His+ in the presence of the inhibitor, the phenotype of the mutant allele, and the mutation is dominant.

CIS-TRANS TESTS

Another use of complementation is to determine whether a mutation is **trans** acting or **cis** acting. These prefixes come from the Latin and mean "on the other side" and "on this side," respectively. A *trans*-acting mutation usually affects a diffusible gene product, either a protein or on RNA. If the mutation affects a protein or RNA product, it can be complemented and it does not matter which DNA has the mutation in a complementation test because the gene product is free to diffuse around in the cell (i.e., the mutation acts in *trans*). A *cis*-acting mutation usually has changed a site on the DNA such as a promoter or an origin of replication. If the mutation affects a site on the DNA, it affects only that DNA and cannot be complemented (i.e., it acts in *cis*). In our example of the histidine synthesis genes, mutations, either recessive or dominant, which affect the enzymes that make histidine would be *trans* acting while a promoter mutation which prevents transcription of the genes for histidine synthesis would be *cis* acting. In subsequent chapters we discuss how *cis-trans* tests have been used to analyze gene expression and other cellular functions.

CLONING BY COMPLEMENTATION

Another very useful application of complementation in bacteria is in cloning. Complementation can be used to identify clones carrying a particular gene by their ability to complement a mutation in the chromosome and restore the normal or wild-type phenotype. Figure 3.24 illustrates using complementation to identify clones of the gene for thymidylate synthetase *(thyA)* in *E. coli*. This enzyme is needed to synthesize dTMP from dump, and so a *thyA* mutant will not be able to replicate its DNA and multiply to form a colony unless thymine is provided in the medium (see chapter 1). A library of the chromosome of wild-type *(thyA+) E. coli* is introduced into a strain of *E. coli* with a *thyA* mutation in its chromosome. The bacteria are then plated on selective plates containing all the necessary growth supplements but without thymine. Any colonies that appear may be due to bacteria containing a clone expressing the *thyA* gene from the clone which is complementing the *thyA* mutation in the chromosome. The clone containing the wild-type copy of the *thyA* gene can then be recovered from the cells in these colonies. For this method to

work, the clone usually must contain the entire gene and the gene must be expressed from the cloning vector. Genes from distantly related organisms will generally not be expressed in a particular host, so this method is generally useful only for identifying clones in the "host of origin," for example a bacterial gene in the bacterium it came from.

Recombination Tests

Complementation tests can tell us how many genes can be mutated to give a particular phenotype and something about the nature of the mutations that give the phenotype. However, recombination tests can give us better information about where mutations are located in the chromosome. **Recombination** is generally defined as the breakage and rejoining of two DNA molecules in new combinations. The breakage and rejoining is called a **crossover**. Recombination can be either specialized or general. We have mentioned some specialized recombination systems, such as the Xer-*dif* system for resolving chromosome dimmers, in chapter 1. However, the most

Figure 3.24 Identification of clones of the *thyA* gene of *E. coli* by complementation. (A) A *thyA* mutant of *E. coli* is transformed by a library of DNA from *thyA+ E. coli,* and the transformants are selected directly on plates lacking thymidine but containing the antibiotic to which the cloning vector confers resistance. (B) Thy+ transformants that contain a clone of the *thyA* gene synthesize the thymidylate synthetase, thus complementing the mutation in the chromosome.

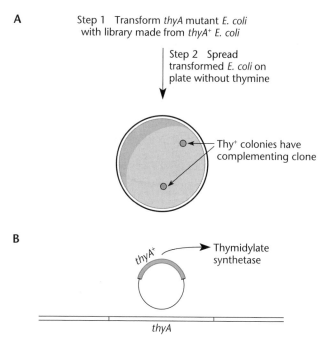

common type of recombination is **generalized recombination,** sometimes called **homologous recombination** because it can occur anywhere but occurs only between two DNA regions that have the same or homologous sequences ("homo-logos" means "same-word" in Greek). Restricting recombination to regions of homology between two DNAs helps ensure that the order of genes in the DNA of a species will not be scrambled each time recombination occurs. In fact, we have seen earlier in the chapter how recombination between the wrong places in DNA (ectopic recombination) can cause deletion, duplication, and inversion mutations. Homologous recombination probably occurs naturally in all organisms and serves the purpose of increasing genetic diversity within a species and/or repairing damage to DNA by restarting replication forks and making one good DNA molecule out of two damaged ones.

We can use recombination for genetic analysis without completely understanding its molecular basis. The detailed mechanisms by which recombination occurs are still under investigation, and the cell obviously uses different pathways for recombination in different situations (see chapter 10). However, geneticists have used recombination for genetic analysis for at least 80 years without knowing the actual molecular mechanisms involved. The simplified model of recombination shown in Figure 3.25 will suffice for our discussion. First the two DNAs pair in the same region where their nucleotide sequences are homologous. In the next step, staggered breaks are made in the two DNA molecules at exactly the same positions. This allows the strands of the two DNAs to cross over and pair with each other by complementary base pairing, as shown. Then the broken ends of one DNA are joined to the broken ends of the other DNA to make two new DNA molecules.

If the breaks and rejoining are occurring in the same place on two identical DNAs, the two new DNA molecules created by the recombination will have the same sequences as the original two DNA molecules and there would be no way of knowing that a recom-

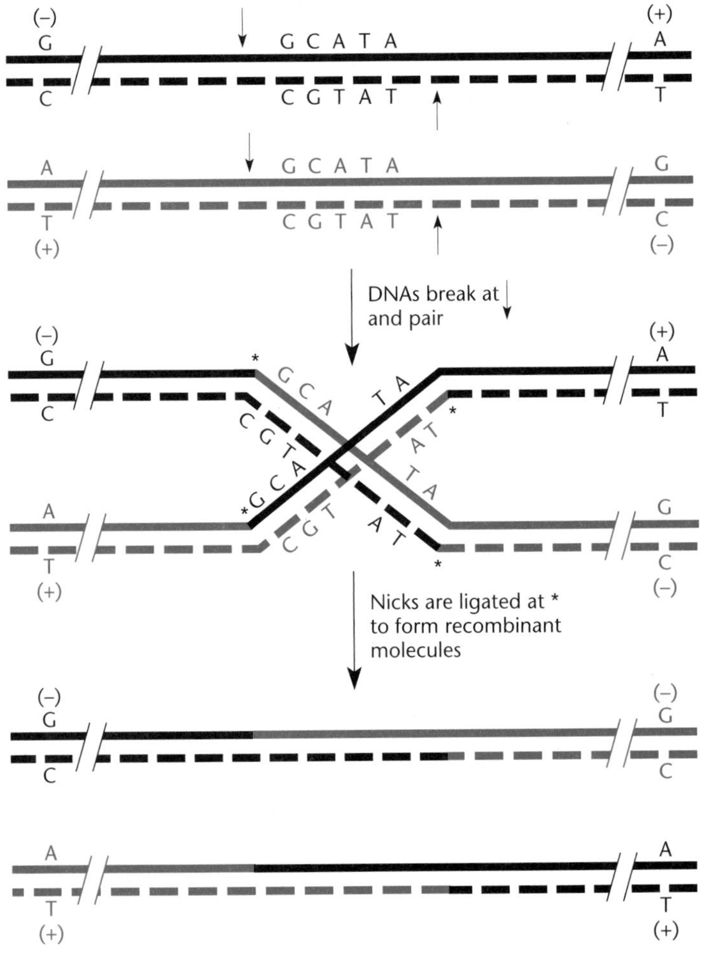

Figure 3.25 A simplified diagram of recombination. Staggered breaks are made in the two DNAs at the same sites on both DNAs. The strands then cross over, and the ends are joined to form two new DNA molecules. The positions of bases are shown only where the mutations have occurred and where the crossing over occurs. (–), mutation; (+), wild type.

bination event has occurred between them. However, as in the example, if the two DNAs have different mutations on either side of where the recombination occurred, there will be sequence differences at the sites of the mutations. A crossover between the regions of the two mutations will then yield two DNAs with different sequences from the original DNAs, one with neither mutation and the other with both mutations. If these two DNAs then **segregate** into progeny, these progeny will be genetically different from either parent. Progeny that are genetically different from either parent as the result of a recombination are called **recombinant types.** Progeny that are genetically identical to one or the other of the two parents are **parental types.** We usually detect recombination by the appearance of recombinant-type progeny.

Bacteria and phages have an advantage over most other organisms in that recombinant types often can be detected even if they are very rare. The process is much like detecting mutants with a positive selection. Conditions are established where only a recombinant type, but neither of the parental types, can multiply. To illustrate the selection of recombinant types, return to the example in Figure 3.25. In this example, both mutations inactivate a gene whose product is required for growth under some conditions, indicated by (–) next to the mutation. Therefore, neither parent is able to grow under the selective conditions. However, if the two parents are mated and a crossover occurs between the regions of the two mutations, recombinant-type progeny can arise which lack both mutations and so can multiply under the selective conditions. These progeny may be only a small percentage of the total progeny, but they can be detected by plating the cross under the selective conditions. Revertants of the mutations in either parent will also form a colony on the selective plates, but reversion requires a specific mutation and so the frequency of reversion is generally much lower than the frequency of recombination.

There are many applications of recombination in genetic analysis. We list a few such applications here.

RECOMBINATION FREQUENCIES

One use of recombination is to locate the region of one mutation or sequence difference based on the recombination frequency between the region of the mutation and the regions of other known mutations in the DNA. The **recombination frequency** is defined as the number of recombinant-type progeny over the total progeny. This frequency multiplied by 100 is the **map distance.** The closer the regions of two mutations are to each other, the smaller the chance that a crossover will occur between them and the lower the recombination frequency.

Therefore, the recombination frequency can be used to estimate how close together the regions of mutations are or the **linkage** between two regions on the DNA. Recombination frequency is only a relative measure of distance because different organisms vary greatly in their frequency of recombination and some regions of the DNA are more apt to have crossovers (i.e., to be recombinogenic) than others. Also, measures of recombination frequencies are more useful in genetic mapping in higher organisms and viruses than they are in bacteria, where recombination usually occurs between only a small part of the chromosome of one parent and the entire chromosome of the other parent. Interpreting recombination data from bacterial crosses varies, depending on the method of genetic exchange which is used. We discuss the analysis of recombination data in genetic mapping in phages and bacteria in subsequent chapters.

REVERSION VERSUS SUPPRESSION

Recombination can also be used to distinguish revertants from strains with suppressor mutations. As mentioned earlier in this chapter, the phenotypic change due to these two types of mutations can be very similar even though their molecular basis is very different. Returning to the example of a *his* mutation, if large numbers of the mutant strain are plated on medium without histidine, a few colonies may arise, apparently due to His$^+$ revertants. However, the *his* mutation may not have reverted but may have acquired a suppressor mutation somewhere else in the chromosome. For example, what if the original *his* mutation were a nonsense mutation? Then a mutation in a tRNA gene elsewhere in the chromosome could create a nonsense suppressor and suppress the mutation, leading to the His$^+$ phenotype even though the original *his* mutation is still present.

Using recombination to distinguish revertants from suppressors is illustrated in Figure 3.26. In the example, the test is being applied to an apparent His$^+$ revertant. The apparent His$^+$ revertant is crossed with the wild type and the progeny of the cross to see if any are His$^-$. If the original *his* mutation had reverted, the mutation no longer exists, so none of the progeny of the cross will be His$^-$. However, if the *his* mutation is being suppressed by another mutation elsewhere in the chromosome, called *supX* in the figure, a crossover could occur in the region between the original mutation and the *supX* mutation, giving rise to progeny that have the original mutation by itself and are therefore His$^-$. The other recombinant type will have only the suppressor mutation, and any phenotypes of this recombinant type will depend upon whether the suppressor by itself confers a phenotype.

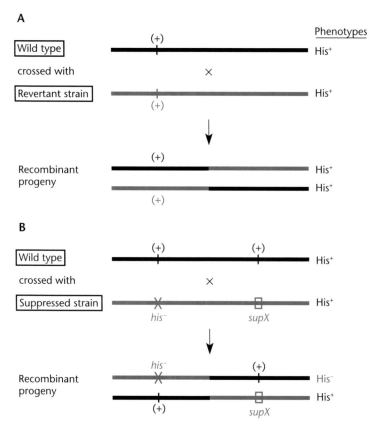

Figure 3.26 Test for reversion versus suppression. (A) The mutation had reverted, giving the His⁺ phenotype. Gold (+) shows the site of the reversion mutation. When the revertant strain is crossed with the wild type (in black), no His⁻ recombinants appear in the progeny. (B) A suppressor mutation, *supX* has suppressed the mutation, giving the His⁺ phenotype. When the suppressed strain is crossed with the wild type (in black), it gives some His⁻ recombinants. The site on the DNA with the *his* mutation is shown as the gold ×, and the site of the suppressor mutation is shown as the gold box.

MARKER RESCUE

Another application of recombination is in identifying clones by **marker rescue.** Recombination between a piece of DNA introduced into the cell and the corresponding region in the chromosome containing a mutation can "rescue" the mutation in the chromosome, restoring the wild-type phenotype. Figure 3.27 illustrates the use of marker rescue to identify a clone containing at least part of the *thyA* gene of *E. coli.* As in complementation cloning, a library of wild-type *E. coli* DNA is introduced into a *thyA* mutant strain of *E. coli.* However, now the cells containing the various clones are plated on permissive plates containing thymine and the plates are incubated to allow the colonies to develop. This plate is then replicated onto a selective plate containing all the necessary growth supplements but lacking thymine. If a particular clone includes the part of the *thyA* gene containing the site of the mutation, recombination between the clone and the chromosome can give rise to some Thy⁺ recombinants within the original colony. These Thy⁺ recombinants can grow to give small microcolonies on the selective plate, as shown. Cloning by marker rescue has advantages over cloning by complementation in that the clone need not contain the entire gene and the gene does not need to be expressed. However, it has the disadvantage that the clone is irretriev-

ably altered by the marker rescue recombination, so that clones carrying the gene cannot be selected directly and there must be some way of returning to the original unaltered clone that showed marker rescue, necessitating the replica plating.

Another use of marker rescue is in mapping the sites of mutations within a gene. Figure 3.28 shows an example of using this method to map mutations within the *thyA* gene of *E. coli.* Clones of the *thyA* gene were constructed with deletions extending different distances into the gene from one side (i.e., nested deletions). If a deleted clone can give Thy⁺ recombinants when introduced into a cell containing a particular *thyA* mutation in the chromosome, then the deleted clone must retain the region of the gene containing the mutation. The pattern of deletions that still show maker rescue localizes the mutation to a particular site in the gene.

Gene Replacements and Transgenics

One of the most useful current technologies involving recombination is its use in introducing foreign DNA into the chromosome of an organism. In some applications, recombination is used to replace the normal gene of an organism with a particular mutated allele of the same gene, often containing a mutation that we have made by

A

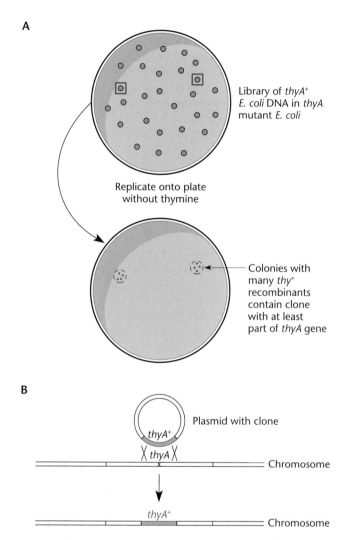

Library of *thyA⁺*
E. coli DNA in *thyA*
mutant *E. coli*

Replicate onto plate
without thymine

Colonies with
many *thy⁺*
recombinants
contain clone
with at least
part of *thyA* gene

B

thyA⁺
Plasmid with clone

thyA

Chromosome

thyA⁺

Chromosome

Figure 3.27 Use of marker rescue to identify a clone containing at least part of the *thyA* gene of *E. coli*. (A) The boxed colonies correspond to Thy⁺ recombinants grown on a replicated plate. See the text for details. (B) The *thyA⁺* gene on the cloning vector recombines with the *thyA* mutant gene on the chromosome to produce Thy⁺ recombinants.

A

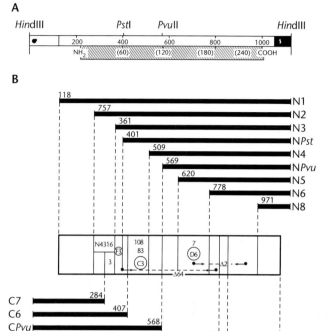

Figure 3.28 (A) Map of the *thyA* gene of *E. coli*. (B) Mutations in the chromosome were mapped by being crossed with deletions extending various distances into the cloned *thyA* gene of *E. coli*. Solid bars show the regions deleted in each of the constructs. N is the amino terminus of the gene, and C is the carboxyl terminus. Adapted from M. Belfort and J. Pedersen-Lane, *J. Bacteriol.* **160**:371–378, 1984.

site-specific mutagenesis (see chapter 1). Once the mutated DNA sequence has replaced the normal sequence in the chromosome, we can determine the effect of the specfic mutation on the phenotypes of the organism. This process of introducing a predetermined mutation into the DNA of an organism is sometimes called "reverse genetics" because it is essentially the reverse of normal genetic analysis. In reverse genetics, first we make the mutation and only afterward do we see the effect of the mutation on the organism. In classical genetics, we know the mutation has occurred only because of its effect on the organism, and afterward we clone and sequence the DNA to determine what kind of mutation caused the phenotype.

Recombination can also be used to introduce new genes into the chromosome of an organism, for example a gene for antibiotic resistance. An organism with foreign DNA in its chromosome is sometimes called a **transgenic organism** and this process is called **transgenics.**

The process of gene replacement is illustrated in Figure 3.29. In the illustration, a short piece of DNA which is homologous to part of the chromosome is introduced into a cell. This is a similar situation to the recombination following bacterial crosses, where generally only a small part of the DNA of one parent is introduced into the other parent. Gene replacement occurs through homologous recombination between the introduced DNA and homologous sequences in the chromosome.

The number of crossovers required to introduce foreign DNA into the chromosome depends upon whether the introduced DNA is linear or circular. As illustrated in Figure 3.29, a single crossover between a short linear DNA and the much longer chromosome will lead to breakage of the DNA, usually a lethal event. Two crossovers are required to replace the chromosomal sequence between

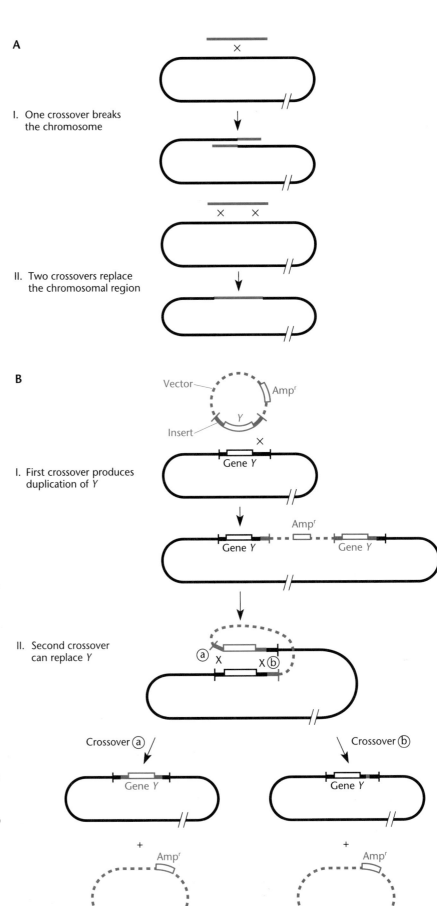

A

I. One crossover breaks
   the chromosome

II. Two crossovers replace
    the chromosomal region

B

Vector — Amp^r

Insert

Y

I. First crossover produces
   duplication of *Y*

Gene *Y*

Amp^r

Gene *Y*   Gene *Y*

II. Second crossover
    can replace *Y*

ⓐ X   X ⓑ

Crossover ⓐ   Crossover ⓑ

Gene *Y*   Gene *Y*

+   +

Amp^r   Amp^r

Y   *Y*

Figure 3.29   Gene replacements.
(A) The introduced DNA is a linear piece
of the chromosome with a slightly
altered sequence. (I) A single crossover
between the short linear piece of DNA
(shown in gold) and the corresponding
homologous region in the chromosome
(shown in black) will break the chromosome and be potentially lethal. (II) Two
crossovers are required to replace the sequence in the chromosome with the altered sequence in the introduced linear
DNA (shown in gold). (B) The
introduced piece of chromosomal DNA
containing gene *Y* with a specific mutation is cloned in a circular plasmid carrying a gene for resistance to ampicillin
(shown in gold). (I) A single crossover
between the cloned DNA and the corresponding homologous region in the
chromosome will insert the plasmid,
bracketing it with the chromosomal region containing a normal gene *Y* (shown
in black) and the plasmid clone containing the mutated copy of gene *Y* (shown
in gold). (II) A second crossover can loop
out the plasmid, leaving only one copy
of gene *Y* in the chromosome. Depending upon where this second crossover
occurs, the copy of gene *Y* left in the
chromosome can be either the mutant
copy (shown in gold; crossover a) or the
original, wild-type copy (shown in black;
crossover b).

150

the regions of the two crossovers with the sequence of the introduced DNA. The sequences of the linear introduced DNA and the chromosome need not be homologous over their entire lengths, only over the regions where the two crossovers occur. If two homologous sequences are on either side of a foreign DNA sequence, recombination between the two homologous sequences called **flanking sequences** and corresponding sequences in the chromosome can insert the foreign DNA into the chromosome and make a transgenic organism, as shown.

A very different situation prevails if the introduced DNA is circular (Figure 3.29). This situation is analogous to a region of the chromosome cloned into a circular plasmid cloning vector and then altered by site-specific mutagenesis in vitro, which is often the starting point for a gene replacement. When the plasmid containing the cloned DNA is then introduced into the cell, a single crossover between the slightly altered cloned DNA and the corresponding homologous sequence in the chromosome will integrate the circular plasmid into the chromosome, as shown. However, the altered cloned DNA sequence does not replace the corresponding sequence in the chromosome. Rather, the homologous sequences on the cloned DNA and the chromosome where the crossover occurred now bracket the integrated plasmid vector as shown, leading to a duplication of these sequences. A second crossover between the duplicated flanking sequences can excise the circular plasmid DNA. Depending upon where it occurs, this second crossover can either restore the original sequence in the chromosome or replace it with the altered cloned sequence in the plasmid.

In practice, gene replacements are not as straightforward as presented above and there are technical difficulties which must be overcome. First, there must be some way of selecting the cells in which the gene replacement has occurred. The recombination events that lead to gene replacement are relatively rare, and only a few cells in a population ever have their normal gene replaced by the mutated copy. Gene replacements are relatively easy to select if the mutation that alters the cloned gene is an insertion of a selectable gene such as a gene for antibiotic resistance. Cloning an antibiotic resistance cassette into the gene on the clone will both disrupt the gene in the clone and almost certainly inactivate the gene product; however, it will also introduce a selectable marker, which can be used later to select the few cells in which the replacement has occurred. In the example shown in Figure 3.30, a kanamycin resistance cassette (Kan^r) has been introduced into a cloned gene. The cells are then plated on medium containing kanamycin to select cells in which

the clone containing the kanaymcin resistance cassette has recombined into the chromosome. In the example, the plasmid cloning vector has a gene for ampicillin resistance (Amp^r). If the plasmid has integrated into the chromosome by a single crossover, the cells will become ampicillin resistant in addtion to kanamycin resistant. However, if a second crossover excises the plasmid from the chromosome and replaces the normal gene in the chromosome with the gene disrupted by the

Figure 3.30 Gene replacement. (A) With a single crossover, the cloning vector will integrate into the chromosome and the cells will become ampicillin resistant (Amp^r) and kanamycin resistant (Kan^r). (B) With a second crossover, the cells lose the cloning vector. Depending on where this second crossover occurs, the cells may be left with only the cloned sequence with the kanamycin resistance cassette and be resistant to kanamycin alone. See the text for details. Plasmid cloning vector sequences are in gold. The cloned region of the chromosome is in black.

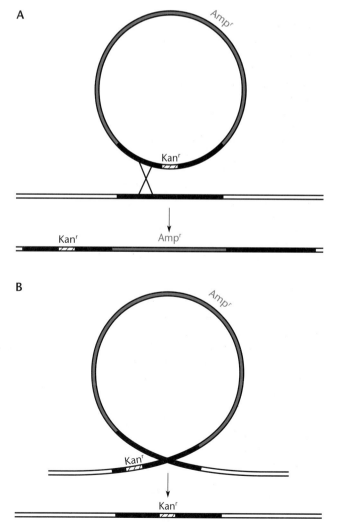

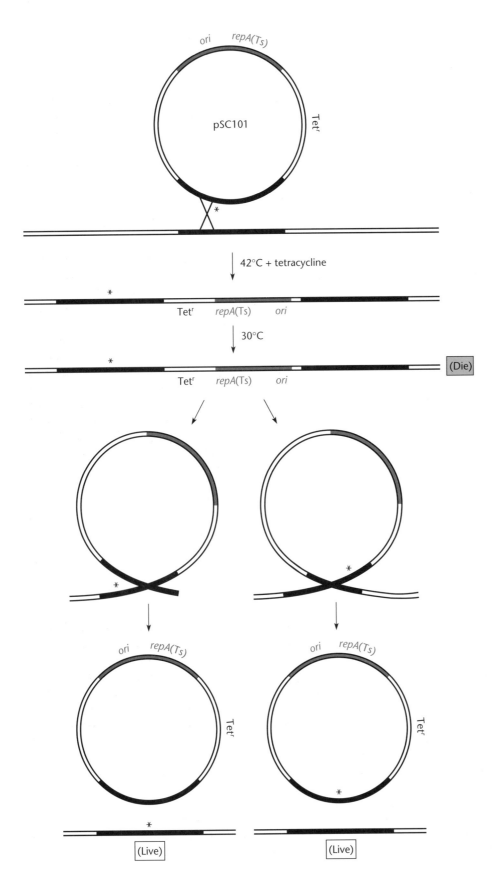

Figure 3.31 A method of selecting for the second crossover that removes the cloning vector and replaces the normal sequence with the sequence mutated in vitro. See the text for details. The asterisk shows the position of the mutation; *ori* denotes the plasmid origin of replication.

kanamycin resistance cassette, the cell will be kanamycin resistant but ampicillin sensitive, allowing detection of the cells in which the second crossover has occurred.

Another problem arises if the DNA is introduced into the cell in a vector that can replicate autonomously in the cell, i.e., is a replicon. This is often the case if the DNA is introduced in a plasmid cloning vector which has its own origin of replication and so can replicate autonomously. Then the cells will retain the antibiotic resistance gene carried on the plasmid, even if the introduced DNA does not recombine with the chromosome since the plasmid can maintain itself independently of the chromosome. This is not usually a problem if the DNA which is introduced into the cell is linear, because linear DNAs do not replicate in bacteria even if they contain an origin of replication. However, using linear DNA for gene replacements in bacteria has its own limitations since linear DNA is degraded in the cell by the RecBCD nuclease (see chapter 10). Methods have been devised to allow gene replacements with linear DNA in *E. coli* in which the RecBCD nuclease is somehow inactivated, and we discuss some of these methods in subsequent chapters. Alternatively, if the cloned DNA for the gene replacement is introduced in a circular plasmid, it is necessary that the cloning vector be somehow converted into a form in which it cannot replicate, i.e., a **suicide vector.** A protein required for replication of the plasmid may have been inactivated by a mutation or the plasmid might be from an unrelated host and may be unable to replicate in the cells. Then the plasmid containing the clone will be lost from the cell unless the cloned gene recombines with the chromosome. Suicide vectors are also used for transposon mutagenesis, as we discuss in chapter 9.

SELECTING GENE REPLACEMENTS BY USING THE LETHAL EFFECT OF A SECOND REPLICATION ORIGIN

As discussed above, most methods for gene replacement require that a selectable gene cassette, such as for antibiotic resistance, be introduced into the cloned gene. Otherwise, it is difficult to select the few cells in which a second crossover has replaced the chromosomal gene with the mutant cloned copy. However, sometimes we want to introduce another, more subtle, type of mutation into the chromosome, for example one which changes a single amino acid in a protein.

An alternative gene replacement method (see Hamilton et al., Suggested Reading) is based on the fact that *E. coli* cells are killed if their chromosome contains two active replication origins. This method is illustrated in Fig-

ure 3.31. The mutant gene is first cloned into a derivative of the plasmid vector pSC101 that confers tetracycline resistance (Tet^r) and has a temperature-sensitive mutation in its *repA* gene [*repA*(Ts)]. Because of the *repA*(Ts) mutation, the pSC101 *ori* region is not active at higher temperatures (around 42°C), so that at these temperatures the plasmid cannot replicate and will be lost from the cells. Only at lower temperatures (around 30°C) will the plasmid origin be active and the plasmid be able to replicate itself.

To use this method, the plasmid with the *repA*(Ts) mutation and the cloned mutant gene is transformed into *E. coli* and transformants are selected on tetracycline-containing plates at the high temperature of 42°C. Since the plasmid cannot replicate at this temperature, the only cells that become Tet^r are those in which the plasmid has integrated into the chromosome by a single crossover between the cloned mutant gene and the normal gene in the chromosome as shown. These cells can be selected by plating on tetracycline-containing medium. The chromosome in these transformants will have two potential origins of replication, its own *oriC* and the origin on the integrated plasmid, but the cells survive because the plasmid *ori* is inactive owing to the high temperature. However, when these Tet^r cells are shifted to 30°C, the temperature at which the pSC101 origin is active, the only cells that survive are those in which a second crossover has occurred, excising the plasmid from the chromosome (as shown in Figure 3.31) and leaving only the chromosomal origin of replication. Therefore, to select the few cells in which a second crossover has excised the plasmid, the cells need only be plated at 30°C. As in the previous example, some of these cells will have the normal gene restored in the chromosome and some will now have the mutant copy instead of the normal gene, depending on where the second crossover occurred.

This method has an advantage over most methods in that the excised plasmid remains in the cells because it is a replicon at 30°C. This feature allows us to determine the cells in which the normal gene has been replaced with the mutant copy because these will be the cells in which the excised plasmid contains the normal copy which it picked up when it left the mutant copy behind in the chromosome. Hence, plasmids are purified from the bacteria in a few of the surviving colonies and the inserted DNA is sequenced to find one which contains the normal sequence. As with all gene replacement techniques, the presence of the gene replacement in the chromosome should still be confirmed by some method, for example by direct PCR sequencing of the gene in the chromosome (see chapter 1).

SUMMARY

1. A mutation is any heritable change in the sequence of DNA of an organism. The organism with a mutation is called a mutant, and that organism's mutant phenotype includes all of the characteristics of the mutant organism that are different from the wild-type, or normal, organism.

2. The mutation rate is the chance of occurrence of a mutation to a particular phenotype each time a cell divides. The mutation rate offers clues to the molecular basis of the phenotype. Quantitative determination of mutation rates is often difficult.

3. One important conclusion from an analysis of population genetics is that the fraction of mutants increases as the population grows. This causes practical problems in genetics, which can be partially overcome by storing organisms without growth or by periodically isolating a single or very few organisms.

4. The type of mutation causing a phenotype can often be ascertained from the properties of the mutation. Base pair changes revert and are often leaky. Frameshift mutations also revert but are seldom leaky. Deletion mutations do not revert and are seldom leaky. They also often inactivate more than one gene simultaneously and can fuse one gene to another. Tandem duplication mutations revert at a high frequency and often have no observable phenotypes, except that they can fuse one gene to another. Inversion mutations also often have no observable phenotype and often revert. Insertion mutations seldom revert and are usually not leaky.

5. Deletions, inversions, and tandem duplication mutations can be caused by recombination between different regions in the DNA. Deletion and tandem duplication mutations are caused by recombination between directly repeated sequences in the DNA and inversions by recombination between inverted repeats. Deletions and tandem duplications arise as reciprocal recombination products between directly repeated sequences in different DNA molecules.

6. If a secondary mutation returns the DNA to its original sequence, we say that the mutation has reverted. If a mutation somewhere else in the DNA restores function, we say that the mutation was suppressed. Suppressors can be either intragenic or intergenic depending upon whether they occur in the same gene or in a different gene from the original mutation, respectively. An example of an intragenic suppressor is a frameshift mutation within the same gene which restores the reading frame shifted by another frameshift mutation. An example of an intergenic suppressor is a mutation in a tRNA gene which changes the tRNA so that it recognizes one of the nonsense codons, allowing translation of a nonsense mutation in another gene.

7. To isolate a mutant means to separate the mutant strain from the many other members of the population which are normal for the phenotype. Bacteria and phages and other viruses have advantages in genetic analysis because of the ease of isolating mutants. They multiply asexually, they are usually haploid, and large numbers can multiply on a single petri plate.

8. Mutations can be either spontaneous or induced. Spontaneous mutations occur as mistakes while the DNA is replicating, while induced mutations are caused by mutagenic chemicals or irradiation. Induction of mutations with mutagens has the advantage that mutations are more frequent and that a specific mutagen will often cause only a specific type of mutation. To ensure that mutations are as representative as possible, different mutagens should be used and the isolation of siblings is to be avoided.

9. Screening for a mutant means devising a way to distinguish the mutant from the normal or wild type. Selecting a mutant means devising conditions in which either only the mutant or the wild type can multiply. Selections can be either positive or negative. In a positive selection, conditions are devised under which the mutant but not the wild type can multiply; in a negative selection, the wild type but not the mutant can multiply under the selective conditions. Most type of mutants can only be selected by a negative selection. However, enrichment can sometimes be used to increase the frequency of the mutant in a population by killing the normal, multiplying cells under the negative selection conditions.

10. Complementation tests reveal whether different mutations inactivate different gene products. To do complementation tests, we introduce two copies of the region of DNA containing the two different mutations into a cell and ask whether the wild-type phenotype is restored. Complementation or allelism tests can be used to determine how many separate genes or, more precisely, regions encoding different gene products are represented in a collection of mutations exhibiting the same phenotype. It can also be used to determine if a mutation is dominant or recessive or *cis* acting or *trans* acting.

11. In recombination, two DNAs are broken and rejoined in new combinations. Generalized or homologous recombination occurs only between two DNAs with the same sequence. Progeny organisms that are different genetically from either parent as a result of recombination are called recombinant types, while progeny that are the same as one of the parents are parental types. Recombination can be used to locate mutations in the chromosome. The frequency of recombinant types, called the recombination frequency, is a measure of how close two regions of the chromosome are to each other. Recombination can also be used to distinguish suppressors from revertants and, through marker rescue, to determine whether a piece of DNA from the chromosome contains the region of a mutation. It can also be used to replace DNA in the chromosome with DNA that has been manipulated in vitro.

12. In reverse genetics, a cloned gene is mutated in vitro and then used to replace the normal gene in the chromosome in a process called gene replacement. Any changes in phenotype are attributed to the mutation introduced in vitro. In transgenics, a foreign DNA is introduced into the chromosome of an organism.

QUESTIONS FOR THOUGHT

1. A single inversion mutation will greatly alter a genetic map, or the order of genes in the DNA. Then why are the genetic maps of *S. enterica* serovar Typhimurium and *E. coli* so similar?

2. Do you suppose that duplication mutations play a role in evolution? If so, why are they not always destroyed as quickly as they form?

3. Why do you suppose that nonsense suppressors are possible in lower organisms but not higher organisms?

4. Can you propose a mechanism by which "directed" mutations might occur?

PROBLEMS

1. In a collection of cultures, all started from a few wild-type bacteria, which cultures are most likely to have had the earliest mutation to a particular mutant phenotype, those with a few mutant bacteria or those with many mutant bacteria?

2. Which phenotype would you expect to have the higher mutation rate, rifampin resistance or Arg$^-$ (arginine auxotrophy)? Rifampin inhibits transcription by binding to RNA polymerase, and Rif^r mutations change the RNA polymerase so that it no longer binds rifampin.

3. Luria and Delbrück grew 100 cultures of 1 ml each to 2×10^9 bacteria per ml. They then measured the number of bacteria resistant to T1 phage in each culture: 20 cultures had no resistant bacteria, 35 had one resistant mutant, 20 had two resistant mutants, and 25 had three or more resistant mutants. Calculate the mutation rate to T1 resistance by using the Poisson distribution.

4. Newcombe spread an equal number of bacteria on each of four plates. After 4 h of incubation, he sprayed plate 1 with T1 and put it back in the incubator. At the same time, he washed the bacteria off of plate 2, diluted them 10^7-fold, and replated them to determine the total number of bacteria. After a further 2 h of incubation, he sprayed plate 3 and washed the bacteria off of plate 4 and diluted them 10^8-fold before replating them. The next morning, he counted the colonies on each plate:

	Plate 1	Plate 2	Plate 3	Plate 4
No. of colonies	10	20	120	22

Calculate the mutation rate to T1 resistance.

5. You have isolated Arg$^-$ auxotrophs of *Klebsiella pneumoniae*.

a. If you plate a few cells with a mutation, *arg-1*, on plates without arginine, they multiply to make tiny colonies. If you plate 10^8 cells, you get some large, rapidly growing colonies. What kind of mutation is *arg-1* likely to be?

b. If you plate another mutant with a different mutation, *arg-2*, you get no growth on plates without arginine. Even if you plate large numbers of mutant bacteria ($>10^8$), you get no colonies. What type of mutation is *arg-2* likely to be?

c. What are some other possible explanations for (a) and (b)?

6. Design an experiment to show that *dam* mutations of *E. coli* are mutagenic, i.e., show higher than normal rates of spontaneous mutations. Assume that you can get a Dam$^-$ mutant in the mail, and do not have to isolate one in the laboratory.

7. Why is it necessary to isolate mutants from different cultures to be certain of getting independent mutations?

8. Design a positive selection for each of the following types of mutants. Discuss what kind of selective plates and/or conditions you would use.

a. Mutants resistant to the antibiotic coumermycin.

b. Revertants of a *trp* mutation that makes cells require tryptophan.

c. Double mutants with a suppressor mutation that relieves the temperature sensitivity due to another mutation in *dnaA*.

d. Mutants with a suppressor in *araA* that relieves the sensitivity to arabinose due to a mutation in *araD*.

e. Mutants with a mutation in *suA* (the gene for the transcription terminator protein Rho) that relieves the polarity of a *hisC* mutation on the *hisB* gene. Hint: make a partial diploid to perform complementation tests.

9. Design an enrichment procedure for reversible, temperature-sensitive mutants with mutations in genes whose products are required for cell growth.

10. Are nonsense suppressor mutations dominant or recessive? Why?

11. You have a strain of *Pseudomonas* that requires arginine, histidine, and serine as a result of nonsense mutations in genes encoding enzymes to make these amino acids. When you plate large amounts of this strain on media lacking all three of these amino acids, a few colonies arise which are Arg$^+$ His$^+$ Ser$^+$. These mutations are almost as frequent as mutations which revert each of the mutations separately. What kind of mutation do you think caused the apparent reversion of all three of these mutations? How would you test your hypothesis?

12. Outline how you would replace the *argH* gene, responsible for making an enzyme for arginine synthesis, with the corresponding gene into which you have inserted a gene for chloramphenicol resistance (Cm^r) in *E. coli*. What would you expect the phenotypes of your mutation to be?

13. You have isolated a nonrevertible *hemA* mutant of *E. coli* that requires δ-aminolevulinic acid to make hemes required for growth on succinate. You wish to use your mutant to clone the *hemA* gene. You do a partial digest of *E. coli* DNA with *Sau*3A, in which the average-sized piece is about 2 kbp, and clone the pieces into plasmid pBR322 cut with *Bam*HI and treated with phosphatase. You than use the ligation mix to transform the *hemA* mutant, selecting the ampicillin resistance gene on the plasmid. You test the colonies for growth on plates containing δ-aminolevulinic acid. How many Amp^r transformants should you have to test to have a reasonable chance of finding a colony that contains bacteria no longer requiring δ-aminolevulinic acid for growth on succinate? There is about 4,500 kbp of DNA in the *E. coli* genome.

14. One of the transformant colonies you test contains a few bacteria that no longer require δ-aminolevulinic acid, but most of the bacteria in the colony still require it. Do you think the clone has all of the *hemA* gene on it? Why or why not? Is the HemA$^+$ phenotype in these few transformants due to recombination or complementation?

SUGGESTED READING

Benzer, S. 1961. On the topography of the genetic fine structure. *Proc. Natl. Acad. Sci. USA* **47**:403–413.

Brenner, S., A. O. W. Stretton, and S. Kaplan. 1965. Genetic code: the nonsense triplets for chain termination and their suppression. *Nature* (London) **206**:994–998.

Crick, F. H. C., L. Barnett, S. Brenner, and R. J. Watts-Tobin. 1961. General nature of the genetic code for proteins. *Nature* (London) **192**:1227–1232.

Hamilton, C. M., M. Aldea, B. K. Washburn, P. Babitzke, and S. R. Kushner. 1989. A new method for generating deletions and gene replacements in *E. coli. J. Bacteriol.* **171**:4617–4622.

Hill, C. W., J. Foulds, L. Soll, and P. Berg. 1969. Instability of a nonsense suppressor resulting from a duplication of genes. *J. Mol. Biol.* **39**:563–581.

Lederberg, J., and E. M. Lederberg. 1952. Replica plating and the indirect selection of bacterial mutants. *J. Bacteriol.* **63**:399.

Luria, S., and M. Delbrück. 1943. Mutations of bacteria from virus sensitivity to virus resistance. *Genetics* **28**:491–511.

Newcombe, H. 1949. Origin of bacterial variants. *Nature* (London) **164**:150–151

Yanofsky, C., B. C. Carlton, J. R. Guest, D. R. Helinski, and U. Henning. 1964. On the colinearity of gene structure and protein structure. *Proc. Natl. Acad. Sci. USA* **51**:266–272.

4

Plasmids

What Is a Plasmid?

In addition to the chromosome, bacterial cells often contain **plasmids**. These DNA molecules are found in essentially all types of bacteria and, as we discuss below, play a significant role in bacterial adaptation and evolution. They also serve as important tools in studies of molecular biology. We address such uses later in the chapter.

Plasmids, which vary widely in size from a few thousand to hundreds of thousands of base pairs (a size comparable to that of the bacterial chromosome), are most often circular molecules of double-stranded DNA. However, some bacteria have linear plasmids, and plasmids from gram-positive bacteria can accumulate single-stranded DNA owing to aberrant rolling-circle replication (discussed below). The number of copies also varies among plasmids, and bacterial cells can harbor more than one type. Thus, a cell can harbor, for example, two different types of plasmids, with hundreds of copies of one plasmid type and only one copy of the other type.

Like chromosomes, plasmids encode proteins and RNA molecules and replicate as the cell grows, and the replicated copies are usually distributed into each daughter cell when the cell divides. However, unlike chromosomes, plasmids generally do not encode functions essential to bacterial growth. They instead provide gene products that can benefit the bacterium under certain circumstances (Table 4.1; Box 4.1). In this chapter we'll discuss the molecular basis for these and other features of plasmids.

Naming Plasmids

Before methods for physical detection of plasmids became available, plasmids made their presence known by conferring phenotypes on the cells harboring

TABLE 4.1	Some naturally occurring plasmids and the traits they carry	
Plasmid	Trait	Original source
ColE1	Bacteriocin which kills *E. coli*	*E. coli*
Tol	Degradation of toluene and benzoic acid	*Pseudomonas putida*
Ti	Tumor initiation in plants	*Agrobacterium tumefaciens*
pJP4	2,4-D (dichlorophenoxyacetic acid) degradation	*Alcaligenes eutrophus*
pSym	Nodulation on roots of legume plants	*Rhizobium meliloti*
SCP1	Antibiotic methylenomycin biosynthesis	*Streptomyces coelicolor*
RK2	Resistance to ampicillin, tetracycline, and kanamycin	*Klebsiella aerogenes*

them. Consequently, many plasmids were named after the genes they carry. For example, R-factor plasmids contain genes for resistance to several antibiotics (hence the name R for *resistance*). These were the first plasmids discovered, when *Shigella* and *Escherichia coli* strains resistant to a number of antibiotics were isolated from the fecal flora of patients in Japan in the late 1950s. The ColE1 plasmid from which many of the cloning vectors were derived carries a gene for the protein colicin E1, a bacteriocin that kills bacteria that do not carry this same plasmid. The Tol plasmid contains genes for the degradation of toluene, and the Ti plasmid of *Agrobacterium tumefaciens* carries genes for *tumor initiation* in plants. This system of nomenclature has led to some confusion, because plasmids carry various genes besides the ones for which they were originally named. Also, we have altered many of these plasmids beyond recognition to make plasmid cloning vectors (see below) and for other purposes.

BOX 4.1

Plasmids and Bacterial Pathogenesis

Plasmids often carry virulence genes required for bacterial pathogenicity. For example, to be pathogenic, some strains of the genus *Escherichia* must carry a large plasmid longer than 200 kb that contains genes for cell invasion and cell adhesion. Interestingly, the genes for regulating the virulence genes on the plasmids are in the chromosome rather than in the plasmid. The genes for the Shiga toxin are also in the chromosome. The Shiga toxin is related structurally to the diphtheria toxin but has an *N*-glycosylase activity that removes a specific adenine base in the 28S rRNA, thus blocking translation. This distribution of virulence genes between the plasmid and the chromosome in strains of *Escherichia* demonstrates the close relationship between plasmids and their hosts.

Most strains of *Salmonella* also require a large plasmid for their pathogenicity. However, the precise virulence genes that are carried on the plasmid are not known, and *Salmonella enterica* serovar Typhi does not require a plasmid to cause typhoid fever.

One of the clearest examples of plasmid virulence genes is in the genus *Yersinia*. The three species of *Yersinia*, *Y. enterocolitica*, *Y. pseudotuberculosis*, and *Y. pestis*, all cause disease, ranging in severity from mild enteritis in the case of *Y. enterocolitica* and *Y. pseudotuberculosis* to bubonic plague in the case of *Y. pestis*. To be pathogenic, all three species must harbor closely related plasmids that are about 70 kb long.

These plasmids encode outer membrane proteins called Yops, which may help the bacteria avoid host phagocytosis. These plasmids also carry genes for Type III secretion systems which injects some of these Yops directly into eukaryotic cells (see Box 2.6). The Yops are synthesized only under conditions of limiting calcium ions and high concentrations of sodium ions—conditions that may mimic the environment inside eukaryotic cells. The "plague bacillus," *Y. pestis*, must also carry two other smaller plasmids to cause disease. One of these encodes a toxin and an antiphagocytic protein similar to Yops, and the other encodes a protease that increases invasiveness. *Y. pestis* also has an iron-scavenging system that allows it to extract iron directly from hemin. The ion-scavenging system is encoded by a large, normally nonessential, chromosomal element that might be an integrated plasmid or prophage. Why so many genes required for pathogenesis are carried on plasmids and other DNA elements is a subject of speculation.

References

Ahmer, B. M., M. Tran, and F. Heffron. 1999. The virulence plasmid of *Salmonella typhimurium* is self-transmissible. *J. Bacteriol.* **181**:1364–1368.

Irarte, M., and G. R. Cornelis. 1999. The 70-kilobase virulence plasmid of yersiniae, p. 91–126. *In* J. B. Kaper and J. Hacker (ed.) *Pathogenicity Islands and Other Mobile Virulence Elements*. American Society for Microbiology, Washington, D.C.

To avoid further confusion, the naming of plasmids is now standarized. Plasmids are given number and letter names much like bacterial strains. A small "p," for *plasmid*, precedes capital letters that describe the plasmid or sometimes give the initials of the person or persons who isolated or constructed it. These letters are often followed by numbers to identify the particular construct. When the plasmid is further altered, a different number is assigned to indicate the change. For example, the plasmid pBR322 was constructed by *B*olivar and *R*odriguez from the ColE1 plasmid and is derivative number 322 of the plasmids they constructed. pBR325 is pBR322 with a chloramphenicol resistance gene inserted. The new number 325 distinguishes this plasmid from pBR322.

Functions Encoded by Plasmids

Depending on their size, plasmids can encode a few or hundreds of different proteins. However, as mentioned above, plasmids rarely encode gene products that are always essential for growth, such as RNA polymerase, ribosomal subunits, or enzymes of the tricarboxylic acid cycle. Instead, plasmid genes usually give bacteria a selective advantage under only some conditions.

Table 4.1 lists a few naturally occurring plasmids and some traits they encode, as well as the host in which they were originally found. Gene products encoded by plasmids include enzymes for the utilization of unusual carbon sources such as toluene, resistance to substances such as heavy metals and antibiotics, synthesis of antibiotics, and synthesis of toxins and proteins that allow the successful infection of higher organisms (Box 4.1). We can use the fact that plasmids generally carry only nonessential genes to distinguish them from the chromosome, particularly when the plasmid is almost as large as the chromosome.

If plasmid genes, such as those for antibiotic resistance and toxin synthesis, were part of the chromosome, all bacteria of the species, not just the ones with the plasmid, would have the benefits of those genes. Consequently, all the members of that species would be more competitive in environments where these traits were desirable. So why are plasmid genes not simply part of the chromosome? Maybe having some genes on plasmids makes the host species able to survive in more environments without the burden of a larger chromosome. Bacteria must be able to multiply very quickly under some conditions to obtain a selective advantage, and smaller bacterial chromosomes can replicate faster than larger ones. Plasmids encoding different traits can then be distributed among different members of the population, where they do not burden any single bacterium too heavily. However, if the environment abruptly changes so that the genes carried on one of the plasmids become essential, the bacteria that carry the plasmid will suddenly have a selective advantage and will survive, thereby ensuring survival of the species. In this way, plasmids allow bacteria to occupy a larger variety of ecological niches and contribute to the evolutionary success of not only the bacterial species but also the plasmids found in that species.

Plasmid Structure

Most plasmids are circular with no free ends, although a few known plasmids are linear (see Box 4.2). In a circular plasmid, all of the nucleotides in each strand are joined to another nucleotide on each side by covalent bonds to form continuous strands that are wrapped around each other. Such DNAs are said to be **covalently closed circular (CCC)**. This structure prevents the strands from separating, and there are no ends to rotate, so that the plasmid can be supercoiled. As discussed in chapter 1, in a DNA that is supercoiled, the two strands are wrapped around each other more or less often than once in about 10.5 bp, as predicted from the Watson-Crick double-helical structure of DNA. If they are wrapped around each other more often than once every

---

BOX 4.2

Linear Plasmids

Not all plasmids are circular. Linear plasmids have been found in many bacteria, including members of the genus *Streptomyces*, which are soil bacteria responsible for making antibiotics, and *Borrelia burgdorferi*, the causative agent of Lyme disease. Linear plasmids, like linear chromosomes, face a "primer problem" because of the properties of DNA polymerases (see chapter 1). No known DNA polymerases can initiate the synthesis of new DNA strands, they can only add deoxynucleotides to a preexisting primer. This means that when replication proceeds to the ends of a linear DNA, the polymerase cannot replicate to the extreme 5′ end of the lagging strand because there is no upstream primer. This is not a problem for circular plasmids because there is always DNA upstream to provide a primer. Different types of linear plasmids may solve the primer problem in different ways. One solution is suggested by the sequence

(continued)

BOX 4.2 (continued)

Linear Plasmids

of the ends of a 16-kb plasmid from *B. burgdorferi* (see Figure). The *B. burgdorferi* linear plasmids solve the "primer problem" by having telomeric sequences on their ends much like some eukaryotic viruses. Such plasmids carry genes for the major surface proteins of the bacteria, breaking the rule that essential genes are never found on plasmids. It is also possible that the chromosomes from these bacteria are linear and the plasmids are directly derived from the chromosome, perhaps through the deletion of large chromosomal segments.

Not only are these plasmids linear, but also their ends are covalently joined, as shown in the figure. If these plasmids are denatured by strand separation, they will form one large, single-stranded, circular molecule. The ends have the same sequence, but they are inverted. Somehow these features allow the plasmids to replicate all the way through without losing DNA from the ends of each replication. If the DNA polymerase could somehow replicate around the end, then this DNA could serve as a primer for replication of the other strand.

Another interesting feature of the ends of these linear plasmids is their similarity to the ends of the linear DNAs of poxviruses and African swine fever virus, as shown in the figure (see Hinnebusch and Barbour). Whether this similarity is significant in terms of the origins of these bacteria and viruses remains to be seen.

Another solution is suggested by the ends of the linear plasmid pSLA2 of *Streptomyces*. This plasmid has a protein covalently attached to its ends. However, this protein does not simply serve as a primer for lagging-strand synthesis as proteins do for some phages (see Box 7.2), because the replication origin for this plasmid is known to be internal to the plasmid and replication proceeds bidirectionally toward both ends. This plasmid contains a number of tandem inverted repeated sequences or palindromes, which may allow hairpin formation by pairing between the inverted repeated sequences. A model has been proposed to explain how these palindromic sequences may allow the leading strand to fold back on itself and form a hairpin with the lagging strand to provide a primer for lagging-strand synthesis. The protein attached to the 5' end of the leading strand may also play a role in this.

References

Casjens, S., N. Palmer, R. van Vugt, W. M. Huang, B. Stevenson, P. Rosa, R. Lathigra, G. Sutton, J. Peterson, R. J. Dodson, D. Haft, E. Hickey, M. Gwinn, O. White, and C.M. Fraser. 2000. A bacterial genome in flux: the twelve linear and nine circular extrachromosomal DNAs in an infectious isolate of the Lyme disease spirochete *Borrelia burgdorferi*. *Mol. Microbiol.* **35**:490–516.

Hinnebusch, J., and A. G. Barbour. 1991. Linear plasmids of *Borrelia burgdorferi* have a telomeric structure and sequence similar to those of a eukaryotic virus. *J. Bacteriol.* **173**:7233–7239.

Qin, Z., and S. N. Cohen. 1998. Replication at the telomeres of the *Streptomyces* linear plasmid pSLA2. *Mol. Microbiol.* **28**:893–903.

Yang, C.-C., C.-H. Huang, C.-Y. Li, Y.-G. Tsay, S.-C. Lee, and C. W. Chen. 2002. The terminal proteins of linear *Streptomyces* chromosomes and plasmids: a novel class of replication priming proteins. *Mol. Microbiol.* **43**:297–305.

A comparison of the telomeres from a linear plasmid of *B. burgdorferi* (A) and from African swine fever virus (B) and vaccinia virus. The ends of the DNAs are covalently joined to each other as shown.

10.5 bp, the DNA is positively supercoiled; if they are wrapped around each other less often, the DNA is negatively supercoiled. Like the chromosome, covalently closed circular plasmid DNAs are usually negatively supercoiled (see chapter 1). Because DNA is stiff, the negative supercoiling introduces stress, and this stress is partially relieved by the plasmid wrapping up on itself, as illustrated in Figure 4.1. In the cell, the DNA wraps around proteins, which relieves some of the stress. The remaining stress facilitates some reactions involving the plasmid, such as separation of the two DNA strands for replication or transcription.

PURIFYING PLASMIDS

The structure of plasmids can be used to purify them away from the chromosomal and other DNA in the cell. Cloning manuals usually give detailed protocols for these methods (see chapter 1, Suggested Reading), but we review them briefly in this section. Many purification procedures are based on the relatively small size of most plasmids. They will often not precipitate at the salt concentrations at which the chromosome will precipitate. Therefore, if extracts of cells are treated at high salt concentrations, the chromosome will often precipitate and

can be removed by centrifugation while the much smaller plasmids stay in the supernatant.

Other purification steps take advantage of the fact that most plasmids are circular and covalently closed and supercoiled. One such purification involving the acridine dye ethidium bromide (EtBr) is illustrated in Figure 4.2. This procedure is based on the fact that covalently closed circular DNAs bind less EtBr than do linear or nicked circular DNAs. EtBr intercalates (inserts itself) between DNA bases, pushing the bases apart and rotating the two strands of the DNA around each other. If the two strands of the DNA are not free to rotate, as in a covalently closed circular plasmid, the binding of EtBr will eventually introduce postive supercoils and increase the stress on the DNA until no more EtBr can bind. EtBr bound to DNA makes it *less* dense in salt solutions made with heavy atoms such as cesium chloride (CsCl). As a consequence, if the DNA is mixed with a solution of CsCl and EtBr and centrifuged to establish a gradient of CsCl concentration, the covalently closed circular plasmid DNAs will band lower, at a position where the solution is more dense (Figure 4.3).

The methods discussed above work well with plasmids that have many copies per cell and are not too large. However, large, low-copy-number plasmids are much more difficult to detect. Most methods for detecting large plasmids involve separating them from the chromosome directly by electrophoresis on agarose gels (see chapter 1). The cells are often broken open directly on the agarose gel to avoid breaking the large plasmid DNA. The plasmid, because of its unique size, makes a sharp band on the gel, distinct from that due to chromosomal DNA, which is usually broken and so gives a more diffuse band. Also, methods such as pulsed-field gel electrophoresis (PFGE) have been devised to allow the separation of long pieces of DNA based on size. These methods depend on periodic changes in the direction of the electric field. The molecules will attempt to reorient themselves each time the field shifts, and the longer molecules will reorient themselves more slowly than the shorter ones and so will move more slowly. Such methods have allowed the separation of DNA molecules hundreds of thousands of base pairs long and the detection of very large plasmids.

Properties of Plasmids

Replication

To exist free of the chromosome, plasmids must have the ability to replicate independently. DNA molecules that can replicate autonomously in the cell are called **replicons**. Plasmids, phage DNA, and the chromosomes are all replicons, at least in some types of cells.

Figure 4.1 Supercoiling of a covalently closed circular plasmid. A break in one strand relaxes the DNA, eliminating the supercoiling.

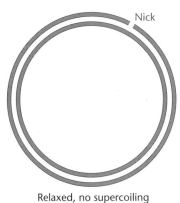

Relaxed, no supercoiling

Supercoiled, covalently closed circular DNA

Figure 4.2 Less EtBr can bind to a covalently closed circular DNA than to a linear or nicked circular DNA. Also, progressively higher EtBr concentrations shift DNA supercoiling from negative to positive. At about 2 μg of EtBr per ml, most DNAs are completely relaxed. The arrow indicates the free rotation of linear DNA.

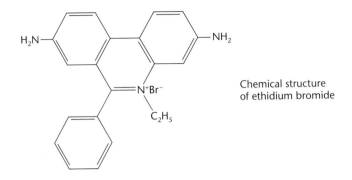

Chemical structure of ethidium bromide

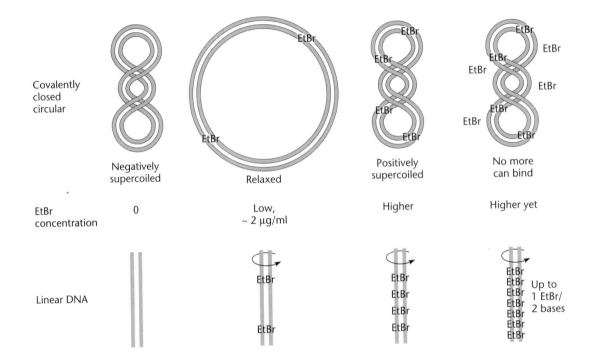

Figure 4.3 Separation of covalently closed circular plasmid DNA from linear and nicked circular DNAs on EtBr-CsCl gradients. After centrifugation, the plasmid DNA will band below the other DNAs because it has a higher buoyant density as a result of less binding of EtBr.

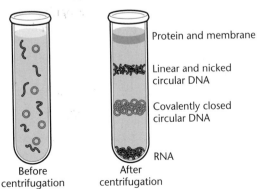

To be a replicon in a particular type of cell, a DNA molecule must have at least one origin of replication, or *ori* site, where replication begins (see chapter 1). In addition, the cell must contain the proteins that enable replication to initiate at this site. Plasmids encode only a few of the proteins required for their own replication. In fact, many encode only one of the proteins needed for initiation at the *ori* site. All of the other required proteins, including DNA polymerases, ligases, primases, helicases, and so on, are borrowed from the host.

Each type of plasmid replicates by one of two general mechanisms, which is determined along with other properties by its *ori* region (see the section on the *ori* region, below). The plasmid replication origin is often named *oriV* for *ori v*egetative, to distinguish it from *oriT*, which is the site at which DNA transfer initiates in plasmid conjugation (see chapter 5). Most of the evidence for the mechanisms described below came from observations of replicating plasmid DNA with the electron microscope.

THETA REPLICATION
Some plasmids begin replication by opening the two strands of DNA at the *ori* region, creating a structure that looks like the Greek letter θ—hence the name **theta replication** (Figure 4.4A and B). In this process, an RNA primer begins replication, which can proceed in one or both directions around the plasmid. In the first case, a single replication fork moves around the molecule until it returns to the origin and then the two daughter DNAs separate. In the other case (bidirectional replication), two replication forks move out from the *ori* region, one in either direction, and replication is complete (and the two daughter DNAs separate) when the two forks meet somewhere on the other side of the molecule.

The theta mechanism is the most common form of DNA replication, especially in gram-negative bacteria. It is used not only by most plasmids, including ColE1, RK2, F, and P1, but also by the chromosome (see chapter 1).

ROLLING-CIRCLE REPLICATION
Other types of plasmids replicate by very different mechanisms. One type of replication is called **rolling-circle replication** because it was first discovered in a type of phage where the template circle seems to roll like a signet ring which has been dipped in ink and rolled on paper, making a copy of its design on the paper. Plasmids which replicate by this mechanism are called **RC plasmids**. This type of plasmid is widespread, being found in both gram-negative and gram-positive eubacteria as well as archaea.

In an RC plasmid, the replication occurs in two stages. In the first stage, the double-stranded circular plasmid DNA replicates to form another double-stranded circular DNA and a single-stranded circular DNA. This stage is analogous to the replication of the DNA of some single-stranded DNA phages (see chapter 7) and to DNA transfer during plasmid conjugation (see chapter 5). In the second stage, the complementary strand is synthesized on the single-stranded DNA to make another double-stranded DNA.

The details of the rolling-circle mechanism of plasmid replication are shown in Figure 4.4C. First the Rep protein recognizes a sequence called the double-strand origin (DSO) on the DNA and makes a single-stranded break in this sequence. The DSO is a palindromic sequence which might form a cruciform structure by base pairing between the inverted-repeated sequences, as shown in the figure, although this has not been experimentally verified. The Rep protein is also known to function as a dimer in some plasmids, as shown in Figure

4.4C. After the Rep protein has made a break in the DSO sequence, it remains covalently attached through one of its tyrosines to the phosphate at the 5′ end at the break. The DNA polymerase III (the replicative polymerase [see chapter 1]) uses the free 3′ hydroxyl end at the break as a primer to replicate around the circle, displacing one of the strands. It may use a host helicase to help separate the strands, or the Rep protein itself may have the helicase activity, depending upon the plasmid. When DNA polymerase III has completed its circle, the Rep protein makes another break, releasing the displaced strand. The ends of the displaced single strand are then joined together by the Rep protein in a process called transesterification, in which the phosphate is transferred from the tyrosine of the Rep protein to the 3′ hydroxyl of the displaced strand. The newly synthesized strand is also ligated by the host cell DNA ligase, creating a covalently closed double-stranded circle, resulting in two circular DNA molecules, one which is single stranded and one which is double stranded.

The displaced circular single-stranded DNA now replicates by a completely different mechanism using only host-encoded proteins. The RNA polymerase first makes a primer at a different origin, the single-stranded origin (SSO), and this RNA then primes replication around the circle by DNA polymerase III. However, the RNA polymerase does not make this primer until the single-stranded DNA is completely displaced during the first stage of replication. This delay is accomplished by locating the SSO immediately counterclockwise of the DSO (Figure 4.4C) so that the SSO will not appear in the displaced DNA until the displacement of the single-stranded DNA is almost complete. After the entire complementary strand has been synthesized, the 5′ exonuclease activity of DNA polymerase I removes the RNA primer, replacing it with DNA, and host DNA ligase joins the ends to make another double-stranded plasmid. The net result is two new double-stranded plasmids synthesized from the original double-stranded plasmid.

In order for the complementary strand of the displaced single-stranded DNA to be synthesized, the RNA polymerase of the host cell must recognize the SSO on the DNA. In some hosts the SSO is not well recognized and single-stranded DNA accumulates. For this reason, some RC plasmids were originally called single-stranded DNA plasmids, although we now know that this is not their normal state. Broad-host-range RC plasmids presumably have an SSO which is recognized by the RNA polymerases of a wide variety of hosts, which allows them to make the complementary strand of the displaced single-stranded DNA in a variety of hosts.

The Rep protein is used only once for every round of plasmid DNA replication and is destroyed after the

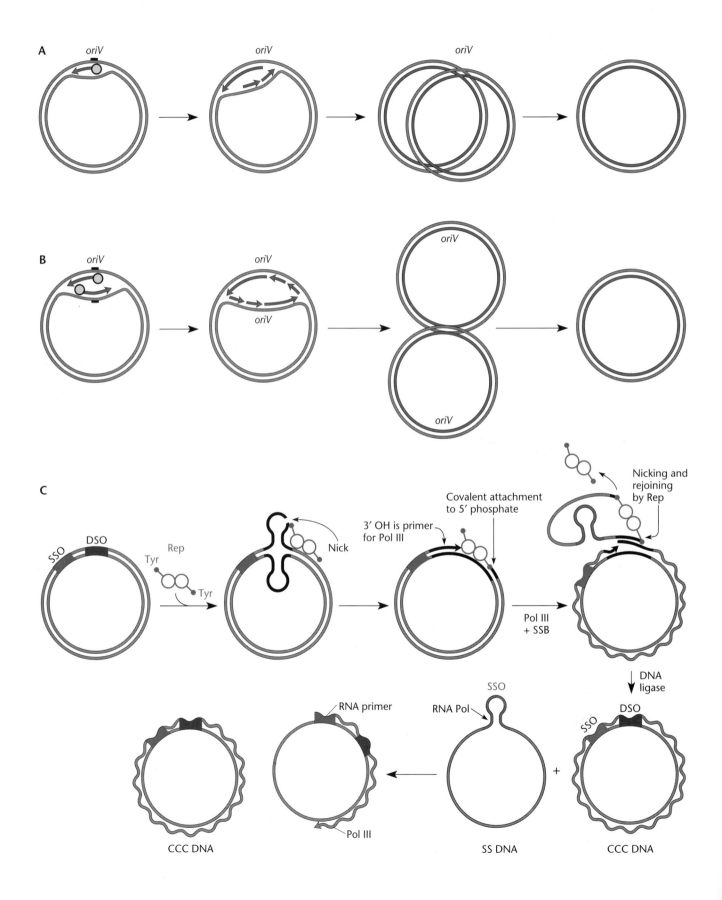

round is completed. This allows the replication of the plasmid to be controlled by the amount of Rep protein in the cell and keeps the total number of plasmid molecules in the cell within narrow limits called the copy number. A little later in the chapter, we discuss how the copy number of other types of plasmids is controlled.

Functions of the *ori* Region

In most plasmids, the genes for proteins required for replication are located very close to the *ori* sequences at which they act. Thus, only a very small region surrounding the plasmid *ori* site is required for replication. As a consequence, the plasmid will still replicate if most of its DNA is removed, provided the *ori* region remains and the plasmid DNA is still circular. Smaller plasmids are easier to use as cloning vectors, as discussed later in the chapter.

In addition, the genes in the *ori* region often determine many other properties of the plasmid. Therefore, any DNA molecule with the *ori* region of a particular plasmid will have most of the characteristics of that plasmid. The following sections describe the major plasmid properties determined by the *ori* region.

HOST RANGE

The **host range** of a plasmid includes all the types of bacteria in which the plasmid can replicate, and the host range is usually determined by the *ori* region. Some plasmids, such as those with *ori* regions of the ColE1 plasmid type, including pBR322 and pET and pUC, have **narrow host ranges**. These plasmids will replicate only in *E. coli* and some other closely related bacteria such as *Salmonella* and *Klebsiella* species. In contrast, plasmids with a **broad host range** include the RK2 and RSF1010 plasmids, as well as the rolling-circle plasmids from gram-postive bacteria mentioned above. The host range of these plasmids is truly remarkable. Plasmids with the *ori* region of RK2 will replicate in most types of gram-negative bacteria, and

RSF1010-derived plasmids will even replicate in some types of gram-positive bacteria. Many of the plasmids isolated from gram-positive bacteria also have quite broad host ranges. For example, pUB110, which was first isolated from the gram-positive *S. aureus*, will replicate in many other gram-positive bacteria, including *Bacillus subtilis*. However, most plasmids isolated from gram-negative bacteria will not replicate in gram-positive bacteria and vice versa, which reflects the evolutionary divergence of these groups (see the introductory chapter).

It is perhaps surprising that the same plasmid can replicate in bacteria which are so distantly related to each other. Broad-host-range plasmids must encode all of their own proteins required for initiation of replication, and so they do not have to depend on the host cell for any of these functions. They also must be able to express these genes in many types of bacteria. Apparently, the promoters and ribosome initiation sites for the replication genes of broad-host-range plasmids have evolved so that they will be recognized in a wide variety of bacteria.

Determining the Host Range

The actual host range of most plasmids is unknown because it is sometimes difficult to determine if a plasmid can replicate in other hosts. First, we must have a way of introducing the plasmid into other bacteria. Transformation systems (see chapter 6) have been developed for some but not all types of bacteria, and if one is available, it can be used to introduce plasmids into the bacterium. Plasmids that are self-transmissible or mobilizable (see chapter 5) can sometimes be introduced into other types of bacteria by conjugation, a process in which DNA is transferred from one cell to another.

Even if we can introduce the plasmid into other types of bacteria, we still must be able to select cells that have received the plasmid. Most plasmids, as isolated from nature, are often not known to carry a convenient selectable

Figure 4.4 Some common schemes of plasmid replication. (A) Unidirectional replication. The origin region is designated *oriV*. Replication terminates when the replication fork gets back to the origin. (B) Bidirectional replication. Replication terminates when the replication forks meet somewhere on the DNA molecule opposite the origin. (C) Rolling-circle replication. A nick is made at the double-strand origin (DSO) by the plasmid-encoded Rep protein, which remains bound to the 5′ phosphate end at the nick. The free 3′ OH end then serves as a primer for the DNA polymerase III (Pol III) which replicates around the circle, displacing one of the old strands as a single-stranded DNA. The Rep protein then makes another nick, releasing the single-stranded circle. DNA ligase then joins the ends to form double-stranded and single-stranded circular DNAs. The host RNA polymerase then makes a primer on the single-stranded DNA origin (SSO), and Pol III replicates the single-stranded (SS) DNA to make another double-stranded circle. DNA Pol I removes the primer, replacing it with DNA, and ligase joins the ends to make another double-stranded circular DNA. CCC, covalently closed circular; SSB, single-strand-DNA-binding protein. Details are given in the text.

gene such as for resistance to an antibiotic, and even if they do, the selectable gene may not be expressed in the other bacterium since most genes will not be expressed in bacteria only distantly related to those in which they were originally found. Sometimes we can introduce a selectable gene into the plasmid that is expressed in many hosts. For example, the kanamycin resistance gene, first found in the Tn5 transposon, will be expressed in most gram-negative bacteria, making them resistant to the antibiotic kanamycin. We can either clone a marker gene into the plasmid or introduce a transposon carrying a selectable marker into the plasmid by methods we shall discuss in chapter 9.

If all goes well and we have a way to introduce the plasmid into other bacteria and the plasmid carries a marker that is likely to be expressed in other bacteria, we can see if the plasmid can replicate in bacteria other than its original host. Clearly, this could be a laborious process, since the mechanisms for introducing DNA into different types of bacteria differ and there are many barriers to plasmid transfer between species. You can see why the host range of plasmids is often extrapolated from only a few examples.

REGULATION OF COPY NUMBER

Another characteristic of plasmids that is determined mostly by their *ori* region is their **copy number,** or the average number of that particular plasmid per cell. More precisely, we can define the copy number as the number of copies of the plasmid in a newborn cell immediately after cell division. All plasmids must regulate their replication; otherwise they would fill up the cell and become too great a burden for the host, or their replication would not keep up with the cell replication and they would be progressively lost during cell division. Some plasmids, such as pIJ101 of *Streptomyces coelicolor,* replicate enough to populate the cell with hundreds of copies. However, others, such as the F plasmid of *E. coli,* replicate only once or a few times during the cell cycle. Table 4.2 lists the copy numbers of these and other plasmids.

The regulation mechanisms used by plasmids with higher copy numbers often differ greatly from those used

by plasmids with lower copy numbers. Plasmids that have high copy numbers, such as ColE1 plasmids, need only have a mechanism that inhibits the initiation of plasmid replication when the number of plasmids in the cell reaches a certain level. Consequently, these molecules are called **relaxed plasmids.** By contrast, low-copy-number plasmids such as F must replicate only once or very few times during each cell cycle and so must have a tighter mechanism for regulating their replication. Hence, these are called **stringent plasmids.** Much more is understood about the regulation of replication of relaxed plasmids than about the regulation of replication of stringent plasmids.

The regulation of relaxed plasmids falls into three general categories. Some plasmids are regulated by an antisense RNA, sometimes called a countertranscribed (ctRNA) because it is transcribed from the same region of the plasmid but from the opposite strand of an RNA essential for plasmid replication. Because the ctRNA is transcribed from the opposite strand from the essential RNA it is able to hybridize to the essential RNA and inhibit its function. The ctRNA of these plasmids is often assisted in its inhibitory role by a protein. Other plasmids are regulated by a ctRNA alone, which inhibits the translation of a protein essential for replication. Yet others are regulated by a protein alone, which binds to repeated sequences in the plasmid DNA called iterons, thereby inhibiting plasmid replication. We consider examples of these three types of regulation in the following sections.

ColE1-Derived Plasmids: Regulation by a ctRNA and a Protein

Plasmids related to the ColE1 plasmid, from which many of the plasmid cloning vectors are derived, are examples of the first type of plasmid regulation. Figure 4.5 shows the genetic map of the original ColE1 plasmid. This plasmid has been put to use in numerous molecular biology studies, and many vectors have been derived from it and the closely related pMB1, including pBR322, the pUC plasmids, and the pET series of plasmids discussed below and in chapters 2 and 7. Although the genetic maps of these cloning vectors (discussed in more detail below), have changed beyond recognition, they all have the *ori* region of the original ColE1-like plasmid and hence they share many of its properties, including the mechanism of replication regulation.

The mechanism of regulation of these plasmids is shown in Figure 4.6. Replication is regulated mostly through the effects of a small plasmid-encoded RNA called RNA I. This small RNA inhibits plasmid replication by interfering with the processing of another RNA called RNA II, which forms the primer for plasmid DNA replication. In the absence of RNAI, RNA II forms an RNA-DNA hybrid at

TABLE 4.2	Copy numbers of some plasmids
Plasmid	Approximate copy number
F	1
P1 prophage	1
RK2	4–7 (in *E. coli*)
pBR322	16
pUC18	~30–50
pIJ101	40–300

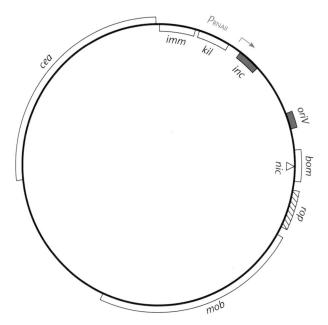

Figure 4.5 Genetic map of plasmid ColE1. The plasmid is 6,646 bp long. On the molecule, *oriV* is the origin of replication; p_{RNAII} is the promoter for the primer RNA II, *inc* encodes RNA I, *rop* encodes a protein that helps regulate copy number, *bom* is a site that is nicked at *nic, cea* encodes the colicin ColE1, and *mob* encodes functions required for mobilization (discussed in chapter 5). Adapted from F. C. Neidhardt, J. L. Ingraham, K. B. Low, B. Magasanik, M. Schaechter, and H. E. Umbarger (ed.), Escherichia coli *and* Salmonella typhimurium: *Cellular and Molecular Biology,* p. 1620, American Society for Microbiology, Washington, D.C., 1987.

the replication origin. RNA II is cleaved by the RNA endonuclease RNase H, releasing a 3′ hydroxyl group that serves as the primer for replication first catalyzed by DNA polymerase I. Unless RNA II is processed properly, it will not function as a primer and replication will not ensue.

RNA I inhibits DNA replication by interfering with primer formation by RNA II by forming a double-stranded RNA with it, as illustrated in Figure 4.6. It can do this because the two RNAs are complementary, as mentioned above. Initially, the pairing between RNA I and RNA II occurs through short exposed regions on the two RNAs that are not occluded by being part of secondary structures. This initial pairing is very weak and therefore has been called a "kissing complex." A protein named Rop helps stabilize the kissing complex, although it is not essential. The kissing complex can then extend into a "hug," with the formation of the double-stranded RNA as shown. Formation of the double-stranded RNA prevents the RNA II from forming the secondary structure required for it to hybridize to the DNA before being processed by RNase H to form the mature primer.

A protein called Rop (sometimes called Rom) helps RNA I to pair with RNA II and therefore helps inhibit plasmid replication. However, it is not clear how Rop works, nor is it essential to maintain the copy number. Mutations that inactivate the Rop protein cause only a moderate increase in plasmid copy number.

This mechanism provides an explanation for how the copy number of ColE1 plasmids is maintained. Since RNA I is synthesized from the plasmid, more RNA I will be made when the concentration of the plasmid is high. A high concentration of RNA I interferes with the processing of most of the RNA II, and replication is inhibited. The inhibition of replication is almost complete when the concentration of the plasmid reaches about 16 copies per cell, the copy number of the ColE1 plasmid.

Formation of the kissing complex involves pairing between very small regions of RNA I and RNA II. However, these regions must be completely complementary for this pairing to occur and plasmid replication to be inhibited. Changing even a single base pair in this short sequence will make the mutated RNA I no longer complementary to the RNA II of the original nonmutant ColE1 plasmid, so it will no longer be able to "kiss" it and regulate its replication. However, note that a mutation in the region of the plasmid DNA encoding RNA I will also change the sequence of RNA II in a complementary way, since they are encoded in the same region of the DNA but from the opposite strands. Therefore, the mutated RNA I can still form a complex with the mutated RNA II made from the same mutated plasmid and prevent its processing, it just cannot interfere with the processing of RNA II from the original nonmutant plasmid. Plasmids which regulate each other's replication are said to be members of the same **incompatability (Inc) group** (see below). A single-base-pair mutation in the RNA I coding region of the plasmid has effectively changed the Inc group of the plasmid to a new Inc group, of which the mutated plasmid is conceivably the sole member! In fact, the naturally occurring plasmids ColE1 and its close relative p15A, from which other cloning vectors such as pACYC184 have been derived, are members of different Inc groups even though they differ by only 1 base in the kissing regions of their RNA I and RNA II. Clever genetic experiments demonstrated Inc group switching by single-base-pair changes in ColE1 plasmids (see Lacatena and Cesareni, Suggested Reading, and chapter 14).

R1 and Col1B-P9 Plasmids: Regulation of Rep Protein by Antisense RNA

The ColE1 plasmids are unusual in that they do not require a plasmid-encoded protein to initiate DNA replication at their *oriV* region, only an RNA primer synthesized from the plasmid. Most plasmids require a plasmid-encoded

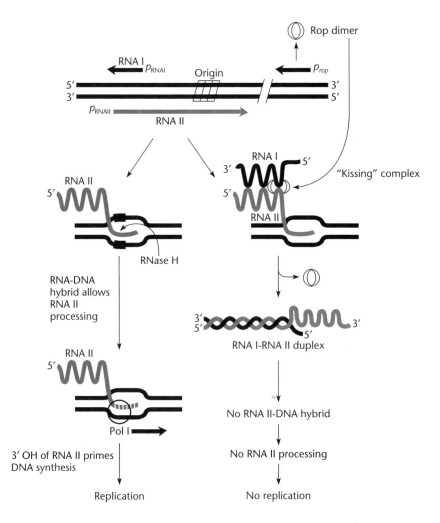

Figure 4.6 Regulation of replication of ColE1-derived plasmids. RNA II must be processed by RNase H before it can prime replication. "Origin" indicates the transition point between the RNA primer and DNA. RNA I binds to RNA II and inhibits the processing, thereby regulating the copy number. p_{RNAI} and p_{RNAII} are the promoters for RNA I and RNA II transcription, respectively. RNA II is shown in gold. The Rop protein dimer enhances the initial pairing of RNA I and RNA II.

protein, often called Rep, to initiate replication. The Rep protein is required to separate the strands of DNA at the *oriV* region, a necessary first step that allows the replication apparatus to assemble at the origin. The Rep proteins are very specific in that they will bind only to the *oriV* of the same type of plasmid because they bind to certain specific DNA sequences within *oriV*. The amount of Rep protein is limiting for replication, meaning that there is never more than is needed to initiate replication. Therefore, the copy number of the plasmid can be controlled, at least partially, by controlling the synthesis of this Rep protein.

The R1 plasmid. One type of plasmid which regulates its copy number by regulating the amount of a Rep protein is the R1 plasmid, a member of the IncFII family of plasmids (see del Solar and Espinosa, Suggested Reading). Like ColE1 plasmids, this plasmid uses a small antisense RNA to regulate its copy number. Also like ColE1 plasmids, the more copies of the plasmid in the cell, the more of this antisense RNA is made and the

more plasmid replication will be inhibited. However, this plasmid uses its antisense RNA to inhibit the synthesis of Rep protein and thereby inhibit the replication of the plasmid DNA.

Figure 4.7 illustrates the regulation of R1 plasmid replication. The plasmid encodes a protein called RepA, which is required for the initiation of replication. The repA gene can be transcribed from two promoters. One of these promoters, called p_{copB} in Figure 4.7, transcribes both the repA and copB genes, making an mRNA that can be translated into the proteins RepA and CopB. The second promoter, p_{repA}, is in the copB gene and so makes an RNA that can encode only the RepA protein. Because the p_{repA} promoter is repressed by the CopB protein, it is only turned on immediately after the plasmid enters a cell and before any CopB protein is made. The short burst of synthesis of RepA from p_{repA} after the plasmid enters a cell causes the plasmid to replicate until it attains its copy number. Then the p_{repA} promoter is repressed by CopB protein, and the repA gene can be transcribed only from the p_{copB} promoter.

A   Plasmid genetic organization

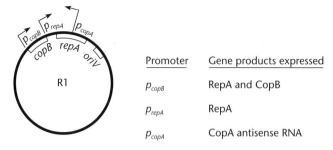

Promoter	Gene products expressed
p_{copB}	RepA and CopB
p_{repA}	RepA
p_{copA}	CopA antisense RNA

B   Replication occurs after plasmid enters cells

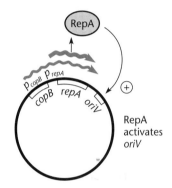

RepA
activates
oriV

C   Replication shutdown

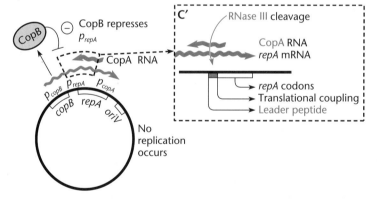

CopB represses
p_{repA}

CopA RNA

No
replication
occurs

C′

RNase III cleavage

CopA RNA
repA mRNA

repA codons
Translational coupling
Leader peptide

Figure 4.7   Regulation of replication of the IncFII plasmid R1. (A) The locations of promoters and genes and gene products involved in the regulation. (B) Immediately after the plasmid enters the cell, most of the *repA* mRNA is made from promoter P_{repA} until the plasmid reaches its copy number. (C) Once the plasmid reaches its copy number, CopB protein represses transcription from P_{repA}. Now *repA* is transcribed only from P_{copB}. (C′) The antisense RNA CopA hybridizes to the leader peptide coding sequence in the *repA* mRNA, and the double-stranded RNA is cleaved by RNase III. This prevents translation of RepA, which is translationally coupled to the translation of the leader peptide.

Once the plasmid has attained its copy number, the regulation of synthesis of RepA, and therefore the replication of the plasmid, is regulated by an antisense RNA called CopA in Figure 4.7. The *copA* gene is transcribed from its own promoter and the RNA product affects the stability of the mRNA made from the p_{copB} promoter. Because the CopA RNA is made from the same region encoding the translation initiation region for the *repA* gene, but from the other strand of the DNA, the two mRNAs will be complementary and can pair to make double-stranded RNA. Then an RNase called RNase III, a chromosomally encoded enzyme

that cleaves some double-stranded RNAs, will cleave the CopA-RepA mRNA.

The reasons why cleavage of this RNA prevents the synthesis of RepA are a little complicated. The 5′ leader region of the mRNA, upstream of where the RepA protein is encoded, actually encodes a short leader polypeptide, and the translation of RepA is coupled to the translation of this leader polypeptide (see chapter 2 for an explanation of translational coupling). Cleavage of the mRNA by RNase III in the leader region will interfere with the translation of this leader polypeptide and, by blocking its translation, will also block translation of the downstream RepA. Therefore, by having the CopA RNA activate cleavage of the mRNA for the RepA protein upstream of the RepA coding sequence, the plasmid copy number is controlled by the amount of CopA RNA in the cell, which in turn depends on the concentration of the plasmid. The higher the concentration of the plasmid, the more CopA RNA will be made and the less RepA protein will be synthesized, maintaining the concentration of plasmid around the plasmid copy number.

The Col1B-P9 plasmid. Yet another level of complexity of the regulation of the copy number by an antisense RNA is provided by Col1B-P9 plasmid (see del Solar and Espinosa, Suggested Reading). As in the R1 plasmid, the *repA* gene (called *repZ* in this case) is translated downstream of the leader peptide gene, called *repY,* and the two are also translationally coupled. The translation of *repY* opens an RNA secondary structure which normally occludes the Shine-Dalgarno (S-D) sequence of the TIR of *repZ*. A sequence in the secondary structure then can pair with the loop of a hairpin upstream of *repY*, forming a pseudoknot (see Figure 2.2 for the definition of a pseudoknot), thus permanently disrupting the secondary structure and leaving the S-D sequence for *repZ* exposed. A ribosome can then bind to the TIR for *repZ* and translate the initiator protein. The Inc RNA pairs with the loop of the upstream hairpin and prevents the hairpin formation, leaving the S-D sequence of the *repZ* coding sequence blocked and preventing translation of the initiator protein.

The Iteron Plasmids: Regulation by Coupling

Many commonly studied plasmids use a very different mechanism to regulate their replication. These plasmids are called iteron plasmids because their *oriV* region contains several repeats of a certain set of DNA bases called an **iteron sequence**. The iteron plasmids include pSC101, F, R6K, P1, and the RK2-related plasmids. The iteron sequences of these plasmids are typically 17 to 22 bp long and exist in about three to seven copies in the *ori* region. In addition, there are usually additional copies of these repeated sequences a short distance away.

One of the simplest of the iteron plasmids is pSC101. For our purposes, the essential features of this plasmid's *ori* region (Figure 4.8) are the gene *repA*, which encodes the RepA protein required for initiation of replication, and three repeated iteron sequences, R1, R2, and R3, through which RepA regulates the copy number. The RepA protein is the only plasmid-encoded protein required for the replication of the pSC101 plasmid and many other iteron plasmids. It serves as a positive activator of replication, much like the RepA protein of the R1 plasmid. The host chromosome encodes the other pro-

teins that bind to this region to allow initiation of replication, which include DnaA, DnaB, DnaC, and DnaG (see chapter 1).

Iteron plasmid replication is regulated by two superimposed mechanisms. First, the RepA protein represses its own synthesis by binding to its own promoter region and blocking transcription of its own gene. Therefore, the higher the concentration of plasmid, the more RepA protein will be made and the more it will repress its own synthesis. Thus, the concentration of RepA protein is maintained within narrow limits and the initiation of replication is strictly regulated. This type of regulation, known as **transcriptional autoregulation**, was discussed in chapter 2.

However, this mechanism of regulation by itself is not enough to regulate the copy number of the plasmid, especially low-copy-number stringent plasmids like F and P1. Iteron plasmids have another mechanism called **coupling** to regulate their copy number within narrow limits. The coupling mechanism of regulation of plasmid replication is illustrated in Figure 4.9. When the concentration of plasmids is high enough, they become coupled to each other through the RepA protein, thereby inhibiting the replication of both coupled plasmids. The coupling mechanism allows plasmid replication to be controlled not only by how much RepA protein is present in the cell but also by the concentration of the plasmid itself or more precisely, the concentration of the iteron sequences on the plasmid.

The coupling mechanism was proposed to explain some puzzling results of experiments designed to study the control of replication of iteron plasmids. The experiments which inspired the coupling model are illustrated in Figure 4.10 (see McEachern et al., Suggested Read-

Figure 4.8 The *ori* region of pSC101. R1, R2, and R3 are the three iteron sequences (CAAAGGTCTAGCAGCAGAATT-TACAGA for R3) to which RepA binds to handcuff two plasmids. RepA autoregulates its own synthesis by binding to the inverted repeats IR1 and IR2. The location of the partitioning site *par* (see the section on Partioning) and the binding sites for the host protein DnaA are also shown.

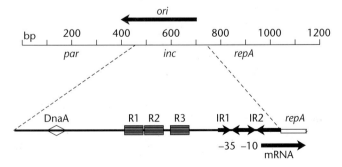

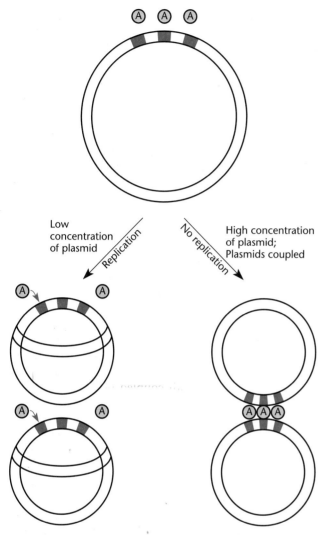

Figure 4.9 The "handcuffing" or "coupling" model for regulation of iteron plasmids. At low concentrations of plasmids, the RepA protein only binds to one plasmid at a time, initiating replication. At high plasmid concentrations, the RepA protein binds to two plasmids simultaneously, handcuffing them and inhibiting replication.

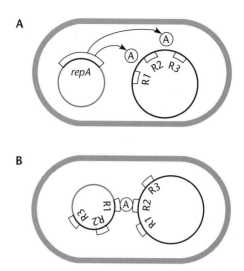

Figure 4.10 Molecular genetic analysis of the regulation of iteron plasmids. (A) The RepA protein is expressed from a clone of the *repA* gene in an unrelated plasmid cloning vector. (B) Additional iteron sequences in an unrelated plasmid can lower the copy number of an iteron plasmid. R1, R2, and R3 are iteron sequences.

ing). One puzzling observation came from experiments to determine the effect of vastly increasing the concentration of RepA protein on the copy number of the plasmid (Figure 4.10A). These experiments were designed to test the hypothesis that the function of the iteron sequences is simply to bind excess RepA protein to limit replication. If iteron sequences are merely binding excess RepA protein, as the concentration of RepA protein increases, eventually all of the iteron sequences should have RepA protein bound to them (i.e., be titrated) and the copy number should increase. To make the experiment easier to interpret, the *repA* gene was expressed from a very strong promoter on another plasmid rather

than the plasmid whose copy number was being measured. This second plasmid did not have the same iteron sequences, so that the effects of increasing the RepA concentration could be determined without the added complication of the amount of RepA also depending on the number of plasmid copies. The researchers observed that increasing the concentration of the RepA protein even by factors of hundreds led to only a modest increase in the copy number of the original plasmid, consistent with the coupling hypothesis.

Other experiments studied the effects of increasing the number of copies of the iteron sequence in the cell, as illustrated in Figure 4.10B. In these experiments, another plasmid that contains a few copies of the iteron sequence but which does not depend on the RepA protein for its own replication was introduced into the same cell with the original plasmid. The extra copies of the interon sequence caused the copy number of the original plasmid to decrease. This effect is not merely due to the iteron sequences on the second plasmid "soaking up" RepA protein, since overproduction of the RepA protein did not overcome the negative effect on plasmid copy number of the extra iteron sequences (not shown in the figure). Apparently, the second plasmid is inhibiting the replication of the first plasmid by being coupled to it through its iteron sequences, as shown.

The coupling model also explains the existence of mutations in the *repA* gene called "copy-up" mutations.

These mutations are thought to increase the copy number of the plasmid by weakening the binding of RepA protein to the iteron sequences, thereby preventing coupling, without affecting the function of RepA protein in initiation of replication. As predicted by the model, increasing the concentration of copy-up mutant RepA protein does overcome the negative effect of extra iterons on the plasmid copy number, unlike increasing the concentration of wild-type RepA protein. Apparently, the extra copies of the mutant RepA protein do not couple the plasmids and inhibit replication, they only cause new initiation of replication and and an increase in the copy number.

Direct support for the coupling model in the replication control of iteron plasmids has come from electron microscopic pictures of purified iteron plasmids mixed with the purified RepA protein for that plasmid. In these pictures, two plasmid molecules can often be seen linked by RepA protein.

HOST FUNCTIONS INVOLVED IN REGULATING PLASMID REPLICATION

In addition to Rep, many plasmids require host proteins to initiate replication. For example, some plasmids require the DnaA protein, which is normally involved in initiating replication of the chromosome, and have *dnaA* boxes in their *oriV* region to which DnaA binds (Figure 4.8). The DnaA protein may also directly interact with Rep proteins of some plasmids. The DnaA protein is involved in coordinating replication of the chromosome with cell division (see chapter 1); making their own replication dependent upon DnaA may allow plasmids to better coordinate their own replication with cell division. Like the chromosome origin *(oriC)*, some *E. coli* plasmids also have Dam methylation sites close to their *oriV*. These methylation sites presumably help to further coordinate their replication with cell division. Both strands of DNA at these sites must be fully methylated for initiation to occur. Immediately after initiation, only one strand of these sites is methylated (hemimethylation), delaying new initiations at these sites (see chapter 1 on sequestration of chromosome origins). In spite of substantial progress, however, the method by which the replication of very stringent plasmids, such as P1 and F with a copy number of only 1, is controlled to within such narrow limits is still something of a mystery and the object of current research.

Mechanisms To Prevent Curing of Plasmids

Cells that have lost a plasmid during cell division are said to be **cured** of the plasmid. Several mechanisms prevent curing, including plasmid addiction sytems (see Box 4.3), site-specific recombinases that resolve multimers, and partitioning systems.

BOX 4.3

Plasmid Addiction

Even with their partitioning functions, plasmids are often lost from multiplying cells. In a kind of ironic revenge, some plasmids encode proteins that kill cells cured of their type. Such functions have been described in many plasmids including the F plasmid, the R1 plasmid, and the P1 prophage, which replicates as a plasmid. These plasmids encode a toxic protein that will kill the cell if it is expressed. The toxic protein kills the cell by various mechanisms, depending on the source. For example, the toxic protein of the F plasmid, Ccd, kills the cells by altering DNA gyrase, so it causes double-stranded breaks in the DNA. The killer protein, Hok, of the plasmid R1 destroys the cellular membrane potential, causing loss of cellular energy. The mechanism of killing by the P1-encoded killer protein, Doc, is not known, but it seems to work indirectly (see below).

Why does the toxic protein kill cells only immediately after they have been cured of the plasmid? When cells contain the plasmid, other proteins or RNAs of the addiction system act as an antidote and either prevent the synthesis of the toxic protein or bind to it and prevent its activity. These other gene products are much more unstable than the toxic protein, however, so that they will be more rapidly inactivated. If a cell is cured of the plasmid, neither the toxic protein nor the antidote will be synthesized, but since the antidote is more unstable, it will not be long before the cured cells have only the toxic protein and are killed. These systems make the cell addicted to the plasmid once it has been acquired, and they prevent cured cells from accumulating.

A recently uncovered twist on plasmid addiction makes it seem even more bizarre. In some cases at least, it is not the plasmid gene products that kill a cell cured of the plasmid, but gene products of the cell itself. It is as if the cured cell is killing itself, overcome by grief over the loss of the plasmid. The bacterium *E. coli* contains the *mazE* and *mazF* genes which somehow kill the cell if translation is severely inhibited. The Doc protein of *E. coli* inhibits protein synthesis after P1 curing, in the absence of its antidote, and this causes *mazEF* to kill the cell (see Hazen et al.).

References
Greenfield T. J., E. Ehli, T. Kirshenmann, T. Franch, K. Gerdes, and K. E. Weaver. 2000. The antisense RNA of the *par* locus of pAD1 regulates the expression of a 33-amino-acid toxic peptide by an unusual mechanism. *Mol. Microbiol.* **37:**652–660.

Hazan, R., B. Sat, M. Reches, and H. Engelberg-Kulka. 2001. Postsegregational killing by the P1 phage addiction module *phd-doc* requires the *Escherichia coli* programmed cell death system *mazEF. J. Bacteriol.* **183:**2046–2050.

RESOLUTION OF MULTIMERIC PLASMIDS

The possibility of losing a plasmid during cell division is increased if the plasmids form multimers during replication. A multimer consists of individual copies of the plasmid molecules linked to each other. The multimers probably occur as a result of defects in replication termination or by recombination of monomers. The formation of multimers effectively lowers the effective copy number because each multimer will segregate into the daughter cells as a single plasmid. Therefore, multimers greatly increase the chance of a plasmid being lost during cell division.

To avoid this problem, many plasmids have site-specific recombination systems that resolve multimers. These systems promote recombination between specific sites on the plasmid if the same site occurs more than once in the molecule, as it would in a multimer. This recombination has the effect of resolving multimers into separate molecules.

The best known example of this mechanism is the *cer*-Xer site-specific recombination system of the ColE1 plasmid (see Guhathakurta et al., Suggested Reading). We have already discussed this system in chapter 1 in connection with segregation of the chromosome of *E. coli*. The XerC and XerD proteins are part of a site-specific recombinase system that acts on a site, *dif,* close to the terminus of replication of the chromosome to resolve chromosome dimers created during chromosome replication. These proteins can also act on a site, *cer,* on plasmid ColE1 to resolve dimers of the plasmid (Figure 4.11). This is yet another example of a case where plasmids commandeer host functions for their own purposes, in this case a site-specific recombination system normally used for resolving chromosome dimers. We discuss site-specific recombinases in more detail in chapter 9.

PARTITIONING

Plasmids also avoid being lost from dividing cells by carrying **partitioning** systems, which ensure at least one copy of the plasmid segregates into each daughter cell during cell division. The functions involved in these systems are called *par* **functions**.

The best evidence for *par* systems comes from calculations, using combinatorial probability, of how often cells will be cured of a plasmid if the plasmid has no *par* system. In the simple example shown in Figure 4.12, the copy number of the plasmid is 4. Immediately after cell division, a cell contains four copies, and immediately before the cell divides, it contains eight copies. If the plasmids are equally divided into the two daughter cells, each will get four plasmids. However, the plasmids usually will not be equally distributed between the two daughter cells, and one daughter cell will get more than

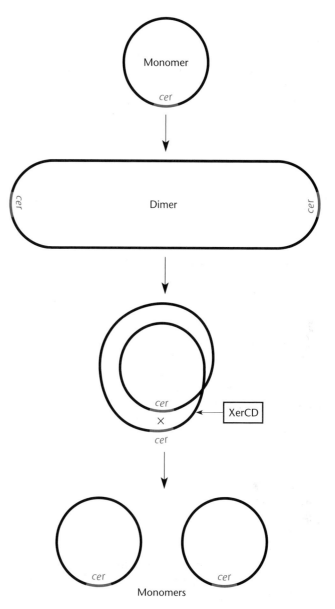

Figure 4.11 The Xer functions of *E. coli* catalyze site-specific recombination at the *cer* site to resolve plasmid dimers.

the other. In fact, with a certain probability, one cell will get all of the plasmids and the other cell will be cured. Since each plasmid can go into either one cell or the other, the probability that one daughter cell will be cured of the plasmid is the same as the probability of tossing eight heads (or tails) of a coin in a row. Thus, the probability that the first plasmid will go into one cell is 1/2, and the probability that the first two plasmids will go into the same cell is 1/2 times 1/2 = 1/4, and so on. The probability that all eight will go into one cell is therefore $(1/2)^8$ or 1/256. Since it is irrelevant which of the two cells is cured, the frequency of curing

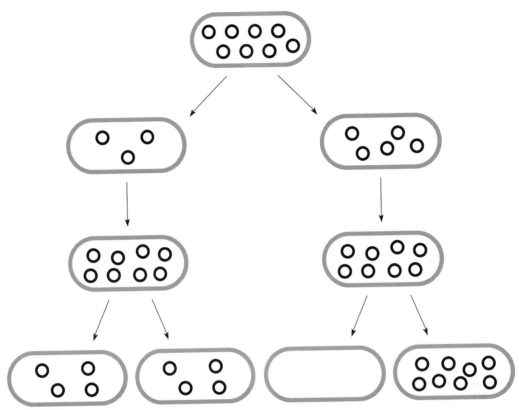

Figure 4.12 Random curing of a plasmid with no *par* site. Each plasmid will have an equal chance of going to one of the daughter cells when the cell divides, and the occasional cell will inherit no plasmids.

will be twice this value, so $2(1/2)^8$ or 1/128 of the cells will be cured each time the cells divide. In general, for a plasmid with a copy number n, the frequency of curing is $2(1/2)^{2n}$, since the number of plasmids at the time of division is twice the copy number. Also, as the cells divide once every generation time, the frequency of cured cells in the population will be roughly equal to the number of generation times that have elapsed times this number. This is the frequency of curing if the sorting of the plasmids into daughter cells is completely random. Therefore, if the fraction of the cells that are cured of the plasmid is less than $2(1/2)^{2n}$ times the number of generation times that have elapsed, the plasmid must have some sort of partitioning function.

This calculation indicates that few cells would be cured of a high-copy-number plasmid each generation, even without a partitioning mechanism. However, a significant fraction of cells would be cured of a low-copy-number plasmid each generation. In fact, with a plasmid with a copy number of only 1, such as F or P1, $2(1/2)^2$ or 1/2 of the cells would be cured each generation. Since cells are seldom cured of even low-copy-number plasmids, some mechanism must ensure that plasmids, especially those with low copy numbers, will be partitioned

faithfully into the daughter cells each time the cell divides.

The *par* Systems

The *par* systems of different single-copy plasmids including P1, F, and R1 are remarkably similar to each other. They consist of a *cis*-acting site, often called *parS*, and two genes, often called *parA* and *parB*, which encode *trans*-acting proteins. The *parS* site is the site at which the two copies of the plasmid are pulled apart during partitioning, and the ParA and ParB proteins act at this site. For historical reasons, the names of these genes are different in different plasmids. For example, in the F plasmid they are called *sop*, for "stability of plasmid," but their functions are probably very similar. The genes for related functions occur in the chromosomes of some bacteria, where they may perform a similar role in chromosome partitioning (see chapter 1), but it was in plasmids where the *par* systems were first discovered.

It is not clear exactly how the ParA and ParB proteins function in plasmid partitioning, and this is an area of active research. It is known that many copies of the ParB protein bind at or around the *parS* site; in fact, there are so many that they may block or silence any genes around the site. The ParA protein does not bind to the *parS* site by itself, but it does bind to the complex of *parS* and ParB. In addition, the ParA protein has an ATPase activity which cleaves ATP to ADP when it is activated by binding to the ParB-*parS* complex.

LOCATION OF Par FUNCTIONS

Clues to how the Par proteins function in plasmid partitioning may come from their location in the cell during plasmid partitioning. This has been made possible by the development of technologies involving fluorescent probes that allow sufficient resolution to localize cellular components within even very small bacterial cells, the same technologies (discussed in chapter 1) that are being used to localize the site of chromosome replication and Min functions. A protein such as the green fluorescent protein is translationally fused either to one of the Par proteins or to a protein such as the LacI repressor that binds to the plasmid DNA. The position of the fluorescing protein is then determined by fluorescence microscopy during the cell cycle in the presence and absence of the other Par functions.

This technology reveals that single copy plasmids such as F or P1 are located in the middle of the cell while they replicate; this is the same place where the chromosome replicates (see chapter 1). After replication, both copies of the plasmid quickly move to one of the quarter positions of the cell, the regions that will become the new center of the cell after it divides. The ParA and ParB proteins follow the plasmid to the quarter cell positions. However, ParB cannot follow the plasmid in the absence of ParA. One interpretation of these results is that ParB initially forms a complex with the *parS* site on the plasmid. The ParA protein then directs the binding of this complex to a site at the quarter position on the membrane. Alternatively, the ParA protein may mediate the release of the ParB-*parS* complex from a site at the middle of the cell, thereby freeing it to bind to new positions at the quarter cell. However, more research is needed to distinguish this and other models of how ParA and ParB function.

WHAT DO THE Par FUNCTIONS BIND TO?

It seems reasonable to assume that the *parS* site helps plasmids partition properly by binding to some sites in the cell that separate during cell division. However, left unanswered is the nature of the sites in the cell to which the *parS* site might bind. One model to explain plasmid partitioning is shown in Figure 4.13A. According to this model, the *parS* sites of the two daughter plasmids bind to unique sites on some cellular structure which pulls apart as the cells grow and divide. But what is the nature of these sites and how many different types of unique sites are there? A vast number of different types of plasmids can coexist in the cell and partition properly. This does not seem possible if their *par* systems are all competing for the same limited number of unique sites. However, if

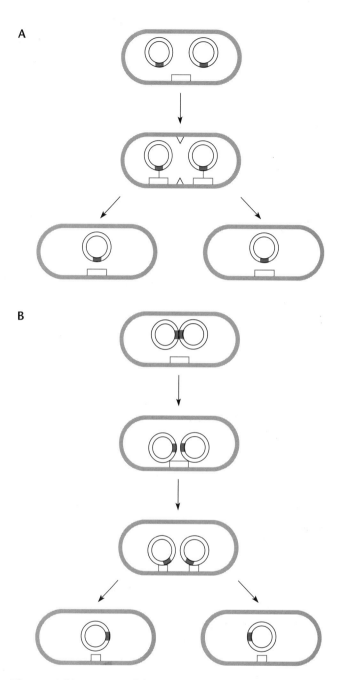

Figure 4.13 Two models for the function of *par* sites. In model A, the plasmid binds to a unique site on the bacterial membrane and the two copies of the plasmid are pulled apart as the site on the membrane divides. In model B, the two copies of the plasmid bind to each other before they bind to a site on the membrane. One then associates with each site when the site on the membrane divides.

they each bind to unique sites, there would have to be a vast number of different types of sites.

A way out of this dilemma is to have the *parS* sites of daughter plasmids obligatorily pair with each other before they can bind to a cellular structure. This obligate pairing strategy is illustrated in Figure 4.13B. If the *parS* sites of the two daughter plasmids *must* bind to each other *prior to* their binding to the site, the site need not be unique. Each of the plasmids then remains bound to one half of the site as it divides during the cell cycle and the two halves move toward opposite ends of the cell. In some ways, this model is analogous to the obligate pairing of homologous chromosomes during the mitotic cycle in eukaryotes. More experiments are needed to distinguish these and other models of plasmid partitioning.

High-copy-number plasmids often have sites that increase proper partitioning of the plasmid, but these regions do not have the same structure and may not work by the same mechanism. For example, the *par* region of pSC101 increases supercoiling of this plasmid. How increased supercoiling aids in proper partitioning is not clear, but perhaps it enhances replication of plasmids immediately after division so that they are less apt to be cured in the next division.

Incompatibility

Many bacteria, as they are isolated from nature, contain more than one type of plasmid. These plasmid types stably coexist in the bacterial cell and remain there even after many cell generations. In fact, bacterial cells containing multiple types of plasmids will not be cured of each plasmid any more frequently than if the other plasmids were not there.

However, not all types of plasmids can stably coexist in the cells of a bacterial culture. Some types will interfere with each other's replication or partitioning so that if two such plasmids are introduced into the same cell, one or the other will be lost at a higher than normal rate when the cell divides. This phenomenon is called **plasmid incompatibility**, and two plasmids that cannot stably coexist are members of the same **incompatibility (Inc) group**. If two plasmids can stably coexist, they belong to different Inc groups. There may be hundreds of different Inc groups, and plasmids are usually classified by their group. For example, RP4, also called RK2, is an IncP (incompatibility group P) plasmid. In contrast, RSF1010 is an IncQ plasmid, so can be stably maintained with RP4 because it belongs to a different Inc group.

Plasmids of the same Inc group can be incompatible because they share the same replication control mechanism and/or because they share the same partitioning (*par*) functions. The following sections address each of these reasons.

INCOMPATIBILITY DUE TO REPLICATION CONTROL

Two plasmids that share the same mechanism of replication control will be incompatible. The replication control system will not recognize the two as different, and so either may be randomly selected for replication.

Figure 4.14A and B contrast distribution at cell division of plasmids of the same Inc group with plasmids of different Inc groups. Figure 4.14A shows a cell containing two types of plasmids that belong to different Inc groups and use different replication control systems. In the illustration, both plasmids exist in equal numbers before cell division, but after division, the two daughter cells are not likely to get the same number of each plasmid. However, in the new cells, each plasmid will replicate to reach its copy number, so that at the time of the next division, both cells will again have the same numbers of the plasmids. This process will be repeated each generation, so very few cells will be cured of either plasmid.

Now consider Figure 4.14B, in which the cell has two plasmids in the same Inc group, both of which use the same replication control system. As in the first example, both plasmids originally exist in equal numbers, and when the cell divides, chances still are that the two daughter cells will not receive the same number of the two plasmids. Note that in the original cell, the copy number of each plasmid is only half its normal number; both plasmids contribute to the total copy number since they both have the same *ori* region and inhibit each other's replication. After cell division, the two plasmids will replicate until the *total* number of plasmids in each cell equals the copy number. The underrepresented plasmid (recall that the daughters may not receive the same number of plasmids) will not necessarily replicate more than the other plasmid, so that the imbalance of plasmid number might remain or become even worse. At the next cell division, the underrepresented plasmid has less chance of being distributed to both daughter cells since there are fewer copies of it. Consequently, in subsequent cell divisions, the daughter cells are much more likely to be cured of one of the two plasmid types.

Incompatibility due to copy number control is probably more detrimental to low-copy-number plasmids than to high-copy-number plasmids. If the copy number is only 1, then only one of the two plasmids can replicate, and each time the cell divides, a daughter will be cured of one of the two types of plasmids.

INCOMPATIBILITY DUE TO PARTITIONING

Two plasmids can also be incompatible if they share the same *par* function. When coexisting plasmids share the same *par* function, one or the other will always be distributed into the daughter cells during division. However, sometimes one daughter cell will receive one type of plasmid and the other cell will get the other plasmid type, producing cells cured of one or the other plasmid.

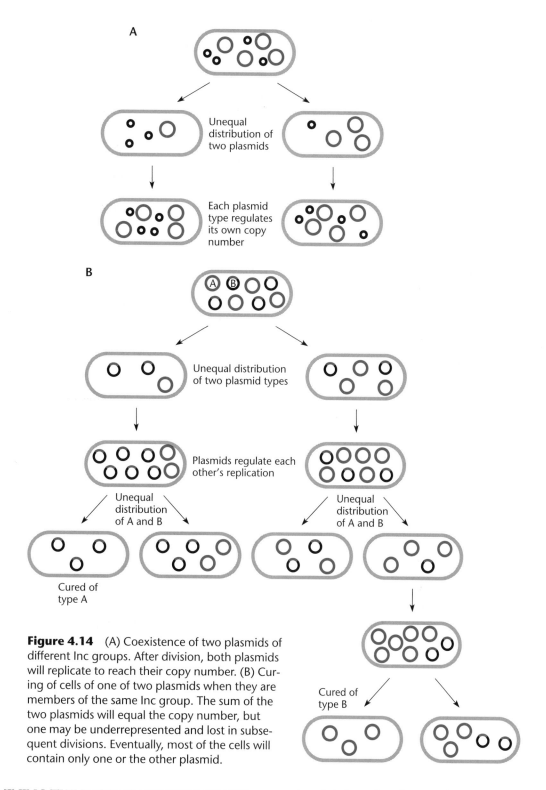

Figure 4.14 (A) Coexistence of two plasmids of different Inc groups. After division, both plasmids will replicate to reach their copy number. (B) Curing of cells of one of two plasmids when they are members of the same Inc group. The sum of the two plasmids will equal the copy number, but one may be underrepresented and lost in subsequent divisions. Eventually, most of the cells will contain only one or the other plasmid.

DETERMINING THE INCOMPATIBILITY GROUP OF A PLASMID

To classify a plasmid by its incompatibility group, we must determine if it can coexist with other plasmids of known incompatibility groups. In other words, we must measure how frequently cells are cured of the plasmid when it is introduced into cells carrying another plasmid of a known incompatibility group. But we can know that cells have been cured of a plasmid only when it encodes a selectable trait, such as resistance to an antibiotic. Then the cells will become sensitive to the antibiotic if the plasmid is lost.

The experiment shown in Figure 4.15 is designed to measure the curing rate of a plasmid that contains the Cam^r gene, which makes cells resistant to the antibiotic chloramphenicol. To measure the frequency of plasmid curing, we grow the cells containing the plasmid in medium containing all the growth supplements and *no* chloramphenicol. At different times, we take a sample of the cells, dilute it, and plate the dilutions on agar containing the same growth supplements but again *no* chloramphenicol. After incubation of the plates, we pick the individual colonies with a loop or toothpick and transfer them to the same type of plate *with* chloramphenicol. To streamline this process, we might replicate the plate onto another plate containing chloramphenicol (see chapter 3). If we do not observe any growth for a colony, the bacteria in that colony must all have been sensitive to the antibiotic and hence the original bacterium that had multiplied to form the colony must have been cured of the plasmid. The percentage of colonies that contain no resistant bacteria is the percentage of bacteria that were cured of the plasmid at the time of plating.

To apply this test to determine if two plasmids are members of the same Inc group, both plasmids must contain different selectable genes, for example, genes encoding resistance to different antibiotics. Then one plasmid is introduced into cells containing the other plasmid. Resistance to both antibiotics is selected for. Then cells containing both plasmids are incubated without either antibiotic and finally grown on antibiotic-containing plates, as above. The only difference is that the colonies are transferred onto two plates, each containing one or the other antibiotic. If the percentage of cells cured of one or the other plasmid is no higher than the percentage cured of either plasmid when it was alone, the plasmids are members of different Inc groups. We continue to apply this test until we find a known plasmid, if any, which is a member of the same Inc group as our unknown plasmid.

Maintaining Plasmids Belonging to the Same Incompatibility Group

We can also use selectable markers such as antibiotic resistance to maintain a population of cells in which most retain two high-copy-number relaxed plasmids, even if the two plasmids belong to the same Inc group. Two relaxed plasmids that are members of the same Inc group will coexist for a limited number of generations during growth of a culture, although one or the other will be cured with a higher than normal frequency (see above). Yet if the plasmids carry genes for resistance to different antibiotics, we can reduce the curing rates of both. For example, consider a cell in which one plasmid contains a gene for tetracycline resistance and the other, which lacks that gene, has a gene for chloramphenicol resistance. By growing the cells in media containing both antibiotics, we can maintain both plasmids. Loss of one plasmid or the other will occur at a high frequency during growth of the culture, but a cell that has lost one of the plasmids will have a limited life span, and it, or its immediate progeny, will die. Therefore, most of the surviving cells at any time will have both plasmids.

Figure 4.15 Measuring curing of a plasmid carrying resistance to chloramphenicol. See the text for details.

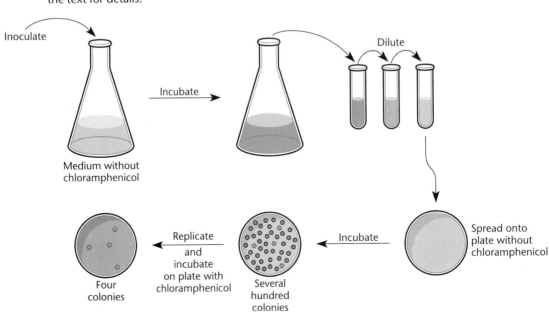

Constructing a Plasmid Cloning Vector

As discussed in chapter 1, a cloning vector is an autonomously replicating DNA (replicon) into which other DNAs can be inserted. Any DNA inserted into the cloning vector will replicate passively with the vector so that many copies (clones) of the original piece of DNA can be obtained.

Plasmids offer many advantages as cloning vectors, and many plasmids have been engineered to serve as plasmid cloning vectors. They generally do not kill the host cell and are relatively easy to purify to obtain the cloned DNA. They also can be made relatively small because few plasmid-encoded functions are required for their replication. In fact, in one of the first cloning experiments, a frog gene was cloned into plasmid pSC101 (see Cohen et al., Suggested Reading).

Most plasmids, as they are isolated from nature, are too large to be convenient for this purpose and/or often do not contain easily selectable genes that can be used to move them from one host to another. In this section, we shall outline some of the steps in making a plasmid cloning vector from a wild-type plasmid and some of the desirable features that have been engineered into plasmid cloning vectors.

Finding the Plasmid *ori* Region

The first step in making a plasmid cloning vector from a wild-type plasmid we have isolated is to find the *ori* region of the plasmid. As discussed above, the *ori* region is usually responsible for most of the properties of the plamid, including replication, copy number control, and partitioning. The origins of some commonly used cloning vectors are listed in Table 4.3.

Recombinant DNA techniques are particularly suited for locating and studying these regions. As shown in Figure 4.16, the plasmid is cut into several pieces with a restriction endonuclease (arrows in the figure) and the pieces are ligated (joined) to another piece of DNA that has a selectable marker, such as resistance to ampicillin (Amp^r). For the experiment to work, the second piece of DNA cannot have a functional origin of replication. The ligated mixture is then used to transform bacteria, and the antibiotic-resistant transformants are selected by plating the mixture on agar plates containing the antibiotic. The only DNA molecules able to replicate and also confer antibiotic resistance on the cells will be hybrids with both the *ori* region of the plasmid and the piece of DNA with the antibiotic resistance gene. Therefore, only cells harboring these hybrid molecules will grow on the antibiotic-containing medium. We can determine which plasmid fragment these have by using methods described in chapter 1, and this will tell us approximately where the *ori* region is located on the plasmid.

TABLE 4.3	Replication origins of several *E. coli* plasmid vectors	
Plasmid	***ori***	**Copy number**
pBR322	pMB1	15–20
pUC vectors	pMB1 mutant	100s
pET vectors	pMB1 mutant	100s
pBluescript	pMB1 mutant	100s
pACYC184	p15A	10–12
pSC101	pSC101	5

We can further localize the *ori* region by cutting the fragment known to contain it into even smaller pieces and repeating the process above until the smallest piece from the plasmid that can function as an origin has been identified.

This same type of analysis also can be used to identify the partitioning region and the region responsible for incompatibility. To locate the *par* sequences, smaller and smaller pieces of DNA that retain origin function are

Figure 4.16 Finding the origin of replication in a plasmid. Random pieces of the plasmid are ligated to a piece of DNA containing a selectable gene but no origin of replication and introduced into cells. Cells that can form a colony on the selective plates contain the selectable gene ligated to the piece of DNA containing the origin.

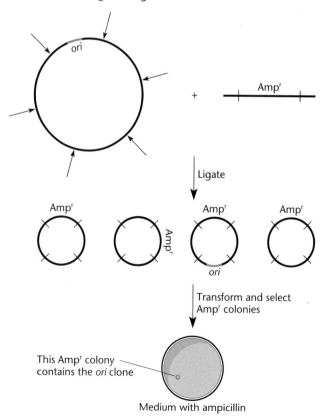

tested to see if they also confer the ability to partition properly. If the plasmid containing the smaller origin region cures at a higher frequency, it has lost its *par* region. Once we identify a plasmid that lacks a *par* region, we can clone pieces of the same type of plasmid back into it to find a piece that restores proper partitioning. In fact, the *par* regions of plasmids were discovered by this means.

INTRODUCING SELECTABLE GENES

Once we have identified the *oriV* of the plasmid, the next step is to introduce genes into the cloning vector that confer selectable properties on cells containing the plasmid. This will allow us to select cells which contain the plasmid vector and to move it from one host cell to another. Genes which confer resistance to an antibiotic make convenient selectable genes. Cells containing the plasmid can then be selected by plating on media containing the antibiotic, and only cells containing the plasmid will multiply to form a colony. Antibiotic resistance genes are often taken from transposons and other plasmids. Some antibiotic resistance genes which have been introduced into cloning vectors are the chloramphenicol resistance gene, Cam^r, of transposon Tn9; the tetracycline resistance gene, Tet^r, of plasmid pSC101; the ampicillin resistance gene, Amp^r, of transposon Tn3; and the kanamycin resistance gene, Kan^r, of the transposon Tn5. The antibiotic resistance gene that is chosen depends upon the uses to which the cloning vector will be put. Some antibiotic resistance genes such as the Tet^r gene from pSC101 are expressed only in some types of bacteria closely related to *E. coli*, while others such as Kan^r from Tn5 are expressed in most gram-negative bacteria.

INTRODUCING UNIQUE RESTRICTION SITES

Since many applications of plasmid cloning vectors require that we introduce clones into restriction sites, it is necessary that our cloning vector have some sites that are unique. If a site is unique, the cognate restriction endonuclease will cut the vector at only that one site when it is used to cut the plasmid. We can then clone pieces of foreign DNA into the unique site, and the cloning vector will remain intact. Restriction sites for six-hitters occur on average about once every 1,000 bp (see chapter 1), so that a cloning vector of 3,000 to 4,000 bp is apt to have more than one site for any particular six-hitter restriction endonuclease. There are many tricks for removing the extra sites. We can change the sequence at some of the sites by site-specific mutagenesis so that they are no longer recognized by the restriction endonuclease. Alternatively, we can make a partial digest of the plasmid so that it is cut at only one of the sites. The site at which it was cut can then be removed by filling in the overhang with DNA polymerase, if the restriction endonuclease leaves a 5' overhang, or by removing the overhang with a single-strand-specific DNase before religating.

Not only do we want to have unique sites for some restriction nucleases in our cloning vector, but also these sites should be located in the plasmid in a way which makes them most useful for cloning. In many plasmid cloning vectors, the unique sites are located in a selectable gene, so that insertion of a foreign piece of DNA in the site will inactivate the selectable gene. This is called **insertional inactivation** and is discussed below. Normally, during a cloning operation, only a small percentage of the cloning vectors will have picked up a foreign DNA insert. If those that have picked up an insertion no longer confer the selectable trait, for example resistance to an antibiotic, they can be more easily identified. Many cloning vectors also have the unique restriction sites for many different restriction endonucleases all grouped in one small region on the plasmid called a polyclonal or a multiple restriction site. This offers the convenience of choosing among a variety of restriction endonucleases for cloning, and the cloned DNA will always be inserted at the same site, independent of the restriction endonuclease used. Polyclonal sites can also be used for **directional cloning.** If the cloning vector is cut by two different incompatible restriction endonucleases with unique sites within the polyclonal site, the resulting overhangs cannot pair to recyclize the plasmid. The plasmid can recyclize only if it picks up a piece of foreign DNA. If the piece of DNA to be cloned has overhangs for the two different sites at its ends, it will usually be cloned in only one orientation into the polyclonal site.

The unique restriction sites can also be placed so that genes cloned into them will be expressed from promoters and translation initiation regions on the plasmid. These are called expression vectors and can be used to express foreign genes in *E. coli* and other convenient hosts. Such vectors can also be used to attach affinity tags to proteins to aid in their purification. We discussed expression vectors and affinity tags in connection with translational and transcriptional fusions in chapter 2.

Examples of Plasmid Cloning Vectors

A number of plasmid cloning vectors have been engineered for special purposes. Almost all of these plasmids have at least some of the features mentioned above for a desirable cloning vector. To reiterate:

1. They are small, so that it can be easily isolated and introduced into various bacteria.
2. They have a relatively high copy number, so that it can be easily purified in sufficient quantities.

3. They carry easily selectable traits, such as a gene conferring resistance to an antibiotic, which can be used to select cells that contain the plasmid.
4. They have one or a few sites for restriction endonucleases, which cut DNA and allow the insertion of foreign DNAs. Also, these sites usually occur in selectable genes to facilitate the detection of plasmids that have foreign DNA inserts by a process called insertional inactivation.

Many plasmid cloning vectors have other special properties that aid in particular experiments. For example, some contain the sequences recognized by phage packaging systems (*pac* or *cos* sites), so that they can be packaged into phage heads (see chapter 7). Expression vectors can be used to make foreign proteins in bacteria. Mobilizable plasmids have mobilization (*mob*) sites and so can be transferred by conjugation to other cells (see chapter 5). Some broad-host-range vectors have *ori* regions that allow them to replicate in many types of bacteria or even in organisms from different kingdoms. Shuttle vectors contain more than one type of replication origin and so can replicate in unrelated organisms. We shall deal with these and some other types of speciality plasmid cloning vectors in more detail below and in later chapters.

CLONING VECTOR pBR322
Figure 4.17 shows a map of pBR322, which embodies many of the desirable traits of a cloning vector. This plasmid is fairly small (only 4,360 bp) and has a relatively high copy number (~16 copies per cell), making it easy to isolate. The vector was constructed by removing all but the essential *ori* region from pMB1, a ColE1-like plasmid, and adding two resistance genes for the antibiotics

tetracycline and ampicillin, which were taken from plasmid pSC101 and transposon Tn*3*, respectively. The plasmid also has several unique sites for restriction endonucleases, including *Bam*HI, *Eco*RI, and *Pst*I. These enzymes cut DNA at specific sites, allowing a piece of foreign DNA to be inserted into the plasmid and then studied.

Insertional inactivation offers a simple genetic test for determining whether a cell contains a pBR322 plasmid with a foreign DNA insert. To illustrate, Figure 4.18 shows a piece of foreign DNA inserted into the *Bam*HI site in the tetracycline resistance (Tet^r) gene in pBR322. The piece of foreign DNA in the *Bam*HI site will disrupt the Tet^r gene and cause the plasmid to lose the ability to confer tetracycline resistance on a bacterium that carries it. The plasmid will still confer ampicillin resistance, however, since the Amp^r gene remains intact. Therefore, cells containing a plasmid with a foreign DNA insert will be ampicillin resistant but tetracycline sensitive, which is easy to test on agar plates containing one or the other antibiotic.

pUC PLASMIDS
Some of the most commonly used plasmid cloning vectors are the pUC vectors and vectors derived from them. One pUC vector, pUC18, is shown in Figure 4.19. The pUC plasmids are very small (with only 2,700 bp of DNA) and have a very high copy number of 30 to 50, making them relatively easy to purify. They also have the easily selectable ampicillin resistance (Amp^r) gene. One of the most useful features of these plasmids is the ease with which they can be used for insertional inactivation. They encode the N-terminal region of the *lacZ* gene product, called the α-peptide, which is not active in the cell by itself but will complement the C-terminal portion of the protein called the *lacZ* β-polypeptide, to make active *lacZ* polypeptide, which turns colonies blue on 5-bromo-4-chloro-3-indolyl-β-D-galactopyranoside (X-Gal) plates. Some host strains such as *E. coli* JM109 have been engineered to make the β-polypeptide of *lacZ*. As a consequence, *E. coli* JM109 containing a pUC plasmid will make blue colonies on X-Gal plates. The pUC plasmids have a multicloning site containing the recognition sequences for many different restriction endonucleases in the coding region for the α-peptide (Figure 4.19). If a foreign DNA is cloned into any one of these sites, the bacterium will not make the α-peptide and the colonies will be colorless on X-Gal plates, making bacteria containing plasmids with inserts easy to identify. These plasmids are also transcription vectors (see chapter 2) because a gene in a piece of DNA directionally cloned into the multicloning site on the plasmid will be immediately downstream of the strong *lac* promoter on the plasmid, called p_{lac} in the figure, and so it

Figure 4.17 The plasmid cloning vector pBR322. Unique restriction sites are highlighted. Amp^r, ampicillin resistance; Tet^r, tetracycline resistance.

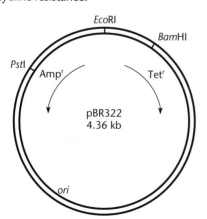

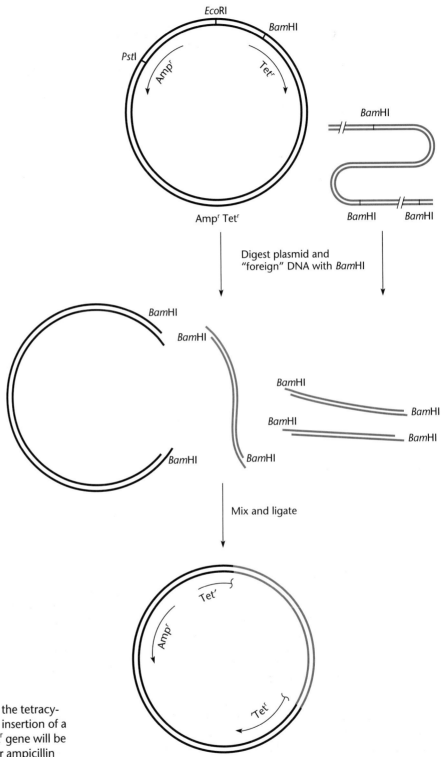

Figure 4.18 Insertional inactivation of the tetracycline resistance (Tet^r) gene of pBR322 by insertion of a foreign DNA into the *Bam*HI site. The Tet^r gene will be disrupted, but the plasmid will still confer ampicillin resistance because the Amp^r gene will still be active.

will be transcribed from the *lac* promoter. The *lac* promoter is also inducible and will be turned on only if an inducer, such as isopropyl-β-D-thiogalactopyranoside (IPTG) or lactose, is added. Thus, the cells can be propagated before the synthesis of the gene product is induced, a feature that is particularly desirable if the gene product is toxic to the cell. Genes cloned into one of the multicloning sites in the *lacZ* gene can also be translated from

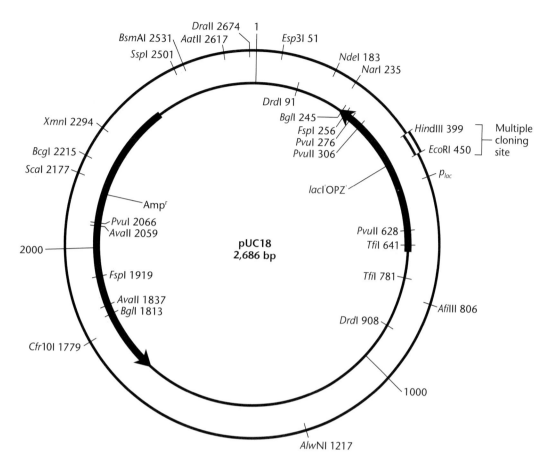

pUC18 multiple cloning site and primer binding regions: 364–480

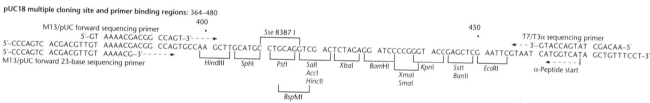

Figure 4.19 A pUC expression vector. A gene cloned into one of the restriction sites in the multiple-cloning site will almost invariably disrupt the coding sequence for the *lacZ* α-peptide. If it is inserted in the correct orientation, the gene will be transcribed from the *lac* promoter called *P_lac_* in the figure. If the ORF for the gene is in the same reading frame as the ORF for the *lacZ* coding sequence, the gene will also be translated from the *lacZ* TIR, and the N-terminal amino acids of *lacZ* will be fused to the polypeptide product of the gene. Adapted with permission from the *BRL-Gibco Catalog*, p. 766. Copyright 2002 Invitrogen Corporation, all rights reserved.

the *lacZ* translational initiation region (TIR) on the plasmid, provided that there are no intervening nonsense codons and the gene is cloned in the same reading frame as the upstream *lacZ* sequences.

Broad-Host-Range Cloning Vectors

Many of the common *E. coli* cloning vectors such as pBR322, the pUC plasmids, and the pET plasmids have been constructed with the pMB1 *ori* region and thus are very narrow in their host range. They will replicate only in *E. coli* and a few of its close relatives. However, some cloning applications require a plasmid cloning vector that will replicate in other gram-negative bacteria, and so cloning vectors have been derived from the broad-host-range plasmids RSF1010 and RK2, which will replicate in most gram-negative bacteria. In addition to the broad-host-range *ori* region, these cloning vectors sometimes contain a *mob* site called an *oriT* site, which can allow them to be transferred into other bacteria (see chapter 5). This latter trait is very useful, because other ways of introducing DNA have not been developed for many types of bacteria.

SHUTTLE VECTORS

Sometimes, a plasmid cloning vector from one organism must be transferred into another organism. If the two organisms are not related, the same plasmid *ori* region is not likely to function in both organisms. Such situations require the use of **shuttle vectors**, so named because they can be used to "shuttle" genes between the two organisms. A shuttle vector has two origins of replication, one that functions in each organism. Shuttle vectors also must contain selectable genes that can be expressed in both organisms.

In most cases, one of the organisms in which the shuttle vector can replicate is *E. coli*. The genetic tests can be performed in the other organism, but the plasmid can be purified and otherwise manipulated by the refined methods developed for *E. coli*.

Some shuttle vectors can replicate in gram-positive bacteria and *E. coli*, whereas others can be used in lower or even higher eukaryotes. For example, plasmid YEpl3 (Figure 4.20) has the replication origin of the 2μm circle, a plasmid found in the yeast *Saccharomyces cerevisiae*, so it will replicate in *S. cerevisiae*. It also has

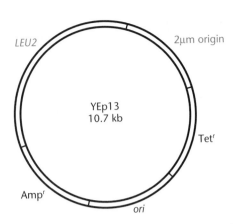

Figure 4.20 The shuttle plasmid YEp13. The plasmid contains origins of replication that will function in *S. cerevisiae* and *E. coli*. It also contains genes that can be selected in *S. cerevisiae* and *E. coli*.

the pBR322 *ori* region and so will replicate in *E. coli*. In addition, the plasmid contains the yeast gene *LEU2*, which can be selected in yeast, as well as Amp^r, which confers ampicillin resistance in *E. coli*. Similar shuttle vectors that can replicate in mammalian or insect cells and *E. coli* have been constructed. Some of these plasmids have the replication origin of the animal virus simian virus 40 and the ColE1 origin of replication.

SUMMARY

1. Plasmids are DNA molecules that exist free of the chromosome in the cell. Most plasmids are circular, but some are linear. The sizes of plasmids range from a few thousand base pairs to almost the length of the chromosome itself.

2. Plasmids usually carry genes for proteins that are necessary or beneficial to the host under some situations but are not essential under all conditions. By carrying nonessential genes on plasmids, bacteria are able to keep their chromosome small but still respond quickly to changes in the environment.

3. Plasmids replicate from a unique origin of replication, or *oriV* region. Many of the characteristics of a given plasmid derive from this *ori* region. These include the mechanism for replication, copy number control, partitioning, and incompatibility. If other genes are added or deleted from the plasmid, it will retain most of its original characteristics, provided that the *ori* region remains.

4. The copy number of a plasmid is the number of copies of the plasmid per cell immediately after cell division. Different types of plasmids use different mechanisms to regulate their initiation of replication and therefore their copy number.

5. Some plasmids use antisense or countertranscribed ctRNAs to regulate their copy number. In ColE1-derived plasmids,

the ctRNA, called RNA I, interferes with the processing of the primer for leading-strand replication, called RNA II. In other cases, including the R1 and Col1B-P9 plasmids, the ctRNA interferes with the expression of the Rep protein required to initiate plasmid DNA replication.

6. Other plasmids called iteron plasmids regulate their copy number by two interacting mechanisms. They control the amount of the Rep protein required to initiate plasmid replication, and the Rep protein also couples plasmids through their iteron sequences.

7. Some plasmids have a special partitioning mechanism to ensure that each daughter cell will get one copy of the plasmid as the cells divide. These partitioning systems are remarkably similar to each other and consist of two genes for proteins usually called ParA and ParB and a site called *parS*. The ParB protein may bind to *parS*, and the ParA protein may then allow this complex to bind to a unique cellular structure that pulls the plasmids apart as the cell divides. Alternatively, the daughter plasmids may have to pair with each other through their *par* sites prior to binding to the cellular structure, which then does not need to be unique.

(continued)

SUMMARY (continued)

8. If two plasmids cannot stably coexist in the cells of a culture, they are said to be incompatible or members of the same Inc group. They can be incompatible if they share the same copy number control system or the same partitioning functions.

9. The host range for replication of a plasmid is all the different organisms in which the plasmid can replicate. Some plasmids are very broad in their host range and can replicate in a wide variety of bacteria. Others are very narrow in their host range and can replicate in only very closely related bacteria.

10. Many plasmids have been engineered for use as cloning vectors. They make particularly desirable cloning vectors for some applications because they do not kill the host, can be small, and are easy to isolate.

QUESTIONS FOR THOUGHT

1. Why are genes whose products are required for normal growth not carried on plasmids? List some genes which you would not expect to find on a plasmid and some genes you might expect to find on a plasmid.

2. Why do you suppose some plasmids are broad host range for replication? Why are all plasmids not broad host range?

3. How do you imagine a partitioning system for a single-copy plasmid like F could work? How might a copy number control mechanism work?

4. How would you find the genes required for replication of the plasmid if they are not all closely linked to the *ori* site?

5. How would you determine which of the replication genes of the host *E. coli* (e.g., *dnaA* and *dnaC*) are required for replication of a plasmid you have discovered?

6. The R1 plasmid has a leader polypeptide translated upstream of the gene for RepA, and cleavage of the mRNA by RNase III occurs in the coding sequence for this leader polypeptide. This blocks the translation of the leader polypeptide and also the translation of the downstream *repA* gene to which it is translationally coupled. Do you think it would have been easier just to have the cleavage occur in the coding sequence for the RepA protein itself? Why or why not?

7. Recent evidence suggests that stringent plasmids replicate in the center of the cell and then quickly move to the quarter positions of the cell before it divides. How would you reconcile this evidence with the model for plasmid partitioning in which the plasmids must pair prior to partitioning?

PROBLEMS

1. The IncQ plasmid RSF1010 carries resistance to the antibiotics streptomycin and sulfonamide. Suppose you have isolated a plasmid that carries resistance to kanamycin. Outline how you would determine whether your new plasmid is an IncQ plasmid.

2. A plasmid has a copy number of 6. What fraction of the cells will be cured of the plasmid each time the cells divide if the plasmid has no partitioning mechanism?

3. You wish to clone pieces of human DNA cut with the restriction endonuclease *Bam*HI into the *Bam*HI site of pBR322. You cut both human DNA and pBR322 with *Bam*HI and ligate them. You transform the ligation mix into *E. coli*, selecting for

Amp^r. Outline how you would determine which of the transformants probably contain a plasmid with a human DNA insert.

4. The ampicillin resistance gene of plasmid RK2 is unregulated. The more copies of the gene a bacterium has, the more gene product will be made. In this case, the resistance of the cell to ampicillin will be higher the more of these genes it has. Use this fact to devise a method to isolate mutants of RK2 that have a higher than normal copy number (copy-up mutations). Determine whether your mutants have mutations in the Rep-encoding gene.

5. Outline how you would determine whether a plasmid has a partitioning system.

SUGGESTED READING

Bagdasarian, M., R. Lurz, B. Ruckert, F. C. H. Franklin, M. M. Bagdasarian, J. Frey, and K. N. Timmis. 1981. Specific purpose plasmid cloning vectors. II. Broad host, high copy number, RSF1010-derived vectors, and a host vector system for cloning in *Pseudomonas*. *Gene* **16:**237–247

Chattoraj, D. K. 2000. Control of plasmid DNA replication by iterons: no longer paradoxical. *Mol. Microbiol.* **37:**467–476.

Cohen, S. N., A. C. Y. Chang, H. W. Boyer, and R. B. Helling. 1973. Construction of biologically functional bacterial plasmids *in vitro*. *Proc. Natl. Acad. Sci. USA* **70:**3240-3244

Del Solar, G., and M. Espinosa. 2000. Plasmid copy number control: an ever-growing story. *Mol. Microbiol.* **37:**492–500. (Microreview.)

Gedes, K., J. Moller-Jensen, and R.B. Jensen. 2000. Plasmid and chromosome partitioning: surprises from phylogeny. *Mol. Microbiol.* 37:455–466. (Microreview.)

Khan, S. A. 2000. Plasmid rolling circle replication: recent developments. *Mol. Microbiol.* 37:477–484. (Microreview.)

Lacatena, R. M., and G. Cesareni. 1981. Base pairing of RNA I with its complementary sequence in the primer precursor inhibits ColE1 replication. *Nature* (London) 294:623–626

McEachern, M. J., M. A. Bott, P. A. Tooker, and D. R. Helinski. 1989. Negative control of plasmid R6K replication: possible role of intermolecular coupling of replication origins. *Proc. Natl. Acad. Sci. USA* 86:7942–7946.

Meacock, P. A., and S. N. Cohen. 1980. Partitioning of bacterial plasmids during cell division: a *cis*-acting locus that accomplishes stable plasmid maintenance. *Cell* 20:529–542.

Miller, C. A., S. L. Beaucage, and S. N. Cohen. 1990. Role of DNA superhelicity in partitioning of the pSC101 plasmid. *Cell* 62:127–133.

Novick, R. P., and F. C. Hoppensteadt. 1978. On plasmid incompatibility. *Plasmid* 1:421–434.

Pogliano, J., T. Q. Ho, Z. Zhong, and D. R. Helinski. 2001. Multicopy plasmids are clustered and localized in *Escherichia coli*. *Proc. Natl. Acad. Sci. USA* 98:4486–4491.

Radloff, R., W. Bauer, and J. Vinograd. 1967. A dye-buoyant-density method for the detection and isolation of closed circular duplex DNA: the closed circular DNA in HeLa cells. *Proc. Natl. Acad. Sci. USA* 57:1514–1521.

5

Conjugation

A REMARKABLE FEATURE OF MANY PLASMIDS is the ability to transfer themselves and other DNA elements from one cell to another in a process called **conjugation.** Joshua Lederberg and Edward Tatum first observed this process in 1947, when they found that mixing some strains of *Escherichia coli* with others resulted in strains that were genetically unlike either of the originals. As discussed later in this chapter, Lederberg and Tatum suspected that bacteria of the two strains exchanged DNA—that is, two parental strains mated to produce progeny unlike themselves but with characteristics of both parents. At that time, however, plasmids were unknown, and it was not until later that the basis for the mating was understood.

Overview

During conjugation, the two strands of a plasmid separate in a process resembling rolling-circle replication (see the section on mechanism of DNA transfer during conjugation in gram-negative bacteria, below), and one strand moves from the bacterium originally containing the plasmid—the **donor**—into a **recipient** bacterium. Then the two single strands serve as templates for the replication of complete double-stranded DNA molecules in both the donor cell and the recipient cell. A recipient cell that has received DNA as a result of conjugation is called a **transconjugant.**

Most naturally occurring plasmids are either **self-transmissible** or **mobilizable,** a fact that suggests that plasmid conjugation is advantageous for plasmids and their hosts. Self-transmissible plasmids encode all the functions they need to move among cells, and sometimes they also aid in the transfer of chromosomal DNA and mobilizable plasmids. The latter encode some but not all of the proteins required for transfer and consequently need the help of self-transmissible plasmids to move.

Any bacterium harboring a self-transmissible plasmid is a potential donor, because it can transfer DNA to other bacteria. In gram-negative bacteria, such cells produce a structure, called a **sex pilus**, that facilitates conjugation (discussed in a later section). Bacteria that lack self-transmissible plasmids are potential recipients, and conjugating bacteria are known as **parents**. Donors are sometimes referred to as **male**.

Self-transmissible plasmids probably exist in all types of bacteria, but those which have been studied most extensively are from the gram-negative genera *Escherichia* and *Pseudomonas* and the gram-positive genera *Enterococcus, Streptococcus, Bacillus, Staphylococcus,* and *Streptomyces*. The best-known transfer systems are those of plasmids isolated from *Escherichia* and *Pseudomonas* species, and so we focus our attention on these gram-negative systems and do not address conjugation in gram-positive bacteria until the end of the chapter.

Classification of Self-Transmissible Plasmids

Bacterial plasmids have many different types of transfer systems, which are encoded by the plasmid *tra* genes (see the section on transfer genes, below). But as discussed in chapter 4, plasmids are usually classified by their incompat-ibility (Inc) group. Accordingly, the F-type plasmids use a transfer system known as the Tra system of IncF plasmids, and the RP4 plasmid uses the Tra system of IncP plasmids.

Despite this nomenclatural link, transfer systems bear little relation to the replication and partitioning functions of a plasmid, the characteristics that determine its Inc group. In fact, the genes for these functions and the transfer genes are located in different regions of the plasmid, and there is no a priori reason for any correlation between them. Nevertheless, there is a high degree of correlation between the type of transfer system and Inc group. There may be a good reason for this. Some products of plasmid transfer genes inhibit the entry of plasmids with the same Tra functions (see below). If the *tra* genes did not correlate with the Inc group, a plasmid would sometimes transfer into a cell that already had a plasmid of the same Inc group, and one of the two plasmids would subsequently be lost.

Interspecies Transfer of Plasmids

Many plasmids have transfer systems that enable them to transfer DNA between unrelated species. These are known as **promiscuous plasmids** and include the IncW plasmids, represented by R388; the IncP plasmids, represented by RP4; and the IncN plasmids, represented by pKM101 (Figure 5.1). The IncP plasmids can transfer

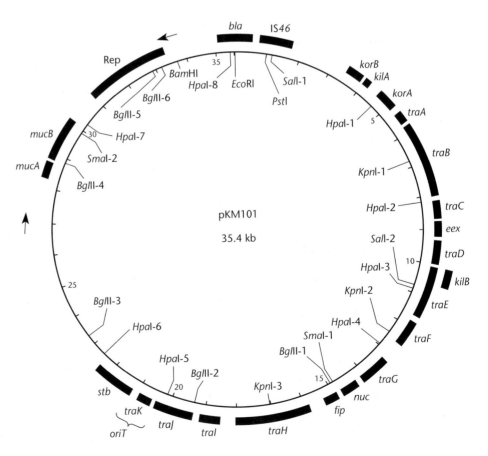

Figure 5.1 Physical and genetic map of the self-transmissible plasmid pKM101. *oriT* is the origin of plasmid transfer. The *tra* genes encode the many transfer functions discussed in this chapter. The *mucA* and *mucB* genes enhance UV mutagenesis by coding for analogs of the *E. coli umuD* and *umuC* genes (see chapter 11). From S. Winans and G. Walker, *J. Bacteriol.* **161**:402–410, 1985.

themselves or mobilize other plasmids from *E. coli* into essentially any gram-negative bacterium. Recent studies showed that plasmids of this group transfer at a low frequency into cyanobacteria, gram-positive bacteria such as *Streptomyces* species, and even plant cells! The F plasmid, which was not known to be particularly promiscuous, can transfer itself from *E. coli* into yeast cells (see Box 5.1).

Transfer of DNA by promiscuous plasmids probably plays an important role in evolution. Such transfer could explain why genes with related functions are often very similar to each other regardless of the organism that harbors them. These genes could have been transferred by promiscuous plasmids fairly recently in evolution, which would account for their similarity relative to the other genes of the two organisms.

The interspecies transfer of plasmids also has important consequences for the use of antibiotics in treating human and animal diseases. Many of the most promiscuous plasmids, including those of the IncP group, such as RP4, and the IncW plasmid R388 were isolated in hospital settings. These large plasmids (commonly called R-plasmids because they carry genes for antibiotic resistance) presumably have become prevalent in recent years in response to the indiscriminate use of antibiotics in medicine and agriculture. The source of the resistance genes may be soil bacteria, such as actinomycetes, that are the producers of antibiotics. In chapter 9, we discuss how transposons might have helped assemble these promiscuous R-plasmids.

Whatever their source, the emergence of R-plasmids indicates why antibiotics should be used only when they are absolutely necessary. In humans or animals treated indiscriminately with antibiotics, bacteria that carry R-plasmids will be selected from the normal flora. R-plasmids can be quickly transferred into an invading pathogenic bacterium, making it antibiotic resistant. Consequently the infection will be difficult to treat.

Mechanism of DNA Transfer during Conjugation in Gram-Negative Bacteria

Much has been learned in the last few years about the detailed mechanism of plasmid conjugation, especially in gram-negative bacteria. An overview of DNA transfer is

---

BOX 5.1

Gene Exchange between Kingdoms

Not only can some plasmids transfer themselves into other types of bacteria, but also they can sometimes transfer themselves into eukaryotes, that is, into organisms of a different kingdom.

Agrobacterium tumefaciens and Crown Gall Tumors in Plants

The first known example of transfer of bacterial plasmids into eukaryotes occurs in the plant disease crown gall, in which a tumor appears on the plant, usually where the roots join the stem (the crown). Crown gall disease is caused by *Agrobacterium tumefaciens*, and the interaction between this bacterium and the plant was the first known example of gene exchange between bacteria and a eukaryote. Virulent strains of *A. tumefaciens* contain a plasmid called the Ti plasmid, for tumor initiation. The Ti plasmids of *Agrobacterium tumefaciens* are in most respects normal bacterial self-transmissible plasmids. A typical Ti plasmid is shown in the Figure, panel A. Like other self-transmissible plasmids, Ti plasmids encode Tra functions that enable them to transfer themselves into other bacteria. What makes these plasmids remarkable is that they can also transfer part of themselves, called the T-DNA region, into plants. This discovery, made in the 1970s, has allowed the construction of transgenic plants because any foreign genes cloned into one of the T-DNA regions of the Ti plasmid will be transferred into

the plant along with the T-DNA and integrated into the plant DNA. The integrated foreign genes can alter the plant, provided they are transcribed and translated in the plant. Panel B of the figure shows the general procedure which is followed. A gene for kanamycin resistance has been inserted into the T-DNA of a Ti plasmid in such a way that it will be expressed in the plant. A piece of the plant leaf is floated in a bath of the bacterium containing the engineered Ti plasmid. Transgenic plants with the T-DNA inserted in their chromosome can be regenerated and selected on plates containing kanamycin. A whole industry has developed around this technology, and agrobacteria have been used to genetically engineer plants to make their own insecticides, to be more nutritious, and to survive more severe growing conditions.

The functions required for transfer of the T-DNA into plants are encoded by a region called the *vir* region (see Figure, panel A). This region is distinct from the *tra* region, which is required for the transfer of the plasmid into other bacteria, but its functions are remarkably similar both to other Tra functions and to other type IV secretion systems. Panel C of the Figure shows the structure of the type IV secretion system encoded by the *vir* region of the Ti plasmid. Like the *tra* region, the *vir* region encodes both a mating-pair

(continued)

BOX 5.1 (continued)

Gene Exchange between Kingdoms

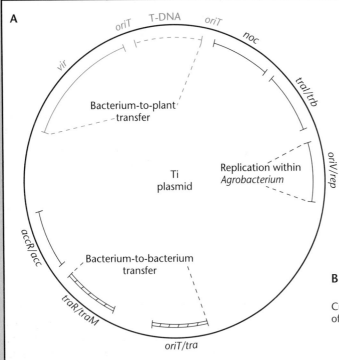

A

oriT T-DNA oriT
vir
noc
traI/trb
Bacterium-to-plant transfer
oriV/rep
Ti plasmid
Replication within Agrobacterium
accR/acc
Bacterium-to-bacterium transfer
traR/traM
oriT/tra

The Ti plasmid. (A) The structure of a Ti plasmid, showing the various regions discussed clockwise from the top. The T-DNA is bordered by the *oriT* sequence which is transferred into plants. The T-DNA contains the genes that are expressed in the plant to make opines and plant hormones (not shown); the *noc* genes, encoding enzymes for the catabolism of the opine nopaline in the bacterium; some *tra* genes, for transfer into other bacteria; the *oriV* region, for replication of the plasmid in the bacterium; *oriT* and *tra* function genes, for transfer into other bacteria; *acc* genes, for catabolism of another opine; and *vir* genes, for transfer into plants. (B) A procedure for making a transgenic plant (for details, see the text). (C) Structure of the type IV secretion system for transfer of T-DNA into plant (for details, see the text).

B

Cut out a piece of the leaf

Float in plate containing an *Agrobacterium* strain with engineered Ti plasmid

Incubate on plate containing plant regeneration medium — Regenerating plants

Excise germinating shoots and transplant to plates containing kanamycin — Kanamycin-resistant shoots

— Transgenic plant

formation (Mpf) system, which elaborates a pilus, and a Dtr system, which processes the DNA for transfer. The pilus is composed of the pilin protein, which is the product of the *virB2* gene and is cyclized, like the pilins of the pili of other type IV secretion systems. The Mpf system also includes a coupling protein, the product of the *virD4* gene, which communicates with the relaxosome, which includes the specific relaxase. The relaxase, the product of the *virD2* gene, cleaves the sequences bordering the T-DNA in the plasmid and remains covalently attached to the 5′ ends of the single-stranded T-DNAs during transfer into the plant. The sequences at which the relaxase cuts these border sequences are similar to the *oriT* sequences of IncP plasmids, and the relaxase makes a cut in exactly the same place in the sequences where the relaxases of IncP plasmids cut their *oriT*.

This is where the T-DNA transfer system begins to differ from normal *tra* DNA transfer systems. In addition to its role as a relaxase, the VirD2 protein contains amino acid sequences that will target it to the plant cell nucleus once it is in the plant. These sequences, called nuclear localization signals, are essentially "passwords" that tell the plant that a particular plant nuclear protein should be transported into the nucleus after it has been translated in the cytoplasm. By imitating the password, the VirD2 protein tricks the plant into transporting it into the nucleus, dragging the attached T-DNA with it. Once in the nucleus, the T-DNA can enter the plant DNA by recombination. Once integrated into the

BOX 5.1 (continued)

Gene Exchange between Kingdoms

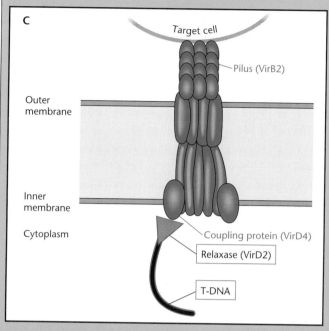

C

Target cell

Pilus (VirB2)

Outer membrane

Inner membrane

Cytoplasm

Coupling protein (VirD4)

Relaxase (VirD2)

T-DNA

The interaction of *A. tumefaciens* with the host plant presents some interesting points of ecology. When a bacterium containing a Ti plasmid encounters a plant, phenolic residues and monosaccharides given off by the plant activate a two-component regulatory system, VirA-VirG, which in turn activates the transcription of the *vir* genes, which transport the T-DNA into the plant. As if to share its good fortune in finding a susceptible plant, the plasmid also induces the VirA-VirB system to activate the transcription of the *tra* genes, which transfer the Ti plasmid into any other surrounding agrobacteria which may lack it and so cannot use that particular opine as a carbon, nitrogen, and energy source. These other surrounding bacteria can then be recruited to aid in the infection process. We shall present a more detailed discussion of two-component systems and how they activate genes in response to extracellular signals in chapter 13.

Transfer of Broad-Host-Range Plasmids into Eukaryotes

The Ti plasmid is obviously designed to transfer part of itself into plant cells. The surprising result of recent studies is that other bacterial plasmids can also transfer themselves or mobilize other plasmids into eukaryotic cells. One striking example is the mobilization of other plasmids into plant cells by the Ti plasmid. As mentioned, the sequences bracketing the T DNA in the Ti plasmid can be thought of as *oriT* sites, and the Tra functions of the Ti plasmid can be thought of as mobilizing the T DNA into plant cells. Plasmid RSF1010, and plasmids derived from it, can also be mobilized into plant cells by the Ti plasmid, provided that they contain the correct *mob* sequence.

Not only do plasmids transfer into plants, but they also transfer into lower fungi. This observation is very surprising because the cell surfaces of bacteria and eukaryotes are very different. So are the surfaces of plant cells, but in the case of the Ti plasmid, we can assume that the transfer functions have evolved to recognize plant cells. Moreover, there is no apparent reason why bacterial plasmids should have evolved to transfer into other Kingdoms. Whatever the reason, the transfer of genes between eukaryotes and bacteria may play an important role in evolution.

plant DNA, the T-DNA of the plasmid encodes the synthesis of plant hormones which induce the plant cells to multiply and form tumors (galls) on the plant, hence the name "crown gall tumors." The T-DNA also encodes enzymes which synthesize unusual small molecules composed of an amino acid such as arginine joined to a carbohydrate such as pyruvate. These compounds, called opines, are excreted from the tumor. The plant is able to express the genes on the T-DNA and make these compounds because the genes on the T-DNA are essentially plant genes with plant promoters and plant TIRs, so they can be expressed once they are in the plant. Meanwhile, back in the bacterium, the Ti plasmid carries genes that allow it to use the particular opine made by that strain as a carbon and nitrogen source. Very few types of bacteria can degrade opines, which gives the *Agrobacterium* species containing this particular Ti plasmid an advantage. In this way, the bacterium creates its own special "ecological niche" at the expense of the plant.

Another plasmid-encoded protein, VirE2, is also transported into the plant cytoplasm. This protein might not enter the nucleus but, rather, might form the channel in the plant membrane through which the T-DNA enters. If other *tra* systems encode such proteins, it would help explain the extreme promiscuity of some conjugation systems (see below). If other transport systems also secrete proteins to form their own channel in the recipient cell membrane through which the DNA must pass, the *tra* system becomes less restricted in its choice of recipient cell. The membrane channel may be able to assemble in almost any type of cell, since bipolar lipid membranes are similar in all organisms.

References

Bates, S., A. M. Cashmore, and B. M. Wilkins. 1998. IncP plasmids are unusually effective in mediating conjugation of *Escherichia coli* and *Saccharomyces cerevisiae*: involvement of the Tra2 mating system. *J. Bacteriol.* **180:**6538–6543.

Buchanan-Wollasten, U., J. E. Passiatore, and F. Cannon. 1987. The *mob* and *oriT* mobilization functions of a bacterial plasmid promote its transfer to plants. *Nature* (London) **328:**172–175.

Zhu, J., P. M. Oger, B. Schrammeijer, P. J. J. Hooykaas, S. K. Farrand, and S. C. Winans. 2000. The basis of crown gall tumorigenesis. *J. Bacteriol.* **182:**3885–3895.

shown in Figure 5.2. Some of this progress has come from the convergence of two seemingly different fields: conjugation and some types of protein secretion. The practical applications of plasmid conjugation systems, especially in plant biotechnology, have also inspired some of this work (see Box 5.1). We outline the process of conjugation in some detail in this section.

Transfer (*tra*) Genes

Conjugation is a complicated process that requires the products of many genes. As mentioned earlier, the genes required for transfer are called the **tra** genes. The products of the *tra* genes are *trans* acting and can act on another plasmid in the same cell. The map of pKM101 (Figure 5.1) shows that this self-transmissible plasmid contains at least 11 of these genes, as well as genes such as *eex* (entry exclusion) that are not required for transfer but play related roles, in this case preventing the entry of other plasmids with the same Tra functions. In addition to the *tra* genes, a *cis*-acting site called *oriT* is required for transfer. With so many genes and sites involved, a large portion of a self-transmissible plasmid is taken up with transfer-related functions, and by necessity, these plasmids are quite large.

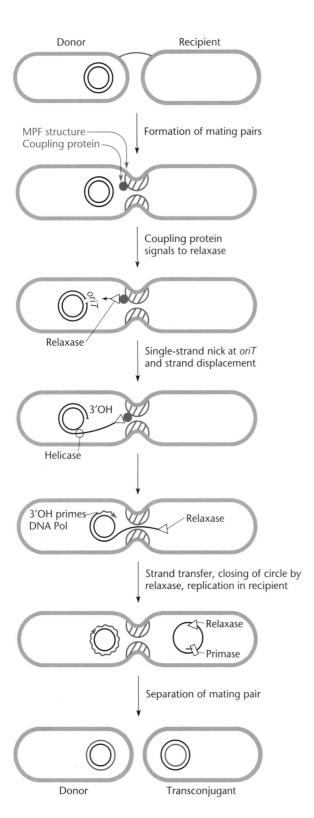

Figure 5.2 Mechanism of DNA transfer during conjugation showing the mating-pair formation (Mpf functions) in gold. The donor cell produces a pilus which forms on the cell surface and which may contact a potential recipient cell and bring it into close contact or may help hold the cells in close proximity after contact has been made, depending on the type of pilus. A pore then forms in the adjoining cell membranes. Upon receiving a signal from the coupling protein that contact with a recipient has been made, the relaxase protein makes a single-stranded cut at the *oriT* site in the plasmid. A plasmid-encoded helicase then separates the strands of the plasmid DNA. The relaxase protein, which has remained attached to the 5′ end of the single-stranded DNA, is then transported out of the donor cell through the channel directly into the recipient cell, dragging the single-stranded attached DNA along with it. Once in the recipient, the relaxase protein helps recyclize the single-stranded DNA. A primase, either that of the host or plasmid encoded and injected with the DNA, then primes replication of the complementary strand to make the double-stranded circular plasmid DNA in the recipient. The 3′ end at the nick made by the relaxase in the donor can also serve as a primer, making a complementary copy of the single-stranded plasmid DNA remaining in the donor. Therefore, after transfer, both the donor and the recipient bacterium end up with a double-stranded circular copy of the plasmid. Details are given in the text.

The *tra* genes of a self-transmissible plasmid required for plasmid transfer can be divided into two components. Some of the *tra* genes encode proteins involved in the processing of the plasmid DNA to prepare it for transfer. This is called the **Dtr component** (for "DNA transfer and replication"). The *tra* genes encoding the Dtr component tend to cluster around the *oriT* site. The bulk of the *tra* genes encode proteins of the **Mpf component** (for "mating-pair formation"). This large membrane-associated structure includes the pilus that is responsible for holding the mating cells together and the channel between the mating cells through which the plasmid is transferred. Below, we outline what is known of these two components.

THE Dtr COMPONENT

The Dtr (or DNA-processing) component of a self-transmissible plasmid is involved in preparing the plasmid DNA for transfer. There are a number of proteins which make up this component and the function of many of these is known.

Relaxase

A central part of the Dtr component of plasmids is the **relaxase.** This is a specific DNA endonuclease which makes a single-strand break or "nick" at the specific site called the *nic* site in the *oriT* sequence (see below) to initiate the transfer process. It also recyclizes the plasmid after transfer. The way in which the relaxase works is similar to the action of Rep protein in rolling-circle plasmid replication and is illustrated in Figure 5.3. The relaxase breaks a phosphodiester bond at the *nic* site by transferring the bond from a deoxynucleotide to one of its own tyrosines. Such a reaction is called a transesterification reaction and requires very little energy because there is no net breakage or formation of new chemical bonds. This transfer leaves the relaxase protein bound to the 5' end of the nick through its tyrosine, and the relaxase protein is transferred into the recipient cell along with the DNA. In fact, it is probably the relaxase protein itself which is transferred and the DNA just goes along for the ride (see Box 5.2).

Once in the recipient cell, the relaxase recyclizes the plasmid by doing essentially the reverse of what it had done in the donor cell. It binds to the two halves of the cleaved *oriT* sequence, holding them together while it transfers the phosphate bond from its tyrosine back to the 3' hydroxyl deoxynucleotide in the DNA (Figure 5.3). This transesterification reaction reseals the nick in the DNA and releases the relaxase, which has done its job and is degraded. The transferred plasmid is now a single-stranded circle in the recipient cell.

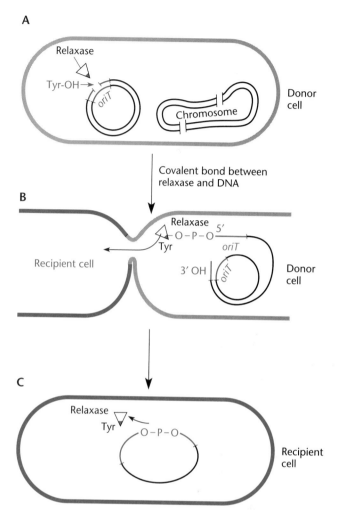

Figure 5.3 Reactions performed by the relaxase. (A) The relaxase nicks the DNA at a specific site in *oriT,* and the 5' phosphate is transferred to one of its tyrosines in a transesterification reaction. (B) The relaxase is transferred to the recipient cell, dragging the DNA along with it. (C) In a reversal of the original transesterification reaction, the phosphate is transferred back to the 3' hydroxyl of the other end of the DNA, recyclizing the DNA and releasing the relaxase.

Relaxosome

The relaxase protein in the donor cell is part of a larger structure called the **relaxosome,** which is made up of a number of proteins that are normally bound to the *oriT* sequence of the plasmid. The function of most of the proteins of the relaxosome is unclear. They might help the relaxase bind to the *oriT* sequence or help separate the strands at the *oriT* sequence to initiate transfer. They might also help in the communication with the coupling protein of the Mpf system, which tells the relaxase when to cut the plamid DNA at the *oriT* site (see below). In some plasmids, one of the other proteins of

BOX 5.2

Conjugation and Type IV Protein Secretion Systems

One of the more intriguing discoveries in modern cell biology is the extent to which systems developed for one purpose have been reassembled and adapted to serve other purposes. We have already seen examples of such molecular "battlebots" that are assembled from the parts of other molecular machines. One example is the relatedness between the syringe-like type III secretion systems, which secrete proteins directly into eukaryotic cells as part of the disease-causing process, and the bacterial flagellum systems, which help bacteria swim (see Box 2.6). Such discoveries have become almost routine since the development of computer technology for searching databases for related sequences. Databases, such as GenBank, that contain the sequences of hundreds of thousands of genes have been assembled over the years. Once you have sequenced a gene, you can ask the computer to tell which other genes in these databases have related sequences. When the sequences of the *tra* genes of many conjugation systems are compared to the sequences of the *vir* genes of type IV secretion systems, it is obvious that many of them have a common ancestry. Apparently, the basic machinery developed to transfer macromolecules from one cell to another has been adapted to many different specialized functions.

Conjugation and type IV protein secretion (see chapter 2) do have much in common. In conjugation, the Tra functions transfer DNA as well as some accompanying proteins from one bacterial cell to another cell, which can be either bacterial or, in some cases such as the Ti plasmid, a eukaryotic plant cell. Type IV protein secretion systems do something similar. They transfer proteins from a bacterial cell directly into a eukaryotic cell as part of the disease-causing mechanism. Both types of systems require pili or another adhesin to hold the cells together during the transfer and special membrane structures through which the macromolecules must pass. Both processes are very specific, and only some types of proteins or plasmid DNAs can be transferred. Nevertheless, it came as a surprise how closely related these two types of systems can be. In fact, they may simply be different manifestations of the same process.

The relatedness of type IV secretion to conjugation is dramatically illustrated by comparisons of the T-DNA transfer system of *Agrobacterium tumefaciens* to some type IV secretion systems. A diagram of the Mpf of the Ti plasmid system is shown in the Figure in Box 5.1. As discussed in Box 5.1, the T-DNA transfer system transfers the T-DNA part of a plasmid from bacterial to plant cells. It also transfers a protein that forms a channel in the plant membrane and the relaxase protein, which doubles as a protein that can target

the T-DNA to the plant nucleus, where it can enter the plant DNA. This transfer system shares features with other plasmid conjugation systems in that it encodes a pilus, relaxase, coupling proteins, and chaperones, as well as many other proteins involved in the transfer process. In fact, most of the proteins of the Tra systems of the pKM101 and R388 plasmids can be assigned homologues in the T-DNA transfer system. However, some pathogenic bacteria that transfer proteins into eukaryotic cells as part of the disease-causing process also have analogous functions. One of the most striking similarities is to the CagA toxin-secreting system of *Helicobacter pylori,* implicated in some types of gastric ulcers. This type IV toxin-secreting system shares at least six protein homologues with the T-DNA system of the Ti plasmid of *Agrobacterium*. The system delivers a toxin directly through the bacterial membranes and into the eukaryotic cell, where the toxin is phosphorylated on one of its tyrosines. In the phosphorylated state, the toxin causes many changes in the cell including alterations in its actin cytoskeleton. Another pathogenic bacterium, *Bordetella pertussis,* the causative agent of whooping cough, also has a type IV secretion system which shares nine homologous proteins with the T DNA system. This system secretes the pertussis toxin through the outer membrane of the bacterial cell. Once outside the cell, the pertussis toxin assembles into a form that can then enter the eukaryotic cell, where it can ADP-ribosylate G proteins, thereby interfering with signaling pathways and causing disease symptoms.

However, the most striking evidence that conjugation is related to type IV secretion has come from the virulence system of *Legionella pneumophila,* which causes Legionnaires disease. Like many pathogenic bacteria, this bacterium can multiply in macrophages, specialized white blood cells that are designed to kill them (see Vogel et al., below). These bacteria are taken up by the macrophage but then secrete proteins which disarm the phagosomes that have engulfed them. The components of this type IV secretion system are analogous to the Tra functions of some self-transmissible plasmids, and this type IV system can even mobilize the plasmid RSF1010 at a low frequency!

References

Christie, P. J. 2001. Type IV secretion: intercellular transfer of macromolecules by systems ancestrally related to conjugation machines. *Mol. Microbiol.* **40:**294–305.

Covacci, A., J. L. Telford, G. Del Giudice, J. Parsonnet, and R. Rappuoli. 1999. *Helicobacter pylori* virulence and genetic geography. *Science* **284:**1328–1333.

Vogel, J. P., H. L. Andrews, S. K. Wong, and R. R. Isberg. 1998. Conjugative transfer by the virulence system of *Legionella pneumophilia. Science* **279:**873–876.

the relaxosome may be the helicase, which helps separate the strands of DNA beginning at the *oriT* sequence, while in others, the helicase seems to be part of the relaxase protein itself. Whatever their function, the other proteins of the relaxosome are not transferred to the recipient, perhaps because the transferred DNA is already single-stranded and so their function is not needed.

Primase

Another component of the Dtr system made in the donor is the **primase.** Primases are needed for chromosomal DNA replication to make an RNA primer to prime the synthesis of the lagging strand of DNA replication (see chapter 1) and to prime plasmid replication (see chapter 4). However, at first the role a primase would play in the donor was not clear. A primase should not be necessary to prime replication in the donor cell since the free 3′ hydroxyl end of DNA created at the nick in *oriT* is the primer for replication during transfer, similar to the priming of the first stage of replication of rolling-circle plasmids (see chapter 4). Synthesis of the complementary strand of DNA in the recipient after DNA transfer should require synthesis of RNA primers, but the primase is made in the donor cell, not in the recipient cell.

One way the plasmid could escape the dilemma of the misplaced primase would be to transfer its own primase into the recipient cell along with the DNA. Clever experiments showed that at least some types of plasmids do just that (see Wilkins and Thomas, Suggested Reading). The researchers reasoned that transfer of the primase would have been difficult to detect biochemically since only a few molecules are transferred and not all of the cells are involved in the transfer. However, if the plasmid primase is transferred to the recipient cell, it might substitute for the host primase in replication in these cells. Therefore they used a recipient cell in which the primase gene has a temperature-sensitive (Ts) mutation (see chapter 3). When the mutant strain is raised to its nonpermissive temperature, chromosomal DNA replication stops for want of a primase. However, if the bacterium has just mated with a self-transmissible plasmid, replication continues in some of the cells, presumably using the newly transferred plasmid primase in lieu of its own inactive primase.

But why would a plasmid bother to make its own primase and transfer it into the recipient cell if it can use the host cell primase instead? The answer may be to make the plasmid more promiscuous and able to transfer into a wider variety of bacterial species. Sometimes a promiscuous plasmid may find it has transferred itself into a type of bacterium which is so distantly related to its original host that the primase in this bacterium does not recognize the sequences on the plasmid DNA necessary to prime the replication of the complementary strand. The plasmid faces a "Catch 22" because it cannot make its own primase in this cell (since the plasmid DNA that has been transferred is single stranded and unable to serve as a template for transcription) and it also cannot make the double-stranded DNA to use as template without its own primase. By transferring its own primase with the DNA, it avoids this problem.

In addition to primases, conjugation systems may secrete other proteins including the proteins that form a channel in the recipient cell membrane. This could explain the extreme promiscuity of some transmissible plasmids that can even transfer themselves into eukaryotic cells (see Box 5.1). Host proteins including RecA may also be transferred by some plasmids. However, it is important to realize that only certain proteins are transferred by a particular transfer system, and these are proteins that are recognized by the coupling protein so that they can "dock" on the channel and be transported (see below).

THE Mpf SYSTEM

The other *tra* genes of a self-transmissible plasmid encode proteins of the Mpf system. This system holds the donor and recipient cells together during the mating and forms the channel through which proteins and DNA are transferred during the mating. It also includes the protein which communicates news of mating-pair formation to the Dtr system, beginning the transfer of plasmid DNA. As mentioned, the Mpf structures involved in conjugal DNA transfer are remarkably similar to type IV protein secretion systems that transfer virulence proteins from pathogenic bacteria directly into eukaryotic cells. They are also related to the DNA uptake systems in naturally transformable bacteria (see Box 5.2 and chapter 6).

The Pilus

The most dramatic feature of the Mpf structure is the **pilus,** a tube-like structure that sticks out of the cell surface. The assembly of a pilus is shown in Figure 5.4. These pili are 10 nm or more in diameter with a central channel. Each pilus is constructed of many copies of a single protein called the **pilin** protein. This protein is synthesized with a long signal sequence that is removed as it passes through the membrane to assemble on the cell surface. The pilin protein is also cyclized, with its head attached to its tail, which is unusual among proteins (see Eisenbrandt et al., Suggested Reading).

The structure of the pilus differs markedly among plasmid transfer systems. For example, the F plasmid encodes a long, thin, flexible pilus; the pKM101 plasmid makes a long, rigid pilus; and IncP plasmids such as RP4 make a short, thick, rigid pilus. The structure of the pilus of an Mpf system can determine the efficiency of transfer

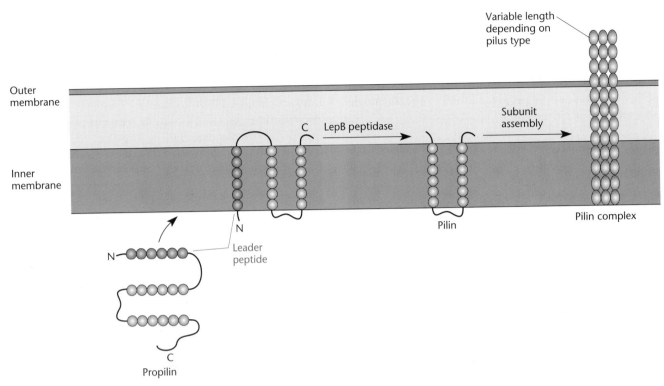

Figure 5.4 Assembly of the pilus on the cell surface. The propilin is processed by LepB peptidase as it passes through the membrane, and then it assembles between the inner and outer membranes (details are given in text).

under various conditions. For example, the long, flexible pilus of the F plasmid allows transfer in liquid medium while the cells are suspended in broth, while the thick, short, rigid pilus of RP4 allows efficient transfer only when the mating cells are fixed to a solid surface like a membrane, where they are less free to move. A long, flexible pilus may facilitate mating in liquid by helping bring more widely dispersed cells together, while a short, rigid pilus may be able to hold mating cells together only when they are concentrated on a solid surface. To make them more versatile, IncI plasmids such as Col1B-P9 make two pili, a long thin one and a short rigid one, the former increasing the frequency of mating in liquid medium and the latter increasing the frequency of mating on a solid surface.

Even though the male-cell pilus was observed a long time ago, the function of pili in conjugation is still unclear. They may only hold cells together during mating, or they may play a more direct role in the DNA transfer. What is clear, however, is that the early assumption, doubtless inspired by anthropomorphic considerations, that DNA passes through the pilus during the mating seems not to be true.

The Channel

In addition to a pilus, the Mpf system encodes a channel or pore through which DNA passes during conjugation.

Some of the *tra* gene products making up this pore are known, but the pore itself has so far escaped detection, and so little is known of its exact structure (see Samuels et al., Suggested Reading, and the Figure in Box 5.1).

Coupling Proteins

The Mpf component is the first to make contact with a recipient cell. However, this information must be communicated to the Dtr component for DNA transfer to occur. The communication between the Dtr and Mpf systems is provided by proteins called **coupling proteins,** which are part of the Mpf system. These coupling proteins provide the specificity for the transport process, so that only some proteins will be transferred (see Hamilton et al., Suggested Reading). The coupling protein is bound to the membrane channel (Figure 5.2) and specifically recognizes the relaxase of the Dtr components of certain plasmids as well as any other proteins to be transferred. Information that the Mpf system has encountered a recipient cell is somehow communicated to the coupling protein, which in turn activates the relaxase to nick the plasmid DNA to initiate the transport process. Coupling proteins are sometimes called docking proteins because they bind to or "dock" proteins on the membrane channel that are to be transported. In order to be "docked," a protein must contain certain amino acid sequences that identify it as a protein to be transported by the system.

The *oriT* Sequence

The *oriT* site is not only the site at which plasmid transfer initiates but also the site at which the DNA ends rejoin to recyclize the plasmid after transfer. Plasmid transfer initiates specifically at the *oriT* site because the specific relaxase encoded by one of the *tra* genes will cut DNA only at this sequence. Also, presumably, the plasmid-encoded helicase will enter DNA only at this sequence to separate the strands. Moreover, after transfer, the two ends of the DNA are probably held together at the *oriT* sequence so that they can be religated. Therefore, to be transferred, the plasmid must have this specific *oriT* sequence. In fact, a self-transmissible plasmid will mobilize any DNA that contains its *oriT* sequence, as we discuss below.

The essential features of *oriT* sequences are currently being investigated. The *oriT* sequence of the F plasmid is known to be less than 300 bp long and contains inverted repeated sequences and a region rich in AT base pairs. The importance of these sequences for *oriT* function is under investigation.

MALE-SPECIFIC PHAGES

Some types of phages can only infect cells that express a certain type of pilus on their cell surface. All phages adsorb to specific sites on the cell surface to initiate infection (see chapter 7), and the these phages use the pilus of the self-transmissible plasmid as their adsorbtion site. Phages that adsorb to the sex pilus of a self-transmissible plasmid are called **male-specific phages** because they infect only donor or "male" cells capable of DNA transfer. Examples of male-specific phages are M13 and R17, which infect only cells carrying the F plasmid, and Pf3 and PRR1, which infect only cells containing an IncP plasmid such as RP4.

Because male-specfic phages infect only cells expressing a pilus, mutations in any *tra* gene required for pilus assembly will prevent infection by the phage. This offers a convenient way to determine which of the *tra* genes of a plasmid encode proteins required to express a pilus on the cell surface and which *tra* genes encode other functions required for DNA transfer. To apply this test to a particular *tra* gene, the phages are used to infect cells containing the plasmid with a mutation in the *tra* gene. If the phage can multiply in the host cell, the *tra* gene which has been mutated must not be one of those that encode a protein required for pilus expression.

Incidentally, the susceptibility of pilus-expressing cells to some phages may explain why the *tra* genes of plasmids are usually tightly regulated. Most self-transmissible plasmids express a pilus only immediately after entering a cell and then only intermittently thereafter (see "Regulation of *tra* genes in IncF plasmids" below). If cells containing the plasmid always expressed the pilus, a male-specific phage could spread quickly through the population, destroying many of the cells and, with them, the plasmid they contain. By only intermittently expressing a pilus, cells containing a self-transmissible plasmid limit their susceptibility to phages that use their pilus as an adsorbtion site.

Efficiency of Transfer

One of the striking features of many transfer systems is their efficiency. Under optimal conditions, some plasmids can transfer themselves into other cells in almost 100% of cell contacts. This high efficiency has been exploited in the development of methods for transferring cloned genes between bacteria and in transposon mutagenesis, both of which require highly efficient transfer of DNA. We shall discuss such methods in subsequent chapters.

REGULATION OF THE *tra* GENES

Many naturally occurring plasmids transfer with a high efficiency for only a short time after they are introduced into cells and then transfer only sporadically thereafter. The rest of the time, the *tra* genes are repressed, and without the synthesis of pilin and other *tra* gene functions, the pilus is lost. For unknown reasons, the repression is relieved occasionally in some of the cells, allowing this small percentage of cells to transfer their plasmid at a given time.

As mentioned, plasmids may normally repress their *tra* genes to prevent infection by some types of phages. The pilus serves as the adsorption site for some phages. If all the cells in a population had a pilus all the time, such a phage could multiply quickly, infecting and killing all the bacteria carrying the plasmid.

This property of only periodically expressing their *tra* genes probably does not prevent the plasmids from spreading quickly through a population of bacteria that does not contain them. When a plasmid-containing population of cells encounters a population that does not contain the plasmid, the plasmid *tra* genes in one of the plasmid-containing cells will eventually be expressed and the plasmid will transfer to another cell. Then when the plasmid first enters a new cell, efficient expression of the *tra* genes leads to a cascade of plasmid transfer from one cell to another. As a result, the plasmid will eventually occupy most of the cells in the population.

AN EXAMPLE: REGULATION OF *tra* GENES IN IncF PLASMIDS

Regulation of the *tra* genes of IncF plasmids has been studied more extensively than that of other types. This regulation is illustrated in Figure 5.5. Transfer of these

A Genetic organization of *tra* region

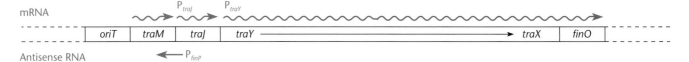

B Immediately after entry into cell

C After plasmid establishment

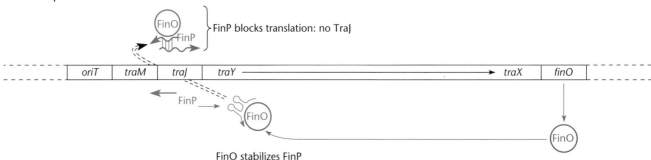

FinO stabilizes FinP

Figure 5.5 Fertility inhibition of the F plasmid. Only the relevant *tra* genes discussed in the text are shown. (A) Genetic organization of the *tra* region. (B) The *traJ* gene product is a transcriptional activator that is required for transcription of the other *tra* genes, Y-X, and *finO* from promoter P*traY*. (C) The translation of the *traJ* mRNA is blocked by hybridization of an antisense RNA, FinP, which is transcribed in the same region from the complementary strand. A protein, FinO, stabilizes the FinP RNA. Details are given in text.

plasmids depends on TraJ, a transcriptional activator. A **transcriptional activator** is a protein required for initiation of RNA synthesis at a particular promoter (see chapter 2). If TraJ were always made, the other *tra* gene products would always be made and the cell would always have a pilus. However, the translation of TraJ is normally blocked by the concerted action of the products of two plasmid genes, *finP* and *finO*, which encode an RNA and a protein, respectively. The FinP RNA is an antisense RNA that is transcribed constitutively from a promoter within and in opposite orientation to the *traJ* gene. Complementary pairing of the FinP RNA and the *traJ* transcript prevents translation of TraJ. The FinO protein stabilizes the FinP antisense RNA. When the plasmid first enters a cell, neither FinP RNA nor FinO protein is present, so TraJ and the other *tra* gene products are made. Consequently, a pilus appears on the cell, and the plasmid can be transferred. After the plasmid has become established in the double-stranded state, however, the FinP RNA and FinO protein will be synthesized so that the

plasmid can no longer transfer. The *tra* genes will later be expressed only intermittently.

The F plasmid was the first transmissible plasmid discovered (Figure 5.6; Table 5.1), and its discovery may have resulted from a happy coincidence involving its *finO* gene. Because of an insertion mutation in this gene (IS3 in Figure 5.6), the F plasmid is itself a mutant that always expresses the *tra* genes. Consequently, a sex pilus almost always extends from the surface of cells harboring this F plasmid, and the F plasmid can always transfer, provided that recipient cells are available, increasing the efficiency of transfer and facilitating the discovery of conjugation. Mutations that increase the efficiency of plasmid transfer, thereby increasing their usefulness in gene cloning and other applications, have been isolated in other commonly used transfer systems.

Mobilizable Plasmids

Some plasmids are not self-transmissible but can be transferred by another self-transmissible plasmid sharing

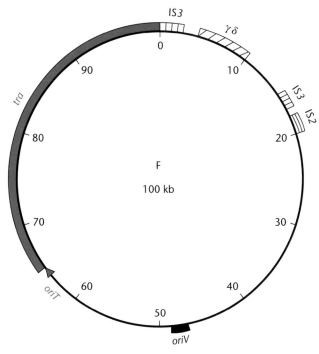

Figure 5.6 Partial genetic and physical map of the 100-kbp self-transmissible plasmid F. The regions IS3 and IS2 are insertion sequences; γδ is also known as transposon Tn1000. oriV is the origin of replication; oriT is the origin of conjugal transfer; the *tra* region encodes numerous *tra* functions. Adapted from N. Willetts and R. Skurray, Structure and function of the F factor and mechanism of conjugation, p. 1110–1133, *in* F. C. Neidhardt, J. L. Ingraham, K. B. Low, B. Magasanik, M. Schaechter, and H. E. Umbarger (ed.), *Escherichia coli and Salmonella typhimurium: Cellular and Molecular Biology,* American Society for Microbiology, Washington, D.C., 1987.

TABLE 5.1	Some F-plasmid genes and sites
Symbol	**Function**
ccdAB	Inhibition of host cell division
incBCE	Incompatibility
oriT	Site of initiation of conjugal DNA transfer
oriV	Origin of bidirectional replication
parABCL	Partitioning
traABCEFGHKLQUVW	Pilus biosynthesis/assembly
traGN	Mating-pair stabilization
traI	Relaxase
traY	Accessory for nickase
traJ, finOP	Regulation of transfer
traST	Entry exclusion

DNA insert including the *oriT* sequence. Transposons and plasmid cloning vectors containing the *oriT* site of a self-transmissible plasmid have many applications in molecular genetics because they can be mobilized into other cells.

While we can construct such a plasmid and they are mobilizable, minimal mobilizable plasmids containing only the *oriT* site of a self-transmissible plasmid do not seem to occur naturally. All mobilizable plasmids isolated so far encode their own Dtr systems, including their own relaxase and helicase. For historical reasons, the *tra* genes of the Dtr system of mobilizable plasmids are called the ***mob* genes** and the region required for mobilization is called the ***mob* region.** The function of the *mob* gene products of mobilizable plasmids seems to be to expand the range of self-transmissible plasmids by which they can be mobilized (see below). A plasmid containing only the *oriT* sequence of a self-transmissible plasmid can be mobilized only by the *tra* system of that self-transmissible plasmid and not by that of other self-transmissible plasmids which do not share the same *oriT* site, while naturally occurring mobilizable plasmids can often be mobilized by a number of *tra* systems.

The process of mobilization of a plasmid by a self-transmissible plasmid is illustrated in Figure 5.8. The process is identical to the transfer of a self-transmissible plasmid, except that the Mpf system of the self-transmissible plasmid is acting not only on its own Dtr system but also on the Dtr system of the mobilizable plasmid. The self-transmissible plasmid forms a mating bridge with a recipient cell and communicates this information via its coupling protein not only to its own relaxase but also to the relaxase of a mobilizable plasmid that happens to be in the same cell. The relaxase of the mobilizable plasmid then makes a single-stranded break at its *oriT* site, and the helicase separates the strands. The relaxase remains bound to the 5′ end of the single-stranded DNA and is

the same cell. Plasmids that cannot transfer themselves but can be transferred by other plasmids are said to be **mobilizable,** and the process by which they are transferred is called **mobilization.** The simplest mobilizable plasmids merely contain the *oriT* sequence of a self-transmissible plasmid, since any plasmid that contains the *oriT* sequence of a self-transmissible plasmid can be mobilized by that plasmid. Expressed in genetic terminology, the Mpf and Dtr systems of the self-transmissible plasmid can act in *trans* on the *cis*-acting *oriT* site of the plasmid and mobilize it.

The ability to allow mobilization is the mechanism by which *oriT* sequences are first located in a plasmid (Figure 5.7). Random clones of the DNA of the self-transmissible plasmid are introduced into a nonmobilizable cloning vector, and the mixture is introduced into a cell containing the self-transmissible plasmid. Any vector plasmids that are mobilized into recipient cells probably contain a

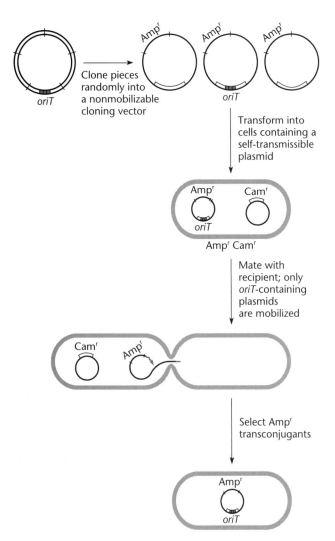

Figure 5.7 Identifying the *oriT* site on a plasmid. Pieces of the plasmid are cloned randomly into a nonmobilizable cloning vector. The mixture is transformed into cells containing the self-transmissible plasmid and mixed with a proper recipient. Pieces of DNA that allow the cloning vector to be mobilized contain the *oriT* site of the plasmid.

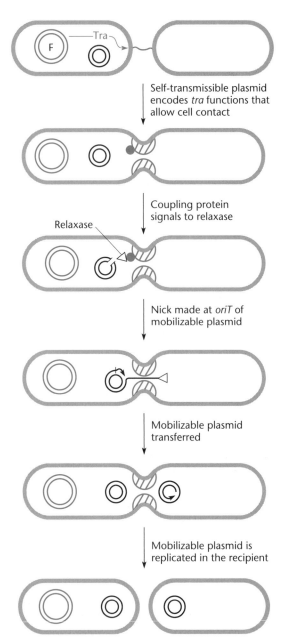

Figure 5.8 Mechanism of plasmid mobilization. The donor cell carries two plasmids, a self-transmissible plasmid, F, which encodes the *tra* functions that promote cell contact and plasmid transfer, and a mobilizable plasmid (gold). The *mob* functions encoded by the mobilizable plasmid make a single-stranded nick at *oriT* in the *mob* region. Transfer and replication of the mobilizable plasmid then occur. The self-transmissible plasmid may also transfer (details are given in text).

transferred into the recipient cell, dragging the single-stranded DNA of the mobilizable plasmid with it. The self-transmissible plasmid is also often transferred at the same time that it mobilizes other plasmids.

The secret of being mobilized by another plasmid is to be recognized by the coupling protein of the other plasmid (see Hamilton et al., Suggested Reading). Any plasmid encoding its own Dtr system can be mobilized by a coresident self-transmissible plasmid, provided that its relaxase can communicate with the coupling protein of the Mpf system of the coresident plasmid. Accordingly, the relaxases of mobilizable plasmids are designed to communicate with a broader range of coupling proteins of self-transmissible plasmids so that they can take advantage

of a number of different Mpf systems, unlike the relaxases of self-transmissible plasmids, which seem to be more specific. This suggests that mobilizable plasmids are designed to be parasitic for transfer on self-transmissible plasmids rather than just being erstwhile self-transmissible plasmids that have lost their own Mpf system. Moreover, some mobilizable plasmids overlap the functions of replication and mobilization so that they can use the same helicase and primase for both processes and their *oriV* site is often placed close to their *oriT* site, again unlike self-transmissible plasmids.

In spite of their versatility, however, not all mobilizable plasmids can be mobilized by all self-transmissible plasmids. For example, the IncQ plasmid RSF1010 can be mobilized by the IncP plasmid RP4 but not by the IncF plasmid F. Apparently, the coupling protein of RP4 can communicate with the relaxase of RSF1010 while the coupling protein of the F plasmid cannot. However, the F plasmid can mobilize the ColE1 plasmid, and so its coupling protein can communicate with the relaxase of this mobilizable plasmid. This complicated interplay between the Mpf and Dtr systems is particularly dramatic for the Ti plasmid and RSF1010. The *tra* system of the Ti plasmid will not mobilize the RSF1010 plasmid into other bacteria, but the *vir* system of the Ti plasmid will mobilize RSF1010 into plants (see Buchanan-Wollasten et al., Box 5.1).

PLASMID MOBILIZATION IN BIOTECHNOLOGY

Mobilization plays an important role in biotechnology because it can be used to efficiently introduce foreign DNA into bacteria. A *mob* site is often introduced into cloning vectors so that they can be efficiently transferred into cells (see Bagdasarian et al., Suggested Reading). Smaller is better in cloning vectors (see chapter 4), and a mobilizable plasmid can be much smaller than a self-transmissible plasmid because it does not need the 15 or so Mpf genes required to assemble the mating bridge, only the 4 or so Dtr genes required for DNA processing. Once foreign DNA has been cloned into such a cloning vector, it can be introduced into even distantly related bacteria by the Mpf system of a larger, promiscuous, self-transmissible plasmid. In addition, some self-transmissible plasmids have been crippled so that they cannot transfer themselves but can transfer only mobilizable plasmids. Then the recipient cell will receive only the mobilized cloning vector in such a transfer and not the self-transmissible plasmid which mobilized it.

A common application of plasmid mobilization technology is in transposon mutagenesis. These methods are most highly developed for gram-negative bacteria. Some plasmids such as RP4 are so promiscuous that they can transfer themselves into essentially any gram-negative bacterium. If such a plasmid is used to introduce a small mobilizable plasmid which has a narrow-host-range origin of replication like the ColE1 origin (see chapter 4), the smaller plasmid will be mobilized into the bacterium but will probably be unable to replicate there and will eventually be lost (e.g., will be a suicide vector). If the smaller plasmid also contains a transposon such as Tn5,

containing the selectable kanamycin resistance gene and with a broad host range for transposition, the only way the recipient cell can become resistant to kanamycin is if the transposon hops into the chromosome of the recipient strain, causing random insertion mutations. The transposon insertion mutants facilitate cloning and can even be used for Hfr mapping (see below). We shall return to such methods of transposon mutagenesis in chapter 9.

However, mobilizable plasmids also present regulatory complications. To meet regulatory requirements or for other reasons, genetic engineers often have to prove that a plasmid containing recombinant genes that confer desirable properties on one bacterium will not be mobilized into another, unknown, bacterium, where the genes might be harmful. But how do we really know that a plasmid does not contain an *oriT* site that will be recognized by some set of Tra functions? Unfortunately, negative evidence of mobilization by all the known self-transmissible plasmids does not mean that a given plasmid cannot be mobilized by *some* plasmid.

TRIPARENTAL MATINGS

Mobilization of a plasmid into a recipient cell is often used for cloning, transposon mutagenesis, or other procedures. Mobilizable plasmids have an advantage over self-transmissible plasmids in their smaller size. Nevertheless, difficulties can be encountered before these plasmids can be mobilized. For example, the self-transmissible plasmid and the plasmid to be mobilized may be members of the same Inc group and so will not stably coexist in the same cell. Also, the self-transmissible plasmid may express its *tra* genes only for a short time after entering a recipient cell.

Triparental matings help overcome some of the barriers to efficient plasmid mobilization. Figure 5.9 illustrates the general method. As the name implies, three bacterial strains participate in the mating mixture. The first strain contains a self-transmissible plasmid, the second contains the plasmid to be mobilized, and the third is the eventual recipient. After the cells are mixed, some of the self-transmissible plasmids in the first strain will transfer into the strain carrying the plasmid to be mobilized. The smaller plasmid will then immediately be mobilized into the final recipient strain, where it is needed for our cloning application. Even if the two plasmids are members of the same Inc group, they will coexist long enough for the mobilization to occur.

Chromosome Transfer by Plasmids

Usually during conjugation, only a plasmid is transferred to another cell. However, plasmids sometimes transfer the chromosomal DNA of their bacterial host, a fact

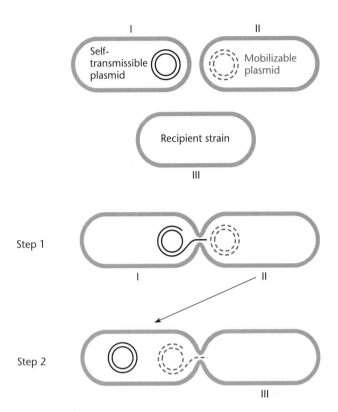

Figure 5.9 Triparental matings. In step 1, a self-transmissible plasmid from parent I transfers into parent II. In step 2, the self-transmissible plasmid transfers the mobilizable plasmid into parent III. This method will work even if the self-transmissible plasmid and the mobilizable plasmid are members of the same Inc group (see the text) and if the self-transmissible plasmid cannot replicate in parent II.

which has been put to good use in bacterial genetics. Without the transfer of genes, bacterial genetics is not possible, and conjugation is one of only three ways in which chromosomal and plasmid genes can be exchanged among bacteria (transduction and transformation are the others).

Formation of Hfr Strains

Sometimes plasmids integrate into chromosomes, and when such plasmids attempt to transfer, they take the chromosome with them. Bacteria that can transfer their chromosome because of an integrated plasmid are called **Hfr strains**, where Hfr stands for *high-frequency recombination*. As we discuss, the name derives from the fact that many recombinants can appear when such a strain is mixed with another strain of the same bacterium.

The integration of plasmids into the chromosome can occur by several different mechanisms, including recombination between sequences on the plasmid and sequences on the chromosome. For normal recombination

to occur, the two DNAs must share a sequence (see chapter 10). Most plasmid sequences are unique to the plasmid, but sometimes the plasmid and the chromosome share an **insertion element** (**IS element;** see chapter 3). These small transposons often exist in several copies in the chromosome and may appear in plasmids (see chapter 9); recombination between these common sequences can result in integration of the plasmid.

Figure 5.10 shows how recombination between the IS2 element in the F plasmid and an IS2 in the chromosome of *E. coli* can lead to integration of the F plasmid. Once integrated, the F plasmid will be bracketed by two copies of the IS2 element. This bacterium is now an Hfr strain. The *E. coli* chromosome contains 20 sites for IS-mediated Hfr formation by the F plasmid.

Transposition can also lead to the integration of plasmids into the chromosome. Plasmids often carry transposons, and integration of a plasmid through transposition can occur even if there are no sequences common to both the plasmid and the chromosome. We defer a discussion of transposons and transposition until chapter 9.

Transfer of Chromosomal DNA by Integrated Plasmids

We mentioned at the start of the chapter that self-transmissible plasmids were first detected in 1947 by Joshua Lederberg and Edward Tatum. These scientists observed recombinant types after mixing some strains of *E. coli* with other strains. **Recombinant types** differ from the

Figure 5.10 Integration of the F plasmid by homologous recombination between IS2 elements in the plasmid and in the chromosome, forming an Hfr cell. Integration can also occur through recombination of the IS3 or γδ sequences on the F plasmid (see Figure 5.6).

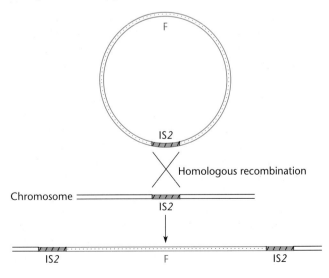

two original, or parental, strains and in this case resulted from the transfer of chromsomal DNA from one strain to another by an F plasmid integrated into the chromosome. In retrospect, it was fortuitous that the strains used by Lederberg and Tatum included some with an integrated F plasmid. Such strains are not common, and, as mentioned, the F plasmid is a mutant that is always ready to transfer, so that recombinant types are more likely to be produced. Also, recall that in 1947, plasmids had yet to be discovered.

Figure 5.11 illustrates the process by which chromosomal DNA is transferred in an Hfr strain. The integrated plasmid will express its *tra* genes, and a pilus will be synthesized. Upon contact with a recipient cell, the DNA will be nicked at the *oriT* site in the integrated plasmid, and one strand will be displaced into the recipient cell. Now, however, after transfer of the portion of the *oriT* sequence and plasmid on one side of the nick, chromosomal DNA will also be transferred into the new cell. If the transfer continues long enough (in *E. coli* approximately 100 min at 37°C), the entire bacterial chromosome will eventually be transferred, ending with the remaining plasmid *oriT* sequences. However, transfer of the entire chromosome (and thus the whole integrated plasmid) is rare, perhaps because the union between the cells is frequently broken, or because the DNA is often broken during conjugation.

FORMATION OF RECOMBINANT TYPES

Because the entire chromosome is seldom transferred, the entire plasmid is usually not transferred, and the DNA will not be able to recyclize after it enters the recipient cell. Hence, the transferred DNA will be lost unless it recombines with the chromosome in the recipient cell. In fact, it would not be possible to tell that an Hfr transfer had occurred unless the transferred chromosomal DNA recombines with the recipient cell chromosome. If this happens and if the donor and recipient bacteria are different in their genotypes, recombinant types might arise that can be identified because they are different from both the Hfr donor bacteria and the recipient bacteria.

In the example shown in Figure 5.12, the donor Hfr strain has an *arg* mutation and therefore will not form colonies on agar plates containing all the growth supplements except arginine, while the recipient strain has a *trp* mutation and therefore will not form colonies on agar plates lacking tryptophan. When the two are mixed, the Hfr strain can transfer DNA into the recipient cell, and sometimes this DNA will replace the recipient DNA at the *trp* allele, replacing it with the wild-type *trp*⁺ allele, as shown. The recipient bacteria will then require neither arginine nor tryptophan and will multiply to form colonies on minimal plates containing neither of these

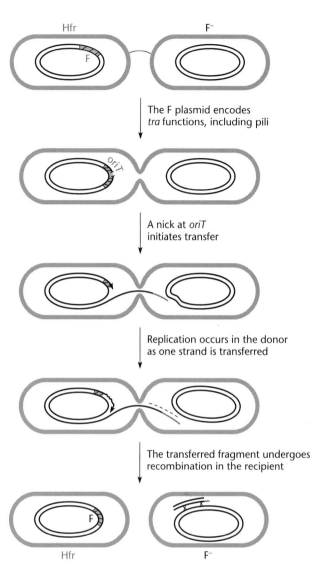

Figure 5.11 Transfer of chromosomal DNA by an integrated plasmid. Formation of mating pairs, nicking of the F *oriT* sequence, and transfer of the 5′ end of a single strand of F DNA proceed as in transfer of the F plasmid. Transfer of the covalently linked chromosomal DNA will also occur as long as the mating pair is stable. Complete chromosome transfer rarely occurs, and so the recipient cell remains F⁻, even after mating. Replication in the donor usually accompanies DNA transfer. Some replication of the transferred single strand may also occur. Once in the recipient cell, the transferred DNA may recombine with homologous sequences in the recipient chromosome.

growth supplements. These Trp⁺ Arg⁺ bacteria are recombinant types because they are genetically unlike either parent. This is basically the experiment that allowed Lederberg and Tatum to discover conjugation. They mixed different strains of *E. coli* that required different growth supplements and showed that some

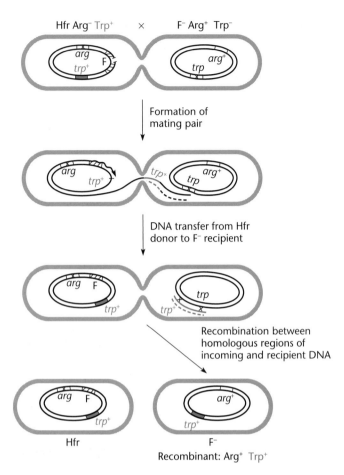

Hfr Arg⁻ Trp⁺   ×   F⁻ Arg⁺ Trp⁻

Formation of
mating pair

DNA transfer from Hfr
donor to F⁻ recipient

Recombination between
homologous regions of
incoming and recipient DNA

Hfr

F⁻
Recombinant: Arg⁺ Trp⁺

Figure 5.12 Formation of recombinant types after DNA transfer by an Hfr strain. If the *trp* region is transferred from the donor to the recipient, it can recombine to replace the homologous region in the recipient, giving rise to a Trp⁺ Arg⁺ recombinant.

mixtures gave rise to recombinant types with neither growth requirement.

As mentioned above, the donor chromosome is almost never transferred in its entirety to the recipient cell during an Hfr mating. Because of premature interruption of mating, genes that are transferred first are transferred at a much higher frequency than genes on the opposite side of the nick. Consequently, the genes exhibit a **gradient of transfer** from one side of the integrated plasmid around the bacterial genome. This gradient of transfer can be used for genetic mapping in bacteria and will be discussed next.

Genetic Mapping with Hfr Crosses

Hfr crosses offer one of the most convenient methods of genetic mapping in bacteria. However, because of the unusual nature of the genetic exchange, data are interpreted differently from the data for other types of genetic crosses. Because this is the first time we have discussed

the details of genetic mapping in bacteria, we'll introduce some important words and concepts here.

Genetic Markers

The DNAs of two individuals of the same species are usually almost identical in sequence over their entire length. However, sequence differences sometimes occur at specific sites in the DNA. These differences could occur if one of the individuals carries a mutation or has a transposon or another DNA element inserted in its DNA. Differences in sequence could also occur because of natural strain variation. If these differences in sequence cause phenotypic differences in the two individuals, the position of the difference in the DNA sequence can be mapped genetically. A DNA sequence variation that can be used for genetic mapping is called a **genetic marker.** What is being mapped is the position on the DNA of the species where a difference in DNA sequence can cause a specific phenotype.

DONOR AND RECIPIENT
In crosses with most organisms, viruses and humans included, both parents participate equally in the cross. The progeny organisms of the cross are as likely to have DNA from one parent as the other. However, in crosses between two bacteria, whether due to transformation, transduction, or conjugation, only part of the DNA of one bacterium, the **donor,** is transferred to the other bacterium, the **recipient.** Consequently, in a bacterial cross, only the recipient can become a recombinant, and in a recombinant type, one or more of the recipient sequences has been replaced by the corresponding donor sequence.

As an example, take the cross shown in Figure 5.12. The recipient in this cross has a *trp* mutation, making it require tryptophan for growth (Trp⁻), and the donor has an *arg* mutation, making it require arginine (Arg⁻). After the cross, the Trp⁻ recipient bacteria may become recombinant for the *trp* marker. In the recombinant, the wild-type *trp⁺* allele of the donor has replaced the mutant *trp* allele of the recipient by a double crossover, as shown. A recombinant type can thus appear which is both Arg⁺ and Trp⁺ and so can grow without any growth supplements. In contrast, if the donor *arg* allele replaces the *arg⁺* allele of the recipient, the recombinant type will be Arg⁻ Trp⁻. Note again that only the recipient bacterium can become a recombinant, because the donor bacterium contributes only a piece of its DNA to the cross and the DNA does not go both ways. The recipient strain becomes recombinant for a particular marker when it obtains the allele of the donor.

SELECTED AND UNSELECTED MARKERS
Another useful concept in bacterial genetics is the concept of **selected markers** and **unselected markers.** Recombinants are usually rare in bacterial crosses, so they

must often first be isolated by positive selection, much like mutants are selected. We first select progeny that are recombinant for one of the markers and then test them to determine how many are also recombinant for other markers. The marker chosen to be the selected one is called the selected marker, and the other markers that are tested for afterward are the unselected markers.

Convenience dictates which marker will be the selected one and whether the wild-type or mutant phenotype will be selected. In general, we want to use a positive selection for the selected marker since it is much easier to detect markers this way. Returning to the example in Figure 5.12, it is easier to use the *trp* marker as our selected marker than the *arg* marker. Recombinants for the *trp* marker have the *trp*⁺ allele of the donor and so are Trp⁺ and can be selected merely by plating the cross on media lacking tryptophan, while recombinants for the *arg* marker have the *arg* mutant allele of the donor and so are Arg⁻ and can be detected only by a negative selection. Once the donor allele for one marker has been selected, the recombinants can be tested for the other, unselected markers. It makes less difference for the unselected markers which allele is in the donor and which is in the recipient, since recombinants for the selected marker must be individually tested for the unselected markers anyway. However, strains recombinant for the unselected markers are still those that received the allele of the donor.

Analysis of Hfr Crosses

Most methods of mapping by Hfr crosses depends on the fact that the DNA is transferred from the donor to the recipient in order, beginning with the DNA on one side and closest to the site of integration of the plasmid, continuing around the chromosome, and ending where the plasmid is integrated but on the other side. Therefore, it is possible to tell the order of markers in the chromosome by determining when particular markers enter the recipient cell during conjugation. This can be done either directly, by periodically disrupting the mating and seeing which markers have entered by that time, or indirectly, by taking advantage of the fact that the frequency of transfer of a marker drops off the farther the marker is from the origin of transfer (the gradient of transfer). We discuss only the gradient-of-transfer method here.

MAPPING BY THE GRADIENT OF TRANSFER
Genetic mapping by the **gradient of transfer** depends upon the fact that the entire donor chromosome is seldom transferred during conjugation. Because the chromosome is enormously long and the entire transfer takes more than an hour, the mating bridge usually breaks long before the transfer is complete. The breakage could be induced by constant bouncing around of the cells

TABLE 5.2	Typical results of an Hfr cross			
Selected marker	Percent recombinant for unselected markers			
	hisG	*trpA*	*argH*	*rif*
hisG		1	7	6
trpA	33		29	31
argH	28	12		89

owing to Brownian motion or to single-strand breaks in the DNA during the transfer. In any case, these periodic interruptions of mating lead to an exponential decay in the transfer frequency of markers as a function of their distance from the origin of transfer.

Table 5.2 lists some data for a typical Hfr cross, and Figures 5.13 through 5.15 illustrate how these data are analyzed. In this example, the marker with the unknown position is due to a mutation, *rif-8,* that confers resistance to the antibiotic rifampin. The Hfr strain used for the cross is PK191, which requires proline because of a small deletion including the *proC* gene Δ(lac-pro). It also has an integrated F plasmid at 42 min, which transfers the chromosome in the clockwise direction (Figure 5.14). The recipient strain has mutations that make it require histidine (*hisG1*), arginine (*argH5*), and tryptophan

Figure 5.13 Mapping by Hfr crosses. The phenotypes and positions of the mutations in the genetic maps of the donor and recipient bacteria are shown. The chromosome will be transferred from the donor to the recipient, starting at the position of the integrated self-transmissible plasmid (arrowhead). The plating media used to select the markers are also shown.

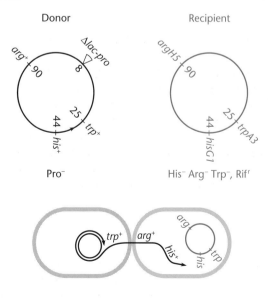

Plate: Arginine plus tryptophan: His⁺
Tryptophan plus histidine: Arg⁺
Arginine plus histidine: Trp⁺

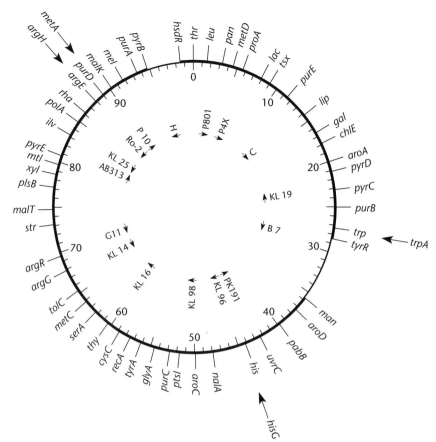

Figure 5.14 Partial genetic linkage map of *E. coli* showing the positions (large arrows) of the known markers used for the Hfr gradient of transfer in Figure 5.15. The small arrows indicate the position of integration of the F plasmid in some Hfr strains, including PK191 (located near the position of *hisG* at 44 min). In each of these Hfr strains, the chromosomal DNA will be transferred beginning from the tip of the arrow. Adapted from B. J. Bachmann, K. B. Low, and A. L. Taylor, *Bacteriol. Rev.* **40**:116–167, 1976.

(*trpA3*). It also has the *rif-8* mutation being mapped, so it is resistant to rifampin. The positions of all the markers except the *rif-8* marker are known (the *Escherichia coli* genetic map is shown in Figure 5.14).

The Hfr strain is mixed with the recipient strain, and the mixture is incubated for a sufficient time to permit transfer of the entire chromosome (more than 100 min for *E. coli* at 37°C). The mating mixture is then plated under conditions in which *neither* the donor *nor* the recipient can grow, *only* the recombinants being selected; otherwise the parent bacteria would grow up and cover the plates, making the detection of recombinants impossible. Plating under conditions where the donor cannot multiply is known as counterselecting the donor. In this case, the donor can be counterselected by omitting proline from the plates, since the donor requires proline for growth. To plate under conditions where only recipients that are recombinant for a particular marker can multiply, the selective plates should contain two of the three amino acids required by the recipient but lack the one that corresponds to the selected marker (e.g., if the selected marker is the region of the *hisG mutation*, the medium should lack histidine). Then only recipient bacteria that are recombinant for the *hisG* marker or His⁺

will be able to multiply and form a colony on the plates. The plates should also contain arginine and tryptophan and lack rifampin so that the His⁺ recombinants will form colonies whether or not they are also Arg⁺ or Arg⁻, Trp⁺ or Trp⁻, or Rif^r or Rif^s. Similarly, to select the region of the *argH* marker, the mating mixture is plated on minimal plates plus tryptophan and histidine. To select the *trpA* marker, it is plated on arginine plus histidine.

TESTING FOR UNSELECTED MARKERS

After transconjugants recombinant for one of the markers have been selected, they are tested to determine if they are also recombinant for one or more of the unselected markers. Table 5.2 shows some representative data in which one of the markers has been selected; the percentage of these recombinants that are also recombinant for each of the unselected markers is given. Remember that the recipient is recombinant for a marker when it has the allele of the donor. Thus, for example, the recombinants for the region of the *argH* marker are Arg⁺ but the recombinants for the region of the *rif-8* marker are Rif^s.

To map by gradient of transfer, the selected marker must come in before all the other markers. Any marker that comes in before the selected marker must have en-

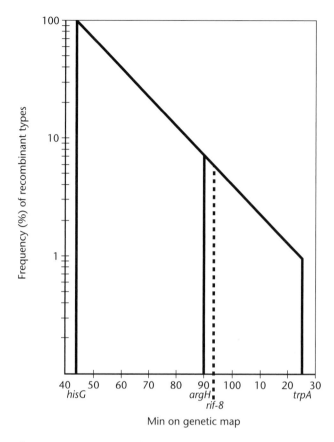

Figure 5.15 Mapping by gradient of transfer during an Hfr cross. The ordinate shows the frequency of each unselected marker with *his* as the selected marker. The abscissa is the distance in minutes from the selected marker. The dashed line shows an estimate of the position of *rif-8* based on the percentage of Rif^s recombinants from the data in Table 5.2.

should appear as a more or less straight line on semilog paper since, as mentioned, the frequency of transmission of markers falls off exponentially the farther they are from the selected marker. The frequency of recombinants of the unknown *rif* marker (Rif^s) is then placed on this line, reading down to determine what map position would give rise to this transmission frequency. These data place the site of the *rif-8* mutation at approximately 90 min, close to the *argH* marker.

The placement of the site of the *rif-8* mutation close to *argH5* is also supported by the results obtained when *argH5* was the selected marker. A very high percentage (89%) of the recombinants selected for being *Arg*$^+$ are also Rif^s and so are recombinant for the *rif* marker. Apparently, few crossovers occurred between the regions of the *argH* and *rif* markers when the *argH* region of the donor replaced the *argH* region of the recipient, indicating that the two markers are very closely linked. If markers are much farther apart than this, so many crossovers occur between the two markers that such genetic linkage is not apparent.

A CAVEAT

The interpretation of mapping data from Hfr crosses can be complicated if the marker being mapped is too close to the marker used to counterselect the donor. In the example, if the *rif-8* mutation were very close to the *proC* mutation of the Hfr donor, there would have been very few crossovers between the two mutations, so that most of the transconjugants recombinant for the *rif* marker will also be Pro$^-$ and so will not grow on the selective plates. Accordingly, to get a reliable map position for an unknown marker, it is best to counterselect the donor with a marker that comes in very late, so that it will not interfere with the frequency of recombinants for the other markers. Detailed protocols for use of Hfr strains for mapping are given in Low, Suggested Reading.

Chromosome Mobilization

The Tra functions of self-transmissible plasmids can also mobilize the chromosome, provided that the chromosome has the *oriT* sequence of the plasmid. Chromosome transfer will begin at the *oriT* sequence. This has allowed the mapping of genes by gradient of transfer in many genera of bacteria by introducing the *oriT* site of a plasmid into the chromosome on a transposon.

Prime Factors

Chromosomal genes can also be transferred when they are incorporated into plasmids. When such a plasmid, called a **prime factor**, transfers itself, it of course takes

tered the recipient cell in recombinants for the selected marker. Once they enter the recipient cell, such unselected markers will show about the same frequency of recombination into the chromosome (about 30% in *E. coli* K-12), so all the unselected markers that come in before the selected marker show a recombination frequency of about 30% and are of little use in mapping. To determine which marker will come in first, we refer to the *E. coli* genetic map shown in Figure 5.14. From the picture, we can see that PK191 transfers the *hisG* marker very early, probably before the unknown *rif* marker, and so we choose *hisG* as our selected marker. After selecting recombinants for the *hisG* marker, we determine what percentage of them are also recombinant for the other markers. We then construct a standard curve on semilog paper on which we plot the frequency of recombinants for each known unselected marker versus its distance in minutes from the *hisG* marker on the genetic map. Such a plot is shown in Figure 5.15. This standard curve

the chromosomal genes with it. Prime factors are usually designated by the name of the plasmid followed by a prime symbol, for example, F′ factor. An R-plasmid such as RK2 carrying bacterial chromosomal DNA is an R′ factor.

Creation of Prime Factors

Like Hfr strains, prime factors can be created through either transposition or homologous recombination. An illustration of the latter process appears in Figure 5.16. A prime factor forms from the chromosome of an Hfr strain, in which the plasmid is bracketed by two copies of an IS element (Figure 5.10). The repeated IS sequences make the Hfr chromosome somewhat unstable, and the plasmid sometimes excises by looping out as a result of

Figure 5.16   Creation of a prime factor by homologous recombination. Recombination may occur between homologous sequences, such as IS sequences, in the chromosomal DNA outside the F factor. The F factor will then contain chromosomal sequences, and the chromosome will carry a deletion.

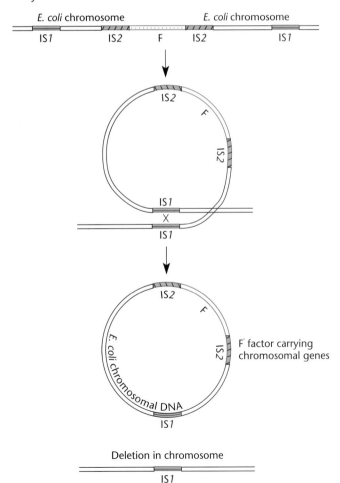

recombination between the flanking IS elements. However, sometimes the recombination occurs not between the IS elements immediately flanking the plasmid but between other DNA sequences repeated elsewhere in the chromosome, as shown in Figure 5.16. This excision will create a larger plasmid—the prime factor—that carries the chromosomal DNA that lay between the recombining DNA sequences. A prime factor can form from recombination between any repeated sequences, including identical IS elements or genes for rRNA, which often exist in more than one copy in bacteria.

Note that a deletion forms in the chromosome when the prime factor loops out. Some of the genes deleted from the chromosome may have been essential for the growth of the bacterium. Nevertheless, the cells will not die, because the prime factor still contains the essential genes, which should be passed on to daughter cells when the plasmid replicates. However, cells that lose the prime factor will die.

Prime factors can be very large, almost as large as the chromosome itself. In general, the larger a prime factor is, the less stable it is. Maintaining large prime factors in the laboratory requires selection procedures designed so that cells will die if they lose some or all of the prime factor. However, most prime factors are small enough to be transferred in their entirety. Because prime factors contain an entire self-transmissible plasmid, a cell receiving a prime factor becomes a donor and can transfer this DNA into other bacteria. Moreover, because prime factors are replicons with their own plasmid origin of replication, they can replicate in any new bacterium that falls within the plasmid host range (see chapter 4). These properties can be used to select cells containing prime factors, as we discuss below.

Complementation Tests Using Prime Factors

As we discussed in chapter 3, complementation tests can reveal whether two mutations are in the same or different genes and how many genes are represented by a collection of mutations that give the same phenotype. Complementation tests can also provide needed information about the type of mutation being studied, whether the mutations are dominant or recessive, and whether they affect a *trans*-acting function or a *cis*-acting site on the DNA. However, complementation tests require that two different alleles of the same gene be introduced into the same cell, and bacteria are normally haploid, with only one allele for each gene in the cell.

Because prime factors contain a region of the chromosome, they can be used to create cells that are stable diploids for the region they carry. However, they contain only a short region of the chromosome and so are diploid for only part of the chromosome. Organisms diploid for only a region of their chromosome are called **partial diploids** or **merodiploids**.

SELECTION OF PRIME FACTORS

To perform complementation tests with prime factors, we must first select a prime factor containing a particular region of the chromosome. Prime factors usually arise from Hfr strains by recombination between repeated flanking sequences in the chromosome (Figure 5.16), but such recombination events occur only infrequently. Somehow cells containing a prime factor must be selected from among the myriad of cells that are still Hfr, with the plasmid still integrated in the chromosome. We discuss two of these selection procedures here.

Selection of Prime Factors Based on the Early Transfer of Distal Markers

One way of selecting prime factors is based on their early transfer of distal markers. The selection depends on the fact that since the Hfr transfers the entire chromosome during a mating, the genes on one side of the integrated plasmid will be transferred very efficiently but the genes on the other side, called the distal markers, will be transferred only after a considerable delay and at a very low frequency. However, in some Hfr donor bacteria, a prime factor has excised from the chromosome, picking up chromosomal genes on both sides of where it was integrated in the Hfr strain (Figure 5.16). These prime factors will transfer markers on both sides of the integrated plasmid early. Therefore, most recombinants for the distal marker that appear early are probably due to the transfer of a prime factor rather than to Hfr transfer.

Figure 5.17 shows how we can take advantage of this property of prime factors to select transconjugants containing a prime factor carrying the *proA* region of the chromosome. In the first step, an Hfr strain which transfers the *proA* marker late because of where the plasmid is integrated and how it is oriented is crossed with a recipient strain, and Pro+ apparent recombinants for the *proA* marker are selected. If the mating is allowed to proceed for only a short time, most of the apparent recombinants that become Pro+ will have picked up a prime factor rather than being true recombinants for the marker. Instead of being true recombinants, they are partial diploids containing two copies of the *proA* gene, one of which is complementing the other, as shown. The partial diploid cells can be distinguished from true recombinants caused by an Hfr cross because the partial diploids contain the entire plasmid sequence and so will themselves be able to produce pili and will be sensitive to male-specific phages (see above). They will also transfer markers on the F′ factor, including the *proA* marker, into other bacteria with high efficiency.

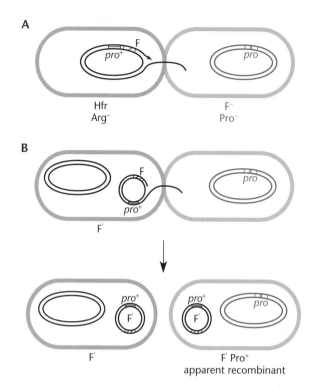

Figure 5.17 Selecting prime factors on the basis of early transfer of a marker. The hatched region indicates F DNA. In B, a prime factor transfers the *pro* marker early. See the text for details.

Selection of Prime Factors Because They Are Replicons

Another selection for prime factors is based on the fact that they are part of the plasmid replicon and so are capable of replication independent of the chromosome. In contrast, genes transferred during an Hfr mating do not contain the entire plasmid and so will be lost unless they recombine with the chromosome. This method is illustrated in Figure 5.18. In the example, an Hfr strain, with the plasmid integrated close to a *his* gene, is mated with a *his* mutant recipient strain made deficient in recombination by a *recA* mutation. Cells with a *rec* mutation are deficient in recombination because most pathways for recombination require the product of the *recA* gene, the RecA protein (see chapter 10). No His+ recombinants should be obtained in the cross with the Hfr strain because the incoming DNA can neither replicate nor recombine with the chromosome and so is lost. However, as in the example above, in a few of the donor cells the plasmid had excised to form a prime factor carrying the *his* marker. When this prime factor is transferred into the recipient cell, it can survive, being capable of independent replication. Its *his+* gene can complement the *his* mutation in the chromosome, giving rise to an apparent His+ recombinant. This second method of isolating prime factors

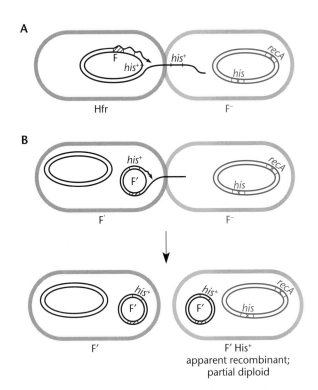

Figure 5.18 Selecting a prime factor by mating into a RecA⁻ recipient. In B, the prime factor is a replicon and so does not need to recombine with the chromosome to be maintained. Any apparent recombinants presumably contain a prime factor. The hatched region indicates F DNA.

is generally preferred because after transfer, the prime factors will be in *recA* mutant cells, where they are more stable and will not be destroyed by recombination with the chromosome. Of course, this latter method is applicable only if *recA* mutants are available for the type of bacteria being studied.

As mentioned, one problem with prime factors, especially if they are very large, is that they tend to be unstable and often suffer large deletions or the cells lose them periodically (i.e., are cured of the prime factor) as they multiply. Consequently, cells containing prime factors usually must be grown under conditions that select for the genes on the prime factor.

Once a prime factor carrying the region of interest has been isolated, it can be mated into other strains to obtain a partial diploid for complementation tests. This was how the complementation tests used to analyze the regulation of the *lac* operon (see chapter 12), as well as many other early genetic analyses, were performed. Now, however, cloning methods have largely replaced the use of prime factors for complementation tests.

Role of Prime Factors in Evolution

Prime factors formed with promiscuous, broad-host-range plasmids probably play an important role in bacte-

rial evolution. Once chromosomal genes are on a broad-host-range plasmid, they can be transferred into distantly related bacteria, where they will be maintained as part of the broad-host-range plasmid replicon. They may then become integrated into the chromosome through recombination or transposition. The similarities between some types of genes, even in distantly related bacteria, suggest that certain genes have been exchanged fairly recently in evolution, and prime factors may have been one of the mechanisms.

Transfer Systems of Gram-Positive Bacteria

Self-transmissible plasmids have also been found in many types of gram-positive bacteria, including species of *Bacillus*, *Streptococcus*, *Staphylococcus*, and *Streptomyces*. In some cases, these plasmids are known to transfer by systems which are similar to the gram-negative transfer systems discussed above. They are transferred complete with their own relaxases and *oriT* sequences. In fact, the *oriT* sequences of gram-positive plasmids are sometimes closely related to those of gram-negative bacteria (see Kurenbach et al., Suggested Reading). Other plasmids found in the gram-positive *Streptomyces* and presumably other related bacteria are very different and are discussed in Box 5.3.

The major differences between plasmids from the two bacterial groups come in the mating-pair formation (Mpf) systems, which can be simpler in gram-positive bacteria because of the lack of an outer membrane. We only discuss plasmids from *Enterococcus* in this section because of their interesting method of attracting mating cells and their medical importance (for a review, see Clewell, Suggested Reading).

Plasmid-Attracting Pheromones

Some strains of *E. faecalis* excrete pheromone-like compounds that can stimulate mating with donor cells. These pheromones are small peptides, each of which stimulates mating with cells containing a particular plasmid. The pheromone-like peptides act by specifically stimulating the expression of *tra* genes in the plasmids of neighboring bacteria, thereby inducing aggregation and mating. Once a cell has acquired a plasmid, it no longer excretes the specific peptide, but it continues to excrete other peptides that will stimulate mating with cells containing other plasmid types. By inducing their Tra functions only when a potential recipient is nearby, the plasmid saves energy and prevents the expression of surface antigens that may provide targets for the host immune system and receptors for male-specific phages. There are many families of transmissible plasmids in the enterococci. All of these plasmids share certain properties and the essential mechanisms for conjugation.

BOX 5.3

Conjugation in Streptomycetes

The mechanism for conjugation in the gram-positive bacteria known as streptomycetes differs greatly from conjugation mechanisms that we have discussed so far. For example, stable mating-pair formation requires numerous *tra* gene products in most bacteria. However, the streptomycete morphology and life cycle facilitate cell contact without plasmid-encoded genes. Commonly found in soil, streptomycetes grow from spores into branching, filamentous hyphae that form an intertwined network. Specialized hyphae within this network eventually differentiate to form haploid spores (see Figure).

As shown in the Figure, the growth of hyphae toward and around each other evidently provides stable contact between different parent strains, since no plasmid-encoded genes involved in establishing or maintaining cell contact have been discovered. Streptomycete transmissible plasmids also lack genes encoding the elaborate proteinaceous structures involved in DNA transfer across cell membranes and walls. Partial fusion of the hyphae may occur, creating an opportunity for DNA transfer.

The single plasmid-encoded *tra* gene product required for conjugation of the well-studied plasmid pIJ101 resembles gene products known to function in DNA partitioning in *E. coli* and *Bacillus subtilis* (see the discussion of *ftsK* in chapter 1). The precise role of this protein is unknown, but it could be speculated that it functions to move DNA between partially fused hyphae.

The efficiency of transfer of plasmid pIJ101 is extremely high—even 100% following the plating of a mixture of donor and recipient spores on agar plates. Moreover, dissemination of the plasmid throughout the recipient mycelium is also highly efficient. The streptomycete hyphae within a mycelial network have few cross walls, and cell compartments are multinucleoid. Both high- and low-copy-number transmissible plasmids are able to spread throughout mycelia and become incorporated in spores. A plasmid genetic locus named *spd* is responsible for efficient spread through hyphae, but its mechanism of action is poorly understood.

Another major difference between gene transfer in streptomycetes and most other bacteria is that chromosome transfer occurs at the same time as plasmid transfer even if the plasmid is not integrated. Also, recombination (and apparently chromosome transfer) is not limited to one parent. Rather, the genotypes of both parents can be changed through recombination, so that as many as 1% of the progeny genomes found in haploid spores can be recombinant. Usually, there is no covalent association of the plasmid with the chromosome; therefore, no gradient of transfer is created. However, if the

(continued)

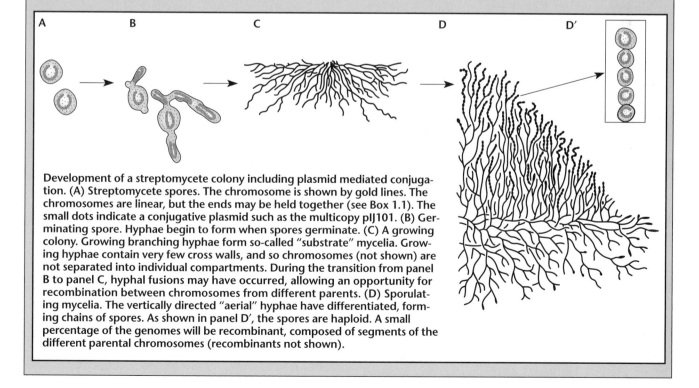

A   B   C   D   D′

Development of a streptomycete colony including plasmid mediated conjugation. (A) Streptomycete spores. The chromosome is shown by gold lines. The chromosomes are linear, but the ends may be held together (see Box 1.1). The small dots indicate a conjugative plasmid such as the multicopy pIJ101. (B) Germinating spore. Hyphae begin to form when spores germinate. (C) A growing colony. Growing branching hyphae form so-called "substrate" mycelia. Growing hyphae contain very few cross walls, and so chromosomes (not shown) are not separated into individual compartments. During the transition from panel B to panel C, hyphal fusions may have occurred, allowing an opportunity for recombination between chromosomes from different parents. (D) Sporulating mycelia. The vertically directed "aerial" hyphae have differentiated, forming chains of spores. As shown in panel D′, the spores are haploid. A small percentage of the genomes will be recombinant, composed of segments of the different parental chromosomes (recombinants not shown).

BOX 5.3 (continued)

Conjugation in Streptomycetes

conjugative plasmid contains chromosomal sequences, it can integrate through homologous recombination, and then a gradient of transfer is apparent.

The mechanisms of generation of recombinant chromosomes are clearly different in many ways from those in Hfr strains of gram-negative bacteria. It is not known whether transfer of either plasmid or chromosomal DNA requires nicking at an *oriT* sequence and subsequent transfer of a single strand from the donor parent. The *Streptomyces* plasmid pIJ101 does contain an *oriT*-like site, which is required for

plasmid transfer, but there is no evidence for nicking at this site. It is notable that this site is not required for generation of recombinant chromosomes in a mating culture.

References
Kieser, T., M. J. Bibb, M. J. Buttner, K. F. Chater, and D. A. Hopwood. 2000. *Practical Streptomyces Genetics.* Norwich, United Kingdom: The John Innes Foundation.

Pettis, G. S., and S. N. Cohen. 2000. Mutational analysis of the *tra* locus of the broad-host-range *Streptomyces* plasmid pIJ101. *J. Bacteriol.* **182:**4500–4504.

The mechanism of pheromone sensing of recipient cells is outlined in Figure 5.19. All of the genes that encode the peptide pheromones are located on the *Enterococcus* chromosome and are cut from the signal peptides of normal cell lipoproteins (see De Boever et al., Suggested Reading). After the signal sequences are cut from the lipoprotein being transported, the active pheromones are produced by proteolysis of the C-terminal 7 or 8 amino acids. Processing occurs as the pheromone is excreted from the cell. By cutting the pheromones from the signal sequences of many different lipoproteins, each cell is able to excrete a number of different pheromones and attract a number of different plasmids.

The pheromone-sensing mechanism is also conserved in the various plasmid families. Sensing of the pheromones by plasmid-containing potential donor cells requires specific proteins located on the cell surface of a plasmid-containing donor cell. The sensing proteins are encoded by the transmissible plasmids. Each type of transmissible plasmid expresses a protein specific for one type of pheromone, and the pheromone is named after

the plasmid it attracts. For example, in the well-studied plasmid pAD1, the plasmid-encoded pheromone-binding protein is called TraC and the pheromone is called cAD1. A given pheromone-binding protein which has bound a specific pheromone signals the cell's peptide uptake system, called the oligopeptide permease, to take up the peptide. Once inside the cell, the pheromone can induce the expression of plasmid genes involved in plasmid transfer, including aggregration substance. This protein coats the donor cell surface and initiates contact with the recipient cell. Following cell-cell contact, the plasmid transfers through a mating channel, much as in gram-negative plasmid transfer.

Once the plasmid has entered the recipient, the new transconjugant will no longer function as a recipient, to acquire additional plasmids of the same family. The limitation of plasmid uptake by a donor cell involves three mechanisms. One mechanism involves the expression of surface exclusion proteins much like the entry exclusion *eex* systems of gram-negative bacteria. The second mechanism involves the shutdown of pheromone sensing

Figure 5.19 Role of pheromones in plasmid transfer. (A) The recipient cell. The pheromone genes are located on the enterococcal chromosome; several examples are shown. The propheromone peptides are the signal sequences cut off of normal cellular proteins. The pheromones are processed from the propheromones when exported. (B) The donor cell. The plasmid-carrying cell expresses TraA, which represses transcription of the other *tra* genes except *traC*, which encodes a cell surface protein that can sense the pheromone. Also shown is TraB, which is discussed in panel E. (C) Mating induction. The pheromone binds TraC on the cell surface of a donor cell in close proximity and enters the cell via the oligopeptide permease system. The pheromone binds to the repressor TraA, releasing it from the DNA and derepressing the synthesis of TraE, which activates the expression of the *tra* genes including those encoding the aggregation substance. (D) Plasmid transfer. The donor cell establishes contact with the recipient cell, and the plasmid transfers, producing a transconjugant. (E) Pheromone shutdown in the transconjugant. Once the cell has become a transconjugant, the inhibitor peptide iAD1 binds to TraC and prevents autoinduction or pheromone stimulation of mating with other donor cells. Also, TraB is an inhibitor protein that somehow functions to shut down induction by preventing excretion of pheromone.

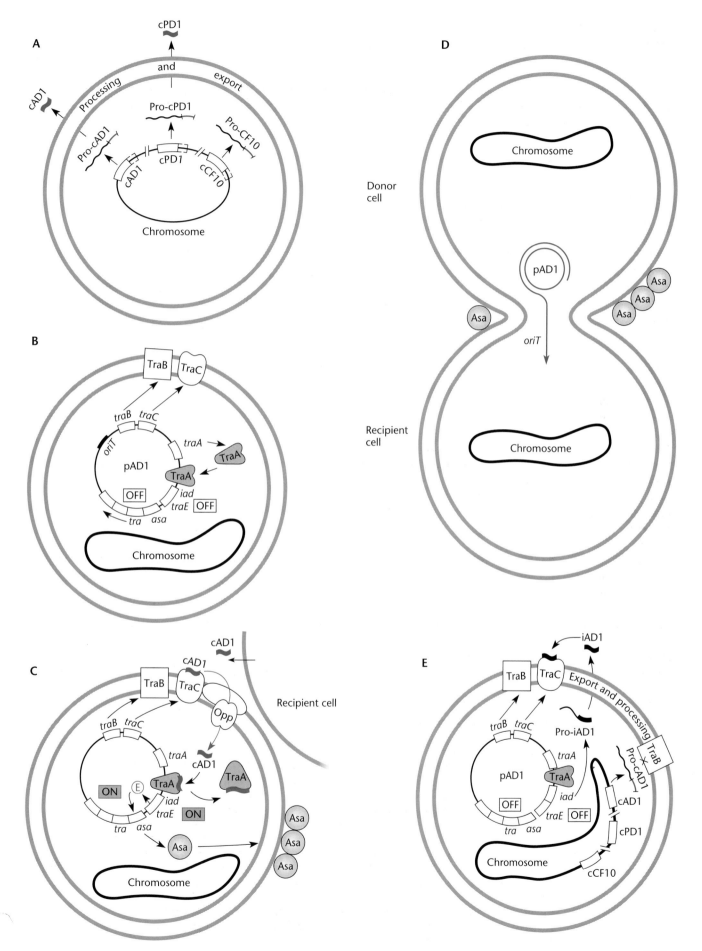

due to synthesis of an inhibitor encoded by a gene of the plasmid. The inhibitor gene product is a peptide of 7 or 8 amino acids, much like the pheromone itself, but differs from the pheromone by only one or a few amino acids, which allows it to bind to the specific pheromone-binding protein. This allows it to competitively inhibit the secretion of the pheromone by the donor cell and prevent autoinduction (i.e., self induction) of its own mating system and that of other potential donors. Yet a third system involves a shutdown protein called TraB in pAD1 and PrgY in a related plasmid. The function of these proteins is unclear, but they are membrane proteins which somehow interfere with the processing or distribution of the pheromone to further prevent autoinduction and to avoid attracting other donors (see Buttaro et al., Suggested Reading).

In contrast to the similarities in pheromone production and sensing, the mechanisms for pheromone induction of *tra* gene expression differ from one type of plasmid to another. For example, in the case of pAD1, the *traA* gene encodes a negative regulator (repressor) (see chapter 2) that represses the transcription of genes encoding aggregation substance and other transfer gene products. The pheromone cAD1 binds to TraA, releasing it from DNA to allow transcription. It is clear that other plasmids use different mechanisms but these mechanisms are less well understood.

The enterococci and their plasmids are of especial importance because they are important hospital-acquired (nosocomial) pathogens. Their diverse plasmids carry both genes that enhance virulence and genes that confer resistance to multiple antibiotics. The enterococcal plasmids can also transfer their genes to other gram-positive bacteria including the very dangerous pathogen *Staphylcoccus aureus*. This is because *S. aureus* can produce pheromones to attract enterococcus plasmid (see De Boever et al., Suggested Reading). This type of transfer is of particular medical importance because enterococcal plasmids can confer resistance to vancomycin, which is often a "last-resort" antibiotic in the treatment of *S. aureus* infections.

Other Types of Transmissible Elements

Plasmids are not the only DNA elements in bacteria that are capable of transferring themselves. Some transposons, called **conjugative transposons**, also encode Tra functions to promote their own transfer.

An example, Tn*916* from *E. faecalis*, is not a replicon capable of autonomous replication, but it can transfer itself from the chromosome of one bacterium to the chromosome of another bacterium without transferring chromosomal genes. Presumably, it transiently excises from one DNA, transfers itself into another cell, and then transposes into the DNA of the recipient bacterium upon entry (see Marra et al., Suggested Reading).

The Tn*916* conjugative transposon and its relatives are known to be promiscuous and will transfer into many types of gram-positive bacteria and even into some gram-negative bacteria. The antibiotic resistance gene they carry, *tetM*, has also been found in many types of gram-positive and gram-negative bacteria. It is tempting to speculate that conjugative transposons such as Tn*916* are responsible for the widespread dissemination of the *tetM* gene. We shall discuss the transposition of conjugative transposons in more detail in chapter 9.

Other elements similar to conjugative transposons have been found in the genus *Bacteroides*. These elements will transfer not only themselves but also other small DNA elements in the chromosome (see Qi et al., Suggested Reading).

SUMMARY

1. Self-transmissible plasmids can transfer themselves to other bacterial cells, a process called conjugation. Some plasmids can transfer themselves into a wide variety of bacteria from different genera. Such plasmids are said to be promiscuous.

2. The plasmid genes whose products are involved in transfer are called the *tra* genes. The site on the plasmid DNA at which transfer initiates is called the origin of transfer *(oriT)*. The *tra* genes can be divided into two groups: those whose products are involved in mating-pair formation (Mpf) and those whose products are involved in processing the plasmid DNA for tranfer (Dtr).

3. The Mpf component includes a sex pilus that extrudes from the cell and holds mating cells together. The pilus is the site to which male-specific phages adsorb. The Mpf sys-tem also includes the channel in the membrane through which DNA and proteins pass, as well as a coupling protein that lies on the channel and docks the relaxase of the Dtr component and other proteins to be transferred on the channel.

4. The Dtr component includes the relaxase, which makes a cut at the *nic* site within the *oriT* sequence and rejoins the ends of the plasmid in the recipient cell. The relaxase also often contains a helicase activity which separates the strands of DNA during transfer. The Dtr component also includes proteins that bind to the *oriT* sequence to form the multiprotein complex called the relaxosome and a primase which primes replication in the recipient cell and is transferred along with the DNA.

(continued)

SUMMARY (continued)

5. Most plasmids only transiently express their Mpf *tra* genes immediately after transfer to a recipient cell and then only intermittently thereafter. Presumably, this helps them avoid infection by male-specific phages that use the sex pilus as their adsorbtion site.

6. Mobilizable plasmids cannot transfer themselves but can be transferred by other plasmids. Mobilizable plasmids encode only a Dtr component; they lack genes to encode an Mpf component. The *tra* genes of a mobilizable plasmid encoding its Dtr component are called the *mob* genes. A mobilizable plasmid can be mobilized by a self-transmissible plasmid only if the coupling protein of the self-transmissible plasmid can dock the relaxase of the mobilizable plasmid. Because they lack an Mpf component, mobilizable plasmids can be much smaller than self-transmissible plasmids, which makes them very useful in molecular genetics and biotechnology.

7. Hfr strains of bacteria have a self-transmissible plasmid integrated into their chromosome. Hfr strains are useful for genetic mapping in bacteria because they transfer chromosomal DNA in a gradient, beginning at the site of integra-tion of the plasmid. Hfr crosses are particularly useful for locating genetic markers on the entire genome.

8. Prime factors are self-transmissible plasmids that have picked up part of the bacterial chromosome. They can be used to make partial diploids for complementation tests. If a prime factor is transferred into a cell, the cell will be a partial diploid for the region of the chromosome carried on the prime factor, making it useful for complementation experiments.

9. Self-transmissible plasmids also exist in gram-positive bacteria. However, these plasmids do not encode a sex pilus. Some gram-positive bacteria have the interesting property of excreting small pheromone-like compounds that stimulate mating with certain plasmids, presumably by inducing their *tra* genes. The existence of such systems em-phasizes the importance of plasmid exchange to bacteria.

10. Some transposons of gram-positive bacteria are also self-transmissible. These so-called conjugative transposons can transfer themselves into other cells even though they are not replicons.

QUESTIONS FOR THOUGHT

1. Why do you suppose the Tra functions, including the *eex* genes, which exclude other plasmids of the same type, are usu-ally the same for all the members of the same Inc group?

2. Why are the *tra* genes whose products are directly involved in DNA transfer usually adjacent to the *oriT* site?

3. Why do you suppose plasmids with a certain *mob* site are mobilized by only certain types of self-transmissible plasmids?

4. Why do self-transmissible plasmids usually encode their own primase function?

5. What do you think is different about the cell surfaces of gram-positive and gram-negative bacteria that causes only the self-transmissble plasmids of gram-negative bacteria to encode a pilus?

6. Why do so many types of phages use the sex pilus of plas-mids as their adsorption site?

7. Why are so many plasmids either self-transmissible or mo-bilizable? Promiscuous?

8. What is the evidence that mobilizable plasmids are not just self-transmissible plasmids that have lost their Mpf?

PROBLEMS

1. After mixing two strains of a bacterial species, you observe some recombinant types that are unlike either parent. These re-combinant types seem to be the result of conjugation, because they appear only if the cells are in contact with each other. How would you determine which is the donor strain and which is the recipient? Whether the transfer is due to an Hfr strain or to a prime factor?

2. How would you determine which of the *tra* genes of a self-transmissible plasmid encodes the pilin protein? The site-spe-cific DNA endonuclease that cuts at *oriT*? The helicase?

3. How would you show that only one strand of the plasmid DNA enters the recipient cell during plasmid transfer?

4. You have discovered that tetracycline resistance can be transferred from one strain of a bacterial species to another.

How would you determine whether the tetracycline resistance gene being transferred is on a self-transmissible plasmid or on a conjugative transposon?

5. Can a male-specific phage infect bacteria containing only a mobilizable plasmid? Why or why not?

6. An Hfr strain that is His⁺ Trp⁺ but has an *argH* mutation is crossed with a recipient that is Arg⁺ but has *hisG* and *trpA* mu-tations, and the cross is plated on minimal plates containing histidine and arginine but no tryptophan. Which is the selected marker, and which are the unselected markers?

7. You have crossed an *E. coli* Hfr strain that has *hisB4 and recA1* mutations with a strain with the *thyA8* mutation and plated the cross on minimal plates with no growth supplements. Almost 80% of the *Thy*⁺ recombinants are very UV sensitive.

Where is the *recA1* marker located in the chromosome? Note that a *recA* mutation will make the cell very sensitive to UV light (see chapter 11).

8. An *E. coli* strain has *metB1* (90 min) and *leuA5* (2 min) mutations, which make it require methionine and leucine, respectively. It also has an *strA7* (73 min) mutation, which makes it resistant to streptomycin, and a Tn5 transposon which confers kanamycin resistance, inserted somewhere in its chromosome. You want to know where the transposon is inserted. You cross the mutant strain with an Hfr strain that is streptomycin sensitive and that transfers counterclockwise from 0 min (Figure 5.14) and has a *hisG2* mutation (44 min) that makes it require histidine. After incubation for 100 min, you plate the cells on minimal plates plus leucine and histidine, to select the *metB* marker. The plates also contain streptomycin to counterselect the donor. After purifying 100 of the Met+ transconjugants,

you test them for the other markers. You find that 15 of them are His−, only 2 are Leu+, and 12 are kanamycin sensitive. Which is the selected and which are the unselected markers? Where is the transposon probably inserted?

9. You do the same cross, incubating for only 10 min and plate to select Leu+ transconjugants. Do you expect these transconjugants to be sensitive to the male-specific phage M13? Why or why not?

10. You have isolated a His− mutant of *E. coli* that you suspect has a nonsense mutation because it is suppressed by intergenic suppressors. You want to map one of the suppressors. To do this, you use an Hfr strain that has the original mutation and transfers clockwise from 30 min on the *E. coli* map. As recipient, you use a strain that has the suppressor as well as the *his* mutation and an *argG* mutation. You find that 80% of the Arg+ recombinants are His−. Where is the suppressor mutation located?

SUGGESTED READING

Bagdasarian, M., R. Lurz, B. Rukert, F. C. H. Franklin, M. M. Bagdasarian, J. Frey, and K. N. Timmis. 1981. Specific-purpose plasmid cloning vectors. II. Broad host range, high copy number, RSF1010-derived vectors, and a host-vector system for gene cloning in *Pseudomonas*. *Gene* 16:237–247.

Buttaro, B. A., M. H. Antiporta, and G. M. Dunny. 2000. Cell associated pheromone peptide (cCF10) production and pheromone inhibition in *Enterococcus faecalis*. *J. Bacteriol.* 182:4926–4933

Clewell, D. 1999. Sex pheromone systems in enterococci, p. 47–65. *In* G. M. Dunny and S.C. Winans (ed.), *Cell-Cell Signaling in Bacteria*. ASM Press, Washington, D.C.

De Boever, E. H., D. B. Clewell, and C. M. Fraser. 2000. *Enterococcus faecalis* conjugative plasmid pAM373: complete nucleotide sequence and genetic analyses of sex pheromone response. *Mol. Microbiol.* 37:1327–1341.

Derbyshire, K. M., G. Hatfull, and N. Willets. 1987. Mobilization of the nonconjugative plasmid RSF1010: a genetic and DNA sequence analysis of the mobilization region. *Mol. Gen. Genet.* 206:161–168.

Eisenbrandt, R., M. Kalkum, R. Lurz, and E. Lanka. 2000. Maturation of IncP Pilin prescursors resembles the catalytic dyad-like mechanism of leader peptidases. *J. Bacteriol.* 182:6751–6761.

Firth, N., K. Ippen-Ihler, and R. A. Skurray. 1996. Structure and function of the factor and mechanism of conjugation, p. 2377–2401. *In* F. C. Neidhardt, R. C. Curtiss III, J. L. Ingraham, E. C. C. Lin, K. B. Low, B. Magasanik, W. S. Reznikoff, M. Riley, M. Schaechter, and H. E. Umbarger (ed.), Escherichia coli *and* Salmonella: *Cellular and Molecular Biology*, 2nd ed. ASM Press, Washington, D.C.

Grahn, A. M., J. Haase, D. H. Bamford, and E. Lanka. 2000. Components of the RP4 conjugative transfer apparatus form an envelope structure bridging inner and outer membranes of donor cells: implications for related macromolecule transport systems. *J. Bacteriol.* 182:1564–1574.

Hamilton, C. M., H. Lee, P.-L. Li, D. M. Cook, K. R. Piper, S. B. von Bodman. E. Lanka, W. Ream, and S. K. Farrand. 2000. TraG from RP4 and VirD4 from Ti plasmids confer re-

laxosome specificity to the conjugal transfer system of pTiC58. *J. Bacteriol.* 182:1541–1548.

Kurenbach, B., D. Grothe, M. E. Farias, U. Szewzyk, and E. Grohmann. 2002. The *tra* region of the conjugative plasmid pIP501 is organized in an operon with the first gene encoding the relaxase. *J. Bacteriol.* 184:1801–1805.

Lederberg, J., and E. L. Tatum. 1946. Gene recombination in *E. coli. Nature* (London) 158:558.

Low, K. B. 1968. Formation of merodiploids in matings with a class of Rec− recipient strains of *E. coli* K12. *Proc. Natl. Acad. Sci. USA* 60:160–167.

Low, K. B. 1996. Genetic mapping, p. 2511–2517. *In* F. C. Neidhardt, R. Curtiss III, J. L. Ingraham, E. C. C. Lin, K. B. Low, B. Magasanik, W. S. Reznikoff, M. Riley, M. Schaechter, and H. E. Umbarger (ed.), Escherichia coli *and* Salmonella: *Cellular and Molecular Biology*, 2nd ed. ASM Press, Washington, D.C.

Manchak, J., K. G. Anthony, and L. S. Frost. 2002. Mutational analysis of F-pilin reveals domains for pilus assembly, phage infection, and DNA transfer. *Mol. Microbiol.* 43:195–205.

Marra, D., B. Pethel, G. G. Churchward, and J. R. Scott. 1999. The frequency of conjugative transposition of Tn916 is not determined by the frequency of excision. *J. Bacteriol.* 181:5414–5418

Matson, S. W., J. K. Sampson, and D. R. N. Byrd. 2001. F plasmid conjugative DNA transfer The TraI helicase activity is essential for DNA strand transfer. *J. Biol. Chem.* 276:2372–2379.

Samuels, A. L., E. Lanka, and J. E. Davies. 2000. Conjugative junctions in RP-4-mediated mating of *Escherichia coli. J. Bacteriol.* 182:2709–2715.

Qi, C., B. J. Paszkiet, N. B. Shoemaker, J. F. Gardner, and A. A. Salyers. 2000. Integration and excision of a *Bacteroides* conjugative transposon, CTnDOT. *J. Bacteriol.* 182:4035–4043.

Watanabe, T., and T. Fukasawa. 1961. Episome-mediated transfer of drug resistance in *Enterobacteriaceae*. 1. Transfer of resistance factors by conjugation. *J. Bacteriol.* 81:669–678.

Wilkins, B. M., and A. T. Thomas. 2000. DNA transfer independent transport of plasmid primase protein between bacteria by the I1 conjugation system. *Mol. Microbiol.* 38:650–657.

6

Transformation

D NA CAN BE EXCHANGED among bacteria in three ways: conjugation, transduction, and transformation. Chapter 5 covered conjugation, in which a plasmid or other self-transmissible DNA element transfers itself and sometimes other DNA into another bacterial cell. In transduction, a subject of chapter 7, a phage carries DNA from one bacterium to another. In this chapter, we discuss transformation, in which cells take up free DNA directly from their environment.

Transformation is one of the cornerstones of molecular genetics because it is often the best way to reintroduce experimentally altered DNA into cells. Since transformation was first discovered in bacteria, ways have been devised to transform many types of animal and plant cells as well.

Discussions of transformation use terms similar to those used in discussions of conjugation. DNA is derived from a **donor bacterium** and taken up by a **recipient bacterium**, which is then called a **transformant**.

Natural Transformation

Most types of bacteria will not take up DNA efficiently unless they have been exposed to special chemical or electrical treatments to make them more permeable. However, **naturally transformable** bacteria can take up DNA from their environment without special treatment. The state of bacteria in which they can take up DNA is called **natural competence**. Naturally competent transformable bacteria are found in several genera, including gram-positive bacteria such as *Bacillus subtilis*, a soil bacterium, and *Streptococcus pneumoniae*, which causes throat infections; and gram-negative bacteria such as *Haemophilus influenzae*, the causative agent of spinal meningitis, and *Neisseria gonorrhoeae*, which causes gonorrhea. Some species of cyanobacteria from the genus *Synechococcus* are also highly naturally transformable.

Discovery of Transformation

Transformation was the first mechanism of gene exchange in bacteria to be discovered. In 1928, Fred Griffith found that one form of the pathogenic pneumococci (now called *Streptococcus pneumoniae*) could be mysteriously "transformed" into another form. Griffith's experiments were based on the fact that *S. pneumoniae* makes two different types of colonies, one pathogenic and the other nonpathogenic. Because they excrete a polysaccharide capsule, the pathogenic strains make colonies that appear smooth on agar plates. Apparently because the capsule allows them to survive in a vertebrate host, these bacteria can infect and kill mice. However, rough-colony-forming mutants that cannot make the capsule and are nonpathogenic in mice sometimes arise from the smooth-colony formers.

In his experiment, Griffith mixed dead *S. pneumoniae* cells that made smooth colonies with live nonpathogenic cells that made rough colonies and injected the mixture into mice (Figure 6.1). Mice given injections of only the rough-colony-forming bacteria survived, but mice that received a mixture of dead smooth-colony formers and live rough-colony formers died. Furthermore, Griffith isolated live smooth-colony-forming bacteria from the blood of the dead mice. Concluding that the dead pathogenic bacteria gave off a "transforming principle" that changed the live nonpathogenic rough-colony-forming bacteria into the pathogenic smooth-colony form, he speculated that this transforming principle was the poly-

saccharide itself. Later, other researchers obtained transformation of rough-colony formers into smooth-colony formers by mixing the rough forms with extracts of the smooth forms in a test tube. Then, about 16 years after Griffith did his experiments with mice, Oswald Avery and his collaborators used the in vitro system to purify the "transforming principle" and showed that it is DNA (see Avery et al., Suggested Reading). The work of Avery and colleagues helped show that DNA, not proteins or other factors in the cell, is the hereditary material (see the introductory chapter).

Competence

The term "competence" refers to the state which some bacteria can enter, in which they can take up naked DNA from their environment. This capability is genetically programmed, and the process of DNA uptake is often called "natural transformation," to distinguish it from transformation induced by electroporation, heat shock, Ca^{2+} treatment of cells, or protoplast uptake of DNA. The genetic programming of competence is widespread but not universal. Generally, more than a dozen genes are involved, encoding both regulatory and structural components of the transformation process.

The general steps that occur in natural transformation are (i) binding of double-stranded DNA to the outer cell surface of the bacterium, (ii) movement of the DNA across the membranes and cell wall of the bacterium, (iii) degradation of one of the DNA strands, (iv) translocation of the remaining single strand of DNA into the cytoplasm of the cell across the inner membrane, and (v) stable integration by homologous recombination of the translocated single strand into the recipient chromosome. We discuss the uptake of DNA in more detail below.

COMPETENCE IN GRAM-POSITIVE BACTERIA

The DNA uptake systems of gram-positive and gram-negative bacteria seem to differ in certain important respects, so we discuss them separately. Two gram-positive species that have been particularly well studied are *B. subtilis* and *S. pneumoniae*. The proteins involved in transformation in these bacteria were discovered on the basis of isolation of mutants that are completely lacking in the ability to take up DNA. The genes affected in the mutants were named *com*, for *com*petence defective.

In *B. subtilis*, the *com* genes are organized into several operons. Several of these, including *comA* and *comK*, are involved in regulation of competence (see below) and others, including *comE*, *comF*, and *comG* encode structural proteins for uptake. The first gene of the *comE* operon, *comEA*, encodes the protein that directly binds extracellular double-stranded DNA. The *comF* genes encode proteins that translocate the DNA into the cell. The

Figure 6.1 The Griffith experiment. Heat-killed pathogenic encapsulated bacteria can convert nonpathogenic noncapsulated bacteria to the pathogenic capsulated form. Type R indicates rough-colony formers and type S indicates smooth-colony formers.

	Bacterial type	Effect in mouse	Bacteria recovered
A	Live type R	Nonpathogenic	None
B	Live type S	Pathogenic	Live type S
C	Heat-killed type S	Nonpathogenic	None
D	Mixture of live type R and heat-killed type S	Pathogenic	Live type S

genes in the *comG* operon encode proteins that provide pore- or channel-like structures that allow DNA to move through the thick peptidoglycan cell wall and across the cytoplasmic membrane. Specific activities of ComG proteins include forming a pore or channel that allows pas-

sage of the DNA (Figure 6.2A). The ComG proteins also function in allowing ComEA to be accessible to the extracellular environment, since ComEA is an integral membrane protein; these ComG proteins either form a channel for ComEA or otherwise modify the cell wall.

Figure 6.2 Structure of DNA uptake competence systems in gram-positive (A) and gram-negative (B) bacteria. Shown are the proteins involved and the channels they form. The nomenclature in panel A is based on *B. subtilis,* and that in panel B is based on *N. gonorrhoeae.* Some of the *B. subtilis* ComG proteins are analogous to the *Neisseria* PilE protein (shaded boxes). The *B. subtilis* ComEC protein is an orthology of the *Neisseria* ComA protein (unshaded boxes). Ss, single-stranded. Details are given in text. Adapted with permission from D. Dubnau, *Annu. Rev. Microbiol.* **53:**217–244, 1999. Copyright 1999 by Annual Reviews www.annualreviews.org.

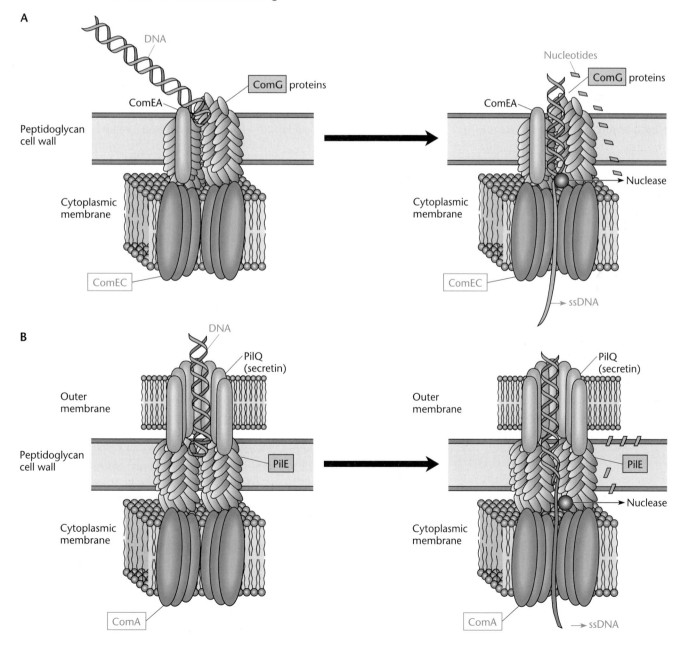

The *comE*, *comF* and *comG* operons are all under the transcriptional control of ComK, a transcription factor that is itself regulated by ComA, as discussed below.

Some of the genes involved in the transformation process are not designated *com*, because such genes were first discovered on the basis of their involvement in another process. For example, one of the nuclease activities which makes double-strand breaks in extracellular DNA is the *nucA* gene product (see Provedi et al., Suggested Reading). These free DNA ends become the substrates for the competence proteins. Other examples of proteins with multiple roles include SSB (single-stranded-DNA-binding protein) and RecA, which functions in the recombination of transforming DNA with the chromosome as well as generally in recombination (see chapter 10).

The lengths of single-stranded DNA incorporated into the recipient chromosome are about 8.5 to 12 kb, as shown by cotransformation of genetic markers; the incorporation takes only a few minutes to be completed.

Transformation in *S. pneumoniae* utilizes similar proteins and mechanisms to those in *B. subtilis* (see Havarstein et al., Suggested Reading).

COMPETENCE IN GRAM-NEGATIVE BACTERIA

As mentioned above, a variety of gram-negative bacteria are also capable of acquiring competence. Some examples are the bacterium *Acinetobacter calcoaceticus* as well as the pathogens *Helicobacter pylori*, *Neisseria spp.*, and *Haemophilus spp.* In the last two, specific uptake sequences are required for the binding of DNA and so these species will take up DNA only of the same species (see below). This differs from the gram-positive bacteria and also from the other gram-negative bacteria (discussed below), which do not require sequence specificity for uptake.

Gram-negative bacteria utilize two different types of DNA uptake systems. One of these involves proteins with structural similarities to the proteins used in the gram-positive bacterial pathway discussed above (Figure 6.2B). The second uses proteins related to some of the type IV secretion-conjugation proteins discussed in chapter 5. We shall discuss each of these pathways in more detail below, but first it is important to comment on the general relatedness of the proteins involved in DNA uptake. As we examine the amino acid sequence similarities of these proteins, it becomes evident that structurally similar proteins occur not only in DNA uptake in both gram-positive and gram-negative bacteria but also in conjugation and protein secretion, using both the so-called type II and type IV pathways and motility of the sort referred to as "twitching" motility, which occurs in *Neisseria* spp.as the result of pilus formation and retraction. The term "PSTC" has been applied to some of the proteins, indicating their multiple roles in *p*ilus formation, *s*ecretion, *t*witching motility, and *c*ompetence. Overall, these re-lated proteins are involved in forming the structures needed to transport large molecules, i.e., DNA or proteins, across the cell wall and membranes.

The PSTC Transformation Pathway in Gram-Negative Bacteria

Many of the proteins that function in DNA uptake in *Haemophilus*, *Neisseria*, and *Acinetobacter* species are related to the *B. subtilis* ComG proteins and their relatives in *S. pneumoniae*. These proteins are called Pil rather than Com because they were discovered on the basis of their involvement in pilus formation. However, they also function in competence, providing structural and translocation functions that allow the passage of DNA through the peptidoglycan and cytoplasmic membrane layers. Besides these obstacles to DNA passage, which are shared by gram-positive bacteria, gram-negative bacteria present an additional obstacle to macromolecular uptake: the outer membrane. Hence, proteins of the secretin class form pores to allow the passage of DNA, much as they do in the processes of protein secretion and even release of filamentous phage.

The Type IV Secretion-Related Pathway in Gram-Negative Bacteria

The second type of mechanism is found in *H. pylori*, an opportunistic pathogen involved in gastrointestinal diseases. In these bacteria, DNA translocation through the membranes and cell walls utilizes proteins related to conjugation in *Agrobacterium* (see Hofreuter et al., Suggested Reading). These proteins are discussed in Box 5.2. Thus this system can function as a two-way DNA transfer system, moving DNA both into and out of the cell.

REGULATION OF COMPETENCE IN *B. SUBTILIS*

The regulation of competence in *B. subtilis* is achieved through a **two-component regulatory system** analogous to those used to regulate many other systems in bacteria (see chapter 13). First, information that the cell is running out of nutrients and the population is reaching a high density is registered by ComP, a **sensor protein** in the membrane (Figure 6.3). The high cell density causes this sensor-kinase protein to phosphorylate itself. The phosphate is then transferred from ComP to ComA, a **response regulator protein**. In the phosphorylated state, the ComA protein is a transcriptional activator (see chapter 2) for several genes, including the operon *srfA*, which is required for competence. At least one of the genes in this operon affects the expression of other genes required for competence.

COMPETENCE PHEROMONES

High cell density is required for competence of *B. subtilis* because of small peptides called **competence pheromones** that are excreted by the bacteria as they multiply (see

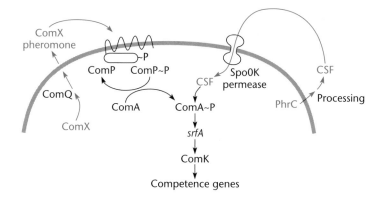

Figure 6.3 Regulation of competence development in *Bacillus subtilis*. The ComP protein in the membrane senses high cell density and is phosphorylated. The phosphate is then transferred to ComA, which allows transcription of competence genes. The ComA protein phosphorylation involves two competence pheromone peptides (shown in gold), one of which is imported by an oligopeptide permease, the product of the *spo0K* gene. CSF, competence-stimulating factor.

Lazazzera et al., Suggested Reading). Cells become competent only in the presence of high concentrations of these peptides, which are reached only when the concentration of cells giving them off is high. The requirement for competence pheromones ensures that cells will be able to take up DNA only when there are other *B. subtilis* cells nearby that are giving off DNA to be taken up.

One competence pheromone peptide is cut out of a longer polypeptide, the product of the *comX* gene. Another gene, *comQ*, which is immediately upstream of *comX*, is also required for synthesis of the competence pheromone, and its product may be the protease enzyme that cuts the competence pheromone out of the longer polypeptide. Once the peptide has been cut out of the longer molecule, it can trigger the phosphorylation of ComP, although the mechanism remains unknown.

B. subtilis cells produce a second competence peptide called competence-stimulating factor (CSF), which is processed from the signal sequence of a longer polypeptide, the product of the *phrC* gene. The CSF peptide is transported into the cell by the oligopeptide permease Spo0K (see below).

RELATIONSHIP BETWEEN SPORULATION AND COMPETENCE

At about the same time as *B. subtilis* reaches stationary phase, some cells acquire competence and some cells sporulate (see chapter 14). Sporulation, a developmental process common to many bacteria, allows a bacterium to enter a dormant state and survive adverse conditions such as starvation, irradiation, and heat. During sporulation, the bacterial chromosome is packaged into a resistant spore, where it remains viable until conditions improve and the spore can germinate into an actively growing bacterium.

Some of the regulatory genes required for sporulation are also required for the development of competence. The *spo0K* gene is an example. This gene was first discovered because of its role in sporulation. A *spo0K* mutant is blocked in the first stage, the "0" stage, of sporulation. The *K* means that it was the 11th gene (as K

is the 11th letter in the alphabet) involved in sporulation to be discovered in that collection. The regulation of sporulation seems to be even more complex than the regulation of competence.

Uptake of DNA during Natural Transformation

Experiments directed toward an understanding of DNA uptake during natural transformation have sought to answer three obvious questions. (i) How efficient is DNA uptake? (ii) Can only DNA of the same species enter a given cell? (iii) Are both of the complementary DNA (cDNA) strands taken up and incorporated into the cellular DNA?

EFFICIENCY OF DNA UPTAKE

The efficiency of uptake is fairly easy to measure biochemically. For example, Figure 6.4 shows an experiment based on the fact that transport of free DNA into the cell will make those strands insensitive to DNases, which

Figure 6.4 Determining the efficiency of DNA uptake during transformation. DNA in the cell is insensitive to DNase. Degraded DNA will pass through a filter. The asterisk refers to radioactively labeled DNA.

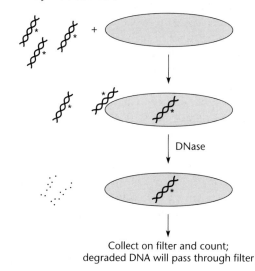

Collect on filter and count;
degraded DNA will pass through filter

cannot enter the cell, because competent cells permit only DNA, not proteins, to enter. In the experiment, labeled DNA is mixed with competent cells and the mixture is then treated with DNase. Any DNA that is not degraded and survives intact must have been taken up by the cells. Such DNA can be distinguished from degraded DNA because intact DNA can be precipitated with acid and collected on a filter whereas degraded DNA will not precipitate and will pass through the filter. Therefore, if the radioactivity on the filter is counted and compared with the total radioactivity of the DNA that was added to the cells, the percentage of DNA that is taken up, or the efficiency of DNA uptake, can be calculated. Experiments such as these have shown that some competent bacteria take up DNA very efficiently.

SPECIFICITY OF DNA UPTAKE

The second question, i.e., whether DNA from only the same species is taken up, is also fairly easy to answer. By using the same assay of resistance to DNases, it has been determined that some types of bacteria will take up DNA from only their own species whereas others can take up DNA from any source. The first group includes *Neisseria gonorrhoeae* and *Haemophilus influenzae*.

A specific **uptake sequence** is required by bacteria that take up DNA only from their own species. Figure 6.5 shows the minimal sequences for *H. influenzae* and *N. gonorrhoeae*. Uptake sequences are long enough that they almost never occur by chance in other DNAs. In contrast, bacteria such as *B. subtilis* seem to take up any DNA. Possible reasons why some bacteria should preferentially take up DNA from their own species while others take up any DNA are subjects of speculation and will be discussed later.

Mechanism of DNA Uptake during Transformation

Although the genetic requirements for transformation are best known for *B. subtilis,* the more efficient uptake of DNA by some other naturally transformable bacteria has allowed biochemical experiments on the uptake of DNA by these species. The general pathway was first worked out for *Streptococcus pneumoniae* but is probably similar in most other naturally transformable bacteria.

Figure 6.5 The uptake sequences on DNA for some types of bacteria. Only DNA with these sequences will be taken up by the bacteria indicated. Only one strand of the DNA is shown.

Haemophilus influenzae	5′ AAGTGCGGTCA 3′
Neisseria gonorrhoeae	5′ GCCGTCTCAA 3′

TRANSFORMATION IN *S. PNEUMONIAE*

Figure 6.6 shows a general scheme for DNA uptake during the transformation of *S. pneumoniae*. In the first step, double-stranded DNA released by lysis of the donor bacteria is bound to specific receptors on the cell surface of the recipient bacterium. The bound DNA is then broken into smaller pieces by endonucleases; one of the two complementary strands is degraded by an exonuclease; and the remaining strand is transported into the cell. The transforming DNA integrates into the cellular DNA in a homologous region by means of strand displacement, a mechanism in which the new strand invades the double helix and displaces an old strand with the same sequence. The old strand is then degraded. If the donor DNA and recipient DNA sequences differ slightly in this region, recombinant types can appear.

Figure 6.6 Transformation in *Streptococcus pneumoniae.* Competence factors accumulate as the cells reach a high density. Double-stranded DNA binds to the cell, and one strand is degraded. The remaining single strand replaces the strand of the same sequence in the chromosome, creating a "heteroduplex" in which one strand comes from the donor and one comes from the recipient.

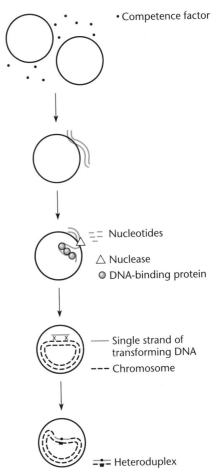

• Competence factor

== Nucleotides
△ Nuclease
◉ DNA-binding protein

— Single strand of transforming DNA
--- Chromosome

=•= Heteroduplex

Evidence for this model comes from several different experiments, some of which are discussed below. Also, the gene for a membrane-bound DNase that may be involved in degrading one of the two strands of the incoming DNA has been found in *S. pneumoniae* (see Puyet et al., Suggested Reading).

TRANSFORMASOMES

The basic scheme described above probably differs among different types of naturally competent bacteria. For example, *Haemophilus influenzae* may first take up double-stranded DNA in subcellular compartments called **transformasomes** (Figure 6.7). The new DNA may not become single stranded until it enters the cytoplasm. However, the basic process of all natural transformation is the same. Only one strand of the DNA enters the interior of the cell and integrates with the cellular DNA to produce recombinant types.

Genetic Evidence for Single-Strand Uptake

Genetic experiments taking advantage of the molecular requirements for transformation can be used to study the molecular basis for transformation; in other words, transformation can be used to study itself. Evidence that

Figure 6.7 Transformation in *Haemophilus influenzae*. Double-stranded DNA is first taken up in transformasomes. One strand is degraded, and the other strand invades the chromosome, displacing one chromosome strand.

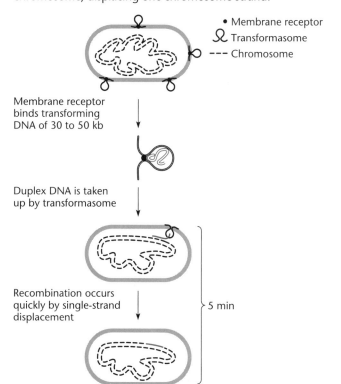

• Membrane receptor
Ω Transformasome
--- Chromosome

Membrane receptor binds transforming DNA of 30 to 50 kb

Duplex DNA is taken up by transformasome

Recombination occurs quickly by single-strand displacement

5 min

DNA has transformed cells is usually based on the appearance of recombinant types after transformation. A recombinant type can form only if the donor and recipient bacteria differ in their genotypes and if the incoming DNA from the donor bacterium changes the genetic composition of the recipient bacterium. The chromosome of a recombinant type will have the DNA sequence of the donor bacterium in the region of the transforming DNA.

Experiments have shown that only double-stranded DNA can bind to the specific receptors on the cell surface, so that only double-stranded—not single-stranded—DNA can transform cells and yield recombinant types. However, we can also conclude from these experiments that the cells actually take up only *single*-stranded DNA, because the DNA enters an "eclipse" phase in which it cannot transform. For example, in the experiment shown in Figure 6.8, an Arg$^-$ mutant requiring arginine for growth is used as the recipient strain, and the corresponding Arg$^+$ prototroph is the source of donor DNA. At various times after the donor DNA has been mixed with the recipient cells, the recipients are treated with DNase, which cannot enter cells but will destroy any DNA remaining in the medium. The surviving DNA in the recipient cells is then extracted and used for retransformation of more auxotrophic recipients, and Arg$^+$ transformants are selected on agar plates without the growth supplement arginine. Any Arg$^+$ transformants must have been due to double-stranded donor DNA in the recipient cells.

Whether or not transformants were observed depends on the time the DNA was extracted from the cells. When the DNA is extracted at time 1, while it is still outside the cells and accessible to the DNase, no Arg$^+$ transformants are observed because the Arg$^+$ donor DNA is all destroyed by the DNase. At time 2, some of the DNA is now inside the cells, where it cannot be degraded by the DNase, but this DNA is single stranded. It has not yet recombined with the chromosome, and so Arg$^+$ transformants are still not observed in step 4. Only at time 3, when some of the DNA has recombined with the chromosomal DNA and so is again double stranded, will Arg$^+$ transformants appear in step 4. Thus, the transforming DNA enters the eclipse period for a short time after it is added to competent cells, as expected if it enters the cell in a single-stranded state.

Plasmid Transformation and Transfection of Naturally Competent Bacteria

Chromosomal DNA can efficiently transform any bacterial cells from the same species that are naturally competent. However, neither plasmids nor phage DNAs can be efficiently introduced into naturally competent cells, because they must recyclize to replicate. Natural transformation requires breakage of the double-stranded DNA

Step 1
Mix Arg⁺ DNA
and recipient
cells

Step 2
Treat mixture
with DNase at
various times

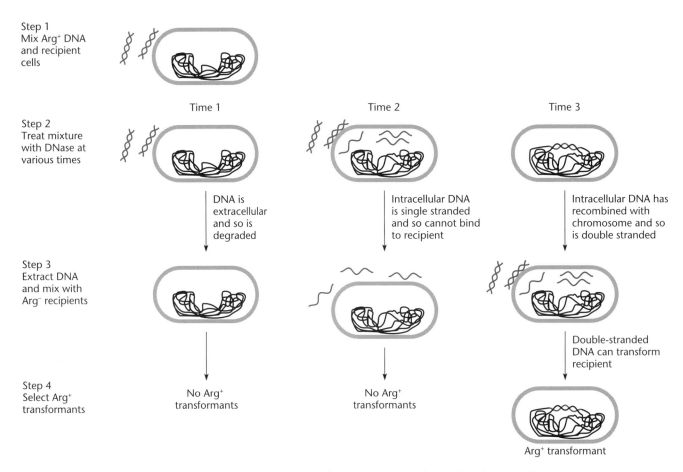

Time 1

DNA is
extracellular
and so is
degraded

Time 2

Intracellular DNA
is single stranded
and so cannot bind
to recipient

Time 3

Intracellular DNA has
recombined with
chromosome and so
is double stranded

Step 3
Extract DNA
and mix with
Arg⁻ recipients

Double-stranded
DNA can transform
recipient

Step 4
Select Arg⁺
transformants

No Arg⁺
transformants

No Arg⁺
transformants

Arg⁺ transformant

Figure 6.8 Genetic assay for the state of DNA during transformation. Only double-stranded DNA will bind to the cell to initiate transformation. The appearance of transformants in step 4 indicates that the transforming DNA was double stranded at the time of DNase treatment.

and degradation of one of the two strands so that a linear single strand can enter the cell. Plasmid and phage DNAs are usually double stranded and must be double stranded to replicate autonomously. However, pieces of single-stranded plasmid or phage DNAs cannot recyclize or make the complementary strand if there are no repeated or complementary sequences at their ends.

Plasmid or phage DNAs often do not have repeated DNA at their ends, and so transformation of naturally competent bacteria with plasmid or phage DNA usually occurs only with DNAs that are **dimerized**. A dimerized DNA is one in which two copies of the molecule are linked head to tail as illustrated in Figure 6.9. If a dimerized plasmid or phage DNA is cut only once, it will still have complementary sequences at its ends that can recombine to recyclize the plasmid, as illustrated in the figure. Such dimers form naturally while plasmid or phage DNA is replicating, so that most preparations of plasmid or phage DNAs contain some dimers. The fact that only

dimerized plasmid or phage DNAs can transform naturally competent bacteria supports the model of transformation described earlier.

The Role of Natural Transformation

The fact that so many gene products play a direct role in competence indicates that the ability to take up DNA from the environment is advantageous. Below, we discuss three possible advantages and the arguments for and against them.

NUTRITION

Organisms may take up DNA for use as a carbon and nitrogen source (see Redfield, Suggested Reading). One argument against this hypothesis is that taking up whole DNA strands for degradation inside the cell may be more difficult than degrading the DNA outside the cell and then taking up the nucleotides. In fact, *B. subtilis* excretes a DNase, which degrades DNA so that it can be

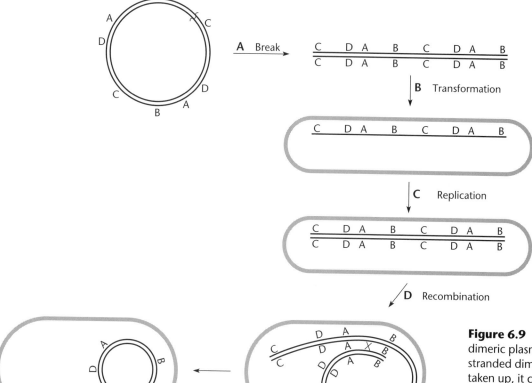

Figure 6.9 Transformation by dimeric plasmids. After the single-stranded dimeric plasmid DNA is taken up, it can serve as a template to make the double-stranded DNA. The repeated ends can recombine with each other to form a circular plasmid.

taken up more easily. The major argument against this hypothesis as a general explanation for transformation in all bacteria is that some bacteria take up only DNA of their own species since DNA from other organisms should offer the same nutritional benefits. Moreover, the fact that competence develops only in a minority of the population, at least in *B. subtilis*, argues against the nutrition hypothesis, since all the bacteria in the population would presumably need the nutrients.

These arguments are attractive but do not disprove the nutrition hypothesis. The bacteria may consume DNA of only their own species because of the danger inherent in taking up foreign DNAs, which might contain prophages, transposons, or other elements that could become parasites of the organism. Furthermore, consumption of DNA from the same species may be a normal part of colony development; cell death and cannibalism are thought to be part of some prokaryotic developmental processes. These processes would require that only some of the cells in the population become DNA consumers. The special circumstances of laboratory cultures may also explain the fact that only a minority of cells become competent. In some natural environments, all of the bacteria may become competent at the same time in response to starvation.

REPAIR

Cells may take up DNA from other cells to repair damage to their own DNA (see Mongold, Suggested Reading). Figure 6.10 illustrates this hypothesis, in which a population of cells is exposed to UV irradiation. The radiation damages the DNA, causing pyrimidine dimers and other lesions (see chapter 11). DNA leaks out of some of the dead cells and enters other bacteria. Because the damage to the DNA will not have occurred at exactly the same places, undamaged incoming DNA sequences can replace the damaged regions in the recipient, allowing at least some of the bacteria to survive. This scenario explains why some bacteria will take up DNA of only the same species, since, in general, this is the only DNA that can recombine and thereby participate in the repair.

If natural transformation helps in DNA repair, we might expect that repair genes will be induced in response to developing competence and that competence will develop in response to UV irradiation or other types of DNA damage. In fact, in some bacteria, including *B. subtilis* and

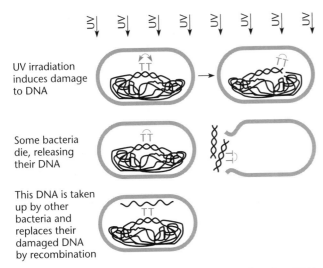

Figure 6.10 Repair of DNA damage by transforming DNA. A region containing a thymine dimer (TT) induced by UV irradiation is replaced by the same, but undamaged, sequence from the DNA of a neighbor killed by the radiation. This mechanism could allow survival of the species.

S. pneumoniae, the *recA* gene required for recombination repair is induced in response to the development of competence (see Haijema et al., and Raymond-Denise and Guillen, Suggested Reading). However, in other bacteria, such as *H. influenzae*, the *recA* gene is not induced in response to competence. There is also no evidence that competence genes are induced in response to DNA damage. Nevertheless, the need for DNA repair is an attractive explanation for why at least some types of bacteria develop competence.

RECOMBINATION

The possibility that transformation allows recombination between individual members of the species is also an attactive hypothesis that is difficult to prove. According to this hypothesis, transformation serves the same function that sex serves in higher organisms: it allows the assembly of new combinations of genes and thereby increases diversity and speeds up evolution. Bacteria do not have an obligatory sexual cycle; therefore, without some means of genetic exchange, any genetic changes that a bacterium accumulates during its lifetime will not be exchanged with other members of the species.

The gene exchange function of transformation is supported by the fact that cells of some naturally transformable bacteria leak DNA as they grow. It is hard to imagine what function this leakage could perform unless the leaked DNA is to be taken up by other bacteria.

In several *Neisseria* species, including *N. gonorrhoeae*, transformation may enhance antigenic variability, allowing the organism to avoid the host immune system (see Box 6.1). In mixed laboratory cultures, transformation does contribute substantially to the antigenic diversity in this species. However, under natural conditions, it is debatable whether most of this antigenic diversity results from recombination between DNAs brought together by transformation or simply from recombination between sequences within the chromosomal DNA of the bacterium.

We still do not know why some types of bacteria are naturally transformable. In fact, transformation may serve different purposes in different organisms. Perhaps transformation is used for DNA repair in soil bacteria such as *B. subtilis*, but it is used to increase genetic variability in obligate parasites such as *N. gonorrhoeae*. Moreover, most types of bacteria may be naturally transformable at low levels.

Whatever its purpose for individual bacterial species, natural transformation has many uses in molecular genetics. It has been used to map genetic markers in chromosomes and to reintroduce DNA into cells after the DNA has been manipulated in the test tube. The interpretation of genetic data obtained by transformation is similar to that of data obtained by transduction. We discuss the interpretion of genetic data obtained by transduction in chapter 7.

Artificially Induced Competence

Most types of bacteria are not naturally transformable, at least not at easily detectable levels. Left to their own devices, these bacteria will not take up DNA from the environment. However, even these bacteria can sometimes be made competent by certain chemical treatments, or DNA can be forced into them by a strong electric field in a process called electroporation.

Calcium Ion Induction

Treatment with calcium ions (see Cohen et al., Suggested Reading) can make some bacteria competent, including *Escherichia coli* and *Salmonella* spp. as well as some species of *Pseudomonas*, although the reason is not understood.

Chemically induced transformation is usually inefficient, and only a small percentage of the cells are ever transformed. Accordingly, the cells must be plated under conditions selective for the transformed cells. Therefore, the DNA used for the transformation should contain a selectable gene such as one encoding resistance to an antibiotic.

TRANSFORMATION BY PLASMIDS

In contrast to naturally competent cells, cells made permeable to DNA by calcium ion treatment will take up both single-stranded and double-stranded DNA. Therefore,

BOX 6.1

Antigenic Variation in *Neisseria gonorrhoeae*

Many types of pathogenic microorganisms avoid the host immune system by changing the antigens on their cell surface. Well-studied examples include trypanosomes, which cause sleeping sickness, and *Neisseria gonorrhoeae*, which causes a sexually transmitted disease.

The pili of *N. gonorrhoeae* are involved in attaching the bacteria to the host epithelial cells. These pili can undergo spontaneous alterations that can change the specificity of binding and confound the host immune system. *N. gonorrhoeae* appears to be capable of making millions of different pili.

The mechanism of pilin variation in *N. gonorrhoeae* is understood in some detail. The major protein subunit of the pilus is encoded by the *pilE* gene. In addition, silent copies of *pilE*, called *pilS*, lack promoters or have various parts deleted. These silent copies share some conserved sequences with each other and with *pilE* but differ in the so-called variable regions. Pilin protein is usually not expressed from these silent copies. However, recombination between a *pilS* gene and *pilE* can change the *pilE* gene and result in a somewhat different pilin protein. This recombination is a type of gene conversion because reciprocal recombinants are not formed (see chapter 10).

Because *N. gonorrhoeae* is naturally transformable, not only could recombination occur between a *pilS* gene and the *pilE* gene in the same organism, but also transformation could allow even more variation through the exchange of *pilS* genes with other strains. Experiments indicate that pilin variation is affected by the presence of DNase. Also experiments with marked *pil* genes indicate that transformation can result in the exchange of *pil* genes between bacteria. These experiments suggest but do not prove that transformation plays an important role in pilin variation during infection.

References

Gibbs, C. P., B.-Y. Reimann, E. Schultz, A. Kaufmann, R. Haas, and T. F. Meyer. 1989. Reassortment of pilin genes in *Neisseria gonorrhoeae* occurs by two distinct mechanisms. *Nature* (London) **338**:651–652.

Mehr, I. J., and H. S. Seifert. 1998. Differential roles of homolgous recombination pathways in *Neisseria gonorrhoeae* pilin antigenic variation, DNA transformation and DNA repair. *Mol. Microbiol.* **30**:697–710.

Meyer, T. F., C. P. Gibbs, and R. Haas. 1990. Variation and control of protein expression in *Neisseria*. *Annu. Rev. Microbiol.* **44**:451–477.

Seifert, H. S., R. S. Ajioka, C. Marchal, P. F. Sparling, and M. So. 1988. DNA transformation leads to pilin antigenic variation in *Neisseria gonorrhoeae*. *Nature* (London) **336**:392–395.

both linear and double-stranded circular plasmid DNAs can be efficiently introduced into chemically treated cells. This fact has made calcium ion-induced competence very useful for cloning and other applications that require the introduction of plasmid and phage DNAs into cells.

TRANSFECTION

In addition to plasmid DNAs, viral DNAs or genomic RNAs can often be introduced into cells by transformation, thereby initiating a viral infection. This process is called **transfection** rather than transformation, although the principle is the same. To detect transfection, the potentially transfected cells are usually mixed with indicator bacteria and plated (see the introductory chapter). If the transfection is successful, a plaque will form where the transfected cells had produced phage, which then infected the indicator bacteria.

Some viral infections cannot be initiated merely by transfection with the viral DNA. These viruses cannot transfect cells, because in a natural infection, proteins in the viral head are normally injected along with the DNA, and these proteins are required to initiate the infection.

For example, the *E. coli* phage N4 carries a phage-specific RNA polymerase in its head that is injected with the DNA and used to transcribe the early genes. Transfection with the purified phage DNA will not initiate an infection, because the early genes will not be transcribed without this RNA polymerase. Another example of a phage in which the infection cannot be initiated by the nucleic acid alone is the phage φ6 (see Box 7.1). This phage has RNA instead of DNA in the phage head and must inject an RNA replicase to initiate the infection; therefore, the cells cannot be infected by the RNA alone. Such examples of phages that inject required proteins are rather rare, and for most phages, the infection can be initiated by transfection.

As an aside, many animal viruses do inject proteins required for multiplication, and these proteins cannot be made after injection of the naked DNA or RNA. For example, a retrovirus such as human immunodeficiency virus (HIV), which causes AIDS, injects a reverse transcriptase required to make a DNA copy of the incoming RNA before it can be transcribed to make viral proteins. Therefore, human cells cannot be transfected with HIV RNA alone.

TRANSFORMATION OF CELLS WITH CHROMOSOMAL GENES

Transformation with linear DNA is one method used to replace endogenous genes with genes altered in vitro. However, most types of bacteria made competent by calcium ion treatment are transformed poorly by chromosomal DNA because the linear pieces of double-stranded DNA entering the cell are degraded by an enzyme called the RecBCD nuclease. This nuclease degrades DNA from the ends; therefore, it does not degrade circular plasmid and phage DNAs. However, inactivating the RecBCD nuclease by a mutation would preclude recombination between the incoming DNA and the chromosome, because the enzyme is required for normal recombination in *E. coli* and other bacteria (see chapter 10).

Nevertheless, methods have been devised to transform competent *E. coli* with linear DNA. One way is to use a mutant *E. coli* lacking the D subunit of the RecBCD nuclease. These *recD* mutants are still capable of recombination, but because they lack the nuclease activity that degrades linear DNA, they can be transformed with linear double-stranded DNAs. We discuss such procedures further in chapter 10.

Electroporation

Another way in which DNA can be introduced into bacterial cells is by **electroporation**. In the electroporation process, the bacteria are mixed with DNA and briefly exposed to a strong electric field. The brief electric shock seems to open the cells up, and the DNA moves into them much as it moves along a gel during electrophoresis. Electroporation works with most types of cells, including most bacteria, unlike the methods mentioned above, which are very specific for certain species. Also, electroporation can be used to introduce linear chromosomal and circular plasmid DNAs into cells. However, electroporation requires specialized equipment.

SUMMARY

1. In transformation, DNA is taken up directly by cells. Transformation was the first form of genetic exchange to be discovered in bacteria, and the demonstration that DNA is the transforming principle was the first direct evidence that DNA is the hereditary material. The bacteria from which the DNA was taken are called the donors, and the bacteria to which the DNA has been added are called the recipients. Bacteria that have taken up DNA are called transformants.

2. Bacteria that are capable of taking up DNA are said to be competent.

3. Some types of bacteria can naturally take up DNA during part of their life cycle. A number of genes whose products are required for competence have been identified. Some of these encode proteins related to proteins of conjugation and the protein secretion systems.

4. The fate of the DNA during natural transformation is fairly well understood. The double-stranded DNA first binds to the cell surface and then is broken into smaller pieces by endonucleases. Then one strand of the DNA is degraded by an exonuclease. The single-stranded pieces of DNA then invade the chromosome in homologous regions, displacing one strand of the chromosome at these sites. Through repair or subsequent replication, the sequence of the incoming DNA may replace the original chromosomal sequence in these regions.

5. Naturally competent cells can be transformed with linear chromosomal DNA but usually not with circular plasmid or circular phage DNAs. Transformation by plasmid DNA usually occurs with dimers or higher multimers of the plasmid that can recyclize by recombination between the repeated sequences at the ends.

6. Some types of bacteria, including *Haemophilus influenzae* and *Neisseria gonorrhoeae*, will take up DNA of only the same species. Their DNA contains short uptake sequences that are required for uptake of DNA into the cells. Other types of bacteria, including *Bacillus subtilis* and *Streptococcus pneumoniae*, seem to be capable of taking up any DNA.

7. There are three possible roles for natural competence: a nutritional function allowing competent cells to use DNA as a carbon, energy, and nitrogen source; a repair function in which cells use DNA from neighboring bacteria to repair damage to their own chromosomes, thus ensuring survival of the species; and a recombination function in which bacteria exchange genetic material among members of their species, increasing diversity and accelerating evolution.

8. Some types of bacteria that do not show natural competence can nevertheless be transformed after some types of chemical treatment or by electroporation. The standard method for making *E. coli* permeable to DNA involves treatment with calcium ions. Cells made competent by calcium treatment can be transformed with plasmid and phage DNAs, making this method one of the cornerstones of molecular genetics.

9. If the cell is transformed with viral DNA to initiate an infection, the process is called transfection.

QUESTIONS FOR THOUGHT

1. Why do you think some types of bacteria are capable of developing competence? What is the real function of competence?

2. How would you determine if the competence genes of *Bacillus subtilis* are turned on by UV irradiation and other types of DNA damage?

3. How would you determine whether antigenic variation in *Neisseria gonorrhoeae* is due to transformation between bacteria or to recombination within the same bacterium?

PROBLEMS

1. How would you determine if a type of bacterium you have isolated is naturally competent? Outline the steps you would use.

2. Outline how you would isolate mutants of your bacterium that are defective in transformation. Distinguish those that are defective in recombination from those that are defective in the uptake of DNA.

3. How would you determine if a naturally transformable bacterium can take up DNA of only its own species or can take up any DNA?

4. How would you determine if a piece of DNA contains the uptake sequence for that species?

5. How would you determine if the DNA of a phage can be used to transfect *E. coli*?

SUGGESTED READING

Avery, O. T., C. M. MacLeod, and M. McCarty. 1944. Studies on the chemical nature of the substance inducing transformation of pneumococcal types. Induction of transformation by a deoxyribonucleic acid fraction isolated from pneumococcus type III. *J. Exp. Med.* **79:**137–159.

Cohen, S. N., A. C. Y. Chang, and L. Hsu. 1972. Nonchromosomal antibiotic resistance in bacteria: genetic transformation of *Escherichia coli* by R-factor DNA. *Proc. Natl. Acad. Sci. USA* **69:**2110

Dubnau, D. 1999. DNA uptake in bacteria. *Annu. Rev. Microbiol.* **53:**217–244.

Haijema, B. J., D. van Sinderen, K. Winterling, J. Kooistra, G. Venema, and L. W. Hamoen. 1996. Regulated expression of the *dinR* and *recA* genes during competence development and SOS induction in *Bacillus subtilis*. *Mol. Microbiol.* **22:**75–86.

Havarstein, L. S., and D. A. Morrison. 1999. Quorum sensing and peptide pheromones in streptococcal competence for genetic transformation, p. 9–26. *In* G. M. Dunny and S. C. Winans (ed.), *Cell-Cell Signaling in Bacteria*. ASM Press, Washington, D.C.

Hofreuter, D., S. Odenbreit, and R. Hass. 2001. Natural transformation competence in *Helicobacter pylori* is mediated by basic components of a type IV secretion system. *Mol. Microbiol.* **41:**379–391.

Lazazzera, B. A., T. Palmer, J. Quisel, and A. D. Grossman. 1999. Cell density control of gene expression and development in *Bacillus subtilis*, p. 27–46. *In* G. M. Dunny and S. C. Winans (ed.), *Cell-Cell Signaling in Bacteria*. ASM Press, Washington, D.C.

Mongold, J. A. 1992. DNA repair and the evolution of competence in *Haemophilus influenzae*. *Genetics* **132:**893–898.

Provedi, R., I. Chen, and D. Dubnau. 2001. NucA is required for DNA cleavage during transformation of *Bacillus subtilis*. *Mol. Microbiol* **40:**634–644.

Puyet, A., B. Greenberg, and S. A. Lacks. 1990. Genetic and structural characterization of *endA*, a membrane-bound nuclease required for transformation of *Streptococcus pneumoniae*. *J. Mol. Biol.* **213:**727–738.

Raymond-Denise, A., and N. Guillen. 1992. Expression of the *Bacillus subtilis dinR* and *recA* genes after DNA damage and during competence. *J. Bacteriol.* **174:**3171–3176.

Redfield, R. J. 1993. Genes for breakfast: the have your cake and eat it too of bacterial transformation. *J. Hered.* **84:**400–404.

7

Lytic Bacteriophages: Genetic Analysis and Use in Transduction

P ROBABLY ALL ORGANISMS ON EARTH are parasitized by viruses, and bacteria are no exception. For purely historical reasons, viruses that infect bacteria are usually not called viruses but are called **bacteriophages** (**phages** for short), even though they have lifestyles similar to those of plant and animal viruses. As mentioned in the introduction, the name *phage* derives from the Greek verb "to eat," and it describes the eaten-out places, or **plaques,** that are formed on bacterial lawns. The plural of phage is phage, but we add an s (phages) when we are discussing more than one type of phage.

Like all viruses, phages are so small that they can be seen only under the electron microscope. As shown in Figure 7.1A, phages are often spectacular in appearance, with **capsids,** or isocahedral heads, and elaborate tail structures that make them resemble interplanetary landing modules. The tail structures allow them to penetrate bacterial membranes and cell walls to inject their DNA into the cell. Animal and plant viruses have much simpler shapes because they do not need such elaborate tail structures. They are either engulfed by the cell, in the case of animal viruses, or enter through wounds, in the case of plant viruses.

Phages differ greatly in their complexity. Smaller phages, such as MS2, usually have no tail, and their heads may consist of as few as two different types of proteins. The heads and tails of some of the larger phages such as T4 have up to 20 different proteins, each of which can exist in as few as 1 copy to as many as 1,000 copies depending on the structural role they play in the phage particle. Phages also infect specific bacterial hosts, and different phages have very different host ranges, as discussed below.

Like all viruses, phages are not live organisms but merely a nucleic acid—either DNA or RNA depending on the type of phage—wrapped in a

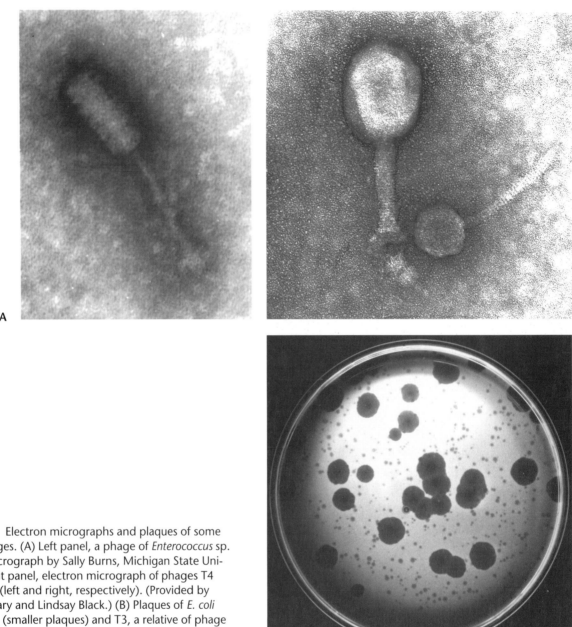

Figure 7.1 Electron micrographs and plaques of some bacteriophages. (A) Left panel, a phage of *Enterococcus* sp. (Electron micrograph by Sally Burns, Michigan State University.) Right panel, electron micrograph of phages T4 and lambda (left and right, respectively). (Provided by Arthur Zachary and Lindsay Black.) (B) Plaques of *E. coli* phages M13 (smaller plaques) and T3, a relative of phage T7 (larger plaques). (Photograph by Kurt Stepnitz, Michigan State University.)

protein and/or membrane coat for protection. This nucleic acid carries genes that direct the synthesis of more phage. In phages, either type of nucleic acid carried in the head is called the **phage genome**. These molecules can be very long, because the genome must be long enough to have at least one copy of each of the phage genes. The length of the DNA or RNA genome therefore reflects the size and complexity of the phage. For instance, the small phage MS2 has only four genes and a rather small RNA genome, whereas phage T4 has more than 200 genes and a DNA genome that is almost 10 μm

long. Long genomes, which can be as much as 1,000 times longer than the head, must be very tightly packed into the head of the phage.

Because phage are so small, they are usually only detected by the "holes" or plaques (Figure 7.1B) they form on **lawns** of susceptible host bacteria (see Introduction). Each type of phage will make plaques on only certain host bacteria, which define its **host range**. Mutations in the phage DNA can alter the host range of a phage or the conditions under which the phage can form a plaque, which is usually how mutations are detected. In this chapter, we discuss what is known about how some representative phages multiply and some of the genetic experiments that have contributed to this knowledge. We also discuss how some phages can be used for genetic mapping and strain construction in bacteria in a process called transduction. First, however, we review some general features of phage development.

Bacteriophage Lytic Development Cycle

Because phages, like all viruses, are essentially genes wrapped in a protein or membrane coat, they cannot multiply without benefit of a host cell. The virus injects its genes into a cell, and the cell furnishes some or all of the means to express those genes and make more viruses.

Figure 7.2 illustrates the multiplication process for a typical large DNA phage. To start the infection, a phage adsorbs to an actively growing bacterial cell by binding to a specific receptor on the cell surface. In the next step, the phage injects its entire DNA into the cell, where transcription of RNA, usually by the host RNA polymerase, begins almost immediately. However, not all the genes of a phage are transcribed into mRNA when the DNA first enters the cell. Only some of the genes of the phage have promoters that mimic those of the host cell DNA and so are recognized by the host RNA polymerase. Those transcribed soon after infection are called the **early genes** of the phage and encode mostly enzymes involved in DNA synthesis such as DNA polymerase, primase, DNA ligase, and helicase. With the help of these enzymes, the phage DNA begins to replicate and many copies accumulate in the cell.

Next, mRNA is transcribed from the rest of the phage genes, the **late genes**, which may or may not be intermingled with the early genes in the phage DNA, depending on the phage. These genes have promoters that are unlike those of the host cell and so are not recognized by the host RNA polymerase alone. Most of these genes encode proteins involved in assembly of the head and tail. After the phage particle is completed, the DNA is taken up by the heads. Finally, the cells break open, or **lyse**, and the new phage are released to infect another sensitive cell. This whole process, known as the **lytic cycle**, takes less than 1 h for many phages, and hundreds of progeny phage can be produced from a single infecting phage.

Actual phage development is usually much more complex than this basic process, proceeding through several

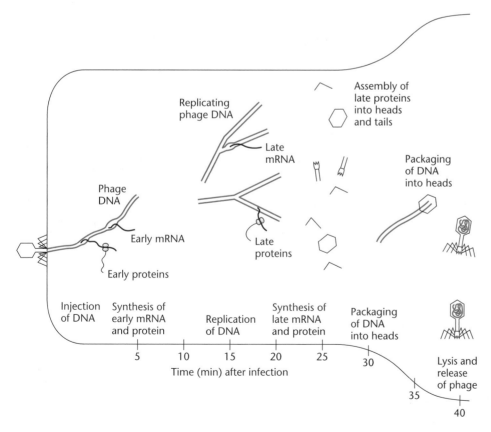

Figure 7.2 A typical bacteriophage multiplication cycle. After the phage injects its DNA, the early genes, most of which encode products involved in DNA replication, are transcribed and translated. Then DNA replication begins, and the late genes are transcribed and translated to form the head and tail of the phage. The DNA is packaged into the heads; the tails are attached; and the cells lyse, releasing the phage to infect other cells.

Replicating phage DNA

Late mRNA

Assembly of late proteins into heads and tails

Packaging of DNA into heads

Phage DNA

Early mRNA

Late proteins

Early proteins

Injection of DNA

Synthesis of early mRNA and protein

Replication of DNA

Synthesis of late mRNA and protein

Packaging of DNA into heads

Lysis and release of phage

5   10   15   20   25   30   35   40

Time (min) after infection

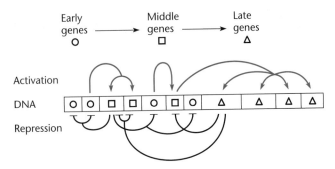

Figure 7.3 Transcriptional regulation during development of a typical large DNA phage. The arrows indicate activation of gene expression; the bars indicate repression of gene expression.

intermediate stages in which the expression of different genes is regulated by specific mechanisms. Most of the regulation is achieved by having genes be transcribed into mRNA only at certain times; this type of regulation is called **transcriptional regulation** (see chapter 2). However, some genes undergo **posttranscriptional regulation**, which occurs after the mRNA has been made. For example, regulation may operate at the level of whether the mRNAs are translated; this is known as **translational regulation**. Other types of posttranscriptional regulation involve the stability of certain RNAs that quickly degrade unless they are synthesized at the right stage of development.

Figure 7.3 shows the basic process of phage gene transcriptional regulation, in which one or more of the gene products synthesized during each stage of development turns on the transcription of the genes in the next stage of development. The gene products synthesized during each stage can also be responsible for turning off the transcription of genes expressed in the preceding stage. Genes whose products are responsible for regulating the transcription of other genes are called **regulatory genes,** and this type of regulation is called a **regulatory cascade** because each step triggers the next step and stops the preceding step. By having such a cascade of gene expression, all the information for the step-by-step development of the phage can be preprogrammed into the DNA of the phage.

Regulatory genes can usually be easily identified by mutations. Mutations in most genes affect only the product of the mutated gene. However, mutations in regulatory genes can affect the expression of many other genes. This fact has been used to identify the regulatory genes of many phages. Below, we discuss some of these genes and their functions, selecting our examples either because they are the basis for cloning technologies or because of the impact they have had on our understanding of regulatory mechanisms in general. All the phages we present in this chapter contain DNA; however, Box 7.1 briefly describes the properties of some RNA phages.

Phage T7: a Phage-Encoded RNA Polymerase

Compared with some of the larger phages, phage T7 has a relatively simple program of gene expression after infection, with only two major classes of genes, the early and late genes. The phage has about 50 genes, many of which are shown on the genome map in Figure 7.4. After infection, expression of the T7 genes proceeds from left to right, with the genes on the extreme left of the genetic map, with genes up to and including gene *1.3*, expressed first. These are the early genes. The genes to the right of *1.3*—the middle DNA metabolism and late phage assembly genes—are transcribed after a few minutes' delay.

Nonsense and temperature-sensitive mutations were used to identify which of the early-gene products is responsible for turning on the late genes. Under nonpermissive conditions, in which the mutated genes were inoperable, amber and temperature-sensitive mutations in gene *1* prevented transcription of the late genes, and so gene *1* was a candidate for the regulatory gene. Later work showed that the product of gene *1* is an RNA polymerase that recognizes the promoters used to transcribe the late genes. The sequence of these promoters differs greatly from those recognized by bacterial RNA polymerases, so that these phage promotors will be recognized only by this T7-specific RNA polymerase. Other phages, including T7's close relative T3 and phage φ29 *of Bacillus subtilis* (Box 7.2), also synthesize their own RNA polymerase, which exclusively recognizes their own promoters. The specificity of phage RNA polymerases for their own promoters has been exploited in many applications in molecular genetics, some of which we discuss below.

Figure 7.4 Genetic map of phage T7.

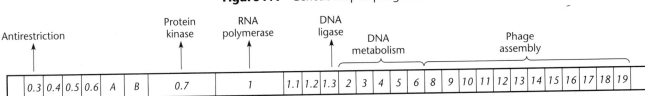

RNA Phages

The capsids (i.e., heads) of many viruses contain RNA instead of DNA. Some of these viruses, the so-called retroviruses, use enzymes called reverse transcriptases to transcribe the RNA into DNA, and these enzymes, because they are essentially DNA polymerases, need primers. In contrast, other RNA viruses, such as the influenza viruses, which cause flu, and the reoviruses, which cause colds, replicate their RNA by using RNA replicases and need no DNA intermediate. Because these RNA replicases have no need for primers, the genomes of RNA viruses can be linear without repeated ends. As we might expect, RNA viruses seem to have higher spontaneous mutation rates during replication, probably because their RNA replicases have no editing functions.

Some phages also have RNA as their genome. Examples include Qβ, MS2, R17, f2, and φ6. The *E. coli* RNA phages Qβ, MS2, R17, and f2 are similar to each other. All have a single-stranded RNA genome that encodes only four proteins: a replicase, two head proteins, and a lysin. Immediately after the RNA enters the cell, it serves as an mRNA and is translated into the replicase. This enzyme replicates the RNA, first by making complementary minus strands and then by using these as a template to synthesize more plus strands. The phage genomic RNA must serve as an mRNA to synthesize the replicase, because no such enzyme exists in *E. coli*. Interestingly, the phage Qβ replicase has four subunits, only one of which is encoded by the phage. The other three are components of the host translational machinery: two of the elongation factors for translation, EF-Tu and EF-Ts, and a ribosomal protein, S1. Because the genomes of these RNA phages also function as an mRNA,

they have served as a convenient source of a single species of mRNA in studies of translation.

Another RNA phage, φ6, was isolated from the bean pathogen *Pseudomonas syringae* subsp. *phaseolicola*. The RNA genome of this phage is double stranded and exists in three segments in the phage capsid, much like the reoviruses of mammals. Also like animal viruses, this phage is surrounded by membrane material derived from the host cell, i.e., an envelope, and the phage enters its host cells in much the same way that animal viruses enter their hosts. However, unlike most animal viruses, φ6 is released by lysis.

The replication, transcription, and translation of the double-stranded RNA of a virus such as φ6 present special problems. Not only must the phage replicate its double-stranded RNA, but it must also transcribe it into single-stranded mRNA since double-stranded RNA cannot be translated. The uninfected host cell will contain neither of the enzymes required for these functions, which therefore must be virus encoded and packaged into the phage head so that they enter the cell with the RNA. Otherwise, neither the transcriptase nor the replicase could be made. Another interesting question with this phage is how three separate RNAs are encapsidated in the phage head. This question has not yet been answered and is common to all viruses with segmented genomes.

References

Blumenthal, T., and G. G. Carmichael. 1979. RNA replication: function and structure of Qβ replicase. *Annu. Rev. Biochem.* **48:**525–548.

Ewen, M. E., and H. R. Revel. 1990. RNA-protein complexes responsible for replication and transcription of the double-stranded RNA bacteriophage φ6. *Virology* **178:**509–519.

T7 PHAGE-BASED EXPRESSION VECTORS

Some of the most useful expression vectors use the T7 RNA polymerase to express foreign genes in *Escherichia coli*. The pET vectors (*p*lasmid *e*xpression *T*7) are a family of plasmid expression vectors that use the T7 phage RNA polymerase and T7 gene 10 promoter to express foreign genes in *E. coli* (see Figure 2.36 for an illustration of one such vector, pET15b). The pET expression vectors use the promoter of the head protein gene (gene 10) of T7, which is very strong. Hundreds of thousands of copies of T7 head protein must be synthesized in a few minutes after infection, making this one of the strongest known promoters. Downstream of the T7 promoter are a number of restriction sites into which foreign genes can be cloned. Any foreign gene cloned downstream of this T7 promoter will be transcribed at

very high rates by the T7 RNA polymerase. A number of variations of the vector shown in the figure have been designed. Some of the pET vectors have strong TIRs for making translational fusions to affinity tags such as His tag, which makes the protein easy to detect and purify on nickel columns (see chapter 2 for a discussion of translational fusions and affinity tags). The T7 promoter can also be made inducible by providing the T7 RNA polymerase only when the foreign gene is to be expressed. This is important in cases where the fusion protein is toxic to the cell.

One general strategy for using a pET vector is illustrated in Figure 7.5. To provide a source of inducible phage T7 RNA polymerase, *E. coli* strains have been constructed in which the phage gene *1* for RNA polymerase is cloned downstream of the inducible *lac* promoter and

Protein Priming

Some viruses, including the adenoviruses and the *Bacillus subtilis* phage φ29, have solved the primer problem by using proteins, rather than RNA, to prime their DNA replication. In the virus head, a protein is covalently attached to the 5' end of the virus DNA. After infection, the DNA grows from this protein, with the first nucleotide attached to a specific serine on the protein. Thus, the virus DNA does not need to form circles or concatemers. The phage DNA polymerase uses this protein to prime its replication by an unusual "sliding back" mechanism. First the DNA polymerase adds a dAMP to the hydroxyl group of a specific serine on the protein. The incorporation of this dAMP is directed by a T in the template DNA. However, the T is the second nucleotide from the 3' end of the template, not the first deoxynucleotide. The DNA polymerase then backs up to recapture the information in the 3' deoxy-nucleotide before replication continues. After replication, the extra dAMP at the 5' end of the newly synthesized strand is removed, the protein is transferred to the 5' end of the newly replicated strand, and replication continues. In this way, no information is lost during replication.

Phage φ29 has also been an important model system to study phage maturation because the phage DNA can be packaged very efficiently into phage heads in a test tube.

References

Escarmis, C., D. Guirao, and M. Salas. 1989. Replication of recombinant φ29 DNA molecules in *Bacillus subtilis* protoplasts. *Virology* **169**:152–160.

Meijer, W. J., J. A. Horcajadas, and M. Salas. 2001. φ29 family of phages. *Microbiol. Mol. Biol. Rev.* **65**:261–287.

Figure 7.5 Strategy for regulating the expression of genes cloned into a pET vector. The gene for T7 RNA polymerase (gene 1) is inserted into the chromosome of *E. coli* and transcribed from the *lac* promoter; therefore, it will be expressed only if the inducer IPTG is added. The T7 RNA polymerase will then transcribe the gene cloned into the pET vector. If the protein product of the cloned gene is toxic, it may be necessary to further reduce the transcription of the cloned gene before induction. The T7 lysozyme encoded by a compatible plasmid, pLysS, will bind to any residual T7 RNA polymerase made in the absence of induction and inactivate it. Also, the presence of *lac* operators between the T7 promoter and the cloned gene will further reduce transcription of the cloned gene in the absence of the inducer IPTG. Reprinted with permission from the *Novagen Catalog*, p. 88, Novagen, Madison, Wis., 2003.

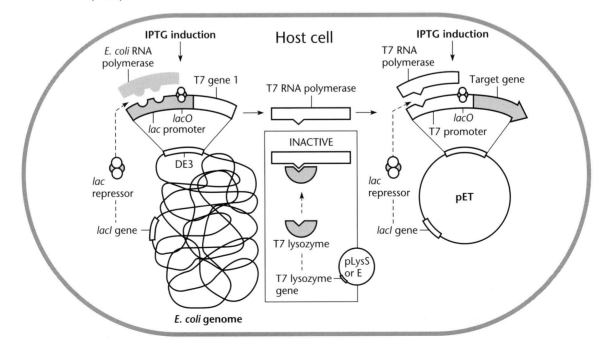

integrated into their chromosome. In these strains, which often have the DE3 suffix, the phage polymerase gene will be transcribed only from the *lac* promoter, so that the T7 RNA polymerase will be synthesized only if an inducer of the *lac* promoter, such as isopropyl-β-D-thiogalactopyranoside (IPTG), is added. The newly synthesized T7 RNA polymerase will then make large amounts of mRNA on the foreign gene cloned into the pET plasmid and large amounts of its protein product.

RIBOPROBES AND PROCESSING SUBSTRATES

Another application of phage RNA polymerases is in making specific RNAs in vitro. Specific RNAs made from a single gene are useful as probes for hybridization experiments (**riboprobes**) or as RNA substrates for processing reactions, such as splicing. This technology is also based on the fact that the phage RNA polymerases will transcribe RNA only from their own promoters. Vectors have been designed with multiple restriction sites bracketed by promoters for phage RNA polymerases. The gene on which RNA is to be made is cloned into one of the restriction sites on the vector, and the vector DNA is purified and cut with a restriction endonuclease on the other side of the cloned gene from the phage promoter. When purified phage RNA polymerase is added along with the other ingredients including the ribonucleoside triphosphates needed for RNA synthesis, the only RNA that is made is complementary to the transcribed strand of the cloned gene. To make RNA complementary to the other strand, the gene can be cloned in the opposite orientation in the cloning vector or a special vector can be used that has the promoter for another phage on the other side of the cloning site. For example, in one such vector, the cloning site is bracketed by a T7 promoter on one side and a φ29 promoter on the other. One strand of the cloned gene DNA will be transcribed into RNA if purified T7 RNA polymerase is added, but the other strand will be transcribed into RNA if φ29 RNA polymerase is added. Purified RNA polymerase of T7 and other phages are available from biochemical supply companies.

PHAGE DISPLAY

Another application of phage T7 is in phage display (see Box 7.3). This is a way of detecting peptides which bind to another molecule such as a hormone or a specific antibody. To use this technology, a randomized peptide-coding sequence is fused to the T7 head protein so that different phages display different versions of the peptide on their surface. The phage that display a version of the peptide which binds to the other molecule are "panned for" and isolated. The cloned DNA can then be sequenced to determine the particular peptide sequence which binds to the molecule.

Phage T4: Transcriptional Activators, Antitermination, a New Sigma Factor, and Replication-Coupled Transcription

Bacteriophage T4 is one of the largest known viruses with a complex structure reminiscent of a lunar landing module (Figure 7.1A). Experiments with this phage have been very important in the development of molecular genetics, so important in fact, that this phage deserves equal status with Mendel's peas (see the introductory chapter). The function of ribosomes, the existence of mRNA, the nature of the genetic code, the confirmation of codon assignments, and many other basic insights originally came from studies with this phage.

Phage T4 is much larger than T7, with over 200 genes (the T4 genome map is shown in Figure 7.6), and the regulation of its gene expression is predictably more complex. In fact, T4 uses many of the known mechanisms of regulation of gene expression at some stage of its life cycle.

Figure 7.7 shows the time course of T4 protein synthesis after infection. Each band in the figure is the polypeptide product of a single phage T4 gene, and some of the gene products are identified by the gene that encodes them (for details on how the bands were obtained, see the legend to Figure 7.7). For example, p37 is the *p*roduct of gene *37* (see the T4 map in Figure 7.6). Because of the way the polypeptides were labeled, the time at which a band first appears is the time at which that gene begins to be expressed, and the time a band disappears is the time that gene is shut off. Clearly, some genes of T4 are expressed immediately after infection. These are called the immediate-early genes. Other genes, called the delayed-early and middle genes, are expressed only a few minutes after infection. Later, expression begins of the true-late genes, so called to distinguish them from some of the delayed-early and middle genes that continue to be expressed throughout infection. Overall, the regulation of protein synthesis during T4 phage development is very complex, as might be expected from such a large virus.

The assigments of the polypeptide products to genes were made by using amber mutations in the genes. If nonsuppressor cells (i.e., cells that lack a nonsense suppressor; see chapter 3) are infected by a phage with an amber mutation in a gene, the band corresponding to the product of that gene will be missing. Translation of the gene will stop at the amber mutation in a nonsuppressor host, leading to the synthesis of a shorter polypeptide, which can sometimes be detected elsewhere on the gel. Sometimes two or more bands are missing as the result of a single amber mutation, such as the two bands missing as a result of an amber mutation in gene *23*, identified as p23 and p23* in Figure 7.7. Several factors could

BOX 7.3

Phage Display

One current application of phages is in phage display. This technology allows the identification and synthesis of proteins that bind tightly to other proteins. An example of such a protein sequence would be a specific antigen which binds tightly to antibodies directed against it. Phage display technology depends upon the fact that phage are made up of DNA coated with protein encoded by that DNA. If a particular protein-coding sequence is translationally fused to the coding sequence for one of the head proteins of the phage, all the progeny of that phage will display the particular pro-

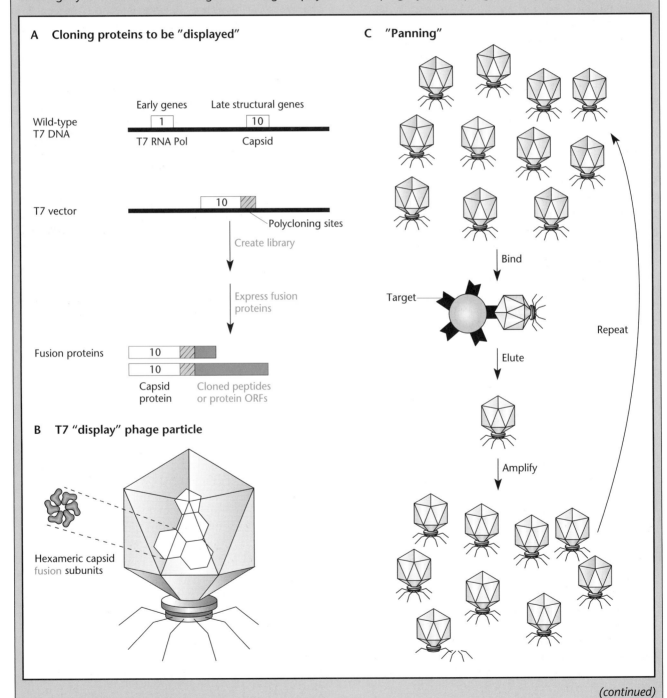

A Cloning proteins to be "displayed"

Wild-type
T7 DNA

Early genes   Late structural genes

1   10

T7 RNA Pol   Capsid

T7 vector   10

Polycloning sites

Create library

Express fusion
proteins

Fusion proteins   10

10

Capsid   Cloned peptides
protein   or protein ORFs

B T7 "display" phage particle

Hexameric capsid
fusion subunits

C "Panning"

Bind

Target

Elute

Repeat

Amplify

(continued)

BOX 7.3 (continued)

Phage Display

tein sequence on their surface. Commercially available kits have been developed which use either M13 or T7 for phage display. The figure outlines the steps in phage display with T7. In panel A, the protein-coding sequence to be displayed (shown in gold) is cloned into the phage DNA to make a translational fusion to the head protein-coding region, gene *10*, of the phage. The DNA is then introduced into cells by transfection or by packaging the DNA into a phage head and using the phage to infect cells. When the phage multiply, their progeny will all display that particular protein sequence on their surface along with the head proteins (panel B). Therefore, if a way can be found to identify and isolate a phage particle displaying the protein sequence on its surface, all the progeny of that phage will also display that particular protein sequence on their surface.

One way to isolate phage displaying a particular protein sequence which binds to another molecule is by a process called panning, so named because of its analogy to panning for gold and other minerals. Panel C illustrates the principle behind panning. A ligand molecule to which the desired protein sequence binds is fixed to a solid surface such as a well in a microtiter plate. Ligands are molecules which bind to proteins and can be either RNA, DNA, another protein, or even a small molecule such as a hormone. A mixture of phages, some of which presumably display the protein se-

quence to which the ligand binds, are put in the well of the plate, and the excess phages are washed out. Any phages which bind to the ligand in the well will be preferentially retained. The remaining phages are then washed out and propagated, and the process is repeated several times to enrich for phages displaying the desired ligand-binding protein sequence. The DNA of the phage displaying the desired protein sequence can then be sequenced to determine the sequence of the peptide which binds to the ligand or target molecule.

One particularly powerful application of phage display is in selecting unknown protein sequences which bind tightly to a particular ligand. Such peptides may be important therapeutic agents because they could block essential bacterial and viral functions. Any phage could in principle be used for phage display. All that is required is that one of the phage coat proteins exposed to the surface can be fused to other proteins without inactivating the phage. In practice, the lengths of the protein sequence that can be fused to the head protein are short so that they do not disrupt the phage structure.

Reference
Harrison, J. L., S. C. Williams, G. Winter, and A. Nissim. 1996. Screening phage antibody libraries. *Methods Enzymol.* **267**:83–109.

---

account for the absence of multiple bands as a result of a single mutation. In this case, the polypeptide product of gene *23*, which makes up the phage head, is cleaved after it is synthesized. Normally, approximately 1,000 copies of this polypeptide are used to build every phage head. The head is first assembled with the p23 polypeptide, and then part of the N terminus of p23 is cut off to form the shorter polypeptide p23* as the head matures into its final form before DNA is encapsidated (i.e., put inside the head). Thus, by disrupting the synthesis of p23, the mutation also prevents the appearance of p23*. The T4 gene products are often referred to as gp23, etc., for *gene product of 23*.

Experiments like those described in the legend to Figure 7.7 were also used to identify the regulatory genes of phage T4. Mutations in these genes prevent the expression of many other genes and so cause the disppearance of many bands from the gel. Mutations in genes named *motA* and *asiA* prevent the appearance of the middle gene products. Mutations in genes *33* and *55*, as well as mutations in many of the genes whose products are re-

quired for T4 DNA replication, prevent the appearance of the true-late gene products. Therefore, *motA*, *asiA*, *33*, and *55*, as well as some genes whose products are required for DNA replication, are predicted to be regulatory genes. Many genetic and biochemical experiments have been directed toward understanding how these T4 regulatory gene products turn on the synthesis of other proteins.

MIDDLE-MODE TRANSCRIPTION

The genes that are turned on a few minutes after infection can be divided into two classes, the "delayed-earlies" and the "middles." The delayed-early genes of T4 are transcribed from the same normal σ^{70} promoters as the immediate-early genes but are regulated by an antitermination mechanism. Without some T4 regulatory gene products, the RNA polymerase which has initiated at an immediate early promoter will stop before it gets to the delayed-early genes. Therefore, the transcription of these genes must await the synthesis of antitermination factors encoded by the phage. We shall defer discussion

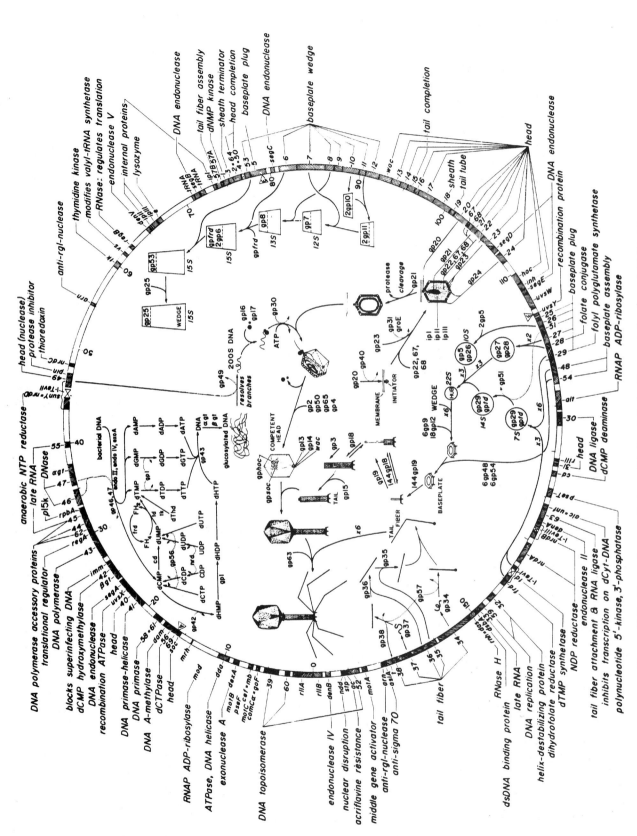

Figure 7.6 Genomic map of phage T4. Reprinted with permission from J. D. Karam (ed.), *Molecular Biology of Bacteriophage T4*, ASM Press, Washington, D.C., 1994.

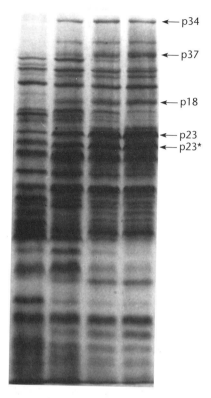

Figure 7.7 Polyacrylamide gel electrophoresis of proteins synthesized during the development of phage T4. The proteins were labeled by adding amino acids containing the ^{14}C radioisotope of carbon at various times after the phage were added to the bacteria. Because phage T4 stops host protein synthesis after infection, radioactive amino acids will be incorporated only into phage proteins, and therefore only the phage proteins will become radioactive. Moreover, only those phage proteins made when the radioactive amino acids were added will be labeled. Hence, we can tell which phage proteins are made at any given time. To separate the proteins, the cells were broken open with the detergent sodium dodecyl sulfate, which also separates the polypeptides that are part of the multimeric proteins. Electrophoresis caused the polypeptides to migrate on an acrylamide gel, forming bands in columns. The smaller polypeptides move faster on these gels, so that the polypeptides are arranged by size from the smallest at the bottom to the largest at the top. The gel was subsequently dried on a piece of filter paper and then laid next to a photographic film, which was exposed to the high-energy light waves given off from the radioactive disintegrations. Each band represents the polypeptide product of a single T4 gene. Lane 1 from the left, proteins synthesized between 5 and 10 min after infection; lane 2, 10 to 15 min; lane 3, 15 to 20 min; lane 4, 30 to 35 min.

of regulation by antitermination of transcription until the next chapter on phage λ, where the mechanism is explained. The middle gene products, however, are transcribed from their own promoters: the so-called "middle-mode" promoters, which look somewhat different from normal σ^{70} promoters in their −35 sequence (Figure 7.8). Because they are somewhat different, transcription from these promoters requires the MotA protein, the product of the *motA* gene, which is a transcriptional activator that allows the bacterial RNA polymerase to transcribe from the middle-mode promoters. We shall discuss transcriptional activators in more detail in chapter 12.

The function of the AsiA protein in stimulating transcription from middle-mode promoters is more obscure. By itself, it binds to a region of σ^{70} and inhibits transcription from normal σ^{70} promoters. However, in combination with MotA, it stimulates transcription from middle-mode promoters (see Colland et al., Suggested Reading).

TRUE-LATE TRANSCRIPTION

The latest genes of T4 to be transcribed are the true-late genes. The products of these genes are mostly the head, tail, and tail fiber components of the phage and enzymes and proteins needed to lyse the cell and release the phage. The initiation of transcription of the true-late genes of T4 has received especial attention because it is coupled to the replication of the DNA. Other viruses, including some human viruses, are known also to couple their late transcription to the replication of their DNA, although it is not known if they use a similar mechanism. This type of transcription regulation would also have obvious advantages in coordinating gene expression and cell division with replication during the normal cell cycle, since genes would be transcribed only when they were replicated. Thus, a detailed understanding of the mechanism of true-late transcription in T4 phage may point to similar mechanisms in other systems, contributing yet another universal basic cellular regulatory mechanism to the list of those first discovered with this phage.

Figure 7.8 Sequence of T4 middle-mode and late promoters. Only the important sequences for recognition by RNA polymerase are shown.

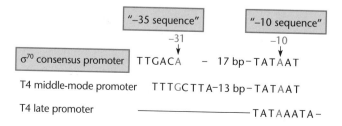

Like the middle genes, the true-late genes of T4 are transcribed from promoters that are different from those of its host (Figure 7.8). They have the sequence TATAAATA, rather than the characteristic −35 and −10 sequences of a bacterial σ^{70} promoter (see chapter 2). Because of this difference, the host RNA polymerase will not normally recognize the T4 promoters. However, the product of the regulatory gene *55* is an alternate sigma factor that binds to the host RNA polymerase, changing its specificity so that it recognizes only the promoters for the T4 true-late genes (see Kassavetis and Geiduschek, Suggested Reading).

Phage T4 and its close relatives are not the only phages to use alternate sigma factors to activate the transcription of their late genes. For example, the *Bacillus subtilis* phage SPO1 also uses this regulatory mechanism. The SPO1 late promoters are very unlike the normal bacterial promoters, but they are also quite unlike the T4 late promoters. In fact, host RNA polymerases can be adapted to recognize a wide variety of promoter sequences merely through the attachment of an alternate sigma factor. This general strategy is also used during many bacterial developmental processes, as we discuss in later chapters.

REPLICATION-COUPLED TRANSCRIPTION

In addition to the alternate sigma factor encoded by gene *55*, many other gene products are required to turn on the transcription of the late genes. Most, notably those of genes *44*, *62*, and *45*, are required for replication of the phage DNA. This observation has led to the conclusion that T4 DNA replication is also required for the expression of the late genes and that the complex of the host RNA polymerase and the gene *55* protein will initiate RNA synthesis efficiently only if the T4 DNA is replicating. Coupling the transcription of the true-late genes to the replication of the phage DNA makes sense from a mechanistic standpoint. Many of the true-late genes encode parts of the phage particle. These proteins are not needed until phage DNA is available to be packaged inside them.

Figure 7.9 shows features of a model to explain the requirement of T4 DNA replication for the activation of true-late gene transcription and related observations (see Herendeen et al., Suggested Reading). According to the model, the promoters for the true-late genes are activated when the gp33 protein, which is bound to RNA polymerase, makes contact with the gp45 protein, which is normally part of the replication apparatus. The gp45 protein is a DNA polymerase accessory protein that acts like the β protein in *E. coli* and wraps around the DNA to form a "sliding clamp," which moves with the DNA polymerase and helps prevent it from falling off the

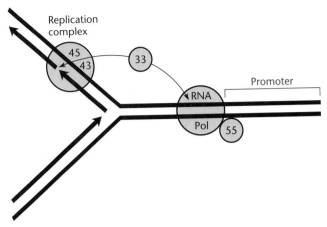

Figure 7.9 Model for replication activation of a T4 late-gene promoter in which the promoter is not activated until a replication fork approaches. The proteins are identified by the genes that encode them. The replication complex includes the DNA polymerase gp43 and the accessory protein gp45. gp33 binds both to the clamp formed by the gp45 on the DNA at the replication fork and to the RNA polymerase at the promoter, thereby activating transcription.

DNA during replication (see chapter 1). The gp44 and gp62 proteins help clamp the gp45 protein onto the DNA. According to the model for how the true-late promoters are activated, the gp45 protein sliding clamp can load on the DNA either at a nick or because it is associated with the replication apparatus. It can then slide along the DNA in either direction. If it makes contact with a gp33 protein bound to the RNA polymerase at a true-late promoter, it will allow the RNA polymerase to initiate transcription from the true-late promoter. This explains why the gp33 and gp45 proteins are required for optimal true-late-gene transcription, as well as gp44 and gp62, since the latter are required for gp45 protein to load on the DNA to form the sliding clamp.

Phage DNA Replication

Unlike the replication of chromosomal or plasmid DNAs, which must be coordinated with cell division, phage DNA replication is governed by only one purpose: to make the greatest number of copies of the phage genome in the shortest possible time. Phage replication can be truly impressive. The one or a few phage DNA molecules that initially enter the cell can replicate to make hundreds or even thousands of copies to be packaged into phage heads in as little as 10 or 20 min. This unchecked replication often makes phage DNA replication easier to study than replication in other systems. Nevertheless, phage replication has many of the same requirements as cellular replication in living organisms, and phage DNA replication has served as a model system to understand DNA replication in bacteria and even in humans.

The structures of many phage genomes present special problems for replication. Some phage genomes, for example, that of phage M13, are single-stranded circles. The DNA of most larger phages is not circular like that of M13 but, rather, is linear. As discussed in chapter 1, linearity presents problems for replication of the extreme 5′ ends of DNA, because DNA polymerases require a primer. Even if the primers for the 5′ ends are synthesized as RNA, once the RNA primer is removed there is no DNA upstream to serve as a primer for the primer's replacement as DNA. Because of this priming problem, a linear DNA would get smaller each time it replicated until essential genes were lost. Eukaryotic chromosomes have telomeres at their ends, but phages with linear genomes solve this primer problem in different ways. One is to use protein primers (see Box 7.2). Other phages have repeated sequences at the ends of their genomic DNA, as we discuss below. Phages use a surprising variety of mechanisms to solve their replication problems; some of these are described in this section.

Phages with Single-Stranded Circular DNA

The genome of some small phages consists of circular single-stranded DNA. The small *E. coli* phages that fit into this category can be separated into two groups. The representative phage of one group, φX174, has a spherical capsid with spikes sticking out, like the ball portion of the medieval weapon known as a morning star. These are called icosahedral phages because their capsid is an icosahedron like a geodesic dome with mostly six-sided building blocks and an occasional five-sided block. In the other group, represented by M13 and f1, a single layer of protein covers the extended DNA molecule, making the phage filamentous in appearance.

The different shapes of phages of these two families determine how they infect and leave the infected cell. The icoshedral phages enter and leave cells much like other phages. They bind to the cell surface, and only the DNA enters the cell; then they lyse or break open the cell to exit after they have developed. In contrast, M13 and other **filamentous phages** are **male-specific** phages because they specifically adsorb to the sex pilus encoded by certain plasmids and so will infect only "male" strains of bacteria (see chapter 5). Unlike most other phages, filamentous phages do not inject their DNA. Instead, the entire phage is ingested by the cell, and the protein coat is removed from the DNA as the phage passes through the inner cytoplasmic membrane of the bacterium. After the phage DNA has replicated, it is again coated with protein as it leaks back out through the cytoplasmic membrane. These phage do not lyse infected cells and leak out only slowly. Consequently, cells infected with M13 or other filamentous phages are "chronically" rather than "acutely" infected. Nevertheless, the filamentous phages form visible plaques, because chronically infected cells grow more slowly than uninfected cells.

The process of infection of the cell by a filamentous single-stranded DNA phage and its release from the cell has been studied in some detail because it serves as a model system for the ability of a large particle like a virus to get through a membrane (see Rakonjac et al., Suggested Reading). Most of these studies have been performed with phage f1, but related phages probably use a similar mechanism.

Phage f1 particle has only five proteins; one of which, the major head protein, pVIII, exists in about 2,700 copies and coats the DNA. The other four proteins are on the ends of the phage and exist in only four or five copies per phage particle. Two of these proteins, pVII and pIX, are located on one end of the phage, and the other two, pVI and pIII, are on the other end. To start the infection, the pIII protein on one end of the phage first makes contact with the end of the sex pilus (Figure 7.10). The sex pilus makes a good first contact point because it sticks out of the cell and hence is very accessible. The pilus retracts when the phage binds to it, drawing the phage to the cell surface. A different region of the same pIII protein then contacts an host inner membrane protein called TolA. How this contact is made is somewhat unclear. The TolA protein sticks into the periplasmic space and might make contact with the outer membrane since it is part of a larger structure whose role has something to do with keeping the outer membrane intact. The phage DNA then enters the cytoplasm, while the major coat protein is stripped off into the host inner membrane.

Release of the phage from the cell uses a different process. This process is also illustrated in Figure 7.10. Unlike injection of the phage DNA, which must rely exclusively on host proteins, secretion of the phage from infected cells can use newly synthesized phage proteins synthesized during the course of the infection. After the phage DNA has replicated a few times to produce the RF DNA form (see below), it enters the rolling-circle stage of replication. As it rolls off the circle, the newly synthesized single-stranded DNA is coated by another protein, pV. The proteins which make up the phage coat are waiting in the membrane, and the major head protein, pVIII, replaces the pV protein on the DNA as it enters the membrane. Only DNA containing the sequence of deoxynucleotides of the *pac* site of the phage will be packaged. The other phage proteins are then added to the particle. Meanwhile, the phage-encoded secretin protein, pIV, has formed a channel in the outer membrane,

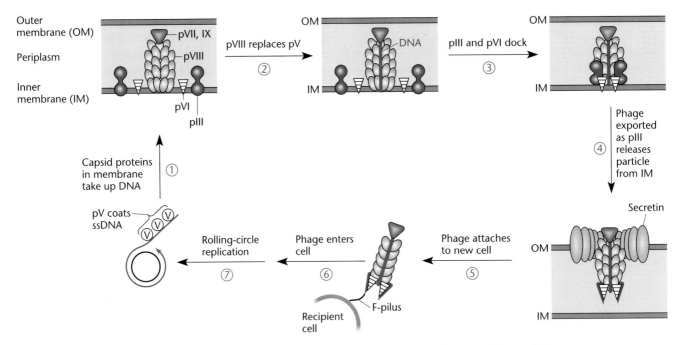

Figure 7.10 Infection cycle of the single-stranded DNA phage f1. Steps 1 through 7 show the encapsidation of phage DNA as it is secreted through the membrane pore formed by the pIV secretin to release the phage and infect a new cell. Details are given in the text. ssDNA, single-stranded DNA.

through which the assembled phage can pass. This channel is related to the channels formed by type II and III secretion systems to inject virulence proteins into eukaryotic cells in bacterial pathogenesis (see Box 2.6).

REPLICATION OF SINGLE-STRANDED PHAGE DNA

Studies of the replication of single-stranded phage DNA have contributed much to our understanding of replication in general. It was with these phages that rolling circle replication was discovered as well as many proteins required for host DNA replication including the proteins PriA, PriB, PriC, and DnaT which are now known to be involved in restarting chromosomal replication forks after they have dissociated upon encountering DNA damage (see Box 1.2 Restarting Replication Forks). Many of the genes for these host proteins were found by isolating host mutants that cannot support the development of these phages and by reconstituting replication systems in vitro by adding host DNA replication proteins to phage DNA until replication was achieved.

The groups working on the secretion of single-stranded DNA phage from the cell have been different from those working on phage DNA replication, and these groups have used different phages. Much of the work on phage DNA replication has been done with M13 and φX174, and the genes for these phages are named somewhat differently from those for phage f1.

First we talk about the replication of the DNA of the filamentous phage M13, and then we compare it to the replication of the icosahedral phage φX174, which turns out to be surprisingly different.

Synthesis of the Complementary Strand To Form the First RF

The steps in the replication of phage M13 DNA are outlined in Figure 7.11. The DNA strand of the phage encapsidated in the phage head is called the + strand. Immediately after the single-stranded DNA enters the cell, a complementary – strand is synthesized on the + strand to form a double-stranded DNA called the **replicative form (RF)**. The formation of this first RF is dependent entirely on host functions, as it must be since no phage proteins enter the cytoplasm with the phage DNA and phage proteins cannot be synthesized from single-stranded DNA. The synthesis of the complementary strand is primed by an RNA made by the normal host RNA polymerase. Once the RNA primer is synthesized, the DNA polymerase III and accessory proteins load on the DNA and synthesize the complementary strand until they have circumvented the DNA and encountered the RNA primer, which they displace. The 5′exonuclease activity of DNA polymerase I then removes the RNA primer, and the nick is sealed by DNA ligase to leave a double-stranded covalently closed RF, which can then be supercoiled (see chapter 1). The repli-

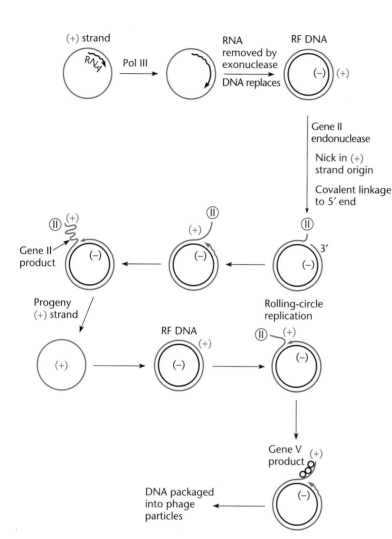

Figure 7.11   Replication of the circular single-stranded DNA phage M13. First, an RNA primer is used to synthesize the complementary minus strand (in black) to form double-stranded RF DNA. The product of gene II, an endonuclease, nicks the plus strand of the RF and remains attached to the 5′ phosphate at the nick. Then more + strands are synthesized via rolling-circle replication, and their – strands are synthesized to make more RFs. Later, the gene V product binds to the plus strands as they are synthesized, preventing them from being used as templates for more RF synthesis and helping package them into phage heads.

cation also occurs on a specific site on the membrane of the bacterium, sometimes called the reduction sequence. This binding may direct the phage DNA to sites where the replication machinery of the host is located.

The icosahedral phages like φX174 use a much more complicated mechanism to initiate synthesis of the first RF. Rather than use just the host RNA polymerase, they assemble a large primosome at a unique site on the single-stranded phage DNA. This primosome is composed of many copies of seven different proteins including DnaB (the replicative helicase), DnaC (which loads the helicase on the DNA) and DnaG, the primase. Most of these proteins are required to initiate replication at the chromsomal origin *oriC* and they have just been commandeered by the phage for the same purpose. However, the primosome also includes the proteins PriA, PriB, PriC, and DnaT, which are not required for initiation of chromosomal replication at *oriC* but, rather, are required to restart chromsomosomal replication after it has been blocked due to encountering damage in the

DNA template. The role of the PriA, PriB, PriC, and DnaT proteins in the initiation of replication of φX174 DNA is not yet clear. Some of these proteins are helicases and may be required to open up a hairpin at the unique origin of replication of the single-stranded DNA and to allow the replication apparatus to load on the DNA and the primosome to move on the DNA. Studies of the replication of φX174 DNA in vitro originally led to the discovery of the Pri proteins (see Box 1.2). Presumably, eukaryotes, including humans, use a similar mechanism to restart replication forks. This is yet another case where studies with phage have led to the discovery of universal phenomena.

Synthesis of More RFs and Phage DNA

Once the first RF of M13 is synthesized, more RFs are made by semiconservative replication. This process requires phage proteins that are synthesized from the first RF. The two strands of the RF are replicated separately and by very different mechanisms. The + strand is replicated by rolling-circle replication from a different origin by a process that is similar to the replication of DNA during plasmid transfer (see chapter 5). First, a nick is made in the RF, at the origin of + strand synthesis, by a specific endonuclease called GpII in M13. A host protein called Rep, a helicase, helps unwind the DNA at the nick. The GpII protein remains attached to the 5′ end of the DNA at the nick, and the DNA polymerase III, with its accessory proteins, extends the 3′ end to synthesize more + strand, displacing the old + strand. The GpII protein bound to the 5′ end of the old displaced strand then reseals the ends of the displaced strand by a transesterification reaction (see chapter 5), re-creating the old circular + strand. This + strand can then serve as the template for − strand synthesis to create another RF.

This process of accumulating RFs continues until a phage protein called GpV begins to accumulate. This protein coats the single-stranded + strand of DNA, probably with the help of the attached GpII, and prevents the synthesis of more RF by a complicated process that is only incompletely understood. The single-stranded viral DNA is then encapsulated in the head, and the cell is lysed (for icosahedral phages like φX174) or transferred to the assembling viral particle in the membrane and leaked out of the cell (for filamentous phages like f1 and M13) as described above.

M13 CLONING VECTORS

Because M13 and related phages have only one DNA strand, these phages provide a convenient vehicle for cloned DNA that we might want to sequence, use as a probe, or use in other applications. Also, because filamentous phages such as M13 have no fixed length and the phage particle will be as long as its DNA, foreign DNA of variable lengths can be cloned into the phage DNA, producing a molecule longer than normal without disrupting the phage's functionality of the phage.

Figure 7.12 shows the M13 cloning vector M13mp18 (see Yanisch-Perron et al., Suggested Reading). Like pUC plasmid vectors, the mp series of M13 phage vectors

Figure 7.12 Map of the M13mp18 cloning vector. The positions of the genes of M13 and the polycloning site containing multiple restriction sites are shown below the map. Cloning into one of these sites will inactivate the portion of the *lacZ* gene on the cloning vector, a process called insertional inactivation. The cloning vector also contains the *lacI* gene, whose product represses transcription from the *p*~LAC~ promoter (see chapter 11).

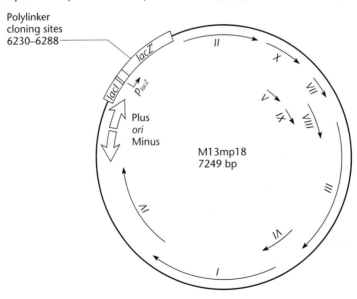

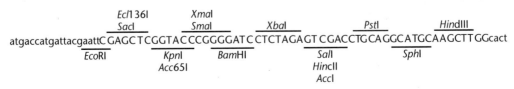

contain the α-fragment-coding portion of the *E. coli lacZ* gene, into which has been introduced some convenient restriction sites (shown as the polylinker cloning site in Figure 7.12, with the multiple restriction sites shown at the bottom of the figure). Phage with a foreign DNA insert in one of these sites can be identified easily by insertional inactivation (see chapter 4), because they will make colorless instead of blue plaques on plates containing X-Gal (5-bromo-4-chloro-3-indolyl-β-D-galactopyranoside).

To use a single-stranded DNA phage vector, the double-stranded RF must be isolated from infected cells, since most restriction endonucleases and DNA ligase require double-stranded DNA. A piece of foreign DNA is cloned into the RF by using restriction endonucleases, and the recombinant DNA is used to **transfect** competent bacterial cells. The term "transfection" refers to the artificial initiation of a viral infection by viral DNA (see chapter 6). When the RF containing the clone replicates to form single-stranded progeny DNA, a single strand of the cloned DNA will be packaged into the phage head. The phage plaques obtained when these phage are plated are a convenient source of one strand of the cloned DNA. Which strand of the cloned DNA is represented in the single-stranded DNA will depend upon the orientation of the cloned DNA in the cloning vector. If it is cloned in one orientation, you get one of the strands; if it is cloned in the other orientation, you get the other strand.

Some plasmid cloning vectors have also been engineered to contain the *pac* site of a single-stranded DNA phage. If cells containing such a plasmid are infected with the phage, the plasmid will be packaged into the phage head in a single-stranded form. Such phages are called phasmids, indicationg that they are a cross between a plasmid and a phage.

Phage Display

Single-stranded filamentous phages are also useful for phage display (see Box 7.3). In fact, these were the first phages to be adapted for such uses. In phage display, a peptide sequence is fused to one of the head proteins of the phage in such a way that it is exposed or displayed on the surface of the phage particle. Phage displaying a desired peptide can be isolated by panning with another molecule that binds to the peptide. These phage have the advantage for phage display that they are small, making panning easier. However, only short peptides and ones which can be transported through the channel with the head protein can be displayed on their surface, since the phage must be secreted through the cell membranes.

SITE-SPECIFIC MUTAGENESIS OF M13 CLONES

Because single-stranded DNA phages like M13 offer a convenient source of only one of the two strands of a cloned DNA, they were used in the first applications of site-specific mutagenesis. As discussed in chapter 1, site-specific mutagenesis involves making a predetermined change in the DNA sequence, unlike random mutagenesis, where the change is made by chance. The standard method of using M13 for site-specific mutagenesis is diagrammed in Figure 7.13. The gene to be mutated has been cloned into a cloning vector such as M13mp18. An oligonucleotide complementary to the region to be mutagenized, except for the change to be made, is synthesized and then hybridized to the single-stranded DNA. DNA polymerase is added, which uses the oligonucleotide as a primer to synthesize the complementary (or −) strand of the M13 DNA, including the clone. The double-stranded RF DNA

Figure 7.13 Site-specific mutagenesis with M13 and a mismatched oligonucleotide primer. See the text for more details.

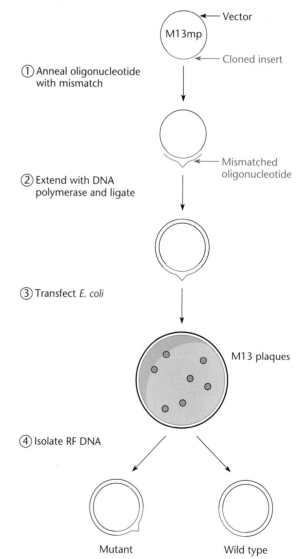

is then ligated to give the covalently closed RF. When this DNA is transfected into cells, some of the plaques will contain progeny phage which have the altered sequence of the oligonucleotide primer rather than the original sequence in the clone, depending upon whether they are descended from the original template strand or the mutagenized complementary strand in the RF.

This method can also be adapted to allow random mutagenesis of a DNA region so that every possible base pair change will be represented in the population of molecules. Instead of a well-defined primer, a mixture of "spiked," or contaminated, oligonucleotide primers is used to mutagenize the DNA. These oligonucleotide primers are synthesized with the deliberate intention of making mistakes. The nucleotide added at each step of the synthesis is deliberately contaminated with a low concentration of the other three nucleotides; the concentration of the contaminating nucleotides is adjusted to make one mistake, on average, in each of the oligonucleotides. When these misfits are used as primers for the synthesis of complementary strands, as above, the DNA synthesized will have a random collection of changes in the region being mutagenized. This method has the advantage that it can be used to make all of the possible base pair changes in a region without preferentially mutagenizing hot spots like chemical mutagens do.

The major difficulty with most methods of site-specific mutagensis for M13 lies in finding which of the plaques contain the few mutated M13 clones among the majority of plaques containing phage which have clones with the original wild-type sequence. A number of methods have been devised to eliminate phages with the original sequence. One of these is illustrated in Figure 7.14. In this method, the thymines in the M13 phage cloning vector are replaced with uracils by propagating the phage in a dUTPase- and uracil-N-glycosylase-deficient (Dut⁻ Ung⁻) host (see chapter 1). When the double-stranded RF molecules are synthesized from such a template, the newly synthesized strand will contain thymines while the template strand contains uracils. When these RFs are transfected into Ung⁺ cells, the uracil-containing template strands will be preferentially degraded, so that most of the phage that survive will be descended from the mutated complementary strands. We discussed another method for eliminating unmutated DNA in chapter 1. This method uses two primers, one that alters a restriction site in the cloning vector and one that makes the desired change in the sequence of the cloned gene (see Figure 1.32). Most of the complementary strands are made using both primers so that they will not contain the restriction site. The RFs are transfected into cells, and the phage are allowed to multiply. The double-stranded RF phage DNAs are isolated from the infected

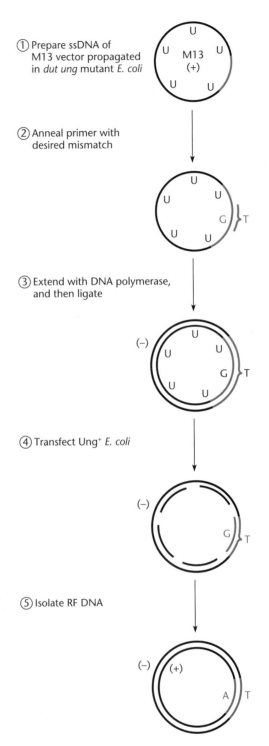

Figure 7.14 Use of uracil-N-glycosylase to eliminate the wild-type sequence after site-specific mutagenesis. See the text for details. ssDNA, single-stranded DNA; DNA Pol, DNA polymerase.

cells and cut with the restriction endonuclease. The RF molecules with the restriction site intact will all be cut, leaving mostly RFs that are descended from the newly synthesized strand with the desired mutation.

Methods for site-specific mutagenesis of M13 clones have been largely replaced by methods using PCR, which do not require additional cloning steps. However, the principles are similar.

Phage T7: Linear DNA That Forms Concatemers

Phage like M13 and φX174 solve the primer problem by having circular DNA, so that there is always DNA upstream to prime the synthesis of new DNA. Other phages, like λ and P22, have cohesive ends at the ends of their DNA that can pair to form circles after infection (see chapter 8). However, some phages, like T7 and T4, never cyclize their DNA but form **concatemers** composed of individual genome-length DNAs linked end to end. The phage DNA can then be cut out of these concatemers so that no information is lost when the phage DNA is packaged.

As shown in Figure 7.15, T7DNA replication begins at a unique *ori* site and proceeds toward both ends of the molecule, leaving the extreme 3′ ends single stranded because there is no way to prime replication at these ends.

However, because T7 has the same sequence at both ends, these single strands are complementary to each other and so can pair, forming a concatemer with the genomes linked end to end. Consequently, the information missing as a result of incomplete replication of the 3′ ends is provided by the complete information at the 5′ end of the other daughter DNA molecule. Individual molecules are then cut out of the concatemers at the unique *pac* sites at the ends of the T7 DNA and packaged into phage heads.

GENETIC REQUIREMENTS FOR T7 DNA REPLICATION

In contrast to single-strand DNA phages, which encode only two of their own replication proteins and otherwise depend on the host replication machinery, T7 encodes many of its own replication functions, including DNA polymerase, DNA ligase, DNA helicase, and primase. The phage T7 RNA polymerase is also required to synthesize the initial primer for phage T7 DNA synthesis. In addition to these proteins, the phage encodes a DNA endonuclease and exonuclease that degrade host DNA to mononucleotides, thereby providing a source of deoxynucleotides for phage DNA replication. Analogous host enzymes can substitute for some of these T7-encoded gene products, so they are not absolutely required for T7

Figure 7.15 Replication of phage T7 DNA. Replication is initiated bidirectionally at the origin (*ori*). The replicated DNAs pair at their terminally repeated ends (TR) to give long concatemers.

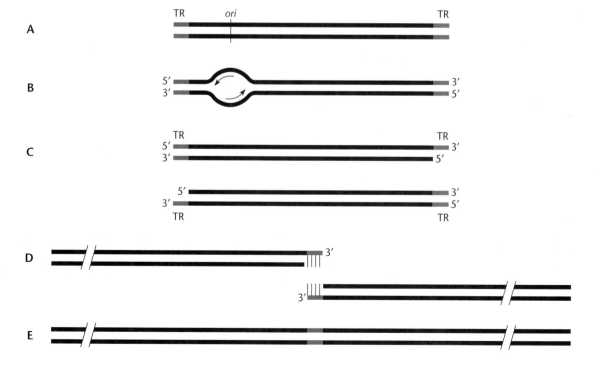

DNA replication. For example, the T7 DNA ligase is not required because the host ligase can act in its stead. T7 DNA replication is a remarkably simple process that requires fewer gene products overall than the replication of bacterial chromosomes and of many other large DNA phages.

Phage T4: Another Linear DNA That Forms Concatemers

Phage T4 also has linear DNA in its head that never cyclizes. It forms concatemers like T7, except that it forms them by recombination between repeated sequences at its ends rather than by pairing between complementary single-stranded ends. However, T4 and T7 differ greatly in how the DNA replicates and is packaged. Also, befitting its larger size, T4 has many more gene products involved in replication than T7 does. As many as 30 T4 gene products participate in replication (Figure 7.6; Table 7.1). In fact, one of the advantages of studying replication with T4 has been the fact that it encodes many of its own replication proteins rather than just using those of its host. It encodes its own DNA polymerase, sliding clamp, clamp-loading proteins, primase, replicative helicase DNA ligase, etc. These proteins are analogous to the replication proteins of bacteria and even eukaryotes (Table 7.1) and it was often in T4 phage where these functions were first discovered; only later were their analogous functions found in uninfected bacteria and eukaryotes. Interestingly, the replication functions of T4 phage are often more similar to those of eukaryotes than to those of the bacteria they infect. For example, the sliding clamp of T4 (the product of gene 45) is more structurally similar to the sliding clamp of eukaryotes, called proliferating-cell nuclear antigen (PCNA), then it is to the sliding clamp of *Escherichia coli*, the bacterium it infects.

OVERVIEW OF T4 PHAGE DNA REPLICATION AND PACKAGING

Phage T4 replication occurs in two stages. These two stages are illustrated in Figure 7.16. In the first stage, T4 DNA replicates from a number of well-defined origins around the DNA. This type of replication is analogous to the replication of bacterial chromosomes and leads to the accumulation of single-genome-length molecules. However, these two daughter molecules will have single-stranded 3' ends because of the inability of DNA polymerase to completely replicate the ends. They lose no information, however, because the sequences at the ends of T4 DNA are repeated, i.e., are terminally redundant (see below). Somewhat later, this type of replication ceases and an entirely new type of replication ensues. The single-stranded repeated sequences at the ends of the genome-length molecules (called terminal redundancies) can invade the same sequence at the ends of other daughter DNAs, forming D-loops, which prime replication to

TABLE 7.1	T4 gene products involved in replication and their homologs in *E. coli* and eukaryotes	
T4 gene product	***E. coli* function**	**Eukaryotic function**[b]
Origin-specific replication		
gp43	Pol III α and ε	DNA pol α, β, γ
gp45	β sliding clamp	PCNA
gp44, gp62	Clamp loader	—[a]
gp41	Replicative helicase	MCM
gp59	DnaC	Edc6
gp61	Primase	—
gp39, gp52, gp60	GyrAB, TopoIV	TopoII
gp30	DNA ligase	—
Rnh	RNase H	—
Recombination-dependent replication		
UvsW	RecG	—
UvsX	RecA	Rad51p (yeast)
UvsY	—	Rad52 (yeast)
gp46, gp47	RecBCD, SbcCD	Rad50p, Mrellp (yeast)
gp32	SSB	RPA
gp59	PriA	—
gp49	RuvABC	—

[a]—, not identified.
[b] PCNA, proliferating-cell nuclear antigen; RPA, SV40 replication protein A.

Stage 1:
specific
origins

Stage 2:
no specific
origins;
recombination
dependent

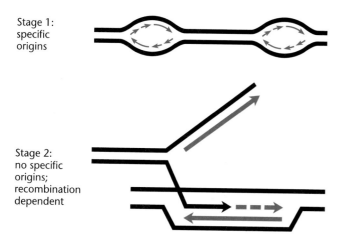

Figure 7.16 Replication of phage T4 DNA. In stage 1, replication initiates at specific origins, using RNA primers. In stage 2, recombinational intermediates furnish the primers for initiation.

form large branched concatemers. This replication (called recombination-dependent replication [RDR]) is analogous to the "replication restarts" discussed in chapter 1 (see Box 1.2) and is now known to occur in all organisms. However it was in T4 where this type of replication was first discovered. We discuss the two stages of T4 DNA replication in more detail below.

Like T7, T4 DNA is packaged into the phage head from concatemers. Periodic cycles of RDR lead to the synthesis of very large, branched concatemers from which individual genome-length DNAs are cut out and packaged into phage heads. However, unlike phage T7, where the DNA is cut at unique *pac* sites, T4 DNA is cut out of the concatemers in phage "headfuls," as illustrated in Figure 7.17. It is like sucking a very long strand of spaghetti into your mouth until your mouth is full and then biting it off—not very polite, but effective. This is also how the terminal redundancies or repeated sequences at the ends of T4 DNA are created. The head of phage T4 holds about 3% more DNA than a single length of the T4 genome, so that each molecule that is cut off includes some sequences from the next genome sequence in the concatemer. These sequences will then be repeated at both ends of the DNA molecule in the head. Also, because it does not cut at unique *pac* sites, each T4 DNA that is packaged will have different se-

quences at its ends. In other words the genomes of the phages that come out of the infection are cyclic permutations of each other. This explains why the genetic map of T4 is circular (Figure 7.6) even though T4 DNA itself never forms a circle. We discuss how different modes of replication and packaging give rise to the various different genetic maps of phages later in the Chapter (see "Phage Genetic Analysis").

DETAILS OF T4 STAGE 1 REPLICATION: REPLICATION FROM DEFINED ORIGINS

As mentioned, the first stage of T4 DNA replication, from defined origins, is analogous to chromosome replication in other organisms including bacteria and eukaryotes. Befitting its large size, T4 uses a number of defined origins around the chromosome to initiate replication. This is more like the replication of eukaryotic chromosomes, which use a number of replication origins, and unlike that of most bacteria and other phages including T7, which use only one origin. The first step in initiating replication from the T4 origins is to synthesize RNA on these origins, using the host RNA polymerase. These primer RNAs also double as mRNAs for the synthesis of middle-mode proteins and are made from middle-mode-type promoters that are first turned on a few minutes after infection and require the MotA protein and RNA polymerase modified by the binding of the AsiA polypeptide (see above). In their role as primers, these short RNAs invade the double-stranded DNA at the origin and hybridize to the strand of the DNA to which they are complementary, displacing the other strand to create a structure called an R-loop. The invading RNA can then prime the leading strand of DNA replication from the origin. Another protein, called gp59, helps load the replicative helicase (gp41) on the DNA to separate the strands of DNA ahead of the replication fork. The gp41 helicase is also associated with a primase (gp61), which primes replication of the lagging strand. Once replication is under way, the DNA polymerase (gp43) is held on the DNA by a sliding clamp (gp45), which has been loaded on the DNA by the clamp-loading proteins (gp44 and gp62).Once synthesis of Okazaki fragments is complete, a DNA ligase (gp30) joins the pieces together, although the host DNA ligase can substitute for this function. Later in infection, a helicase called Usw can

Headful

Figure 7.17 T4 DNA headful packaging. Packaging of DNA longer than a single genome equivalent gives rise to repeated terminally redundant ends and circularly permuted genomes.

displace these R-loops, suppressing origin-specific replication in favor of recombination-dependent replication (see below).

STAGE 2: RECOMBINATION-DEPENDENT REPLICATION

In the second stage of replication, the leading strand of T4 replication is primed by recombination intermediates rather than by primer RNAs synthesized by RNA polymerase (see Mosig, Suggested Reading). The first step in recombination-dependent replication is the invasion of a complementary double-stranded DNA by a single-stranded 3′ end to form a three-stranded **D-loop** (Figure 7.16). This invading single-stranded 3′ end could have been created during an earlier round of origin-specific replication by the inability of the DNA polymerase to replicate to the end of the molecule, or it could have been created by the action of exonucleases such as gp46 and gp47 on the ends of the molecule. In this respect, gp46 and 47 are analogous to the RecBCD protein of *E. coli* (Table 7.1) (see chapter 10). The complementary sequence which the free 3′ end invades could be in the terminally redundant sequence at one end of a sister DNA created earlier by origin-specific replication or could even be a complementary sequence in the middle of the DNA of a coinfecting phage, since T4 DNAs are cyclically permuted (see above). This pairing of the invading strand with the complementary strand in the invaded DNA is promoted by the T4 *uvsX* gene product (Table 7.1), which is analogous in function to the RecA protein, which forms synapses in uninfected cells (see chapter 10). Normally, single-stranded T4 DNA is coated with a single-stranded-DNA-binding protein, gp32, and the UvsX protein might need the help of another T4 protein, UvsY, to displace the gp32 protein. Once the D-loop has formed, the invading 3′ end can serve as a primer for new leading-strand DNA replication. The primase gp62 can then be loaded on the displaced strand, with the help of the replicative helicase gp41, for lagging-strand replication. Repeated rounds of strand invasion and replication lead to very long branched concatemers which can then be packaged into phage heads.

However, if the concatamers from which T4 packages DNA into its heads later in infection are branched, what keeps the phage head from "choking" on the branches when it tries to package a concatemer? This is where the gp49 protein, a branch-specific DNA endonuclease analogous to RuvABC (Table 7.1) comes into play. By cleaving branches in the concatemers, it allows the phage to package past branches to fill its head with DNA.

This is a simplified version of what must be a much more complicated mechanism of replication-dependent replication. It ignores some known features of RDR such as its bidirectionality, as well as details about the roles of some of the helicases and exonucleases, among other enzymes, known to be required for this process. The details of RDR in this and other systems are still being uncovered.

The discovery of RDR is an interesting example of how progress in basic cell biology often occurs. First, a basic cellular mechanism is discovered and characterized in a relatively simple organism, such as a phage, which is more accessible to experimentation than higher organisms. Once the basic cellular function is characterized in this type of organism, it is found to exist in all organisms. The number of developments that have come this way are too numerous to mention but include essential features of how proteins are synthesized, how DNA replicates, and how proteins fold.

The discovery of RDR follows the same pattern. It has been known since the 1960s that T4 DNA replication ceases prematurely in the absence of recombination. As evidence, when cells are infected by T4 with mutations in any of the recombination function genes, genes *46, 47, 49, 59, uvsX*, etc. (see chapter 10) replication begins normally but soon ceases. This was named the DNA-arrested phenotype. In the early 1980s, Gisela Mosig proposed that recombination functions were required for T4 DNA replication later in infection because replication at later times is dependent on recombination intermediates. She proposed a model whereby D-loops formed by strand invasion could prime the leading strand of DNA replication. It was some time before this model was generally accepted and even longer before it was thought to be anything but an arcane mechanism confined to this type of phage. Only in the last couple of years has it come to be recognized that ubiquitous phenomena such as double-strand break repair, replication restart, and intron and intein mobility, processes common to all organisms, use basically the same mechanism as RDR (see Kruezer, Suggested Reading).

Genetic Analysis of Phages

Phages are ideal for genetic analysis (see the introductory chapter). They have short generation times and are haploid. Mutant strains can be stored for long periods and resurrected only when needed. Also, phages multiply as clones in plaques, and large numbers can be propagated on plates or in small volumes of liquid media. Different phage mutants can be easily crossed with each other, and the progeny can be readily analyzed. Because of these advantages, phages were central to the development of molecular genetics, and important genetic principles such as recombination, complementation, suppression, and *cis*- and *trans*-acting mutations are most easily demonstrated with phages. In this section, we discuss the general principles of genetic analysis with phages. However, most of the genetic principles presented here are the same for all organisms, including humans. Only the details of how genetic experiments are performed differ from organism to organism.

Infection of Cells

The first step in doing a genetic analysis of phages or any other virus for that matter is to infect cells with the phage. Phages can infect only cells which are sensitive or susceptible to them and they can only multiply in cells which are permissive for their development. To multiply in a cell, not only must the phage adsorb to the cell surface of the bacterium and inject its nucleic acid, either DNA or RNA, but also a permissive host cell must provide all of the functions needed for multiplication of the phage. Therefore, most phages can infect and multiply in only a very limited number of types of bacteria. The types of cells which a phage can multiply are called its host range. Sometimes a normally permissive type of cell can become a **nonpermissive host** for the phage as a result of a single mutation or other genetic change. Alternatively, a virus or a mutant of the phage may be able to multiply in a particular type of host cell under one set of conditions, for example at lower temperatures, but not under a different set of conditions, for example at higher temperatures. The conditions under which it can multiply are **permissive conditions,** while the conditions under which it cannot multiply are **nonpermissive conditions.**

MULTIPLICITY OF INFECTION

Infecting permissive cells with a phage is simple enough in principle. The phage and potential bacterial host need only be mixed with each other; and some bacteria and phage will collide at random, leading to phage infection. However, the percentage of the cells that will be infected depends upon the concentration of phage and bacteria. If the phage and bacteria are very concentrated, they will collide with each other to initiate an infection more often than if they are more dilute.

The efficiency of infection is affected not only by the concentration of phage and bacteria but also by the ratio of phage to bacteria, the **multiplicity of infection (MOI).** For example, if 2.5×10^9 phage are added to 5×10^8 bacteria, there will be $2.5 \times 10^9 / 5 \times 10^8 = 5$ phage for every cell, and the MOI is 5. If only 2.5×10^8 phage had been added to the same number of bacteria, the MOI would have been 0.5.

The MOI can be either high or low. If the number of phage greatly exceeds the number of cells to infect, the cells are infected at a **high MOI.** Conversely, a **low MOI** indicates that the cells outnumber the phages. To illus-

trate, an MOI of 5 is considered high; there are five times as many phage as bacteria. An MOI of 0.5 is low; there is only one phage for every two bacteria. Whether a high or low MOI is used depends upon the nature of the experiment. At a high enough MOI, most of the cells will be infected by at least one phage; at a low MOI, many of the cells will remain uninfected but each cell which is infected will usually be infected by only one phage.

Even at very high MOI, not all the cells will be infected. There are two reasons for this. First, infection by phage is never 100% efficient. The surface of each cell may have only one or a very few receptors for the phage, and a phage can infect a cell only if it happens to bind to one of these receptors. There is also the statistical variation in the number of phages which bind to each cell. Because the chance of each phage binding to a cell is random, the number of phages infecting each cell will follow a normal distribution. Some cells will be infected by five phages—the average MOI—but some will be infected by six phages, some by four, some by three, and so on. Even at the highest MOIs, some cells by chance will receive no phage and will remain uninfected.

The minimum fraction of cells that will escape infection due to statistical variation can be calculated by using the Poisson distribution, which is an approximation to the normal distribution. In chapter 3 we discussed how Luria and Delbrück used the Poisson distribution to estimate mutation rates. According to the Poisson distribution, the probability of a cell receiving no phages and remaining uninfected (P_0) is at least e^{-MOI}, since the MOI is the average number of phages per cell. If MOI = 5, then $P_0 = e^{-5} = \sim 0.0067$; i.e., at least 0.67% of the cells will remain uninfected. At an MOI of only 1, e^{-1}, or at least ~37%, of the cells will remain uninfected. In other words, at most ~63% of the cells will be infected at an MOI of 1. Even this is an overestimation of the fraction of cells that will be infected, since, as mentioned, some of the viruses will never actually infect a cell.

Phage Crosses

Once the mutations to be tested are chosen for a genetic analysis, the mutated DNAs of two members of the same species must be put together into the same cell. This is called **crossing.** If the DNAs of the two different organisms are in a cell at the same time, the genes of both mutant strains can be expressed and the two DNAs can recombine with each other. In cellular organisms, crosses are usually performed by mating the two organisms to form zygotes that can develop into the mature organism. In phage and other viruses, crosses are performed by infecting the same cell with different strains of the virus at the same time.

To be certain that many of cells in a culture are simultaneously infected by both strains of a phage, we must use a high MOI of both phages. The Poisson distribution can again be used to calculate the maximum fraction of cells that will be infected by both mutant phages at a given MOI of each. If an MOI of 1 for each mutant phage is used for the infection, then at least e^{-1} or ~0.37 (37%) of the cells will be uninfected by each phage strain and at most $1 - 0.37 = 0.63$ (63%) of the cells will be infected with each mutant strain of the phage. Since the chance of being infected with one strain is independent of the chance of being infected by the other strain, at most $0.63 \times 0.63 \approx 0.40$ (40%), of the cells will be infected by *both* phage strains at an MOI of 1. You can see that only when both phage strains have a high MOI will most of the bacteria be infected by both strains.

Recombination and Complementation Tests with Phages

As discussed in chapter 3, two basic concepts in classical genetic analysis are recombination and complementation. The types of information derived from these tests are completely different. In recombination, the DNA of the two parent organisms is assembled in new combinations, so that the progeny have DNA sequences from both parents. In complementation, the gene products synthesized from two different DNAs interact in the same cell to produce a phenotype.

RECOMBINATION TESTS

The principles of recombination are the same for all organisms, but they are most easily illustrated with phage. Figure 7.18 gives a simplified view of what happens when two DNA molecules from different strains of the same phage recombine. The two mutant phage strains infecting the cell are almost identical, except that one has a mutation at one end of the DNA and the other has a mutation at the other end. The sequences of the two DNAs therefore differ only at the ends, the sites of the two mutations. Recombination occurs by means of a crossover between the two DNA molecules, where the two DNAs are broken at the same place and the ends of one DNA created by the break are joined to the ends of the other DNA to create two new molecules which are identical in sequence to the original molecules except that one part now comes from one of the original DNA molecules and the other part comes from the other. What effect, if any, the crossover will have depends upon where it occurs. If the crossover occurs between the sites of the two mutations, two new types of recombinant DNA molecules will appear: one has neither mutation, and the other has both mutations. Progeny phage that have packaged the DNAs with these new DNA sequences are **recombinant types** because they are unlike either parent (see chapter 3). Progeny phage that have

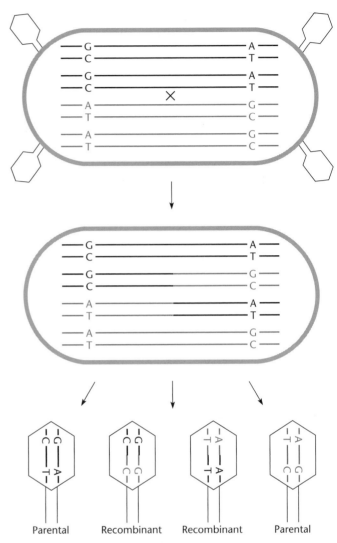

Figure 7.18 Recombination between two phage mutations. The two different mutant parent phages infect the same permissive host cell and their DNA replicates. Crossovers occur in the region between the two mutations, giving rise to recombinant types that are unlike either parent phage. Only the positions of the mutated base pairs are shown. The DNA of one parent phage is shown in black, and the other is shown in gold.

packaged a DNA molecule with one and only one of the mutations are called **parental types** because they are like the original phages that infected the cell. The appearance of recombinant types tells us that recombination has occurred. Note that the decision about what will be a recombinant type and what will be a parental type depends upon how the mutations were distributed between the parent phages. In Figure 7.18, one parent strain had one mutation and the other parent strain had the other mutation. However, one strain could have had both mutations and the other strain could have had neither mutation. In that case, the recombinant types would have had only one or the other of the two mutations and the parental types would have had either both mutations or neither mutation.

RECOMBINATION FREQUENCY

The closer together the regions of sequence difference are to each other in the DNA, the less room there is between them for a crossover to occur. Therefore, the frequency of recombinant-type progeny is a measure of how far apart the mutations are in the DNA of the phage. This number is usually expressed as the recombination frequency. In general, independent of the type of organism involved in the cross, the recombination frequency is defined as the number of recombinant progeny divided by the total number of progeny produced in the cross (see chapter 3). When the recombination frequency is expressed as a percentage, it is called the map unit. For example, the regions of two mutations in the DNA give a recombination frequency of 0.01 if 1 in 100 of the progeny are recombinant types. The regions of the two mutations are then $0.01 \times 100 = 1$ map unit apart.

Different organisms differ greatly in their recombination activity; therefore, map distance is only a relative measure and a map unit represents a different physical length of DNA for different organisms. Also, recombination frequency can indicate the proximity of two mutated regions only when the mutations are not too far apart. If they are far apart, two crossovers will often occur between them, reducing the apparent recombination frequency. Note that while one crossover between the regions of two mutations will create recombinant types, two crossovers will re-create the parental types. In general, odd numbers of crossovers will produce recombinant types and even numbers of crossovers will re-create parental types.

COMPLEMENTATION TESTS

Complementation is also most easy to demonstrate with phages and other viruses, although the nomenclature and concepts are the same for all organisms. As with recombination tests, to perform a complementation test with phages, cells are infected simultaneously by different strains of a particular phage. However, rather than measure the frequency with which recombination occurs between the regions of mutations, complementation measures the interaction between gene products synthesized from different DNAs in the same cell (see chapter 3). Usually, with phages, we are asking whether the two mutations complement each other to allow the phage to multiply under conditions that are nonpermissive for either of them alone. If they complement each other, both mutant phages will multiply; if they do not complement

each other, neither will multiply. If the two mutations complement each other, they are probably in different genes (see Table 3.5).

Figure 7.19 illustrates complementation tests with phages. In the example, two different mutant strains of a phage infect the same host cell. This host cell is normally a permissive host for the wild-type phage but cannot propagate either of the mutant phages by itself because they cannot make a gene product required for multiplication on that host. The outcome of the infection depends upon whether the mutations are in the same or different genes. If they are in different genes (left side of the figure), each DNA furnishes one of the needed gene products and so the two mutations will complement each other and both mutant viruses can multiply. If, however, the two mutations are in the same gene (right side of figure), neither DNA will furnish that gene product and so the mutations do not complement each other

Figure 7.19 Complementation tests between phage mutations. Phages with different mutations infect the same host cell, in which neither mutant phage can mulitiply. (Left) The mutations, represented by the minus signs, in different genes (*M* and *N*). Each mutant phage will synthesize the gene product that the other one cannot make; complementation will occur, and new phage will be produced. (Right) The mutations (minus signs) prevent the synthesis of the *M* gene product. There is no complementation, the mutants cannot help each other multiply, and no phages are produced.

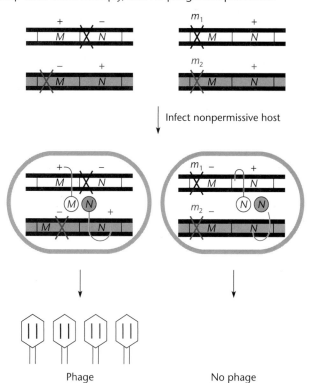

and neither mutant strain will multiply. Usually the interpretation of complementation tests follows this simple rule. However, as discussed in chapter 3, the interpretation of complementation experiments can be complicated by intragenic complementation and by polarity and translational coupling. Complementation tests can also be used to determine whether a mutation affects a gene product (i.e., is trans acting) or whether it affects a site on the DNA such as a promoter or origin of replication (i.e., is cis acting).

Note the difference in the hosts used for recombination and complementation tests with phage or other viruses. In a recombination test, the permissive host cells are infected with the two strains and the phage are allowed to multiply before the genotypes of the progeny phages are tested for recombinant types. However, in the complementation test, the host cells are infected with both mutant strains under nonpermissive conditions, and only if the mutations complement each other will the phages multiply. If complementation occurs, most of these progeny phages will still be the parental types, unable to multiply alone in subsequent infections under nonpermissive conditions.

Experiments with the rII Genes of Phage T4

We illustrate the basic principles of phage genetics by using the rII genes of phage T4. Experiments with these genes were responsible for many early developments in molecular genetics, including the discovery of nonsense codons, the definition of the nature of the genetic code, and the discovery of gene divisibility. Considering their historical importance, it is ironic that we still do not know what these gene products do for the phage. However, as discussed in chapter 3, one of the advantages of genetic analysis is that one can perform a genetic analysis without knowledge of the functions of the genes involved.

The name rII means "rapid-lysis mutants type II." Many genes of this phage were named before the current three-letter names for genes became conventional. Phage with an r-type mutation cause the infected cells to lyse more quickly than the normal (r^+) phage, a property that rII mutant phage share with the other rapid-lysis mutants types I and III. Phage with a rapid-lysis mutation can be distinguished by the appearance of their plaques on *E. coli* B indicator bacteria. The plaques formed by wild-type r^+ phage have fuzzy edges because of a phenomenon called lysis inhibition, which delays lysis of the infected cells. However r-type mutants do not show lysis inhibition, which causes them to form hard-edged, clear plaques. The hard-edge, clear-plaque phenotype makes it easy to distinguish rapid-lysis mutants from the wild type (Figure 7.20).

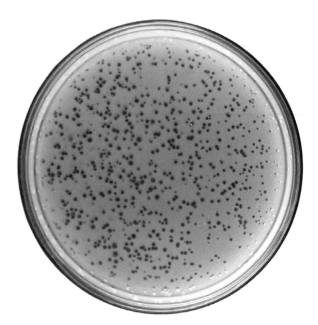

Figure 7.20 Plaques of phage T4. Most plaques are fuzzy edged, but some due to rII or other r-type mutants have hard edges because of rapid lysis of the host cells.

The property of rII mutants that distinguishes them from the other types of rapid-lysis mutants is that they cannot multiply in strains of *E. coli* that carry the λ prophage; these strains are designated *E. coli* K-12λ or Kλ (see chapter 8). This property also greatly facilitates complementation tests with rII mutants and makes possible the detection of even very rare recombinant types.

COMPLEMENTATION TESTS WITH rII MUTANTS
In about 1950, Seymour Benzer and others realized the potential of using the rII genes of T4 to determine the detailed structure of genes. The first question asked was that of how many genes, or **complementation groups,** are represented by rII mutations. To obtain an answer, numerous phage with rII mutations were isolated and pairwise complementation tests were performed in which two different rII mutants infected the nonpermissive host, *E. coli* Kλ. When the two rII mutations complemented each other, phage were produced. These complementation tests revealed that all rII mutations could be sorted into two complementation groups, or genes, which were named rIIA and rIIB. Presumably, the rIIA and rIIB genes encode different polypeptides, both of which are required for multiplication in *E. coli* Kλ.

RECOMBINATION TESTS WITH rII MUTANTS
The next step in the genetic analysis was to perform recombination tests between rII mutations to determine the location of the rIIA and rIIB genes with respect to

each other and to order the mutations within these genes. Recombination between rII mutations was measured by infecting permissive *E. coli* B with two different rII mutants and allowing the phage to multiply. The progeny phage were then plated to measure the frequency of recombinant types.

If recombination can occur between the two rII mutations, two different recombinant-type progeny would appear: double mutants with both rII mutations, and wild-type, or r^+, recombinants with neither mutation. The recombinant types with both mutations are difficult to distinguish from the parental types. However, the r^+ recombinants with neither rII mutation are easy to detect because they can multiply and form plaques on *E. coli* Kλ. Therefore, when the progeny of the cross are plated on *E. coli* Kλ, any plaques that appear are due to r^+ recombinants. As discussed in chapter 3, the recombination frequency equals the *total* number of recombinant types divided by the total progeny of the cross. We can assume that about half of the recombinants are not being detected because they are double-mutant recombinants, and so the total number of recombinant types is the number of r^+ recombinants multiplied by 2. All the progeny should form plaques on *E. coli* B, so this is a measure of the total progeny. Hence, the recombination frequency between the two rII mutations is twice the number of phage that form plaques on *E. coli* Kλ divided by the number of phage that form plaques on *E. coli* B. There will be many fewer r^+ recombinant progeny than total progeny, so that if the cross is plated directly on the two types of bacteria, there may be only a few plaques on *E. coli* Kλ and millions of plaques on *E. coli* B—too many to count. Therefore the phage progeny of the cross must be serially diluted before they are plated on the two types of bacteria, and the final recombination frequency must take these differences in dilution into account.

To illustrate, let us cross an rII mutant that has the mutation r168 with another rII mutant that has the mutation r131. We infect *E. coli* B with the two mutants and incubate the infected cells to allow the phage to multiply. To determine the number of r^+ recombinants, we dilute the phage by a factor of 10^5 and plate on the *E. coli* Kλ indicator bacteria. To determine the total number of progeny, we dilute the phage by a factor of 10^7 and plate on the *E. coli* B indicator bacteria. After incubating the plates overnight, we observe 108 plaques on the *E. coli* Kλ plate and 144 plaques on the *E. coli* B plate. The total number of recombinant-type plaques is twice the number of plaques on *E. coli* K-12λ or 2 × 108 = 216, since we are counting only the r^+ recombinants and not the double-mutant recombinants, which are produced in equal numbers. This number must be multiplied by the dilution 10^5 to get the total number of recombinant-type

phage, which is 2.16×10^7. The total number of progeny phage is $144 \times 10^7 = 1.44 \times 10^9$.

From the equation for recombination frequency (RF) in chapter 3,

$$RF = \text{total recombinant progeny/total progeny}$$

the recombination frequency between the two mutations is 0.015. If we want to express this in map units, we multiply the recombination frequency by 100 which gives 1.5. The two mutations r168 and r131 are therefore 1.5 map units apart.

Crosses such as these revealed that rIIA mutations are close to rIIB mutations, suggesting that the rIIA and rIIB genes are adjacent in the T4 DNA.

ORDERING rII MUTATIONS BY THREE-FACTOR CROSSES

Measuring the recombination frequency between two mutations can give an estimate of how close together the two mutations are in the DNA. However, to determine the relative order of mutations from such crosses, it is necessary to measure recombination frequencies very accurately. For example, let us say we have three rIIA mutations, r21, r3, and r12, and we want to use recombination frequencies to determine their order in the rIIA gene. When we cross r12 with r3, we obtain a recombination frequency of approximately 0.01. When we cross r21 with r12, we also obtain a recombination frequency of approximately 0.01. When we cross r3 with r21, we get a recombination frequency of approximately 0.02. From these data alone, we suspect that r12 is between r3 and r21 and that the order of the three mutations is r3-r12-r21.

Three-factor crosses offer a less ambiguous method for ordering mutations. The principle behind a three-factor cross is illustrated in Figure 7.21. In this method, a mutant strain that has two mutations is crossed with another strain that has the third mutation. The number of wild-type recombinants is then determined. In such a cross, the number of crossovers required to make a wild-type recombinant will depend on the order of the three mutations, and the more crossovers required to make a wild-type recombinant, the less frequent that recombinant type will be.

In Figure 7.21, a three-factor cross is being used to order the three rII mutations. First, a double mutant is constructed with the r21 and r3 mutations. Then this double mutant is crossed with a mutant that had only the r12 mutation. If the order is r3-r21-r12, only one crossover between r21 and r12 is required to make an r^+ recombinant, and the frequency of r^+ recombinants should be about 0.01, the frequency of recombination

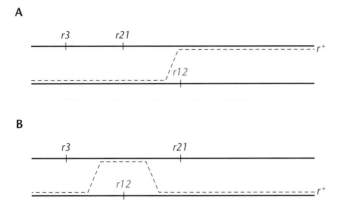

Figure 7.21 A three-factor cross to map *r*II mutations. See the text for details.

between the two single mutations *r21* and *r12*. The same is true if the order is *r21-r3-r12,* where, again, only one crossover is required to make *r*⁺ recombinants (data not shown). If, however, the order is *r3-r12-r21,* with the *r12* mutation in the middle, as we suspected from the earlier two-factor crosses, two crossovers are required and the frequency of *r*⁺ recombinants will be much lower than 0.01.

Theoretically, if the two crossovers were independent, the frequency of double crossovers should be the product of the frequencies of each of the single crossovers, or about $0.01 \times 0.01 = 0.0001$, which is only 1/100 of the frequency of the single crossover. However, because of high negative interference, also called gene conversion (see chapter 10), crossovers close to each other in the DNA are not truly independent, and one crossover will greatly increase the likelihood of what appears to be a second crossover nearby, making the frequency of apparent double crossovers much higher than predicted. Nevertheless, the frequency of double crossovers will generally be much lower than the frequencies of the individual crossovers, permitting the use of three-factor crosses to unambiguously order mutations.

ORDERING LARGE NUMBERS OF *r*II MUTATIONS BY DELETION MAPPING

One of the early contributions of *r*II genetics was the ordering by Benzer of large numbers of mutations in the *r*II genes, both spontaneous mutations and mutations induced by mutagens (see Benzer, Suggested Reading). As part of his genetic analysis of the structure of the *r*II genes, Benzer wanted to determine how many sites there are for mutations in *r*IIA and *r*IIB and to determine whether all sites within these genes are equally mutable or whether some are preferred. To find these answers, he turned to **deletion mapping**. The principle behind this

method is that if a phage with a point mutation is crossed with another phage with a deletion mutation, no wild-type recombinants will appear when the point mutation lies within the deleted region. It is much easier to determine whether there are any *r*⁺ recombinants at all than it is to carefully measure recombination frequencies. Therefore, deletion mapping offers a convenient way to map large numbers of mutations.

For this approach, Benzer needed deletion mutations extending known distances into the *r*II genes. Some of the *r*II mutants he had already isolated had the properties of deletion mutations (see chapter 3). First, these *r*II mutations did not revert, as indicated by the lack of plaques due to *r*⁺ revertant phage, even when very large numbers of *r*II mutant phage were plated on *E. coli* Kλ. Second, these mutations did not map at a single position, or point, as would base pair changes or frameshift mutations. They did not give *r*⁺ recombinants when crossed with many different *r*II mutations, some of which at least gave *r*⁺ recombinants when crossed with each other and so must have been at different positions in the *r*II genes.

Figure 7.22 shows a set of Benzer deletions that are particularly useful for mapping *r*IIA mutations. These lengthy deletions begin somewhere outside of *r*IIB and remove all of that gene, extending various distances into *r*IIA. One deletion, *r1272,* extends through the entire *r*II region, completely removing both *r*IIA and *r*IIB.

Armed with deletions with known endpoints, Benzer quickly localized the position of any new *r*II point mutation by crossing the mutant phage with other phage that each had one of these deletions. For example, if a point mutation gives *r*⁺ recombinants when crossed with the deletion *rA105* but not when crossed with *rpB242;* this point mutation must lie in the short region between the end of *rA105* and the end of *rpB242*. Therefore, with no

Figure 7.22 Some of Benzer's *r*II deletions in phage T4. The deletions remove all of *r*IIB and extend various distances into *r*IIA. The shaded bars show the region deleted in each of the mutations.

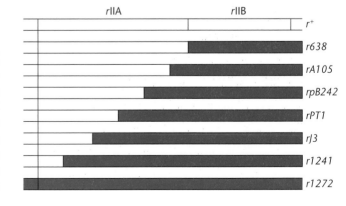

more than seven crosses, a mutation could be localized to one of seven segments of the *r*IIA gene. The position of the mutation could be located more precisely through additional crosses with other mutations located within this smaller segment.

MUTATIONAL SPECTRA

The numerous point mutations within the *r*IIA and *r*IIB genes that Benzer found by deletion mapping included spontaneous as well as induced mutations. Figure 7.23 illustrates the map locations of some of the spontaneous mutations. Spontaneous mutations can occur everywhere in the *r*IIA gene, but Benzer noted that some sites are "hot spots," in which many more mutations occur than at other sites. Mutagen-induced mutations also have hot spots, which differ from each other and from those of spontaneous mutations (data not shown).

The tendency of different mutagens to mutate some sites much more frequently than others has practical consequences in genetic analysis (see chapter 3). It is apparent from Figure 7.23 that if Benzer had studied only spontaneous mutations in the *r*II genes, almost 30% of these would have been at A6c, a major hot spot for spontaneous mutations. Therefore, to obtain a random collection of mutations in a gene requires isolating not only spontaneous mutations but also ones induced with different mutagens.

Methods are now available that allow essentially random mutagenesis of selected regions of DNA. These methods involve the use of special oligonucleotide primers for site-specific mutagenesis and PCR mutagenesis. We discussed some of these methods in chapter 1.

THE *r*II GENES AND THE NATURE OF THE GENETIC CODE

Of all the early experiments with the T4 *r*II genes, some of the most elegant were those that revealed the nature of the genetic code. These were conducted by Francis Crick and his collaborators (see Crick et al., Suggested Reading). These experiments not only have great historical importance but also are a good illustration of classical genetic principles and analysis.

At the time Crick and his collaborators began these experiments, he and Watson had used the X-ray diffraction data of Rosalind Franklin and Maurice Wilkins and the biochemical data of Erwin Chargaff and others to solve the structure of DNA (see the introductory chapter). This structure indicated that the sequence of bases in DNA determines the sequence of amino acids in protein. However, the question remained of how the sequence of bases is read. For example, how many bases in DNA encode each amino acid? Does every possible sequence of bases encode an amino acid? Is the code "punctuated," with each code word demarcated, or does the cell merely begin reading at the beginning of a gene and continue to the end, reading a certain number of bases each time? The ease with which the *r*II genes of T4 could be manipulated made this system the obvious choice for use in experiments to answer these questions.

The experiments of Crick et al. were successful for two reasons. First, the extreme N-terminal region of the *r*IIB polypeptide, the so-called B1 region, is nonessential for activity of the *r*IIB protein. The B1 region can be deleted or all the amino acids it encodes can be changed without affecting the activity of the polypeptide. Note that this is not normal for proteins. Most proteins cannot tolerate such extensive amino acid changes in any region.

The second reason for the success of these experiments is that acridine dyes specifically induce frameshift mutations by causing the removal or addition of a base pair in DNA (see chapter 11). This conclusion required a leap of faith at the time. The mutations caused by acridine dyes are usually not leaky but are obviously not deletions because they map as point mutations and revert. Also, the frequency of revertants of mutations due to acridine dyes increases when the mutant phage are

Figure 7.23 Mutational spectrum for spontaneous mutations in a short region of the *r*I genes. Each small box indicates one mutation observed at that site. Large numbers of boxes at a site indicate hot spots, where spontaneous mutations often occur. Adapted with permission from S. Benzer, *Proc. Natl. Acad. Sci. USA* **47**:403–415, 1961.

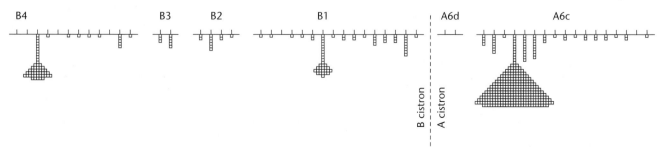

again propagated in the presence of acridine dyes but not when they are propagated in the presence of base analogs, which at the time were suspected to cause only base pair changes. It was reasoned that if acridine dye-induced mutations could not be reverted by base pair changes, the mutations induced by acridine dyes themselves could not be base pair changes. This evidence that acridine dyes cause frameshift mutations may seem flimsy in retrospect, yet it was convincing enough to Crick et al. that they proceeded with their experiments on the nature of the genetic code.

Intragenic Suppressors of a Frameshift Mutation in *r*II B1

The first step in their analysis was to induce a frameshift mutation in the *r*II B1 region by propagating cells infected with the phage in the presence of the acridine dye proflavin. Crick and his colleagues named their first *r*II mutation FC0 for "Francis Crick Zero." The FC0 mutation prevents T4 multiplication in *E. coli* Kλ because it inactivates the *r*IIB polypeptide. That an acridine-induced mutation in the region encoding the nonessential B1 portion of the B polypeptide can inactivate the *r*IIB polypeptide in itself suggests that the mutations are frameshifts, since, as mentioned above, merely changing an amino acid in the B1 region should not inactivate the gene.

Selecting Suppressor Mutations of FC0

The next step in the Crick et al. analysis was to select suppressor mutations of FC0. As discussed in chapter 3, a suppressor mutation restores the wild-type phenotype by altering the DNA sequence somewhere else than the site of the original mutation. To select suppressors of FC0, Crick et al. merely needed to plate large numbers of FC0 mutant phage on *E. coli* Kλ. A few plaques due to phenotypically *r*+ phage appeared. These phage could either have been revertants of the original FC0 mutation or have had two mutations, the original FC0 mutation plus a suppressor.

To determine which of the *r*+ phage had suppressor mutations, Crick et al. applied the classic genetic test for suppression (see chapter 3). In such a test, an apparent revertant is crossed with the wild type. If any of the progeny are recombinant types with the mutant phenotype, the interpretation is that the mutation had not reverted but had been suppressed by a second-site mutation that restored the wild-type phenotype.

Figure 7.24 illustrates the principle behind this test as applied to the apparent revertants of the FC0 *r*II mutation of T4. The apparent wild-type revertant is crossed with the wild-type phage. If the mutation has reverted, all of the progeny will be *r*+ and there will be no *r*II mutant recombinants. In contrast, if the mutation has been

A Reversion

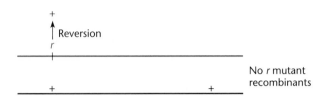

B Suppression

Figure 7.24 Classical genetic test for suppression. Phages that have apparently reverted to wild type (*r*+) are crossed with wild-type *r*+ phage. (A) If the mutation has reverted, all of the progeny will be *r*+. (B) If the mutation has been suppressed by another mutation, x, there will be some *r*II mutant recombinants among the progeny.

suppressed, the suppressing mutation can be crossed away from the FC0 mutation, and *r*II mutant recombinant types will appear that cannot multiply in *E. coli* Kλ. In the test used by Crick et al., most of the apparent *r*+ revertants of FC0 gave some *r*II mutant recombinants when crossed with the wild type; therefore, the FC0 mutation in these apparent revertants was being suppressed rather than reverted. Moreover, there were very few *r*II mutant recombinants, and so the suppressing mutations must have been very close to the original FC0 mutation, presumably also in the B1 region of the *r*IIB gene, since only crossovers between the regions of the FC0 mutation and the suppressing mutation will give rise to *r*II mutant recombinants.

Isolating the Suppressor Mutations

A double mutant with both the FC0 mutation and the suppressor mutation is *r*+ and will multiply in *E. coli* Kλ. But would a mutant with a suppressor mutation alone be phenotypically *r*II or *r*+? If the suppressor mutations by themselves produce a phenotypically *r*II mutant, presumably some of the *r*II mutant recombinants obtained by crossing the suppressed FC0 mutant with wild-type T4 would be single-mutant recombinants with only the suppressor mutation. This can be easily tested. If a recombinant is phenotypically *r*II− because it has the suppressor mutation rather than the FC0 mutation, it should give some *r*+ recombinants when crossed with the FC0 single mutant. Some of the *r*II mutant recombinant phages did give some *r*+ recombinants when crossed with FC0 mutants, indicating that these phages have an *r*II mutation different from the FCO mutation, presumably the suppressor mutation by itself.

Selecting Suppressor-of-Suppressor Mutations

Because the suppressor mutations of FC0 by themselves make the phage *r*IIB⁻ and prevent multiplication on *E. coli* Kλ, the next question was whether the suppressor mutations of FC0 could be suppressed by "suppressor-of-suppressor" mutations. As with the original FC0 mutation, when Crick et al. plated large numbers of T4 with a suppressor mutation on Kλ, they observed a few plaques. Most of these resulted from second-site suppressors. Moreover, these suppressor-of-suppressor mutations were *r*IIB⁻ when isolated by themselves. This process could be continued indefinitely.

Frameshift Mutations and Implications for the Genetic Code

To explain these results, Crick and his collaborators proposed the model shown in Figure 7.25. They proposed that FC0 is a frameshift mutation that alters the reading frame of the *r*IIB gene by adding or removing a base pair so that all the amino acids inserted in the protein from that point on are wrong. This explains how the FC0 mutation can inactivate the *r*IIB polypeptide, even though it occurs in the nonessential N-terminal-encoding B1 region of the gene.

The suppressors of FC0 are also frameshift mutations in the *r*IIB1 region. The suppressors either remove or add a base pair, depending on whether FC0 adds or removes a base pair, respectively. As long as the other mu-

Figure 7.25 Frameshift mutations and suppression. The FC0 frameshift is caused by the addition of 1 bp, which alters the reading frame and makes an *r*II mutant phenotype. FC0 can be suppressed by another mutation, FC7, that deletes 1 bp and restores the proper reading frame and the *r*⁺ phenotype. The FC7 mutation by itself confers an *r*II mutant phenotype. Regions translated in the wrong frame are underlined.

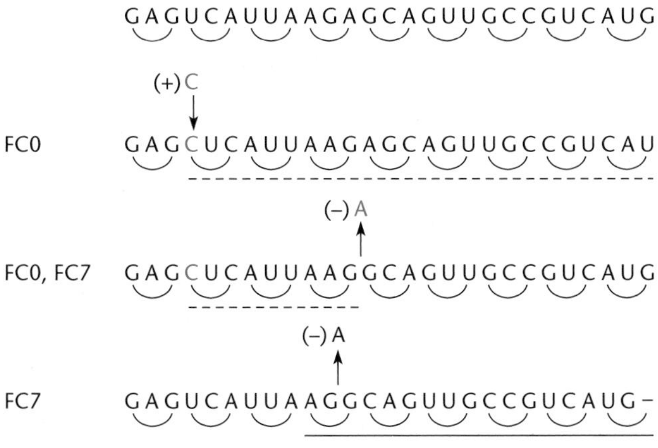

tation has the opposite effect to FC0, an active *r*IIB polypeptide will often be synthesized. The results of these experiments had several implications for the genetic code.

The code is unpunctuated. At the time, it was not known if something demarcated the point where a code word in the DNA begins and ends. Consider a language in which all the words have the same number of letters. If there were spaces between the words, we could always read the words of a sentence correctly because we would know where one word ended and the next word began. However, if the words did not have spaces between them, the only way we would know where the words began and ended would be to count the letters. If a letter were left out or added to a word, we would read all the following words wrong. This is what happens when a base pair is added to or deleted from a gene. The remainder of the gene is read wrong; therefore, the code must be unpunctuated.

The code is three lettered. The experiments of Crick et al. also answered the question of how many letters are in each word of this language; i.e., how many bases in DNA are being read for each amino acid inserted in the protein. At the time, there were theoretical reasons to believe that the number is larger than 2. Since DNA contains four "letters" or bases (A, G, T, and C), only 4 × 4 = 16 possible amino acid code words could be made out of only two of these letters. However, at least 20 amino acids were known to be inserted into proteins (the known number is now 22 [see chapter 2]), and so a two-letter code would not yield enough code words for all the amino acids. However, three bases per code word results in 4 × 4 × 4 = 64 possible code words, plenty to encode all 20 amino acids.

They could test the assumption that the code is three lettered. The reading frame of a three-letter code would not be altered if 3 bp was added to or removed from the *r*IIB1 region. Continuing with the letter analogy, an extra word would then be put in or left out but all the other words would be read correctly. Therefore, in the B1 region, if three suppressors of FC0 or three suppressors of suppressors were combined in the same phage DNA, a complete new code word would be added to or subtracted from the molecule and the correct reading frame would be restored. Thus, the phage should be *r*⁺ and should multiply in *E. coli* Kλ. Experimental results were consistent with this hypothesis, indicating that 3 bp in DNA encodes each amino acid inserted into a protein.

The code is redundant. The results of these experiments also indicated that the code is redundant; that is, more than one word codes for each amino acid. Crick et al.

reasoned that if the code were not redundant, most of the code words, i.e., 64 − 20 = 44, would not encode an amino acid; then a ribosome translating in the wrong frame would almost immediately encounter a code word that does not encode an amino acid, and translation would cease. The fact that most combinations of suppressors with FC0 and with suppressors of suppressors restored the r^+ phenotype indicated that most of the possible code words do encode an amino acid.

Nonsense code words and termination of translation. Although their evidence indicated that most code words encode an amino acid, it also indicated that not all of them do. If all possible words signified an amino acid, the entire *r*IIB1 region should be translatable in any frame and a functional polypeptide would result, provided that the correct frame was restored before the translation mechanism entered the remainder of the *r*IIB gene. However, if not all the words encode an amino acid, a "forbidden" code word that does not encode an amino acid might be encountered during translation in a wrong frame. In this situation, not all combinations of suppressors and suppressors of suppressors would restore the r^+ phenotype. Crick et al. observed that some combinations of suppressors and suppressors of suppressors did cause forbidden code words to be encountered in the *r*IIB region. However, other combinations in the same region resulted in a functional *r*IIB polypeptide. For example, in Figure 7.26, a nonsense codon (UAA) is encountered when a region is translated in the +1 frame because 1 bp was removed. However, no nonsense codons are encountered when 1 bp is added and the same region is translated in the −1 frame.

Figure 7.26 Frameshift suppression and nonsense codons. The nonsense codon UAA is encountered in the +1 frame due to deletion of 1 bp in the DNA. Translation will terminate even if the correct translational frame is restored farther downstream. In the −1 frame, due to addition of 1 bp, no nonsense codons are encountered. A downstream deletion restores the correct reading frame, and the active polypeptide is translated.

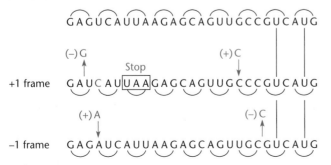

Even more convincing evidence that some code words do not encode an amino acid came from experiments with the deletion *r1589* (see Benzer and Champe, Suggested Reading). The deletion *r1589* removes much of *r*IIA and the nonessential *r*IIB1 region, eliminating the nonsense codon or codons that are normally at the end of the *r*IIA gene and the translation initiation region of *r*IIB. This deletion mutation thereby causes translation initiated at *r*IIA to proceed into *r*IIB, resulting in a fusion protein in which the N terminus comes from *r*IIA and the rest of the protein comes from *r*IIB. Since most of *r*IIA but only the nonessential B1 region of *r*IIB is deleted, this fusion protein has *r*IIB activity but not *r*IIA activity, as can be demonstrated by complementation tests.

Although the fusion protein does not require the *r*IIA portion for *r*IIB activity, Benzer and Champe found that some base pair change mutations in the *r*IIA region prevented *r*IIB activity. These base pair changes presumably caused nonsense codons that stopped translation in the *r*IIA region. Other base pair change mutations that did not disrupt *r*IIB activity were presumably missense mutations, which resulted in insertion of the wrong amino acid in the *r*IIA portion of the fusion protein but did not stop translation.

Benzer and Champe also found that even the presumed nonsense mutations did not prevent *r*IIB activity in some strains of *E. coli* and so were in a sense "ambivalent." We now know that these "permissive" strains of *E. coli* are nonsense suppressor strains with mutations in tRNA genes that allow readthrough of one or more of the nonsense codons (see chapter 3).

Postscript on the Crick et al. Experiments

The experiments of Crick et al. laid the groundwork for the subsequent deciphering of the genetic code by Marshall Nirenberg and his colleagues, who assigned an amino acid to each of the 61 3-base sense codons. Other researchers later used reversion and suppression studies to determine that the nonsense codons are UAG, UAA, and UGA.

ISOLATING DUPLICATION MUTATIONS OF THE *r*II REGION

Our final example of the genetic manipulation of the *r*II genes of T4 is the isolation of tandem duplication mutations (see, for example, Symonds et al., Suggested Reading). These experiments help contrast the differences between complementation and recombination and also illustrate some of the genetic properties of tandem duplication mutations discussed in chapter 3.

The isolation of tandem duplication mutations of the *r*II region depended on the properties of two deletions in this region, the aforementioned *r1589* deletion and an-

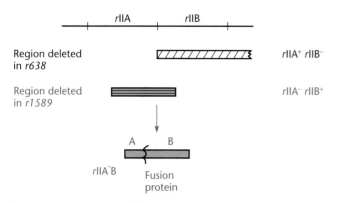

Figure 7.27 The *r*II deletions *r638* and *r1589*. An *r*IIAB fusion protein is made in a strain with *r1589*. See the text for details.

other deletion, *r638* (Figure 7.27). As mentioned, the *r1589* deletion removes the N-terminal-coding B1 region of the *r*IIB gene as well as the C-terminal-coding part of *r*IIA. Phage with this deletion make a fusion protein with the N terminus of the *r*IIA protein fused to most of *r*IIB and are phenotypically *r*IIA⁻ *r*IIB⁺. The deletion mutation *r638* deletes all of *r*IIB but does not enter *r*IIA, so that phage with this deletion are *r*IIA⁺ *r*IIB⁻. Because one deleted DNA makes the product of the *r*IIA gene and the other makes the product of the *r*IIB gene, the two deletion mutations can complement each other. However, they cannot recombine to give *r⁺* recombinants, because they overlap, both deleting the B1 region of the *r*IIB gene.

Even though recombination should not occur between the two deletions to give *r⁺* recombinants, when *E. coli* B is infected simultaneously with the two deletion mutants and the progeny are plated on *E. coli* Kλ, a few rare plaques due to *r⁺* phage arise. These phenotypically *r⁺* phage have tandem duplications of the *r*II region (Figure 7.28A). Each copy of the *r*II region has a different deletion mutation, and these mutations complement each other to give the *r⁺* phenotype.

Figure 7.28A also illustrates how these tandem duplication mutations might arise. Sometimes, while the DNA is replicating, recombination mistakenly occurs between two short directly repeated regions on either side of the *r*II region (ectopic recombination). Such mistaken crossovers are rare because repeated sequences in DNA, when they exist at all, are usually very short. However, once such a crossover occurs, one of the recombinant-type phage will have a duplicate of the *r*II region, both copies of which, in the example, have the *r1589* mutation. If a phage with such a duplication then infects the same cell as a phage with the *r638* deletion, one of the copies of the *r*II region can recombine with the DNA of the other parent and the *r638* deletion will replace the

r1589 deletion in one of the copies. This second recombination will occur very frequently, because of the extensive homology between the duplicated regions. The phage that package this DNA will then have two copies of the *r*II region, one copy with the *r1589* deletion and the other copy with the *r638* deletion. In subsequent infections, the two deletion mutations can complement each other to make the phage phenotypically *r⁺*. These phage are diploid for the *r*II regions and surrounding genes because they have two copies of these genes.

The salient property of tandem duplication mutations is that they are very unstable, because recombination anywhere in the long duplicated region will destroy the duplication (see chapter 3). These *r⁺* phage exhibit this instability. If the *r⁺* phage with the putative duplication are propagated in *E. coli* B, where there is no selection for phage with the duplication, a very high percentage of the progeny phage will be unable to multiply in *E. coli* Kλ.

Figure 7.28B illustrates why the duplications are unstable. A crossover between either of the duplicated segments, x or y, will cause the intervening sequences to be deleted and one copy of the duplication to be lost. The resulting phage, some of which have the *r1589* deletion whereas others have the *r638* deletion, are haploid segregants that now have only one copy of the duplicated region. The term "segregants" is used rather than recombinants because the recombination that destroys the duplication occurs spontaneously while the phage is multiplying and does not require crosses. The x and y regions are usually quite long and identical in sequence, so that haploid segregants appear quite frequently.

The two haploid types segregate at a characteristic frequency for each duplication mutation. As we can see from the duplication shown in Figure 7.28B, a crossover in region x will yield the *r1589* haploid whereas a crossover in region y will yield the *r638* haploid. Therefore, the haploid type that segregates at the highest frequency will depend upon which region, x or y, is longer. If x is longer than y, the *r1589* single mutant will segregate more frequently. However, if x is shorter than y, *r638* will segregate more frequently.

Constructing the Genetic Linkage Map of a Phage

The *r*II genes are only two genes out of the hundreds of genes of phage T4 (Figure 7.6), and these genes do not even exist in other types of phage. In order to obtain the genetic map of a phage, we need a way to identify many more genes of phage. A picture that shows many of the genes of an organism and how they are ordered with respect to each other is known as the **genetic linkage map** of the organism, so named because it shows the proximity or linkage of the genes to each other. This linkage is

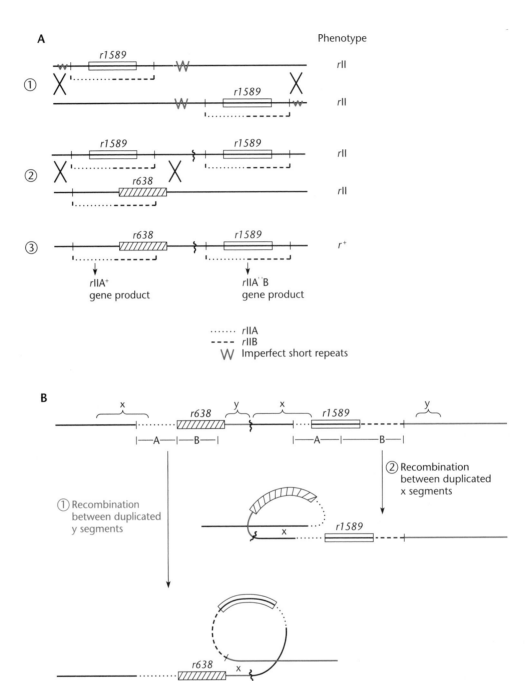

Figure 7.28 (A) Model for how tandem duplications in the *r*II region form. Recombination between short repeated sequences flanking the *r*II region may occur at a low frequency, giving rise to a duplication of one or more genes. This duplication can recombine with the other deletion strain in the region of one of the repeated sequences, giving rise to a duplication in which one copy of *r*IIB has the *r1589* deletion and the other copy has the *r638* deletion. The two deletions complement each other, so that the phage is phenotypically *r*⁺. (B) Tandem duplications are unstable because recombination between the duplicated regions can destroy the duplication. Which deletion mutation remains in the haploid segegant depends upon where the recombination occurs. (1) Recombination between the y duplicated segments gives rise to a haploid segregant with only the *r638* deletion. (2) Recombination between the two duplicated regions designated x gives rise to a haploid segregant with only the *r1589* deletion. In this example, *r1589* haploid segegants will be about three times more frequent than *r638* segregants because x is approximately three times as long as y.

determined by genetic crosses. Physical methods for mapping DNA, discussed in chapter 1, give rise to a physical map, which can often be correlated with the genetic map.

Conditional-lethal mutations, including temperature-sensitive and nonsense mutations, are very useful types of mutations in phages and can be used to identify any gene that is essential for multiplication of the phage. If a phage has a mutation to a nonsense codon (UAA, UAG, or UGA) in an essential gene, it will multiply and form a plaque only on a permissive host with a nonsense suppressor tRNA (see chapter 2). Phage with a temperature-sensitive mutation in an essential gene will multiply and form plaques at a lower (permissive) temperature but not at a higher (nonpermissive) temperature. Because such mutations can be isolated in any essential gene, nonsense and temperature-sensitive mutations can be used to identify many of these genes and to construct a relatively complete genetic linkage map of a phage. Not all genes can be identified this way, however, since the products of some genes of the phage may be nonessential on a particular host, so these genes cannot be identified by conditional-lethal mutations and other methods must be used for these genes.

The first step in constructing the conditional-lethal map of a phage is to isolate a large number of temperature-sensitive and nonsense mutations of the phage by mutagenizing the phage with various mutagens and then plating the surviving phage on suppressing bacteria at the permissive temperature. Then plaques are picked, and the phage is tested for multiplication on nonsuppressing bacteria and at the nonpermissive temperature. Phage that cannot form plaques on the nonsuppressing bacteria or at the nonpermissive temperature have nonsense mutations or temperature-sensitive mutations, respectively, in essential genes.

IDENTIFYING PHAGE GENES BY COMPLEMENTATION TESTS
Once a large collection of mutations of the phage have been assembled, the mutations can be placed into complementation groups or genes. As discussed above, two mutations that do not complement each other are probably in the same complementation group or gene. Complementation tests are done under conditions where neither mutant can multiply. For example, to test for complementation between two different nonsense mutations, nonsuppressing bacteria are infected with the two mutants simultaneously and the progeny are plated on amber-suppressing bacteria to determine how many progeny phage are produced. To test for complementation between a temperature-sensitive mutation and a nonsense mutation, nonsuppressing bacteria are infected at the high (nonpermissive) temperature. The progeny phage are then plated on nonsense-suppressing bacteria at the permissive temperature to see how many phage

are produced. Note that a temperature-sensitive mutation and a nonsense mutation could be in the same gene; in other words, they could be allelic. Even though they are different types of mutations, they will not complement each other if they are allelic.

Each time we find a mutation that complements all the other mutations in the collection, we have found a new gene. Once two mutations have been found that do not complement each other, only one of the two mutations need be used for further complementation tests. If many of the complementation groups are represented by only a single mutation, many other essential genes are probably not yet represented by any mutations. More mutants must then be isolated to identify more of the essential genes. Eventually, more and more of the new mutations will sort into one of the previously identified complementation groups and the collection of mutations in essential genes will be almost complete.

MAPPING PHAGE GENES
Once most of the genes of the phage have been identified, representative mutations in each of the genes can be mapped with respect to each other. To measure the frequency of recombination between two mutations, cells are infected under conditions that are permissive for both mutations. For example, to cross a temperature-sensitive mutation with an amber mutation, amber suppressor cells are infected at the low (permissive) temperature. After phage are produced, the progeny phage are plated under conditions permissive for both mutations, to measure the total progeny, and under conditions that are nonpermissive for both mutations, to measure the number of wild-type recombinants. Hence, if a mutant strain with a nonsense mutation in one gene is crossed with another mutant with a temperature-sensitive mutation in a different gene, the progeny should be plated on nonsuppressing bacteria at the high (nonpermissive) temperature to measure the number of wild-type recombinants and at the low (permissive) temperature on nonsense-suppressing bacteria to measure the total progeny. From the frequency of wild-type recombinants, the map distance between the mutations and therefore the genes they are in can be calculated by using the equation for recombination frequency given above. Genes are said to be **linked** when the recombination frequency between mutations in the genes is lower than random, indicating that fewer than a random number of crossovers are occurring between them.

ORDERING MUTATIONS BY THREE-FACTOR CROSSES
It is often difficult to order mutations in closely linked genes by measuring recombination frequencies alone. The recombination frequencies need to be measured very carefully to be certain of their order. In such cases,

three-factor crosses may be a less ambiguous way to order the mutations.

Figure 7.29 illustrates how a three-factor cross can be used to order two amber mutations with respect to a temperature-sensitive mutation, although the same reasoning could apply to any combination of mutations. In particular, we want to know whether the site of the temperature-sensitive mutation lies between the sites of the two amber mutations or outside them. We cross a double mutant containing both amber mutations with a single mutant containing only the temperature-sensitive mutation. We then plate the progeny on nonsuppressing bacteria at the nonpermissive temperature for the temperature-sensitive mutation, conditions under which only wild-type recombinants that lack all three mutations should form plaques. The frequency of wild-type recombinants will depend upon the map order. If the site of the temperature-sensitive mutation lies outside the sites of the two amber mutations, then a single crossover between the nearest amber mutation and the temperature-sensitive mutation should result in a wild-type recombinant. In fact, the recombination frequency should be comparable to the recombination frequency in a two-factor cross where a mutant with one of the amber mutations is crossed with a mutant with the temperature-sensitive mutation. However, if the site of the temperature-sensitive mutation lies between the sites of the two amber mutations, two crossovers are re-

quired to make the wild-type recombinant. Since two crossovers are less apt to occur than a single crossover, the frequency of wild-type recombinants should be much lower with this map order. As discussed above, the frequency of the double crossover is often much higher than the product of the frequencies of the single crossovers due to high negative interference (gene conversion). We discuss phenomenona associated with recombination such as gene conversion in more detail in chapter 10.

GENETIC LINKAGE MAPS OF SOME PHAGES

Using methods such as those described above, genetic linkage maps have been determined for several commonly used phages. We have shown some of these maps throughout the chapter, along with the function of some of the gene products where known (see, for examples, Figures 7.4, 7.6, and 7.12).

One noticeable feature of most phage genetic maps is that genes whose products must physically interact, such as the products involved in head or tail formation, tend to be clustered. The argument is that this clustering may allow recombination between closely related phages without disruption of gene function. If the genes whose products must physically interact were not close to each other, recombination between two closely related phages could separate the genes and give rise to inviable phage. For example, if the head genes were not clustered, recombination

Figure 7.29 Three-factor cross for ordering conditional-lethal mutations. In a three-factor cross, the frequency of wild-type recombinants depends upon the order of the three mutations. (A) The temperature-sensitive mutation, *ts3*, lies outside the region of the two amber mutations, *am1* and *am2*. A single crossover will give a wild-type recombinant that has none of the mutations. (B) The *ts3* mutation lies between the two amber mutations. Now two crossovers are required to give the wild-type recombinant, making it much rarer. Details are given in the text. See also problem 4 at end of the chapter.

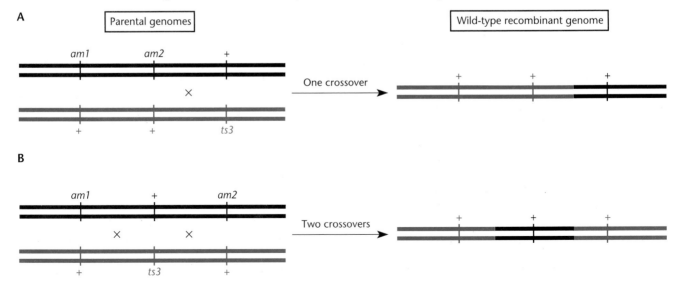

with the DNA of another related phage would often replace some of the head genes of the first phage with the corresponding head genes from the other phage. If the head of the phage could not be assembled from this mixture of head proteins, the phage would be inviable. Therefore, the potential advantages of the new recombinant phage would be lost. This hypothesis has received support recently from the structure of the family of phages related to λ. These phages seem to be modular in construction, made up of regions, or cassettes, assembled from different phages of this large family of phages.

FACTORS THAT DETERMINE THE FORM OF THE LINKAGE MAP

Another easily discernible feature of phage genetic maps is that some are linear while others are circular (compare the T7 map in Figure 7.4 to the T4 map in Figure 7.6). A genetic map is found to be circular when, as the genes are ordered from left to right by genetic crosses, the last gene to be ordered is found to be linked to the first gene so that you have come full circle. The form of the genetic map does not necessarily correlate with the linearity or circularity of the phage DNA itself, however. Some phages with circular DNA in the cell, such as λ (see chapter 8), have a linear linkage map, while some with linear DNA in the cell, such as T4, have a circular map. To understand how the genetic maps arise, we need to review how some phage replicate their DNA and how it is packaged into phage heads.

Phage λ

Phage λ has a linear genetic map, even though the DNA forms a circle after it enters the cell. As we discuss in chapter 8, phage λ has a linear map because its concatemers are cleaved at unique *cos* sites before being packaged into the phage head. The position of these *cos* sites determines the end of the linear genetic map. As an illustration, consider a cross between two phages with mutations in the *A* and *R* genes at the ends of the phage DNA (see the λ map in chapter 8). Even though different parental alleles of the *A* and *R* genes can be next to each other in the concatemers prior to packaging, these alleles will be separated when the DNA is cut at the *cos* site during packaging of the DNA. Therefore, the *A* and *R* genes will appear to be unlinked in genetic crosses and the genetic map will be linear, with the *A* and *R* genes at its ends. For this reason, all types of phages which package DNA from unique *pac* or *cos* sites have linear genetic linkage maps with the ends defined by the position of the *pac* or *cos* site.

Phage T4

Phage T4 has a circular genetic map even though its DNA never forms a circle. The reason for its circular

map is that T4 has no unique *pac* site and the DNA is packaged by a headful mechanism from long concatemers (Figure 7.17). Consequently, any two T4 phage DNAs in different phage heads from the same infection will not have the same ends but will be cyclic permutations of each other. Therefore, genes that are next to each other in the concatemers will still be together in most of the phage heads and so will appear linked in crosses, producing a circular map.

Phage P22

Phage P22 is a phage of *Salmonella*. It is closely related to λ and replicates by a similar mechanism. However, unlike λ, it has a circular linkage map. The difference is that P22 begins packaging at a unique *pac* site or *cos* site, like λ, but then packages a few genomes by a processive headful mechanism, like T4, giving rise to a circular genetic map.

Phage P1

Phage P1 has linear DNA in the head, which forms a circle by recombination between terminally repeated sequences at its ends after infection. The DNA then replicates as a circle and forms concatemers from which the DNA is packaged by a headful mechanism. However, unlike most phages, which package DNA by a headful mechanism, the genetic map of P1 is linear. The genetic map of P1 is linear because it has a very active site-specific recombination system called *cre-lox*, which promotes recombination at a particular site in the DNA. Because recombination at this site is so frequent, genetic markers on either side of the site appear to be unlinked, giving rise to a linear map terminating at the *cre-lox* site. The function of this site-specific recombination system in phage development is unknown. It is nonessential for phage development but may help promote the formation of circles after infection if the site happens to be located in the terminal redundancies. Because of its sensitivity and specificity, the *cre-lox* recombination system of P1 has many uses. We mentioned one of these uses in chapter 5, where the system was used to show that some proteins encoded by the Ti plasmid enter the plant cell nucleus along with the T-DNA during the formation of crown gall tumors.

Generalized Transduction

Bacteriophages not only infect and kill cells but also sometimes transfer bacterial DNA from one cell to another in a process called **transduction**. There are two types of transduction in bacteria: **generalized transduction**, in which essentially any region of the bacterial DNA can be transferred from one bacterium to another, and **specialized transduction**, in which only certain genes close to the attachment site of a lysogenic phage in the chromosome can be transferred. These two types of transduction have fundamentally different mechanisms. In this section, we discuss only generalized transduction; coverage of specialized transduction is deferred until it can be addressed in the context of lysogeny in chapter 8.

Figure 7.30 illustrates the process of generalized transduction. While phages are packaging their own DNA, they sometimes mistakenly package the DNA of the bacterial host instead. These phages are still capable

Figure 7.30 An example of generalized transduction. A phage infects a Trp$^+$ bacterium, and in the course of packaging DNA into heads, the phage mistakenly packages some bacterial DNA containing the *trp* region instead of its own DNA into a head. In the next infection, this transducing phage injects the Trp$^+$ bacterial DNA instead of phage DNA into the Trp$^-$ bacterium. If the incoming DNA recombines with the chromosome, a Trp$^+$ recombinant transductant may arise. Only one strand of the DNA is shown.

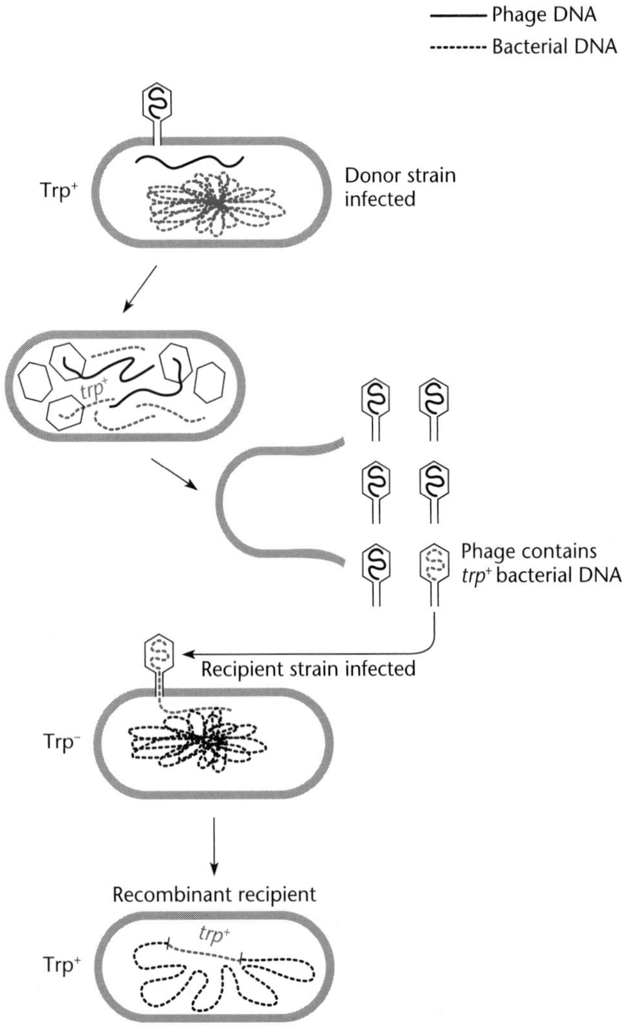

of infecting other cells, but progeny phage will not be produced. If the bacterial DNA is a piece of the bacterial chromosome, it may recombine with the host chromosome if they have sequences in common. If the injected DNA is a plasmid, it may replicate after it enters the cell and thus be maintained. If the incoming DNA contains a transposon, the transposon may hop, or insert itself, into a host plasmid or chromosome (see chapter 9).

The nomenclature of transduction is much like that of transformation and conjugation. Phages capable of transduction are called **transducing phages**. The original bacterial strain in which the transducing phage had multiplied is called the **donor strain**. The infected bacterial strain is called the **recipient strain**. Cells that have received DNA from another bacterium by transduction are called **transductants**.

Transduction occurs very rarely for a number of reasons. First, mistaken packaging of host DNA is itself rare, and transduced DNA must survive in the recipient cell to form a stable transductant. Since each of these steps has a limited probability of success, transductants can be detected only by powerful selection techniques.

What Makes a Transducing Phage?

Not all phages can transduce. The phage must not degrade the host DNA completely after infection, or no host DNA will be available to be packaged into phage heads. The packaging sites, or *pac* sites, of the phage must not be so specific that such sequences will not occur in host DNA. Some examples of transducing phages help illustrate these properties (see Table 7.2).

Phage P1, which infects gram-negative bacteria, is a good transducer because it has less *pac* site specificity

TABLE 7.2	Characteristics of generalized transducing phages P1 and P22	
	Phage	
Characteristic	**P1**	**P22**
Length (kb) of DNA packaged	100	44
Length (%) of chromosome transduced	2	1
Packaging mechanism	Sequential headful	Sequential headful
Specificity of markers transduced	Almost none	Some markers transduced at low frequency
Packaging of host DNA	Packaged from ends	Packaged from *pac*-like sequences
% of transducing particles in lysate	1	2
% of transduced DNA recombined into chromosome	1–2	1–2

than most phages and packages DNA via a headful mechanism; therefore, it will efficiently package host DNA. About 1 in 10^6 phage P1 particles will transduce a particular marker. It also has a very broad host range for adsorption and can transduce DNA from *E. coli* into a wide variety of other gram-negative bacteria including members of the genera *Klebsiella* and *Myxococcus*.

The *Salmonella enterica* serovar Typhimurium phage P22 is also a very good transducer and, in fact, was the first transducing phage to be discovered (see Zinder and Lederberg, Suggested Reading). Like P1, P22 has *pac* sites that are not too specific and packages DNA by a headful mechanism. From a single *pac*-like site, about 10 headfuls of DNA can be packaged before the mechanism requires another *pac* site. Because of even this limited *pac* site specificity, however, some regions of *Salmonella* DNA will be transduced by P22 at a much higher frequency than others.

Other phages are not normally transducing phages but can be converted into them by special treatments. For example, T4 normally degrades the host DNA after infection but works extremely well as a transducing phage if its genes for the degradation of host DNA have been inactivated. Because phage T4 does not have *pac* sites, it packages any DNA including the host DNA with high efficiency.

In contrast, phage λ does not work well for generalized transduction, because it normally packages DNA between two *cos* sites rather than by a headful mechanism. It will very infrequently pick up host DNA by mistake, but then it does not cut the DNA properly when the head is filled unless another *cos*-like sequence lies the right distance along the DNA. Thus, potential transducing particles usually have DNA hanging out of them that must be removed with DNase before the tails can be added. Even with these and other manipulations, λ works poorly as a generalized transducer.

Transducing phages have been isolated for a wide variety of bacteria and have greatly aided genetic analysis of these bacteria. Transduction is particularly useful for moving alleles into different strains of bacteria and making isogenic strains that differ only in a small region of their chromosomes so that they are very useful for strain construction. However, if no transducing phage is known for a particular strain of bacterium, finding one can be very time-consuming. Therefore, generalized transduction has been largely replaced by conjugation, transposon mutagenesis, and DNA cloning, particularly in bacteria that are not well characterized.

Mapping of Bacterial Markers by Transduction

Because bacterial DNA is carried from one cell to another during transduction, it offers a means of genetic mapping in bacteria. In fact, the three major means of gene exchange in bacteria are transduction, transformation, and Hfr conjugation (see the introductory chapter). Hfr crosses are very useful for locating the approximate map position of markers in the entire chromosome. However, for more precise mapping, techniques such as generalized transduction and transformation are generally more useful. The principles behind genetic mapping by transduction and transformation are similar, but we discuss only mapping by transduction in this chapter.

Like Hfr mapping, mapping by transduction is based on genetic markers, with one marker being the selected marker and the other markers being the unselected markers. However, rather than being based on gradient of transfer like Hfr mapping (see chapter 5), mapping by transduction is based on whether the regions of markers can be carried in the same phage head, i.e., can be **cotransduced**. If a transductant which has become recombinant for the selected marker sometimes also becomes recombinant for another unselected marker, the regions of the two markers are **cotransducible** and therefore very close to each other. In fact, they must be less than the length of the phage DNA from each other. A phage head will hold only a rather small length of DNA relative to the length of the chromosome. The actual length of DNA that can be accommodated by the head depends upon the type of phage but is usually only approximately 1 to 2% of the length of the bacterial genome (Table 7.2). Furthermore, if two markers are cotransduced, it is safe to assume that they were carried in on the same piece of DNA and not on different pieces of DNA being introduced by different phage particles. Few phage particles in a lysate carry bacterial DNA rather than their own DNA, so the chances are slight that a cell could have been infected by two of these phage particles simultaneously. Even if the MOI were so high that some cells were infected by two or more transducing particles, they would almost always be infected by normal phages at the same time; then the cells would be killed and would not become transductants.

Not only does the appearance of transductants which are recombinant for two markers mean that the markers must be very close to each other, but also the higher the percentage of the transductants that are recombinant for both markers, the closer together the two markers are likely to be. The percentage of the total transductants selected for one marker that are also recombinant for the other marker is called the **cotransduction frequency** between the two markers. In principle, the cotransduction frequency between two markers is a constant for any two markers and should be independent of which of the two markers is selected and whether it is the mutant or wild-type form of the gene that is selected. A cross with the selected and unselected markers reversed is called a **reciprocal cross.** Note that, as with Hfr crosses, a transductant that has the allele of the donor is recombinant for that marker.

In this section, we illustrate how mapping data from transductional crosses is interpreted by using an actual example. Similar reasoning would apply to transformational crosses.

MAPPING BY COTRANSDUCTION FREQUENCIES

In the example shown in Figure 7.31, phage P1 is being used for transduction to further refine the mapping of a mutation to resistance to the antibiotic rifampin named *rif-8* which we had used as an example in chapter 5. In particular, we want to know the map position of the *rif-8* mutation relative to the known map position of a nearby *argH5* mutation.

The first step is to determine if the regions of the *argH* and *rif* markers are close enough to each other on the DNA to be cotransducible. We select one of the markers and then determine whether any of the recombinants for that marker are also recombinant for the other marker. For practical reasons, it is easier to use the *argH* marker rather than the *rif* marker as the selected marker. It is also much easier to select Arg⁺ transductants than Arg⁻ transductants, since the former need only be plated on minimal plates without arginine and any transductants that form colonies are Arg⁺. Thus, we shall select the

Figure 7.31 Cotransduction of two bacterial genetic markers. The regions of the *argH* and *rif-8* mutations are close enough together that both regions can be carried on a piece of DNA fitting into a phage head. After transduction, some of the Arg⁺ transductants are also rifampin sensitive. See the text for details.

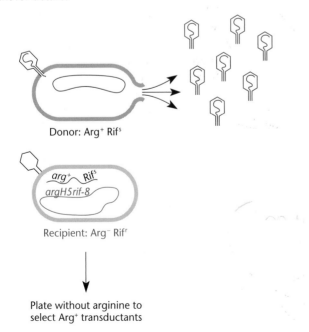

Donor: Arg⁺ Rifˢ

Recipient: Arg⁻ Rifʳ

Plate without arginine to select Arg⁺ transductants

argH marker by using a donor bacterium that is Arg⁺ and a recipient that has the *argH5* mutation. The *rif-8* mutation, the unselected marker, can be in either the donor or the recipient.

In the experiment illustrated in Figure 7.31, the transducing phage are grown on donor cells that are wild type for the *argH* and *rif* genes and so are phenotypically Arg⁺ Rifˢ. These phage are then used to infect recipient cells that have the *argH* and *rif-8* mutations and so are Arg⁻ Rifʳ. The Arg⁺ transductants are then selected by plating the infected cells on minimal plates without arginine, and the purified transductants are tested for the unselected *rif* marker. Note that the transductants which are now rifampin sensitive are recombinant for the *rif-8* marker since that is the allele of the donor. In the Arg⁺ transductants that are also rifampin sensitive, the region of the donor DNA that contains the *rif* marker will have replaced the corresponding region of the recipient. In the example, 33% of the Arg⁺ transductants are also Rifˢ. Thus, the two markers are cotransducible, and the cotransduction frequency is about 33%. Note that if we did the reciprocal cross, grew the phage on a donor with the *rif-8* mutation, and used it to transduce an *argH5* mutant, selecting for rifampin-resistant transductants and then testing for the *argH* marker, the result should be about the same. About 33% of the Rifʳ transductants should be Arg⁺, again indicating that the regions of the *rif* and *argH* markers are 33% cotransducible.

We can estimate how close together on the *E. coli* DNA the *argH* and *rif* markers would have to be so that they could be cotransducible by P1 phage. The chromosome is 100 min long, and the P1 phage head only holds about 2% of that length (Table 7.2); therefore, the two markers must be less than 2 min apart. Translating this distance into base pairs of DNA, the *E. coli* chromosome is about 4.5×10^6 bp long, and the P1 phage head will hold only 0.02 times 4.5×10^6 or 90,000 bp. Thus, to be cotransducible, two markers in the DNA must be less than 90,000 bp apart.

ORDERING THREE MARKERS BY COTRANSDUCTION FREQUENCIES

As mentioned, the closer together two markers are in the DNA, the more likely they are to be carried in the same phage head and the higher their cotransduction frequency. Therefore, cotransduction frequencies can also be used to determine which markers are closest to each other on the DNA and therefore to determine the order of markers. To illustrate, we shall use cotransduction to order the *argH5* and *rif-8* markers with respect to another marker in this region. The third marker is due to a mutation in the *metA* gene, whose product is required to make methionine. We have already determined that the *argH* and *rif* markers are about 33% cotransducible. By doing similar transduction experiments, we determine that the *argH* and *metA* markers are about 13% cotransducible and the *rif* and *metA* markers are almost 80% cotransducible. Hence, the *metA* marker appears to be closer to the *rif* marker than either is to the *argH* marker. We also know that the *rif* marker is somewhat closer to the *argH* marker (33%) than the *metA* marker is to the *argH* marker (13%). From these data, we conclude that the *rif* and *metA* markers must be on the same side of *argH* and that the order of the three markers is *argH–rif-8–metA*.

ORDERING MUTATIONS BY THREE-FACTOR CROSSES

A careful determination of cotransduction frequencies can reveal the order of markers in the DNA. However, three-factor crosses offer a less ambiguous way to determine marker order. This technique can be used in the genetic mapping of any organism in which genetic crosses are possible and was mentioned earlier in the chapter in connection with ordering mutations in phage. The only difference in transductional crosses or other types of crosses in bacteria comes from the fact that in bacterial crosses only a small part of the chromosome is recombining with the entire chromosome. As mentioned in chapter 3 in connection with gene replacements, a single crossover between a short linear piece of the chromosome and the entire chromosome will break the chromosome and be lethal. Therefore, a minimum of two crossovers are required to replace the chromosome sequence with the sequence on the transducing DNA and form a recombinant type. In general, in such a cross, odd numbers of crossovers (one, three, five, etc.) will break the chromosome and be lethal; therefore, any viable recombinant types must have originated from an even number of crossovers (two, four, six, etc.).

To illustrate the ordering of bacterial markers by transductional three-factor crosses, we again use the example of the *argH*, *metA*, and *rif* markers (Figure 7.32). In the example in the figure, the donor has the *metA15*

mutation and the recipient has the *argH5* and *rif8* mutations. After transduction, the *argH* marker was selected and the Arg⁺ transductants were tested for the unselected *metA* and *rif* markers. Four recombinant types are possible in a cross of this type. Three of the four recombinant types require only two crossovers, with any particular order of the three markers. The fourth recombinant type requires four crossovers which should make this recombinant type rarer than the others. Therefore, a determination of which recombinant type is rarest should reveal the order of the three markers.

Table 7.3 shows some representative data for the frequency of each of the four recombinant types in a cross of this type. Only two of the transductants were Arg⁺ Met⁻ Rifʳ, making this the rarest recombinant type. This recombinant type probably required four crossovers. As shown in Figure 7.32, the markers must be ordered *argH–rif-8–metA* to make this the recombinant type which requires four crossovers. In contrast, only two crossovers are required to make this recombinant type if the order were *argH–metA–rif-8* (with the metA marker in the middle). We conclude that the order is probably *argH5–rif-8–metA15*, which is also consistent with the order we obtained based on cotransduction frequencies alone.

Incidentally, resistance to rifampin is due to a mutation in the gene for the β subunit of RNA polymerase (see chapter 2). Therefore, the gene for the β subunit of RNA polymerase must lie between the *argH* and *metA* genes in the chromosome of *E. coli*.

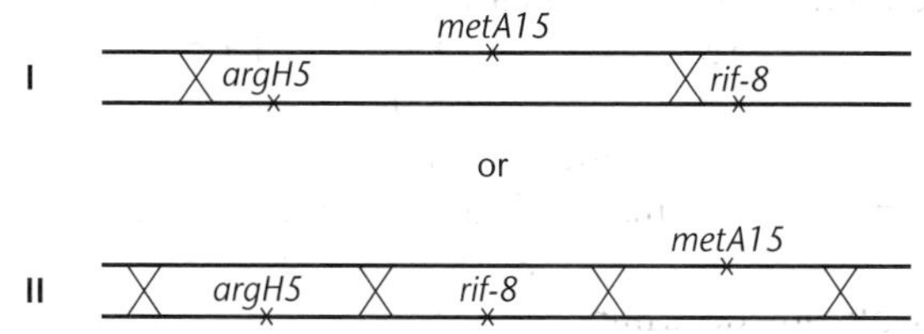

USING TRANSDUCTION FOR STRAIN CONSTRUCTION

One of the major uses of transduction is in constructing isogenic bacterial strains. Even different strains of the same species can differ at a number of genetic loci. An experiment often requires strains which differ by only one mutation or other genetic difference but are otherwise identical. Such strains are said to be **isogenic** (see chapter 3). Only by comparing isogenic strains can we be sure that any phenotypes we observe are due to that genetic difference alone and not to some other differences in the chromsomes of the two strains. Transduction introduces only a small region of the chromosome,

Figure 7.32 The number of crossovers required to make the rarest recombinant type, Arg⁺ Rifʳ Met⁻, with two different orders of the three markers. Since order II, *argH–rif-8–metA*, requires four crossovers, this is probably the order of the three markers. See the text for details.

TABLE 7.3	Typical transductional data from a three-factor cross
Recombinant phenotype	**No. of recombinants**
Arg⁺ Met⁺ Rifʳ	61
Arg⁺ Met⁺ Rifˢ	22
Arg⁺ Met⁻ Rifʳ	2
Arg⁺ Met⁻ Rifˢ	11

and so any differences between the original recipient strain and a transductant must have been carried in on the same piece of DNA. If other genetic differences are contributing to phenotypic differences between the two strains, they must be very closely linked.

Sometimes we can use transduction to move mutations into a strain, even if the mutation has no easily selectable phenotype. For example, we might use a closely linked transposon carrying an antibiotic resistance gene to move such a mutation. For this purpose, collections of *E. coli* strains have been assembled which have transposon insertions around the genome which are all cotransducible with at least one other transposon insertion in the collection. Therefore, the site of any mutation will be cotransducible with at least one of the transposon insertions. We can use the collection as a donor for transduction, selecting the antibiotic resistance gene on the transposon and testing a number of the transductants for the mutation. A recipient strain which has lost the mutation will probably have the transposon integrated close to the site of the mutation. We can then repeat the transduction with just this isolated strain as a donor but this time save a transductant which has retained the mutation, discarding the ones which have lost it due to cotransduction. These transductants will have the transposon inserted close to the mutation so that they can be used to move the mutation into other strains, selecting the antibiotic resistance gene on the closely linked transposon and thereby constructing an isogenic strain with the mutation.

The Role of Transduction in Bacterial Evolution

Phages may play an important role in evolution by promoting the horizontal transfer of genes between individual members of a species, as well as between distantly related bacteria. The DNA in phage heads is usually more stable than naked DNA and so may persist longer in the environment. Also, many phages have a broad host range for adsorption. For example, phage P1, which infects *E. coli*, will also inject its DNA into a number of other gram-negative bacterial species including *Myxococcus xanthus*. The host range of P1 is partially affected by the orientation of an invertible DNA segment encoding the tail fibers (see chapter 9). Although incoming DNA from one species will not recombine with the chromosome of a different species if they share no sequences, stable transduction of genes between distantly related bacteria becomes possible when the transduced DNA is a broad-host-range plasmid that can replicate in the recipient strain or contains a transposon that can hop into the DNA of the recipient cell.

SUMMARY

1. Viruses that infect bacteria are called bacteriophages, or phages for short.

2. The productive developmental cycle of a phage is called the lytic cycle. The larger DNA phages undergo a complex program of gene expression during development.

3. The products of phage regulatory genes regulate the expression of other phage genes during development. One or more regulatory genes in each stage of development turn on the genes in the stage to follow and turn off the genes in the preceding stage, a regulatory cascade. In this way, all the information for the stepwise development of the phage can be preprogrammed into the phage DNA.

4. Phage T7 encodes an RNA polymerase that specifically recognizes the promoters for the late genes of the phage. Because of its specificity and because transcription from the T7 promoters is so strong, this system has been used to make cloning vectors for expressing large amounts of protein from cloned genes.

5. All the genes of phage T4 are transcribed by the host RNA polymerase, which undergoes many changes in the course of infection. Phage T4 goes through a number of steps of gene regulation in its development, and the genes are named based on when their products appear in the infected cell. The immediate-early genes are transcribed immediately after infection from σ^{70}-type promoters. The delayed-early genes are transcribed through antitermination of transcription from immediate-early genes. The transcription of the middle genes is from phage-specific middle-mode promoters and requires a transcriptional activator, MotA, as well as a polypeptide, AsiA, which binds tightly to the RNA polymerase. The true-late genes are also transcribed from phage-specific promoters and require a phage-encoded sigma factor, gp55. In addition, T4 couples the transcription of its late genes to the replication of the phage DNA, so that the late genes will not be transcribed until the phage DNA begins to replicate.

6. All phages encode at least some of the proteins required to replicate their nucleic acids. They borrow others from their hosts.

7. The requirement of DNA polymerases for a primer prevents the replication of the extreme 5′ ends of linear DNAs. Phages solve this replication primer problem in different ways. Some replicate as circular DNA. Others form long concatemers by rolling-circle replication, by linking single DNA molecules end to end by recombination, or by pairing through complementary ends.

8. Bacteriophages are ideal for illustrating the basic principles of classical genetic analysis, including recombination and

(continued)

SUMMARY (continued)

complementation. To perform recombination tests with phages, cells are infected by two different mutants or strains of phage and the progeny are allowed to develop. The progeny are then tested for recombinant types. To perform complementation tests with phages, the cells are infected by two different strains under conditions which are nonpermissive for both mutants. If the mutations complement each other, the phage will multiply. Recombination tests can be used to order mutations with respect to each other. Complementation tests can be used to determine whether two mutations are in the same functional unit or gene.

9. Nonsense and temperature-sensitive mutations are very useful for identifying the genes of a phage whose products are essential for multiplication.

10. The linkage map of a phage shows the relative positions of all of known genes of the phage with respect to each other. Whether the linkage map of a phage is circular or linear depends upon how the DNA of the phage replicates and how it is packaged into phage heads.

11. Transduction occurs when a phage accidentally packages bacterial DNA into a head and carries it from one host to another. In generalized transduction, essentially any region of bacterial DNA can be carried.

12. Not all types of phages make good transducing phages. To be a good transducing phage, the phage must not degrade the host DNA after infection, must not have very specific *pac* sites, and must package DNA by a headful mechanism.

13. Transduction is very useful for genetic mapping in bacteria and for strain construction. Using transduction to map bacterial genetic markers is based on the fact that only a small part of the bacterial genome can be packaged in a phage head. If two markers are close enough together to be packaged in a phage head, they are cotransducible. Since transduction is so infrequent, one marker is selected and the transductants are tested for the cotransduction of the region of the other marker. The frequency of cotransduction increases the closer two markers are to each other.

QUESTIONS FOR THOUGHT

1. Why do you suppose phages regulate their gene expression during development so that genes whose products are involved in DNA replication are transcribed before genes whose products become part of the phage particle? What do you suppose would happen if they did not do this?

2. Why do you suppose some phages encode their own RNA polymerase while others change the host RNA polymerase?

3. Why do phage often encode their own replicative machinery rather than depend upon that of the host?

4. Why do some single-stranded DNA phages such as φX174 use a complicated mechanism to initiate replication that uses

the cellular proteins PriA, PriB, PriC, and DnaT, which are normally used by the cell to reinitiate replication at blocked forks, while others, such as f1, use a much simpler process, involving only the host RNA polymerase, to prime replication?

5. Phage often change the cell to prevent subsequent infection by other phage of the same type once the infection is under way. This is called superinfection exclusion. What do you suppose would be the consequences of another phage of the same type superinfecting a cell that is already in the late state of development of the phage if this mechanism did not exist?

PROBLEMS

1. We have precisely determined the titer of a stock of virus by counting the viruses under the electron microscope. How would you determine the effective MOI of the virus (i.e., the fraction of viruses that actually infect a cell) under a given set of conditions?

2. Phage components (e.g., heads and tails) can often be seen in lysates by electron microscopy, even before they are assembled into phage particles. In studying a newly discovered phage of *Pseudomonas putida*, you have observed that amber mutations in gene *C* of the phage prevent the appearance of phage tails in a nonsuppressor host. Similarly, mutations in gene *T* prevent the appearance of heads. However, mutations in gene *M* prevent the appearance of either heads or tails. Which gene, *C*, *T*, or *M*, is most likely to be a regulatory gene? Why?

3. Would you expect to be able to isolate amber mutations in the *ori* sequence of a phage? Why or why not?

4. To order the three genes *A*, *M*, and *Q* in a previously uncharacterized phage you have isolated, you cross a double mutant having an amber mutation in gene *A* and a temperature-sensitive mutation in gene *M* with a single mutant having an amber mutation in gene *Q*. About 90% of the Am⁺ recombinants that can form plaques on the nonsuppressor host are temperature sensitive. Is the order Q-A-M or A-Q-M? Why?

5. Phage T1 packages DNA from concatemers beginning at a unique *pac* site and then packaging by a processive headful mechanism, cutting about 6% longer than a genome length each time. However, it packages a maximum of only three headfuls from each concatemer. Would you expect T1 to have a linear or a circular genetic map? Draw a hypothetical T1 map.

6. Most types of phage encode lysozyme enzymes which break the cell wall of the infected cell late in infection to release the phage. How would you determine which of the genes in the linkage map of a phage you have isolated encodes the lysozyme? Hint: You can purchase egg white lysozyme from biochemical supply companies.

7. Calculate the cotransduction frequencies of the *argH* marker and the *rif* marker from the three-factor cross data in Table 7.3. Also calculate the cotransduction frequency of the *argH* and *metA* markers. How does this compare to the cotransduction frequencies in the example given in the text?

8. You wish to use transduction to determine if the order of three *E. coli* markers is *metB1–argH5–rif-8* or *argH5–metB1–rif-8*, so you do a three-factor cross. The donor for the transduction has the *rif-8* mutation, and the recipient has the *metB1* and *argH5* mutations. You select the *argH5* marker by plating on minimal

plates plus methionine, purify 100 of the Arg$^+$ transductants, and test for the other markers. You find that 17 are Arg$^+$ Met$^+$ Rif^r, 20 are Arg$^+$ Met$^+$ Rif^s, 60 are Arg$^+$ Met$^-$ Rif^s, and 3 are Arg$^+$ Met$^-$ Rif^r. What is the cotransduction frequency of the *argH* and *metB* markers? Of the *argH* and *rif-8* markers? What is the order of the three markers deduced from the three-factor cross data? Are the results consistent?

9. You have isolated a *hemA* mutant of *E. coli* which requires δ-aminolevulinic acid for growth. You wish to move this mutation into other genetic backgrounds. You obtain a strain of *E. coli* from the Yale Stock Collection that has a Tn*10* transposon conferring resistance to tetracycline inserted only 0.5 min away from the *hemA* gene. Explain the steps involved in using transduction to move your *hemA* mutation into other *E. coli* strains.

SUGGESTED READING

Benzer, S. 1961. On the topography of genetic fine structure. *Proc. Natl. Acad. Sci. USA* **47:**403–415.

Benzer, S., and S. P. Champe. 1962. A change from nonsense to sense in the genetic code. *Proc. Natl. Acad. Sci. USA* **48:** 1114–1121.

Campbell, A. M. 1996. Bacteriophages, p. 2325–2338. *In* F. C. Neidhardt, R. Curtiss III, J. L. Ingraham, E. C. C. Lin, K. B. Low, B. Magasanik, W. S. Reznikoff, M. Riley, M. Schaechter, and H. E. Umbarger (ed.), Escherichia coli *and* Salmonella: *Cellular and Molecular Biology,* 2nd ed. ASM Press, Washington, D.C.

Christensen, A. C., and E. T. Young. 1982. T4 late transcripts are initiated near a conserved DNA sequence. *Nature* (London) **299:**369–371.

Colland, F., G. Orsini, E. N. Brody, H. Buc, and A. Kolb. 1998. The bacteriophage T4 AsiA protein: a molecular switch for σ^{70} dependent promoters. *Mol. Microbiol.* **27:**819–829.

Crick, F. H. C., L. Barnett, S. Brenner, and R. J. Watts-Tobin. 1961. General nature of the genetic code for proteins. *Nature* (London*)* **192:**1227–1232.

Fluck, M. M., and R. H. Epstein. 1980. Isolation and characterization of context mutations affecting the suppressibility of nonsense mutations. *Mol. Gen. Genet.* **177:**615–627.

Herendeen, D. R., G. A. Kassavetis, and E. P. Geiduschek. 1992. A transcriptional enhancer whose function imposes a requirement that proteins track along the DNA. *Science* **256:**1298–1303.

Herendeen, D. R., K. P. Williams, G. A. Kassavetis, and E. P. Geidushek. 1990. An RNA polymerase binding protein that is required for communication between an enhancer and a promoter. *Science* **248:**573–578.

Kassavetis, G. A., and E. P. Geiduschek. 1984. Defining a bacteriophage T4 late promoter: bacteriophage T4 gene 55 protein suffices for directing late promoter recognition. *Proc. Natl. Acad. Sci. USA* **81:**5101–5105.

Kreuzer, K. N. 2000. Recombination-dependent DNA replication in phage T4. *Trends Biochem. Sci.* **25:**165–173.

Marciano, D. K., M. Russel, and S. M. Simon. 2001. Assembling filamentous phage occlude pIV channels. *Proc. Natl. Acad. Sci. USA* **98:**9359–9364.

Masters, M. 1996. Generalized transduction, p. 2421–2441. *In* F. C. Neidhardt, R. Curtiss III, J. L. Ingraham, E. C. C. Lin, K. B. Low, B. Magasanik, W. S. Reznikoff, M. Riley, M. Schaechter, and H. E. Umbarger (ed.), Escherichia coli *and* Salmonella: *Cellular and Molecular Biology,* 2nd ed. ASM Press, Washington, D.C.

Molineux, I. J. 2001. No syringes please, ejection of phage T7 DNA from the virion is enzyme driven. *Mol. Microbiol.* **40:**1–8. (Microreview.)

Mosig, G. 1998. Recombination and recombination-dependent replication in bacteriophage T4. *Annu. Rev. Genet.* **32:**379–413.

Mosig, G., A. Luder, A. Ernst, and N. Canan. 1991. Bypass of a primase requirement for bacteriophage T4 DNA replication *in vivo* by a recombination enzyme, endonuclease VII. *New Biol.* **3:**1195–1205.

Nomura, M., and S. Benzer. 1961. The nature of deletion mutants in the *r*II region of phage T4. *J. Mol. Biol.* **3:**684–692.

Rakonjac J., J. Feng, and P. Model. 1999. Filamentous phage are released from the bacterial membrane by a two-step mechanism involving a short C-terminal fragment of pIII. *J. Mol. Biol.* **289:**1253–1265.

Studier, F. W. 1969. The genetics and physiology of bacteriophage T7. *Virology* **39:**562–574.

Studier, F. W., A. H. Rosenberg, J. J. Dunn, and J. W. Dubendoff. 1990. Use of T7 RNA polymerase to direct expression of cloned genes. *Methods Enzymol.* **185:**60–89.

Symonds, N., P. Vander Ende, A. Dunston, and P. White. 1972. The structure of *r*II diploids of phage T4. *Mol. Gen. Genet.* **116:**223–238.

Yanisch-Perron, C., J. Vieira, and J. Messing. 1985. Improved M13 phage cloning vectors and host strains: nucleotide sequences of the M13mp18 and pUC19 vectors. *Gene* **33:**103–119.

Zinder, N. D., and J. Lederberg. 1952. Genetic exchange in *Salmonella. J. Bacteriol.* **64:**679–699.

8

Phage λ and Lysogeny

IN THE LAST CHAPTER we reviewed the lytic development of some representative phages. During lytic development, the phage infects a cell and multiplies, producing more phage that can then infect other cells. But this is not the only life-style of which phage are capable. Some phage are able to maintain a stable relationship with the host cell in which they neither multiply nor are lost from the cell. Such a phage is called a **lysogenic phage.** In the lysogenic state, the phage DNA is either integrated into the host chromosome or replicates as a plasmid. The phage DNA in the lysogenic state is called a **prophage** and the bacterium harboring a prophage is a **lysogen** for that phage. Thus, a bacterium harboring the prophage P2 would be a P2 lysogen.

In a lysogen, the prophage acts like any good parasite and does not place too great a burden upon its host. The prophage DNA is mostly quiescent; most of the genes expressed are those required to maintain the lysogenic state, and most of the others are turned off. Often the only indication that the host cell carries a prophage is that the cell is immune to superinfection by another phage of the same type. The prophage state can continue almost indefinitely unless the host cell suffers potentially lethal damage to its chromosomal DNA. Then, like a rat leaving a sinking ship, the phage can be **induced** and enter lytic development, producing more phage. The released phage can then infect other cells and develop lytically to produce more phage or lysogenize the new bacterial cell, hopefully one with a brighter future than its original host.

Studies with phage lysogeny were important in forming concepts of how viruses can remain dormant in their hosts and how they can convert cells into cancer cells. Also, many genes of benefit to the host bacterium, including virulence genes, whose products are required to make bacteria pathogenic, are carried on prophages, a topic covered later in the chapter. In addition to prophages that can be induced under some circumstances, bacteria carry many **defective prophages** that no longer can form infective phage due to deletion of some of their essential genes. These DNA elements are suspected of being defective prophages rather than normal parts of the chromosome because they are not common to all the strains of a species of bacterium and they carry genes related to the genes of other known phages. Defective prophages may be important in evolution because they eventually lose their identity and contribute useful genes to the chromosome of their host.

In this chapter, we discuss some examples of lysogenic phages and how the lysogenic state is achieved and maintained. One striking insight to come from these studies is the extent to which bacteria and their lysogenic phages depend upon each other, to the point where the distinction between the bacterial host and the parasitic virus begins to blur.

Phage λ

Phage λ is the lysogenic phage which has been studied most extensively and the one to which all others are compared. Although lysogeny was suspected as early as the 1920s, the first convincing demonstration that bacterial cells could carry phages in a quiescent state was made with λ phage in about 1950 (see Lwoff, Suggested Reading). In this experiment, apparently uninfected *Escherichia coli* cells could be made to produce λ phage by irradiating them with UV light. Since then, phage λ has played a central role in the development of the science of molecular genetics. A large number of very clever and industrious researchers have collaborated to make the interaction of λ with its host, *E. coli*, our best understood biological system (see Gottesman, Suggested Reading). This research has revealed not only the complexity and subtlety of biological systems but also their utility and beauty. First, we discuss how λ replicates lytically, and then we discuss how it can form a lysogen and how these two states are coordinated.

Phage λ: Lytic Development

Phage λ is fairly large, with a genome intermediate in size between those of T7 and T4 (Figure 8.1). Some gene products and sites encoded by λ are listed in Tables 8.1 and 8.2, respectively. Phage λ goes through three major stages during development. The first λ genes to be expressed after infection are *N* and *cro*. Most of the genes expressed next play a role in replication and recombination. Finally, the late genes of the phage are expressed, encoding the head and tail proteins of the phage particle and enzymes involved in cell lysis.

TABLE 8.1	Some λ gene products and their function
Gene product	**Function**
N	Antitermination protein acting at t_L^1, t_R^1, and t_R^2
O, P	Initiation of λ DNA replication
Q	Antitermination protein acting at t_R'
CI	Repressor; protein inhibitor of transcription from p_L and p_R
CII	Activator of transcription of *cI* and *int*
CIII	Stabilizer of CII
Cro	Protein inhibitor of CI synthesis
Gam	Protein required for rolling-circle replication
Red	Proteins involved in λ recombination
Int	Integrase; protein required for site-specific recombination with chromosome
Xis	Excisase; protein forms complex with Int and functions in excision of prophage

TRANSCRIPTION ANTITERMINATION

Many mechanisms of gene regulation now known to be universal were first discovered in phage λ. One of these is **transcription antitermination** (see chapter 2). In regulation by transcription antitermination, transcription begins at the promoter but then soon terminates unless certain conditions are met. Then the transcription continues into other downstream genes, hence the name "antitermination" because the mechanism works against

TABLE 8.2	Some sites involved in phage λ transcription and replication[a]
Sites	**Function**
p_L	Left promoter
p_R, p_R'	Right promoters
o_L	Operator for leftward transcription; binding sites for CI and Cro repressors
o_R	Operator for rightward transcription; binding sites for CI and Cro repressors
t_L^1, t_L^2	Termination sites of leftward transcription
t_R^1, t_R^2, t_R'	Termination sites of rightward transcription
nutL	N utilization site for leftward transcribing RNA Pol (i.e., the site at which N binds to RNA Pol)
nutR	N utilization site for rightward transcribing RNA Pol
qut	Q utilization site for antitermination at p_R'
p_{RE}	Promoter for repressor establishment; activated by CII
p_{RM}	Promoter for repressor maintenance; activated by CI
p_I	Promoter for *int* transcription; activated by CII
POP'	Attachment site (*attλ*)
cos	Cohesive ends of λ genome (12-bp single-stranded ends in linear genome anneal to form circular genome after infection)

[a]In λ, essential genes have single-letter names while nonessential genes have more conventional three-letter names.

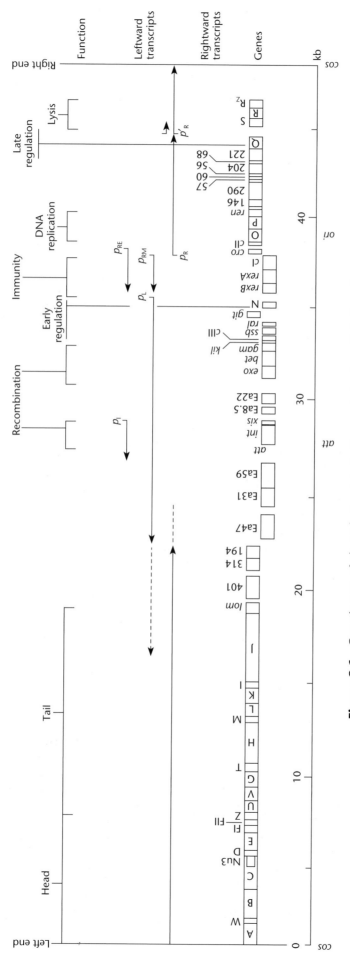

Figure 8.1 Genetic map of phage λ. Reprinted with permission from D. L. Daniels, J. L. Schroeder, W. Szybalski, F. Sanger, and F. R. Blattner, A molecular map of coliphage lambda, p. 473, *in* R. W. Hendrix, J. W. Roberts, F. W. Stahl, and R. A. Weisberg (ed.), *Lambda II*, Cold Spring Harbor Laboratory Press, Cold Spring Harbor, N.Y., 1983.

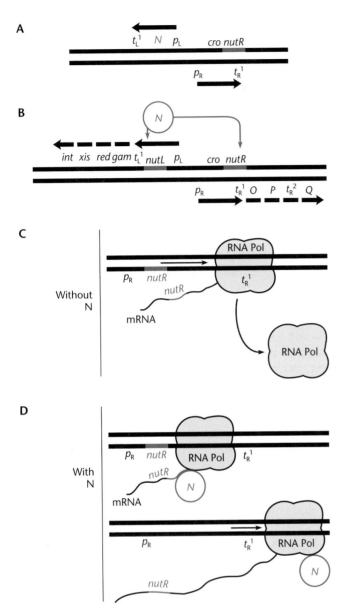

Figure 8.2 Antitermination of transcription in phage λ. (A) Before the N protein is synthesized, transcription starts at promoters p_L and p_R and stops at transcription terminators t_L^1 and t_R^1. (B) The N protein causes transcription to continue past t_L^1 and t_R^1 into *gam-red-xis-int* and *O-P-Q*, respectively. (C and D) Mechanism of antitermination by N, showing rightward transcription only. (C) In the absence of N, transcription initiated at p_R terminates at the terminator t_R^1. (D) If N has been made, it will bind to RNA polymerase as the polymerase transcribes the *nutR* site, possibly because N undergoes a conformational change when it binds to the *nut* sequence in the RNA. This change is required before N can bind to the RNA polymerase. With N bound, the RNA polymerase does not stop transcription at t_R^1. The sites and gene products shown here are defined in Tables 8.1 and 8.2.

or "anti to" termination. After being discovered in λ, this type of regulation was found in many other systems including in the regulation of transcription of human immunodeficiency virus (see below).

Phage λ uses antitermination of regulation at two stages in its development. At an early stage, it uses antitermination protein N to regulate the synthesis of its recombination and replication functions. At a later stage, it uses antitermination protein Q to regulate the synthesis of its late proteins including the head and tail proteins. We cover the early regulation involving the N antitermination protein first.

The N Protein

The N protein is responsible for the first stage of λ antitermination. When the λ DNA first enters the cell, transcription begins normally and terminates at transcription termination sites after two short RNAs are synthesized (Figure 8.2A). One of these RNAs encodes the Cro protein, which is an inhibitor of repressor synthesis, as discussed in the section on lysogeny (see below). However, the other encodes the N protein, the antitermination factor that permits the RNA polymerase to continue along the DNA, as shown in Figure 8.2B to D.

Figure 8.2C and D outlines the current picture for how N protein antiterminates, showing only rightward transcription. Initially, transcription initiated at the rightward p_R promoter terminates at the transcription terminator designated t_R^1. One of the sequences transcribed into RNA is *nutR* (for N utilization rightward). When the N protein appears, it will bind to the RNA polymerase after the *nutR* region on DNA has been transcribed (Figure 8.2D). The RNA polymerase with N protein riding on it does not stop for any termination site, transcribing past the t_R^1 terminator and any other terminator it encounters.

Similar events are occurring during leftward transcription. When the *nut* site on the other side, called *nutL* (for N utilization leftward), is transcribed, the N protein binds to the RNA polymerase and prevents any further termination, allowing transcription to continue into other genes, including *gam* and *red*.

This model for how N regulates the expression of the early genes by antiterminating transcription was first proposed on the basis of indirect genetic experiments. First it was shown that transcription initiated at the λ promoters p_R and p_L (see the genetic map in Figure 8.1) soon terminates unless N protein is present. This led to the conclusion that N was acting as an antiterminator by allowing transcription through downstream transcription terminators. However, surprisingly, N could antiterminate only the transcription that had initiated at the p_R

and p_L promoters and not the transcription initiating from other promoters closer to the terminators. This led to the conclusion that N does not act only at the terminators to prevent termination but that some sites upstream of the terminators are required for N action, possibly the p_R and p_L promoters themselves. Then it was shown that the sites required for N antitermination were actually not the p_R and p_L promoters but sites somewhat downstream of the promoters. It was hypothesized that N must bind to these sites to allow transcription through the downstream termination sites, and these were named the N utilization sites or *nut* sites.

Identifying the hypothetical *nut* sites on mRNA involved some clever selectional genetics (see Salstrom and Syzbalski, Suggested Reading, and "Genetic experiments with phage λ" below). This quest must have required a leap of faith, since all the evidence for the existence of the *nut* sites up to this point was indirect. However, the investigators were able to isolate *nutL* mutations that had all the predicted characteristics of *nut* mutations. These mutations prevent the expression of genes downstream of the terminator site; hence, they no longer allow antitermination, are *cis* acting, and affect only mRNA from the DNA which has the mutation. They also map in approximately the right place for a predicted *nutL* mutation, just downstream of the p_L promoter and just upstream of the N gene. Once such mutations had been isolated, the exact base pair change in the mutations was determined by comparing the DNA sequence of the *nutL* mutants to the DNA sequence of wild-type λ in this region. Once the site of *nutL* mutations was identified by DNA sequencing, a region with similar sequences was found just to the right of the *cro* gene and was assumed to be *nutR*.

The location of the *nut* sites supported another element of the model, i.e., that the N protein binds to the *nut* site sequence in the mRNA rather than in the DNA. Because of their location between genes, the *nut* sequences are not normally translated into protein. In fact, translating the *nut* sites interferes with antitermination. Apparently, ribosomes translating the mRNA can interfere with the binding of N to a *nut* site and thereby interfere with antitermination.

Figure 8.3 shows the sequence of the *nut* sites of λ as well as the *nut* sequences of some other related phages. The *nut* sites consist of a sequence, called BoxB, that forms a "hairpin" secondary structure in the mRNA because it is encoded by a region of the DNA with a twofold rotational symmetry (see chapter 2). The original *nutL* mutations all change bases in BoxB and disrupt the twofold symmetry of the sequence, preventing formation of the hairpin in the mRNA. Thus, apparently the formation of the BoxB hairpin in the mRNA is important for the binding of the N protein. In fact, structural studies performed more recently have indicated that the N protein can bind to RNA with just the BoxB hairpin and that both N protein and the BoxB secondary structure change upon this binding. This supports the idea that the N protein changes its conformation upon binding to the BoxB sequence and that only N in the changed conformation can bind to RNA polymerase and prevent termination.

The function of the BoxA and BoxC sequences in *nut* sites is more obscure. These sequences are common to the *nut* sites of all phages, and BoxA-like sequences occur in some bacterial genes including the genes for rRNA, where they play an important role in preventing premature termination of transcription (see chapter 13).

Figure 8.3 The sequences of some antitermination (*nut*) sites in bacteriophages and in bacterial rRNA genes. The sequences consist of similar box A, box B, and sometimes box C sequences. Adapted from D. I. Friedman and M. Gottesman, p. 29, *in* R. W. Hendrix, J. W. Roberts, F. W. Stahl, and R. A. Weisberg (ed.), *Lambda II*, Cold Spring Harbor Laboratory Press, Cold Spring Harbor, N.Y., 1983.

	Box A	Box B	Box C
λ*nutL*	ATGAAGGTGACGCTCTTAAAAATTAAGCCCTGAAGAAGGGCAGCATTCAAAGCAGAAGGCTTTGGGGTGTGTGATAC		
λ*nutR*	TAAATAACCCCGCTCTTACACATTCCAGCCCTGAAAAAGGGCATCAAATTAAACCACACCTATGGTGTATGCATTTAT		
21*nutR*	TAAGCAAATTGCTCTTTAACAGTTCTGGCCTTTCACCTCTAACCGGGTGAGCAAACATCAGCGGCAAATCCATTGGGTGTGCGCTA		
P22*nutL*	AACGCTCTTTAACTTCGATGATGCGCTGACAAAGCGCGAACAAATACCAAACGAGATTGGTTTGGACTGGCGTGTGGT		
λ*qut*	ATGGGTTAATTCGCTCGTTGTGGTAGTGAGATGAAAAGAGGCGGCGCTTACTACCGATTCCGCCTAGTTGGTCACTT		

Adding to the mystery is the fact that point mutations in BoxA of a phage λ *nut* site can prevent antitermination but deleting the entire BoxA sequence does not. It is an attractive idea that the BoxA sequences in antitermination sequences like the λ *nut* sites help bind host Nus proteins to promote antitermination (see below). No one has found a function for BoxC. It could be that these sites are required to regulate antitermination in subtle ways not easily detectable in a laboratory situation.

Host Nus Proteins

The N protein does not act alone to cause antitermination. A number of *E. coli* proteins, some of which are involved in transcription termination and antitermination in the uninfected *E. coli* cell, are also involved in λ antitermination. Host proteins which collaborate with N to cause antitermination are called Nus proteins (for *N utilization substance*). Many of the *nus* genes were first identified by *E. coli* mutations which prevent killing by phage λ after induction (see Friedman et al., Suggested Reading, and "Genetic Experiments with phage λ" below). There are six *nus* genes, *nusA* to *nusG*, that have been identified thus far using this and other types of selections. Four of these, *nusA*, *nusB*, *nusE*, and *nusG*, encode proteins that are involved in transcription termination and/or antitermination in the uninfected host. The products of these genes travel with the N-*nut*-RNA polymerase complex and may help hold the complex together. Surprisingly, *nusE* mutations are in a gene for a ribosomal protein, S10. This is surprising because translation is not thought to be required for antitermination and, in fact, inhibits it (see above). Perhaps the S10 protein plays a dual role in the cell, one in translation and another in transcription antitermination. Other *nus* mutations have a less direct effect on antitermination. For example, *nusD* mutations affect the host ρ factor, which is required for transcription termination at ρ-dependent termination sites (see chapter 2). These mutations may affect N antitermination, by causing stronger termination that cannot be overcome by N. Other *nus* mutations called *nusC* are present in the RNA polymerase β subunit. They may alter the binding of the antitermination complex to the RNA polymerase and thereby reduce antitermination.

The Q Protein

One of the genes under the control of the N antiterminator is gene *Q*, whose product is responsible for the transcription of the late genes of λ including the head and tail genes. Thus, λ marches through a regulatory cascade, with one of the earliest gene products (N) directing the synthesis of another gene product (Q), which in turn directs the transcription of the late genes. Like N, the Q protein of λ is an antiterminator, which allows transcription from the late promoter p_R' to proceed through terminators into downstream genes. The mechanism of antitermination by Q has been studied in great detail and is very different from antitermination by N. Like N, the Q protein loads on RNA polymerase in response to a sequence located close to the promoter, called *qut* (for *Q utilization site*) (Table 8.2; Figure 8.3). But the similarity ends here. The *qut* site is not found exclusively in the mRNA; some of it must be present in the DNA since some of the required *qut* sequence is not even transcribed into mRNA, being upstream of the start site of transcription at p_R'. As a prelude to antitermination, the RNA polymerase transcribes a short RNA of only 16 to 17 nucleotides from the late promoter p_R' before it pauses. The RNA polymerase may pause because it confuses the pause site for a promoter. The σ-factor has not yet cycled off the RNA polymerase (see chapter 2), and the pause site has a –10 sequence similar to the –10 sequence at promoters with which the σ-factor interacts. The Q protein then loads on the paused RNA polymerase, making it oblivious to further transcription termination and allowing it to proceed untrammeled through the late genes of the phage, much as the N protein prevents transcription termination by remaining bound to the RNA polymerase.

The details of Q protein antitermination have been studied in some detail (see Marr and Roberts, Suggested Reading). Apparently, the Q protein can load on the RNA polymerase only when it is positioned just right at the pause site. Sometimes the RNA polymerase overshoots the pause site and makes an RNA 17 nucleotides or more in length. Then the RNA polymerase must backtrack to 16 nucleotides while the GreA and GreB proteins cleave the extra RNA that extrudes from the RNA polymerase as a result of the backtracking (see chapter 2). The Q protein can then load on the RNA polymerase and send it on its way at last.

ANTITERMINATION IN OTHER SYSTEMS

Like many regulatory systems, antitermination was first discovered in phage, but it is used in many other systems. This mechanism operates not only in related phages such as P22 but also for many bacterial genes. As mentioned, antitermination regulates transcription of the rRNA genes of all bacteria. It is also known to regulate the *bgl* operon of *E. coli* and the aminoacyl-tRNA synthetase genes of *B. subtilis* (see Box 12.3). Often, where regulation by antitermination of transcription occurs, sequences similar to λ *nut* sequences also are present (Figure 8.3). Antitermination may also be used to regulate eukaryotic genes. For example, the *myc* oncogene of mammals and, as mentioned, the transcription of human immunodeficiency virus, which causes AIDS, may also be regulated through antitermination.

Phage λ: Linear DNA That Replicates as a Circle

The replication of λ DNA has also been studied extensively and has been one of the major model systems for understanding replication in general. The λ DNA is linear in the phage head but cyclizes, that is, forms a circular molecule, after it enters the cell through pairing between its cohesive ends, or *cos* sites. These sites are single stranded and complementary to each other for 12 bases and so can join by complementary base pairing, which is why they are cohesive or "sticky." Once the cohesive ends are paired, DNA ligase can join the two ends to form covalently closed circular λ DNA molecules. Circular DNA molecules can replicate because there will always be DNA upstream to serve as a primer (see chapter 1).

CIRCLE-TO-CIRCLE OR θ REPLICATION OF λ DNA

Once circular λ DNA molecules have formed in the cell, they can replicate by a mechanism similar to the θ replication described for the chromosome in chapter 1 and for plasmids in chapter 4. Replication initiates at the *ori* site in gene O (Figure 8.1) and proceeds in both directions, with both leading- and lagging-strand synthesis in the replication fork (Figure 8.4). When the two replication forks meet somewhere on the other side of the circle, the two daughter molecules separate.

ROLLING-CIRCLE REPLICATION OF λ DNA

After a few circular λ DNA molecules have accumulated in the cell by θ replication, the rolling-circle type of replication ensues. The initiation of λ rolling-circle replication is similar to that of M13 in that one strand of the circular DNA is cut and the free 3′ end serves as a primer to initiate the synthesis of a new strand of DNA that displaces the old strand. DNA complementary to the displaced strand is also synthesized to make a new double-stranded DNA. The λ process differs in that the displaced individual single-stranded molecules are not released when replication around the circle is completed. Rather, the circle keeps rolling, giving rise to long tandem repeats of individual λ DNA molecules called **concatemers** (Figure 8.4). The concatemers are like the picture obtained by rolling an engraved ring, dipped in ink, across a piece of paper. The pattern on the ring will be repeated over and over again on the paper.

In the final step, the long concatemers are cut at the *cos* sites into λ-genome length pieces as they are packaged into phage heads. Phage λ will only package DNA from concatemers, and at least two λ genomes must be linked end to end in a concatemer, because the packaging system in the λ head recognizes one *cos* site on the concatemeric DNA and takes up DNA until it arrives at the next *cos* site, which it cleaves to complete the packaging.

GENETIC REQUIREMENTS FOR λ DNA REPLICATION

Unlike in T4 and T7, which encode many proteins for replication, the products of only two λ genes, O and P, are required for λ DNA replication. As Figure 8.4 illustrates,

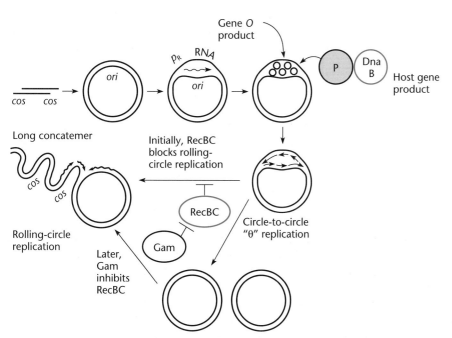

Figure 8.4 Overview of replication of phage λ. See the text for details.

both of these proteins are required for priming DNA replication at the *ori* site. The O protein is thought to bend the DNA at this site by binding to repeated sequences, similar to the mechanism by which DnaA protein initiates chromosome replication (see chapter 1), and the P protein binds to the O protein and to components of the host replication machinery, thus commandeering them for λ DNA replication; it therefore acts like DnaC.

RNA synthesis must also occur in the *ori* region for λ replication to initiate. It normally initiates at the p_R promoter and may be required to separate the DNA strands at the origin or may serve as a primer for rightward replication.

A third λ protein, the product of the *gam* gene, is required for the shift to the rolling-circle type of replication, albeit indirectly. The RecBCD nuclease (an enzyme that facilitates recombination) of *E. coli* somehow inhibits the switch to rolling-circle replication, but Gam inhibits the RecBCD nuclease. Therefore, a *gam* mutant of λ is restricted to the θ mode of replication, and concatemers can form from individual circular λ DNA molecules only by recombination. Because λ requires concatemers for packaging, a *gam* mutant of λ will not multiply in the absence of a functional recombination system, either its own or the RecBCD pathway of its host. This fact led to the detection of *chi* sequences, which are important in RecBCD recombination and will be discussed further in chapter 10.

PHAGE λ CLONING VECTORS

The many cloning vectors derived from phage λ offer numerous advantages. The phage multiply to a high copy number, allowing for the synthesis of large amounts of DNA and protein. Also, it matters less if the cloned gene encodes a toxic protein than with plasmid cloning vectors. The toxic protein will not be synthesized until the phage infects the cell, and the infected cell is destined to die anyway. It is also relatively easy to store libraries in the relatively stable λ phage head.

COSMIDS

As mentioned, λ packages DNA into its head by recognizing *cos* sites in concatemeric DNA; therefore, any DNA containing a *cos* sequence will be packaged into phage heads. In particular, plasmids containing *cos* sites can be packaged into λ phage heads. Such plasmid cloning vectors, called **cosmids**, also offer many advantages for genetic engineering, including **in vitro packaging**. In this procedure, plasmid DNA is mixed with extracts of λ-infected cells containing heads and tails of the phage. The DNA will be taken up by the heads, and because λ particles will self-assemble in the test tube, the tails will be attached to the heads to make infectious λ

particles, which can then introduce the cosmid into cells by infection. Any λ cloning vector will serve in this method, and infection is a more efficient way of introducing DNA into bacteria than is transfection or transformation.

Another major advantage of cosmids is that the size of the cloned DNA is limited by the size of the phage head. If the piece of DNA cloned into a cosmid is too large, the *cos* sites will be too far apart and the DNA will be too long to fit into a phage head. However, if the cloned DNA is too small, the *cos* sites will be too close to each other and the phage heads will have too little DNA and be unstable. Therefore, the use of cosmids ensures that the pieces of DNA cloned into a vector will all be approximately the same size, which is sometimes important for making libraries (see chapter 1).

LYSOGENIC VECTORS

As mentioned, phage λ sometimes infects cells and then becomes quiescent (see the section on lysogeny, below), and some λ cloning vectors retain this ability. If a gene is cloned into λ and the phage is then allowed to lysogenize a host, the cloned gene will exist in only one copy in the chromosome. This can be useful for complementation studies, which are often complicated if a gene is present in too many copies.

Lysogeny

Phage λ is the classical example of a phage that can form lysogens. In the lysogenic state, very few λ genes are expressed, and essentially the only evidence that the cell harbors a prophage is that the lysogenic cells are immune to superinfection by more λ. The growth of the immune lysogens in the plaque is what gives the λ plaque its characteristic fried-egg appearance, with the lysogens forming the "yolk" in the middle of the plaque (see Figure 8.5).

Some phage λ mutants form plaques that are clear because they do not contain immune lysogens. These phage have mutations in the *c*I, *c*II, or *c*III gene, where the "c" stands for *c*lear plaque. These mutations prevent the formation of lysogens. Understanding the regulation of the λ lysogenic pathway and the function of the *c*I, *c*II, and *c*III gene products in forming lysogens required the concerted effort of many people. Their findings illustrate the complexity and subtlety of biological regulatory pathways and serve as a model for other systems (see Gottesman, Suggested Reading).

The *c*II Gene Product

Figure 8.6 illustrates the process of forming a lysogen after λ infection and how the *c*I, *c*II, and *c*III gene products are involved. After λ infects a cell, the decision

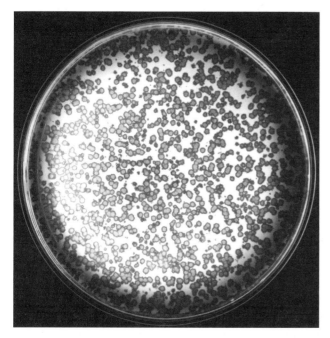

Figure 8.5 Phage λ plaques with a typical fried-egg appearance. (Photograph by Kurt Stepnitz.)

Figure 8.6 Formation of lysogens after λ infection. (A) the cII and cIII genes are transcribed from promoters p_R and p_L, respectively. (B) CII activates transcription from promoters p_{RE} and p_I, leading to the synthesis of CI repressor and the integrase Int, respectively. (C) The repressor shuts off transcription from p_L and p_R by binding to o_R and o_L. Finally, the Int protein integrates the λ DNA into the chromosome (see Figure 8.7).

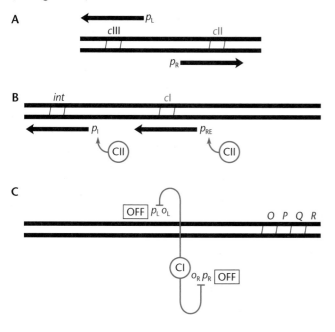

whether the phage enters the lytic cycle and makes more phage or forms a lysogen depends on the outcome of a competition between the product of the *c*II gene, which acts to form lysogens, and the products of genes in the lytic cycle that replicate the DNA and make more phage particles. Most of the time, the lytic cycle wins, the λ DNA replicates, and more phage are produced. However, about 1% of the time, depending on environmental factors such as the richness of the medium, the *c*II gene product wins the race and a lysogen is formed.

The CII protein promotes lysogeny by activating the RNA polymerase to begin transcribing at two promoters, which are otherwise inactive (Figure 8.6). Proteins that enable RNA polymerase to begin transcription at certain promoters are called **transcriptional activators** (see chapter 2). One of the promoters activated by CII is p_{RE}, which allows transcription of the *c*I gene. The product of this gene, the **CI repressor**, prevents transcription from the promoters p_R and p_L, which service many of the remaining genes of λ. We discuss the CI repressor in more detail below. The other promoter activated by the CII protein, p_I, allows transcription of the integrase (*int*) gene. The Int enzyme integrates the λ DNA into the bacterial DNA to form the lysogen.

The role of the *c*III gene product in lysogeny is less direct. CIII inhibits a cellular protease that degrades CII. Therefore, in the absence of CIII, the CII protein will be rapidly degraded and no lysogens will form.

Phage λ Integration

As discussed above, the λ DNA forms a circle immediately after infection by pairing between the *cos* sequences at its ends. The Int protein can then promote the integration of the circular λ DNA into the chromosome, as illustrated in Figure 8.7. Int is an integrase that specifically promotes recombination between the attachment sequence (called *attP* for *att*achment *p*hage) on the phage DNA and a site on the bacterial DNA (called *attB* for *att*achment *b*acteria) that lies between the galactose (*gal*) and biotin (*bio*) operons in the chromosome of *E. coli*. This is a nonessential region of the *E. coli* chromosome, but some phages integrate into essential genes of the bacterium (see Box 8.1). Because the Int-promoted recombination does not occur at the ends of λ DNA but, rather, at the internal *attP* site, the prophage map is different from the map of DNA found in the phage head. In the phage head, the λ DNA has the *A* gene at one end and the *R* gene at the other end (see the λ map in Figure 8.1). In contrast, in the prophage, the *int* gene is on one end and the *J* gene is on the other (Figure 8.7). The relative order of the genes on the maps is still the same, but the prophage and phage maps are cyclic permutations of each other.

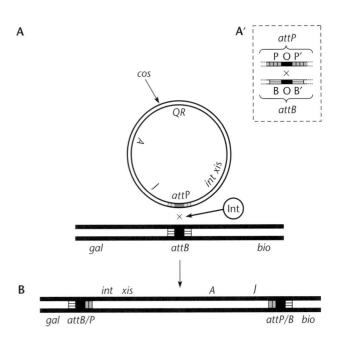

Figure 8.7 Integration of λ DNA into the chromosome of *E. coli.* (A) The Int protein promotes recombination between the *attP* sequence in the λ DNA and the *attB* sequence in the chromosome. The inset (A′) shows the region in more detail, with sequences POP′ and BOB′. The common core sequence of the two sites is shown in black. (B) The gene order in the prophage. The *cos* site is where the λ DNA is cut for packaging and recircularization after infection. The location of the *int, xis, A,* and *J* genes in the prophage is shown (refer to λ map in Figure 8.1). The *E. coli gal* and *bio* operons are on either side of the prophage DNA in the chromosome.

BOX 8.1

Phages That Integrate into Essential Genes of the Host

Until fairly recently, it seemed axiomatic that lysogenic phage could integrate only into sites on the bacterial DNA between genes; otherwise, they would inactivate bacterial genes and be deleterious to their host. The archetypal phage λ, which integrates into a nonessential region between the *gal* and *bio* operons of *E. coli*, supported this contention. However, we now know that some phages and other DNA elements such as pathogenicity islands (see chapter 9) integrate directly into genes, often into genes for tRNAs. Examples include the *Salmonella* phage P22, which integrates into a threonine tRNA gene; the *E. coli* phage P4, which also integrates into a leucine tRNA gene; the *Haemophilus influenzae* phage HPc1, which integrates into a leucine tRNA gene; and the virus-like element SSV1 of the archebacterium *Sulfolobus* sp., which integrates into an arginine tRNA gene. It is not known why so many phages use tRNA genes as their attachment sites. Perhaps it is because tRNA genes are relatively highly conserved in evolution. A phage could lysogenize a different species of bacterium if the sequence of its attachment site were highly conserved, and thus it could be found in the new chromosome. Another possible explanation is that phages seem to prefer sequences with twofold rotational symmetry for their attachment sites. The sequences of tRNA genes have such symmetry, since the tRNA products of the genes can form hairpin loops, and in fact, most phage seem to integrate into the region of the tRNA gene that encodes the anticodon loop.

Not all phages that integrate into genes use tRNA genes for their attachment sites, however. For example, the phage, φ21, a close relative of λ, and e14, a defective prophage, both integrate into the isocitrate dehydrogenase (*icd*) gene of *E. coli*. The product of the *icd* gene is an enzyme of the tricarboxylic acid cycle and is required for optimal utilization of most energy sources, as well as for the production of precursors for some biosynthetic reactions. Inactivation of the *icd* gene would cause the cell to grow poorly on most carbon sources.

How can phages integrate into genes and not inactivate the gene, thereby compromising the host? The answer is that they duplicate part of the gene in their *attP* site. The 3′ end of the gene is repeated in the phage *attP* site with very few changes, so that when the phage integrates, the normal 3′ end of the gene will be replaced by the very similar phage-encoded sequence. It is an interesting question in evolution how the bacterial sequence could have arisen in the phage. Perhaps the phage first arose as a specialized transducing particle, which then adapted to using the substituted bacterial genes as their normal attachment site in the chromosome.

References

Hill, C. W., J. A. Gray, and H. Brody. 1989. Use of the isocitrate dehydrogenase structural gene for attachment of e14 in *Escherichia coli* K-12. *J. Bacteriol.* **171:**4083–4084.

Reiter, W. D., P. Palm, and S. Yeats. 1989. Transfer RNA genes frequently serve as integration sites for prokaryotic genetic elements. *Nucleic Acids Res.* **17:**1907–1914.

The recombination promoted by Int is called **site-specific recombination** because it occurs between specific sites, one on the the bacterial DNA and another on the phage DNA. This site-specific recombination is not normal homologous recombination but, rather, a type of nonhomologous recombination because the sequences of the phage and bacterial *att* sites are mostly dissimilar. They have a common core sequence, O, of only 15 bp— GCTTT(TTTATAC)TAA—flanked by two dissimilar sequences, B and B′, in *attB*, and P and P′ in *attP* (Figure 8.7, inset). The recombination always occurs within the bracketed 7-bp sequence. Because the region of homology is so short, this recombination would not occur without the Int protein, which recognizes both *attP* and *attB*. We discuss other examples of site-specific recombination in chapter 9.

Maintenance of Lysogeny

After the lysogen has formed, the *cI* repressor gene is one of the few λ genes to be transcribed. The CI repressor binds to two regions, called **operators**. These operators, o_R and o_L, are close to promoters p_R and p_L, respectively, and, by binding the repressor, prevent transcription of most of the other genes of λ. Operators are discussed further in chapter 12. In the prophage state, the *cI* gene is transcribed from the p_{RM} promoter (for *re*pression *m*aintenance), which is immediately upstream of the *cI* gene, rather than from the p_{RE} promoter used immediately after infection. The p_{RM} promoter is not used immediately after infection because its activation requires the CI repressor. We discuss the regulation of CI synthesis in more detail below.

Regulation of Repressor Synthesis in the Lysogenic State

The CI repressor is the major protein required to maintain the lysogenic state; therefore, its synthesis must be regulated even after a lysogen has formed. If the amount of repressor drops below a certain level, the transcription of the lytic genes will begin and the prophage will be induced to produce phage. However, if the amount of repressor increases beyond optimal levels, cellular energy is wasted making excess repressor. The mechanism of regulation of repressor synthesis in lysogenic cells is well understood and has served as a model for gene regulation in other systems (Figure 8.8).

To function, the CI repressor protein must be a homodimer (see chapter 2) composed of two identical polypeptides encoded by the *cI* gene. At very low concentrations of CI polypeptide, the dimers will not form and the repressor will not be active.

Each CI polypeptide consists of two parts, or domains. In fact, the λ repressor was probably the first protein shown to have separable domains (see the discussion

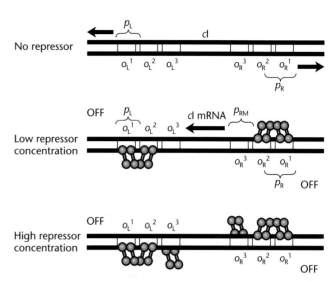

Figure 8.8 Regulation of repressor synthesis in the lysogenic state. The dumbbell shape represents the two domains of the repressor. The dimeric repressor, shown as two dumbbells, binds cooperatively to o_R^1 and o_R^2, (and o_L^1 and o_L^2), repressing transcription from p_R (and p_L) and activating transcription from p_{RM}. At higher repressor concentrations, it also binds to o_R^3 (and o_L^3), preventing transcription from p_{RM}. The relative affinity of the repressor for the sites is $o_R^1 > o_R^2 > o_R^3$ and $o_L^1 > o_L^2 > o_L^3$, respectively.

of the Lieb experiments on the λ repressor in "Genetic experiments with phage λ" below). We now know that many proteins have a modular construction, with different functions of the protein separated into different domains. One of these domains promotes the formation of a dimer by binding to another polypeptide. The other domain on each polypeptide binds to an operator sequence on the DNA. To illustrate this structure, the CI polypeptides are traditionally drawn as dumbbells, with the ends of each dumbbell indicating the two domains, and the complete dimer repressor protein is presented as two dumbbells bound to each other (Figure 8.8).

The repressor regulates its own synthesis by binding to the operator sequences and by being both a repressor and an activator of transcription. The important operator for regulating repressor synthesis is o_R, to the right of the repressor gene, and so only this operator is shown, although both operators have the same structure. The operator o_R can be divided into three repressor-binding sites, o_R^1, o_R^2, and o_R^3. If the concentration of repressor is low, only the o_R^1 site will be occupied, which is sufficient to repress transcription from promoter p_R, which overlaps the operator site (Figure 8.8). However, as the repressor concentration increases, o_R^2 will also be occupied. Only repressor bound at o_R^2 can activate transcription from the promoter p_{RM}, which is why p_{RM} is used to transcribe the repressor gene in the lysogen only when there is some repressor in the cell. With very high concentrations of repressor, o_R^3 will also be occupied, and transcription from p_{RM} will be blocked, blocking synthesis of more repressor, although there is less direct experimental evidence for this part of the model. By synthesizing more repressor when there is less in the cell and less when there is more, the cell maintains the levels of repressor within narrow limits.

Part of the regulation is based on the fact that repressor can bind to o_R^2 only if o_R^1 is already occupied by repressor and can bind to o_R^3 only if o_R^2 is already occupied. Apparently, the repressor protein bound at one site makes contact with another repressor molecule, allowing it to bind at the adjacent site. This is called **cooperative binding**, because protein bound at one site cooperates in the binding of a protein to an adjacent site. Cooperative binding is used in many regulatory systems, some of which we discuss in later chapters.

Immunity to Superinfection

The CI repressor in the cell of a lysogen will prevent not only the transcription of the other prophage genes by binding to operators o_L and o_R but also the transcription of the genes of any other λ phage infecting the lysogenic cell by binding to the operators of that phage. Thus, bacteria lysogenic for λ are immune to λ superinfection. However, λ lysogens can still be infected by any relative of λ phage that has different operator sequences. The λ CI repressor could not bind to such sequences. Any two phages that differ in their operator sequences are said to be **heteroimmune**.

The Induction of λ

Phage λ will remain in the prophage state until the host cell DNA is damaged. Figure 8.9 outlines the process of induction of λ. When the cell attempts to repair the damage to its DNA, short pieces of single-stranded DNA accumulate and bind to the RecA protein of the host. The RecA protein with single-stranded DNA attached then binds to the λ CI repressor, causing it to cleave itself. The DNA-binding domain is separated from the domain involved in dimer formation. Without the dimerization domain, the DNA-binding domains can no longer bind tightly to the operators. As the repressors drop off the operators, transcription will initiate from the promoters p_R and p_L, and the lytic cycle will begin.

THE Cro PROTEIN

Early during induction, more repressor could be made to interfere with later stages in lytic development or even reestablish lysogeny. However, the *cro* gene product, which is one of the first λ proteins to be made after induction, prevents the synthesis of more repressor. Cro

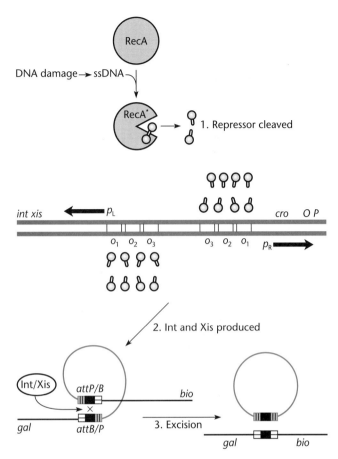

Figure 8.9 Induction of λ. Accumulation of single-stranded ssDNA due to damage to the DNA results in activation of the RecA protein, which promotes the autocleavage of the CI repressor protein, separating the dimerization domain of the protein from the DNA-binding domain, so that the repressor can no longer form dimers and bind to DNA. Transcription of *int-xis* and *O* and *P* ensues, and the phage DNA (in gold) excises from the chromosome and replicates. ssDNA, single-stranded DNA.

does this by binding to the operator sequences, although in reverse order of repressor binding. This binding is illustrated in Figure 8.10. Cro binds first to the o_R^3 site and then to the o_R^2 site, thereby preventing the CI repressor from binding to the o_R^2 site and activating its own synthesis from the p_{RM} promoter. Cro also binds to o_L^3, thereby preventing CI repressor binding to o_L. Thus, p_L will no longer be repressed.

EXCISION

Once the repressor is out of the way, transcription from p_L and p_R can begin in earnest. Among the genes transcribed from p_L are *int* and *xis*. Unlike after infection, when only Int is synthesized, after induction both the integrase (Int) and excisase (Xis) proteins are synthesized

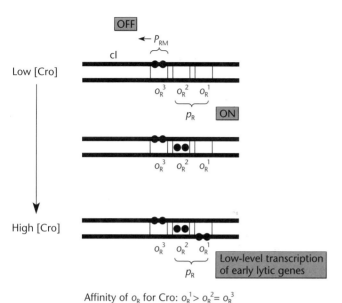

Affinity of o_R for Cro: $o_R^1 > o_R^2 = o_R^3$

Figure 8.10 Cro prevents repressor binding and synthesis by binding to the operator sites in reverse order from the repressor. By binding to o_R^3, Cro prevents repressor activation of transcription from p_{RM} while allowing transcription from p_R. Eventually Cro accumulates to the point where it binds to o_R^1 and o_R^2 and blocks transcription of early RNA.

(see below). Excision requires both of these proteins, because the recombination occurs between different sequences from those used for integration. Integration requires recombination between the *attP* sequence on the phage and the *attB* sequence in the chromosome (see the section on phage λ integration above); however, excision requires recombination between hybrid *attP-attB* sequences that exist at the junctions between the prophage DNA and the chromosomal DNA. These sequences are

different from either *attB* or *attP*. Therefore, Int alone is not capable of excising the prophage, and by making just Int after infection, the phage ensures that a newly integrated prophage will not be excised. The prophage will be excised only after induction, when both Int and Xis are made. Box 8.2 describes the molecular basis for the timing of Int and Xis synthesis.

While the Int and Xis proteins are excising λ DNA from the chromosome, the O and P genes are being transcribed from p_R. These proteins will promote the replication of the excised λ DNA. Therefore, a few minutes after the cellular DNA is damaged, the phage DNA is replicating, repressor levels are dropping, and the phage is irreversibly committed to lytic development. In about 1 h, depending on the medium and the temperature, the cell will lyse, spilling about 100 phage into the medium.

Competition between the Lytic and Lysogenic Cycles

As mentioned above, some cells infected by λ follow the lytic pathway, while others become lysogens. Figure 8.11 and Table 8.3 review the competition for entry into the lysogenic cycle versus the lytic cycle. After infection, when there is no CI repressor in the cell, the N and *cro* genes are transcribed. As discussed earlier in this chapter, the N gene product acts as an antiterminator and allows the transcription of many genes, including *cII* and *cIII*, as well as the genes encoding the replication proteins O and P.

The phage will enter the lytic or the lysogenic cycle according to the outcome of a race between the CII activator protein and the Cro protein, which may be determined by chance or the host cell's metabolic state. If the CII protein wins, it will activate the synthesis of the CI repressor from the p_{RE} promoter and the integrase from the p_I promoter. The CI repressor will bind to the

TABLE 8.3	Steps leading to lytic growth and lysogeny	
Steps leading to lytic growth	**Steps leading to lysogeny**	
1. Transcription from p_L and P_R	1. Same as for lytic growth	
2. N and Cro are made	2. Same as for lytic growth	
3. N allows CII expression	3. Same as for lytic growth	
4. **CII degraded**	4. **CII stable**	
5. Low CII concentration means that little CI is made	5a. High CII concentration activates p_I, and so Int is made and λ DNA integrates	
	5b. High CII concentration activates p_{RE}, and so CI is made	
6. Cro binds at o_R^3 and o_L^3, blocking binding by any low level of CI that is made	6. CI outcompetes Cro, and so CI binding at o_R and o_L both represses p_L and p_R and positively autoregulates at p_{RM}, maintaining lysogeny	
7. Meanwhile, N allows O and P replication gene transcription		
8. A second antiterminator, Q, allows late-gene transcription, and so λ phage particles are made		

BOX 8.2

Retroregulation

The term **retroregulation** means that the expression of a gene is regulated by sequences downstream of it rather than upstream, such as at the promoter. How phage λ ensures that only Int will be made after infection but that both Int and Xis will be made after prophage induction is an example of retroregulation. Initially, after infection, both the *int* and *xis* genes are transcribed from the promoter p_L. However, the *int* and *xis* coding parts of this RNA do not survive long enough to be translated, because the RNA contains an RNase III cleavage site downstream of the *int* and *xis* coding sequences. The RNA is cleaved at this site and degraded past the *int* and *xis* coding sequences by a 3′ exonuclease, probably RNase II. This regulation was named retroregulation because mutations downstream of the *int* gene changed the RNase III cleavage site, preventing enzyme recognition, and thus stabilized the RNA and allowed both *int* and *xis* to be translated from the transcript initiated at p_L.

As discussed in the text, the *int* gene will also be transcribed from the promoter p_I, which is actually located in the *xis* gene (see Figure 8.1). This resulting RNA does not contain a *nut* site, and so the N protein cannot bind to the RNA polymerase and allow it to proceed past termination signals as far as the coding sequence for the RNase III cleavage site; therefore, the RNA will be stable. Moreover, the *int* RNA contains all of the *int* sequence but only part of the *xis* sequence, so that only Int can be made from this RNA.

Immediately after induction, however, both Int and Xis can be made from the RNA produced from the p_L promoter, because *xis-int* RNA produced from this promoter will now be stable. As shown in the figure, during integration of the phage, the coding region for the RNase III cleavage site has been separated from the *xis-int* coding region, since this region is on the other side of the *attP* site, which is split during integration of the phage DNA. Therefore, the long RNA transcript initiated at p_L will no longer contain the RNase III cleavage site at its 3′ end and so will be stable. Both Int and Xis can be translated from this RNA after induction. This is an example of posttranscriptional regulation, because it occurs after the RNA synthesis has occurred on the gene (see chapter 12). It is interesting how the phage has taken advantage of the reorganization of its genes during integration to regulate the expression of its genes.

References

Guarneros, G., C. Montanez, T. Hernandez, and D. Court. 1982. Posttranscriptional control of bacteriophage λ *int* gene expression from a site distal to the gene. *Proc. Natl. Acad. Sci. USA* **79:**238–242.

Schmeissner, U., K. M. McKenny, M. Rosenberg, and D. Court. 1984. Removal of a terminator structure by RNA processing regulates *int* gene expression. *J. Mol. Biol.* **176:**39–53.

A After λ infection, *int* expressed from p_I

B After induction, *int* and *xis* expressed from p_L

(A) After infection, the *xis* and *int* genes cannot be expressed from the p_L promoter. Because of *N*, transcription from p_L will continue past *int* into an RNase III cleavage site. The RNA will be cleaved and digested back into *xis* and *int*, removing them from the RNA. Xis also cannot be expressed from p_I because the p_I promoter is in the *xis* gene. The RNA from p_I is stable, however, because this transcript does not contain a *nut* site and so will not continue through *int* to the RNase III cleavage site. The gold region indicates the location of the coding information for the RNase III site, but RNase III cleaves only the mRNA transcript. (B) In the prophage, however, the sequence encoding the RNase III cleavage site is separated from the *xis-int* coding sequence, so that the RNA made from p_L is stable.

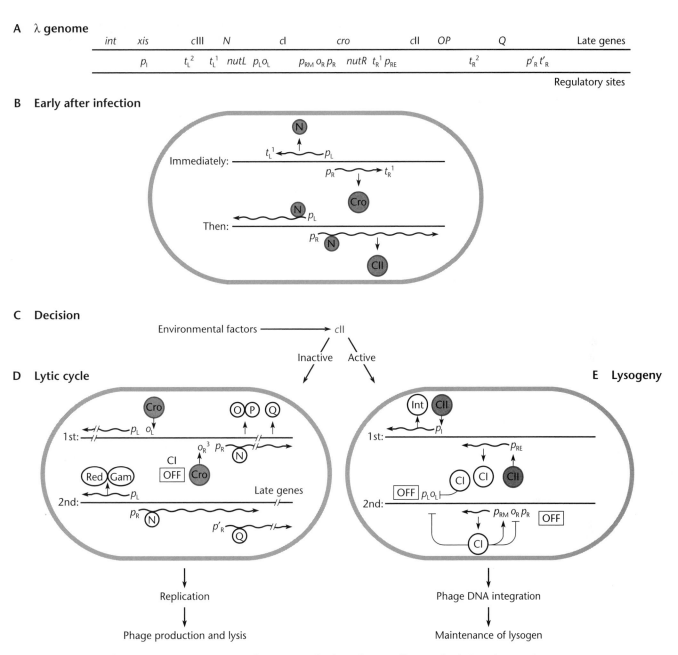

A   λ genome

int	xis	clII	N	cI	cro	clII	OP	Q	Late genes

| | p_I | t_L^2 t_L^1 nutL $p_L o_L$ | | p_{RM} o_R p_R nutR t_R^1 p_{RE} | | t_R^2 | p'_R t'_R | | |

Regulatory sites

B   Early after infection

Immediately: t_L^1 ← p_L N
p_R → t_R^1 Cro

Then: N ← p_L
p_R N   CII

C   Decision

Environmental factors ⟶ cII

Inactive   Active

D   Lytic cycle   **E   Lysogeny**

1st: Cro p_L o_L O P Q
o_R^3 p_R N

CI OFF Cro

Red Gam p_L Late genes

2nd: p_R N
p'_R Q

1st: Int CII p_I
p_{RE}

CI CI CII

OFF $p_L o_L$

2nd: p_{RM} o_R p_R OFF
CI

Replication   Phage DNA integration

Phage production and lysis   Maintenance of lysogen

Figure 8.11   Competition determing whether phage will enter the lytic or lysogenic cycle. (A) Key genes (top line) and sites (bottom line) are shown. (B) Gene expression early after infection. (C) The abundance of active CII protein determines whether the phage enters the lytic or lysogenic cycle. (D) The synthesis of Cro will promote lytic development by repressing the synthesis of CI repressor. Once O and P are synthesized, the replication of the λ DNA will dilute out the CI repressor. (E) The synthesis of CII will promote lysogeny.

operators o_L and o_R and repress the synthesis of more Cro, as well as O and P, the DNA will integrate, and the lysogen will form. However, if the Cro protein wins, it will prevent the synthesis of more CI repressor. Then, without more CI repressor, some transcription will occur from genes O and P and replication of the λ DNA will begin. Eventually, there will be too much DNA for the repressor to bind to all of it, and transcription of O and P will increase further, followed by yet more DNA replication. The Q protein, which is also an antiterminator, will be synthesized next, allowing transcription of the head, tail, and lysis genes. Phage will then be assembled, the cells will lyse, and newly minted phage will spill out into the medium.

Specialized Transduction

Lysogenic phages are capable of another type of transduction called specialized transduction. Two properties distinguish specialized from generalized transduction. In generalized transduction, essentially any gene of the donor bacterium can be transduced into the recipient bacterium. However, in specialized transduction, only bacterial genes close to the attachment site of the prophage can be transduced. Also, the specialized transducing phage carries *both* bacterial genes and phage genes instead of only bacterial genes, like a generalized transducing phage.

Figure 8.12 illustrates how specialized transduction occurs in phage λ. In a λ lysogen of *E. coli*, the λ prophage is integrated close to and between the *gal* and *bio* genes in the chromosome. The *gal* gene products degrade galactose for use as a carbon and energy source, and the *bio* gene products make the vitamin biotin.

Specialized transduction can occur when a phage picks up neighboring bacterial genes during induction of the prophage. As shown in Figure 8.12, a specialized transducing phage carrying the *gal* genes, called λd*gal*, forms as the result of a mistake during the excision recombination. As mentioned in the preceding section, when the phage DNA is excised from the bacterial DNA, recombination occurs between the hybrid *attP-attB* sites at the junction between the prophage and host DNA. However, recombination sometimes occurs by mistake between the prophage DNA and a neighboring site in the bacterial DNA. The DNA later packaged into the head will include some bacterial sequences, as shown. Such transducing phage are very rare because the erroneous recombination that gives rise to them is extremely infrequent, occurring at one-millionth the frequency of normal excision. Furthermore, the recombination must, by chance, occur between two sites that are approximately a λ genome length apart, or the DNA would not fit into a phage head.

Because of the rarity of these transducing phage, powerful selection techniques are required to detect them. To select λ phage carrying *gal* genes of the host, induced phage are used to infect Gal⁻ recipient bacteria, and Gal⁺ transductants are selected on plates with galactose as the sole carbon source. In the rare Gal⁺ transductants, a λ phage carrying *gal* genes may have integrated into the chromosome, providing the *gal* gene product that the mutant lacks. If such a Gal⁺ lysogen is colony purified and the prophage is induced from it, all of the resultant phage progeny will carry the *gal* genes. Such a lysogenic strain produces an **HFT lysate** (for *h*igh-*f*requency *t*ransduction) because it produces phage that can transduce bacterial genes at a very high frequency.

Because the λ-phage head can hold DNA of only a certain length, the transducing particles will of necessity have lost some phage genes to make room for the bacterial genes. Which phage genes are lost depends on where the mistaken recombination occurs. If the recombination occurs to the left of the prophage, some of the head and tail genes will be replaced by *gal* genes of the host, as in Figure 8.12. However, if the mistake in recombination occurs to the right, the *int* and *xis* genes will be replaced by the *bio* genes of the host (not shown).

Clearly, the properties of the transducing particle will be determined by the lost phage genes. For example, the λd*gal* phage shown in Figure 8.12 lack essential head and tail genes and so cannot multiply without a wild-type λ **helper phage** to provide the missing head and tail proteins (Figure 8.13). These phage are therefore called λd*gal*, where the "d" stands for *d*efective.

Transducing phage in which the *int* and *xis* genes have been replaced by *bio* will be able to multiply, since the genes on this side of *attP* are not required for multiplication. However, they cannot form a lysogen without the help of a wild-type phage. Because they can multiply and form plaques, *bio* transducing phages are called λp*bio*, in which the "p" stands for *p*laque forming.

Specialized transducing phage particles played a major role in the development of microbial molecular genetics, including the first isolation of genes and the discovery of IS elements in bacteria. They can also be used to map

Figure 8.12 Formation of a λd*gal* transducing particle. A rare mistake in recombination between a site in the prophage DNA (in this case between *A* and *J*) and a bacterial site to the left of the prophage results in excision of a DNA particle in which some bacterial DNA has replaced phage DNA.

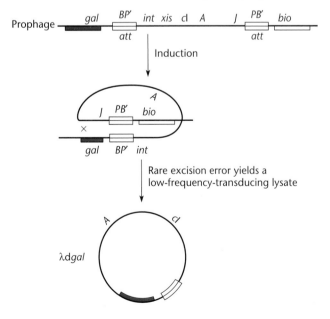

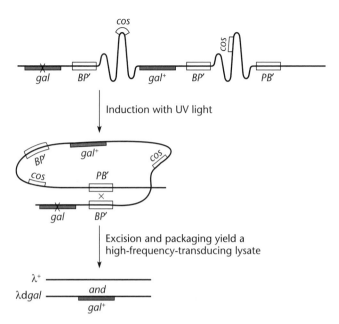

Figure 8.13 Induction of the λdgal phage from a dilysogen containing both λdgal and a wild-type λ in tandem. Recombination between the hybrid *PB′* and *BP′* sites at the ends excises both phages. The wild-type "helper" phage helps the λdgal phage to form phage particles. See the text for details.

phage genes (see below). Although their general use has been largely supplanted by recombinant DNA techniques, they continue to have special applications.

Other Lysogenic Phages

Phage λ was the first lysogenic phage to be extensively studied and thus serves as the archetypal lysogenic phage. However, many other types of lysogenic phages are known. Many use somewhat different strategies to achieve and maintain the prophage state. We briefly describe some of them here.

Phage P2

Phage P2 is another lysogenic phage of *E. coli*. The phage DNA is linear in the phage head but has cohesive ends like λ, which cause the DNA to cyclize immediately after infection. The phage replicates as a circle, and the DNA is packaged from these circles instead of from concatamers as λ does normally. Also, like λ, the genetic map of P2 phage is linear because it has a unique *cos* site at which the circles are cut during packaging.

One way in which P2 differs significantly from λ, which almost always integrates into the same site in the *E. coli* chromosome, is that P2 can integrate into many sites in the bacterial DNA, although it uses some sites

more than others. Like λ, P2 requires one gene product to integrate and two gene products to excise. P2 prophage is much more difficult to induce than λ, however. It is not inducible by UV light, and even temperature-sensitive repressor mutations cannot efficiently induce it. The only known ways to induce it are to infect with another P2 or P4 (see below).

Phage P4

Even viruses can have parasites! Phage P4 depends on phage P2 for its lytic development (see Kahn et al., Suggested Reading). Thus, it is a representative of a group called **satellite viruses**, which need other viruses to multiply. Phage P4 does not encode its own head and tail proteins but, rather, uses those of P2. Thus, P4 can multiply only in a cell that is lysogenic for P2 or that has been simultaneously infected with a P2 phage. When P4 multiplies in bacteria lysogenic for P2, it induces transcription of the head and tail genes of the P2 prophage, which are normally not transcribed in the P2 lysogen. This is illustrated in Figure 8-14A. P4 uses two mechanisms to induce transcription of the late genes of P2. It induces the P2 lysogen because it makes an inhibitor of the P2 repressor protein, which binds to the P2 repressor, inactivating it and inducing P2 to enter the lytic cycle. However, even though the P2 DNA replicates after induction by P4 and all the P2 proteins are made, the phage which are made will contain mostly P4 DNA. This is because P4 makes a protein called Sid, which causes the P2 proteins to assemble into heads that are smaller than normal, with only one-third the volume of a normal P2 head. These heads are too small to hold P2 DNA but large enough to hold P4 DNA, which is only about one-third the length of P2 DNA, so that the heads will be filled with P4 DNA instead. Nevertheless, a P4 *sid* mutant, which cannot make Sid protein, can still multiply in a P2 lysogen. The heads in the lysate, which are now the larger P2 size, contain either P2 DNA or P4 DNA. However, those which contain P4 DNA have two or three copies of the P4 DNA to fill the larger heads.

P4 can still multiply in P2 lysogens even if it cannot induce the P2 prophage, which remains in the chromosome. P4 accomplishes this by *trans* activating the transcription of the head and tail genes of P2. At first it was not obvious how P4 could induce the transcription of the late genes of P2, since, like T4 phage, the transcription of the late genes of P2 is normally coupled to replication and the P2 DNA will not replicate if the prophage is not induced. To bypass this replication dependence, P4 makes a protein called δ which activates the transcription of the P2 late genes without P2 replication, even though the transcription seems to occur from the same promoters as the normal P2 replication-dependent transcription.

Because it wears the protein coat of P2, phage P4 particle looks similar to P2. Only the head of P4 is smaller, to accommodate the shorter DNA. While the DNAs of P2 and P4 have otherwise very different sequences, the *cos* sites at the ends of the DNA are the same, so that the head proteins of P2 can package either DNA.

Phage P4 can also form a lysogen; when it does so, it usually integrates into a unique site on the chromosome.

Not only can P4 infection induce a P2 prophage, but also P2 infection can induce a P4 prophage (Figure 8.14B). Apparently P4, which cannot multiply by itself, does not want to be caught sleeping as a prophage if the cell happens to be infected by P2. Not only would it die along with its host, but also it would miss the opportunity to multiply and infect new hosts. Again, most of the phage which emerge from the infection after P4 is in-

Figure 8.14 P2 can't win for losing. (A) A P2 lysogen is infected with P4. P4 makes a protein which binds to and inhibits the P2 repressor inducing the P2 prophage. The P4 protein Sid makes the P2 head proteins form a smaller head, which packages the shorter P4 DNA rather than P2 DNA. Therefore, mostly P4 phage are released when the cells lyse. (B) A P4 lysogen is infected with P2. Now the P4 prophage is induced, and its replicating DNA is packaged by head proteins made by the infecting P2. Again, mostly P4 phage are released from the lysed cell even though it was a P2 phage that infected the cell. Details are given in the text.

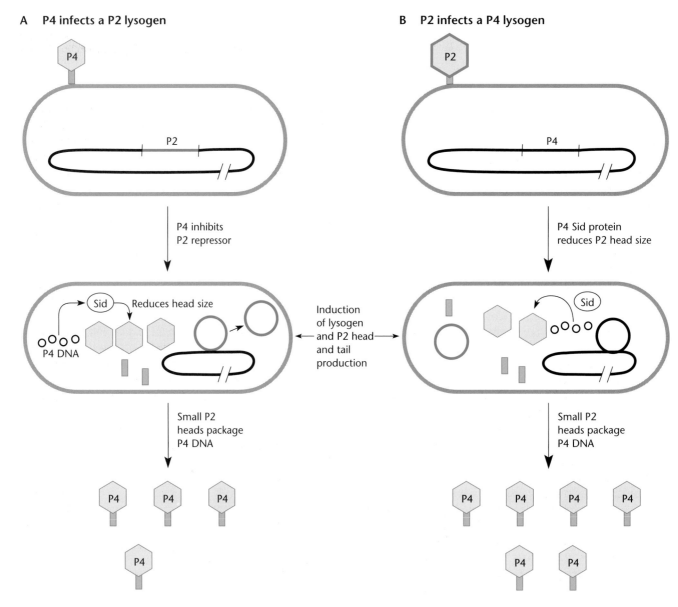

duced have P4 DNA wrapped in a smaller-than-normal P2 coat, even though it was a P2 phage that infected the cell. One phage enters the cell and emerges as a different phage. No matter who starts the infection, P2 comes out the loser.

Not only can phage P4 integrate into the chromosome, it can also replicate autonomously as a circle in the prophage state, as does P1 (see below). Because of this ability to maintain itself as a circle, phage P4 has been engineered for use as a cloning vector.

Phages P2 and P4, as well as their many relatives, have a very broad host range and infect many members of the *Enterobacteriaceae* including *Salmonella* and *Klebsiella* spp. as well as some *Pseudomonas* spp. They are also related to phage P1, although their lifestyles and strategies for lytic development and lysogeny are very different.

As is often the case, once the interaction of P2 and P4 had been discovered and characterized, other examples of DNAs that parasitize phages were discovered. One of the more intriguing is the parasitization of a *Staphylococcus aureus* phage by a pathogenicity island carrying the toxin gene for toxic shock syndrome. The phage gives the pathogenicity island its mobility, allowing it to move between strains of *S. aureus* (see Box 8.3).

Phage P1: a Plasmid Prophage

Not all prophages integrate into the chromosome of the host to form a lysogen. Some, represented by P1, form a prophage that replicates autonomously as a plasmid. Other phages are known to sometimes exist as plasmids in the prophage state, including P4 (see above) and some mutants of λ. In these cases, partial repression of the gene expression of the phage limits replication and keeps

---

BOX 8.3

How a Pathogenicity Island Gets Around

Many bacteria have large DNA elements integrated in their chromosome; these elements are called genetic islands, and they contain genes which confer special properties on the bacteria that carry them. Like prophages and integrated plasmids, genetic islands are not normal regions of the chromosome. They are not carried by all the strains of a particular species and also often precisely excise from the chromosome, deleting no chromosomal sequences and leaving no part of themselves behind. They also carry genes which allow a bacterium carrying them to occupy special ecological niches. However, they are neither integrated plasmids nor prophages, even defective ones. They are not capable of autonomous replication, nor do they encode any gene products required to make a phage upon induction. Nevertheless, they seem to be mobile and able to move from one strain of bacterium to another, because identical genetic islands are sometimes found in otherwise less closely related bacterial strains. They carry an integrase (*int*) gene, whose product allows them to integrate specifically into a region of the host DNA, often a tRNA gene. They also carry inverted repeated sequences at their ends, which are presumably involved in their integration. However, very few genetic islands have actually been demonstrated to move in a laboratory situation.

Pathogenicity islands are a type of genetic island which carry genes required for pathogenicity. For example, *Yersinia* species have a pathogenicity island which carries genes for iron scavenging in the animal host, and the *cag* pathogenicity island of *Helicobacter pylori* encodes a type IV protein secretion system to secrete a toxin required for

pathogenicity (see chapter 5). Because they are not found in all strains of a species, pathogenicity islands are assumed to be able to move from one host to another. However, the only pathogenicity island whose movement has been demonstrated is SaP11, which is found in some strains of *Staphylococcus aureus*. This pathogenicity island carries the gene for the toxin which causes toxic shock syndrome, and only *S. aureus* organisms which carry the island are capable of causing the disease. The SaP11 pathogenicity island moves by specifically parasitizing a *S. aureus* phage called 80α, in a process strikingly similar to the way in which P4 parasitizes P2 phage (see above). When phage 80α infects a *S. aureus* bacterium carrying SaP11, the pathogenicity island excises from the chromosome and replicates, apparently with the help of phage replication proteins. Like P4, the pathogenicity island directs the phage to make smaller heads, which then package the pathogenicity island rather than the phage DNA. When such a phage infects another cell, the pathogenicity island is injected and can integrate into the chromosome of its new host, using its Int protein. Thus, the major difference between phage P4 and the SaP11 pathogenicity island is that P4 encodes all the proteins to replicate its own DNA while SaP11 depends upon the infecting phage for its replication proteins. Maybe P4 should be called a genetic island rather than a phage and you should never judge a DNA element by its coat alone.

Reference

Ruzin, A., J. Lindsay, and R. P. Novick. 2001. Molecular genetics of SaP11, a mobile pathogenicity island in *Staphylococcus aureus*. *Mol. Microbiol.* **41**:365–377.

the copy number of the plasmid low but very variable. However, P1 in the prophage state is a bona fide plasmid. The P1 plasmid prophage maintains a copy number of 1 and combines many of the other features of true plasmids including a partitioning system and plasmid addiction system. Because it is a true plasmid, combined with the convenience of having a phage cycle which facilitates the isolation of DNA, etc., plasmid P1 is one of the major model systems for studying plasmid copy number control, segregation, and partitioning (see chapter 4). Another interesting aspect of this phage is that it has an invertible segment (see chapters 7 and 9). A region of the phage DNA encoding the tail fibers frequently inverts, and the host range of the phage depends upon the orientation of this invertible segment.

Phage Mu

Another phage which forms lysogens is Mu, which integrates randomly into the chromosome. Because it integrates randomly, it often integrates into genes and causes random insertion mutations, hence its name Mu (for *mu*tator phage). This phage is essentially a transposon wrapped in a phage coat, and it integrates and replicates by transposition. For this reason, we defer a discussion of phage Mu until chapter 9.

Use of Lysogenic Phages as Cloning Vectors

Lysogenic phages offer some distinct advantages as cloning vectors. Because they can multiply as a phage, obtaining large amounts of the cloned DNA is relatively easy. Since they can also integrate into the DNA of the host, a cloned gene will exist in only two copies, one in the prophage and one at the normal site, which is important in complementation studies with bacteria.

Cloning into lysogenic phage can also facilitate genetic mapping or gene replacements. In such experiments, a lysogenic phage that lacks its *att* site but has an easily selectable marker (such as resistance to an antibiotic) introduced into it is used. The gene of interest is cloned into the phage DNA, which is introduced into the cell either by in vitro packaging and infection or by transfection. Lysogens will form by recombination between the cloned gene in the phage and its counterpart in the chromosome (Figure 8.15). The phage will be integrated at this site in the chromosome rather than at the *attB* site because the phage *attP* site is deleted. When the antibiotic resistance gene on the phage DNA is mapped genetically by methods discussed in chapters 5 and 7, the original location in the chromosome of the gene will be known. This may be the preferred way to map a cloned gene for which no convenient phenotype is available, since antibiotic resistance markers are relatively easy to map genetically. We discuss gene replacement in chapter 3.

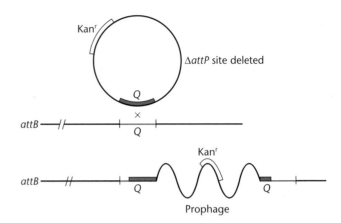

Figure 8.15 Use of a phage cloning vector lacking its *attP* site to mark a cloned chromosomal gene for mapping or for gene replacement. In the example, a phage lysate containing a DNA clone of the bacterial gene *Q* is used to infect the bacterium. Because the phage has its *attP* site deleted, it can only integrate by recombination between the *Q* gene in the phage and in the chromosome. The location of the *Q* gene in the chromosome can then be ascertained by mapping the kanamycin resistance (Kan^r) gene on the prophage.

Lysogenic Phages and Bacterial Pathogenesis

In a surprising number of instances, prophages carry genes for virulence factors or toxins required by the pathogenic bacteria they lysogenize. For example, the bacteria that cause diphtheria, scarlet fever, botulism, tetanus, and cholera all require lysogenic phages for virulence. Even λ phage carries genes that confer on its *E. coli* host serum resistance and the ability to survive in macrophages. As mentioned earlier, the process by which a prophage converts a nonpathogenic bacterium to a pathogen is an example of lysogenic conversion.

E. coli and Dysentery: Shiga Toxins

Pathogenic strains of *E. coli* are prime examples of bacteria that are not pathogenic unless they harbor prophages or other DNA elements carrying virulence genes. These bacteria are part of the normal intestinal flora unless they carry a certain DNA element. Then they can cause severe diseases, including bacterial dysentery with symptoms such as bloody diarrhea. The famous *E. coli* strain O157, which has caused many outbreaks of bacterial dysentery worldwide, is one example of such a lysogenic *E. coli*. In fact, bacterial dysentery due to these bacteria is the major cause of infant mortality worldwide.

In one particularly clear example, a group of prophages very closely related to λ can make *E. coli* pathogenic by encoding toxins called Shiga toxins, so named because they were first discovered in *Shigella dysenteriae*, a very close relative of *E. coli* which should be moved into the same genus. Like cholera toxin and many other toxins, the Shiga toxin is composed of two subunits, A and B. The B subunit helps the A subunit enter an endothelial cell of the host. The A subunit is an *N*-glycosidase, a type of enzyme that cleaves the bond between the base and the sugar in nucleotides, removing the base from RNA or DNA. There are many *N*-glycosidases known, including uracil-*N*-glycosidase, which removes uracil from DNA (see chapter 1), and the *N*-glycosidases that remove other damaged bases from DNA to avoid mutagenesis (see chapters 1 and 11). However, the Shiga toxin A subunit is very specific in that it removes only a certain adenine base from the 28S rRNA. Removal of this adenine from the 28S rRNA in a ribosome blocks translation by interfering with binding of the translation factor EF-1α to the ribosome. Interestingly, this adenine in the 28S rRNA seems to be the "Achilles heel" of the ribosome and is a popular target of translation-blocking systems. The rRNAs are highly conserved, and an adenine occurs in this position in the rRNAs of all eukaryotes. Plant enzymes such as ricin that protect the plant by blocking translation in cells infected by virus are also *N*-glycosidases that remove this same adenine from their own 28S rRNA, killing the cell and preventing multiplication of the virus. Yeast also make an enzyme called saracin, which has the same target, although the function of this enzyme is unknown.

As mentioned, the Shiga toxins are encoded by genes on prophages rather than being normal chromosomal genes. A prophage genetic map of phage φ361 encoding Shiga toxin 2 (Stx2), is shown in Figure 8.16. Note the remarkable similarity between the genetic map of this prophage and the genetic map of the λ prophage shown in Figure 8.1. Obviously, they are very close relatives. The toxin genes *stx2A* and *stx2B* lie just downstream of the Q gene and upstream of the lysis genes S and R. The genes in this region of phage λ are late genes, transcribed with the other late genes from the p_R' promoter, and so they would be transcribed only if the lysogen is induced. However, the toxin genes in φ361 have their own weak promoter, p_{Stx}, and so they are weakly expressed, even in the lysogen (see Wagner et al., Suggested Reading).

Diphtheria

Diphtheria is the classic example of a disease caused in part by the product of a gene carried by a lysogenic phage. Pathogenic strains of *Cornyebacterium diphtheriae* differ from nonpathogenic strains in that they are lysogenic for phage β (see Freeman, Suggested Reading). The β prophage carries a gene, *tox*, for the diphtheria toxin. The diphtheria toxin is an enzyme that kills eukaryotic cells by ADP-ribosylating (attaching adenosine diphosphate to) the EF-2 translation factor, thereby inactivating it and blocking translation in the cell. The *tox* gene of the β prophage is transcribed only when *C. diphtheriae* infects its eukaryotic host or under conditions that mimic this environment. We discuss the mechanism of action of the toxin and the regulation of the *tox* gene in chapter 13 under Global Regulation.

Cholera

Another recently discovered example of toxin genes carried by a lysogenic phage involves the bacterium that causes cholera, *Vibrio cholerae* (see Waldor and Mekalanos, Suggested Reading). In this case, the toxin genes are carried on a single-stranded filamentous phage called CTXφ for *cholera toxin phage*. In many ways this phage is like any other filamentous single-stranded DNA phage like f1 (see Chapter 7). It infects the cell by attaching to a pilus and then enters the cell through the TolA, TolQ, and TolR channel. The CTXφ phage can also form a lysogen and exist as a prophage. The prophage integrates at a specific bacterial attachment site in the *Vibrio cholerae* chromosome. If the host lacks this site in its chromosome, the prophage can maintain itself as a double-stranded

Figure 8.16 Close relatives of λ encode Shiga toxins. Shown is the prophage genetic map of phage φ361 indicating the positions of the toxin genes. The gold indicates that the repressor and toxin genes are expressed in the lysogen. Details are given in the text. Reprinted with permission from P. L. Wagner, M. N. Neely, X. Zhang, D. W. K. Acheson, M. K. Watdor, and D. L. Friedman, *J. Bacteriol.* **183**:2081–2085, 2001.

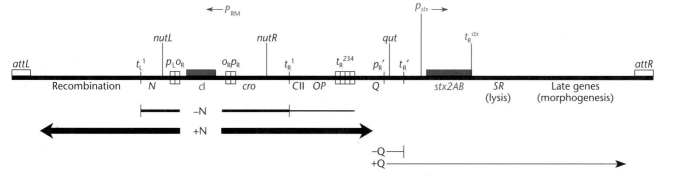

plasmid which replicates by a rolling-circle mechanism analogous to the rolling-circle mechanism by which other single-stranded phages replicate their replicative form (RF) (chapter 7). When it is found integrated in the chromosome, the prophage often exists in tandem repeats, with two or more phage genomes linked head to foot. Only lysogens in which the prophage exists in tandem repeats, can be induced to make more phage (see Moyer et al., Suggested Reading). The phage does not excise itself from the chromosome like λ and many other phages. Instead the prophage replicates by a rolling-circle mechanism to spin off more + strands, which can be packaged into phage heads, much like + strands are made from circular RFs in other phages. The rolling-circle replication initiates at the origin for + strand synthesis in one copy of the prophage in the chromosome and continues into the adjacent prophage DNA, terminating at the origin in the second prophage to make a complete phage genome. This explains why the prophage must exist in multiple tandem repeats in the chromosome. If only a single copy of the prophage existed in the chromosome, the phage would be able to replicate only part of its genome when it is induced.

The cholera toxin genes, *ctxA* and *ctxB,* encode another AB-type toxin in which the B subunit helps the A subunit, an enzyme, into the eukaryotic cell. The apparatus which secretes the subunits of the cholera toxin from the *V. cholerae* cell is an example of a type II secretion system and was mentioned in chapter 2. It is interesting that the cholera toxin genes on the prophage are regulated by a chromosomal gene, *toxR,* that also regulates the synthesis of the pili that serve as the receptor sites for the phage. These pili are also important virulence determinants because they enable the bacteria to adhere to the intestinal mucosa. We also discuss the regulation of cholera toxin genes in chapter 13.

Botulism and Tetanus

Other striking examples of diseases caused by toxins encoded by lysogenic phages are botulism and (probably) tetanus. Botulinum toxin causes a flaccid paralysis, in which the muscles are unable to contract, while tetanus toxin causes a rigid paralysis, in which the muscles remain flexed. Because of its ability to relax muscles, botulinum toxin (Botox) is currently used to treat involuntary muscle spasms such as facial "tics" and for the cosmetic purpose of removing facial lines associated, for example, with aging. Botox is made by recombinant DNA techniques from the gene cloned from the prophage.

The two very different diseases botulism poisoning and tetanus are caused by different toxins, but both toxins are encoded by lysogenic phages of the gram-positive genus *Clostridium.* Recent evidence indicates that these two toxins work by a common mechanism. They both

cleave the same neuronal protein, synaptobrevin, in exactly the same position in the amino acid sequence, although the routes of entry of the toxins into the host and the symptoms they cause are very different (see Schiavo et al., Suggested Reading).

Synopsis

It is becoming increasingly apparent that some of the genes which make bacteria virulent are carried on prophages and other DNA elements such as plasmids and transposons. We have listed only a few examples here; however, many of the other classical diseases which have plagued humankind since recorded history are turning out to have prophage involvement. But why are toxin genes and other virulence factors often encoded by prophages instead of being normal genes in the chromosome? The argument is similar to that used to explain why genes are carried on plasmids (see chapter 4). Having toxin and other virulence genes on a movable DNA element like a phage may allow the bacterium to adapt to being pathogenic without all the members of the population having to carry extra genes. Furthermore, many virulence proteins are also strong antigens, thus allowing a nonlysogenic bacterium to colonize the host without alerting the host immune system and to become pathogenic only if it is infected by the phage.

Genetic Experiments with Phage λ

Earlier we discussed how the interaction of phage λ with its host has been one of the major contributors to our present concepts of how cells function, but we have not gone into detail about the types of experiments which contributed to these concepts. These are great examples of the uses of selectional genetics and genetic analysis in general. The following are some examples of how selectional genetics has been used to analyze the interaction of λ with its host, *E. coli,* and has led to many of the conceptual advances we have described. We do not always credit individuals for these experiments but, rather, just review the types of experiments that were done.

Genetics of λ Lysogeny

Our understanding of how λ forms lysogens was formed by a genetic analysis. Because phage λ is capable of lysogeny, the plaques of λ are cloudy in the middle due to the growth of immune lysogens in the plaque (Figure 8.5). Mutants of λ that cannot form lysogens do not contain these immune lysogens and so are easily identified by their clear plaques. These "clear-plaque mutants," called C-type mutants, have mutations in λ genes whose products are required for the phage to form lysogens.

Complementation tests can reveal how many genes are represented by clear-plaque mutants. Now, however, rather than asking whether two mutants can help each other to multiply, we are asking whether two mutants can help each other to form a lysogen, since this is the function of the genes represented by clear-plaque mutants. Cells are infected by two different clear-plaque mutants simultaneously, and the appearance of lysogens is monitored. Lysogens can be recognized by their immunity to infection by the phage, which allows them to form colonies in the presence of the phage. One way to perform this test is to mix one of the mutant phages with the bacteria and streak the mixture on a plate. The other mutant is then streaked at right angles to the first streak. Very few bacterial colonies will grow in the streaks because the individual mutants cannot form lysogens. However, if bacterial colonies grow in the region where the two streaks cross, immune lysogens are forming due to infection of some cells by both mutant phages and complemention between the two mutations to form a lysogen. Such complementation tests revealed three complementation groups or genes to which the clear-plaque mutations of λ belonged: *cI*, *cII*, and *cIII*. In addition to mutations in the clear-plaque genes, mutations in the *int* gene can prevent the formation of stable lysogens, although Int⁻ mutants make somewhat cloudy plaques. In this case, the λ DNA, while not integrated into the chromosome, may make the cells transiently immune.

Further genetic tests revealed different roles for CI, CII, and CIII in lysogeny. Mutations in the *cII* and *cIII* genes can be complemented to form lysogens, and these lysogens can harbor a single prophage with a *cII* or *cIII* mutation. However, lysogens harboring a single prophage with a *cI* mutation are never seen. Apparently, the *cI* mutation in the phage must be complemented by another mutation to maintain the phage in the lysogenic state. This observation led to the idea that CII and CIII are required to form lysogens but are not necessary to maintain the lysogenic state once a lysogen has formed. The CI protein, on the other hand, is required to form a lysogen and to maintain the lysogenic state.

Because they can be complemented, the *cI*, *cII*, and *cIII* mutations must affect *trans*-acting functions, either proteins or RNAs required to form lysogens. Another set of mutations, called the *vir* mutations, also prevent lysogeny and cause clear-plaque formation but cannot be complemented and so are *cis* acting (see chapter 3). These *vir* mutations allow the mutant phage to multiply and form clear plaques even on λ lysogens. DNA sequencing has revealed that phage with *vir* mutations are multiple mutants with mutations in the o_R^1 and o_R^2 sequences as well as o_L^1. These mutations change the operators so that they can no longer bind the CI repressor, thereby preventing lysogeny (see "Regulation of repressor synthesis in the lysogenic state" above).

Genetics of the CI Repressor

Many proteins are now known to be assembled from "modules" or domains with separable functions. One of the goals of modern proteomics is to identify the domains in proteins in an attempt to guess the function of the protein. The λ repressor product of the *cI* gene was the first protein shown to have separable domains. One region of the protein binds to the operator sequences on the DNA and the other region binds to another repressor monomer to form an active repressor dimer. The first indication that the CI repressor has separable domains came from genetic experiments that demonstrated intragenic complementation between temperature-sensitive mutations in the *cI* gene (see Lieb, Suggested Reading). As discussed in chapter 3, complementation usually occurs only between mutations in different genes and intragenic complementation is possible only if the protein product of the gene is a multimer composed of more than one identical polypeptide encoded by that gene.

Figure 8.17 illustrates the experiments that demonstrated intragenic complementation by some temperature-sensitive mutations in the *cI* gene. Lysogenic cells containing a prophage with one *cI* temperature-sensitive mutation were heated to the nonpermissive temperature and infected with a phage carrying a different *cI* temperature-sensitive mutation. At this temperature, infection by one or the other mutant phage alone will invariably kill the cell because the repressor will be inactivated so that λ cannot form a lysogen. However, if the two mutations complement each other to form an active repressor, a few cells may become lysogens and survive.

The results clearly demonstrated intragenic complementation between some of the mutations in the *cI* gene. In particular, some mutations in the amino (N)-terminal part of the polypeptide, which we now know to be involved in DNA binding, complement some mutations in the carboxyl (C)-terminal part, which we now know to be involved in dimer formation. Apparently, in some cases, dimers can form if only one of the two polypeptides has a mutation in the C-terminal domain. Furthermore, the dimer can sometimes bind to DNA if only one of the two polypeptides in the dimer has a mutation in the N-terminal domain. The ability to form active repressor out of two mutant polypeptides is what leads to intragenic complementation.

Isolation of λ *nut* Mutations

Some of the most elegant genetic experiments with phage λ involved the isolation and mapping of mutations in the λ *nut* sites (see Salstrom and Syzbalski, Suggested Reading). These experiments also illustrate some basic principles of genetic selections and analysis, and so we go into them in some detail. As discussed earlier in the chapter,

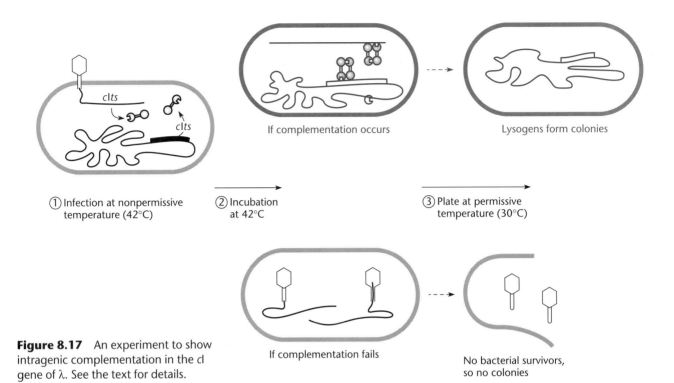

① Infection at nonpermissive temperature (42°C)

② Incubation at 42°C

③ Plate at permissive temperature (30°C)

If complementation occurs

Lysogens form colonies

If complementation fails

No bacterial survivors, so no colonies

Figure 8.17 An experiment to show intragenic complementation in the cI gene of λ. See the text for details.

the existence of the *nut* sites was predicted from a model for how N antiterminates transcription. The *nut* sites (for *N utilization sites*) were the sites on the mRNA to which N must bind before it could bind to the RNA polymerase and prevent termination (Figure 8.2), allowing transcription to proceed into the downstream genes. Mutations in the DNA coding sequence for one of these *nut* sites could prevent the binding of the N protein to the mRNA, thereby causing transcription to stop at the next transcription termination site and preventing transcription of downstream genes.

The first decision was whether to try to isolate *nutL* or *nutR* mutations first. The investigators wisely decided that it would be easier to isolate *nutL* mutants than to isolate *nutR* mutants because *nutL* mutations would prevent the transcription of nonessential genes to the left of *cI*, including the *gam* and *red* genes, while *nutR* mutations would prevent the transcription of essential genes to the right of the *cI* gene, including the *O* and *P* genes required for replication (see the λ map in Figure 8.1). Therefore, phages with mutations that completely inactivate *nutL* should still be viable but mutations that inactivate *nutR* should be lethal and preclude isolation of the mutant phage. However, even though *nutL* mutations should not be lethal, they might be very rare. For all they knew, the *nut* sequences in DNA may be very short, consisting of only a few base pairs, and only mutations that changed one of these base pairs would inactivate the *nut* site. Selecting rare mutations requires a positive selection. It meant finding conditions under which phages

with a mutation that inactivates the *nutL* site can form plaques whereas wild-type λ cannot.

The positive selection used to isolate *nutL* mutations is illustrated in Figure 8.18. The selection is based on the observation that for unknown reasons, wild-type λ cannot multiply in *E. coli* lysogenized by another phage, P2, because the products of the *gam* and *red* genes (the *red* gene later turned out to be two genes, *exo* and *bet* [see chapter 10]) of the infecting λ interact somehow with the *old* gene product of the P2 prophage and kill the cell. Mutations in the predicted *nutL* site which prevent the transcription of both the *gam* and *red* genes should result in phage able to form plaques on *E. coli* lysogenic for P2. Isolating *nutL* mutants of λ should therefore be easy: just plate millions of mutagenized λ on a P2 lysogen, and some of the plaques that form may be due to λ with *nutL* mutations.

Unfortunately, *nutL* mutants are not the only type of mutant that can form plaques under these conditions. As shown in the figure, double mutants of λ with point mutations in both the *gam* and *red* genes or with a deletion mutation that simultaneously inactivates both the *red* and *gam* genes will also multiply and form plaques on a P2 lysogen. Fortunately, double mutants should not be much more common than *nutL* single mutants since the chance of getting two mutations is the product of the chances of getting either single mutation. Deletion mutations that include both the *red* and *gam* genes would be more frequent. However, by inducing mutations using a specific mutagen that causes only point mutations, it is

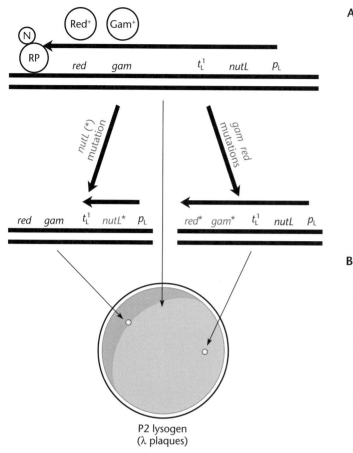

Figure 8.18 Positive selection for λ *nutL* mutations. See the text for details. RP, RNA polymerase.

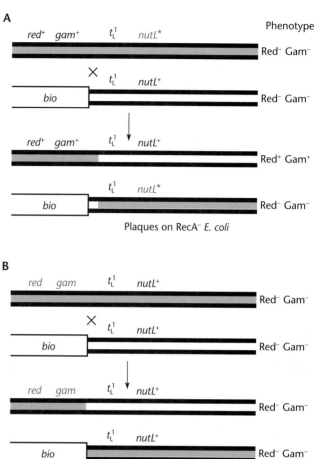

Figure 8.19 Using λp*bio*-substituted phage to map *nutL* mutations. If the mutation that makes λ phenotypically Gam⁻ Red⁻ maps outside the substituted region, Gam⁺ Red⁺ recombinants will arise that can multiply in RecA⁻ *E. coli*. Only the region of the *nutL* region and the *gam* and *red* genes is shown.

possible to lower the percentage of all the mutations that will be deletions (see chapters 7 and 11). Nevertheless, in spite of all precautions, some of the λ mutants that multiply on P2 lysogens will be *gam* and *red* mutants, and these must be distinguished from any *nutL* mutants.

One way to distinguish *nutL* mutants from the other types of mutants that can form plaques on P2 lysogens is by genetic mapping since *gam* and *red* mutations should map to the left of the t_L^1 terminator while *nutL* mutations should map to the right of the t_L^1 terminator (see Figure 8.1). Genetic mapping in phage λ is facilitated by the collection of λp*bio* specialized transducing phage in which *E. coli* genes have replaced some of the λ genes. The endpoints of the substitutions in many of these phages have been precisely mapped. The way in which they can be used to map mutations is illustrated in Figure 8.19. In the example, a Red⁻ Gam⁻ mutant of λ is crossed with a phage λ in which the *bio* substitution includes the *red* and *gam* genes. The appearance of Red⁺ Gam⁺ recombinants indicates that the mutation that makes the phage Red⁻ Gam⁻ must lie outside the substituted region. As will be discussed in chapter 10 in con-

nection with the discovery of *chi* (χ) sites, only Red⁺ Gam⁺ recombinants of λ can form concatemers and hence form plaques on RecA⁻ *E. coli*. Therefore, even very rare Red⁺ Gam⁺ recombinants can be detected by plating the progeny of the cross on RecA⁻ *E. coli*.

Even if the mutation which allows the phage to multiply in a P2 lysogen maps to the right of the t_L^1 terminator, it is not necessarily a *nutL* mutation. Other types of mutations have the potential to reduce *gam* and *red* transcription. For example, leaky *N* mutations might reduce the transcription of *gam* and *red* enough to allow plaques to form on a P2 lysogen and might also allow sufficient *O* and *P* transcription for a plaque to form. However, such *N* mutations, as well as many other such types of mutations, can be distinguished from *nutL* mutations because they are *trans* acting rather than *cis* acting. To determine if the mutation is *cis* acting, a P2 lysogen is infected simultaneously with two different mutant λ phages, one which has a potential *nutL* mutation and the other which has a *gam red* deletion mutation. If the potential *nutL* mutation is *cis* acting, the phage should still multiply on the P2 lysogen, since the *nutL* mutant phage cannot make the *gam* and *red* gene products even if it is furnished with all the other λ gene products in *trans*. Note that *nut* mutations should behave as though they are *cis* acting even though they affect a site on the RNA (a diffusible molecule) rather than a site on the DNA, because the mutation affects transcription termination only from the same DNA.

As mentioned earlier, once the *nutL* mutations had been genetically mapped, investigators could locate the base pair change in the mutation through DNA sequencing and comparison of the sequence of the mutant DNA to the known sequence of wild-type λ DNA in this region. They then found the identical *nutR* sequence to the right of the *cI* gene and similar antiterminator sequences in other DNAs (Figure 8.3).

Isolation of Host *nus* Mutations

The last genetic analysis with λ phage we discuss is the isolation of host *nus* mutations, because they illustrate some additional important concepts in selectional genetics. As mentioned, host *nus* mutations affect host gene products that are required for N antitermination (see Friedman et al., Suggested Reading). The λ N gene product does not act alone and host proteins are required for efficient N antitermination. A host chromosomal mutation which affects one of these proteins may reduce antitermination by the λ N gene product. However, host *nus* mutations are rare, and a positive selection was required to isolate them. Also, it was necessary to reduce the frequency of other types of mutants which might be much more frequent than *nus* mutants.

The selection for *nus* mutations was based on the fact that while induction of a wild-type λ prophage will invariably kill the host, the cell can survive induction of an N⁻ mutant prophage. The reason they survive is that antitermination by N is required to synthesize the P gene product and other λ gene products that kill the cell. A *nus* mutant of the host should also survive induction of the prophage since the *nus* mutation should also prevent N antitermination. Therefore, induction of the prophage should provide a positive selection for *nus* mutations. If an *E. coli* lysogen containing a λ prophage with a temperature-sensitive mutation in its *cI* gene is induced by raising the temperature and the culture is later plated at the lower (permissive) temperature, at least some of the bacteria which survive and grow up to form a colony should have a *nus* mutation in their chromosome. However, as with the isolation of *nut* mutations in the phage, it is necessary to do the selection in a way which will reduce the frequency of other types of mutants which will be much more frequent than *nus* mutants. For example, cells cured of the prophage would also survive the heat treatment. To reduce the frequency of cured cells, the investigators used a deleted prophage that contained the P gene and could kill the cell but could not be induced to excise from the chromosome. There are also a myriad of mutations in the prophage itself that will allow the cell to survive the induction, for example N mutations. The frequency of surviving mutants with N mutations or other types of mutations in the prophage could be reduced by using a double lysogen, with two copies of the prophage in the chromosome. They reasoned that a mutation in the N gene of just one of the two copies of the prophage would not save the cell and that two N mutations, one in each copy of the prophage, would be required, greatly reducing the frequency of this type of survivor among the mutants. Once surviving mutants were selected, they were mapped by Hfr crosses and transduction, using methods described in chapters 5 and 7. The investigators were not interested in mutations that mapped to where the prophage was inserted in the chromosome and were presumably in the prophage; they were interested only in mutations elsewhere in the *E. coli* genome, and these mutations defined the chromosomal *nus* genes. In this way, they found the *E. coli* genes *nusA*, *nusB*, etc., whose products are involved in transcription termination and antitermination in the uninfected host.

SUMMARY

1. Some phages are capable of lysogeny, in which they persist in the cell as prophages. Such phages are called lysogenic phages. A bacterium harboring a prophage is called a lysogen. In the lysogen, most of phage gene products made are involved in maintaining the prophage state. The prophage DNA can either be integrated into the chromosome or replicate autonomously as a plasmid.

2. The *E. coli* phage λ (lambda) is the prototype of a lysogenic phage. It was the first lysogenic phage to be discovered and the one to which all others are compared.

3. Phage λ regulates its early transcription through antitermination proteins N and Q. These proteins bind to the

(continued)

SUMMARY (continued)

RNA polymerase and allow it to transcribe through transcription termination sites.

4. The N protein must first bind to a sequence *nut* in the mRNA before it can bind to the RNA polymerase. With the N protein riding on it, the RNA polymerase can transcribe through transcription terminators into the O, P, and Q genes on the right and the *red, gam,* and *int* genes on the left. *E. coli* proteins called Nus are involved in N antitermination. At least some of the Nus proteins are involved in transcription termination antitermination in the host.

5. The Q protein allows the RNA polymerase to transcribe through a termination site into the late genes, including the head, tail, and lysis genes of the phage. The RNA polymerase first makes a short RNA and then stops. The Q gene product must bind to a *qut* sequence in the DNA close to the promoter before it can bind to the stalled RNA polymerase and allow it to transcribe into the late genes of the phage.

6. Phage λ DNA is linear in the phage head, with short complementary single-stranded 5′ ends called the *cos* or cohesive ends. Because they have complementary sequences, the *cos* ends can base pair with each other after the DNA enters the cell to form a circle. The phage DNA then replicates as a circle a few times before it enters the rolling-circle mode of replication, which leads to the formation of long concatemers in which many genome length DNAs are linked end to end.

7. Phage λ DNA can be packaged only from concatamers and not from unit-length genomes. This is because λ begins filling the head at one *cos* site and stops only when it gets to the next *cos* site in the concatemer. The concatemers from which λ DNA is packaged can be formed either by rolling-circle replication or by recombination between single-length circles.

8. Whether or not λ enters the lysogenic state depends upon the outcome of a race between the *c*II gene product and the products of the lytic genes of the phage. The product of the *c*II gene is a transcriptional activator that activates the transcription of the *c*I and *int* genes after infection.

The *c*I gene product is the repressor that blocks transcription of most of the genes of λ in the prophage state, and the *int* gene product is an enzyme that is required to integrate λ DNA into the bacterial chromosome.

9. The CI repressor protein of λ is a homodimer made up of two identical polypeptides encoded by the *c*I gene. The repressor blocks transcription by binding to operators o_R and o_L on both sides of the *c*I gene, preventing the utilization of two promoters, p_R and p_L, which are responsible for the transcription of genes to the right and to the left of the *c*I gene, respectively.

10. The synthesis of repressor is regulated in the lysogenic state through its binding to three repressor binding sites within o_R. These sites are named o_R^1, o_R^2, and o_R^3. The regulation of repressor synthesis has served as a model for regulation in other systems.

11. Damage to the DNA of its host can cause the λ prophage to be induced and produce phage. The activated RecA protein of the host causes the λ repressor protein to cleave itself so that it can no longer form dimers and be active. The process of excision of λ DNA is essentially the reverse of integration, except that excision requires both the *int* and *xis* gene products of λ.

12. Very rarely, when λ DNA excises, it picks up neighboring bacterial DNA and becomes a transducing particle. This type of transduction is called specialized transduction, because only bacterial genes close to the insertion site of the prophage can be transduced.

13. Lysogenic phages are often useful as cloning vectors. A bacterial gene cloned into a prophage will exist in only two copies, one in its normal site in the chromosome and another in the prophage. This can be important in complementation tests or in other applications. If the prophage is induced, the cloned DNA can be recovered in large amounts.

14. Lysogenic phages often carry genes for bacterial toxins. Examples include the toxins that cause diphtheria, botulism, scarlet fever, cholera, and, possibly, toxic shock syndrome.

QUESTIONS FOR THOUGHT

1. Why do you suppose the λ prophage uses different promoters to transcribe the *c*I repressor gene immediately after infection and in the lysogenic state?

2. Why is only one protein, Int, required to integrate the phage DNA into the chromosome while two proteins, Int and Xis, are required to excise it? Why not just make one different Int-like protein that excises the prophage?

3. Can you propose an alternative explanation for why toxins

and other proteins that make bacteria pathogenic are often encoded by phages?

4. Why do you suppose some types of prophage can be induced only if another phage of the same type infects the lysogenic cell containing them? What purpose does this serve?

5. Is P4 a phage or a genetic island? What distinguishes these two types of DNA elements?

PROBLEMS

1. Lambda (λ) *vir* mutations cause clear plaques because they change the operator sequences so that they no longer bind repressor. How would you determine if a clear mutant you have isolated has a *vir* mutation instead of a mutation in any one of the three genes *cI*, *cII*, or *cIII*?

2. Lambda (λ) integrates into the bacterial chromosome in the region between the galactose utilization (*gal*) genes and the biotin biosynthetic (*bio*) genes on the other side. Outline how you would isolate an HFT strain carrying the *bio* operon of *E. coli*. Would you expect your transducing phage to form plaques? Why or why not?

3. A *vir* mutation changes the operator sequences. Would you expect λ with *vir* mutations in the o_R^1 and o_L^1 sites to form plaques on λ lysogens? Why or why not?

4. The DNA of a λ specialized transducing particle usually integrates next to a preexisting prophage. Would you expect just Int or both Int and Xis to be required for this integration? Why?

5. Why can you sometimes get intragenic complementation between two temperature-sensitive mutations in the *cI* gene of λ phage but never between two amber (UAG) mutations?

6. You have isolated a relative of P2 phage from sewage, using *E. coli* as the indicator bacterium. How would you determine if P4 phage can parasitize (i.e., can be a satellite virus of) your P2-like phage?

7. You have isolated a phage from a strain of *Staphylococcus aureus* known to cause food poisoning owing to production of a toxin. Outline how you would go about determining if the toxin is encoded by a prophage. Assume that you can detect the toxin by its ability to kill human cells in culture.

SUGGESTED READING

Cairns, J., M. Delbrück, G.S. Stent, and J. D. Watson. 1966. *Phage and the Origins of Molecular biology.* Cold Spring Harbor Laboratory Press, Cold Spring Harbor, N.Y.

Costa, J. J., J. L. Michel, R. Rappuoli, and J. R. Murphy. 1981. Restriction maps of corynebacteriophages βc and βvir and physical location of the diphtheria *tox* operon. *J. Bacteriol.* **148:**124–130.

Echols, H., and H. Murialdo. 1978. Genetic map of bacteriophage lambda. *Microbiol. Rev.* **42:**577–591.

Freeman, V. J. 1951. Studies on the virulence of bacteriophage-infected strains of *Corneybacterium diphtheriae*. *J. Bacteriol.* **61:**675–688.

Friedman, D. I., M. F. Baumann, and L. S. Baron. 1976. Cooperative effects of bacterial mutations affecting λ N gene expression. *Virology* **73:**119–127.

Gottesman, M. 1999. Bacteriophage λ: the untold story. *J. Mol. Biol.* **293:**177–180.

Hendrix, R. W., J. W. Roberts, F. W. Stahl, and R. A. Weisberg (ed.). 1983. *Lambda II.* Cold Spring Harbor Laboratory Press, Cold Spring Harbor, N.Y.

Johnson, A. D., A. R. Poteete, G. Lauer, R. T. Sauer, G. K. Akers, and M. Ptashne. 1980. λ repressor and cro⁻ components of an efficient molecular switch. *Nature* (London) **294:**217–223.

Johnson, L. P., M. A. Tomai, and P. M. Schlievert. 1986. Bacteriophage involvement in group A streptococcal pyrogenic exotoxin A production. *J. Bacteriol.* **166:**623–627.

Juhala, R. J., M. E. Ford, R. L. Duda, A. Youlton, G. F. Hatfull, and R. W. Hendrix. 2000. Genomic sequences of bacteriophages HK97 and HK022: pervasive mosaicism in the Lambdoid phages. *J. Mol. Biol.* **299:**27–51.

Kahn, M. L., R. Ziermann, G. Deho, D. W. Ow, M. G. Sunshine, and R. Calendar. 1991. Bacteriophage P2 and P4. *Methods Enzymol.* **204:**264–280.

Li, J., R. Horwitz, S. McCracken, and J. Greenblatt. 1992. NusG, a new *Escherichia coli* elongation factor involved in transcriptional antitermination by the N protein of phage λ. *J. Biol. Chem.* **267:**6012–6019.

Lieb, M. 1976. Lambda *cI* mutants intragenic complementation with a *cI* promotes mutant. *Mol. Gen. Genet.* **146:**291–297.

Lwoff, A. 1953. Lysogeny. *Bacteriol. Rev.* **17:**269–337.

Marr, M. T., and J. W. Roberts. 2000. Function of transcription cleavage factors GreA and GreB at a regulatory pause site. *Mol. Cell* **6:**1275–1285.

Moyer, K. E., H. H. Kimsey, and M. K. Waldor. 2001. Evidence for a rolling circle mechanism of phage DNA synthesis from both replicative and integrated forms of CTXφ. *Mol. Microbiol.* **41:**311–323.

Ptashne, M. 1992. *A Genetic Switch,* 2nd ed. Blackwell Scientific Publications, Cambridge, Mass.

Salstrom, J. S., and W. Szybalski. 1978. Coliphage λ *nutL*: a unique class of mutations defective in the site of N product utilization for antitermination of leftward transcription. *J. Mol. Biol.* **124:**195–222.

Schiavo, G., F. Benfenati, B. Poulain, O. Rossetto, P. Polverino de Laureto, B. R. dasGupta, and C. Montecucco. 1992. Tetanus and botulinum-B neurotoxins block neurotransmitter release by proteolytic cleavage of synaptobrevin. *Nature* (London) **359:**832–835.

Skalka, A. 1977. DNA replication—bacteriophage lambda. *Curr. Top. Microbiol. Immunol.* **78:**201–211.

Wagner, P. L., M. N. Neely, X. Zhang, D. W. K. Acheson, M. K. Watdor, and D. L. Friedman. 2001. Role for a phage promoter in Shiga toxin 2 expression from a pathogenic *Escherichia coli* strain. *J. Bacteriol.* **183:**2081–2085.

Waldor, M. K., and J. J. Mekalanos. 1996. Lysogenic conversion by a filamentous phage encoding cholera toxin. *Science* **272:**1910–1914.

Transposition and Site-Specific Recombination

RECOMBINATION IS THE BREAKING AND REJOINING OF DNA in new combinations. In homologous recombination, which accounts for most recombination in the cell, the breaking and rejoining occurs only between regions of two DNA molecules that have similar or identical sequences. Homologous recombination requires that the two DNAs pair through complementary base pairing (see chapter 10). However, other types of recombination, known as **nonhomologous recombination**, sometimes occur in cells. They do not require that the two DNAs have the same sequence of nucleotides. Rather, nonhomologous recombination depends on enzymes that recognize specific regions in DNA, which may or may not have sequences in common, and promote recombination between those regions. This chapter addresses some examples of nonhomologous recombination in bacteria and the mechanisms involved, including transposition by transposons and site-specific recombination that occurs during the integration and excision of prophages and other DNA elements, the inversion of invertible sequences, and resolution of cointegrates by resolvases.

Transposition

Transposons are DNA elements that can hop, or **transpose**, from one place in DNA to another. Transposable DNA elements were first discovered in corn by Barbara McClintock in the early 1950s and about 20 years later in bacteria by others. Transposons are now known to exist in all organisms on Earth, including humans. In fact, from the human genome project it is apparent that almost half of our DNA may be transposons! The movement by a transposon is called **transposition**, and the enzymes that promote transposition are called **transposases**. The transposon itself usually encodes its own transposases, so that it carries with it the ability to hop each time it moves. For this reason, transposons have been called "jumping genes."

Not all DNA elements that can move are true transposons. For example, "homing" DNA elements, which include some types of moveable RNA and protein introns, move by means of endonucleases that make a specific double-stranded break in the DNA at a given site. Then, through homologous recombination aimed at repairing the double-stranded break, the DNA element is inserted at that site. No specific transposases are required for the movement of homing DNA, and these DNA elements can move only into the same sequence in other DNA molecules that lack them. We discuss homing endonucleases in more detail in chapter 10 (see Box 10.1).

True transposons should also be distinguished from **retrotransposons**, so named because they behave like RNA retroviruses with a DNA intermediate. An RNA copy of a region is made and then copied into DNA by a reverse transcriptase. The DNA intermediate then integrates elsewhere by various mechanisms that may or may not be analogous to transposition. Although only a few examples of retrotransposons are known in bacteria, such elements are well known in fungi.

Although transposons probably exist in all organisms on Earth, they are best understood in bacteria, where they obviously play an important role in evolution. Transposons found in different bacterial genera may be more closely related to each other than are the bacteria in which they are found. This suggests that transposons move among different genera of bacteria with some regularity. As mentioned in previous chapters, transposons may enter other genera of bacteria during transfer of promiscuous plasmids or via transducing phage. Some transposons are themselves conjugative or can be induced to form phage, as discussed later in this chapter. Transposons may offer a way of introducing genes from one bacterium into the chromosome of another bacterium to which it has little DNA sequence homology.

Overview of Transposition

The net result of transposition is that the transposon appears at a place in DNA different from where it was originally. Many transposons are essentially cut out of one DNA and inserted into another (Figure 9.1), whereas other transposons are copied and then inserted elsewhere. Regardless of the type of transposon, however, the DNA from which the transposon originated is called the **donor DNA** and the DNA into which it hops is called the **target DNA**.

In all transposition events, the transposase enzyme cuts the donor DNA at the ends of the transposon and then inserts the transposon into the target DNA. How-

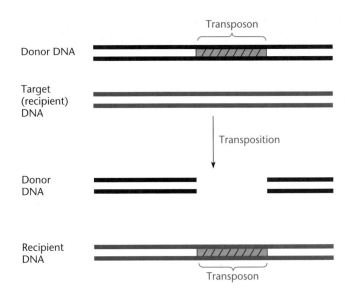

Figure 9.1 Overview of transposition. See the text for details.

ever, the details of the mechanism can vary. Some types of transposons may exist free of other DNA during the act of transposition, but many transposons, before and after they hop, remain contiguous with other DNA molecules. Later in this chapter, we discuss more detailed models for transposition.

Transposition must be tightly regulated and occur only rarely; otherwise the cellular DNA would become riddled with the transposon, which would have many deleterious effects. Transposons have evolved elaborate mechanisms to regulate their transposition so that they hop very infrequently and do not often kill the host cell. The frequency of transposition varies from about once in every 10^3 cell divisions to about once in every 10^8 cell divisions, depending upon the type of transposon. Thus, the chance of a transposon hopping into a gene and inactivating it is not much higher than the chance that a gene will be inactivated by other types of mutations (see chapter 3).

Structure of Bacterial Transposons

There are many different types of bacterial transposons. Some of the smaller ones are about 1,000 bp long and carry only the genes for the transposases that promote their movement in DNA and the genes that regulate this movement. Larger transposons may also contain one or more other genes, such as those for resistance to an antibiotic.

One distinguishing feature of bacterial transposons is that all those identified so far contain repeats at their ends, which are usually **inverted repeats** (Figure 9.2). As discussed in chapter 1, two regions of DNA are inverted repeats if the sequence of nucleotides on one strand in

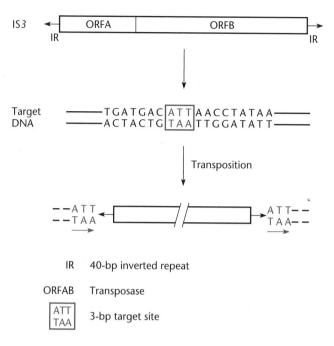

IR   40-bp inverted repeat

ORFAB   Transposase

ATT / TAA   3-bp target site

Figure 9.2 Structure of the insertion sequence element IS3 and its related family members. The inverted repeats are shown as arrows, and the 3-bp target sequence that will be duplicated after transposition is boxed.

one region, when read in the 5′-to-3′ direction, is the same or almost the same as the 5′-to-3′ sequence of the opposite strand in the other region.

Another common feature is the presence of short **direct repeats** of the target DNA that bracket the transposon (Figure 9.2). Direct repeats have the same or almost the same 5′-to-3′ sequence of nucleotides on the same strand. As shown in Figure 9.2, the target DNA originally contains only one copy of the sequence at the place where the transposons insert. During the insertion of the transposon, this sequence is duplicated. Most transposons can insert into many places in DNA and so have little or no target specificity. Thus, the duplicated sequence will vary with the sequence at the site in the target DNA into which the transposon inserted. However, even though the duplicated sequences differ, the number of duplicated base pairs is characteristic of each transposon. Some duplicate as few as 3 bp and others as many as 9 bp. The molecular models for transposition discussed later in the chapter offer an explanation for the duplicated sequences.

Types of Bacterial Transposons

Each type of bacterium carries its own unique transposons, although many transposons are related across species as though they were recently exchanged. In this section, we describe some of the common types of transposons.

INSERTION SEQUENCE ELEMENTS

The smallest bacterial transposons are called **insertion sequence elements** (IS elements). These transposons are usually only about 750 to 2,000 bp long and encode little more than the transposase enzymes that promote their transposition.

Insertion sequence elements carry no selectable genes, and they were discovered only because they inactivate a gene if they happen to hop into it. The first IS elements were detected as a type of *gal* mutation that was unlike any other known mutations. This type of mutation resembled deletion mutations in that it was nonleaky; however, unlike deletion mutations, it could revert, albeit at a lower frequency than base pair changes or frameshifts. Such anomalous *gal* mutations were also very polar and could prevent the transcription of downstream genes (see chapter 2). Later work showed that these mutations resulted from insertion of about 1,000 bp of DNA into a *gal* gene. Moreover, they were due to insertion of not just any piece of DNA but one of very few sequences.

Originally, four different IS elements were found in *Escherichia coli*: IS1, IS2, IS3, and IS4. Most strains of *E. coli* K-12 contain approximately six copies of IS1, seven copies of IS2, and fewer copies of the others. Almost all bacteria carry IS elements, with each species harboring its own characteristic IS elements, although sometimes related IS elements can be found in different bacteria. To date, more than 700 IS elements have been found in bacteria, although most of these can be placed in about 20 families (see Mahillon and Chandler, Suggested Reading). Plasmids also often carry IS elements, which are important in the formation of Hfr strains (see chapter 5).

Figure 9.2 shows the structure of the IS element IS3. In addition to the inverted repeats at its ends, it consists of two open reading frames (ORFA and ORFB). The reading frame of ORFB is shifted −1 relative to the reading frame of ORFA, but a programmed −1 frameshift (see Box 2.4) causes the synthesis of a fusion protein ORFAB, which is the active transposase. The smaller protein made from ORFA when the frameshifting does not occur regulates transcription of the transposase gene. The target site sequence that is duplicated in the target DNA upon insertion of IS3 is 3 bp long. As mentioned above, the length of such direct repeats is characteristic of each type of transposon.

Although the original IS elements were discovered only because they had hopped into a gene, causing a detectable phenotype, IS elements are now more often discovered during hybridization experiments with cloned regions of bacterial DNA as probes or in genomic sequencing.

COMPOSITE TRANSPOSONS

Sometimes two IS elements of the same type form a larger transposon, called a **composite transposon**, by bracketing other genes. Figure 9.3 shows the structures of three composite transposons, Tn5, Tn9, and Tn10. Tn5 consists of genes for kanamycin resistance (Kan^r) and streptomycin resistance (Str^r) bracketed by copies of an IS element called IS50. Tn9 has two copies of IS1 bracketing a chloramphenicol resistance gene (Cam^r). In Tn10, two copies of IS10 flank a gene for tetracycline resistance (Tet^r). Some composite transposons, such as Tn9, have the bracketing IS elements in the same orientation, whereas others, including Tn5 and Tn10, have them in opposite orientations.

Outside-End Transposition

Figure 9.4 illustrates transposition of a composite transposon. Each IS element can transpose independently as long as the transposase acts on both of its ends. However, because all the ends of the IS elements in a composite transposon are the same, a transposase encoded by one of the IS elements can recognize the ends of either. When such a transposase acts on the inverted repeats at the farthest ends of a composite transposon, the two IS elements will transpose as a unit, bringing along the genes between them. These two inverted repeats are called the "outside ends" of the two IS elements because they are the *farthest* from each other.

The two IS elements that form composite transposons are often not completely autonomous, because of mutation in the transposase gene of one of the elements. Thus, only one of the IS elements encodes an active transposase. However, this transposase can act on the outside ends to promote transposition of the composite transposon.

Inside-End Transposition

The transposase encoded by one IS element in a composite transposon can also act on the "inside ends" of both IS elements, that is, the two ends that are closest to each other. Inside-end transposition presumably happens as

Figure 9.4 Two IS elements can transpose any DNA between them. (A) Action of the transposase at the ends of an isolated IS element causes it to transpose. (B) Two IS elements of the same type are close to each other in the DNA. Action of the transposase on their outside ends causes them to transpose together, carrying along the DNA between them. *A* denotes the DNA between the IS elements (hatched bar), and arrows indicate the inverted repeated (IR) sequences at the ends of the IS elements. Dashed lines represent the target DNA.

A   IS element

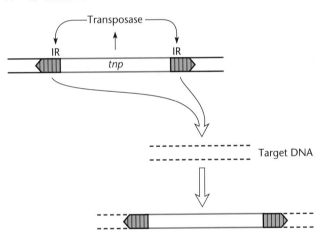

B   Composite: 2 IS + gene *A*

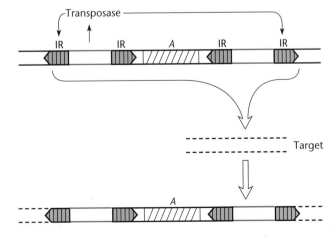

Figure 9.3 Structures of some composite transposons. The commonly used genes for kanamycin resistance, Kan^r, and the gene for chloramphenicol resistance, Cam^r, come from Tn5 and Tn9, respectively. The active transposase gene is in one of the two IS elements. Note that the IS elements can be in either the same or opposite orientation (arrows). Str^r, gene encoding streptomycin resistance; Tet^r, gene encoding tetracycline resistance; Ble^r, gene encoding bleomycin resistance.

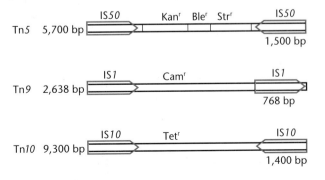

often as outside-end transposition but has very different consequences.

One possible outcome of inside-end transposition is the creation of a new composite transposon, which was first demonstrated with Tn*10* (see Foster et al., Suggested Reading). In these experiments, Tn*10* was inserted into a small plasmid with an origin of replication (*ori*) and an ampicillin resistance gene (Amp^r). The plasmid served as the donor DNA. As shown in Figure 9.5, transposition with the outside ends of the IS*10* element would move Tn*10*, with the tetracycline resistance gene, Tet^r, to another DNA. However, transposition from the inside ends would create a new composite transposon carrying the Amp^r gene and the plasmid origin of replication (*ori*) to another DNA. If this new composite transposon hops into a target DNA that does not have a functional origin of replication, it may confer on that DNA the ability to replicate. In the experiment, a λ phage with amber mutations in its replication genes *O* and *P* was used to infect amber-suppressing cells containing the donor plasmid with the transposon. The progeny phage were then plated on a non-amber-suppressor host. In this host, the phage could not replicate from the λ origin of replication because of the amber mutations in their replication genes. However, any phage into which the new composite transposon had hopped would be able to replicate by using the plasmid origin of replication. As expected, those few phages that formed plaques did contain the new composite transposon. The inside ends of the IS*10* elements of Tn*10* must have been used for its transposition.

Deletions and inversions can also be caused by inside-end transposition of a composite transposon to a nearby

target on the same DNA (Figure 9.6). The neighboring sequences between the orginal site of insertion of the transposon and the site into which it is trying to transpose will be either deleted or inverted. Whether a deletion or inversion is created depends on how the inside ends of the IS elements in the transposon are attached to the target DNA. If the inside ends cross over each other before they attach, the neighboring sequences will be inverted. If they do not cross over each other, the neighboring sequences will be deleted. As shown in Figure 9.6, the DNA between the two IS elements in the composite transposon will also be deleted, independent of what

Figure 9.6 Rearrangements of DNA caused by composite transposons. Attempts to transpose by the inside ends of a composite transposon to a neighboring target sequence can cause either a deletion or an inversion of the intervening sequences, depending upon how the ends are attached. For an explanation of steps 1 and 2, see Figure 9.14.

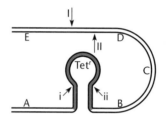

① Double-strand breaks at inside ends (i and ii) of transposon Tn*10*

② Staggered break (I and II) in target DNA

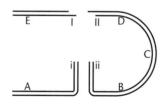

③ Tet^r gene is destroyed

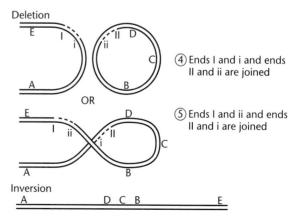

Deletion

④ Ends I and i and ends II and ii are joined

OR

⑤ Ends I and ii and ends II and i are joined

Inversion

Figure 9.5 Either the outside or inside ends of the IS elements in a composite transposon can be used for transposition. Outside-end (a) transposition will transpose Tn*10* (gold), including the gene encoding tetracycline resistance (Tet^r), whereas inside-end (b) transposition will transpose the plasmid, including the origin of replication and the gene for ampicillin resistance (Amp^r).

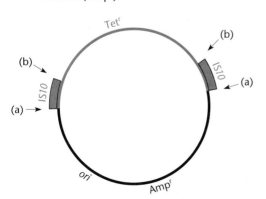

happens to the neighboring DNA. Therefore, these re-arrangements are usually accompanied by the loss of any resistance gene on the composite transposon, which is how they are usually selected. For example, methods have been developed to select tetracycline-sensitive derivatives of *E. coli* harboring the Tn*10* transposon. Most of these tetracycline-sensitive derivatives will have deletions or inversions of DNA next to the site of insertion of the Tn*10* element. Presumably, inside-end transposition is responsible for most of the often-observed instability of DNA caused by composite transposons.

Assembly of Plasmids by IS Elements

Any time two IS elements of the same type happen to hop close to each other on the same DNA, a composite transposon is born. These transposons have not yet evolved a defined structure such as the named transposons (e.g., Tn*10*) described above. Nevertheless, the two IS elements can transpose any DNA between them. In this way, "cassettes" of genes bracketed by IS elements can be moved from one DNA molecule to another.

Many plasmids seem to have been assembled from such cassettes. Figure 9.7 shows a naturally occurring plasmid carrying genes for resistance to many antibiotics. Such plasmids are historically called R-factors, because they confer resistance to so many different antibiotics (see chapter 4). Notice that many of the resistance genes

Figure 9.7 R-factors, or plasmids containing many resistance genes, may have been assembled, in part, by IS elements. The tetracycline resistance (Tet^r) gene is bracketed by IS*3* elements, and the region containing the other resistance genes (the r determinant, gold) is bracketed by IS*1* elements. Reprinted with permission from H. Saedler, p. 66, in A. I. Bukhari, J. A. Shapiro, and S. Adhya (ed.), *DNA Insertion Elements, Plasmids and Episomes*, Cold Spring Harbor Laboratory Press, Cold Spring Harbor, N.Y., 1977.

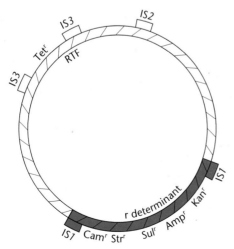

on the plasmid are bracketed by IS elements. In particular, IS*3* flanks the tetracycline resistance gene and IS*1* brackets the genes for resistance to many other antibiotics. Apparently, the plasmid was assembled in nature by resistance genes hopping onto the plasmid from some other DNA via the bracketing IS elements. In principle, any two transposons of the same type can move other DNA lying between them by a similar mechanism, but because IS elements are the most common transposons and often exist in more than one copy per cell, they probably play the major role in the assembly of plasmids.

NONCOMPOSITE TRANSPOSONS

Composite transposons are not the only ones to carry resistance genes. Such genes can also be an integral part of transposons known as **noncomposite transposons** (Figure 9.8). They are bracketed by short inverted repeats, but the resistance gene is part of the minimum transposable unit.

Noncomposite transposons seem to belong to a number of families in which the members are related to each other by sequence and structure. Interestingly, different members of transposon families, notably the Tn*21* family, often carry different resistance genes, even though they are almost identical otherwise (see Figure 9.8). Apparently, the resistance genes can integrate into the transposon by means of specific integrase enzymes, much like lysogenic phages can integrate into the chromosome of bacteria (see the section on integrases of transposon integrons, below).

Assays of Transposition

To study transposition, we must have assays for it. As mentioned above, insertion elements were discovered because they create mutations when they hop into a gene. However, this is usually not a convenient way to assay transposition, because transposition is infrequent and it is laborious to distinguish insertion mutations from the myriad other mutations that can occur. If the transposon carries a resistance gene, the job of assaying transposition is easier. But how do we know if a transposon has hopped in the cell? The cells are resistant to the antibiotic no matter where the transposon is inserted in the cellular DNA, and so hopping from one place to another makes no difference in the level of resistance of the cell. Obviously, detecting transposition requires special methods.

SUICIDE VECTORS

One way to assay transposition is with **suicide vectors**. Any DNA, including plasmid or phage DNA, that cannot replicate (i.e., is not a replicon) in a particular host can be used as a suicide vector. These DNAs are called suicide vectors because, by entering cells in which they

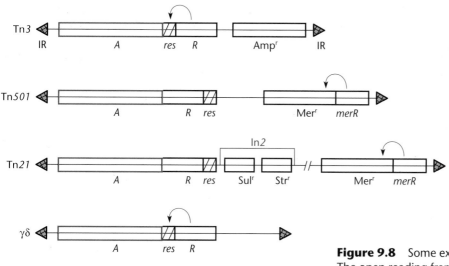

◀ ▶	Terminal inverted repeats
A	Transposase
R	Resolvase and repressor of *A* and *R* transcription
res	Recombination site
merR	Regulator of Mer^r and transposase
Mer^r	Mercury resistance
Amp^r	Ampicillin resistance
IR	Inverted repeat

Figure 9.8 Some examples of noncomposite transposons. The open reading frames encoding the proteins are boxed. The terminal inverted repeats ends are shown as hatched arrows. *A* is the transposase; *R* is the resolvase and repressor of *A* transcription; *res* is the site at which resolvase acts; Mer^r is the mercury resistance region; and *merR* is the regulator of mercury resistance gene transcription In*2* is described in Figure 9.25. Tn*3* was originally found on the broad-host-range plasmid pR1drd19, Tn*501* was found on the *Pseudomonas* plasmid pUS1, Tn*21* was found on the *Shigella flexneri* plasmid R100, and γδ was found on the chromosome of *E. coli* and on the F plasmid.

cannot replicate, they essentially kill themselves. To assay transposition with a suicide vector, we use one to introduce a transposon carrying an antibiotic resistance gene into an appropriate host. How the suicide vector itself is introduced into the cells will depend upon its source. If it is a phage, the cells could be infected with that phage. If it is a plasmid, it could be introduced into the cells through conjugation.

Once in the cell, the suicide vector will remain unreplicated and will eventually be lost. The only way the transposon can survive and confer antibiotic resistance on the cells is by hopping to another DNA molecule that is capable of autonomous replication in those cells, for example a plasmid or the chromosome. Therefore, when the cells under study are plated on antibiotic-containing agar and incubated, the appearance of colonies, as a result of the multiplication of antibiotic-resistant bacteria, is evidence for transposition. These cells have been mutagenized by the transposon, since the transposon has hopped into a cellular DNA molecule—either the chromosome or a plasmid. This is called **transposon mutagenesis**.

Phage Suicide Vectors

Some derivatives of the phage λ are designed to be used as suicide vectors. These phage have been rendered inca-

pable of replication in nonsuppressing hosts by the presence of nonsense codons in their replication genes *O* and *P* (see chapter 8). They have also been rendered incapable of integrating into the host DNA by deletion of their attachment region, *attP*. Such a λ phage can be propagated on an *E. coli* strain carrying a nonsense suppressor. However, in a nonsuppressor *E. coli*, it cannot replicate or integrate. Because of the narrow host range of λ, these suicide vectors can normally only be used in strains of *E. coli* K-12.

Plasmid Suicide Vectors

Plasmid cloning vectors can also be used as suicide vectors. The plasmid containing the transposon with a gene for antibiotic resistance can be propagated in a host in which it can replicate and is then introduced into a cell in which it cannot replicate. In principle, any plasmid with a conditional-lethal mutation, nonsense or temperature sensitive, in a gene required for plasmid replication can be used as a suicide vector. The plasmid could be propagated in the permissive host or under permissive conditions and then introduced into a nonpermissive host or into the same host under nonpermissive conditions, depending upon the type of mutation. Alternatively, a narrow host range plasmid could be used. It can be propagated in a host in which it can replicate and introduced into a different species in which it cannot. The most efficient way to introduce a plasmid into cells is by conjugation. If the plasmid contains a *mob* region, it can be mobilized into the recipient cell by using the Tra functions of a self-transmissible plasmid (see chapter 5). Taking advantage of the extreme promiscuity of some transmissible plasmids, methods have been developed which allow transposon mutagenesis of essentially any gram-negative bacterium. This means of introducing suicide vectors into cells can be used to mutagenize almost any bacterium, including those about which little to nothing is known. If the self-transmissible plasmid is promiscuous for transfer, it can be used to mobilize a narrow-range plasmid into an unrelated bacterium, where the mobilized plasmid will be a suicide vector since it is unable to replicate. Some types of transposons are also very broad in their host range for transposition, so that the transposon may hop in the recipient bacterium. If the transposon also carries a selectable gene which is expressed in the bacterium, insertion mutants in which the transposon has hopped can be selected. The gene into which the transposon has hopped can then be relatively easily cloned and sequenced. This method is particularly useful for gram-negative bacteria because promiscuous plasmids such as the IncP plasmid RP4 are known to transfer efficiently into essentially any gram-negative bacterium and the Tn5 transposon will transpose in almost any gram-negative bacterium. ColE1-derived plasmids are often the suicide vectors of choice in such applications because they can only replicate some in enteric bacteria including *E. coli*. We discuss such methods in more detail later in the chapter (see "Transposon Mutagenesis").

THE MATING-OUT ASSAY FOR TRANSPOSITION

Transposition can also be assayed by using the "mating-out" assay. In this assay, a transposon in a nontransferable plasmid will not be transferred into other cells unless it hops into a plasmid that is transferable. Figure 9.9 shows a specific example of a mating-out assay with *E. coli*. In the example shown, the transposon Tn10 carrying tetracycline resistance has been inserted into a small plasmid that is neither self-transmissible nor mobilizable. This small plasmid is used to transform cells containing F, a larger, self-transmissible plasmid. While the cells are growing, the transposon may hop from the smaller plasmid into the F plasmid in a few of the cells. Later, when these cells are mixed with streptomycin-resistant recipient cells, any F plasmid into which the transposon hopped will carry the transposon when it transfers to a new cell, thus conferring tetracycline resistance on that cell. Transposition can be detected by plating the mating mixture on

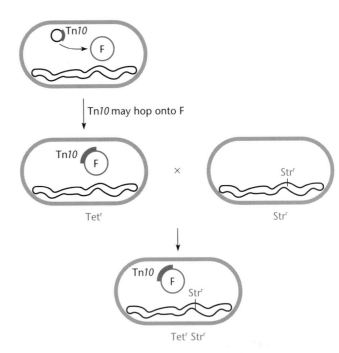

Figure 9.9 Example of a mating-out assay for transposition. See text for details.

agar containing tetracycline and counterselecting the donor with streptomycin.

The appearance of antibiotic-resistant transconjugants in a mating-out assay is not absolute proof of transposition. Some transconjugants could become antibiotic resistant by means other than transposition of the transposon into the larger, self-transmissible plasmid. The smaller plasmid containing the transposon could have been somehow mobilized by the larger plasmid, or the smaller plasmid could have been fused to the larger plasmid by recombination or by cointegrate formation (see below). A few representative transconjugants should be tested, for example, by using restriction digests (see chapter 1) to verify that they contain only the larger plasmid with the transposon inserted.

Genetic Requirements for Transposition

Once assays for transposition had been developed, it was possible to determine the genetic requirements for transposition. The first analysis of the genetic requirements for transposition used transposon Tn3, which transposes by a replicative mechanism (see Gill et al., Suggested Reading, and Figure 9.8 for a diagram of Tn3). The questions are essentially the same as for any other genetic analysis. How many gene products are required for transposition of Tn3, and where are the sites at which they act? Where do the genes for these gene products lie on the transposon? Do any intermediates of transposition accumulate when one or more of these gene products is inactivated? Obtaining answers to these questions was the first step in developing a molecular model for transposition of Tn3.

ISOLATION OF MUTATIONS IN THE TRANSPOSON

As in any genetic analysis, the first step in analyzing the genetic requirements for transposition was to isolate mutations in the transposon. A plasmid containing the transposon was cut randomly with DNase, and then DNA linkers containing the recognition sequence for a restriction endonuclease were ligated into the cut plasmid. This creates random insertion mutations around the plasmid including the transposon, and the presence of the restriction site on the inserted DNA makes the site

of the mutation easy to map physically (see chapter 1). Insertions of a DNA linker will also disrupt the DNA sequence and, if inserted into a translated open reading frame (ORF), will usually cause a frameshift or create an in-frame nonsense codon.

Insertion mutations scattered in various places around the transposon were tested for their effects on transposition by the mating-out assay. As illustrated in Figure 9.10, cells containing both a small, nonmobilizable plasmid carrying the mutant Tn3 and a larger, self-transmissible plasmid were mixed with recipient cells and tranconjugants resistant to ampicillin were selected. Because the smaller plasmid was not mobilizable, ampicillin-resistant transconjugants could be produced only by donor cells in which the transposon had hopped from the smaller plasmid into the larger, self-transmissible

Figure 9.10 Molecular genetic analysis of transportation of the replicative transposon Tn3. The transposon is in gold. The asterisk marks the position of the mutation. Amp^r, ampicillin resistance. See the text for details.

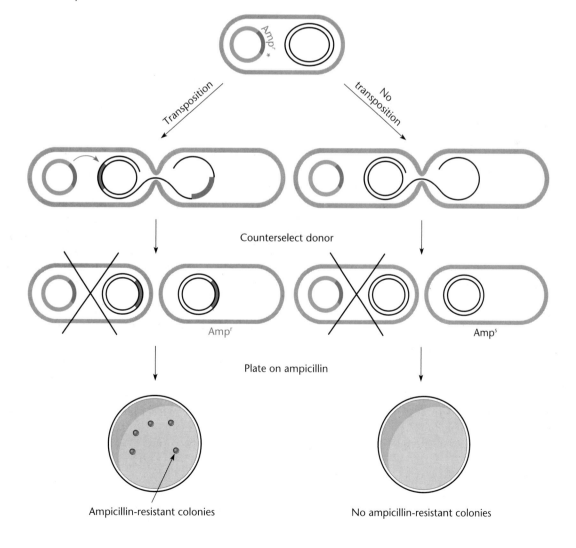

one, which was then transferred into the recipient cell. When no ampicillin-resistant transconjugants were observed, the mutation in the transposon must have prevented transposition into the self-transmissible plasmid. When larger than normal numbers of ampicillin-resistant transconjugants were observed, the mutation must have increased the frequency of transposition.

The effect of a mutation depends on its position in the transposon. As shown in Figure 9.11, mutations in the inverted repeat (IR) sequences and mutations that disrupt the *tnpA* ORF can totally prevent transposition. In contrast, mutations that disrupt the *tnpR* ORF result in higher than normal rates of transposition and the formation of cointegrates, in which the self-transmissible plasmid and the smaller plasmid, which originally contained the transposon, are now joined and are transferred together into the recipient strain. The cointegrate contains two copies of the transposon bracketing the smaller plasmid as shown. Mutations in the short sequence called *res* (for "resolution sequence") also give rise to cointegrates, but unlike *tnpR* mutations, they result in normal, not elevated, rates of transposition.

COMPLEMENTATION TESTS WITH TRANSPOSON MUTATIONS

Complementation tests were used to determine which mutations in transposon Tn3 disrupt *trans*-acting functions and which disrupt *cis*-acting sequences or sites. The complementation tests used the same mating-out assay illustrated in Figure 9.10, except that the cell in which the transposition was to occur also contained another Tn3-related transposon inserted into its chromosome (Figure 9.12). This other transposon is capable of transposition but lacks an ampicillin resistance gene so that its own transposition will not confuse the analysis. The data are interpreted as in any other complementation test. If the mutation in the transposon in the plasmid inactivates a *trans*-acting function, it will be complemented by the corresponding gene in the transposon in the chromosome and the transposon should now be able to transpose. However, if the mutation in the plasmid transposon inactivates a *cis*-acting site, it will not be complemented and will not transpose properly, even in the presence of the chromosomal transposon, since mutations that inactivate *cis*-acting sites cannot be complemented. The comple-

Figure 9.11 Effects of mutations in different genes required for transposition of Tn3. In the left-hand pathway, a *tpnA* or *IR* mutation prevents transposition, and so no Amp^r transconjugants form. In the right-hand pathway, transposition by a *tpnR* or *res* mutant leads to the formation of Amp^r transconjugants that contain the mobilizable and self-transmissible plasmids fused to each other in a cointegrate.

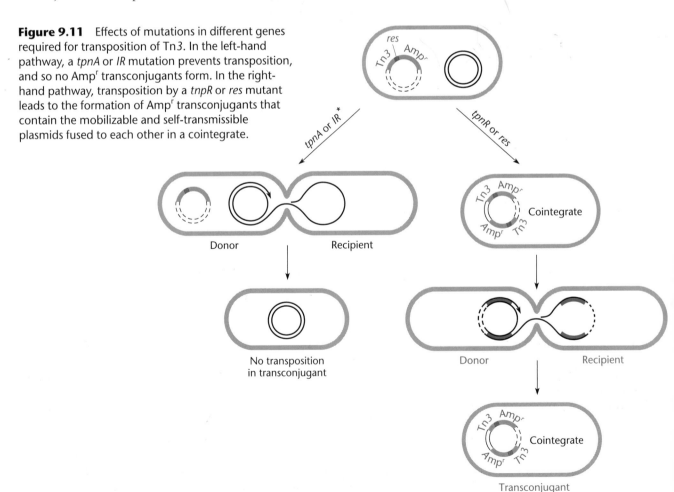

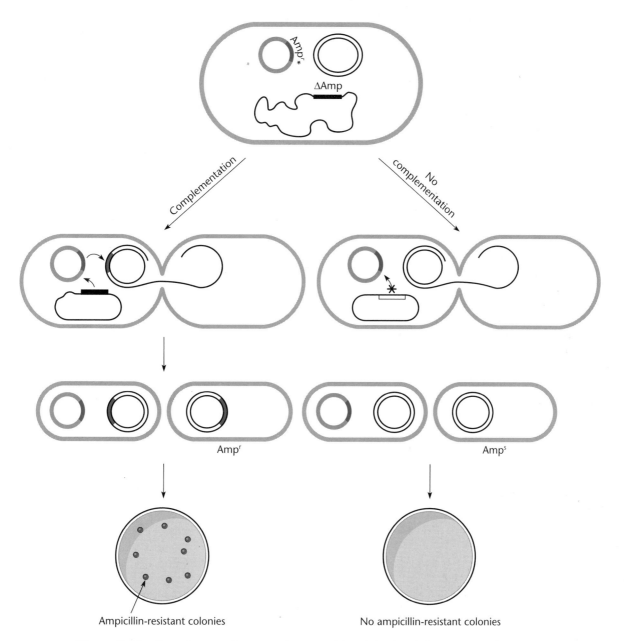

Figure 9.12 Complementation tests of transposition-defective Tn*3* mutations. The mutant Tn*3* transposon being complemented is in gold, and the asterisk indicates the mutation. See the text for details.

mentation tests revealed that mutations that inactivate either the ORF called *A* or the ORF called *R* (Figure 9.8) could be complemented to give normal transposition. However, neither mutations in the IR sequences at the ends of the transposon nor those in the sequence called *res* could be complemented to give normal transposition. Mutations in an IR sequence prevented transposition altogether, even in the presence of the complementing copy of Tn*3*, while mutations in *res* permitted transposition but still gave rise to cointegrates. The investigators con-

cluded that *tnpA* and *tnpR* encode *trans*-acting proteins while IR and *res* are *cis*-acting sites on the DNA.

These genetic data prompted formulation of a model for replicative transposition of Tn*3* and other Tn*3*-like transposons. Briefly, mutations in the *tnpA* gene prevent transposition because the *tnpA* gene encodes the transposase TnpA, which promotes transposition. Mutations in the IR elements at the ends of the transposon also prevent transposition, because these are the sites at which the TnpA transposase acts to promote transposition.

The behavior of mutations in *tnpR* and *res* was more difficult to explain. To reiterate, *tnpR* mutations not only cause higher than normal rates of transposition but also cause the formation of cointegrates. Mutations in *res* also cause cointegrates to form but do not affect the frequency of transposition and cannot be complemented. To explain these results, the investigators proposed that *tnpR* encodes a protein with two functions. First, the TnpR protein acts as a repressor (see chapter 2), which represses the transcription of the *tnpA* gene for the transposase. By inactivating the repressor, *tnpR* mutations cause higher rates of transposition by allowing more TnpA synthesis. Second, the TnpR protein acts as a recombinase that resolves cointegrates by promoting site-specific recombination between the *res* sequences in the two copies of the transposon in the cointegrate (see Figure 9.13). This explains why both *tnpR* and *res* mutations cause the accumulation of cointegrates. Either type will prevent the site-specific recombination that resolves the cointegrates causing cointegrates to accumulate, but only *tnpR* mutations can be complemented because only *tnpR* encodes a diffusible gene product.

Molecular Models for Transposition

Most bacterial transposons seem to transpose by one of two mechanisms. Some transposons, represented by Tn*3* and phage Mu, use a **replicative mechanism** of transposition, in which the entire transposon replicates during transposition, resulting in two copies of the transposon. Other transposons, represented by many of the IS elements and composite transposons such as Tn*10*, Tn*5*, and Tn*9*, transpose by a **cut-and-paste**, or conservative, mechanism, in which the transposon is removed from one place and inserted into another.

REPLICATIVE TRANSPOSITION
The first detailed model to be developed for transposition attempted to explain all of what was known about Tn*3* transposition and the transposition of other transposons such as phage Mu (see Shapiro, Suggested Reading, and Box 9.1). This model continues to be generally accepted for those types of transposons. The model incorporates the following observations, some of which we have already mentioned.

1. Whenever a transposon such as Tn*3* hops into a site, a short sequence of the target DNA is duplicated. The number of bases duplicated is characteristic of each transposon. (For Tn*3*, 5 bp is duplicated; for IS*1*, 9 bp is duplicated.)
2. The formation of a **cointegrate,** in which the donor and target DNAs have become fused and encode two copies of the transposon, is an intermediate step in the transposition process.

3. Once the cointegrate has formed, it can be resolved into separate donor and target DNA molecules either by the host recombination functions or by a transposon-encoded **resolvase** that promotes recombination at internal *res* sequences.
4. The donor DNA and target DNA molecules both have a copy of the transposon after resolution of the cointegrate. Therefore, the transposon does not actually move but duplicates itself, and a new copy appears somewhere else; hence the name "replicative transposition."
5. Neither transposition nor, for some transposons, resolution of cointegrates requires the normal recombination enzymes or extensive homology between the transposon and the target DNA.

Figure 9.13 shows the molecular details of the model for replicative transposition. In the first step, the transposase makes single-stranded breaks at each junction between the transposon and the donor DNA and a double-stranded break in the target DNA. The break in the target DNA is staggered so that the nicks in the two strands are separated by the same number of base pairs as will be duplicated in the target DNA during insertion of the transposon, as explained below. The cutting leaves two 5′ ends and two 3′ ends in the target DNA and a 5′ end and a 3′ end at each junction between the transposon and the donor DNA. The 5′ ends in the target DNA are then joined (ligated) to the 3′ ends of the transposon (Figure 9.13). Replication then proceeds in both directions over the transposon, with the free 3′ ends of the target DNA serving as primers. After replication over the transposon, the 3′ ends of the newly synthesized strands are ligated to the remaining free 5′ ends of the donor DNA to form the cointegrate. The last step, **resolution** of the cointegrate, results from recombination between the two *res* sites in the cointegrate promoted by the resolvase of the transposon (see section on resolvases, below). Resolution of the cointegrate gives rise to two copies of the transposon, one at the former, or donor site, and a new one at the target site.

This model explains why cointegrates are obligate intermediates in replicative transposition. After replication has proceeded over the transposon in both directions, the donor DNA and the target DNA will be fused to each other, separated by copies of the transposon, as shown.

This model also explains why, after transposition, a short target DNA sequence of defined length has been duplicated at each end of the transposon. Because it makes a staggered break in the target DNA, the transposase causes a short region of the target DNA to be duplicated when replication proceeds from this staggered break over the transposon. The number of base pairs of

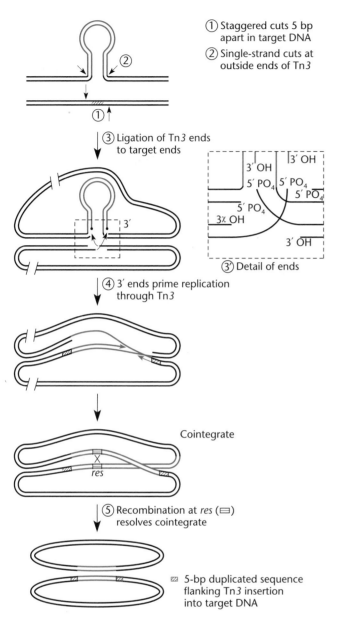

① Staggered cuts 5 bp apart in target DNA

② Single-strand cuts at outside ends of Tn*3*

③ Ligation of Tn*3* ends to target ends

③ Detail of ends

④ 3′ ends prime replication through Tn*3*

Cointegrate

res

⑤ Recombination at *res* (▱) resolves cointegrate

▱ 5-bp duplicated sequence flanking Tn*3* insertion into target DNA

Figure 9.13 Replicative transposition of Tn*3* (gold) and the formation and resolution of cointegrates. At sites labeled 1 and 2, breaks are made in the target DNA and at the ends of the transposon, respectively. (3) The 3′ OH ends of the transposon (dots) are ligated to 5′ PO₄ ends of the target DNA. The inset (3′) shows details of the ends. (4) The free 3′ ends of target DNA prime replication in both directions over the transposon to form the cointegrate. (5) The cointegrate is resolved by recombination promoted by the resolvase TnpR at the *res* sites.

target DNA duplicated at the ends of the transposon will be the same as the number of base pairs between the nicks in the two strands in the staggered break and will be characteristic of the transposase enzyme for each type of transposon.

Finally, this model explains why replicative transposition is independent of most host functions including DNA ligase and the normal recombination functions such as RecA (see chapter 10). The transposase cuts the target and donor DNAs and promotes ligation of the ends. Also, the normal recombination system is not needed to resolve the cointegrate into the original replicons, because the resolvase specifically promotes recombination between the *res* elements in the cointegrates. Although cointegrates can also be resolved by homologous recombination anywhere within the repeated copies of the transposon, the resolvase greatly increases the rate of resolution by actively promoting recombination between the *res* sequences.

CUT-AND-PASTE TRANSPOSITION

Some bacterial transposons, including many of the IS elements and composite transposons, do not transpose by a replicative mechanism but, rather, by the simpler cut-and-paste mechanism illustrated in Figure 9.14. In this mechanism, the transposase makes double-stranded breaks at the ends of the transposon, cutting it out of the donor DNA, and then pastes it into the target DNA at the site of a staggered break. When the single-stranded gaps created by the nicks in the target DNA are filled in, a short sequence will be duplicated. For most types of transposons, removal of the transposon from the donor DNA probably leaves breaks in the donor DNA, which is consequently degraded, as shown in Figure 9.14.

The molecular details of how cut-and-paste transposition works are more complicated than this and may differ between transposons. For example, some IS elements are thought to go through a single-stranded circular DNA intermediate. Others are known to replicate by a rolling-circle mechanism similar to that used by some phages and plasmids. However, the net result is the same: the transposon is cut out of one DNA and inserted into another.

Genetic and biochemical experiments have contributed the evidence for cut-and-paste transposition of some transposons. Some of the strongest evidence supporting this mechanism has come from recently developed in vitro transposition systems. In the next few sections, we describe in detail some of the ways in which cut-and-paste transposition differs from replicative transposition and the evidence for each of these features.

No Cointegrate Intermediate

In cut-and-paste transposition, cointegrates do not form as a necessary intermediate, as they do in the replicative mechanism. This conclusion is supported by indirect evidence suggesting that IS elements and composite transposons do not form these intermediates. For example,

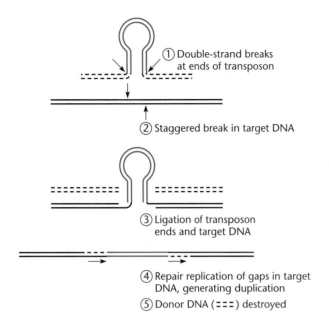

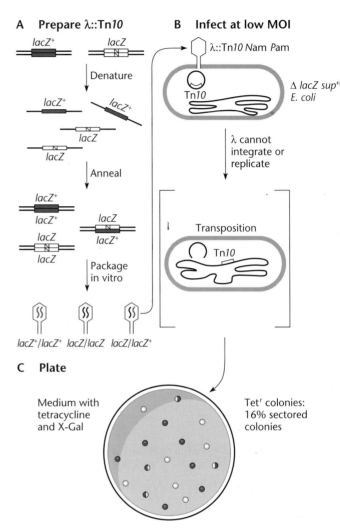

Figure 9.14 Cut-and-paste transposition. (1) Double-strand breaks are made at the ends of the transposon (arrows). (2) Staggered breaks are made in the target DNA (arrows). (3) The free 3' ends of the transposon are ligated to the 5' ends of the target DNA. The dashed lines represent the donor DNA, which will be degraded. (4) DNA polymerase fills in the gaps of the target DNA, producing the short duplication of target DNA at the ends of the transposon. (5) Donor DNA is destroyed.

Figure 9.15 Genetic evidence for nonreplicative transposition by Tn*10*. (A) Preparation of λ::Tn*10 lacZ/lacZ*+ heteroduplex DNA. (B) The λ::Tn*10* infects a nonpermissive *E. coli* host. Because the λ contains N*am* and P*am* mutations, it cannot integrate or replicate. If the transposon replicates during transposition, the bacteria in the Tet' colonies will get only one or the other strand of DNA in the heteroduplex, and the colonies will be all blue or all colorless. If both strands are transferred, some colonies will be sectored, part blue and part white. See the text for further details. MOI, multiplicity of infection. Note that blue is represented by gold.

there are no mutants of the transposons Tn*10* and Tn*5* that accumulate cointegrates as there are for Tn*3*. Moreover, even if cointegrates are formed artificially by recombinant DNA techniques, there is no evidence that the cointegrates can be resolved except by the normal host recombination system. Therefore, the cut-and-paste transposons do not seem to encode their own resolvases, which they would be likely to do if cointegrates were a normal intermediate in their transposition process.

Both Strands of the Transposon Transpose

The primary difference between replicative and cut-and-paste transposition is that in the latter, both strands of the transposon move to the target DNA. The results of genetic experiments with Tn*10* (outlined in Figures 9.15 and 9.16) support this conclusion (see Bender and Kleckner, Suggested Reading). The first step in these experiments was to introduce different versions of transposon Tn*10* into a λ suicide vector. Both of the two Tn*10* derivatives contained a copy of the *lacZ* gene as well as the Tet' gene Tn*10* usually carries. However, one of the Tn*10* derivatives carried three missense mutations in the *lacZ* gene to inactivate it. The strands of the two λ

DNAs were separated and reannealed to make heteroduplex DNA, in which each of the strands came from a λ phage carrying a different derivative of Tn*10*. Consequently, the heteroduplex DNAs had one strand with a good copy of *lacZ* and another strand with the mutated copy of *lacZ*.

In the next step, the heteroduplex DNA was packaged into λ heads in vitro (see chapter 8) and used to infect

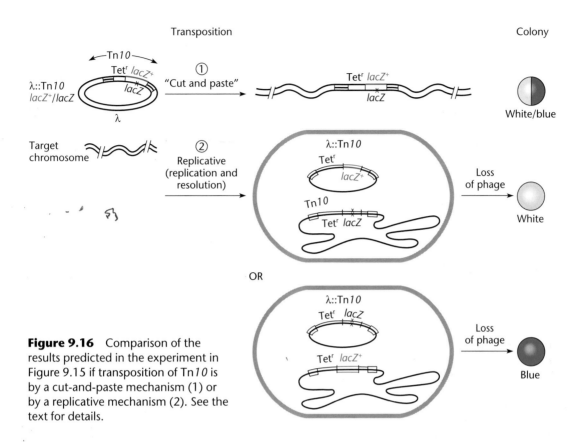

Figure 9.16 Comparison of the results predicted in the experiment in Figure 9.15 if transposition of Tn*10* is by a cut-and-paste mechanism (1) or by a replicative mechanism (2). See the text for details.

Lac⁻ cells. Because this λ was a suicide vector, the only cells that became Tet^r were ones in which the Tn*10* derivatives had hopped into the chromosome. If the transposition had occurred by a replicative mechanism, the Tet^r colonies would have contained either Lac⁺ or Lac⁻ bacteria (Figure 9.16), since the information in only one of the two strands could have been transferred. If, however, the transposition had occurred by a nonreplicative cut-and-paste mechanism, the cells in at least some Tet^r colonies due to transposition would have received both the Lac⁺ and the Lac⁻ strands of the heteroduplex DNA. When these heteroduplex DNAs replicated, they would give rise to both Lac⁺ and Lac⁻ bacteria in the same colony, making "sectored" blue-and-white colonies on X-Gal plates (see chapter 12), as shown in Figure 9.15. In the experiment, sectored blue-and-white colonies were observed, supporting the conclusion that both strands were transferred.

Transposon Leaves the Donor DNA

A major difference between replicative and cut-and-paste transposition is the number of copies of the transposon after transposition. The replicative mechanism leaves the cell with two copies, and the cut-and-paste mechanism leaves it with one. It might seem easy to determine whether a copy of the transposon still exists in the donor DNA after transposition; however, it is usually difficult. For example, if the donor DNA containing the transposon is a multicopy plasmid and the transposition occurs by a cut-and-paste mechanism, only one copy of the plasmid will lose its transposon, leaving many plasmids intact with the transposon. Even if the transposition occurs from a DNA that normally exists in a single-copy, such as the chromosome or a single copy plasmid, a copy of the transposon remaining in the donor DNA could be attributed to replication of the donor DNA prior to transposition.

For much the same reasons, it is difficult to determine whether the donor DNA is resealed after the transposon is cut out of it during cut-and-paste transposition. It might seem that the donor DNA must be resealed after the transposon hops; otherwise, transposition would leave a double-stranded break in the donor DNA, which would be a lethal event in many cases. However, cells usually carry more than one copy of a plasmid, and even the chromosome is usually in a partial state of replication, so that many regions exist in more than one copy per cell. After transposition, the double-strand break could then be repaired by recombination with the other daughter DNA in a process called double-strand break repair (see chapter 10). Ironically, since the daughter DNA will still contain the transposon, the double-strand break repair will restore the transposon to its original donor site, making the transposition appear replicative even though it is not. Even if the transposon were to leave an unreplicated region of the chromosome and the donor DNA were left unrepaired and the cell died, this would be difficult to detect. Transposition events are infrequent, and a few dead cells would go undetected in a large population of bacteria.

The best evidence that the donor DNA is not resealed or, at least, is not resealed correctly after transposition by most cut-and-paste transposons comes from reversion studies. For a transposon insertion mutation to revert, the transposon must be completely removed from the DNA in a process called **precise excision**. Not a trace of the transposon can remain, including the duplication of the short target sequence, or the gene would probably remain disrupted and nonfunctional.

If the transposon were precisely excised and the donor DNA were resealed every time a transposon hopped by a cut-and-paste mechanism, then transposon insertion mutations would revert every time the transposon hopped. However, reversion of insertion mutations occurs at a much lower rate than does transposition itself. Moreover, mutations in the transposon itself that inactivate the transposase and render the transposon incapable of transposition do not further lower the reversion frequency, as might be expected if the few revertants that are seen resulted from another transposition event. Presumably, the rare precise excisions that cause transposon insertion mutations to revert are due to homologous recombination between the short duplicated target sequences bracketing the transposon and are unrelated to transposition itself. Therefore, for most cut-and-paste transposons, the donor DNA is apparently left broken after the transposon is cut out of it, as shown in Figure 9.14.

RELATIONSHIP BETWEEN REPLICATIVE AND CUT-AND-PASTE TRANSPOSITION

Even though replicative transposition and cut-and-paste transposition seem different, they are actually mechanistically related. As shown in Figures 9.13 and 9.14, the major difference is in the number of strand cuts that the transposase enzyme makes in the junction between the transposon and the donor DNA. A cut-and-paste transposase makes cuts in both strands in the junction, whereas a replicative transposase cuts only one strand at the junction. Otherwise, the two mechanisms are similar. In both, the cut 5′ ends of the target DNA are joined to the free 3′ ends of the transposon. Then in both mechanisms, the free 3′ ends of the target DNA are used as primers for replication that proceeds until a free 5′ end in the donor DNA, to which the newly replicated DNA can be joined, is reached.

The difference comes from how much DNA must be replicated before this 5′ end is reached. In cut-and-paste transposition, replication proceeds only a short distance owing to the staggered break in the target DNA before the growing 3′ end reaches the free 5′ end at the end of the transposon and the two ends join. As mentioned, this replication gives rise to the short duplications of the target DNA but does not duplicate the transposon itself. In replicative transposition, in contrast, the replication must proceed over the region of the staggered break and then over the entire transposon before the free 5′ end is reached on the other side of the transposon. This process duplicates the target DNA as well as the transposon and gives rise to a cointegrate.

A dramatic confirmation of the similarity between the cut-and-paste and replicative mechanisms of transposition came with the demonstration that a cut-and-paste transposon can be converted into a replicative transposon by a single amino acid change in the transposase (see May and Craig, Suggested Reading). Transposon Tn7 normally transposes by a cut-and-paste mechanism, and different regions of the transposase make the cuts in the opposite strands of DNA at the ends of the transposon. If the region that makes one of the two cuts is inactivated by a mutation, the transposase will cut only one of the two strands, like a replicative transposase. The Tn7 transposon with such an altered transposase then transposes by the replicative mechanism, forming a cointegrate. Apparently, the transposase need only make the appropriate cuts and joinings, and the replication apparatus of the cell does the rest. It is somewhat surprising that the same transposase enzyme can support both types of transposition, since the transposon presumably needs different cellular replication machineries for each. Replication of tens of thousands of base pairs during replicative transposition would be a much more involved process than replication of a few base pairs during cut-and-paste transposition.

TARGET SITE SPECIFICITY

No transposable elements insert completely randomly into target DNA. Most transposable elements show some target specificity, hopping into some sites more often than into others. Even Tn5, which is famous for hopping almost at random, does prefer some sites to others. Usually it is not the sequence of DNA at the target site which matters but, rather, some structure in the DNA such as a bend or the presence of other proteins bound to the DNA, such as transcription factors.

Tn7 is the extreme case of a transposon with target specificity. It transposes with a high frequency into only one site in the E. coli DNA, called attTn7. Recent studies have given insights into how this selectivity is achieved (see Kuduvalli et al., Suggested Reading). The Tn7 transposition machinery consists of five proteins, TnsA, TnsB, TnsC, TnsD and TnsE. Of these, TnsA and TnsB make up the transposase that cuts and joins the DNA strands and the other proteins regulate transposition and direct the transposon to the target sequence. The role of TnsD may be to direct the transposon to its target sequence, *att*Tn7. By binding to the *att*Tn7 sequence, TnsD may induce changes representative of triple-stranded structures in the *att*Tn7 site. This directs TnsC to stimulate transposition into the site. In the absence of TnsD, TnsE will stimulate transposition into other sites in the chromosome. This transposition is inefficient but random. It is possible that this random transposition is occurring when transient triple-stranded structures, created for example by recombination (see chapter 11), appear in the DNA, but this needs to be investigated.

EFFECTS ON GENES ADJACENT TO THE INSERTION SITE

Most insertion element and transposon insertions cause polar effects if they insert into a gene transcribed as a polycistronic mRNA. The inserted element contains transcriptional stop signals and may also contain long stretches of sequence that is transcribed but not translated. The latter may cause Rho-dependent transcriptional termination.

Some insertions may enhance expression of a gene adjacent to the insertion site. This expression can result from transcription that originates within the transposon. For example, both Tn5 and Tn10 contain outward-facing promoters near their termini which can initiate transcription into neighboring genes.

REGULATION OF TRANSPOSITION

Transposition of most transposons occurs rarely, as discussed above, because transposons self-regulate their transposition frequency. The regulatory mechanisms used by various transposons differ greatly. For some, transposition occurs primarily just after a replication fork has passed through the element. Newly replicated DNA is hemimethylated (see chapter 1), and hemimethylated DNA has been shown not only to activate the transposase promoter of Tn10 but also to increase the activity of the transposon ends. Most transposons employ mechanisms to modulate the level of transposase transcription and/or translation (see Craig, Suggested Reading).

TARGET IMMUNITY

Another feature of some transposons such as Tn3 is that they do not hop close to another transposon in the same target DNA. This is called target site immunity and is limited to only some transposons. The immunity can extend over 100,000 bp of DNA. It is not clear what causes target site immunity. In Mu, where it has been most extensively studied, the presence of MuB bound to the ends of the first Mu insertion may preferentially bind MuA, making less of it available for insertion in a new, nearby target site. Also, the presence of two transposons close to each other can cause instability due to mechanisms like those described above in the section on inside-end transposition.

Transposon Mutagenesis

One of the most important uses of transposons is in transposon mutagenesis. This is a particularly effective form of mutagenesis because a gene that has been marked with a transposon is relatively easy to map by genetic crosses or by physical mapping with restriction endonucleases. Furthermore, genes marked with a transposon are also relatively easy to clone by using plate hybridizations or by selecting for selectable genes carried on the transposon.

Not all types of transposons are equally useful for mutagenesis. A transposon used for mutagenesis should have the following properties:

1. It should transpose at a fairly high frequency.
2. It should not be very selective in its target sequence.
3. It should carry an easily selectable gene, such as one for resistance to an antibiotic.
4. It should have a broad host range for transposition, if it is to be used in a variety of different kinds of bacteria.

Transposon Tn5 is ideal for random mutagenesis of gram-negative bacteria because it embodies all of these features. Not only does Tn5 transpose with a relatively high frequency, but also it has almost no target specificity and will transpose in essentially any gram-negative bacterium. It also carries a kanamycin resistance gene that is expressed in most gram-negative bacteria. Figure 9.17A illustrates a popular method for transposon mutagenesis of gram-negative bacteria other than *E. coli* (see Simon et al., Suggested Reading). In addition to the broad host range of Tn5 and the promiscuity of RP4, this method takes advantage of the narrow host range of ColE1-derived plasmids, which will replicate only in *E. coli* and a few other closely related species. Phage Mu is another transposon-like element that can hop in many types of gram-negative bacteria and shows little target specificity (Box 9.1). Equally universal methods are not available for transposon mutagenesis in gram-positive bacteria. No transposons of gram-positive bacteria have been identified that fulfill all the criteria above, although

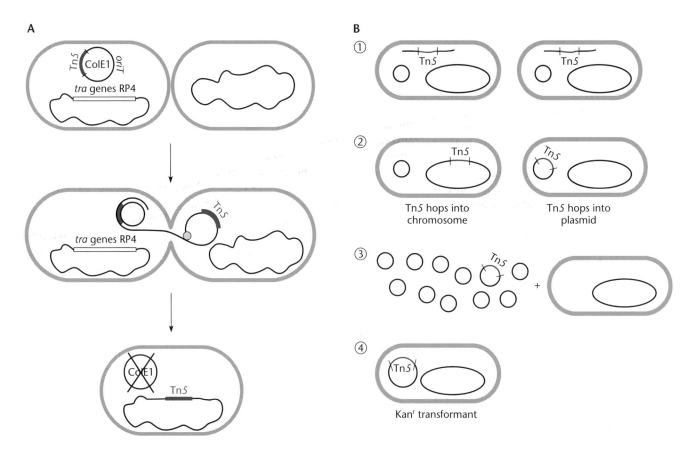

Figure 9.17 Transposon Tn5 mutagenesis. (A) A standard protocol for transposon mutagenesis of gram-negative bacteria. A suicide ColE1-derived plasmid containing the *oriT* sequence of promiscuous plasmid RP4 and transposon Tn5 is mobilized into the bacterium by the products of the RP4 transfer genes, which are inserted in the chromosome. The transposon hops into the chromosome of the recipient cell, and the ColE1 plasmid is lost because it cannot replicate. The Tn5 transposon is shown in gold. (B) Random transposon mutagenesis of a plasmid. In step 1, the transposon Tn5 is introduced into cells on a suicide vector. In step 2, the culture is incubated, allowing the Tn5 time to hop, either into the chromosome (large circle) or into a plasmid (small circle). Plating on kanamycin-containing medium will result in the selection of cells in which a transposition has occurred. In step 3, plasmid DNA is prepared from Kan^r cells and used to transform a Kan^s recipient. In step 4, selection for Kan^r will allow the identification of cells that have acquired a Tn5-carrying plasmid.

Tn917 is a transposon that hops fairly randomly in some gram-positive bacteria. Also Tn916, a conjugative transposon of gram-positive bacteria (Box 9.2), has the advantages that it transfers itself from one cell to another and shows little target specificity although it does not integrate in many gram-positive bacteria. However, with recently developed techniques such as in vitro mutagenesis, it is now possible to perform transposon mutagenesis in many bacteria for which useful transposons are not available (see Box 9.3).

Transposon Mutagenesis of Plasmids

Transposons are particularly useful for the mutagenesis of plasmids. The relatively small size of plasmids makes physical mapping easier, and it is often fairly easy to isolate large numbers of transposon insertions in a plasmid.

Figure 9.17B illustrates the steps in the selection of plasmids with transposon insertions in *E. coli*. A suicide vector containing the transposon, in the example, Tn5, is introduced into cells harboring the plasmid. Cells in which the transposon has hopped into cellular DNA, either the plasmid or the chromosome, are then selected by plating on medium containing the antibiotic to which a transposon gene confers resistance, in this case kanamycin. Only the cells in which the transposon has hopped to another DNA will become resistant to the antibiotic, since the transposons which remain in the suicide vector will be lost with the suicide vector. In most of the antibiotic-resistant bacteria, the transposon will have hopped into the chromosome rather than into the plasmid, simply because the chromosome is the larger target. The plasmids in these bacteria will be normal. However, the plasmids can be isolated from the few bacteria in which the transposon has hopped into the plasmid, by mating the plasmid into another *E. coli* strain and selecting the antibiotic resistance on the transposon if the plasmid being mutagenized is self-transmissible. Alternatively, the antibiotic-resistant colonies which have the transposon either in the plasmid or in the chromosome can be pooled and the plasmids can be isolated from them by one of the procedures outlined in chapter 4. This mixture of plasmids, most of which are normal, is then used to transform another strain of *E. coli,* selecting for the antibiotic resistance gene on the transposon in the example, kanamycin. The antibiotic-resistant transformants should contain the plasmid with the transposon inserted somewhere in it. Voilà, in a few simple steps, plasmids with transposon insertion mutations have been isolated. This method can be used to randomly mutagenize a DNA cloned in a plasmid or to mutagenize the plasmid itself.

PHYSICAL MAPPING OF THE SITE OF TRANSPOSON INSERTIONS IN A PLASMID

Once plasmids with transposon insertion mutations have been isolated, it is relatively easy to map the sites of

the insertions relative to the positions of known restriction sites on the plasmid. This mapping relies on the fact that the size and number of DNA fragments obtained with a restriction endonuclease are dependent on the number and location of the sites for that restriction endonuclease in the DNA being cut. The insertion of the transposon into the plasmid will introduce new restriction sites and change the size of some of the fragments.

As an example, consider the small plasmid pAT153 and the transposon Tn5 (Figure 9.18). Use of a small plasmid like pAT153 makes interpretation of the physical mapping data easier, but the same general methods are applicable to much larger plasmids.

To locate the site of insertion, we need to use two different restriction endonucleases, both of which cut in the transposon and the original plasmid. The restriction endonucleases *Pst*I and *Hind*III fulfill this requirement (Figure 9.18). Plasmid pAT153 is 3,600 bp (3.6 kb) long and has only one *Hind*III site and one *Pst*I site. The transposon Tn5 is 5.6 kb long and has two *Hind*III and four *Pst*I sites. Cutting the original pAT153 plasmid without the transposon insertion with either *Hind*III or *Pst*I should give one band of 3.6 kb. However, the plasmid with the transposon inserted is much larger (3.6 kb + 5.6 kb = 9.2 kb) and also contains restriction sites introduced on the transposon; therefore, cutting this larger molecule with *Hind*III should yield three bands and cutting it with *Pst*I should yield five bands, because of the

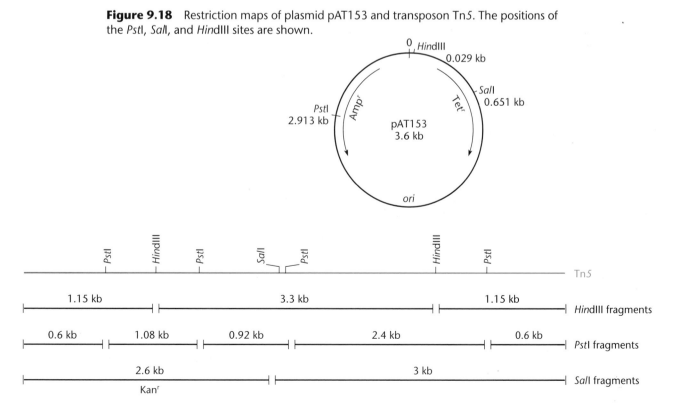

Figure 9.18   Restriction maps of plasmid pAT153 and transposon Tn5. The positions of the *Pst*I, *Sal*I, and *Hind*III sites are shown.

Phage Mu: a Transposon Masquerading as a Phage

Phage Mu is a lysogenic phage that can integrate into the bacterial DNA after infection. However, the phage was known to be different from most lysogenic phages almost from the time it was first discovered. One unusual thing about it is that it causes random mutations when forming lysogens, giving it its name Mu, for *mut*ator phage. The phage causes detectable mutations because, unlike λ and other known phages, it has no unique bacterial attachment site but inserts almost randomly into the chromosome. When it happens to insert into a gene, it will inactivate the gene and can cause a mutant phenotype. In contrast, λ almost always inserts at a unique site in a nonessential region between the *gal* and *bio* genes (see chapter 8), therefore, lysogenization by λ seldom causes a mutant phenotype. Other unusual properties of phage Mu were discovered when the strands of DNA in the phage heads were separated by heating, renatured, and observed under the electron microscope. These experiments were undertaken to determine whether phage Mu DNA has unique ends like T7 and λ or is cyclically permuted like T4 and P22. If the DNA has unique ends, the single-stranded DNAs will find partners that are complementary to them over their entire length, so that the renatured DNAs will be nice double-stranded molecules with no single-stranded ends. However, renatured, cyclically permuted DNA molecules have a very different appearance. Each single-stranded DNA will usually pair with another molecule that has different ends, so that most of the molecules will have single-stranded ends which can pair with complementary regions in the single-stranded ends of other molecules to give very complicated, branched structures. Surprisingly, renatured Mu DNA gives neither of these patterns. Instead, the renatured Mu DNA molecules look like they are having a bad-hair day, with single-stranded "split ends" from 500 to 2,000 bases in length. These single-stranded ends are made up of host DNA from various regions of the chromosome which are not complementary to each other. Mu has random host DNA attached to its ends because of the way it replicates and is packaged (see Ljungquist and Bukhari, below). Mu DNA replicates by replicative transposition without resolution of the concatamers. First, it replicatively transposes to another place in the chromosome, so that it now exists in two copies in the chromosome, the original site and the new site. Rather than be resolved, each of these two Mu DNAs then replicatively transposes to yet other sites and so forth until the entire chromosome of the bacterium is riddled with hundreds of copies of the Mu DNA. These multiple copies of Mu DNA are then packaged into phage heads by making cuts in the sur-

rounding host DNA 500 to 2,000 base pairs from the ends of the Mu DNA, leaving the adjacent host DNA attached to the ends. Since each Mu DNA was inserted at a different place in the chromosome, each of the packaged Mu DNA molecules will have different host DNA sequences at its ends, giving rise to "split ends" after denaturation and renaturation.

Mu uses transposition both to integrate its prophage into the host DNA to form a lysogen and to replicate its DNA during lytic development. However, it uses different transposition mechanisms for these two processes. Integration of the prophage requires a single "cut-and-paste" transposition event, while replication requires repeated rounds of replicative transposition. It is still a mystery how Mu can use these two distinct transposition mechanisms, since both require the same two Mu proteins, MuA and MuB, which make up the transposase (see Figure). Some hypotheses being tested are that the initial DNA which infects the cell has a protein attached to its ends, which allows the DNA to integrate but blocks further transposition. Another is that different regions of the MuB protein might be involved in the two different processes (see Roldan and Baker, below).

Another unusual feature of renatured Mu DNA is that it often has an unpaired region of about 3,000 base pairs in the middle, which forms a "bubble" when the DNA is renatured. This region of Mu DNA, called the G-segment (see Figure), is an invertible sequence which flips around or inverts at a high frequency. If two single-stranded Mu DNAs which have the G-segment in opposite orientations attempt to pair, this region will loop out and form a single-stranded bubble that is clearly visible under the electron microscope. Like phage P1, which has a similar invertible segment (see chapter 7), the host range of phage Mu is partially determined by the orientation of the G-segment, which encodes elements of the tail fibers. Recall from chapter 7 that the tail fibers help determine the host range of a phage, so the ability of the phage to infect a particular host depends upon the orientation of the G-segment. It is ironic that the G-segment may not have been observed if the phage from which the DNA had been prepared had been grown lytically in *Escherichia coli* K-12 for many generations rather than having been induced from a lysogen. Only phage Mu with the G-segment in one orientation is able to infect *E. coli* K-12, so that phage with the G-segment in the other orientation are selected against when the phage is grown in *E. coli* K-12. However, both orientations accumulate when the phage is grown as a prophage in the lysogen. When the prophage is

(continued)

BOX 9.1 (continued)

Phage Mu: a Transposon Masquerading as a Phage

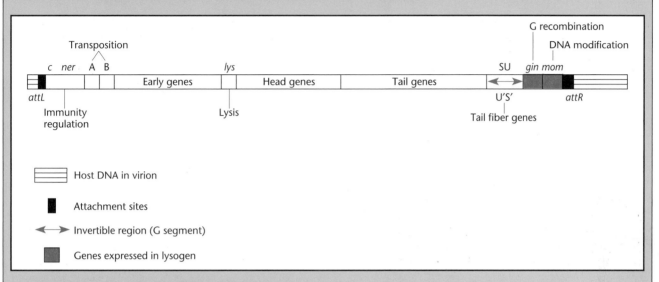

induced, phage with both orientations of the G-segment are produced equally and form the bubble when the DNAs are denatured and renatured.

Because it transposes randomly into the chromosome, Mu is useful for transposon mutagenesis and for creating random gene fusions. Besides its usefulness in bacterial genetics and its general interest as a phage, Mu is an effective tool for biochemical studies of transposition because it trans-poses so frequently. We discuss uses of Mu in making random gene fusions and in in vivo cloning later in this chapter.

References

Ljungquist, E., and A. I. Bukhari. 1977. State of prophage Mu upon induction. *Proc. Natl. Acad. Sci. USA* **74:**3143–3147.

Roldan, L. A. S., and T. A. Baker. 2001. Differential role of MuB protein in phage Mu integration vs replication: mechanistic insights into two transposition pathways. *Mol. Microbiol.* **40:**141–156.

*Hin*dIII and *Pst*I sites in the transposon. With each restriction endonuclease, the fragments should add up to 9.2 kb. Some of these fragments will always be the same size, regardless of where the transposon has inserted in the plasmid. These are called the **internal fragments** because they come from within the transposon and contain only transposon DNA. Cutting with *Hin*dIII leaves one internal fragment of 3.3 kb, while cutting with *Pst*I leaves internal fragments of 1.08, 0.92, and 2.40 kb. However, two fragments in each case extend from a restriction site in the transposon to one in the plasmid and so contain both transposon and plasmid DNA. The sizes of these fragments, called the **junction fragments,** will vary depending upon where the transposon has inserted in the plasmid. From the sizes of the junction fragments, we can determine where the transposon inserted in the plasmid.

The representative data in Figure 9.19 illustrate how the site of insertion of the Tn*5* transposon can be determined for a particular insertion mutation. To obtain Figure 9.19A, the plasmid with the transposon insertion was digested separately with *Hin*dIII and *Pst*I and the fragments were applied to an agarose gel, along with marker DNA fragments of known size. During electrophoresis, smaller fragments move faster than larger fragments and the rate of migration decreases exponentially with size (see chapter 1). Therefore, on semilog paper, a plot of distance versus size for the marker DNAs should give a standard curve approximating a straight line in the region of interest, as shown in Figure 9.19B. From this standard curve, we can estimate the sizes of the unknown fragments by plotting the distance they travel and reading over to obtain their sizes. Thus, according to Figure 9.19B, the three *Hin*dIII fragments are approximately 4.60, 3.45, and 1.40 kb, adding up to 9.45 kb, which is close enough to the expected 9.2 kb. We estimate the five *Pst*I fragments to be 3.40, 2.20, 1.60, 1.00, and 0.87 kb, which add up to 9.07 kb, again close enough to 9.2 kb. Therefore, all the restriction fragments seem to be present and accounted for.

Our next step is to identify the internal fragments in each digestion, since the remaining fragments will be the junction fragments. After *Hin*dIII digestion, the internal fragment of 3.30 kb is probably the *Hin*dIII fragment

A

B

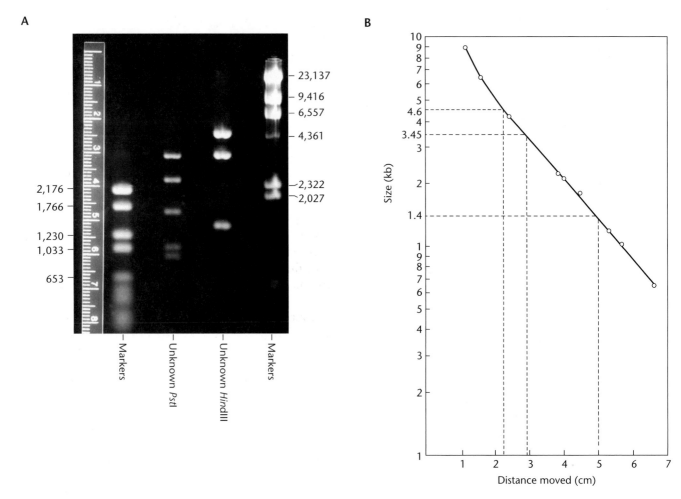

Figure 9.19 (A) Agarose gel electrophoresis of a Tn*5*-mutagenized pAT153 plasmid cut with *Hin*dIII and *Pst*I. The sizes of the marker fragments are also shown. The DNA fragments on the gel were stained with ethidium bromide and photographed under UV illumination. (B) Standard curve of the sizes of the known fragments plotted against the distance moved in the gel shown in panel A. The sizes of the unknown fragments can be estimated from the standard curve derived from the marker fragments by plotting the distances that the unknown fragments moved on the gel (dotted lines). Only the positions of the *Hin*dIII fragments are shown.

that we have estimated to be 3.45 kb. That leaves the 4.60- and 1.40-kb fragments to be the junction fragments. Similarly, the internal *Pst*I fragments that should be 1.08, 0.92, and 2.40 kb are probably the fragments that we have estimated to be 1.00, 0.87, and 2.20 kb, leaving the 3.40- and 1.60-kb fragments to be the *Pst*I junction fragments.

The next step is to examine the junction fragments to determine how the distance between the insertion site for the transposon and one of the restriction sites on the plasmid. Let us start with the *Hin*dIII site on the plasmid. The smaller of the two *Hin*dIII junction fragments is about 1.40 kb, which must be the distance from the

*Hin*dIII site in the original plasmid to the nearest *Hin*dIII site in the transposon. Since 1.15 kb of the junction fragment is taken up by transposon DNA, the actual site of insertion of the transposon is 1.4 − 1.15 = 0.25 kb from the plasmid *Hin*dIII site. However, from the *Hin*dIII data alone, the transposon could be inserted either 0.25 kb clockwise or 0.25 kb counterclockwise of the *Hin*dIII site on the plasmid as the plasmid map is drawn in Figure 9.20. To determine the side of the *Hin*dIII site on which the transposon is inserted, we need to refer to the size of the *Pst*I fragments. From the size of the smallest *Pst*I junction fragment, which we estimated to be 1.60 kb, the transposon must be inserted 1.60 − 0.60 = 1.00

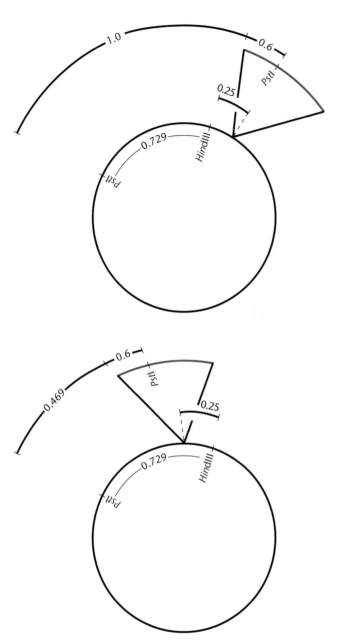

Figure 9.20 Using the size of the smallest junction fragments from Figure 9.19 to determine the site of insertion of the transposon in the plasmid. The size of the smallest *Hin*dIII junction fragment indicates that the transposon must have inserted 0.25 kb from the *Hin*dIII site in the plasmid. The smallest *Pst*I junction fragment indicates that the transposon must be positioned clockwise of the plasmid *Hin*dIII site.

kb from the plasmid *Pst*I site, since 0.60 kb of this junction fragment is taken up with transposon DNA. The only way the transposon could be inserted both approximately 0.25 kb from the *Hin*dIII site and 1.00 kb from the *Pst*I site is to be inserted approximately 0.25 kb clockwise of the *Hin*dIII site, as shown in Figure 9.20.

Transposon Mutagenesis of the Bacterial Chromosome

The same methods used to mutagenize a plasmid with a transposon can also be used to mutagenize the chromosome. A gene with a transposon insertion is much easier to map or clone than a gene with another type of mutation, making this a popular method for mutagenesis of chromosomal genes. Transposons have been engineered to carry a bacterial origin of replication (*oriV*) or an origin of transfer (*oriT*), which facilitate cloning of the region of the transposon insertion or genetic mapping by conjugation, as outlined in chapter 5.

The major limitation of transposon mutagenesis is that transposon insertions will usually inactivate a gene, a lethal event in a haploid bacterium if the gene is essential for growth. Therefore, this method can generally be used only to mutate genes that are nonessential or essential under only some conditions. However, it can still be used to map essential genes by isolating transposon insertion mutations that are not in the gene itself but close to it in the DNA. If the transposon is inserted close enough, it might be used to map or clone the gene. It is also important to remember that insertion of some transposons may increase the expression of genes at the insertion site (see above).

PHYSICAL MAPPING OF CHROMOSOMAL TRANSPOSON INSERTIONS

Physical mapping of transposon insertion mutations in the chromosome is more difficult than mapping of insertions in plasmids, because the chromosome is so large. Usually, eight-hitter restriction nucleases must be used instead of six-hitter enzymes, so that there are larger but fewer fragments. However, the fragments obtained with eight-hitter restriction endonucleases are usually too large to resolve on normal agarose gels, and techniques such as pulsed-field gel electrophoresis (see chapter 1) must be used to separate them.

Even if the fragments can be separated, fragments containing an inserted transposon are often difficult to identify because the few thousand base pairs added by a transposon will usually not make a significant difference in the size of such large fragments. However, if the transposon itself has a restriction site for the eight-hitter enzyme, any fragment containing the transposon will have a new site for the eight-hitter enzyme. Adding the eight-hitter will cut the fragment containing the transposon in two pieces making it easy to identify. Fortuitously, transposon Tn5 has a site for the eight-hitter *Not*I. Wherever transposon Tn5 has inserted in the chromosome, a new *Not*I site is introduced, which can be used in the physical mapping of the transposon insertion. Once the eight-hitter fragment containing the transposon has been identified, the site of the insertion of the transposon in this

BOX 9.2

Conjugative Transposons

The conjugative transposons are transposons combined with transfer functions, such as those of a self-transmissible plasmid (see the text and chapter 5).

Conjugative transposons are very large because they must carry *tra* genes as well as transposition functions. The first conjugative transposon to be discovered was found in *Enterococcus faecailis* and carries tetracycline resistance (see Senghas et al., below, and Figure panel A). Later, other conjugative transposons such as CTnDOT, which also carries tetracycline resistance, were found in *Bacteroides* species (see Cheng et al., below). These *Bacteroides* conjugative transposons also mobilize other smaller elements in the chromosome of *Bacteroides,* much as self-transmissible plasmids mobilize other smaller plasmids (see chapter 5). Interestingly, one of these smaller mobilizable elements, called NBU1, is known to encode an integrase related to the λ integrase and like λ, integrates by recombination between its *att* site and a specific *att* site in the chromosome, rather than by transposition.

In order to transpose from the DNA of one cell to the DNA of another cell, a conjugative transposon must first excise from the DNA of the cell in which it resides, be transferred into the other cell, and then integrate into the DNA of the second cell. This process has been studied extensively in the transposon Tn*916* and is outlined in the Figure panel B. Like phage λ, Tn*916* requires two proteins, Int and Xis, to excise. Excision of the element requires cutting of the DNA next to the ends of the transposon. The integrase first makes a staggered break in the donor host DNA near the ends of the transposon, to leave single-stranded ends 6 nucleotides long. The flanking sequences shown in the Figure are arbitrary because they differ depending upon where the transposon was inserted. Because they are random, these single-stranded ends (called the coupling sequences) are not complementary to each other. Nevertheless, these ends pair to form a circular intermediate of the transposon, including the unpaired region formed by the ends of the transposon which form a heteroduplex coupling sequence. Panel B, part b, shows how the imprecise excision of Tn*916* from the donor chromosome will leave a short deletion in the donor DNA. The circular intermediate cannot replicate but can be transferred into another cell by a process which is much like plasmid transfer. The transposon has its own *oriT* sequence and *tra* genes (see Figure panel A). In fact, this *oriT* sequence is related to those of the self-transmissible plasmids RP4 and F. The *tra* genes are *orf23-orf13,* and the *oriT* sequence lies between *orf20* and *orf21* (see the Tn*916* map panel A, below) To initiate the transfer, a single-stranded break is made in the *oriT* sequence of the excised circular transposon and a single strand of the transposon is transferred to the recipient cell by the transposon-encoded Tra functions. Once in the recipient cell, the ends rejoin and a complementary strand is synthesized to make another double-stranded circular transposon. The Int protein of the transposon then integrates the transposon into the DNA of the recipient cell by making a blunt-end break in the target DNA and at the ends of the transposon and joining the ends of the two DNAs. Because a blunt-end break is made in the target DNA, no target site duplication accompanies the integration.

Conjugative transposons, including Tn*916,* can transfer only after they have excised, not while they are still integrated in the chromosome. The transposon uses a clever mechanism to ensure that it will be transferred only after it is excised. The *tra* genes are arranged so that they can be transcribed only after the transposon has excised and has formed a circle. To accomplish this, the *tra* genes are transcribed from the promoter called p_{orf7} (see Figure). Transcription from this promoter will run off the end when the transposon is integrated but will continue around the circle into the *tra* genes (*orf23-orf13*) when the transposon has excised and formed a circle. The Int protein may also be bound to the *oriT* sequence while the transposon is inserted in the chromosome, blocking access to the *oriT* sequence by the Tra functions and precluding transfer (see Hinerfeld and Churchward, below). The advantage to the transposon of thus regulating its transfer seems obvious. If they could transfer themselves while still integrated in the chromosome, they would mobilize the chromosome, much like self-transmissible plasmids mobilize the chromosome in Hfr strains (see chapter 5). However, then, as occurs during *Hfr* transfer, only the part of the transposon on one side of the *oriT* site would enter the recipient, and the transposon would decapitate itself.

The process of transfer of the *Bacteroides* conjugative transposons such as CTnDOT may be similar, but there are important differences (see Cheng et al., below). Rather than integrate randomly in the recipient chromosome, the transposon has certain preferred integration sites. This may be due to a short complementary sequence of 10 bp in the transposon and the integration site. The transfer of the CTnDOT transposon is also induced by tetracycline in the medium, much like opines induce transfer of the Ti plasmid, unlike transfer by the Tn*916* transposon, which is not induced by tetracycline.

The conjugative transposons related to Tn*916* are remarkably broad in their host range and have been found to transfer into and transpose in a wide variety of bacteria. They usually carry a gene, *tetM,* that confers resistance to

(continued)

BOX 9.2 (continued)

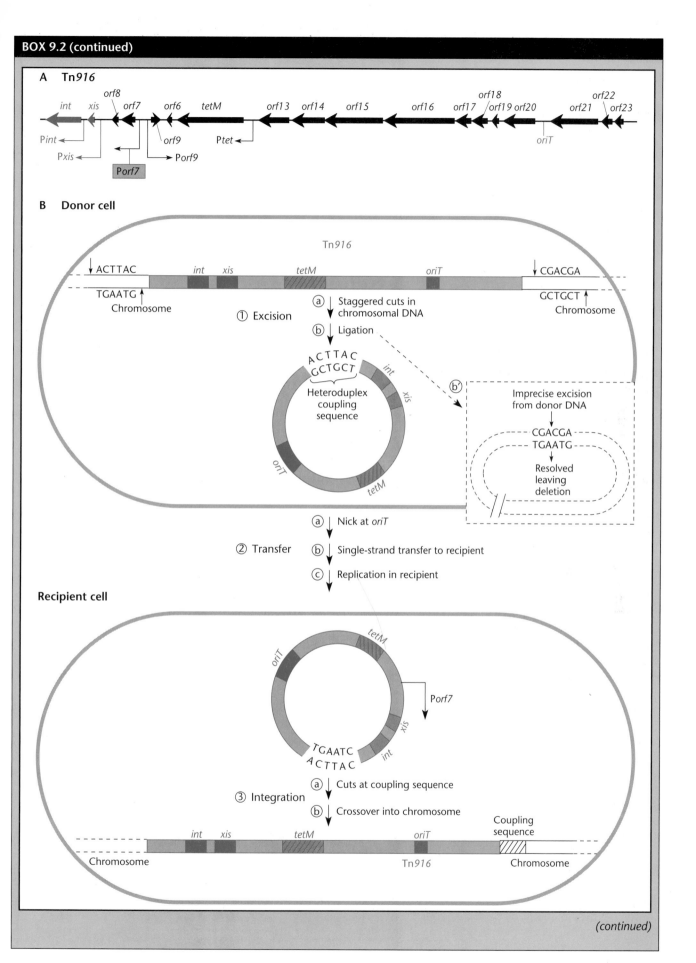

A Tn916

B Donor cell

Recipient cell

(continued)

BOX 9.2 (continued)

Conjugative Transposons

tetracycline. This gene has been found widely in gram-positive as well as gram-negative bacteria, and it is reasonable to suppose that the conjugative transposons have played a role in its dispersal.

References
Cheng, Q., B. J. Paszkiet, N. B. Shoemaker, J. F. Gardner, and A. A. Salyers. 2000. Integration and excisions of a *Bacteroides* conjugative transposon, CTnDOT. *J. Bacteriol.* **182:**4035–4043.

Hinerfeld, D., and G. Churchward. 2001. Specific binding of integrase to the origin of transfer (*oriT*) of the conjugative transposon Tn*916*. *J. Bacteriol.* **183:**2947–2951.

Marra, D., and J. R. Scott. 1999. Regulation of excision of the conjugative transposon Tn*916*. *Mol. Microbiol.* **31:**609–621.

Senghas, E., J. M. Jones, M. Yamamoto, C. Gawron-Burke, and D. B. Clewell. 1988. Genetic organization of the bacterial conjugative transposon, Tn*916*. *J. Bacteriol.* **170:**245–249.

somewhat smaller piece can be further localized by using six-hitters and more standard techniques.

Transposon Mutagenesis of All Bacteria

One of the most useful features of transposon mutagenesis is that it can be applied to many types of bacteria, even ones which have not been extensively characterized. Methods have been developed to perform transposon mutagenesis of almost all gram-negative bacteria as well as many gram-positive bacteria. All that is needed is a way of introducing a transposon into the bacterium, provided that the transposon can hop in the bacterium. The transposon should also carry a gene that can be selected in the bacterium. Some of these methods were mentioned earlier and are outlined in more detail below.

PROMISCUOUS PLASMIDS

One common method of mutagenizing gram-negative bacteria uses the transfer system of self-transmissible plasmids such as the IncP plasmids, which are very promiscuous for transfer and will transfer themselves or mobilize other plasmids into essentially any gram negative bacterium (see Figure 9.17A and Simon et al., Suggested Reading). These plasmids are used to mobilize a smaller plasmid containing a compatible *mob* site and the transposon Tn*5*, which will transpose in most gram-negative bacteria and contains a gene for kanamycin resistance that can be selected in most gram-negative bacteria. This smaller plasmid also has the replication origin of the ColE1 plasmid, which is narrow host range and capable only of replicating in *E. coli* and a few close relatives; this makes it a suicide vector in most gram-negative bacteria. The bacterium to be mutagenized is mixed with *E. coli* strains carrying these two plasmids. The larger IncP plasmid will mobilize the smaller Tn*5*-containing plasmid into the cells, and the transposon will hop. Cells in which the transposon has hopped can be selected on plates containing kanamycin and irgasan, an antibiotic to which *E. coli* is sensitive but most other gram-negative bacteria are resistant.

CLONING GENES MUTATED WITH A TRANSPOSON INSERTION

Genes that have been mutated by transposon insertion are usually relatively easy to clone by cloning the easily identified antibiotic resistance gene in the transposon. Since some antibiotic genes, for example the kanamycin resistance gene in Tn*5*, are expressed in many types of bacteria, this method can even be used to clone genes from one bacterium in a cloning vector from another. This is particularly desirable because most cloning vectors and recombinant DNA techniques have been designed for *E. coli*. To clone a gene mutated by a transposon from a bacterium distantly related to *E. coli*, the DNA from the mutagenized strain is cut with a restriction endonuclease that does not cut in the transposon and ligated into an *E. coli* plasmid cloning vector cut with the same or a compatible enzyme. The ligation mixture is then transformed into *E. coli*, and the transformed cells are spread on a plate containing the antibiotic to which the transposon confers resistance. Only cells containing the mutated, cloned gene will multiply to form a colony.

Transposons can be engineered to make cloning of transposon insertions even more efficient by introducing an origin of replication into the transposon so that the DNA containing the transposon need only cyclize to replicate autonomously in *E. coli*. The use of such a transposon for cloning transposon insertions is illustrated in Figure 9.21. In the example, the transposon carrying a plasmid origin of replication has inserted into the gene to be cloned. The chromosomal DNA is isolated from the mutant cells and cut with a restriction endonuclease that does not cut in the transposon, *Eco*RI in the example. When the cut DNA is religated, the fragment containing the transposon will become a circular replicon with the plasmid origin of replication. If the ligation mixture is

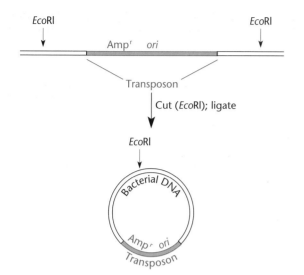

Figure 9.21 Cloning genes mutated by insertion of a transposon. A transposon used for mutagenesis of a chromosome contains a plasmid origin of replication (*ori*), and the chromosome is cut with the restriction endonuclease *Eco*RI and religated. If the ligation mix is used to transform *E. coli*, the resulting plasmid in the Amp^r transformants will contain the sequences that flanked the transposon insertion in the chromosome. Chromosomal sequences are shown in black, and transposon sequences are shown in gold.

used to transform *E. coli* and the ampicillin resistance gene on the transposon is selected, the chromosomal DNA surrounding the transposon will have been cloned. Since the restriction endonuclease cuts outside the transposon, any clones of the transposon cut from the chromosome will also include sequences from the gene of interest into which the transposon had inserted.

Once the gene containing the transposon insertion has been cloned, it can be used in several ways. We may want to directly sequence the gene with primers complementary to the ends of the transposon. Alternatively, we may need a clone of the wild-type gene without the transposon insertion; in this case, we could use the clone with the transposon insertion as a probe to identify the wild-type gene by screening a library of wild-type DNA in a plate hybridization (see chapter 1). This method allows the cloning of genes about which nothing is known except the phenotype of mutations that inactivate the gene, and it can be easily adapted to clone genes from any bacterium in which the transposon can hop to create the original chromosomal mutation. Once mutants are obtained with the transposon inserted in the gene of interest, the remaining manipulations are performed in *E. coli*. This can be of particular advantage if the bacterium being studied is difficult to grow or maintain in a laboratory situation.

Using Transposon Mutagenesis To Make Random Gene Fusions

The ability of some transposons to hop randomly into DNA has made them very useful for making random gene fusions to reporter genes (see chapter 2). Fusing a gene to a reporter gene can make regulation of the gene much easier to study or can be used to identify genes subject to a certain type of regulation or those localized to certain cellular compartments. Once a gene subject to a certain type of regulation has been identified in this way, it can be easily cloned and studied using methods such as those described above.

As discussed in chapter 2, gene fusions can be either transcriptional or translational. In a transcriptional fusion, one gene is fused to the promoter for another gene so that the two genes are transcribed into mRNA together. In a translational fusion, the ORFs for the two proteins are fused to each other in the same reading frame so that translation, initiated at the translational initiation region (TIR) for one protein, will continue into the ORF for the other protein, making a fusion protein.

Transposons engineered to make random gene fusions include Tn*3HoHo1*, Tn*5lac*, and Tn*5lux*. These transposons carry a reporter gene at one end that either has its own TIR, if the transposon is to make transcriptional fusions, or lacks a TIR, if the transposon is to make translational fusions. When the transposon hops into the chromosome, it will fuse its reporter gene to whatever gene it hops into, provided that it has hopped into the gene in the right orientation.

Mu*d*(Amp^r, *lac*)

The prototype transposon for making random gene fusions is the Mu*d*(Amp^r, *lac*) transposon (see Casadaban and Cohen, Suggested Reading). This transposon has been largely supplanted by more elaborate constructions with specialized uses, but we can use it to illustrate the basic principles involved. The Mu*d*(Amp^r, *lac*) transposon is derived from phage Mu, which will transpose with almost no target specificity (see Box 9.1). Phage Mu also has a quite broad host range and will productively infect and transpose in many gram-negative bacteria, including *Erwinia carotovora*, *Citrobacter freundii*, and *E. coli*.

Figure 9.22 shows the essential features of random gene fusions created by the Mu*d*(Amp^r, *lac*) transposon. Most of the phage Mu DNA is removed, except the ends and the transposase, which is the product of Mu genes A and B. The transposase and the ends of phage Mu are sufficient for transposition of the phage DNA into the chromosome after infection. Close to one of its ends, the Mu*d*(Amp^r, *lac*) transposon carries the *lacZ* gene of *E. coli* as its reporter gene. The *lacZ* gene has no promoter of its own and so will not normally be transcribed. However, if the trans-

Figure 9.22 Structure of random gene fusions created by the original Mud(Amp^r, *lac*) transposon. The essential transposon elements are the ends of Mu, the reporter gene *lacZ*, and a gene for ampicillin resistance (Amp^r). The *lacZ* gene lacks its own promoter and so will be transcribed only from a promoter outside the transposon (*p*). The sequences outside the transposon are shown as dashed lines, and the Mu sequences of the transposon are shown in gold.

poson has hopped into a target DNA in such a way that the *lacZ* gene is positioned in the correct orientation downstream of a promoter, the *lacZ* gene will be turned on and will express β-galactosidase, which is easily detected by colorimetric assays with dyes such as X-Gal or ONPG. The Mud(Amp^r, *lac*) transposon also has an ampicillin resistance gene that can be used to select only the cells that have the transposon inserted somewhere in their DNA.

The procedure for using the Mud(Amp^r, *lac*) transposon to make random gene fusions is illustrated in Figure 9.23. The first step is to prepare a lysate by inducing a Mu prophage in cells that also contain the Mud(Amp^r, *lac*) transposon. The cell must also contain the normal phage Mu (a helper phage) to furnish all the proteins needed to make a phage, including the head and tail proteins, which the Mud(Amp^r, *lac*) transposon cannot make for itself. Since Mu cannot distinguish its own DNA from the Mud(Amp^r, *lac*) transposon DNA, which has the same ends, the replicating phage Mu will sometimes package the Mud(Amp^r, *lac*) transposon instead of its own DNA. Therefore, when the cells lyse, some of the phage particles that are released will contain the Mud(Amp^r, *lac*) transposon. This phage lysate is then used to infect cells at a low multiplicity of infection (MOI). Any ampicillin-resistant transductants, selected on ampicillin plates, will have the Mud(Amp^r, *lac*) transposon inserted somewhere in their chromosome. If the Mud(Amp^r, *lac*) transposon has inserted downstream of an active promoter, the lacZ gene on the transposon will be transcribed from the promoter and the cells will make β-galactosidase.

The fact that the Mud(Amp^r, *lac*) transposon expresses *lacZ* only if the gene into which it hops is transcribed has been used to identify genes that are turned on only under certain conditions. A classical example is

Figure 9.23 Isolating random gene fusions with Mud(Amp^r, *lac*). The Mu prophage is induced in a cell containing the Mud(Amp^r, *lac*) transposon (in gold), which will be packaged into some of the phage. All Amp^r transductants have Mud(Amp^r, *lac*) somewhere in their chromosome, and in cells that form blue colonies on X-Gal plates, the Mud(Amp^r, *lac*) transposon has hopped downstream of a promoter in such a way that the *lacZ* gene is transcribed.

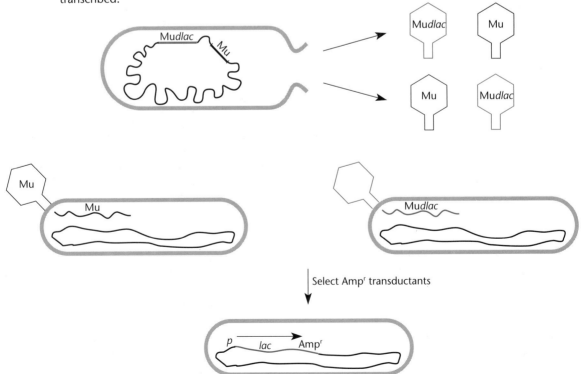

BOX 9.3

Transposon Mutagenesis In Vitro

While in vivo transposon mutagenesis is a very useful technology, it does have some limitations. One of the limitations is that it is necessary to introduce the transposon on a suicide vector, which may give some residual false-positive results for transposon insertion mutants if the suicide vector is capable of limited replication. Another limitation is that it is not very efficient and requires powerful positive selection techniques to isolate the mutants. Another limitation comes if a specific plasmid or other smaller DNA sequence is to be mutated. There is no target specificity to the insertion mutants, and so most of the time the transposon hops into the chromosome; those few with transposon insertions in the smaller target DNA must be found among the myriad of mutants in the chromosome. Yet another limitation is that the target DNA to be mutated must be a replicon in the cell in which the mutagenesis is performed. Finally, useful transposons have not been found for most bacteria.

In vitro transposon mutagenesis avoids many of these limitations. Such techniques have been developed for derivatives of transposon Tn5, Tn552, mariner, and others. This technology is made possible by the fact that the transposase enzyme by itself performs most of the reactions of the "cut-and-paste" transposition reaction. It cuts the DNA at the outside ends of the inverted repeats on the transposon and the target DNA and joins the ends to each other. Therefore, if the target DNA is mixed with a donor DNA containing the transposon and the purified transposase is added, the transposon will insert into the target DNA. The efficiency of transposition can be increased by using mutant transposases with enhanced activity. The transposon can even be engineered to lack a transposase gene so that it will not hop in subsequent generations or cause genetic instability such as deletions once it is in the chromosome. In fact, only the sequences at the ends of the inverted repeats are needed; in Tn5, these are only 19 bp long. Once it is mutagenized, the target DNA can be introduced into cells by whatever means are available and those with transposon insertions can be selected as for in vivo transposon mutagenesis. The target DNA can be either a replicon, such as a plasmid which repli-

cates in the recipient cell, or random linear pieces of the chromosomal DNA of the recipient if it is being introduced into cells which can be transformed with linear DNA (see chapter 6). The linear pieces will recombine with the chromosome and replace the chromosomal sequence with the sequence mutated with the transposon. This offers a way of doing random chromosomal transposon mutagenesis of bacteria for which no transposon mutagenesis system is available (see, for example, Gering et al., below).

Another variation of this method for doing transposon mutagenesis, which can be applied to mutagenize the DNA of almost any bacterium and even eukaryotic cells, is to use "transpososomes" (see Goryshin et al., below). A transpososome is a transposon to which the transposase protein is already attached so that it does not have to be made in the cell. This latter feature is important because the transposase gene on the transposon might not be expressed in a distantly related bacterium and certainly not in a eukaryotic cell. As in other methods of transposon mutagenesis, the transposon should carry a selectable gene that is expressed in the cell to be mutagenized. Transpososomes based on Tn5 are made by running the transposition reaction in vitro in the absence of magnesium ions. Under these conditions, the donor DNA is cut but the transposon remains attached to the cut ends. To use a transpososome for transposon mutagenesis, the DNA-protein complex is electroporated into the cells. The transposase attached to the transposon will then catalyze the transposition of the transposon into the chromosome. The transposase enzyme introduced with the transposon is soon degraded, preventing further transposition.

References

Gering, A. M., J. N. Nodwell, S. M. Beverley, and R. Losick. 2000. Genomewide insertional mutagenesis in *Streptomyces coelicolor* reveals additonal genes involved in morphological differentiation. *Proc. Natl. Acad. Sci. USA* **97**:9642–9647.

Goryshin, I. Y., J. Jendrisak, L. M. Hoffman, R. Mais, and W. S. Reznikoff. 2000. Insertional transposon mutagenesis by electroporation of released Tn5 transposition complexes. *Nature (Biotechnology)* **18**:97–100.

the identification of the *din* (damage-inducible) genes of *E. coli*, which are turned on only after DNA damage (see Kenyon and Walker, Suggested Reading). To use Mu*d*(Amp^r, *lac*) to identify *din* genes, these investigators first used the method described above to isolate ampicillin-resistant (Amp^r) transductants with Mu*d*(Amp^r, *lac*) inserted somewhere in their genome. They then needed to identify transductants with the transposon in-

serted in the correct orientation into a gene that is induced when the DNA is damaged. To accomplish this, they replicated the plates containing the Amp^r transductants onto two other sets of plates, one containing only X-Gal and the other containing X-Gal plus a DNA-damaging agent such as mitomycin. The colonies that were blue on the set of plates containing the DNA-damaging agent but not on the plates with X-Gal alone were known to have the transposon inserted in a *din* gene in the right orientation. By mapping the ampicillin resistance gene on the transposon, the investigators could map a number of *E. coli din* genes that are turned on after DNA damage. We discuss the *din* genes and their induction in more detail in chapter 11.

In Vivo Cloning

Transposons can also be used for in vivo cloning. Like other cloning procedures, in vivo cloning requires a library of recombinant DNA. However, these libraries do not need to be made in vitro with restriction endonucleases and DNA ligase as outlined in chapter 1. Instead, in vivo cloning relies on genetic methods and lets bacteriophages and transposons do most of the work of making the library.

Most in vivo cloning procedures are also based on phage Mu and the fact that Mu replicates by replicative transposition without resolving the cointegrates. After Mu replicates, copies of the phage genome are inserted all over the bacterial chromosome (see Box 9.1). During the normal infection cycle, these phage genomes are then packaged into phage heads by using specific *pac* sites at the ends of the phage DNA.

CONSTRUCTING MINI-MU ELEMENTS
Because the phage Mu packaging system recognizes the ends of phage DNA, it can, in principle, recognize the outside ends of two phage genomes lying close to each other in the chromosome and package both phage Mu DNAs plus any chromosomal DNA between them. This does not normally happen, because the phage Mu head is large enough to hold only a single Mu genome. However, if a shortened version of Mu, called a mini-Mu element, is used, there is extra room in the phage head for chromosomal DNA (see Groisman and Casadaban, Suggested Reading). These mini-Mu elements lack all of the functions of Mu except the ends required for transposition and packaging. In addition, to make them useful for in vivo cloning, the mini-Mu elements have a plasmid origin of replication and an antibiotic resistance gene cloned into them, the significance of which will become apparent later.

Figure 9.24 illustrates how mini-Mu elements may be used for in vivo cloning. A mini-Mu element which has been introduced into a cell on a plasmid is induced to replicate (transpose) by infection of the cells with a helper phage, wild-type Mu, which contributes the transposase functions need for the mini-Mu to transpose. Both the wild-type Mu and the mini-Mu elements will then transpose around the chromosome, making many copies of themselves. Later in the infection, while the normal phage DNAs are being cut out and packaged, some pairs of mini-Mu elements, which happen to be

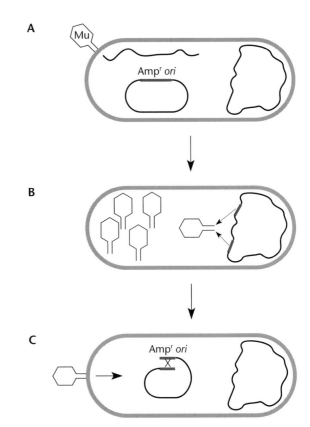

Figure 9.24 In vivo cloning with mini-Mu. In the first cell, a mini-Mu (in gold) on a plasmid is induced to replicate by infection with a helper phage Mu, allowing the mini-Mu to hop randomly around the chromosome as it replicates. In the second cell, pairs of mini-Mu elements, along with chromosomal DNA lying between them, are packaged into some phage Mu heads. The sites of cleavage for packaging the two mini-Mu elements are slightly outside the ends of the mini-Mu elements (arrows). Upon infection of the third cell, the two mini-Mu elements can recombine to form a circular DNA that can replicate because of the origin of replication on the mini-Mu elements. The ampicillin resistance gene also carried by the mini-Mu allows selection of transductants.

separated by a phage Mu length of DNA, will also be cut out and packaged along with the chromosomal DNA between them. When the cells lyse, the released phage will be a mixture of phage containing normal Mu DNA and others containing chromosomal DNA packaged between copies of the mini-Mu element.

The plasmid origin of replication and the selectable gene on the mini-Mu now come into play. When new cells are infected with the phage lysate at a low MOI, some of the cells will be infected by phage containing the mini-Mu elements and chromosomal DNA. The DNA injected into these cells can then cyclize by recombination between the identical mini-Mu sequences at the ends to form a circular DNA that can replicate from the plasmid origin of replication. Because the mini-Mu element also contains the gene for ampicillin resistance, transductants containing autonomously replicating plasmids can be selected by plating on ampicillin-containing plates. If enough phage in the lysate carried chromosomal DNA, every region of the bacterial DNA will be represented in the collection of plasmids, constituting a library of the bacterial DNA. Once a library has been obtained, a clone containing the desired gene can be found in the library through complementation or hybridization techniques, such as those discussed in chapters 1 and 3. This general method has been used in a variety of gram-negative bacteria, including *E. coli*, *Klebsiella* spp., and *Erwinia* strains (for some examples, see Van Gijsegem and Toussaint, and Chen and Maloy, Suggested Reading).

Site-Specific Recombination

Another type of nonhomologous recombination, **site-specific recombination,** occurs only between specific sequences or sites on DNA. It is promoted by enzymes called **site-specific recombinases,** which recognize two specific sites in DNA and promote recombination between them. Even though the two sites generally have short sequences in common, the regions of homology are usually too short for normal homologous recombination to occur efficiently. Therefore, efficient recombination between the two sites requires the presence of a specific recombinase enzyme. We have already mentioned some site-specific recombination systems in connection with the resolution of chromosome and ColE1 plasmid dimers by the XerCD site-specific recombinase. In this section, we discuss some other examples of site-specific recombination in bacteria and phages.

Developmentally Regulated Excision of Intervening DNA

Genes sometimes contain sequences that are not present in the final RNA or protein product of the gene. The extra or intervening sequences in genes are often removed from the RNA or the protein product by splicing after the RNA or protein is synthesized (see Box 2.2). However, sometimes the intervening sequences are cut out of the DNA iself, before the gene is expressed. One function of site-specific recombinases is to remove these intervening DNA sequences from genes. Note that this type of gene rearrangement is possible only in cell lines that do not need to reproduce themselves, that is, those that are terminally differentiated. Otherwise, the intervening DNA sequences would be lost to subsequent generations.

The classic example of DNA rearrangements during terminal differentiation is in the vertebrate immune cells that make antibodies. In this case, site-specific recombination during the differentiation of the immune cells re-

moves different lengths of sequences from the few germ line genes that encode antibodies, thereby creating hundreds of thousands of new genes encoding antibodies of different specificities. The antibody-encoding genes in the germ line cells (eggs and sperm) remain intact, however, so that no DNA sequences are lost to future generations and the genes can undergo similar rearrangements in subsequent generations.

Most bacteria multiply by cell division, and none of the cells undergo terminal differentiation. Therefore, irreversible DNA rearrangements are generally not possible in bacteria. Nevertheless, bacteria do manifest a few examples of terminal differentiation.

Some types of bacteria sporulate by forming endospores, a process during which the cell divides into a mother cell and a spore (see chapter 14). Only the spore need contain the full complement of the bacterial DNA since the mother cell is required only to make the spore and will eventually lyse. Consequently, the mother cell terminally differentiates and therefore can undergo irreversible DNA rearrangements needed for the sporulation process. In *Bacillus subtilis* sporulation, an intervening sequence of 42 kbp is cut out of the gene for the sigma factor σ^K, which is required for transcription of some genes in the mother cell. Unless this intervening sequence is removed, σ^K will not be synthesized, some of the gene products required for sporulation will not be expressed, and sporulation will be blocked (see Kunkel et al., Suggested Reading).

Terminal differentiation also occurs in the formation of highly specialized cells responsible for fixing atmospheric nitrogen in some types of filamentous cyanobacteria (see Golden et al., Suggested Reading). These cells, called heterocysts, appear periodically in the filaments when the bacteria are growing under conditions of nitrogen starvation. The heterocysts never divide to form new cells and disappear when the cells are provided with a good source of nitrogen, so that they do not need a full complement of DNA and can undergo irreversible DNA rearrangements. As occurs in *B. subtilis* sporulation, intervening sequences are cut out of some genes while the cells differentiate. In the cyanobacteria, these sequences are cut out of at least two genes. One is the *nifD* gene, which encodes a product required for nitrogen fixation in the heterocysts after an intervening sequence of 11 kbp is cut out.

The process of removing the intervening sequences seems to be similar in both developmental systems. Recombination between directly oriented sites bracketing the intervening sequence results in its excision. In both cases, the site-specfic recombinase required to promote recombination between the bracketing sites is encoded within the intervening sequence itself and is expressed only during differentiation. Therefore, the recombinase will be expressed only in the cells undergoing terminal differentiation, and the intervening sequence will not be lost from normally replicating cells.

The function of these intervening sequences is unclear. Even though they are removed only during development, their removal seems to play no important regulatory role in either sporulation or heterocyst differentiation. Permanently removing the intervening sequences from the chromosome, so that the σ^K and *nifD* genes are not interrupted, does not adversely affect the ability of the bacteria to sporulate or to develop heterocysts that can fix nitrogen. One possibility is that these intervening sequences are parasitic DNAs like phages or transposons. If they integrate themselves into an essential gene, the cell cannot easily delete them without killing itself. As with transposons, unless the deletion event precisely removes the intervening sequence, the gene will be inactive, disrupting an important cellular function. By excising themselves during differentiation, however, the parasitic DNA elements allow sporulation or nitrogen fixation by their hosts and so have no deleterious effects—the mark of a good parasite.

Integrases

Integrases also recognize two sequences in DNA and promote recombination between them; therefore, they are no different in principle from the site-specific recombinases that resolve cointegrates or remove intervening sequences. However, rather than remove a DNA sequence by promoting recombination between two directly repeated sequences on the *same* DNA, integrases act to integrate one DNA into another by promoting recombination between two sites on *different* DNAs.

PHAGE INTEGRASES
The best-known integrases are the Int enzymes, which are responsible for the integration of circular phage DNA into the DNA of the host to form a prophage (see chapter 8). Briefly, the phage integrase specifically recognizes the *attP* site in the phage DNA and the *attB* site on the bacterial chromosome and promotes recombination between them. Usually, phage integrases are extremely specific. Only the *attP* and *attB* sites will be recognized, so that the phage will only integrate at one or at most a few places in the bacterial chromosome. In a reversal of this reaction performed by the combination of the integrase and excisase, the hybrid *attP-attB* sites flanking the prophage recombine to excise the prophage.

Because of their specificity, phage integrases have a number of potential uses in molecular genetics. For example, the reaction performed by the phage γ integrase and excisase has been capitalized on in cloning technologies (see, for example, Gateway Technology in the Invit-

rogen catalogue). In these applications, a PCR fragment containing the gene of interest is cloned into a plasmid vector called the entry vector so that the clone is flanked by one of the hybrid *attB-attP* sites which flank integrated prophage λ DNA in the chromosome. If this is mixed with a destination vector containing the other *attP-attB* hybrid site and λ integrase and excisase are added, site-specific recombination between the sites will yield the destination vector with the cloned gene inserted flanked by *attB* sites. While this technology does not remove the requirement for the initial cloning, which can be laborious, once a clone is made in the entry vector, it can be transferred quickly into a number of different destination vectors.

INTEGRASES OF TRANSPOSON INTEGRONS
Integrases are also important in the evolution of some types of transposons. The first indication of this importance came from comparisons of members of some families of transposons, such as the Tn*21* family (see Figure 9.25). The three transposons in this figure are similar, except that they have different antibiotic resistance genes inserted at exactly the same site, designated *P* in the figure. The transposon In*0* has no antibiotic resistance gene inserted at the *P* site, whereas In*3* has a gene for trimethoprim resistance inserted at this site. In*2*, which is actually a transposon inserted within the Tn*21* transposon, has a gene for streptomycin and spectinomycin resistance inserted at this site. Yet other members of this family carry different genes at this site, including those for resistance to kanamycin, gentamicin, trimethoprim, chloramphenicol, and oxacillin, as well as one for mercury resistance. Presumably, In*0* is a direct descendant of a primordial transposon that did not have an antibiotic resistance gene at the *P* site. The other transposons then evolved from this primordial transposon as different antibiotic resistance genes were inserted at the *P* site.

But how did those resistance genes come to occupy the same site on the transposon? The answer came from the discovery of a gene, homologous to the phage integrases, lying next to the *P* site on the transposons. Furthermore, the integrase encoded by this gene will promote recombination between two DNAs containing the 59-bp sequence adjacent to the integration site. Presumably, site-specific recombination between the *P* site in the transposon and similar sites in other DNAs carrying resistance genes has resulted in the incorporation of the resistance genes into the site. The advantages of this system are obvious. By carrying different resistance genes, the transposon allows the cell, and thereby itself, to survive environments containing various toxic chemicals. However, where the resistance genes originally came from remains a mystery.

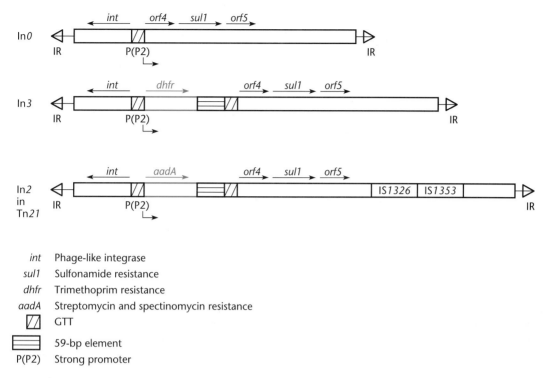

int   Phage-like integrase
sul1   Sulfonamide resistance
dhfr   Trimethoprim resistance
aadA   Streptomycin and spectinomycin resistance
▨   GTT
☰   59-bp element
P(P2)   Strong promoter

Figure 9.25 Integrons of the Tn*21* family. The transposon In*0* contains only the integrase gene and a gene for sulfonamide resistance. IR denotes a 25-bp inverted repeated sequence. In In*3*, the 3-bp GTT sequence has been the site of an integrase-mediated insertion of the gene for resistance to tetrahydrofolate reductase inhibitors, *dhfr*. In the In*2* element found inserted in Tn*21*, the GTT has been the site of insertion of the *aadA* gene, conferring resistance to streptomycin and spectinomycin. The expression of the inserted resistance genes is driven by the integron promoter P(P2). Also shown are 59-bp elements located at the 3′ ends of the inserted resistance-encoding sequences. These 59-bp sequences are thought to be involved in the Int-mediated recombination reactions. In addition, some IS elements have hopped into the transposon, as shown.

PATHOGENICITY ISLANDS

Integrases and nonhomologous recombination also play a role in the integration of at least some types of genetic islands into the chromosome. Like plasmids and prophages, genetic islands often carry clusters of genes which allow the bacterium to occupy specific environmental niches. They can be hundreds of thousands of deoxynucleotides long and carry hundreds of genes. The **pathogenicity island** (PAI) is a type of genetic island that carries genes required for pathogenicity of the bacterium. PAIs carry genes for resistance to multiple antibiotics in *Shigella flexneri*, for alpha-hemolysin and fimbriae in pathogenic *E. coli*, for scavenging and storing iron in *Yersinia*, and for a type III secretion system in *Helicobacter pylori*, to give just some examples. We have already mentioned the PAI SaP11, which carries the gene for the toxin involved in toxic shock syndrome in *Staphlococcus aureus* (see Box 8.3). This PAI has its own integrase and so can integrate itself into the chromosome

of new strains of *S. aureus*, making them pathogenic. It can also move from strain to strain of *S. aureus* by behaving like a satellite virus of phage 80α and allowing itself to be transduced by the phage. If a cell carrying the pathogenicity island is infected by phage 80α, the island will excise and replicate with the help of phage replication proteins. It will then be packaged into phage heads, from where it can be injected into and integrate into the chromosome of another bacterium.

However, the demonstration of movement by SaP11 makes it the exception. Most PAIs have not been demonstrated to move from one bacterium to another or even to integrate into a DNA which lacks them. The only evidence that they have moved recently is that they are not found in all the strains of a bacterium and that the base composition (G+C content) of their DNA and their codon usage are often different from those of the chromosomal DNA as a whole. These characteristics are often taken as evidence of recent horizontal transfer of genes from one strain of bacterium to another. PAIs also often carry vestigial integrase genes that are broken up by nonsense and other types of mutations, suggesting that they once encoded functional integrases, which are no longer functional. The PAI elements are often flanked by short repeated sequences, either direct or inverted, which may be the sites at which the integrase acted to integrate the PAI. Apparently, the PAIs did move into the strain some time ago, perhaps on a promiscuous plasmid, and integrated into the chromosome, but their DNA has mutated over time so that they are no longer capable of integrating. Interestingly, many PAIs are integrated into tRNA genes in the chromosome. Part of the tRNA gene is duplicated on the PAI, so that the gene will still be functional (see Box 8.1). Chromosomal tRNA genes may be preferred sites of integration, because they have almost the same sequence in different species of bacteria. By using a highly conserved tRNA gene as its bacterial attachment (*attB*) site, the PAI can integrate into the chromosome of any bacterial strain in which it finds itself.

Even though most PAIs cannot integrate, some researchers decided to make one that could do so (see Rakin et al., Suggested Reading). More accurately, they constructed a plasmid which could integrate using the integrase of a PAI. They accomplished this by using parts of a PAI from *Yersinia*. As background, different strains of *Yersinia* differ greatly in their pathogenicity depending upon the DNA elements they carry. The most pathogenic species of *Yersinia* are *Y. pestis*, which causes bubonic plague, and *Y. enterocolitica* and *Y. pseudotuberculosis*, which cause mild intestinal upsets. These strains carry a PAI called HPI (high-pathogenicity island). This PAI, of 40 kb, is integrated into one of the asparagine-tRNA genes and encodes enzymes to make small molecules called siderophores, which help scavenge for iron, which is in limited supply in the eukaryotic host. Even though this PAI from different strains has not be demonstrated to move, perhaps the functional parts of each PAI could be assembled into a plasmid cloning vector of *E. coli*, which would then be able to integrate into an asparagine tRNA gene of *E. coli*. As mentioned, the tRNA genes are very similar in different species, and *E. coli* has four asparagine-tRNA genes, all of which are very similar in the region where the PAI is integrated in *Yersinia*.

In order to integrate, the PAI needs both a functional integrase and functional *attP* and *attB* sites, by analogy to phage λ. First, the investigators reasoned that the integrase itself from the PAI in *Y. pestis* might be functional, and so they cloned this integrase gene into a plasmid cloning vector so that it would be transcribed at a high level from a phage T7 promoter (see the discussion of T7-based expression vectors in chapter 7). However, they needed to reconstruct a site that would be recognized by the integrase. The PAI itself would not have such a site. The situation is analogous to what happens when phage λ integrates (see chapter 8). The *attP* site on the phage recombines with the *attB* site in the chromosome, resulting in *attP-attB* hybrid sites flanking the integrated prophage, which are no longer recognized by the integrase. Presumably such hybrid sites exist at the ends of the PAI and are also not recognized. To reconstruct the original *attP* site from the hybrid sites, they needed to know where in the hybrid sites the *attP* sequence ends and the *attB* sequence in the asparagine-tRNA gene begins. To determine this, they compared the sequences at the ends of the PAI to the sequence of asparagine-tRNA genes in strains that do not have the PAI integrated to determine which sequences are due to the *attB* sequence in the tRNA gene. They then constructed the original *attP* sequence using PCR and cloned this into the plasmid which already contained the integrase. This plasmid then integrated into an asparagine-tRNA gene of *E. coli* when the integrase gene on the plasmid was induced, showing that all the features needed to integrate the PAI were present.

Resolvases

The resolvases of transposons discussed above are another type of site-specific recombinase. These enzymes promote the resolution of cointegrates by recognizing the *res* sequences that occur in one copy in the transposon but occur in two copies in direct orientation in cointegrates. Recombination between these two *res* sequences will excise the DNA lying between them, resolving the cointegrate.

DNA Invertases

The DNA invertases promote site-specific recombination between two closely linked sites on DNA. The main difference between the reactions promoted by DNA invertases and those catalyzed by resolvases is that two sites recognized by invertases are in reverse orientation with respect to each other whereas the sites recognized by resolvases are in direct orientation. As discussed in chapter 3 in the section on types of mutations, recombination between two sequences that are in direct orientation will delete the DNA between the two sites whereas recombination between two sites that are in inverse orientation with respect to each other will invert the intervening DNA.

The sequences acted upon by DNA invertases are called **invertible sequences**. These short sequences may carry the gene for the invertase or may be adjacent to it. Therefore, the invertible sequence and its specific invertase form an invertible cassette that sometimes plays an important regulatory role in the cell, some examples of which follow.

PHASE VARIATION IN *SALMONELLA* SPECIES

One type of invertible sequence is responsible for **phase variation** in some strains of *Salmonella*. Phase variation was discovered in the 1940s with the observation that some strains of *Salmonella* can shift from making flagella composed of one flagellin protein, H1, to making flagella composed of a different flagellin protein, H2. The shift can also occur in reverse, i.e., from making H2-type flagella to making H1-type flagella. The flagellar proteins are the strongest antigens on the surface of many bacteria, and periodically changing their flagella may help these bacteria escape detection by the host immune system.

Two features of the *Salmonella* phase variation phenomenon suggested that the shift in flagellar type was not due to normal mutations. First, the shift occurs at a frequency of about 10^{-4} per cell, much higher than normal mutation rates. Second, both phenotypes are completely reversible—the cells switch back and forth between the two types of flagella.

Figure 9.26 outlines the molecular basis for the *Salmonella* antigen phase variation (see Simon et al., Suggested Reading). A DNA invertase called the Hin invertase causes the phase variation by inverting an invertible sequence upstream of the gene for the H2 flagellin. The invertible sequence contains the invertase gene itself and a promoter for two other genes: one encodes the H2 flagellin and another encodes a repressor of H1 gene transcription. With the invertible sequence in one orientation, the promoter will transcribe the H2 gene and the repressor gene, and only the H2-type flagellum will be expressed on the cell surface. When the sequence is in the other orientation, neither the H2 gene nor repressor gene will be transcribed, because the promoter will be facing backward. Now, however, without the repressor, the H1 gene can be transcribed; therefore, in this state, only the H1-type flagellum will be expressed on the cell surface. Clearly, the Hin DNA invertase that is encoded in the invertible sequence is expressed in either orientation, or the inversion would not be reversible.

OTHER INVERTIBLE SEQUENCES

There are a few other known examples of regulation by invertible sequences in bacteria and phages. For example, fimbria synthesis in some pathogenic strains of *E. coli* is regulated by an invertible sequence. Fimbriae are required for the attachment of the bacteria to the eukaryotic cell surface and may also be important targets of the host immune system.

Invertible sequences also exist in some phages. An example is the invertible region of phages Mu discussed in Box 9.1. Phage P1 and the defective prophage e14 shown in Figure 9.26 also have invertible regions. These

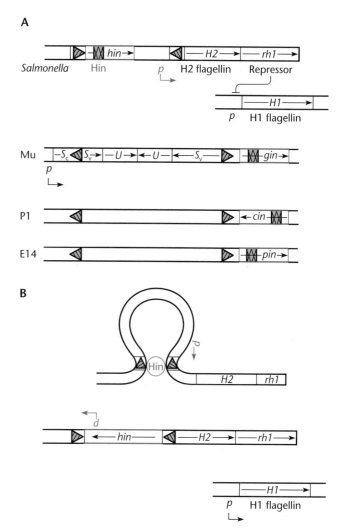

Figure 9.26 Regulation of the *Salmonella* phase variation and some other members of the family of Hin invertases. (A) Invertible sequences, bordered by gold arrows, of *Salmonella* and several phages. (B) Hin-mediated inversion. In one orientation, the H2 flagellin gene, as well as the repressor gene *rh1*, are transcribed from the promoter *p* (in gold). In the other orientation, neither of these genes is transcribed and the H1 flagellin is synthesized instead. The invertase Hin is made constitutively from its own promoter. Redrawn from G. F. Hatfull and N. D. F. Grindley, p. 362, *in* R. Kucherlapati and G. R. Smith (ed.), *Genetic Recombination*, American Society for Microbiology, Washington, D.C., 1988.

phages use invertible sequences to change their tail fibers. The tail fibers made when the invertible sequences are in one orientation differ from those made when the sequences are in the other orientation, broadening the host range of the phage. In phage Mu, the tail fiber genes expressed when the invertible sequence is in one orientation allow the phage to adsorb to *E. coli*, K-12, *Serratia marcescens,* and *Salmonella enterica* serovar Typhi. In the other orientation, the phage is able to absorb to other strains of *E. coli, Citrobacter,* and *Shigella sonnei.*

THE Hin FAMILY OF RECOMBINASES

As mentioned above, many site-specific recombinases, whether they be integrases, resolvases, or invertases, are sometimes closely related to each other. Dramatic evidence for these relationships came from the discovery that the Hin invertase will invert the Mu, P1, and e14 invertible sequences and vice versa (see Van De Putte et al., Suggested Reading). Because they are related, these DNA invertases are called the Hin family of invertases, named after the first member to be discovered. The members of the Hin family of invertases are also related by sequence to the resolvases of Tn3-like transposons. It is tempting to speculate that these recombinases were all derived from the same original enzyme, which has been pressed into service in many different regulatory systems.

Topoisomerases

As discussed in chapter 1, topoisomerases are enzymes that make single- or double-stranded breaks in DNA and allow DNA strands to pass through the breaks before resealing them. Because they catalyze the breakage and joining of DNA strands, topoisomerases essentially perform a nonhomologous recombination reaction. However, since they usually rejoin the original strands after DNA passes through the break, they do not, in general, create new combinations of sequences. Only when they mistakenly join DNA strands from two different regions or different DNA molecules do they cause recombination. Because this recombination is often between sequences that have little or no sequence similarity, the recombination promoted by topoisomerases is also an example of nonhomologous recombination. Topoisomerases are sometimes responsible for mutations, including deletions and frameshifts.

SUMMARY

1. Nonhomologous recombination is the recombination between specific sequences on DNA that occurs even if the sequences are mostly dissimilar.

2. Transposition is the movement of certain DNA sequences, called transposons, from one place in DNA to another. The smallest known bacterial transposons are IS elements, which contain only the genes required for their own transposition. Other transposons carry genes for resistance to substances such as antibiotics and heavy metals. Transposons have played an important role in evolution and are useful for mutagenesis, gene cloning, and random gene fusions.

3. Composite transposons are composed of DNA sequences bracketed by IS elements. Inside-end transposition by composite transposons can cause the deletion and inversion of neighboring DNA sequences.

4. Bacterial transposons transpose by two distinct mechanisms: replicative transposition and cut-and-paste, or conservative, transposition. In replicative transposition, which is used by such transposons as Tn3 and Mu, the entire transposon is replicated, leading to formation of a cointegrate. In cut-and-paste transposition, which is used by many IS elements and composite transposons, the transposon is cut out of the donor DNA and inserted somewhere else.

5. In transposon mutagenesis, a gene is disrupted by insertion of a transposon, which can introduce a selectable marker and additional restriction sites at the site of insertion that can be used for genetic and physical mapping of the gene.

6. Especially engineered transposons carrying reporter genes can be used to make random gene fusions. Insertion of the transposon into a gene can lead to expression of the reporter

gene on the transposon from the promoter or TIR of the disrupted gene, depending on whether the fusion is transcriptional or translational.

7. Genes that have been mutated by insertion of a transposon are often easy to clone in *E. coli* if an antibiotic resistance gene is present on the transposon. Some transposons have been engineered to contain a plasmid origin of replication so that the restriction fragment containing the transposon need not be cloned in another cloning vector but need only be cyclized after ligation to form a replicon in *E. coli.*

8. Site-specific recombinases are enzymes that promote recombination between certain sites on the DNA. Examples of site-specific recombinases include resolvases, integrases, and DNA invertases. The genes for many of these site-specfic recombinases have sequences in common, suggesting they may have been derived from a common ancestor.

9. Resolvases are site-specific recombinases encoded by replicative transposons that resolve cointegrates by promoting recombination between short *res* sequences in the copies of the transposon in the cointegrate.

10. Integrases promote nonhomologous recombination between specific sequences on a DNA element such as a phage DNA and the chromosome, integrating the phage DNA into the chromosome to form lysogens. They also integrate antibiotic resistance gene cassettes into transposons. Transposons that encode an integrase which allows then to accept these antibiotic gene cassettes are called integrons. Integrases also play a role in integrating pathogenicity islands and other

(continued)

SUMMARY (continued)

types of genetic islands into the chromosome. These are large DNA elements (50,000 to 100,000 bp) which carry genes, including genes for pathogenicity, that allow the bacterium to occupy unusual ecological niches.

11. DNA invertases promote nonhomologous recombination between short inverted repeats, thereby changing the orientation of the DNA sequence between them. The sequences they invert, invertible sequences, are known to play an important role in changing the host range of phages and the bacterial surface antigens to avoid host immune defenses.

12. Topoisomerases can also perform nonhomologous recombination. However, instead of creating new combinations of sequences, topoisomerases usually reattach the original strands. Only rarely do they promote an illegitimate recombination event.

QUESTIONS FOR THOUGHT

1. For those transposons that transpose replicatively (e.g., Tn3), why are there not multiple copies of the transposon around the genome?

2. How do you think that transposon Tn3 and its relatives have spread throughout the bacterial kingdom?

3. Do you think that transposons are just parasitic DNAs, or do they serve a useful purpose for the host?

4. Where do you suppose the genes that were inserted into integrons in the evolution of transposons came from?

5. If the DNA invertase enzymes are made continuously, why do the invertible sequences invert so infrequently?

PROBLEMS

1. Outline how you would use an HFT λ strain to show that some *gal* mutations of *E. coli* are due to insertion of an IS element.

2. In the experiments shown in Figures 9.15 and 9.16, what would have been observed if Tn10 transposed by a replicative mechanism?

3. List the advantages and disadvantages of transposon mutagenesis over chemical mutagenesis to obtain mutations.

4. Outline how you would use transposon mutagenesis to mutagenize plasmid pBR322 in *E. coli* with transposon Tn5. Determine what size of junction fragments would be obtained with *Pst*I and *Hin*dIII if the transposon hopped into the 1-kb position clockwise of 0 kb on the plasmid. See chapter 4 for a map of pBR322. The plasmid is 4.36 kb, and the *Hin*dIII and *Pst*1 sites are at 0.029 and 3.673 kb, respectively. See Figure 9.18 for a map of Tn5.

5. How would you determine whether a new transposon you have discovered integrates randomly into DNA?

6. In the example shown in Figures 9.17 to 9.20, digestion with the restriction endonuclease *Sal*I gives two fragments of 2.972 and 6.228 kb. In what orientation has the transposon inserted? Draw a picture of the transposon inserted in the plasmid, showing the position of the kanamycin resistance gene.

7. You have isolated a strain of *Pseudomonas putida* that can grow on the herbicide 2,4-dichlorophenoxyacetic acid (2,4-D) as the sole carbon and energy source. Outline how you would clone the genes for 2,4-D utilization (a) by transposon mutagenesis and (b) by complementation in the original host.

8. Propose a mechanism whereby the same two proteins, MuA and MuB, could integrate Mu by a single cut-and-paste transposition event to form a lysogen after infection but replicatively transpose Mu DNA hundreds of times after induction of the lysogen.

9. The Int protein of conjugative transposons like Tn916 must integrate the transposon into the chromosome of the recipient cell after transfer. How would you show whether the Int protein of the transposon must be synthesized in the recipient cell or can be transferred in with the transposon during conjugation, much like primases are transferred during plasmid conjugation?

10. How would you show whether the G segment of Mu can invert while it is in the prophage state or whether it inverts only after the phage is induced?

11. The defective prophage e14 resides in the *E. coli* chromosome and has an invertible sequence. How would you determine if its invertase can invert the invertible sequences of Mu, P1, and *Salmonella enterica* serovar Typhimurium?

12. The red color of *Serratia marcescens* is reversibly lost with a high frequency. Outline how you would attempt to determine if the change in pigment is due to an invertible sequence.

13. How would you prove that the Mu phage transposon probably does not encode a resolvase?

SUGGESTED READING

Bender, J., and N. Kleckner. 1986. Genetic evidence that Tn*10* transposes by a nonreplicative mechanism. *Cell* **45**:801–815.

Berg, C. M., and D. E. Berg. 1996. Transposable element tools for microbial genetics, p. 2588–2612. *In* F. C. Neidhardt, R. Curtiss III, J. L. Ingraham, E. C. C. Lin, K. B. Low, B. Magasanik, W. S. Reznikoff, M. Riley, M. Schaechter, and H. E. Umbarger (ed.), Escherichia coli *and* Salmonella: *Cellular and Molecular Biology*, 2nd ed. ASM Press, Washington, D.C.

Casadaban, M. J., and S. N. Cohen. 1979. Lactose genes fused to exogenous promoters in one step using a Mu-lac bacteriophage: in vivo probe for transcriptional control sequences. *Proc. Natl. Acad. Sci. USA* **76**:4530–4533.

Chen, L.-M., and S. Maloy. 1991. Regulation of the proline utilization in enteric bacteria: cloning and characterization of the *Klebsiella put* control region. *J. Bacteriol.* **173**:783–790.

Collis, C. M., G. D. Recchia, M.-J. Kim, H. W. Stokes, and R. M. Hall. 2001. Efficiency of recombination reactions catalyzed by class I integron integrase IntI1. *J. Bacteriol.* **183**:2535–2542.

Craig, N. L. 1996. Transposition, p. 2339–2362. *In* F. C. Neidhardt, R. Curtiss III, J. L. Ingraham, E. C. C. Lin, K. B. Low, B. Magasanik, W. S. Reznikoff, M. Riley, M. Schaechter, and H. E. Umbarger (ed.), Escherichia coli *and* Salmonella: *Cellular and Molecular Biology*, 2nd ed. ASM Press, Washington, D.C.

Craig, R., M. Craige, M. Gellert, and A. Lambowitz (ed.). 2002. *Mobile DNA II*. ASM Press, Washington, D.C.

Foster, T. J., M. A. Davis, D. E. Roberts, K. Takeshita, and N. Kleckner. 1981. Genetic organization of transposon Tn*10*. *Cell* **23**:201–213.

Gill, R., F. Heffron, G. Dougan, and S. Falkow. 1978. Analysis of sequences transposed by complementation of two classes of transposition-deficient mutants of Tn*3*. *J. Bacteriol.* **136**:742–756.

Golden, J. W., S. G. Robinson, and R. Haselkorn. 1985. Rearrangement of nitrogen fixation genes during heterocyst differentiation in the cyanobacterium Anabaena. *Nature* (London) **327**:419–423.

Groisman, E. O., and M. J. Casadaban. 1986. Mini-Mu bacteriophage with plasmid replicons for in vivo cloning and *lac* gene fusing. *J. Bacteriol.* **168**:357–364.

Kanaar, R., P. Van de Putte, and N. R. Cozzarelli. 1988. Ginmediated DNA inversion: product structure and the mechanism of strand exchange. *Proc. Natl. Acad. Sci. USA* **85**:752–756.

Kenyon, C. J., and G. C. Walker. 1980. DNA-damaging agents stimulate gene expression at specific loci in *Escherichia coli*. *Proc. Natl. Acad. Sci. USA* **77**:2819–2823

Kuduvalli, P. N., J. E. Rao, and N. L. Craig. 2001. Target DNA structure plays a critical role in Tn7 transposition. *EMBO J.* **20**:924–932.

Kunkel, B., R. Losick, and P. Stragier. 1990. The *Bacillus subtilis* gene for the developmental transcription factor σ^K is generated by excision of a dispensable DNA element containing a sporulation recombinase gene. *Genes Dev.* **4**:525–535.

Mahillon, J., and M. Chandler. 1998. Insertion sequences. *Microbiol. Mol. Biol. Rev.* **82**:725–774.

May, E. W., and N. L. Craig. 1996. Switching from cut-and-paste to replicative Tn7 transposition. *Science* **272**:401–404.

Nash, H. A. 1996. Site-specific recombination: integration, excision, resolution, and inversion of defined DNA segments, p. 2363–2376. *In* F. C. Neidhardt, R. Curtiss III, J. L. Ingraham, E. C. C. Lin, K. B. Low, B. Magasanik, W. S. Reznikoff, M. Riley, M. Schaechter, and H. E. Umbarger (ed.), Escherichia coli *and* Salmonella: *Cellular and Molecular Biology*, 2nd ed. ASM Press, Washington, D.C.

Rakin, A., C. Noelting, P. Schropp and J. Heesemann. 2001. Integrative module of the high-pathogenicity island of *Yersinia*. *Mol. Microbiol.* **39**:407–415.

Shapiro, J. A. 1979. Molecular model for the transposition and replication of bacteriophage Mu and other transposable elements. *Proc. Natl. Acad. Sci. USA* **76**:1933–1937.

Simon, M., J. Zeig, M. Silverman, G. Mandel, and R. Doolittle. 1980. Phase variation: evolution of a controlling element. *Science* **209**:1370–1374.

Simon, R., U. Preifer, and A. Puhler. 1983. A broad host range mobilization system for in vivo genetic engineering: transposon mutagenesis in gram negative bacteria. *Bio/Technology* **1**:784–790.

Van De Putte, P., R. Plasterk, and A. Kuijpers. 1984. A Mu *gin* complementing function and an invertible DNA region in *Escherichia coli* K-12 are situated on the genetic element e14. *J. Bacteriol.* **158**:517–522.

Van Gijsegem, F., and A. Toussaint. 1983. In vivo cloning of *Erwinia carotovora* genes involved in catabolism of hexuronates. *J. Bacteriol.* **154**:1227–1235.

PART III

Genes in Action

Even though they are the simplest organisms on Earth, with only a few thousand genes, the cells of bacteria are nevertheless almost unimaginably complex. The few thousand gene products of bacteria interact to perform many functions required by the living cell. The task of molecular genetics is to identify the genes whose products are involved in the various functions of the cell and to determine what each of these gene products contributes to the function. The cellular functions that we will discuss are recombination, repair, and gene regulation. These functions are central to living cells and involve the DNA itself, so in a sense we are using genetics to study genetics. Also, many of these basic pathways have been found to have features in common with the analogous pathways in other organisms, including humans, so the studies with bacteria have laid the foundation for similar studies in other organisms.

Molecular Basis of Recombination

Even two organisms belonging to a species that must reproduce sexually will not be genetically identical. As chromosomes are assembled into germ cells (sperm and eggs), one chromosome of each pair of homologous chromosomes is chosen at random. Consequently, it is highly unlikely that two siblings will be alike, because each of their parents' germ cells contains a mixture of chromosomes originally derived from the siblings' grandparents. In addition, the chance of two siblings being the same is even lower because of **genetic recombination.** Because of recombination, genetic information that was previously associated with one DNA molecule may become associated with a different DNA molecule, or the order of the genetic information in a single DNA molecule may be altered.

All organisms on Earth probably have some mechanism of recombination, suggesting that recombination is very important for species survival. The new combinations of genes obtained through recombination allow the species to adapt more quickly to the environment and speed up the process of evolution. Recombination can allow an organism to change the order of its own genes or move genes to a different replicon, for example, from the chromosome to a plasmid. Recombination genes also play an important role in repair of damage to DNA and in mutagenesis, topics that will be covered in the next chapter.

As we mentioned at the beginning of chapter 9, the two general types of recombination are nonhomologous, or site-specific, recombination and homologous recombination. Nonhomologous recombination occurs relatively rarely and requires special proteins that recognize specific sequences and promote recombination between them. **Homologous recombination** occurs much more often. This type of recombination can occur between any two DNA sequences that are the same or very similar, and it usually involves the breaking of two DNA molecules in the same region, where the sequences are similar, and the joining of one DNA to the other. The result is called a **crossover**. Depending upon the organism, homology-dependent crossovers can occur between homologies as short as 23 bases, although longer homologies produce more frequent crossovers.

Because of its importance in genetics, we have already mentioned homologous recombination in previous chapters, for example in discussions of deletion and inversion mutations and genetic analysis. Determination of recombination frequencies allows us to measure the distance between mutations and thus can be used to map mutations with respect to each other, as we discussed in chapters 3 and 7, among others. Moreover, the clever use of recombination can take some of the hard work out of cloning genes and making DNA constructs.

When we have discussed the use of recombination for genetic mapping and many other types of applications in previous chapters, we used a simplified description of recombination: the strands of two DNA molecules break at a place where they both have the same sequence of bases and then the strands of the two DNA molecules join with each other to form a new molecule. This model is obviously too simple, but we could use recombination without knowing its molecular mechanisms in any detail. People have been using recombination for 80 years or more without knowing the actual molecular mechanisms involved. In fact, as we'll see, models of recombination are still being debated and recombination can proceed by different mechanisms depending upon the situation. In this chapter, we focus on the actual mechanisms of recombination—what actually happens to the DNA molecules involved—discussing some molecular models and some of the genetic evidence that supports or contradicts those models. We also discuss the proteins involved in recombination mostly in *Escherichia coli*, the bacterium for which recombination is best understood.

Overview of Recombination

Recombination is a remarkable process. Somehow, two enormously long DNA molecules in a cell link up and exchange sequences. Moreover, recombination usually occurs only at homologous regions of two DNAs. Thus, these regions must line up so that they can be broken and rejoined, and the long DNA molecules on either side of the point of recombination must change their configuration with respect to each other. This complicated process clearly involves many functions. But before presenting detailed models for how recombination might occur, and the functions involved, we consider the basic

requirements any recombination model must satisfy and what functions to expect of gene products directly involved in recombination.

Requirement 1: Identical or Very Similar Sequences in the Crossover Region

The distinguishing feature of homologous recombination is that the deoxynucleotide sequence in the two regions of DNA where a crossover occurs must be the same or very similar, and all recombination models must start with this requirement. This prerequisite serves a very practical function in recombination. The sequence of nucleotides in molecules of DNA from different individuals of the same species will usually be almost identical over the entire lengths of the molecules, and by extension, two DNA molecules will generally share identical sequences only in the same regions. Thus, recombination usually occurs only between sites on the two DNAs that are in the same place with respect to the entire molecule. Recombination between different regions in two DNA molecules does sometimes occur because the same or similar sequences occur in more than one place in the DNAs. This type of recombination, sometimes called **ectopic** or **homeologous recombination**, gives rise to deletions, duplications, inversions, and other gross rearrangements of DNA (see chapter 3). Not surprisingly, cells have evolved special mechanisms to discourage ectopic recombination, one of which we discuss in the next chapter in the section on mismatch repair.

Requirement 2: Complementary Base Pairing between Double-Stranded DNA Molecules

Mandatory complementary base pairing between strands of the two DNA molecules ensures that recombination will occur only between sequences at the same **locus**, that is, the same place on the molecules. The point at which two double-stranded DNA molecules are held together by complementary base pairing between their strands is called a **synapse**. All recombination models must involve synapse formation. However, in double-stranded DNA, the bases are usually hidden inside the helix, where they are not available for base pairing. Separating the strands in various regions to expose the bases would be too slow to allow efficient recombination. Therefore, the cell must contain some functions that allow one strand of DNA to locate and pair with its complementary sequence even though the bases are mostly hidden from pairing inside the double helix.

Requirement 3: Recombination Enzymes

For a crossover, or true recombination, to occur between two DNA molecules, the strands of each molecule must be broken and rejoined to the corresponding strands of the other DNA molecule. Therefore, DNA endonucleases and ligases—enzymes that break DNA strands and rejoin them, respectively—are required for recombination.

Requirement 4: Heteroduplex Formation

The regions of complementary base pairing between the two DNA molecules in a synapse are called **heteroduplexes**, because the strands in these regions come from different DNA molecules. In principle, for a synapse to form, heteroduplexes need form only between two of the strands of the DNA molecules being recombined. However, evidence indicates that all strands of the two DNA molecules are involved in heteroduplex formation, an observation that must be explained by the models.

Molecular Models of Recombination

Several models have been proposed to explain recombination at the molecular level. These models include the required features of recombination discussed above and also account for additional experimental evidence. However, no single model of recombination can make an exclusive claim to the truth, and recombination may occur by different pathways in different situations. Nevertheless, these models serve as a framework for forming hypotheses that can be tested through experimentation and will help focus thinking about recombination at the molecular level.

The Holliday Double-Strand Invasion Model

The first widely accepted model for recombination was proposed by Robin Holliday in 1964. Figure 10.1 illustrates the basic steps of the **Holliday model**. According to this model, recombination is initiated by two single-stranded breaks made simultaneously at exactly the same place in the two DNA molecules to be recombined. Then the free ends of the two broken strands cross over each other, each pairing with its complementary sequence in the other DNA molecule to form two heteroduplexes. The ends are then ligated to each other, resulting in a cruciform-like structure called a **Holliday junction** in which the two double-stranded molecules are held together by their crossed-over strands (Figure 10.1). The formation of Holliday junctions is central to all the recombination models discussed here.

Once formed, Holliday junctions can undergo a rearrangement that changes the relationship of the strands to each other. This rearrangement is called an **isomerization** because no bonds are broken. As shown in the figure, isomerization causes the crossed strands in configuration I to uncross. A second isomerization occurs to create configuration II, where the ends of the two double-stranded DNA molecules are in the recombinant configuration with respect to each other. This may seem surprising, but experiments with models show that the two structures I and II are actually equivalent to each other and the Holliday junction can

Figure 10.1 The Holliday model for genetic recombination. One strand of each DNA molecule is cut at the same position and then pairs with the other molecule to form a heteroduplex (region of gold paired with black). The strands are then ligated to form the Holliday junction. This DNA structure can isomerize between forms I and II. Cutting and ligating resolves the Holliday junction. Depending on the conformation of the junction, the flanking markers A, B, a, and b will recombine or remain in their original conformation. The product DNA molecules will contain heteroduplex patches.

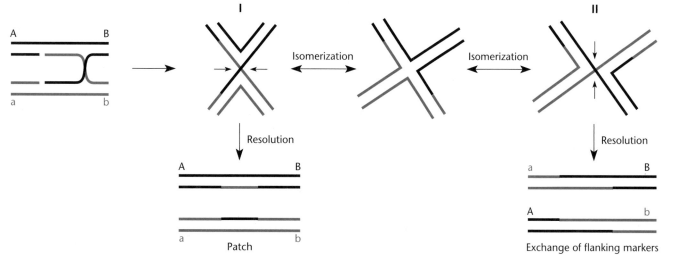

change from one form to the other without breaking any hydrogen bonds between the bases. Hence, flipping from one configuration to the other requires no energy and should occur quickly, so that each configuration should be present approximately 50% of the time.

Once formed, the Holliday junction can be cut and re-ligated, or **resolved**, as shown in Figure 10.1. Whether or not recombination occurs depends on the configuration of the junction at resolution. If the Holliday junction is in configuration I, the flanking DNA sequences will not recombine. However, if the Holliday junction is in the II configuration when it is resolved, DNA sequences will be exchanged, as indicated by the flanking markers shown in Figure 10.1.

Holliday junctions also move up and down the DNA by breaking and re-forming the hydrogen bonds between the bases. This process is called **branch migration** (Figure 10.2). The same number of hydrogen bonds are broken and re-formed as the cross-connection moves, so that no energy is required. However, without the expenditure of some energy, hydrogen bonds may not be broken fast enough for efficient branch migration. Specific ATP-hydrolyzing proteins seem to be required for branch migration (see the section on the Ruv proteins).

As shown in Figure 10.2, branch migration in Holliday junctions can have the effect of increasing the length of the heteroduplex regions. If a heteroduplex extends to include a region of differing sequence, a mismatch will

occur, possibly leading to gene conversion (discussed later in the chapter). The other models we discuss all invoke branch migration of Holliday junctions to explain the experimental evidence concerning gene conversion and the length and distribution of heteroduplexes.

The Holliday model is called a double-strand invasion model because one strand from each DNA molecule invades the other DNA molecule (Figure 10.1), explaining how heteroduplexes can form on both molecules of DNA during recombination. However, one problem with this model is that the two DNA molecules must be simultaneously cut at almost the same place to initiate recombination. But how could the two like DNA molecules line up for pairing before they are cut when the bases are hidden inside the double-stranded DNA helix and so are not free to pair with other DNA molecules? Also, if the two DNA molecules were not aligned, how could they be cut at exactly the same place? To answer these questions, Holliday proposed the existence of certain sites on DNA that are cut by special enzymes to initiate the recombination. However, there is no evidence for such sites, and recombination seems to occur more or less randomly over the entire DNA.

Despite these objections, the Holliday double-strand invasion model has served as the standard against which all other models of recombination are compared. All the most favored models involve the formation of Holliday junctions and branch migration. They mostly differ in the earlier stages, before Holliday junctions have formed.

Single-Strand Invasion Model

One way to overcome the objections to the Holliday model is to modify it with the proposal of a single-strand invasion model such as the one shown in Figure 10.3 (see Meselson and Radding, Suggested Reading). In this model, a strand of one of the two DNA molecules is cut at random and then the exposed end invades another double-stranded DNA until it finds its complementary sequence. If it finds such a sequence, it will displace one strand. The DNA polymerase will then fill the gap left by the invading strand, using the remaining strand as a template. The displaced strand is degraded, and its remaining end is ligated to the newly synthesized DNA strand on the opposite molecule to form a Holliday junction (Figure 10.3). Once the Holliday junction has formed, isomerization and resolution can produce recombinants as in the Holliday model.

By this model, a heteroduplex will at first form on only one of the two DNAs. However, once the Holliday junction forms, branch migration can create a heteroduplex on the other DNA. This accounts for the formation of heteroduplexes on both DNAs.

Figure 10.2 Migration of Holliday junctions. By breaking the hydrogen bonds holding the DNAs together in front of the branch and re-forming them behind, the junction migrates and extends the regions of pairing (i.e., the heteroduplexes) between the two DNAs. The heteroduplex region is hatched. In the example, two mismatches, GA and CT, form in the heteroduplex region because one of the DNA molecules has a mutation in this region.

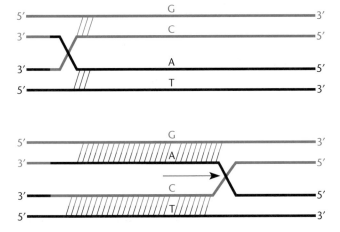

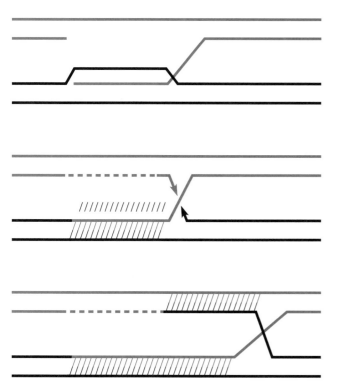

Figure 10.3 The single-strand invasion model. A single-stranded break in one of the two DNA molecules frees a single-stranded end that invades the other DNA molecule. The gap left on the cut DNA is filled by DNA polymerase (dashed line). The displaced strand on the other DNA molecule is degraded, and the two ends are joined (arrows). Initially, a heteroduplex, represented by hatching, will form on only one of the two DNA molecules. Branch migration will cause another heteroduplex to form on the other DNA molecule. Isomerization can recombine the flanking DNA molecules, as in the Holliday model.

Double-Strand Break Repair Model

In both the single-strand and the double-strand invasion models, a single-strand break in one or both of the two DNA molecules, respectively, initiates the recombination event. A priori, it seems unlikely that a double-strand break in DNA could initiate recombination. If a single-strand break is made in DNA, one strand still holds the molecule together. However, if a double-strand break occurred, the two parts of the DNA should fall apart, a lethal event. Thus, double-strand break models were at first ruled out as being counterintuitive. However, it now seems clear that recombination is often initiated by a double-strand break in one of the two DNA molecules, at least in some situations.

The first evidence that double-strand breaks in DNA can initiate recombination came from genetic experiments with *Saccharomyces cerevisiae* (see Szostak et al.,

Suggested Reading). These experiments were aimed at analyzing the ability of recombination between plasmids and chromosomes to repair double-strand breaks and gaps in the plasmids by inserting the corresponding sequence from the chromosomes. However, the initiation of recombination by double-strand breaks is now known to be a general mechanism. For example, homing DNA endonucleases in bacteria, phages, and lower eukaryotes initiate recombination by making a double-strand break (see Box 10.1).

Figure 10.4 illustrates the model for this type of recombination (see Szostak et al., Suggested Reading). Both strands of one of the two DNA molecules participating in the recombination are broken, and the 5' ends from each break are digested by an exonuclease, leaving a gap with exposed single-stranded 3' tails. One of these tails invades the other double-stranded DNA until it finds its

Figure 10.4 The double-strand break repair model. (1) A double-strand break in one of the two DNAs initiates the recombination event. The arrows indicate the degradation of the 5' ends at the break. (2) One 3' end, or tail, then invades the other DNA, displacing one of the strands. (3) This 3' end serves as a primer for DNA polymerase, which extends the tail until it can eventually be joined to a 5' end (black arrow). Meanwhile, the displaced strand (in black) serves as a template to fill the gap left in the first DNA (dashed lines). Two Holliday junctions form and may produce recombinant flanking DNA, depending upon how they are resolved.

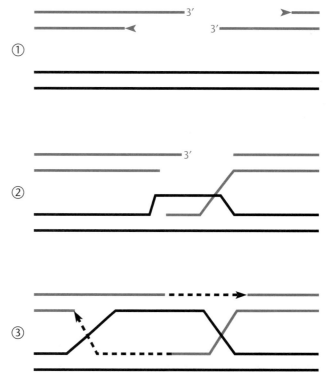

BOX 10.1

Breaking and Entering: Some Introns and Inteins Move by Double-Strand Break Repair

Introns and inteins are parasitic DNA elements that are sometimes found in genes, both in eukaryotes and in prokaryotes (see Box 2.2). Like all good parasites, they have as little effect on the health of their host as possible. This is a matter of self-interest, because if their host dies, they die with it. When they hop into a DNA, they could disrupt a gene, which could inactivate the gene product and be deleterious to their host. As discussed in Box 2.2, they can avoid inactivating the product of the gene in which they reside by splicing their sequences out of the mRNA before it is translated (in the case of introns) or out of the protein product of the gene after it is made (in the case of inteins). Sometimes this splicing requires other gene products of the intron or the host, and sometimes it occurs spontaneously, in a process called self-splicing. Self-splicing introns were one of the first-known examples of RNA enzymes or ribozymes.

Many introns and inteins are able to move from one DNA to another. When they move, they usually move from one gene into exactly the same location in another allele of that same gene which previously lacked them. In that way, they can move through a population until almost all of the individuals in the population have the intron in that location. Because they always return to the same site, this process is called **homing**. There is a good reason why they choose to move into exactly the same position in the same gene rather than to other places in the same gene or even other places in the genome. The sequences in the gene around the intron or intein, called exon or extein sequences, respectively, also participate in the splicing reaction; therefore, if they find themselves somewhere else where these flanking exon sequences are different, they will not be able to splice themselves out of the RNA or protein. Homing allows the intron or intein to spread through a population by parasitizing other DNAs which lack it but never disrupting the product of an essential gene and disabling or killing the new host as it moves.

There are two basic mechanisms by which introns home. Some introns, called the group II introns, move by a process called retrohoming. These introns encode a reverse transcriptase which makes a DNA copy of the intron in the mRNA and then essentially splices the intron into the target DNA by a process analogous to splicing the intron out of the mRNA (see Box 2.2). Other introns, called group I introns, and inteins home by double-strand break repair. To move by double-strand break repair, the intron or intein first makes a double-strand break in the homing site of the target gene into which it must move. To accomplish this, the intron or intein encodes a specific DNA endonuclease called a homing nuclease, which will make a break only in this particular sequence. In the case of group I introns, this homing endonuclease is usually encoded by an ORF on the intron. In the case of inteins, the intein iself becomes the homing endonclease after it is spliced out of the protein. After the double-strand cut is made by the specific endonuclease, the corresponding gene containing the intron repairs the cut by double-strand break repair, replacing the sequence without the intron with the corresponding sequence containing the intron by gene conversion (see Figure 10.4). After repair, both DNAs now contain the intron in exactly the same position. Other DNA elements, including the mating-type loci of yeast, are known to move by a similar mechanism of double-strand break repair.

References

Belfort, M., M. E. Reaban, T. Coetzee, and J. Z. Dalgaard. 1995. Prokaryotic introns and inteins: a panoply of form and function. *J. Bacteriol.* **177**:3897–3903.

Shub, D. A., J. M. Gott, M.-Q. Xu, B. F. Lang, F. Michel, J. Tomaschewski, J. Pedersen-Lane, and M. Belfort. 1988. Structural conservation among three homologous introns of bacteriophage T4 and group I introns of eukaryotes. *Proc. Natl. Acad. Sci. USA* **85**:1151–1155.

complementary sequence. Then DNA polymerase extends the tail along the complementary sequence, displacing the other strand with the same sequence, until it reaches the free 5′ end of the invading strand and is joined to it by DNA ligase. The other free 3′ end can then be used as a primer to fill in the remaining gap by using the displaced strand as a template before being joined to the free 5′ end in its own strand by DNA ligase. This causes two Holliday junctions to form (Figure 10.4).

Whether recombination occurs depends on which configuration the two Holliday junctions are in when resolved. If both are in the same configuration—either I or II (Figure 10.1)—when they are resolved, no crossovers, and thus no recombination, will result. However, if the two junctions are in different configurations at resolution, recombination will occur.

In this model, heteroduplex DNA will initially form in only one DNA molecule, between the strand that does the initial invading and its complementary sequence. However, as in the single-strand invasion models, the migration of the Holliday junctions can lead to heteroduplex formation on both strands.

Molecular Basis for Recombination in *E. coli*

As with many cellular phenomena, much more is known about the molecular basis for recombination in *E. coli* than in any other organism. At least 25 proteins involved in recombination have been identified in *E. coli*, and specific roles have been assigned to many of these (see Table 10.1).

chi (χ) Sites and the RecBCD Nuclease

The first step in recombination in *E. coli* is usually performed by the RecBCD nuclease. The enzyme is called RecBCD because it is composed of three subunits, RecB, RecC, and RecD, encoded by the genes *recB*, *recC*, and *recD*, respectively. RecBCD is a large, complex enzyme that plays a number of important roles in bacteria. Not only is the RecBCD enzyme required for most recombination, but also it is responsible for degrading foreign linear DNA in the cell and is involved in replication restarts when the replication fork stalls at single- or double-strand breaks in the template DNA, as well as in some types of repair (see Box 10.2). We discuss the role of RecBCD in more detail below, but to give an overview, the role of the RecBCD enzyme is to prepare single-stranded ends of DNA that can invade other double-stranded DNAs as the first step in recombination. It also helps the RecA protein load onto these single-stranded

DNAs to form helical nucleoprotein filaments, which can then invade other double-stranded DNAs as the first step in recombination. All recombination pathways require such a strand invasion step, which usually requires loading the RecA protein onto single-stranded DNA to form the helical nucleoprotein filament. An entirely different pathway, called the RecF pathway, is required to load the RecA protein onto single-stranded "gaps" in DNA for recombination. We return to the RecF pathway and the role of the RecA protein in recombination in more detail in later sections.

The *recB* and *recC* genes were among the first *rec* genes found because their products are required for recombination after Hfr crosses (see Clark and Margulies, Suggested Reading, and the section on genetic analysis of recombination below). Their products were later shown to also be required for transductional crosses. The *recD* gene product is not required for recombination after such crosses so this gene had to be found in other ways (see below). RecB and RecC were shown later also to be required for recombination after transduction. We now know that the RecBCD enzyme is required for the recombination that occurs after conjugation and transduction because the small pieces of DNA transferred into cells by Hfr crosses or by transduction are the natural substrates for the RecBCD enzyme. The RecBCD enzyme processes the ends of these pieces to form single-stranded 3′ ends, which can then invade the chromosomal DNA to form recombinants (see below). If a different selection had been used in these original searches which favored another class of *rec* genes, for example those encoding enzymes of the RecF pathway, other *rec* genes would have been found first.

TABLE 10.1	Some genes encoding recombination functions in *E. coli*		
Gene	**Mutant phenotype**	**Enzymatic activity**	**Probable role in recombination**
recA	Recombination deficient	Enhanced pairing of homologous DNAs	Synapse formation
recBC	Reduced recombination	Exonuclease, ATPase, helicase, chi-specific endonuclease	Initiates recombination by separating strands cutting DNA at the *chi* site
recD	Rec⁺ χ independent	Exonuclease	Degrades 3′ ends
recF	Reduced plasmid recombination	Binds ATP and single-stranded DNA	Substitutes for RecBCD at gaps
recJ	Reduced recombination in RecBC⁻	Single-stranded exonuclease	Substitutes for RecBCD at gaps
recN	Reduced recombination in RecBC⁻	ATP binding	Substitutes for RecBCD at gaps
recO	Reduced recombination in RecBC⁻	DNA binding and renaturation	Substitutes for RecBCD at gaps
recQ	Reduced recombination in RecBC⁻	DNA helicase	Substitutes for RecBCD at gaps
recR	Reduced recombination in RecBC⁻	Binds double-stranded DNA	Substitutes for RecBCD at gaps
recG	Reduced Rec in RuvA⁻ B⁻ C⁻	Branch-specific helicase	Migration of Holliday junctions
ruvA	Reduced recombination in RecG⁻	Binds to Holliday junctions	Migration of Holliday junctions
ruvB	Reduced recombination in RecG⁻	Holliday junction-specific helicase	Migration of Holliday junctions
ruvC	Reduced recombination in RecG⁻	Holliday junction-specific nuclease	Resolution of Holliday junctions
priA, priB, priC, dnaT	Reduced recombination	Helicase?	Reload replication forks

BOX 10.2

The Three "R"s: Recombination, Replication, and Repair

One of the most gratifying times in science comes when phenomena that were originally thought to be distinct are discovered to be but different manifestations of the same process. Such a discovery is usually followed by rapid progress as the mass of information accumulated on the different phenomena is combined and reinterpreted. This is true of the fields of recombination, replication, and DNA repair, sometimes called the three "R"s (see Box 1.2). Replication can be initiated in a number of different ways, and many of these involve recombination; recombination often requires normal replication mechanisms, and some types of DNA repair require recombination as well as normal replication.

Appreciation of the role of recombination in replication was slow in coming. It was known for a long time that some phages, such as T4, need the recombination functions to initiate replication (see chapter 7). In these phages, recombination intermediates such as D-loops function to initiate DNA replication later in infection. However, this was thought to be an abnormality of these phages. Normally, initiation of chromosomal replication in bacteria does not require recombination functions. However, after extensive DNA damage due to irradiation or other agents, a new type of initiation, which does require recombination, comes into play in bacteria. This type of initiation was originally called stable DNA replication (SDR) because it continued even after protein synthesis stopped (see Kogoma, below). Normally, initiation of DNA replication at the chromosomal *oriC* site requires new protein synthesis, and so, in the absence of protein synthesis, replication will continue only until all the ongoing rounds of replication are completed, and no new rounds will be initiated (see chapter 1). However, after extensive DNA damage due to irradiation, etc., initiation of new rounds of replication occurs even in the absence of protein synthesis. Interestingly, this SDR often initiates close to the *oriC* site, although other sites can also be used. To initiate SDR, a double-strand break is made in the DNA close to the *oriC* site and recombination functions cause the formation of a D-loop. The PriA protein and other replisome-loading functions, maybe including DnaC, then reload the replication apparatus on the DNA, and replication is under way again. It is not quite clear how PriA reloads the replication apparatus at D-loops. The current view is that it binds to the three-stranded junction at the D-loop and then, in a complex with PriB, DnaC, and DnaT, directs the DnaB helicase to bind. This model was inspired by the discovery that some *dnaC* mutants bypass the need for PriA. The role of

DnaC is to help DnaB helicase to bind at the normal *oriC* origin of replication, but it may also be required to reload DnaB at double-stranded breaks. Therefore, if some mutant DnaC can bypass the need for PriA, PriA might also be directing the binding of DnaB, but to a D-loop rather than to *oriC*. This raises the question of why *E. coli* encodes a PriA protein at all if it could perform the same function by slightly changing its DnaC protein. The answer could be that even though a mutant DnaC can help reload replication forks at D-loops, it is nowhere near as efficient at this loading as is the PriA protein.

Recombination also plays a role in restarting replication forks after the replication apparatus has been derailed as a result of encountering damage in the DNA template in a process analogous to stable DNA replication. In fact, some authors have gone so far as to purpose that this is the primary role of recombination in bacteria, to repair damage to the DNA (see Kuzmimov and Stahl, below). The pathway of recombination that is used depends upon the type of damage encountered. If a single-stranded break is encountered in either the leading or lagging strand of the DNA, a double-strand break will ensue. The RecBCD enzyme then degrades in from the break until it encounters a χ site. The 3' single-stranded end formed as a result of this degradation can then invade the sister DNA to form a D-loop. However, a gap may form if the lagging strand has DNA damage over which the replication fork cannot replicate. Then the RecF pathway is responsible for forming the D-loop. Once a D-loop has formed, the PriA proteins, etc., will help reassemble the replication fork at the D-loop as above and replication will continue. The necessity to form D-loops to repair some types of DNA damage explains why *E. coli recB* and *recC* mutants, as well as *recF*, *recN*, *recQ*, and *recO* mutants are more sensitive to irradiation and other types of DNA damage than are their recombination-proficient parents (see chapter 11).

A role for replication in recombination completes the circle. As described in the text, most models show recombination proceeding through one or more Holliday junctions, which are then resolved by being cut with an X-phile such as RuvC. Depending upon how the resolvase cuts the Holliday junctions, a recombinant can ensue. However, recombinants need not be created this way. If the recombination functions form a D-loop between two different DNAs in the cell and the replication apparatus loads on this D-loop as described above, then the replication appa-

(continued)

The Three "R"s: Recombination, Replication, and Repair

ratus that started out replicating one DNA will have switched to replicating the other DNA and a recombinant will ensue. Such models of recombination, which used to be called "copy choice," have now come back into favor and may account for at least some of the recombination products which are observed.

References

Kogoma, T. 1997. Stable DNA replication: interplay between DNA replication, homologous replication, and transcription *Microbiol. Mol. Biol. Rev.* **61:**212–238.

Kuzimov, A., and F. W. Stahl. 1999. Double-strand end repair via the RecBCD pathway in *Escherichia coli* primes DNA replication. *Genes Dev.* **13:**345–356.

HOW RecBCD WORKS

The RecBCD protein is a remarkable enzyme with many enzyme activities. It has single-stranded DNA endonuclease and exonuclease activities as well as DNA helicase and DNA-dependent ATPase activities. To put all these activities in perspective, it is useful to think of the RecBCD protein as a DNA helicase with associated nuclease activities. Its job is to put single-stranded 3′ tails on DNA that can invade other DNAs to initiate recombination. To perform this job, it loads on one end of a double-stranded DNA and unwinds the DNA, looping out the 3′ to 5′ strand of DNA as it goes (see Figure 10.5). These loops are cut into small pieces by its 3′ to 5′ nuclease activity, leaving the 5′ to 3′ strand mostly intact. This process will continue for up to 30,000 bp or until the RecBCD protein encounters a sequence on the DNA called a *chi* (χ) site, which in *Escherichia coli* has the sequence 5′GCTGGTGG3′ but in other types of bacteria has somewhat different sequences. These sites were first found because they stimulate recombination in phage λ (see "Discovery of χ Sites" below), so they were given the Greek symbol χ, which looks like a crossover. When RecBCD encounters a χ site, its 3′-to-5′ nuclease activity is inhibited, but its 5′-to-3′ nuclease activity may be stimulated, leading to formation of the free 3′ single-stranded tail, as shown in Figure 10.5. Note that the χ sequence does not have twofold rotational symmetry like the sites recognized by many restriction endonucleases. This means that the sequence will be recognized on only one strand of the DNA and the RecBCD nuclease will pass over the sequence if it occurs only on the other strand, making the orientation of the χ site all important, as we discuss below. After a RecBCD enyme has formed a 3′ single-stranded tail on the DNA, it will direct a RecA protein molecule to bind to the DNA next to where it is bound (see below). This is called **cooperative binding**, because one protein is helping another to bind. In more physicochemical terms, the incoming RecA protein makes contact both with the DNA and with the RecBCD protein already on the DNA, helping stabilize its binding

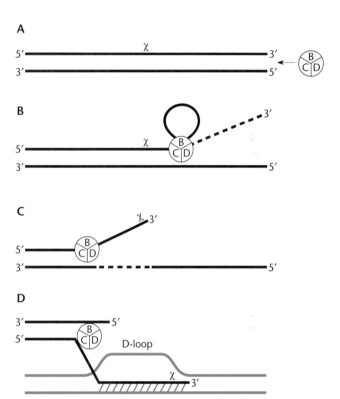

Figure 10.5 Model for promotion of recombination initiation at a *chi* site by the RecBCD enzyme. (A) RecBCD loads onto a double-stranded end. (B) Its helicase activity separates the strands, and the 3′-to-5′ nuclease activity degrades one strand until it encounters a χ site. (C) The χ site inhibits the 3′-to-5′ nuclease and stimulates the 5′-to-3′ nuclease, leaving a single-stranded 3′ end. (D) The single-stranded 3′ end can invade another double-stranded DNA, forming a D-loop.

to the DNA. More RecA proteins then bind cooperatively to the first RecA protein to form a helical nucleoprotein filament to prepare for the next step in recombination, as we discuss in more detail below.

The discovery of χ sites and their role in recombination came as a complete surprise, and it took many years and a lot of clever experimentation to figure out how they work (see "Discovery of χ Sites" below). A detailed model consistent with much of the available information has emerged from this work. According to this detailed model, the RecD subunit of the RecBCD enzyme is responsible for the 3′ to 5′ nuclease activity and χ sites work by inhibiting the RecD subunit and stimulating the 5′ to 3′ nuclease activity. Thus, as the RecBCD enzyme moves along the DNA, opening the strands, its RecD subunit will degrade the 3′-to-5′ strand. This will continue until the sequence 5′GCTGGTGG3′ (the χ sequence) is encountered on the strand being degraded by RecD. The DNA with this sequence can bind to the RecD subunit and inhibit its 3′-to-5′ exonuclease activity but stimulate the 5′-to-3′ activity of the remainder of the enzyme. The RecBCD enzyme continues to move on past the site, degrading the 5′-to-3′ strand but leaving the 3′-to-5′ strand intact. The end result is a single-stranded 3′-ended tail which contains the χ site sequence. The *Bacillus subtilis* enzyme is known to function similarly, except that the 5′-to-3′ strand might be degraded even before a χ site is encountered, and the χ site has a different sequence in *B. subtilis* (see Chedin and Kowalczkowski, Suggested Reading).

Much of the evidence for this detailed model of χ site action is genetic. First, the RecD subunit of the RecBCD enzyme seems to be responsible for its 3′-to-5′ nuclease activity, because *recD* mutants which lack this subunit do not exhibit the nuclease activity. As evidence, a linear DNA can be transformed into a RecD⁻ mutant of *E. coli* and will not be degraded. Another prediction of the model which is fulfilled is that RecD⁻ mutants are proficient for recombination, and this recombination does not require χ sites. This is predicted by the model, since the RecBC enzyme lacking the RecD subunit still has its helicase activity to separate the strands and these single-stranded ends are not degraded even if they do not contain a χ site, so they are available to invade other DNAs and promote recombination. This property of RecD⁻ mutants, of not degrading linear DNA but still being proficient for recombination, is what makes RecD⁻ mutants of *E. coli* useful for gene replacements with linear DNA (see chapter 3).

This model also explains why recombination is only stimulated on the 5′ side of a χ site. Until the RecBCD enzyme reaches a χ site, the displaced strand is degraded and so is not available for recombination. Only after the RecBCD protein passes completely through a χ site will the strand survive to invade another DNA, so that only DNA on the 5′ side of the site will survive. This model is also supported by the known enzymatic activites of the RecBCD protein, as well as by electron microscopic visualization of the RecBCD protein acting on DNA. It has also received experimental support from the results of in vitro experiments with purified RecBCD nuclease and DNA containing a χ site (see Dixon and Kowalcykowski, Suggested Reading).

WHY χ?

One of the hardest questions to answer in biology is why? To answer this question with certainty, we must know everything about the organism and every situation in which it might find itself, both past and present. Nevertheless, it is tempting to ask why *E. coli* and other bacteria use such a complicated mechanism involving χ sites for their major pathway of recombination initiated by ends of DNA and double-strand breaks. Adding to the mystery is the fact that they would not need χ sites at all if they were willing to dispense with the RecD subunit of the RecBCD nuclease. As mentioned, in the absence of RecD, recombination proceeds just fine without χ sites and the cells are viable.

One idea is that the self-inflicted dependency on χ sites allows the RecBCD nuclease to play a dual role in recombination and in defending against phages and other foreign DNAs. Small pieces of foreign DNA such as a phage DNA entering the cell are not apt to have a χ site, since 8-bp sequences like χ occur by chance only once in about 65,000 bp, longer than many phage DNAs. The RecBCD nuclease will degrade a DNA until it encounters a χ site, and if it does not encounter a χ site it will degrade the entire DNA. *E. coli* DNA, by contrast, has many more of these sites than would be predicted by chance. Therefore, *E. coli* DNA is apt to be degraded by RecBCD less extensively than are phage and other foreign DNAs. In support of the idea that RecBCD is designed to help defend against phages is the extent to which phages go to avoid degradation by this enzyme. Many phages avoid degradation by RecBCD by attaching proteins to the ends of their DNA or by making proteins that inhibit RecBCD.

Another possible reason for χ sites is that they might help direct recombination to regions in the DNA which are better for D-loop formation (see below). The χ sequence itself is relatively GT rich (note that seven of the eight bases in the χ sequence are G's or T's) and χ sites are also often surrounded by many other G's and T's. Sequences rich in G's and T's are preferred sites for binding RecA. When χ is used, GT-rich sequences more often end up on the 3′ single-stranded tail where they can enhance the invasion of other DNAs and the formation of D-loops.

It is also possible that the real role of recombination is in recombination repair and that χ sites help in replication restarts. As evidence, most χ sites in *E. coli* are oriented to help with replication restarts (see Box 10.2).

The RecF Pathway

The other major pathway used to prepare single-stranded DNA for D-loop formation in *E. coli* is the RecF pathway. This pathway needs the products of the *recF*, *recQ*, *recO*, *recR*, and *recJ* genes and cannot prepare DNA ends for recombination. It is used mostly to initiate recombination at single-stranded gaps in DNA. These gaps may be created during repair of DNA damage or by the replication fork jumping over a lesion in the DNA, leaving a single-stranded gap. The RecFOR proteins can then use the single-stranded DNA at the gap to form a D-loop by promoting the invasion of a sister DNA. This D-loop can then be used to restart replication, making the RecF pathway important in recombination repair of DNA damage (see chapter 11).

The RecF pathway does not have a "superstar" like RecBCD, and so it needs a number of proteins to perform the tasks the RecBCD nuclease can perform alone. The RecQ protein is a helicase like RecBCD but lacks a nuclease activity to degrade the strands it displaces. RecJ may provide the exonuclease activity that RecQ lacks. The RecQ protein also lacks the ability to load the RecA protein on the single-stranded DNA it creates with its helicase activity. The RecO and RecR proteins may help the RecA protein to bind to and coat the single-stranded DNA created by RecQ. They do this by helping displace the single-stranded-DNA binding protein (SSB) from the single-stranded DNA. SSB creates problems for RecA because it will bind to single-stranded DNA as quickly as it forms, and RecA cannot displace SSB by itself. As we have seen, the RecBCD protein solves this problem by helping load the first RecA protein on the DNA, displacing SSB in the process. The RecO and RecR proteins may help remove SSB, which immediately coats the single-stranded DNA created by RecQ, thereby allowing SSB to be replaced by RecA. The role of the RecF protein is less clear, but it possesses ATP and DNA binding activities and a weak ATPase activity.

Synapse Formation and the RecA Protein

Once a single-stranded DNA is created by the RecBCD pathway or the RecF pathway, it must invade another DNA to form a **D-loop,** so named because it resembles a letter D as usually drawn (Figure 10.6). The joining of two DNAs in this way is called a **synapse,** and the process by which an invading strand can replace one of the two strands in a double-stranded DNA is called **strand exchange.** Searching for complementary DNA is the job of the RecA protein, whose role in recombination is outlined in Figure 10.6. As the single-stranded DNA is created, it is coated by RecA to form a helical nucleoprotein filament. As mentioned above, the RecBCD and RecOR proteins may help RecA bind to the DNA and form this filament. The nucleoprotein filament then goes looking for a homologous double-stranded DNA to invade. This is one of the most amazing and mysterious steps in all of cell biology. Somehow the single-stranded DNA must search through all of the double-stranded

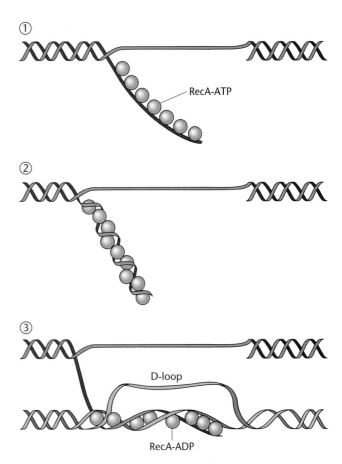

Figure 10.6 Synapse formation between two homologous DNAs by RecA protein. In steps 1 and 2, the RecA protein binds to the single-stranded end and forces it into an extended helical structure. In step 3, the helical single-stranded DNA can pair with a homologous double-stranded DNA in its major groove to form a D-loop. Details in text.

DNA in the cell, which even in a simple bacterium can be more than 1 mm long, looking for its homologous sequence to pair with and invade. But how could the single-stranded DNA know when the sequence is complementary? The bases of the double-stranded DNA are on the inside of the helix, where it is unable to see them, and it seems that it would be too slow for it to go around opening each double-stranded DNA to test for complementarity. Not only is synapse formation remarkably fast, but also it is remarkably efficient. Once an incoming single-stranded DNA enters the cell, for example during an Hfr cross, it finds and recombines with its complementary sequence in the chromosome almost 100% of the time.

While the details of how two DNAs find each other to form a D-loop remain obscure, a number of observations have been made that shed light on this process. One observation is that a single-stranded RecA nucleoprotein filament can somehow change, or "activate," double-stranded DNAs merely by binding to them, even if the two DNAs are not complementary. This activation presumably has something to do with the way in which the RecA nucleoprotein filament scans DNA looking for its complementary sequence and may involve stretching the double-stranded DNA and partially separating its two strands. Once the strands of the double-stranded DNA are activated, even by a noncomplementary RecA nucleoprotein filament, a complementary single-stranded DNA not bound to RecA can invade it and exchange with one of its strands, forming a D-loop. This was named *trans* activation because the RecA nucleoprotein filament which activated the DNA is not necessarily the one which invades it (see Mazin and Kowalczykowski, Suggested Reading).

These researchers demonstrated the *trans* activation of double-stranded DNAs by the RecA nucleoprotein filament by first coating a single-stranded DNA with RecA, removing the extra RecA, and then adding the DNA to a double-stranded DNA to which it was not complementary. No strand exchange occurred, because the two DNAs were not complementary. However, when a second single-stranded DNA, which was complementary to the double-stranded DNA, was added, it invaded the double-stranded DNA and exchanged strands to form a D-loop. Apparently, the first single-stranded DNA coated with RecA had somehow "prepared" the double-stranded DNA to be invaded by the second single-stranded DNA, which was not coated with RecA. However, it was necessary to show that the RecA protein had not come off the first single-stranded DNA and bound to the second single-stranded DNA. This was accomplished by using an analog of ATP which is not hydrolyzable (i.e., in which the γ phosphate cannot be cleaved off). This analog of ATP could bind to RecA and allow it to bind to the first single-stranded DNA. However, without ATP cleavage, the RecA could not cycle off the first DNA and bind to the second DNA (see above). It is not clear what happens to a double-stranded DNA when it is activated by a RecA nucleoprotein filament, but the helix may be transiently extended and the strands may be partially separated to allow the single-stranded DNA to search for its complementary sequence.

Also, *trans* activation may be a mechanism for forming R-loops in the DNA. An R-loop is like a D-loop but is formed by the 3' end of an RNA invading a double-stranded DNA rather than a single-stranded DNA. The 3' end of RNA in the R-loop can then be used as a primer to initiate DNA replication. The RecA protein cannot bind to RNA, only to single-stranded DNA. However, if the RecA protein *trans* activates a double-stranded DNA, the *trans*-activated DNA may allow an RNA to invade it. R-loops may also be formed by RNA polymerase when it separates the strands of DNA during transcription to form a transcription bubble (see chapter 2), and it is difficult to distinguish these two mechanisms.

The Ruv and RecG Proteins and the Migration and Cutting of Holliday Junctions

Once a D-loop has formed as the result of the concerted action of RecBCD and RecA or the RecF pathway and RecA, the two DNAs can join via Holliday junctions, using one of the molecular mechanisms described in the earlier section on molecular models of recombination. As discussed earlier in the chapter, Holliday junctions are remarkable structures that can do many things. The branches on the Holliday junction can **isomerize** so that the crossed strands become the uncrossed strands. This process should occur spontaneously since no hydrogen bonds need to be broken or reformed for isomerization. Holliday junctions can also **migrate** when hydrogen bonds break on one side of the Holliday junction and reform on the other side, so that the Holliday junction moves from one place on the DNA to another. Or two of the strands of the Holliday junction can be cut to resolve the Holliday junction. Depending on which two strands are cut, the two DNAs will return to their original configuration or will take up a new configuration to form recombinants.

In *E. coli,* branch migration and the resolution of Holliday junctions can be performed by the three Ruv proteins, RuvA, RuvB, and RuvC, which are encoded by adjacent genes. Recent work on the crystal structures of these proteins has shown that they form interesting structures that give clues to how they function in the migration and resolution of Holliday junctions (Figure 10.7) (see West, Suggested Reading). The RuvA protein is a specific Holliday junction binding protein whose role is to force the Holliday junction into a certain structure amenable to the subsequent steps of migration and resolution. Four copies of the RuvA protein come together to form a flat structure like a flower with four petals. The Holliday junction lies flat on the flower and thus is forced into a flat (planar) configuration. There is a short region in the middle of the Holliday junction where the strands are not base paired and the single strands form a sort of square. Another tetramer of RuvA may then bind to the first to form a sort of turtle shell, with the four arms of double-stranded DNA emerging from the "leg holes." The RuvB protein then forms a hexameric (six-membered) ring encircling one arm of the DNA, as shown in Figure 10.7. The DNA is then pumped through the RuvB ring, using ATP cleavage to drive the pump, thereby forcing the Holliday junction to migrate. Holliday junction migration should not require any energy since as many hydrogen bonds form on one side of the Holliday junction as are broken on the other side when

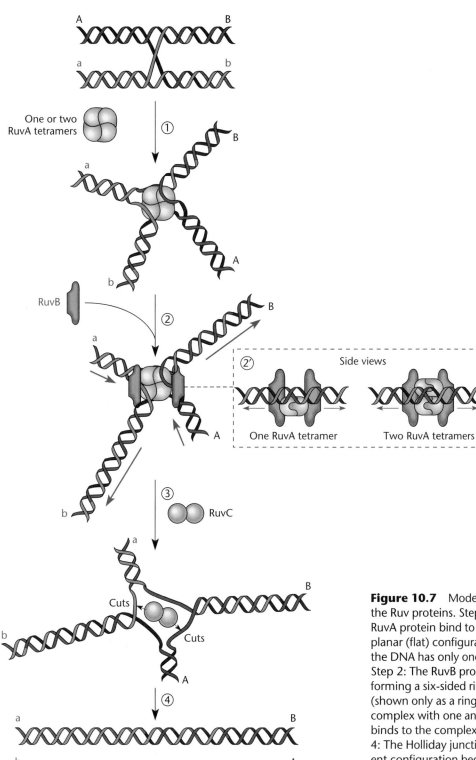

Figure 10.7 Model for the mechanism of action of the Ruv proteins. Step 1: One or two tetramers of the RuvA protein bind to a Holliday junction, holding it in a planar (flat) configuration. Note that at the beginning, the DNA has only one turn of heteroduplex (gold-gray). Step 2: The RuvB protein binds to the RuvA complex, forming a six-sided ring around one strand of the DNA (shown only as a ring). Step 2′: Side view of the complex with one and two tetramers. Step 3: RuvC binds to the complex and cuts two of the strands. Step 4: The Holliday junction has been resolved into a different configuration because of the way the strands were cut. Note that there are now more turns of heteroduplex (gold-gray).

it migrates. Theoretically, any reaction which does not require an input of energy will eventually happen spontaneously. However, spontaneous migration is too slow, and the expenditure of energy by the RuvB pump greatly accelerates the migration and gives it a directionality so that it does not just drift back and forth.

After the RuvA and RuvB proteins have caused Holliday junctions to migrate, they can be cut (resolved) by the RuvC protein. The RuvC protein is a specialized DNA endonuclease which cuts the two crossed strands of a Holliday junction simultaneously. Like many enzymes that make double-stranded breaks in DNA, two identical polypeptides encoded by the *ruvC* gene come together to form a homodimer, which is the active form of the enzyme. Because the enzyme has two copies of the polypeptide, it has two DNA endonuclease active centers, each of which can cut one of the DNA strands to make a double-strand break. RuvC can only cut a Holliday junction that is bound to RuvA and RuvB.

The evidence for the obligatory association of the RuvA, RuvB, and RuvC proteins to function is genetic: mutants with either a *ruvA*, a *ruvB*, or a *ruvC* mutation are indistinguishable in that they are all defective in the resolution of Holliday junctions (see below). To a geneticist, this means that RuvC cannot act to resolve a Holliday junction without RuvA and RuvB being present and bound to the Holliday junction. However, this leads to an apparent conflict with the structural information about RuvA and RuvB discussed above, indicating that RuvA forms a turtle shell-like structure over the Holliday junction. How could RuvC enter the turtle shell formed by RuvA to cleave the crossed DNA strands in the inside? Perhaps a tetramer of RuvA is bound to only one face of the Holliday junction, leaving the other face open for RuvC to bind and cut the crossed strands. However, it seems unlikely that the Holliday junction could be held tightly enough in this way to not be dislodged by the RuvB pump. Another idea is that the RuvA shell opens up somehow to allow RuvC to enter and cut the crossed strands.

The RuvC protein is a member of a large group of specialized DNA endonucleases, called the **X-philes** because the crossed strands of a Holliday junction, as usually drawn, resemble the letter "X" (see Sharples, Suggested Reading). They are also sometimes called cruciform-cutting enzymes because the structure they cut can also be drawn as a cruciform. Some X-philes, such as RuvC, are very specific and will cut only the crossed strands of a Holliday junction, while others will also cut the crossed strands at D-loops. The obligatory association of RuvC with the RuvA shell and the RuvB pump may give it its specificity for Holliday junctions. RuvC is also unusual in that it has some sequence specificity at the site where it cuts. It always cuts next to two T's in the DNA. Note that the crossed strands of DNA are not the

complementary strands but the strands with the same sequence. If one strand has the sequence T-T, the other strand will also have the sequence T-T at that position. Presumably, RuvA and RuvB cause the Holliday junction to migrate until this sequence is encountered and RuvC can cut the crossed strands.

Gram-positive bacteria do not seem to have a RuvC-like X-phile, but they do have another X-phile which is analogous to an alternative X-phile in *E. coli* called YggF. This alternative X-phile is not as specific as RuvC in that it will cut D-loops as well as Holliday junctions. Phages, including T4 and T7, often encode their own X-phile, which will also usually cut D-loops as well as Holliday junctions (Table 10.2). It will obviously take some effort to sort out the contributions of the various X-philes to recombination, replication restarts, and other aspects of DNA metabolism.

Support for this model of the action of RuvA, RuvB, and RuvC has come from observations of purified Ruv proteins acting on artificially synthesized structures that resemble Holliday junctions (see Parsons et al., Suggested Reading). These junctions are constructed by annealing four synthetic single-stranded DNA chains that have pairwise complementarity to each other (Figure 10.8). These synthetic structures are not completely analogous to a real Holliday junction in that they are not made from naturally occurring DNA. Rather, four single strands are synthesized that are complementary to each other in the regions shown and therefore form a cross. A Holliday junction made this way will be much more stable than a natural Holliday junction because the branch cannot migrate. Real Holliday junctions are too unstable for these experiments; they quickly separate into two double-stranded DNA molecules.

Experiments performed with such synthetic Holliday junctions indicated that purified Ruv proteins act sequentially on a synthetic Holliday junction in a manner consistent with the above model. First, RuvA protein

TABLE 10.2	Analogy between phage and host recombination functions
Phage function	**Analogous *E. coli* function**
T4 UvsX	RecA
T4 gene 49	RuvC
T7 gene 3	RuvC
T4 genes 46 and 47	RecBCD
λ ORF in *nin* region	RecO, RecR, RecF
Rac *recE* gene	RecJ, RecQ
λ *gam*	Inhibits RecBCD
λ *exo*	RecBCD, RecJ
λ *bet*	RecA
rusA (DLP12 prophage)	RuvC

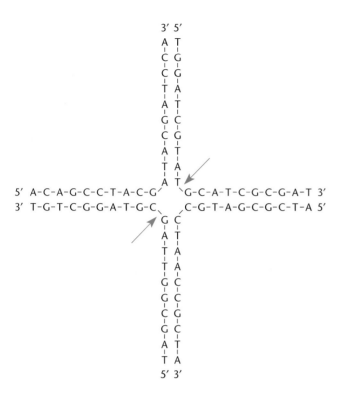

Figure 10.8 A synthetic Holliday junction with four complementary strands. The junction cannot migrate but can be disrupted by RuvA and RuvB. It can also be cut by RuvC (arrows) and other Holliday junction resolvases such as the gene *49* and gene *3* products of phages T4 and T7, respectively.

seems to promote the migration of Holliday junctions only after they have formed, the RecG protein can turn forks into Holliday junctions or bind to three-stranded junctions like those at a D-loop. Therefore, it seems likely that RecG represents a completely different pathway from RuvABC for dealing with DNA branches, but the two pathways can lead to the same result, a recombinant.

One role of RecG might occur when the replication fork encounters an obstacle such as damage in the DNA (see McGlynn and Lloyd, Suggested Reading). The RecG protein may then cause the stalled replication fork to back up, much like a train backs up to allow the track to be repaired, before it moves on (see Box 10.2). This would get the replication apparatus out of the way until the damage to the DNA could be repaired by the repair enzymes, as described in chapter 11. The extra DNA could then be degraded, and the replication fork could proceed, maybe with the help of PriA and DnaC to load the DnaB helicase back on the DNA. Alternatively, if the damage is irreparable, the Holliday junction formed by the backing up of the replication fork by RecG could be cut by RuvC and the cut ends of the Holliday junction could then invade the other daughter DNA to form a D-loop, which could be acted upon by the PriA protein to restart the replication fork. Such mechanisms could increase the survival of cells after DNA damage or could increase the frequency of recombination by a copy choice mechanism as described in Box 10.2; this would explain why *recG* mutants are sensitive to DNA damage and are defective in recombination under some circumstances. Alternatively, the replication fork may encounter another protein, such as RNA polymerase, stalled on the DNA. By backing up the replication fork, RecG may allow another helicase, such as the Rep helicase, to bind to the DNA and displace the obstacle from the DNA, much like the "cowcatcher" on a train sweeps up things ahead of the train. There is some evidence that the Rep helicase may be better than the DnaB helicase at displacing objects from the DNA. Once the obstacle has been displaced, the PriA protein may allow the normal DnaB helicase to reload and permit replication to continue. There are at least 12 different helicases in *E. coli*, and more work is needed to determine the roles of the various helicases and resolvases in recombination and replication. This is an active current area of research.

Recombination in bacteria has some differences from recombination in most other organisms. In bacteria, generally only small pieces of incoming donor DNA recombine with the chromosome, whereas in other organisms, two enormously long double-stranded DNA molecules of equal size usually recombine. Nevertheless, the requirements for recombination in bacteria are likely to be similar to those in other organisms. In fact, accumulating evidence supports the existence of proteins analogous to many of

bound specifically to the synthetic Holliday junctions, and then a combination of RuvA and RuvB caused a disassociation of the synthetic Holliday junctions, simulating branch migration in a natural DNA molecule. The dissocation required the energy in ATP to break the hydrogen bonds holding the Holliday junction together, as predicted. The RuvC protein then specifically cut the synthetic Holliday junction in two of the four strands.

Another helicase in *E. coli* which can help junctions to migrate is RecG (Table 10.1) For a long time, the function of this helicase was unknown. It has very little effect on recombination when RuvABC is present, which suggested that its role is redundant with respect to RuvABC. The idea was that RecG could play the role of RuvAB and promote the migration of Holliday junctions. Then another resolvase could play the role of RuvC and resolve the Holliday junction, providing a backup for RuvABC. However, it now seems that the roles of these proteins are not redundant and they each have their own unique role to play (see McGlynn and Lloyd, Suggested Reading). For one thing, they move in opposite directions on single-stranded DNA. The RuvAB helicase moves in the 5′-to-3′ direction on single-stranded DNA, while the RecG helicase moves in the 3′-to-5′ direction. Also, unlike RuvAB, which

the recombination proteins of bacteria in both eukaryotes and archeae. For example, the Rad51 proteins of yeast and humans are analogous to RecA and can form helical nucleoprotein filaments similar to those formed by the bacterial RecA protein. A protein called CCE1 in yeast specifically cuts Holliday junctions and is related to RuvC.

Phage Recombination Pathways

Many phages encode their own recombination functions, some of which can be important for the multiplication of the phage. For example, as discussed in chapter 7, some phages use recombination to make primers for replication and concatemers for packaging. Also, phage recombination systems may be important for repairing damaged phage DNA and for exchanging DNA between related phages to increase diversity. Phages may encode their own recombination systems to avoid dependence on host systems for these important functions.

Many phage recombination functions have been shown to be analogous to the recombination proteins of the host bacteria (Table 10.2), and in many cases, the phage recombination proteins were discovered before their host counterparts. As a result, studies of bacterial recombination systems have been heavily influenced by simultaneous studies of phage recombination systems.

Rec Proteins of Phages T4 and T7

Phages T4 and T7 depend upon recombination for the formation of DNA concatemers after infection (see chapter 7). Therefore, recombination functions are essential for the multiplication of these phages. Many of the T4 and T7 Rec proteins are analogous to those of their hosts. For example, the gene *49* protein of T4 and the gene *3* protein of T7 are X-phile endonucleases that resolve Holliday junctions and are representative of phage proteins discovered before their host counterparts, in this case RuvC. The gene *46* and *47* products of phage T4 may perform a reaction similar to the RecBCD protein of the host, although no evidence indicates the presence of χ-like sites associated with this enzyme. The UvsX protein of T4 and the Bet protein of λ are analogous to the RecA protein of the host.

The RecE Pathway of the *rac* Prophage

Another classic example of a phage-encoded recombination pathway is the RecE pathway encoded by the *rac* prophage of *E. coli* K-12. The *rac* prophage is integrated at 29 min in the *E. coli* genetic map and is related to λ. This defective prophage cannot be induced to produce infective phage, as it lacks some essential functions for multiplication.

The RecE pathway was discovered by isolating suppressors of *recBCD* mutations, called *sbcA* mutations

(for *suppressor of BC*), that restored recombination in conjugational crosses. The *sbcA* mutations were later found to activate a normally repressed recombination function of the defective prophage *rac*. Apparently, *sbcA* mutations inactivate a repressor that normally prevents the transcription of the *recE* gene, as well as other prophage genes. When the repressor gene is inactivated, the RecE protein will be synthesized and can then substitute for the RecBCD nuclease in recombination.

The Phage λ *red* System

Phage λ also encodes recombination functions. The best characterized is the *red* system, which requires the products of adjacent λ genes *exo* and *bet*. The product of the *exo* gene is an exonuclease that degrades one strand of a double-stranded DNA from the 5′ end to leave a 3′ single-stranded tail. The *bet* gene product is known to help the renaturation of denatured DNA and to bind to the λ exonuclease. Unlike the other recombination systems that we have discussed, the λ *red* recombination pathway does not require the RecA protein. The λ *red* system is the basis for a very useful gene replacment technique (see Box 10.3).

Interestingly, the RecE protein of *rac* and the λ *exo* exonuclease may be similar. The RecE protein needs RecA to promote *E. coli* recombination, but it does not need RecA to promote λ recombination. It is not too surprising that λ and *rac* encode similar recombination functions, since *rac* and λ are related phages.

Besides the *red* system, phage λ encodes another recombination function that can substitute for components of the *E. coli* RecF pathway (see Sawitzke and Stahl, Suggested Reading, and Table 10.2). Apparently, phages can carry components for more than one recombination pathway.

Genetic Analysis of Recombination in Bacteria

One reason we understand so much more about recombination in *E. coli* than in most other organisms is because of the relative ease of doing genetic experiments with this organism. In this section, we'll discuss some of the genetic experiments that have led to our present picture of the mechanisms of recombination in *E. coli*.

Isolating Rec⁻ Mutants of *E. coli*

As in any genetic analysis, the first step in studying recombination in *E. coli* was to isolate mutants defective in recombination. Such mutants are called **Rec⁻ mutants** and have mutations in the *rec* genes, whose products are required for recombination. Two very different approaches were used in the first isolations of Rec⁻ mutants of *E. coli*.

One approach used for selection of Rec⁻ mutants was to select them directly on the basis of their inability to

BOX 10.3

Gene Replacements in *E. coli* with Phage λ Recombination Functions

One of the major advantages of using bacteria and other lower organisms for molecular genetic studies is the relative ease of doing gene replacements with some of these organisms (see the discussion of gene replacements in chapter 3). To perform a gene replacement, a piece of the DNA of an organism is manipulated in the test tube to change its sequence in some desired way. The DNA is then reintroduced into the cell, and the recombination systems of the cell cause the altered sequence of the reintroduced DNA to replace the normal sequence of the corresponding DNA in the chromosome. Because it depends upon homologous recombination, gene replacement requires that the sequence of the reintroduced DNA be homologous to the sequence of the DNA it replaces. However, the homology need not be complete, and minor changes such as base pair changes can be introduced into the chromosome in this way as a type of site-specific mutagenesis. Also, the reintroduced DNA need not be homologous over its entire length, only where the recombination occurs. This makes it possible to use gene replacement to make large alterations such as construction of an in-frame deletion to avoid polarity effects and insertion of an antibiotic resistance gene cassette into the chromosome. If the sequences on both sides of the alteration (the flanking sequences) are homologous to sequences in the chromosome, recombination between these flanking sequences and the chromosome will insert the alteration.

A method has recently been developed for performing gene replacements in *E. coli* by using the phage λ Red pathway (see Table 10.2). This method makes it possible to introduce single-stranded DNAs as short as 30 bases and have the sequence of the introduced DNA replace the corresponding sequence in the chromosome. This is important because the synthesis of single-stranded DNAs of this length has become routine and a single-stranded DNA of this length with any desired sequence can be purchased for as little as $1 per base. This obviates the need for other types of site-specific mutagenesis for making specific changes in a sequence, all of which are tedious and require a certain amount of technical skill.

Methods for gene replacement in *E. coli* have usually relied upon the RecBCD-RecA recombination pathway since this is the major pathway for recombination in *E. coli*. However, the λ Red system has many advantages over the RecBCD-RecA pathway. First, it is more efficient. If the gene replacement system is not highly efficient, only a very low percentage of the organisms will get the replacement. Minor changes, such as single-amino-acid changes in a protein, usually offer no positive selection, and it becomes nec-

essary to screen thousands of individuals to find one with the replacement. It also is generally better to use linear DNA rather than circular DNA for the replacement. A single crossover between a circular DNA, say a DNA cloned in a plasmid, and the chromosome will insert the circular DNA and duplicate the cloned DNA rather than replace it (see the discussion of gene replacements in chapter 3). A second crossover can remove the duplication and replace the chromosomal sequence with the sequence of the cloned DNA, but it is often difficult to find those with a double crossover among the many with only a single crossover. With linear DNA, a single crossover will break the chromosome and kill the cell, and so all of the survivors will have had at least two crossovers. Gene replacements with linear DNA are possible with *recD* mutants since the RecD subunit of the RecBCD enzyme is reponsible for degrading linear DNA. However, RecD⁻ mutants are somewhat sick and hard to handle. Also, they do not maintain plasmids very well, and their recombination is depressed. Finally, the λ Red system can use a short homology, as mentioned; as few as 30 bp is required.

The figure outlines the procedure for using the λ Red system for gene replacements. Panel A shows the structure of the *E. coli* strain which is required. It carries a defective prophage in which most of the λ genes have been deleted except the recombination genes *gam-bet-exo* (see Table 10.2 and Figure 8-1 [the λ genetic map]). Panel B shows the replacement of a sequence in the plasmid by the corresponding region on another plasmid, in which the sequence has been disrupted by introduction of an antibiotic resistance cassette (Abʳ). The region of the plasmid amplified with PCR and the double-stranded DNA fragment carrying the antibiotic resistance cassette flanked by sequences homologous to the sequences to be replaced in the target plasmid is electroporated into the cells. The *gam* gene product, Gam, inhibits the RecBCD enzyme, thereby allowing the use of the linear DNA fragment. The *exo* gene product, Exo, is an exonuclease which plays the role of RecBCD, degrading one strand of a double-stranded DNA, thereby exposing a single strand for strand invasion. The *bet* gene product, Bet, plays the role of RecA, binding to the single-stranded DNA exposed by Exo and promoting synapse formation and strand exchange with another DNA. Panel C shows the replacement if a single stranded oligonucleotide is used to introduce either an in-frame deletion or a single-base-pair change into the target DNA. The procedure is similar, except that if a single-stranded linear DNA is used instead of a double-stranded

(continued)

BOX 10.3 (continued)

Gene Replacements in *E. coli* with Phage λ Recombination Functions

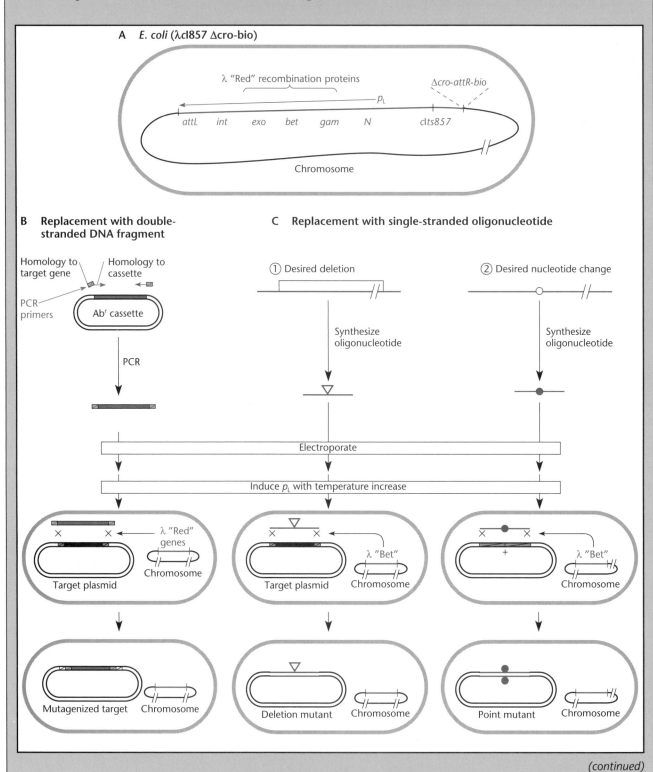

(continued)

BOX 10.3 (continued)

Gene Replacements in *E. coli* with Phage λ Recombination Functions

DNA, the *gam* and *exo* genes can be dispensed with and only *bet* is needed. Without the Gam protein, the RecBCD enzyme will be active and RecBCD can degrade single-stranded DNA to some extent, but the Bet protein might coat the single-stranded DNA, protecting it from degradation by RecBCD. Exo will not act on single-stranded DNA, but it is not needed anyway if the DNA is single-stranded. Only Bet is needed to promote pairing between the introduced single-stranded DNA and the chromosomal DNA and strand exchange. Then replication or repair will replace the normal sequence within the mutant sequence in both strands as shown. It is still necessary to screen the survivors for ones in which the sequence of the introduced DNA has replaced the corresponding sequence in a plasmid or the chromosome.

Interestingly, this method of gene replacement shows a strand bias, meaning that a single-stranded oligonucleotide complementary to one strand is more apt to replace the corresponding chromosomal sequence than is an oligonucleotide complementary to the other strand in any particular region. Which strand is preferred correlates with the direction of replication in the region. The *E. coli* chromosome replicates bidirectionally from the origin, so that on one side of the *oriC* region the replication fork moves in one direction while on the other side it moves in the opposite direction; the sequence which is more easily replaced corresponds to the lagging strand. Presumably the single-stranded gaps which are produced on the lagging strand at the replication fork are sites of strand invasion promoted by Bet, which, unlike RecA, is not able to help a single-stranded DNA invade a completely double-stranded DNA. Apparently, Bet needs a single-stranded gap to get its "foot in the door" by pairing between the two DNAs before it can promote the invasion of adjacent double-stranded DNA.

Reference

Ellis, H. M., D. Yu, T. DiTizio, and D. L. Court. 2001. High efficiency mutagenesis, repair and engineering of chromosomal DNA using single-stranded oligonucleotides. *Proc. Natl. Acad. Sci. USA* **98:**6742–6746.

support recombination (see Clark and Margulies, Suggested Reading). The idea behind this selection was that an *E. coli* strain with a mutation that inactivates a required *rec* gene should not be able to produce recombinant types when crossed with another strain.

In one experiment designed to isolate mutants defective in recombination, a Leu⁻ strain of *E. coli* was mutagenized with nitrosoguanidine. Individual strains, some of which might now also have a *rec* mutation, were then crossed separately with an Hfr strain. A Rec⁻ mutant would have produced no Leu⁺ recombinants when crossed with the Hfr strain.

To cross thousands of the mutagenized bacteria separately with the Hfr strain to find one that gave no recombinants would have been very laborious, but replica plating facilitated the crosses. A few mutant strains gave no Leu⁺ recombinants when crossed with the Hfr strain and were candidates for Rec⁻ mutants. We discuss bacterial techniques such as replica plating and Hfr crosses in chapters 3 and 5.

However, just because the mutants give no recombinants in a cross does not mean that they are necessarily Rec⁻ mutants. For instance, the mutants might have been normal for recombination but defective in the uptake of DNA during conjugation. This possibility was ruled out by crossing the mutants with an F′-containing strain instead of an Hfr strain. As discussed in chapter 5, apparent recombinant types can appear without re-

combination in an F′ cross because the F′ factor can replicate autonomously in the recipient cells; that is, it is a replicon. However, the DNA must still be taken up during transfer of the F′ factor, so that if mutants are defective in DNA uptake, no apparent recombination types would appear in the F′ cross. Normal frequencies of apparent recombinant types appeared when some of the mutants were crossed with F′ strains; therefore, these mutants were not defective in DNA uptake during conjugation but, rather, had defects in recombination. These and other criteria were used to isolate several recombination-deficient Rec⁻ mutants.

The other approach used in the first isolations of recombination-deficient mutants of *E. coli* was less direct (see Howard-Flanders and Theriot, Suggested Reading). These isolations depended on the fact that some recombination functions are also involved in the repair of UV-damaged DNA. Therefore, using methods described in chapter 11, Howard-Flanders and Theriot isolated several repair-deficient mutants and tested them to determine if any were also deficient in recombination. Some, but not all, of these repair-deficient mutants could also be shown to be defective in recombination in crosses with Hfr strains.

COMPLEMENTATION TESTS WITH *rec* MUTATIONS
Once Rec⁻ mutants had been isolated, the number of *rec* genes could be determined by complementation tests. The original *rec* mutations defined three genes of the

bacterium: *recA*, *recB*, and *recC*. The *recB* and *recC* mutants were less defective in recombination and repair than were the *recA* mutants. In fact, the RecA function is the only gene product absolutely required for recombination in *E. coli* and many other bacteria.

MAPPING *rec* GENES

The next step was to map the *rec* genes. This task may seem impossible, as a *rec* mutation causes a deficiency in recombination, but it is the frequency of recombination with known markers that reveals map position (see chapter 3). Crosses with Rec⁻ mutants can be successful only if the donor, but not the recipient, strain has the *rec* mutation. After DNA transfer, even recipient cells which become Rec⁻ will retain recombination activity at least long enough for recombination to occur. Other markers are selected, and the presence or absence of the *rec* mutation is scored as an unselected marker by the UV sensitivity of the recombinants or some other phenotype due to the *rec* mutation.

Using crosses such as those described above, investigators found that *recA* mutations mapped at 51 min and *recB* and *recC* mutations mapped close to each other at 54 min on the *E. coli* genetic map. Later studies showed that the products of the *recB* and *recC* genes, as well as of the adjacent *recD* gene, comprise the RecBCD nuclease, which initiates the major recombination pathway in *E. coli* (see above). The *recD* gene was not found in the original selection, because its product is not essential for recombination under these conditions. In fact, as discussed above, mutations in the *recD* gene can stimulate recombination by preventing degradation of the displaced strand and making recombination independent of χ sites.

Other Recombination Genes

In addition to the *recA* and *recBCD* genes, other genes whose products participate in recombination in *E. coli* have been found (Table 10.1). Many of these genes were not found in the original selections, because inactivating them alone does not sufficiently reduce either recombination or repair in wild-type *E. coli*.

THE RecF PATHWAY

As discussed, the RecF pathway in *E. coli* involves the products of the *recF*, *recJ*, *recN*, *recO*, *recQ*, and *recR* genes. The genes of the RecF pathway of recombination were not discovered in the original selections of *rec* mutations because they are not normally required for recombination after Hfr crosses. By themselves, the *recB* and *recC* mutations reduce recombination after an Hfr cross to about 1% of its normal level. Mutations in any of the genes of the RecF pathway can prevent the residual recombination that occurs in a *recB* or *recC* cell, first, sug-

gesting that these genes are responsible for only a minor pathway of recombination in *E. coli*.

The RecF pathway would be more efficient in Hfr crosses, except that the products of two other *E. coli* genes interfere with it. These genes, named *sbcB* and *sbcC*, were first detected because mutations in them suppress the recombination deficiency in *recB* and *recC* mutants (hence the name *sbc,* for suppressor of *BC*). We know that these suppressor mutations act by allowing stimulation of the RecF pathway, because, for example, the extra recombination in a *recB sbcB* mutant was eliminated by a third mutation in a gene of the RecF pathway, for example, *recQ*. Apparently, the products of the *sbcB* and *sbcC* genes normally inhibit the RecF pathway, perhaps because they have an enzymatic activity that destroys one of the intermediates of this pathway.

But why would the *E. coli* cell seem to invest so many genes in a pathway that is normally responsible for only a small part of the total recombination in the cell, especially when the efficiency of this second pathway is limited by other gene products that the cell itself synthesizes? The answer may lie in the way the original experiments were performed, which may have diminished the role of the RecF pathway. The experiments that revealed the RecF pathway were Hfr crosses in which a large portion of the donor DNA was introduced into the recipient cell by conjugation. These experiments also put no constraints on the time required for recombination, which could have occurred over a long period. However, the RecF pathway may be very important if short pieces of DNA are introduced and recombination is allowed to proceed for only a limited time. Evidence for this hypothesis comes from transduction in *Salmonella enterica* serovar Typhimurium (see Miesel and Roth, Suggested Reading).

THE *ruvABC* AND *recG* GENES

As discussed above, the products of the *ruv* and *recG* genes are involved in the migration and cutting of Holliday junctions. There are three adjacent *ruv* genes: *ruvA*, *ruvB*, and *ruvC*. The *ruvA* and *ruvB* genes are transcribed into a polycistronic mRNA, and the *ruvC* gene is adjacent but independently transcribed. The *recG* gene lies elsewhere in the genome.

The discovery of the role the *ruv* genes play in recombination involved some interesting genetics. The *recG* and *ruvABC* genes were not found in the original selections for recombination-deficient mutants because by themselves, mutations in these genes do not significantly reduce recombination. The *ruv* genes were found only because mutations in them can increase the sensitivity of *E. coli* to killing by UV irradiation. The *recG* gene was found because double mutants with mutations in the *recG* gene and one of the *ruv* genes are severely deficient in recombination (see Lloyd, Suggested Reading). The RecG and Ruv proteins presumably perform overlapping functions in recombination and so can substitute for each other. If one of the *ruv* genes is inactive, the *recG* gene product is still available to migrate Holliday junctions and vice versa. The enzyme that resolves Holliday junctions in a *ruv* mutant is not clear. Redundancy of function is a common explanation in genetics for why some gene products are nonessential. We discuss other examples of redundant functions elsewhere in this book.

Once the Ruv proteins had been shown to be involved in recombination, it took some clever intuition to find that their role is in the migration and resolution of Holliday junctions (see Parsons et al., Suggested Reading). The Ruv proteins were first suspected to be involved in a late stage of recombination because of a puzzling observation: even though *ruv* mutants give normal numbers of transconjugants when crossed with an Hfr strain, they give many fewer transconjugants when crossed with strains containing F′ plasmids. As mentioned above, mating with F′ factors should result in, if anything, more transconjugants than crosses with Hfr strains, because F′ factors are replicons, which can multiply autonomously and do not rely on recombination for their maintenance.

One way to explain this observation is to propose that the Ruv proteins function late in recombination. If the Ruv proteins function late, recombinational intermediates might accumulate in cells with *ruv* mutations, thus having a deleterious effect on the the cell. If so, *recA* mutations, which block an early step in recombination by preventing the formation of synapses, might suppress the deleterious effect of the *ruv* mutations on F′ crosses. This was found to be the case. A *ruv* mutation had no effect on the frequency of transconjugants in F′ crosses if the recipient cells also had a *recA* mutation. Once genetic evidence supported a late role for the Ruv proteins in recombination, the process of Holliday junction resolution became the candidate for that role, since this is the last step in recombination. Biochemical experiments were then used to show that the Ruv proteins help in the migration and resolution of synthetic Holliday junctions, as described above.

DISCOVERY OF χ SITES
The discovery on DNA of χ sites that stimulate recombination by the RecBCD nuclease also required some interesting genetics. Many sites that are subject to single- or double-strand breaks are known to be "hot spots" for recombination. In some cases, such as recombination initiated by homing enzymes (see Box 10.1), it is clear that breaks at specific sites in DNA can initiate recombination. In general, however, the frequency of recombination seems to correlate fairly well with physical distance on DNA, as though recombination occurs fairly uniformly throughout DNA molecules.

It came as some surprise, therefore, to discover that the major recombination pathway for Hfr crosses in *E. coli*, the RecBCD pathway, does occur through specific sites on the DNA—the χ sites. Like many important discoveries in science, the discovery of χ sites started with an astute observation. This observation was made during studies of host recombination functions using λ phage (see Stahl et al., Suggested Reading). The experiments were designed to analyze the recombinant types that formed when the phage *red* recombination genes were deleted. Without its own recombination functions, the phage requires the host RecBCD nuclease. In addition, if the phage is also a *gam* mutant, it will not form a plaque unless it can recombine. Therefore, plaque formation by a *red gam* mutant phage λ is an indication that recombination has taken place.

The reason that *red gam* mutant phage λ cannot multiply to make a plaque without recombination is somewhat complicated. As discussed in chapter 8, phage λ cannot package DNA from genome-length DNA molecules but only from concatemers in which the λ genomes are linked end to end. Normally, the phage makes concatemers by rolling-circle replication. However, if the phage is a *gam* mutant, it cannot switch to the rolling-circle mode of replication because the RecBCD nuclease, which is normally inhibited by Gam, will somehow block the switch. Therefore, the only way a *gam* mutant phage λ can form concatemers is by recombination between the circular λ DNAs formed via θ replication. If the phage is also a *red* mutant (i.e., lacks its own recombination functions), the only way it can form concatemers is by RecBCD recombination, the major host pathway. Therefore, *red gam* mutant phage λ requires RecBCD recombination to form plaques, and the formation of plaques can be used as a measure of RecBCD recombination under these conditions.

χ mutations were discovered when large numbers of *red gam* mutant phage λ were plated on RecBCD⁺ *E. coli*. Very few phage were produced, and the plaques that formed were very tiny. Apparently, very little RecBCD recombination was occurring between the circular phage DNAs produced by θ replication. However, λ mutants that produced much larger plaques sometimes appeared. The circular λ DNAs in these mutants were apparently recombining at a much higher rate. The responsible mutations were named χ **mutations** because they increase the frequency of crossovers (crossed-over chromosomes are called *chi*asma in eukaryotes). Once the mutations were mapped and the DNA was sequenced, χ mutations in λ were found to be base pair changes that created the sequence 5'GCTGGTGG3' somewhere in the λ DNA. The presence of this sequence appears to stimulate recombination by the RecBCD pathway. Since wild-type λ has no such sequence anywhere in its DNA, recombination by RecBCD is very infrequent unless the χ sequence is created by a mutation.

Why wild-type λ has no χ sequences seems obvious. Phage λ normally uses its own recombination system, and so with no need for the RecBCD enzyme, it also has no need for χ sites. Bacterial DNA, by contrast, has many χ sites—many more than would be predicted by chance alone.

Further experimentation with χ sites revealed several interesting properties. For example, they stimulate crossovers only to one side of themselves, the 5' side. Very little stimulation of crossovers occurs on the 3' side. Also, if only one of the two parental phages contains a χ site, most of the recombinant progeny will not have the χ site, so that the χ site itself is preferentially lost during the recombination. These properties of χ sites led to models for χ site action such as the one presented earlier in this chapter.

Gene Conversion and Other Manifestations of Heteroduplex Formation during Recombination

GENE CONVERSION

As discussed earlier in this chapter, models for recombination are based in part on the evidence concerning the formation of heteroduplexes during recombination. The first such evidence came from studies of gene conversion in fungi. Understanding this process requires some knowledge of the sexual cycles of fungi. Some fungi have long been favored organisms for the study of recombination because the spores that are the products of a single meiosis are often contained in the same bag, or ascus (see any general genetics textbook). When two haploid fungal cells mate, the two cells fuse to form a diploid zygote. Then the homologous chromosomes pair and replicate once to form four chromatids that recombine with each other before they are packaged into spores. Therefore, each ascus will contain four spores (or eight in fungi such as *Neurospora crassa,* in which the chromatids replicate once more before spore packaging).

Since both haploid fungi contribute equal numbers of chromosomes to the zygote, their genes should show up in equal numbers in the spores in the ascus. In other words, if the two haploid fungi have different alleles of the same gene, two of the four spores in each ascus should have an allele from one parent and the other two spores should have the allele from the other parent. This is called a 2:2 segegation. However, the two parental alleles sometimes do not appear in equal numbers in the spores. For example, three of the spores in a particular ascus might have the allele from one parent while the remaining spore has the allele from the other parent—a 3:1 segregation instead of the expected 2:2 segregation. In this case, an allele of one of the parents appears to have been converted into the allele of the other parent during meiosis. The term **gene conversion** originates from this phenomenon.

Gene conversion is caused by repair of mismatches created on heteroduplexes during recombination, and Figure 10.9 shows how such mismatch repair can con-

Figure 10.9 Repair of a mismatch in a heteroduplex region formed during recombination can cause gene conversion. A plus sign indicates the wild-type sequence, and a minus sign indicates the mutant sequence. See the text for details.

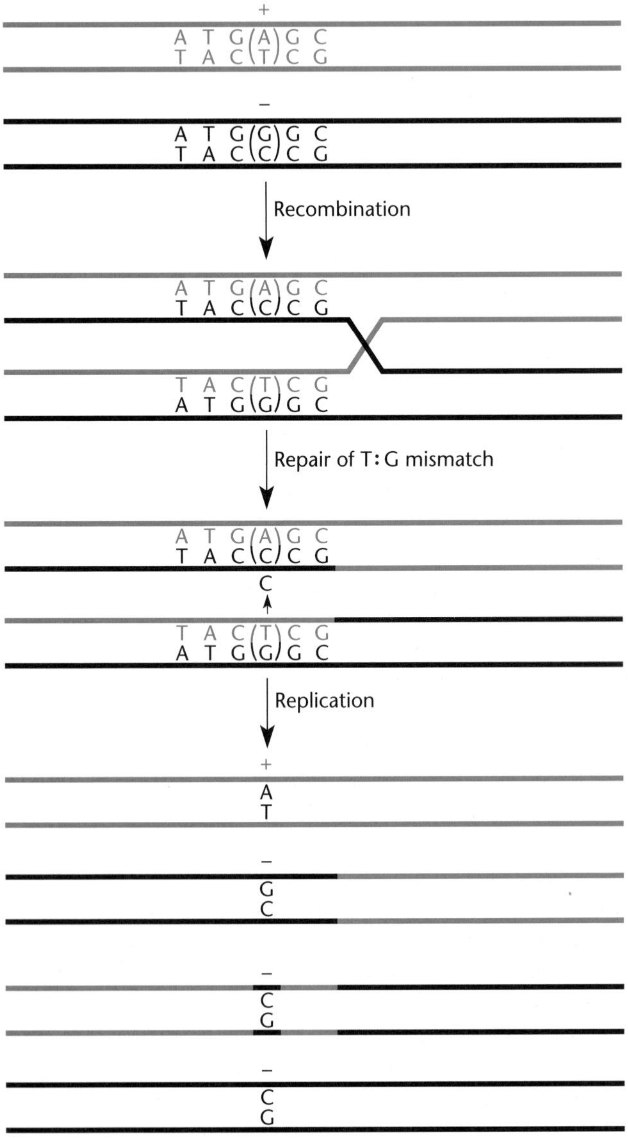

vert one allele into the other when the DNA molecules of the two parents recombine. In the illustration, the DNAs of two parents are identical in the region of the recombination except that a mutation has changed a wild-type AT pair into a GC in one of the DNA molecules. Hence, the parents have different alleles of this gene. When the two individuals mate to form a diploid zygote and their DNAs recombine during meiosis, one strand of each DNA may pair with the complementary strand of the other DNA in this region. A mismatch will result, with a G opposite a T in one DNA and an A opposite a C in the other DNA (Figure 10.9). If a repair system changes the T opposite the G to a C in one of the DNAs, then after meiosis three molecules will carry the mutant allele sequence, with GC at this position, but only one DNA will have the wild-type allele sequence, with AT at this position. Hence, one of the two wild-type alleles has been converted into the mutant allele.

MANIFESTATIONS OF MISMATCH REPAIR IN HETERODUPLEXES IN PHAGES AND BACTERIA

Gene conversion is difficult to detect in crosses with bacteria and phages, since the products of a single recombination event are not contained in an ascus like they are in some fungi. Heteroduplexes do form during recombination in bacteria and phages, but the repair of mismatches in these structures is manifested in other ways.

Map Expansion

In prokaryotes, mismatch repair in heteroduplexes can increase the apparent recombination frequency between two closely linked markers, making the two markers seem farther apart than they really are. This manifestation of mismatch repair in heteroduplexes formed during recombination is called **map expansion** because the genetic map appears to increase in size.

Figure 10.10 shows how mismatch repair can affect the apparent recombination frequency between two markers. In the illustration, the two DNA molecules participating in the recombination have mutations that are very close to each other, so that crossovers between the two mutations to give wild-type recombinants should be very rare. However, a Holliday junction occurs nearby, and the region of one of the two mutations is included in the heteroduplex, creating mismatches that can be repaired. If the G in the GT mismatch in one of the DNA molecules is repaired to an A, the progeny with the DNA will appear to be a wild-type recombinant. Therefore, even though the potential crossover that caused the formation of heteroduplexes did not occur in the region between the two mutations, apparent wild-type recombinants resulted. The apparent recombination due to mismatch repair might occur even when the Holliday

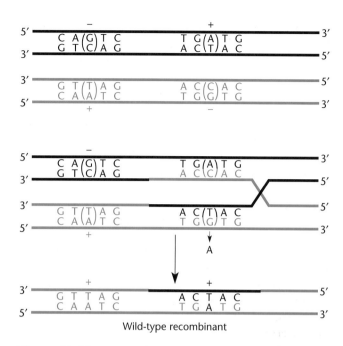

Figure 10.10 Repair of mismatches can give rise to recombinant types between two mutations. A plus sign indicates the wild-type sequence, and a minus sign indicates the mutant sequence. The position of the mutations is shown in parentheses. See the text for details.

junction is resolved so that the flanking sequences are in their original configuration; thus, a true crossover does not result. Therefore, although gene conversion and other manifestations of mismatch repair are generally associated with recombinant DNA molecules, they do not appear only in DNA molecules with obvious crossovers.

Marker Effects

Mismatch repair of heteroduplexes also can cause **marker effects**, phenomena in which two different markers at exactly the same locus show different recombination frequencies when crossed with the same nearby marker. For example, two different transversion mutations might change a UAC codon into UAA and UAG codons in different strains. However, when these two strains are crossed with another strain with a third nearby mutation, the recombination frequency between the ochre mutation and the third mutation might appear to be much lower than the recombination frequency between the amber mutation and the third mutation, even though the amber and ochre mutations are exactly the same distance on DNA from the third mutation. Such a difference between the two recombination frequencies can be explained because the amber and ochre mutations are causing different mismatches to form during recombination and one of these may be recognized and repaired more readily by the mismatch repair system than the other.

Marker effects also occur because the lengths of single DNA strands removed and resynthesized by different mismatch repair systems will vary (see chapter 11), and the chance that mismatch repair will lead to apparent recombination depends upon the length of these sequences, or patches. As is apparent in Figure 10.10, a wild-type recombinant will only occur if mismatches due to both mutations are not removed on the same repair patch. If the patch that is removed in repairing one mismatch also removes the other mismatch, one of the parental DNA sequences will be restored and no apparent recombination will occur.

High Negative Interference

Another manifestation of mismatch repair in heteroduplexes is **high negative interference**. This has the reverse effect of interference in eukaryotes, a phenomenon in which the stiffness of the chromatids can cause one crossover to reduce the chance of another crossover nearby. In high negative interference, one crossover greatly *increases* the apparent frequency of another crossover nearby.

High negative interference is often detected during three-factor crosses with closely linked markers. In chapter 7, we discussed how three-factor crosses can be used to order three closely linked mutations. Briefly, if one parent has two mutations and the other parent a third mutation, the frequency of the different types of recombinants after the cross will depend on the order of the sites of the three mutations in the DNA. Twice as many apparent crossovers are required to produce the rarest recombinant type as are needed for the more frequent recombinant types.

However, owing to mismatch repair of heteroduplexes, the rarest recombinant type can occur much more frequently than expected. Figure 10.11 shows a three-factor cross between markers that are created by three mutations. Wild-type sequences are marked with a plus and mutant sequences with a minus. With the molecules pictured, the formation of a wild-type recombinant should require two crossovers: one between mutations 1 and 2 and another between mutations 2 and 3. Theoretically, if the two crossovers were truly in-

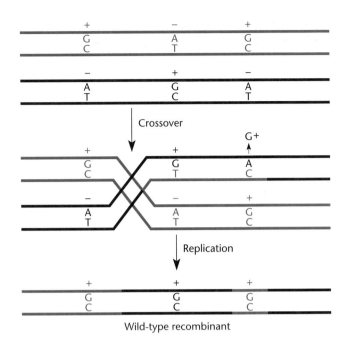

Figure 10.11 High negative interference due to mismatch repair. See the text for details.

dependent, the frequency of wild-type recombinants in the three-factor cross should equal the frequency when mutation 1 is crossed with mutation 2 times the frequency when mutation 2 is crossed with mutation 3, in separate crosses. Instead, the frequency of the wild-type recombinants in the three-factor cross is often much higher than this product. As shown in Figure 10.11, if a crossover occurs between two of the mutations, a heteroduplex formed by the Holliday junction might include the region of the third mutation. Then repair of the third mutant site mismatch can give the appearance of a second nearby crossover, greatly increasing the apparent frequency of that crossover.

It is important to realize that repair of mismatches can occur wherever two DNAs are recombining. However, only if two markers are close enough together that normal crossovers are infrequent will mismatch repair in the heteroduplexes contribute significantly to apparent recombination frequencies and cause effects such as map expansion, high negative interference, and marker effects.

SUMMARY

1. Recombination is the joining of DNA strands in new combinations. Homologous recombination occurs only between two DNA molecules that have the same sequence in the region of the crossover.

2. Models of recombination can be divided into one- or two-strand invasion models and double-strand break repair models.

(continued)

SUMMARY (continued)

3. All recombination models involve the formation of Holliday junctions. The Holliday junctions can migrate and isomerize so that the crossed strands uncross and then recross in a different orientation. Holliday junctions can then be resolved by specific DNA endonucleases to give recombinant DNA products.

4. The region over which two strands, originating from different DNA molecules, are paired in a Holliday junction is called a heteroduplex.

5. Repair of mismatches on the heteroduplex DNA molecules formed as intermediates in recombination can give rise to such phenomena as gene conversion, map expansion, high negative interference, and marker effects.

6. In *E. coli*, the major pathway for recombination during conjugation and transduction is the RecBCD pathway. The RecBCD protein loads on the DNA at a double-stranded break and moves along the DNA, looping out a single strand and degrading one strand from the 3′ end. If a sequence in the DNA called the χ sequence is encountered as the protein moves along the DNA, the 3′-to-5′ nuclease activity on RecBCD is inhibited, and a 5′-to-3′ activity is stimulated, leaving a free 3′ end that can invade other double-stranded DNAs.

7. The RecF pathway is another recombination pathway in *E. coli* which might be required for recombination between single-stranded gaps in DNA and double-stranded DNA. The RecF pathway requires the products of the *recF*, *recJ*, *recN*, *recO*, *recQ*, and *recR* genes. This pathway can function in Hfr crosses only if the *sbcB* and *sbcC* genes have been inactivated. Apparently, the SbcB and SbcC enzymes destroy intermediates created during recombination by the RecF pathway.

8. The RecA protein promotes synapse formation and strand displacement and is required for recombination by both the RecBCD and RecF pathways. The RecA protein forces a single-stranded free end of DNA into a helical nucleoprotein filament, which can then *trans* activate other double-stranded DNAs, looking for its complementary sequence. If it finds its complementary sequence, it invades it, forming a D-loop.

9. Holliday junctions can migrate by two separate pathways in *E. coli*, the RuvABC pathway and the RecG pathway. In the first pathway, the RuvA protein binds to Holliday junctions and then the RuvB protein binds to RuvA and promotes branch migration with the energy derived from cleaving ATP. The RuvC protein is a Holliday junction-specific endonuclease that cleaves Holliday junctions to resolve recombinant products. The RecG protein is also a Holliday junction-specific helicase, but it moves Holliday junctions in the opposite direction from RuvB and can substitute for the RuvABC system.

10. Many phages also encode their own recombination systems. Sometimes, phage recombination functions are analogous to host recombination functions. The gene *49* product of T4 phage and the gene *3* product of T7 resolve Holliday junctions. A cryptic phage, *rac*, found in *E. coli* K-12 strains encodes a recombination system called RecE, which can be activated by *sbcA* mutations and substitute for the RecBCD system in conjugational and transductional crosses. Phage λ also encodes recombination systems, one of which can partially substitute for components of the RecF pathway. Another λ recombination system, the *red* system, is encoded by two genes, *exo* and *bet*. The *exo* gene product is analogous to RecBCD in that it degrades one strand of a double-stranded DNA to make a single-stranded DNA for strand invasion. The *bet* gene product is analogous to RecA in that it promotes synapse formation between two DNAs.

QUESTIONS FOR THOUGHT

1. Why do you suppose essentially all organisms have recombination systems?

2. Why do you suppose the RecBCD protein promotes recombination through such a complicated process?

3. Why are there overlapping pathways of recombination that can substitute for each other?

4. What is the real function of the RecF pathway? Why do you think the cell encodes the *sbcB* and *sbcC* gene products that interfere with the RecF pathway?

5. Why do some phage encode their own recombination systems—why not rely exclusively on the host pathways?

6. Propose a model for how the RecG protein could substitute for RuvABC when it has only a helicase activity and not an X-phile resolvase activity.

PROBLEMS

1. Describe how you would determine if recombinants in an Hfr cross have a *recA* mutation. Note: *recA* mutations make the cells very sensitive to mitomycin and UV irradiation.

2. How would you determine if the products of other genes participate in the *recG* pathway of migration and resolution of Holliday junctions? How would you find such genes?

3. Describe the recombination promoted by homing double-stranded nucleases to insert an intron by using the double-strand break repair model.

4. Design an experiment to determine whether recombination due to the RecBC nuclease without the RecD subunit is still stimulated by χ.

SUGGESTED READING

Arnold, D. A., and S. C. Kowalczykowski. 2000. Facilitated loading of RecA protein is essential to recombination by RecBCD enzyme. *J. Biol. Chem.* **275**:12261–12265

Chedin, F., and S.C. Kowalczykowski. 2002. A novel family of regulated helicases/nucleases from gram-positive bacteria: insights into the initiation of DNA recombination. *Mol. Microbiol.* **43**:823–834. (Microreview.)

Clark, A. J., and A. D. Margulies. 1965. Isolation and characterization of recombination-deficient mutants of *Escherichia coli* K12. *Proc. Natl. Acad. Sci. USA* **62**:451–459.

Dixon, D. A., and S. C. Kowalcykowski. 1993. The recombination hotspot Chi is a regulatory sequence that acts by attenuating the nuclease activity of the *E. coli* RecBCD enzyme. *Cell* **73**:87–96.

Holliday, R. 1964. A mechanism for gene conversion in fungi. *Genet. Res.* **5**:282–304.

Howard-Flanders, P., and L. Theriot. 1966. Mutants of *Escherichia coli* defective in DNA repair and in genetic recombination. *Genetics* **53**:1137–1150.

Jones, M., R. Wagner, and M. Radman. 1987. Mismatch repair and recombination in *E. coli. Cell* **50**:621–626.

Lloyd, R. G. 1991. Conjugal recombination in resolvase-deficient *ruvC* mutants of *Escherichia coli* K-12 depends on *recG. J. Bacteriol.* **173**:5414-5418.

Lloyd, R. G., and C. Buckman. 1985. Identification and genetic analysis of *sbcC* mutations in commonly used *recBC sbcB* strains of *Escherichia coli* K-12. *J. Bacteriol.* **164**:836–844.

Mazin, A. V., and S. C. Kowalczykowski. 1999. A novel property of the RecA nucleoprotein filament: activation of double-stranded DNA for strand exchange in *trans. Genes Dev.* **13**:2005–2016

McGlynn, P., and R. G. Lloyd. 2000. Modulation of RNA polymerase by (p)ppGpp reveals a RecG-dependent mechanism for replication fork progression. *Cell* **101**:35–45

Meselson, M. S., and C. M. Radding. 1975. A general model for genetic recombination. *Proc. Natl. Acad. Sci. USA* **2**:358–361.

Miesel, L., and J. R. Roth. 1996. Evidence that SbcB and RecF pathway functions contribute to RecBCD-dependent transductional recombination. *J. Bacteriol.* **178**:3146–3155.

Parsons, C. A., I. Tsaneva, R. G. Lloyd, and S. C. West. 1992. Interaction of *Escherichia coli* RuvA and RuvB proteins with synthetic Holliday junctions. *Proc. Natl. Acad. Sci. USA* **89**:5452–5456.

Sawitzeke, J. A., and F. W. Stahl. 1992. Phage λ has an analog of *Escherichia coli recO, recR* and *recF* genes. *Genetics* **130**:7–16.

Sharples, G. J. 2001. The X philes: structure-specific endonucleases that resolve Holliday junctions. *Mol. Microbiol.* **39**:823–834. (Microreview.)

Stahl, F. W., M. M. Stahl, R. E. Malone, and J. M. Crasemann. 1980. Directionality and nonreciprocality of Chi-stimulated recombination in phage λ. *Genetics* **94**:235–248.

Szostak, J. W., T. L. Orr-Weaver, R. J. Rothstein, and F. W. Stahl. 1983. The double-strand-break repair model for recombination. *Cell* **33**:25–35.

West, S. C. 1998. RuvA gets X-rayed on Holliday. *Cell* **94**:699–701. (Minireview.)

11

DNA Repair
and Mutagenesis

THE CONTINUITY OF SPECIES from one generation to the next is a tribute to the stability of DNA. If DNA were not so stable and were not reproduced so faithfully, there could be no species. Before DNA was known to be the hereditary material and its structure determined, a lot of speculation centered around what types of materials would be stable enough to ensure the reliable transfer of genetic information over so many generations (see, for example, Schrodinger, Suggested Reading in the introductory chapter). Therefore, the discovery that the hereditary material is DNA—a chemical polymer no more stable than many other chemical polymers—came as a surprise.

Evolution has resulted in a design for the DNA replication apparatus that minimizes mistakes (see chapter 1). However, mistakes during replication are not the only threats to DNA. Since DNA is a chemical, it is constantly damaged by chemical reactions. Many environmental factors can damage this molecule. Heat can speed up spontaneous chemical reactions, leading, for example, to the deamination of bases. Chemicals can react with DNA, adding groups to the bases or sugars, breaking the bonds of the DNA, or fusing parts of the molecule to each other. Irradiation at certain wavelengths can also chemically damage DNA, which can absorb the energy of the photons. Once the molecule is energized, bonds may be broken or parts may be fused. DNA damage can be very deleterious to cells because their DNA may not be able to replicate over the damaged area and so the cells could not multiply. Even if the damage does not block replication, replicating over the damage can cause mutations, many of which may be deleterious or even lethal. Obviously, cells need mechanisms for DNA damage repair.

To describe DNA damage and its repair, we need to first define a few terms. Chemical damage in DNA is called a **lesion**. Chemical compounds or treatments that cause lesions in DNA can kill cells and can also increase the frequency of mutations in DNA. Such treatments that generate mutations are called **mutagenic treatments** or **mutagens**. Some mutagens, known as **in vitro mutagens**, can be used to damage DNA in the test tube and then produce mutations when this DNA is introduced into cells. Other mutagens damage DNA only in the cell, for example by interfering with base pairing during replication. These are called **in vivo mutagens**.

In this chapter, we discuss the types of DNA damage, how each type of damage to DNA might cause mutations, and how bacterial cells repair the damage to their DNA. Many of these mechanisms seem to be universal and are shared by higher organisms including humans.

Evidence for DNA Repair

Before discussing specific types of DNA damage, we should make some general comments on the outward manifestations of DNA damage and its repair. The first question is how we even know a cell has the means to repair a particular type of damage to its DNA. One way is to measure killing by a chemical or by irradiation. The chemical agents and radiation that damage DNA also often damage other cellular constituents, including RNA and proteins. Nevertheless, cells exposed to these agents usually die as a result of chemical damage to the DNA. The other components of the cell can usually be resynthesized and/or exist in many copies, so that even if some molecules are damaged, more of the same type of molecule will be there to substitute for the damaged ones. However, a single chemical change in the enormously long chromosomal DNA of a cell can prevent the replication of that molecule and subsequently cause cell death unless the damage is repaired.

To measure cell killing—and thereby demonstrate that a particular type of cell has DNA repair systems—we can compare the survival of the cells exposed intermittently to small doses of a DNA-damaging agent with that of cells that receive the same amount of treatment continuously. If the cells have DNA repair systems, more will survive the short intervals of treatment because some of the damage will be repaired between treatments. Consequently, at the end of the experiment, fewer intermittently treated cells will have been killed than cells exposed to continuous treatment. In contrast, if DNA is not repaired during the rest periods, whether the treatment occurs at intervals or continuously should make no difference. The cells will accumulate the same amount of damage regardless of the different treatments, and the same fraction of cells should survive both regimes.

Another indication of repair systems comes from the shape of the killing curves. A killing curve is a plot of the number of surviving cells versus the extent of treatment by an agent that damages DNA. The extent of treatment can refer to the length of time the cells are irradiated or exposed to a chemical that damages DNA or to the intensity of irradiation or the concentration of the damaging chemical.

The two curves in Figure 11.1 contrast the shapes of killing curves for cells with and without DNA repair systems. In the curve for cells without a repair system for the DNA-damaging treatment, the fraction of surviving cells drops exponentially, since the probability that each cell will be killed by a lethal "hit" to its DNA is the same during each time interval. This exponential decline gives rise to a straight line when plotted on semilog paper as shown.

The other curve shows what happens if the cell has DNA repair systems. Rather than dropping exponentially with increasing treatment, this curve extends horizontally first, creating a "shoulder." The shoulder appears because repair mechanisms repair lower levels of damage, allowing many of the cells to survive. Only with higher treatment levels, when the damage becomes so extensive that the repair systems can no longer cope with it, will the number of surviving cells drop exponentially with increasing levels of treatment.

Figure 11.1 Survival of cells as a function of the time of treatment with a DNA-damaging agent. The fraction of surviving cells is plotted against the duration of treatment. A shoulder on the survival curve indicates the presence of a repair mechanism.

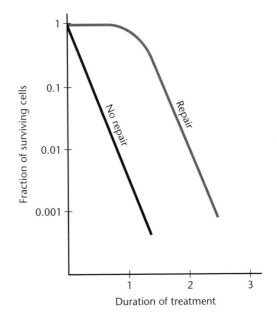

Among the survivors of DNA-damaging agents, there may be many more mutants than before. However, it is very important to distinguish DNA damage from mutagenesis. In particular types of cells, some types of DNA damage are mutagenic while others are not, independent of their effect on cell survival. Recall from chapter 3 that mutations are permanent heritable changes in the sequence of nucleotides in DNA. The damage to DNA caused by a chemical or by irradiation is not by itself a mutation because it is not heritable. A mutation might occur because the damage is not repaired and the replication apparatus must proceed over the damage, making mistakes because complementary base pairing does not occur properly at the site of the damage. Alternatively, mistakes might be made during attempts to repair the damage, causing changes in the sequence of nucleotides at the site of the damage. In the following sections, we discuss some types of DNA damage, how they might cause mutations, and the repair systems that can repair them. Most of what we describe is known for *Escherichia coli,* for which these systems are best understood.

Specific Repair Pathways

Different agents damage DNA in different ways, and different repair pathways operate to repair the various forms of damage. Some of these repair pathways repair only a certain type of damage, whereas others are less specific and repair many types. We first discuss examples of damage repaired by specific repair pathways.

Deamination of Bases

One of the most common types of damage to DNA is the deamination of bases. Some of the amino groups in adenine, cytosine, and guanine are particularly vulnerable and can be removed spontaneously or by many chemical agents (Figure 11.2). When adenine is deaminated, it becomes **hypoxanthine.** When guanine is deaminated, it becomes **xanthine.** When cytosine is deaminated, it becomes **uracil.**

Deamination of DNA bases is mutagenic because it results in base mispairing. As shown in Figure 11.2, hypoxanthine derived from adenine will pair with the base cytosine instead of thymine, and uracil derived from the deamination of cytosine will pair with adenine instead of guanine.

The type of mutation caused by deamination depends on which base is altered. For example, the hypoxanthine that results from deamination of adenine will pair with cytosine during replication, incorporating C instead of T at that position. In a subsequent replication, the C will pair with the correct G, causing an AT-to-GC transition in the DNA. Similarly, a uracil resulting from the deami-

nation of a cytosine will pair with an adenine during replication, causing a GC-to-AT transition.

DEAMINATING AGENTS

Although deamination often occurs spontaneously, especially at higher temperatures, some types of chemicals react with DNA and remove amino groups from the bases. Treatment of cells or DNA with these chemicals, known as **deaminating agents,** can greatly increase the rate of mutations. Which deaminating agents are mutagenic in a particular situation depends upon the properties of the chemical.

Hydroxylamine

Hydroxylamine specifically removes the amino group of cytosine and consequently causes only GC-to-AT transitions in the DNA. However, hydroxylamine, an in vitro mutagen, cannot enter cells, so it can be used only to mutagenize purified DNA or viruses. Mutagenesis by hydroxylamine is particularly effective when the treated DNA is introduced into cells deficient in repair by the uracil-*N*-glycosylase enzyme (see chapter 1), for reasons we discuss below.

Bisulfite

Bisulfite will also deaminate only cytosine, but these cytosines must be in single-stranded DNA. This property of bisulfite has made it useful for **site-directed mutagenesis.** If the region to be mutagenized in a clone is made single stranded, the bisulfite will preferentially limit mutagenesis to this single-stranded region. Use of bisulfite for site-directed mutagenesis of cloned DNAs has been largely supplanted by oligonucleotide-directed and PCR mutagenesis (see chapter 1).

Nitrous Acid

Nitrous acid not only deaminates cytosines but also removes the amino groups of adenine and guanine (Figure 11.2). It also causes other types of damage. Because it is less specific, nitrous acid can cause both GC-to-AT and AT-to-GC transitions as well as deletions. Nitrous acid can enter some types of cells and so can be used as a mutagen both in vivo and in vitro.

REPAIR OF DEAMINATED BASES

Because base deamination is potentially mutagenic, special enzymes have evolved to remove deaminated bases from DNA. These enzymes, **DNA glycosylases,** break the glycosyl bond between the damaged base and the sugar in the nucleotide. A unique DNA glycosylase exists for each type of deaminated base and will remove only that particular base. Specific DNA glycosylases discussed in later chapters remove bases damaged in other ways.

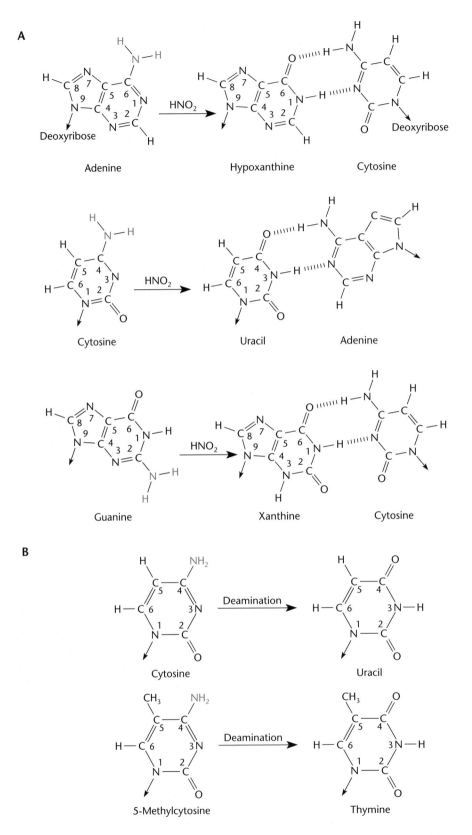

Figure 11.2 (A) Modified bases created by deaminating agents such as nitrous acid (HNO$_2$). Some deaminated bases will pair with the wrong base, causing mutations. (B) Spontaneous deaminations. Deamination of 5-methylcytosine will produce a thymine that is not removable by the uracil-*N*-glycosylase.

Figure 11.3 illustrates the removal of damaged bases from DNA by DNA glycosylases. After the base has been removed by the specific glycosylase, nucleases called **AP endonucleases** cut the sugar-phosphate backbone of the DNA next to the missing base. The AP endonuclease can cut either next to the spot from which a pyrimidine (C or T) has been removed (an *apy*rimidinic site) or next to where a purine (A or G) has been removed (an *ap*urinic site). After the cut is made, the free 3′-hydroxyl end is used as a primer by the repair DNA polymerase (DNA polymerase I in *E. coli*) to synthesize more DNA, while the 5′ exonuclease activity associated with the DNA polymerase degrades the strand ahead of the DNA polymerase. In this way, the entire region of the DNA strand around the deaminated base is resynthesized and the normal base is inserted in place of the damaged one.

VERY SHORT PATCH (VSP) REPAIR OF DEAMINATED 5-METHYLCYTOSINE

Most organisms have some 5-methylcytosine bases instead of cytosines at specific sites in their DNA. These bases are cytosines with a methyl group at the 5 position on the pyrimidine ring instead of the usual hydrogen (Figure 11.2B). Specific enymes called **methyltransferases** transfer the methyl group to this position after the DNA is synthesized. The function of these 5-methylcytosines is often obscure, but we know that they sometimes help protect DNA against cutting by restriction endonucleases and may help regulate gene expression in higher organisms.

The sites of 5-methylcytosine in DNA are often hot spots for mutagenesis, because deamination of 5-methylcytosine yields thymine rather than uracil (Figure 11.2B), and thymine in DNA will not be recognized by the uracil-N-glycosylase since it is a normal base in DNA. These thymines in DNA will be opposite guanines and so could in principle be repaired by the methyl-directed mismatch repair system (see below and chapter 1). However, in a GT mismatch created by a replication mistake, the mistakenly incorporated base can be identified because it is in a newly replicated strand, as yet unmethylated by Dam methylase, whereas the GT mismatches created by the deamination of 5-methylcytosine are generally not in newly synthesized DNA. Repairing the wrong strand will cause a GC-to-AT transition in the DNA.

In *E. coli* K-12 and some other enterics, most of the 5-methylcytosine in the DNA occurs in the second C of the sequence 5′CCWGG3′/3′GGWCC5′, where the middle base pair (W) is generally either AT or TA. The second C in this sequence is methylated by an enzyme called DNA cytosine methylase (Dcm) to give $C^mCAGG/GGTC^mC$. Because of the mutation potential, *E. coli* K-12 has evolved a special repair mechanism for deaminated 5-methylcytosines that occur in this sequence. This repair system specifically removes a thymine whenever it appears as a TG mismatch in this sequence.

Because a small region, or "patch," of the DNA strand containing the T is removed and resynthesized during the repair process, the mechanism is called **very short patch (VSP) repair**. In VSP repair, the Vsr endonuclease, the product of the *vsr* gene, binds to a TG mismatch in the $C^mCWGG/GGWTC$ sequence and makes a break next to the T. The T is then removed and the strand resynthesized by the repair DNA polymerase (DNA polymerase I), which inserts the correct C (see Lieb et al., Suggested Reading).

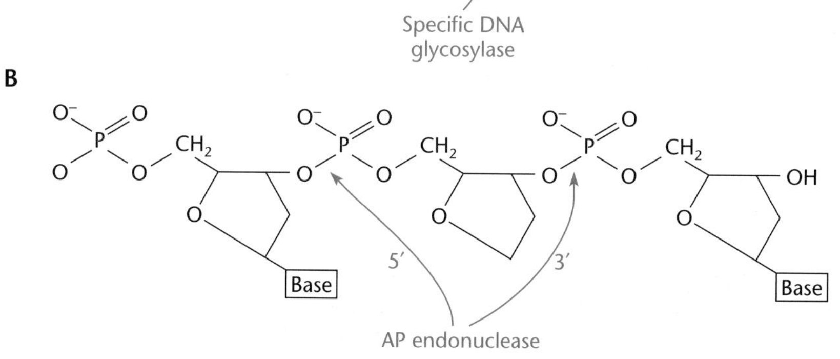

Figure 11.3 Repair of altered bases by DNA glycosylases. (A) The specific DNA glycosylase removes the altered base. (B) Most apurinic or apyrimidinic (AP) endonucleases cut the DNA backbone on the 5′ side of the apurinic or apyrimidinic site. The strand is degraded and resynthesized, and the correct base is restored (not shown).

The Vsr repair system is very specialized and usually only repairs TG mismatches in the sequence shown above. Therefore, this repair system would have only limited usefulness if methylation did not occur in this sequence. The *vsr* gene is immediately downstream of the gene for the Dcm methylase, ensuring that cells that inherit the gene to methylate the C in C^mCWGG/GGWC^mC will also usually inherit the ability to repair the mismatch correctly if it is deaminated. While only enterics like *E. coli* have been shown to have this particular repair system, many other organisms have 5-methylcytosine in

their DNA, and we expect that similar repair systems will be found in these organisms.

Damage Due to Reactive Oxygen

Although molecular oxygen (O_2) is not damaging to DNA and other macromolecules, other, more reactive forms of oxygen are very damaging. The reactive forms of oxygen have more electrons than molecular oxygen and include superoxide radicals, hydrogen peroxide, and hydroxyl radicals. These forms of oxygen may be produced by normal cellular reactions (see Box 11.1). Alter-

---

BOX 11.1

Oxygen: the Enemy Within

To respire, all aerobic organisms, including humans, must take up molecular oxygen (O_2). At normal temperatures, molecular oxygen reacts with very few molecules. However, some of it is converted into more reactive forms such as superoxide radicals (O_2^-), hydrogen peroxide (H_2O_2), and hydroxyl radicals ($\cdot OH$). Hydrogen peroxide is probably formed inadvertently by flavoenzymes during oxidative respiration (see Seaver and Imlay, below). It is also produced deliberately in the liver to help detoxify recalcitrant molecules and by lysozomes to kill invading bacteria. Iron in the cell can then catalyze the conversion of hydrogen peroxide to hydroxyl radicals, which may be the form in which oxygen is most damaging to DNA.

Cells have evolved many enzymes to help reduce this damage. These include catalases that reduce reactive oxygen molecules as they form and repair enzymes such as exonucleases and glycosylases that remove the damaged bases from DNA before they can cause mutations (see text).

Accumulation of DNA damage due to reactive oxygen has been linked to many degenerative diseases such as cancer, arthritis, cataracts, and cardiovascular disease. It has been estimated that a rat at 2 years of age has about 2 million DNA lesions per cell, and some types of human cells have been shown to accumulate DNA damage with age. The synthesis of these active forms of oxygen helps explain why some compounds (such as asbestos) that are not themselves mutagenic or chronic infections can increase the rate of cancer. The reactive forms of oxygen that are synthesized in response to these conditions by macrophages may be the real mutagens.

The importance of internally generated reactive oxygen in cancer has received dramatic confirmation recently in a published report (see Al-Tassan et al., below). These authors report that a genetic disease characterized by increased rates of colon cancer is due to mutations in the human repair gene *MYH*, which is analogous to the *mutY* gene of *E. coli*. Siblings who have inherited this predisposition to can-

cer, called familial adenomatous polyposis, are heterozygous for different mutant alleles of the *MYH* gene. The *mutY* gene product of *E. coli* is a specific *N*-glycosylase which removes adenine bases that have mistakenly paired with 8-oxoG in the DNA (see the text), and the human enzyme is known to have a similar activity. Furthermore, like *mutY* mutants of *E. coli*, human *MYH* mutants show higher than normal rates of GC-to-TA transversion mutations in another gene, *APC*, which is often associated with this type of cancer. Apparently, 8-oxoG also normally accumulates in human DNA and, in the absence of the *N*-glycosylase to remove A:oxoG mismatches during replication, will cause an increased rate of GC-to-TA transversion mutations, some of which convert normal cells into cancer cells.

Obviously, any mechanism for reducing the levels of these active forms of oxygen should increase longevity and reduce the frequency of many degenerative diseases. Fruits and vegetables produce antioxidants, including asorbic acid (vitamin C), tocopherol (vitamin E), and carotenes such as β-carotene (found in large amounts in carrots), that destroy these molecules. Consumption of adequate amounts of fruits and vegetables that contain these compounds may greatly reduce the rate of cancer and many degenerative diseases.

References

Al-Tassan, N., N. H. Chmiel, J. Maynard, N. Fleming, A. L. Livingston, G. T. Williams, A. K. Hodges, D. R. Davies, S. S. David, J. R. Sampson, and J. P. Cheadle. 2002. Inherited variants of MYH associated with somatic G:C to T:A mutations in colorectal tumors. *Nat. Genet.* **30:**227–232.

Ames, B. N., M. K. Shigenaga, and T. M. Hagen. 1993. Oxidants, antioxidants and the degenerative diseases of aging. *Proc. Natl. Acad. Sci. USA* **90:**7915–7922.

Seaver, L. C., and J. A. Imlay. 2001. Alkyl hyperoxide reductase is the primary scavenger of endogenous hydrogen peroxide in *Escherichia coli*. *J. Bacteriol.* **183:**7173–7181.

Slupska, M. M., W. M. Luther, J.-H. Chiang, H. Yang, and J. H. Miller. 1999. Functional expression of hMYH, a human homolog of *Escherichia coli* MutY protein. *J. Bacteriol.* **181:**6210–6213.

natively, they can arise as a result of environmental factors, including UV irradiation and chemicals such as the herbicide paraquat.

Because the reactive forms of oxygen normally appear in cells, all aerobic organisms must contend with the resulting DNA damage and have evolved elaborate mechanisms to remove these chemicals from the cellular environment. In bacteria, some of these systems are induced by the presence of the reactive forms of oxygen, and these genes encode enzymes such as superoxide dismutases, catalases, and peroxide reductases, among others, which help destroy the reactive forms. These systems also include genes that encode repair enzymes that help repair the oxidative damage to DNA caused by the reactive forms. The accumulation of this type of damage may be responsible for the increase in cancer rates with age and for many age-related degenerative diseases (see Box 11.1).

8-oxoG

One of the most mutagenic lesions in DNA caused by reactive oxygen is the oxdized base **7,8-dihydro-8-oxoguanine (8-oxoG or GO)** (Figure 11.4). This base appears frequently in DNA, and DNA Pol III often mispairs it

Figure 11.4 (A) Structure of 8-oxodG. (B) Mechanisms for avoiding mutagenesis due to 8-oxoG (GO). In pathway 1, an A mistakenly incorporated opposite 8-oxodG is removed by a specific glycosylase (MutY), and the strand is degraded and resynthesized with the correct C. In pathway 2, the 8-oxoG is itself removed by a specific glycosylase (MutM), and the strand is degraded and resynthesized with a normal G. In a third pathway, the 8-oxodG is prevented from entering DNA by a specific phosphatase (MutT) that degrades the triphosphate 8-oxodGTP to the monophosphate 8-oxodGMP. Adapted from M. L. Michaels and J. H. Miller, *J. Bacteriol.* **174**:6321–6325, 1992.

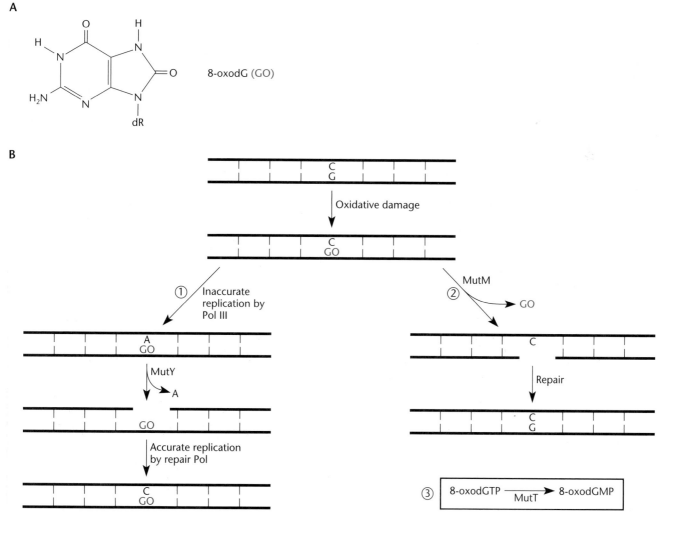

with adenine. Because of the mutagenic potential of 8-oxoG, cells have evolved many mechanisms for avoiding the resultant mutations.

MutM, MutY, AND MutT

Bacteria with a mutation in a *mut* gene suffer higher than normal rates of spontaneous mutagenesis. This is how they are selected, and we discuss the isolation of *mut* mutations below (see "Genetic Analysis of Repair Pathways"). *E. coli* has a number of different *mut* genes, and we discussed some of them in chapter 1 in connection with editing and mismatch repair. The products of three *mut* genes of *E. coli*, MutM, MutY, and MutT, are exclusively devoted to preventing mutations due to 8-oxoG, a tribute to the mutagenic potential of this damaged base. The effects of mutations in these genes are additive in the sense that the rate of spontaneous mutations is higher if two or all three of the *mut* genes are mutated than if only one of them is mutated. Moreover, the generally high rate of spontaneous mutagenesis in *mutM*, *mutT*, and *mutY* mutants supports the conclusions that internal oxidation of DNA is an important source of spontaneous mutations and that 8-oxoG, in particular, is a very mutagenic form of damage to DNA. The way in which the gene products MutM, MutT, and MutY prevent mutations due to 8-oxoG lesions is the result of an excellent study (see Michaels et al., Suggested Reading). We discuss each separately.

MutM

The MutM enzyme is an N-glycosylase that specifically removes the 8-oxoG base from the deoxyribose sugar in DNA (Figure 11.4). This repair pathway functions like other N-glycosylase repair pathways we have discussed except that the depurinated strand is cut by the AP endonuclease activity of MutM itself, degraded by an exonuclease, and resynthesized by DNA polymerase I (Figure 11.3).

The MutM protein is present in larger amounts in cells that have accumulated reactive oxygen, because the *mutM* gene is part of a regulon induced in response to oxidative stress. We discuss regulons in more detail in chapter 13.

MutY

The MutY enzyme is also a specific N-glycosylase. However, rather than removing 8-oxoG directly, the MutY N-glycosylase specifically removes adenine bases that have been mistakenly incorporated opposite an 8-oxoG in DNA (Figure 11.4). Repair synthesis by DNA Pol I then introduces the correct C to prevent a mutation, as with other N-glycosylase-initiated repair pathways.

In vitro, the MutY enzyme will also recognize a mismatch that results from accidental incorporation of an A

opposite a normal G and will remove the A. However, its major role in avoiding mutagenesis in vivo seems to be to prevent mutations due to 8-oxoG. As evidence, mutations that cause the overproduction of MutM completely suppress the mutator phenotype of *mutY* mutants (see Michaels et al., Suggested Reading). The interpretation of this result is as follows. If a significant proportion of all spontaneous mutations in a *mutY* mutant resulted from misincorporation of A's opposite normal G's, excess MutM should have no effect on the mutation rate, because removal of 8-oxoG should not affect this type of mispairing. However, the fact that excess MutM suppresses mutagenesis in *mutY* mutants suggests that it functions to ensure that very little 8-oxoG will persist in the DNA to mispair with A.

MutT

The MutT enzyme operates by a very different mechanism (see Figure 11.4): it prevents 8-oxoG from entering the DNA in the first place. The reactive forms of oxygen can oxidize not only guanine in DNA to 8-oxoG but also the base in dGTP to form 8-oxodGTP. Without MutT, 8-oxodGTP will be incorporated into DNA by DNA polymerase, which cannot distinguish 8-oxodGTP from normal dGTP. The MutT enzyme is a phosphatase that specifically degrades 8-oxodGTP to 8-oxodGMP so that it cannot be used in DNA synthesis.

GENETICS OF 8-oxoG MUTAGENESIS

Genetic evidence obtained with *mutM*, *mutT*, and *mutY* mutants is consistent with these functions for the products of the genes. First, these activities explain why the effects of mutations in these genes are additive. If *mutT* is mutated, more 8-oxodGTP will be present in the cell to be incorporated into DNA, increasing the spontaneous mutation rate. If MutM does not remove these 8-oxoG's from DNA, spontaneous mutation rates will increase even further. If MutY does not remove some of the A's that mistakenly pair with the 8-oxoG's, the spontaneous mutation rate will be higher yet.

Mutations in the *mutM*, *mutY*, and *mutT* genes also increase the frequency of only some types of mutations, which again can be explained by the activities of these enzymes. For example, the enzyme functions explain why *mutM* and *mutY* mutations increase only the frequency of GC-to-TA transversion mutations. If MutM does not remove 8-oxoG from DNA, then mispairing of the 8-oxoG's with A's can lead to these transversions. Moreover, GC-to-TA transversions will occur if MutY does not remove the mispaired A's opposite the 8-oxoG's in the DNA. By contrast, while *mutT* mutations can increase the frequency of GC-to-TA transversion mutations, they can also increase the frequency of TA-to-GC transversions. This is possible because some of the 8-oxodGTP molecules present in the cell owing to the *mutT* mutation may enter the DNA incorrectly by pairing with an A (instead of a C) while others may enter the DNA correctly opposite a C but then pair wrongly with an A.

Alkylation

Alkylation is another common type of damage to DNA. Both the bases and the phosphates in DNA can be alkylated. The responsible chemicals, known as **alkylating agents**, add alkyl groups (CH_3, CH_3CH_2, etc.) to the bases or phosphates. Some examples of alkylating agents are ethyl methanesulfonate (EMS; nitrogen mustard gas), methyl methanesulfonate (MMS), and *N*-methyl-*N'*-nitro-*N*-nitrosoguanidine (nitrosoguanidine, NTG, or MNNG).

Many reactive groups of the bases can be attacked by alkylating agents. The most reactive are the N^7 of guanine and the N^3 of adenine. These nitrogens can be alkylated by EMS or MMS to yield methylated or ethylated bases such as N^7-methylguanine and N^3-methyladenine. Alkylation of the bases at these positions can severely alter their pairing with other bases, causing major distortions in the helix.

Some alkylating agents, such as nitrosoguanidine, can also attack other atoms in the rings, including the O^6 of guanine and the O^4 of thymine. The addition of a methyl group to these atoms makes O^6-methylguanine (Figure 11.5) and O^4-methylthymine, respectively. Altered bases with an alkyl group at these positions are particularly mutagenic because the helix is not significantly distorted, so that the lesions cannot be repaired by the more general repair systems discussed below. However, the altered base will often mispair, producing a mutation. In this section, we discuss the repair systems specific to these types of alkylated bases.

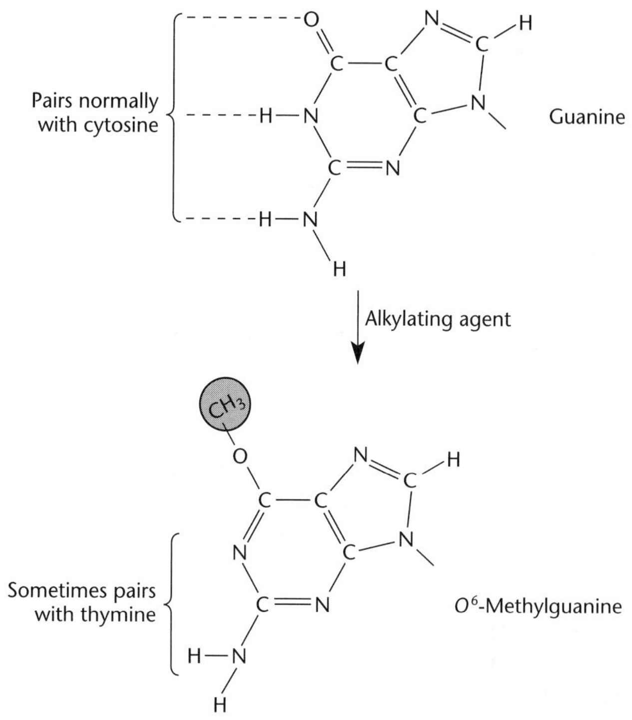

SPECIFIC N-GLYCOSYLASES

Alkylated bases can be removed by specific *N*-glycosylases. The repair pathways involving these enzymes work in the same way as other *N*-glycosylase repair pathways in that first the alkylated base is removed by the specific *N*-glycosylase and then the apurinic or apyrimdinic DNA strand is cut by an AP endonuclease. Exonucleases degrade the cut strand, which is then resynthesized by DNA polymerase I. In *E. coli*, an *N*-glycosylase that removes the bases 3-methyladenine and 3-methylguanine has been identified. This enzyme, which is encoded by the *alkA* gene, is induced as part of the adaptive response (see below).

METHYLTRANSFERASES

The cell can also use special proteins to repair bases damaged from alkylation of the O^6 carbon of guanine and the O^4 carbon of thymine. These proteins, called methyl-

Figure 11.5 Alkylation of guanine to produce O^6-methylguanine. The altered base sometimes pairs with thymine, causing mutations.

transferases, directly remove the alkyl group at these positions by transferring the group from the altered base in the DNA to themselves. The two major methyltransferases in *E. coli* are Ogt and Ada. Ogt plays the major role when the cells are growing actively but when the cells reach stationary phase and quit growing or if the cell is exposed to an external methylating agent, Ada is induced as part of the adaptive response (see below) and becomes the major methyltransferase (see below). They are not true enzymes because they do not catalyze the reaction but, rather, are consumed during it. Once they have transferred a methyl or other alkyl group to themselves, they are inactive and will be eventually degraded. That the cell is willing to sacrifice an entire protein molecule to repair a single O^6-methylguanine or O^4-methylthymine lesion is a tribute to the mutagenic potential of these lesions.

THE ADAPTIVE RESPONSE

Many of the genes whose products, including specific *N*-glycosylases and methyltransferases, repair alkylation damage in *E. coli* are part of the **adaptive response**. The products of these genes are normally synthesized in small amounts but are produced in much greater amounts if the cells are exposed to an alkylating agent. The name "adaptive response" comes from early evidence suggesting that *E. coli* "adapted" to damage caused by alkylating agents.

If *E. coli* cells are briefly treated with an alkylating agent such as nitrosoguanidine (NTG), they will be better able to survive subsequent treatments with alkylating agents. We now know that the cell adapts to the alkylating agent by inducing a number of genes whose products are involved in repairing alkylation damage to DNA. The adaptive-response genes seem to be most important for conferring resistance to alkylating agents that transfer methyl (CH_3) groups to DNA. Resistance to alkylating agents that transfer longer groups, such as ethyl (CH_3CH_2), to DNA seems to be due mostly to excision repair (see below).

Some of the the genes induced as part of the adaptive response include *ada*, *aidB*, *alkA*, and *alkB*. The product of the *ada* gene is a methyltransferase that removes methyl or ethyl groups from O^6-methylguanine or O^4-methylthymine or from the phosphate groups in the backbone of the DNA. The Ada protein is also the regulatory protein that regulates transcription of the other genes of the adaptive response, as we discuss below. The *alkA* gene product is a an N-glycosylase that removes many alkylated bases, including 3-methyladenine and 3-methylguanine, from DNA (see above). The product of the *aidB* gene is unknown, but it increases the resistance of cells to some methylating agents. It is related to an enzyme in human cells which may degrade nitrosoguanidine and related compounds whose products can methylate DNA (see below), causing mutations and therefore being potentially carcinogenic. Therefore, the AidB protein may detoxify alkylating agents before they can act. The product of the *alkB* gene is also unknown but may increase the resistance of cells to some alkylating agents by allowing replication past the alkylated bases in DNA, although how it does this is a mystery.

Regulation of the Adaptive Response

How does treatment with an alkylating agent cause the concentration of some of the enzymes involved in repairing alkylation damage to increase from a few to many thousands of copies? The regulation is achieved through the state of methylation on the Ada protein, a methyltransferase, as shown in Figure 11.6. In addition to its role in repairing alkylation damage to DNA, the Ada protein is a transcriptional activator which, in its methylated form, activates the transcription of its own gene and other genes of the adaptive response. The *ada* gene is part of an operon with *alkB*, while the *aidB* and *alkA* genes are separately transcribed, as shown in Figure 11.6. After alkylation damage occurs, methyl groups can be transferred to amino acids at two different positions on the Ada protein. The amino acid at one position accepts a methyl group from either O^6-methylguanine or O^4-methylthymine, and the amino acid at the other position accepts methyl groups from any phosphate in the

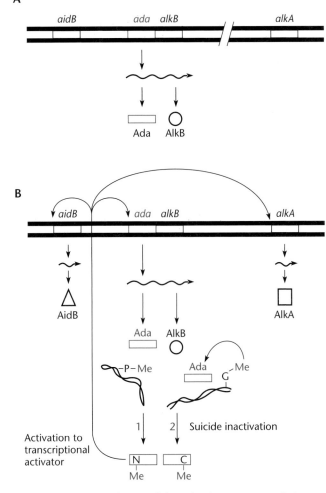

Figure 11.6 Regulation of the adaptive response. Only a few copies of the Ada protein normally exist in the cell. After damage due to alkylation, the Ada protein, a methyltransferase, transfers alkyl groups from damaged DNA phosphates to itself, converting itself into a transcriptional activator (1), or from a damaged base, inactivating itself (2). See the text for details.

DNA backbone. Methylation at either position will prevent further acceptance of methyl groups at that position. Transfer of a methyl group from an alkylated base to the Ada C terminus inactivates the Ada protein for additional acceptance of methyl or other groups (i.e., suicide inactivation in Figure 11.6). However, methyl groups transferred from the phosphate backbone to the Ada N terminus convert the Ada protein into a transcriptional activator. In this state, the Ada protein will activate transcription of its own gene, the *ada* gene, as well as the other genes of the adaptive response. We have discussed transcriptional activators in chapter 8 in connection with the CI and CII proteins of phage λ, and we discuss them again in much more detail in chapter 12.

Interestingly, some of the adaptive-response genes are also turned on during stationary phase when the cells stop growing. The transcription of genes during stationary phase requires σ^s rather than σ^{70} (see chapter 13), and the methylated Ada protein can also activate transcription by RNA polymerase containing σ^s at these promoters. However, transcription only of the *ada-alkB* operon and the *aidB* gene is activated by methylation in stationary phase; *alkA* gene transcription is not activated but instead is actually repressed by methylated Ada protein. This is not unusual, in that many transcriptional activators can also be repressors (see chapter 12). The purpose of inducing *aidB* during stationary phase may be to prevent the natural accumulation of nitrosoamines, which could alkylate DNA while the cells are resting. The situation is analogous to the mutagenic potential of reactive oxygen (see Box 11.1). Nitrosoamines such as nitrosourea can accumulate in cells that have run out of oxygen and are using nitrate as a terminal electron acceptor for anaerobic respiration. Accepting an electron converts nitrate to nitrite, which is chemically reactive and can react with other cellular constituents to form nitrosoamines, which can methylate DNA (see Taverna and Sedgwick, Suggested Reading). It makes sense to head off the damage to DNA during stationary phase by preventing the accumulation of these alkylating agents by AidB instead of waiting until the damage is done, since once the bases in DNA have been alkylated, removing them with the AlkA glycosylase during stationary phase may actually be deleterious to the cell. This type of repair requires DNA replication to finish the job of resynthesizing the deleted stretch of DNA (see Figure 11.3), and very little replication of DNA occurs in stationary-phase cells. As an aside, there are nitrosoamines in cigarette smoke and similar compounds may accumulate in meats preserved with nitrites when they are cooked. These alkylating agents may be carcinogenic in humans, and a repair system analogous to the AidB adaptive response enzyme in *E. coli* may help protect us against the accumulation of such alkylating agents and the subsequent DNA damage, which could lead to cancer.

Pyrimidine Dimers

One of the major sources of natural damage to DNA is UV irradiation due to sun exposure. Every organism that is exposed to sunlight must have mechanisms to repair UV damage to its DNA. The conjugated-ring structure of the bases in DNA causes them to strongly absorb light in the UV wavelengths. The absorbed photons energize the bases, causing their double bonds to react with other nearby atoms and hence forming additional chemical bonds. These chemical bonds result in abnormal linkages between the bases in the DNA and other bases or between bases and the sugars of the nucleotides.

One common type of UV irradiation damage is the **pyrimidine dimer**, in which the rings of two adjacent pyrimidines become fused (Figure 11.7). In one of the two possible dimers, the 5-carbon atoms and 6-carbon atoms of two adjacent pyrimidines are joined to form a **cyclobutane ring**. In the other type of dimer, the 6-carbon of one pyrimidine is joined to the 4-carbon of an adjacent pyrimidine to form a **6-4 lesion**.

PHOTOREACTIVATION OF CYCLOBUTANE DIMERS

Because the cyclobutane-type pyrimidine dimer due to UV irradiation is so common, a special type of repair system called **photoreactivation** has evolved to repair it. The photoreactivation repair system separates the fused bases of the cyclobutane pyrimidine dimers rather than replacing them. This mechanism is named "photoreactivation" because this type of repair occurs only in the presence of visible light.

Figure 11.7 Two common types of pyrimidine dimers caused by UV irradiation. In the top panel, two adjacent thymines are linked through the 5- and 6-carbons of their rings to form a cyclobutane ring. In the bottom panel, a 6-4 dimer is formed between the 6 carbon of a cytosine and the 4 carbon of a thymine 3′ to it.

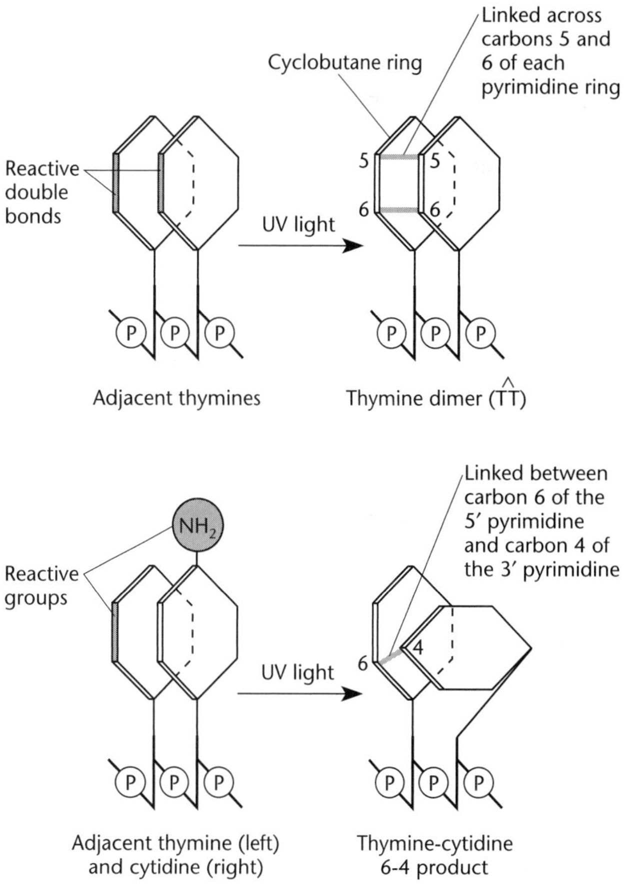

In fact, photoreactivation was the first DNA repair system to be discovered. In the 1940s, Albert Kelner observed that the bacterium *Streptomyces griseus* was more likely to survive UV irradiation in the light than in the dark. Photoreactivation is now known to exist in most organisms on earth, with the important exception of placental mammals like ourselves. Don't think that when you are soaking up rays the damaging effects of UV irradiation are being repaired just because you are also being exposed to visible light. You don't have a photoreactivation system, although you might have a repair system derived from it.

The mechanism of action of photoreactivation is shown in Figure 11.8. The enzyme responsible for the repair is called **photolyase**. This enzyme, which contains a reduced flavin adenine dinucleotide (FADH$_2$) group that will absorb light of wavelengths between 350 and 500 nm, binds to the fused bases. Absorption of light then gives photolyase the energy it needs to separate the fused bases.

There is some evidence that the photoreactivating system may also help repair pyrimidine dimers even in the dark. By binding to pyrimidine dimers, it may help make them more recognizable by the nucleotide excision repair system discussed below.

N-GLYCOSYLASES SPECIFIC TO PYRIMIDINE DIMERS

There are also specific *N*-glycosylases that recognize and remove pyrimidine dimers. This repair mechanism is similar to the mechanisms for deaminated and alkylated bases discussed above and involves AP endonucleases and the removal and resynthesis of strands of DNA containing the dimers.

General Repair Mechanisms

As mentioned, not all repair mechanisms in cells are specific for a certain type of damage to DNA. Some types of repair systems can repair many different types of damage. Rather than recognizing the damage itself, these repair systems recognize distortions in the DNA structure caused by improper base pairing.

The Methyl-Directed Mismatch Repair System

One of the major pathways for avoiding mutations in *E. coli* is the **methyl-directed mismatch repair system**. We have already mentioned the mismatch repair system in chapter 1, in connection with lowering the rate of mistakes made during the replication of DNA, and in chapter 10, in connection with gene conversion. The methyl-directed mismatch repair system is not specific and can repair essentially any damage that causes a slight distortion in the DNA helix, including mismatches, frameshifts, incorporation of base analogs, and some types of alkylation damage that cause only minor distortions in the DNA helix. Alkylation that causes more significant distortions is repaired by other systems including the nucleotide excision repair discussed later. In general, DNA damage that causes only minor distortions of the helix is repaired by the methyl-directed mismatch repair system whereas damage that causes more significant distortions is repaired by other pathways.

BASE ANALOGS

Base analogs are chemicals that resemble the normal bases in DNA. Because they resemble the normal bases, these analogs can sometimes be converted into a deoxynucleoside triphosphate and enter DNA. Incorporation of a base analog can be mutagenic because the

Figure 11.8 Photoreactivation. The photoreactivating enzyme (photolyase) binds to cyclobutane pyrimidine dimers (gold) even in the dark. Absorption of light by the photolyase causes it to cleave the bond between the two pyrimidines, restoring the bases to their original form. (Courtesy of Dr. P. C. Hanawalt.)

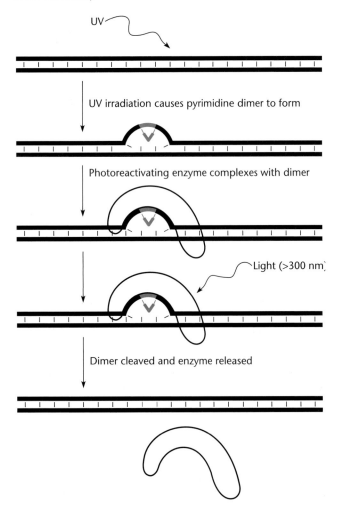

UV

UV irradiation causes pyrimidine dimer to form

Photoreactivating enzyme complexes with dimer

Light (>300 nm)

Dimer cleaved and enzyme released

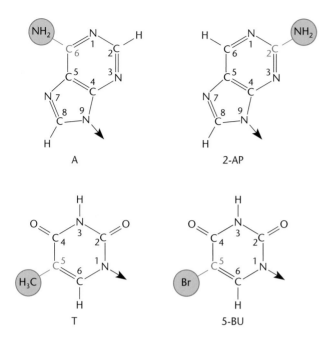

Figure 11.9 Base analogs 2-aminopurine (2-AP) and 5-bromouracil (5-BU). The amino groups that are at different positions in adenine (A) and 2-AP are circled in gold, as are the methyl group in thymine (T) and the bromine group in 5-BU.

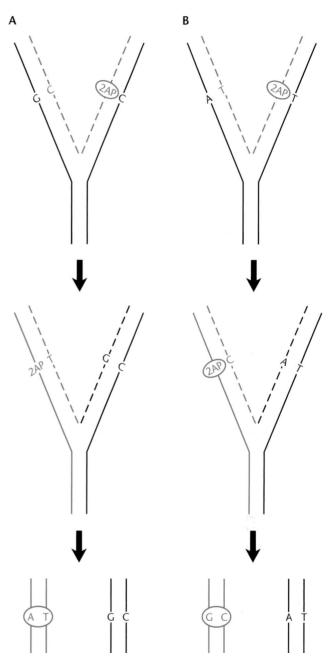

Figure 11.10 Mutagenesis by incorporation of the adenine analog 2-AP into DNA. The 2-AP is first converted to the deoxyribose nucleoside triphosphate and then inserted into DNA. (A) Analog incorrectly pairs with a C during its incorporation into the DNA strand. (B) Analog is incorporated correctly opposite a T but mispairs with C during subsequent replication. The mutation is circled in both panels.

analog often pairs with the wrong base, leading to base pair changes in the DNA. Figure 11.9 shows two base analogs, 2-aminopurine (2-AP) and 5-bromouracil (5-BU). 2-AP resembles adenine, except that it has the amino group at the 2 position rather than at the 6 position. This analog sometimes incorrectly pairs with cytosine. 5-BU, which resembles thymine except for a bromine atom instead of a methyl group at the 5 position, sometimes incorrectly pairs with guanine.

Figure 11.10 shows how mispairing by a base analog might cause a mutation. In the illustration, the base analog 2-AP has entered a cell and been converted into the nucleoside triphosphate. The deoxyribose 2-aminopurine triphosphate is then incorporated during synthesis of DNA and sometimes pairs with cytosine in error. Which type of mutation occurs depends on when the 2-AP mistakenly pairs with cytosine. Even if the 2-AP enters the DNA by mistakenly pairing with C instead of T, in subsequent replications it will usually pair correctly with T. This will cause a GC-to-AT transition mutation. However, if it is incorporated properly by pairing with T but pairs mistakenly with C in subsequent replications, an AT-to-GC transition mutation will result.

FRAMESHIFT MUTAGENS

Another type of damage repaired by the methyl-directed mismatch repair system is the incorporation of frameshift mutagens, which are usually planar molecules of the acridine dye family (see chapters 1 and 7). These chemicals are mutagenic because they intercalate between bases in the same strand of the DNA, thereby increasing the distance between the bases and preventing them from aligning properly with bases on the other strand. The frameshift mutagens include acridine dyes such as 9-aminoacridine, proflavin, and ethidium bromide, as well as some aflatoxins made by fungi.

Figure 11.11 illustrates a model for mutagenesis by a frameshift mutagen. Intercalation of the dye forces two of the bases apart, causing the two strands to slip with respect to each other. One base will thus be paired with the base next to the one with which it previously paired. This slippage is most likely to occur where a base pair in the DNA is repeated, for example in a string of AT or GC base pairs. Whether a deletion or addition of a base pair occurs depends upon which strand slips, as shown in Figure 11.11. If the dye is intercalated into the template DNA prior to replication, the newly synthesized strand might slip and incorporate an extra nucleotide. However, if it is incorporated into the newly synthesized strand, the strand might slip backward, leaving out a base pair in subsequent replication.

Figure 11.11 Mutagenesis by a frameshift mutagen. Intercalation of a planar acridine dye molecule between two bases in a repeated sequence in DNA forces the bases apart and can lead to slippage. (A) The dye inserts itself into the new strand, resulting in deletion of a base pair (–). (B) The dye comes into the old strand, adding a base pair.

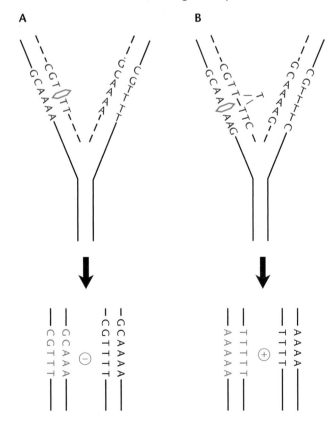

MECHANISM OF METHYL-DIRECTED MISMATCH REPAIR

As discussed briefly in chapter 1, the methyl-directed mismatch repair system requires the products of the *mutS*, *mutL*, and *mutH* genes. Like the *mut* genes whose products are involved in avoiding 8-oxoG, these *mut* genes were found in a search for mutations that increase spontaneous mutation rates (see below). The products of these three genes were predicted to participate in the same repair pathway because of the observation that double mutants did not exhibit higher spontaneous mutation rates than cells with mutations in each of the single genes alone. In other words, the effect of inactivating two of these *mut* genes by mutation is not additive. Generally, mutations that affect steps of the same pathway do not have additive effects.

As discussed in chapter 1, another gene product that participates in the methyl-directed mismatch repair system is the product of the *dam* gene. The *dam* gene encodes the Dam methylase, which methylates adenine in the sequence GATC/CTAG. This enzyme methylates the DNA at this sequence only after it is synthesized, so that the new strand of DNA will be temporarily unmethylated after the replication fork moves on. By having the *mut* gene products repair only the newly synthesized strand, the cell usually avoids mutagenesis, because most mistakes are made during synthesis of new strands. If the old strand were degraded and resynthesized with the new strand as a template, the damage would be fixed as a mutation and would be passed on to future generations.

It is mysterious how the mismatch repair system can use the state of methylation of the GATC/CTAG sequence to direct itself to the newly synthesized strand, even though the nearest GATC/CTAG sequence is probably some distance from the alteration. Figure 11.12 presents a model that may explain this mechanism. First, the MutS protein binds to the alteration in the DNA that is causing a minor distortion in the helix (an AC mismatch in the figure). Then two molecules of the MutH protein bind to MutS, and the DNA slides through the complex in both directions, resulting in a loop containing the alteration and held together by the MutS-MutH complex as shown. When one of the MutH subunits in the complex reaches the nearest hemimethylated GATC/CTAG sequence in either the 5′ or 3′ direction, it cuts the unmethylated strand, which is then degraded back to and through the mismatch. Depending upon the side of the mismatch where the first hemimethylated GATC is encountered, this degradation might have to be in either the 3′-to-5′ or the 5′-to-3′ direction on the other strand. However, all known exonucleases can degrade in only one direction, so how does the cell accomplish this?

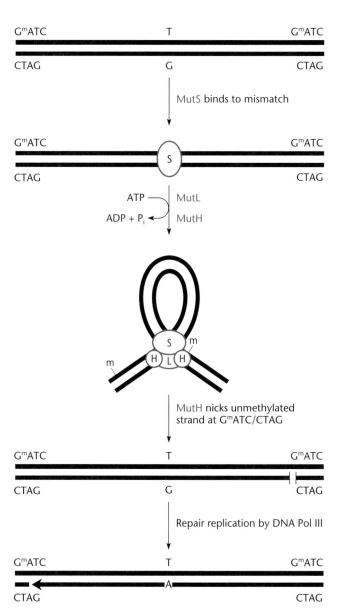

Figure 11.12 Model for repair, by the methyl-directed mismatch repair system, of slight distortions in the helix as a result of mismatches and of incorporation of base analogs and frameshift mutagens into DNA. The MutS protein binds to the distortion (in the example, this is due to incorporation of a C opposite an A). The MutL and MutH proteins then bind together, and the DNA is drawn through the complex in both directions until it reaches a hemimethylated GATC sequence. The MutH protein nicks the DNA in the unmethylated strand, which is then degraded back over the mismatch. Resynthesis of a new strand by DNA polymerase III restores the correct base. Adapted with permission from P. Modrich, *Annu. Rev. Genet.* **25:**229–253, 1991. © 1991 by Annual Reviews www.annualreviews.org

E. coli solves this problem by using four different exonucleases, two of which, ExoVII and RecJ, can degrade in only the 5'-to-3' direction and two of which, ExoI and ExoX, can degrade in only the 3'-to-5' direction. Also, these exonucleases can degrade only single-stranded DNA, and so a helicase is needed to first separate the strands. This work is done by the UvrD helicase, which is a general helicase also used by the excision repair system (see below and Table 11.1). This dual role explains why the *uvrD* gene has two names: *uvrD* and *mutU*. It was named *mutU* because the gene was identified in a search for mutations that led to an increased frequency of spontaneous mutations due to its role in mismatch repair. It was named *uvrD* because mutations in the gene cause an increased sensitivity to UV irradiation and other DNA-damaging agents due to its role in excision repair. The MutL protein is also required for mismatch repair, but its role is less clear. It may recruit the UvrD helicase and maybe also the exonucleases to the MutS-MutH complex. After the strand containing the mismatch has been degraded, DNA polymerase III resynthesizes a new strand, using the other strand as the template and removing the cause of the distortion. It is somewhat surprising that DNA polymerase III, the replication DNA polymerase, is used for this repair. Most other repair reactions use DNA polymerase I or II, the normal repair enzymes.

In addition to their role in general mismatch repair, the MutS and MutL proteins participate in very-short-patch (VSP) repair, which occurs at the site of methylated cytosines in *E. coli* (see above and Lieb et al., Suggested Reading). The MutS protein may bind to the T in the T-G mismatch created by the deamination of the methylated C at this position, thereby attracting the attention of the Vsr endonuclease to the mismatch. The MutL protein may then recruit the UvrD helicase and exonucleases to degrade the strand causing the mismatch, consistent with the roles of these proteins in general mismatch repair.

TABLE 11.1	Genes involved in the UvrABC endonuclease repair pathway
Gene	**Function of gene product**
uvrA	DNA-binding protein
uvrB	Loaded by UvrA to form a DNA complex; nicks DNA 3' of lesion
uvrC	Binds to UvrB-DNA complex; nicks DNA 5' of lesion
uvrD	Helicase II; helps remove damage-containing oligonucleotide
polA	Pol I; fills in single-strand gap
lig	Ligase; seals single-strand nick

GENETIC EVIDENCE FOR METHYL-DIRECTED MISMATCH REPAIR

Genetic experiments support general models of methyl-directed mismatch repair like that presented in Figure 11.12. Probably the most convincing evidence for the role of methylation in directing the repair system came from experiments with heteroduplexes of λ DNA (see Pukkila et al., Suggested Reading). We have discussed the synthesis and use of heteroduplex DNAs for experiments in other chapters. For example, in chapter 9 we showed how they can be used to distinguish replicative from cut-and-paste transposition.

To determine the role of methylation in directing the mismatch repair system, heteroduplex λ DNAs were prepared in which the two strands came from phage with different mutations—we shall call them phage mutant 1 and phage mutant 2. The DNA from phage mutant 2 was unmethylated because it was propagated on *E. coli* with a *dam* mutation. One of the phage was also propagated in medium with heavy isotopes, as in the Meselson and Stahl experiment (see chapter 1), so that the heteroduplex DNAs could be purified later. After the DNA from both mutants was purified, the DNA strands of both were separated by heating and then were renatured in the presence of each other. The heteroduplex DNAs were then purified on density gradients. Because mutant 1 and mutant 2 carried different mutations, the heteroduplex DNA had mismatches at the sites of those mutations. These DNAs were transfected into cells, the phage progeny were plated, and their genotypes were tested to see which mutation they had inherited and which of the two strands had been preferentially repaired. The progeny phage exhibited predominantly the genotype of mutant 1, which had the methylated DNA. In the reverse experiment, mutant 1 was propagated on the *dam* mutant so that it would have the unmethylated DNA. In this case, the progeny phage that arose from the heteroduplex DNA had the genotype of mutant 2; therefore, once again, the methylated DNA sequence was preferentially preserved. These results indicate that the sequence of the unmethylated strand is usually the one that is repaired to match the sequence of the methylated strand.

Other evidence supporting the role of methylation in directing the mismatch repair system came from genetic studies of 2-AP sensitivity in *E. coli* (see Glickman and Radman, Suggested Reading). These experiments were based on the observation that a *dam* mutation makes *E. coli* particularly sensitive to killing by 2-AP, as expected from the model. As mentioned, cells incorporate 2-AP indiscriminately into their DNA because they cannot distinguish it from the normal base adenine. The methylated mismatch repair system repairs DNA containing 2-AP because the incorporated 2-AP causes a slight distortion in the helix. Since the 2-AP is incorporated into the newly synthesized strand during the replication of the DNA, the strand containing the 2-AP will be normally transiently unmethylated, so that the 2-AP-containing strand will be repaired, removing the 2-AP. In a *dam* mutant, however, neither strand will be methylated, and so the mismatch repair system cannot tell which strand was newly synthesized and may try to repair both strands. The rationale was that if two 2-APs are incorporated close to each other, the mismatch repair system may try to simultaneously remove the two mismatches by repairing the opposite strands. However, cutting two strands at sites opposite each other will result in a double-strand break in the DNA, which may kill the cell. However, whether this is the real mechanism of killing is unclear.

One prediction based on this proposed mechanism for the sensitivity of *dam* mutants to 2-AP is that the toxicity of 2-AP in *dam* mutants should be reduced if the cells also have a *mutL*, *mutS*, or *mutH* mutation. Without the products of the mismatch repair enzymes, the DNA will not be cut on either strand, much less on both strands simultaneously. The cells may suffer higher rates of mutations but at least they will survive. This prediction was fulfilled. Double mutants that have both a *dam* mutation and a mutation in one of the three *mut* genes were much less sensitive to 2-AP than were mutants with a *dam* mutation alone. Furthermore, *mutL*, *mutH*, and *mutS* mutations can be isolated as suppressors of *dam* mutations on media containing 2-AP. In other words, if large numbers of *dam* mutant *E. coli* cells are plated on medium containing 2-AP, the bacteria that survive are often double mutants with the original *dam* mutation and a spontaneous *mutL*, *mutS*, or *mutH* mutation.

THE ROLE OF THE MISMATCH REPAIR SYSTEM IN PREVENTING HOMEOLOGOUS AND ECTOPIC RECOMBINATION

As discussed in chapter 3, some DNA rearrangements, such as deletions and inversions, are caused by recombination between similar sequences in different places in the DNA. This is called **ectopic recombination**, or "out-of-place" recombination. Many sequences at which ectopic recombination occurs are similar but not identical. Also, recombination between DNAs from different species often occurs between sequences that are similar but not identical. In general, recombination between similar but not identical sequences is called **homeologous recombination**.

The mismatch repair system reduces ectopic and other types of homeologous recombination. As evidence, recombination between similar but unrelated bacteria such as *E. coli* and *Salmonella enterica* serovar Typhimurium is greatly enhanced if the recipient cell has a *mutL*, *mutH*, or *mutS* mutation. Also, the frequency of deletions and other types of DNA rearrangements is enhanced among bacteria with a *mutL*, *mutH*, or *mutS* mutation. Because the sequences at which the homeologous recombination occurs are not identical, the heteroduplexes formed during recombination between different regions in the DNA or between the same region from different species often contain many mismatches that bind MutS and the other proteins of the mismatch repair system. Evidence suggests that the mismatch repair system may inhibit homeologous and ectopic recombination by interfering with the ability of RecA to form synapses where there are extensive mismatches (see Worth et al., Suggested Reading).

MISMATCH REPAIR SYSTEMS IN OTHER ORGANISMS

All organisms probably have mismatch repair systems that help reduce mutagenesis. In fact, deficiencies in a human mismatch repair system have been linked to human cancer (see Box 11.2). However, unlike bacteria, which have only one MutS and one MutL protein, humans have at least five MutS analogs and four MutL homologs (see Harfe and Jinks-Robertson, Box 11.2). Other eukaryotes, from yeast on up, also have multiple forms of each of these proteins. The MutS and MutL proteins in these organisms seem to assemble in different combinations, and each combination may play a specialized role in repairing a particular type of mismatch. The proteins probably function in a similar way to MutS and MutL, because two of the proteins are known to form a sliding clamp on DNA, which may move away from the mismatch much like the MutS-MutL complex moves away from a mismatch in *E. coli* (see Gradia et al., Box 11.2). However, most eukaryotes and even most types of bacteria do not have a Dam methylase and so cannot use GATC methylation to direct the mismatch repair system to the newly synthesized strand. In fact, some organisms such as yeasts have very little methylation of any type in the bases in their DNA. Some other mechanism is apparently used to direct the mismatch repair system to the newly synthesized strand, perhaps the structure of the replication fork itself.

Nucleotide Excision Repair

One of the most important general repair systems in cells is **nucleotide excision repair**. This type of repair is very efficient and seems to be common to most types of organisms on Earth. It is also relatively nonspecific and will repair many different types of damage. Because of its efficiency and relative lack of specificity, the nu-

---

BOX 11.2

Cancer and Mismatch Repair

Cancer is a multistep process that is initiated by mutations in oncogenes and other tumor-suppressing genes. It is not surprising, therefore, that DNA repair systems form an important line of defense against cancer. There is recent evidence that people with a mutation in a mismatch repair gene are much more likely to develop some types of cancer, including cancer of the colon, ovary, uterus, and kidney. This disease, called HNPCC (for hereditary nonpolyposis colon cancer), results from mutations in a gene called *hMSH2* (for human *mutS* homolog 2). This gene was first suspected to be involved in mismatch repair because people who inherited the mutant gene showed a higher frequency of short insertions and deletions that should normally be repaired by the mismatch repair system (see the text). People with this hereditary condition were found to have inherited a mutant form of a gene homologous to the *mutS* gene of *E. coli*. Moreover, the *hMSH2* gene can cause increased spontaneous mutations when expressed in *E. coli*, probably because it interferes with the normal mismatch repair system. Like its analog, MutS, the *hMSH2* gene product may bind to mismatches but then does not interact properly with MutL and MutH. In any case, the mismatch repair system in humans seems to play an important role in defending against cancer.

References

Fishel, R., M. K. Lescoe, M. R. S. Rao, N. G. Copeland, N. A. Jenkins, J. Garber, M. Kane, and R. Kolodner. 1993. The human mutator gene homolog MSH2 and its association with hereditary nonpolyposis colon cancer. *Cell* **75:**1027–1038.

Gradia, S., D. Subramanian, T. Wilson, S. Acharya, A. Makhov, J. Griffith, and R. Fishel. 1999. hMSH2-hMSH6 forms a hydrolysis-independent sliding clamp on mismatched DNA. *Mol. Cell* **3:**255–261.

Harfe, B. D., and S. Jinks-Robertson. 2000. DNA mismatch repair and genetic instability. *Annu. Rev. Genet.* **34:**359–399.

---

cleotide excision repair system is very important to the ability of the cell to survive damage to its DNA.

The nucleotide excision repair system is relatively nonspecific because, like the mismatch repair system, it recognizes the distortions in the normal DNA helix that results from damage, rather than the chemical structure of the damage itself. However, nucleotide excision repair recognizes only major distortions in the helix; it will not repair lesions such as base mismatches, O^6-methylguanine, O^4-methylthymine, 8-oxoG, or base analogs, all of which result in only minor distortions and must be repaired by other repair systems. Nucleotide excision repair will, however, repair almost all the types of damage caused by UV irradiation, including cyclobutane dimers, 6-4 lesions, and base-sugar cross-links. Nucleotide excision repair will also collaborate with recombination repair to remove cross-links formed between the two strands by some chemical agents such as psoralens, *cis*-diamminedichloroplatinum (cisplatin), and mitomycin (see below).

The excision repair system can be distinguished from most other types because the DNA containing the damage is actually excised from the DNA. For example, after cells are irradiated with UV light, short pieces of DNA (oligodeoxynucleotides) containing pyrimidine dimers and the other types of damage induced by UV appear in the medium.

MECHANISM OF NUCLEOTIDE EXCISION REPAIR

Because nucleotide excision repair is such an important line of defense against some types of DNA-damaging agents, including UV irradiation, mutations in the genes whose products are required for this type of repair can make cells much more sensitive to these agents. In fact, mutants defective in excision repair were identified because they were killed by much lower doses of irradiation than was the wild type. Table 11.1 lists the *E. coli* genes whose products are required for nucleotide excision repair. The products of some of these genes, such as *uvrA*, *uvrB*, and *uvrC*, are involved only in excision repair, while the products of others, including the *polA* and *uvrD* genes, are also required for other types of repair.

How these gene products participate in excision repair is illustrated in Figure 11.13. The products of the *uvrA*, *uvrB*, and *uvrC* genes interact to form what is called the **UvrABC endonuclease**. The function of these gene products is to make a nick close to the damaged nucleotide, causing it to be excised.

In more detail, two copies of the UvrA protein and one copy of the UvrB protein form a complex that binds nonspecifically to DNA even if it is not damaged. This complex then migrates up and down the DNA until it hits a place where the helix is distorted because of DNA damage (in the illustration, because of a thymine dimer). The complex will then stop, the UvrB protein will bind to the damage, and the UvrA protein will leave, being replaced by UvrC. The binding of UvrC protein to UvrB causes UvrB to cut the DNA about 4 nucleotides 3' of the damage. The UvrC protein then cuts the DNA 7 nucleotides 5' of the damage. Once the DNA is cut, the UvrD helicase removes the oligonucleotide containing the damage and the DNA polymerase I resynthesizes the strand that was removed, using the complementary strand as a template.

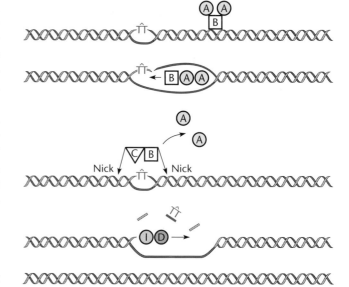

Figure 11.13   Model for nucleotide excision repair by the UvrABC endonuclease. See the text for details. A, UvrA; B, UvrB; C, UvrC; D, UvrD; I, Pol I.

INDUCTION OF NUCLEOTIDE EXCISION REPAIR

Although the genes of the excision repair system are almost always expressed at low levels, *uvrA*, *uvrB*, and *uvrD* are expressed at much higher levels after the DNA has been damaged. This is a survival mechanism that ensures that larger amounts of the repair proteins will be synthesized when they are needed. Because they are inducible by DNA damage, the *uvr* genes that are induced by DNA damage fall into a class of genes known as the *din* genes (for damage inducible), which includes *recF*, *recA*, *umuC*, and *umuD*. Many *din* genes, including *uvrA*, *uvrB*, and *uvrD*, are part of the SOS regulon (see later sections).

CAN EXCISION REPAIR BE DIRECTED?

Damage in some regions of the DNA presents a more immediate problem for the cell than does damage in other regions. For example, pyrimidine dimers in transcribed regions of DNA will block not only replication of the DNA but also transcription of RNA from the DNA. It makes sense for the cell to repair the damage that occurs in transcribed genes first, so that these genes can be transcribed and translated into proteins.

Evidence suggests that a mechanism that directs the nucleotide excision repair system preferentially to transcribed regions of the DNA may exist (see Box 11.3). A protein called transcription repair coupling factor (TRCF) displaces RNA polymerase molecules that are stalled at the site of the DNA damage because they cannot transcribe over the damage. The TRCF protein may

Transcription-Repair Coupling

DNA damage in transcribed regions presents a special problem for the cell because the RNA polymerase can stall at the damage, interfering with both expression of the genes and repair of the damage. Recent evidence indicates that DNA damage that occurs within transcribed regions of the DNA and in the transcribed strand will be repaired preferentially by the nucleotide excision repair system. This is called transcription-repair coupling. Some of the evidence for transcription-repair coupling is that mutations occur more frequently when the nontranscribed strand of DNA in a particular region is the one damaged, as expected if damage in this strand were not repaired by the relatively mistake-free excision repair system. This enhanced repair of transcribed strands requires that the gene be actively transcribed and suggests that the RNA polymerase molecule blocked at the site of the DNA damage helps direct the UvrB protein to the damage. The protein that seems to be directly involved in directing the UvrB protein to stalled RNA polymerase is TRCF (transcription-repair coupling factor), the product of the *mfd* gene in *E. coli*. TRCF may bind to the DNA with a stalled RNA polymerase and remove it to get it out of the way of the UvrABC endonuclease. TRCF may then remain bound to the DNA and direct the UvrAB proteins to the damage, accelerating the rate of repair.

Human cells have a counterpart to the TRCF that may play a similar role. Deficiencies in the human analog of TRCF may be the cause of the human genetic disease known as Cockayne's syndrome.

References

Selby, C. P., and A. Sancar. 1993. Transcription-repair coupling and mutation frequency decline. *J. Bacteriol.* **175**:7509–7514.

Selby, C. P., and A. Sancar. 1993. Molecular mechanism of transcription-repair coupling. *Science* **260**:53–58.

then help the UvrA and UvrB proteins bind to the damage and the UvrA protein to leave, thereby increasing the rate of repair.

Recombination Repair

In all of the repair systems discussed above, the cell removes the damage from the DNA, often using the information in the complementary strand to restore the correct sequence. Hopefully, the damage is repaired before the replication apparatus arrives on the scene and tries to replicate over the damage, causing the chromosome to break or mistakes to be made. But what hap-

pens if the damage occurs on both strands of the DNA at the same place, so there is no undamaged complementary strand to use as a template to resynthesize the correct sequence? And what happens if both strands of the DNA are broken simultaneously immediately opposite each other? In all these cases, there is no good strand of DNA opposite the damage to use as a template to repair the damage. Alternatively, the damage might just be so extensive that the repair systems are overwhelmed and are unable to repair all of the damage before the replication apparatus arrives. In all the instances where the damage is not repaired, the cell has no choice but to tolerate it. Mechanisms that allow the cell to tolerate DNA damage without ever repairing it are called **damage tolerance** mechanisms.

One type of damage tolerance mechanism is **recombination repair.** This type of repair is also sometimes called postreplication repair because it depends upon the damaged region already having been replicated so that it can use the sequence of the sister DNA in the same cell to restore the correct sequence. By using its recombination functions to promote recombination between the sister DNA and the damaged DNA, the cell can put a good strand of DNA opposite the damaged DNA and allow the damage to be repaired or allow the replication apparatus to move past the damage. The situation is analogous to a train being shunted off to a different track to avoid damage ahead on its original track. In fact, some people think that this may be the major function of recombination in bacteria, to tolerate damage to the DNA rather than to confer some long-term evolutionary benefit (see Box 10.2).

Early on, it was recognized that mutations in the *rec* genes of *E. coli* made the cell more sensitive to DNA-damaging agents (see Kato et al., Suggested Reading, and below). Mutations in genes in either the RecBCD or RecFOR pathway can make the cell more sensitive to killing by DNA damage. As discussed in chapter 10, the RecBCD pathway is used at double-stranded breaks or ends of DNA while the RecFOR pathway is used primarily at gaps in the DNA. We can imagine scenarios where each of these situations might occur. One such scenario is when the replication fork encounters a nick. The replication of the lagging strand may continue past the damage, but the leading strand runs off forming a free double-stranded end. Free double-stranded ends are the substrate for the RecBCD enzyme which can bind to the free end and separate the strands, degrading the 3′ ended strand until it encounters a χ site. It might be significant that in *E. coli* most χ sites are arranged so that the RecBCD enzyme will soon encounter one with the right polarity when it degrades back from a blocked replication fork (see chapter 10). When the RecBCD nuclease encounters a χ site, it will stop degrading the 3′-ended strand, giving rise to a 3′ single-stranded tail which can load on RecA to form a nucleoprotein filament and invade the other sister DNA to form a D-loop (see Figure 10.5). For reasons which are not completely understood, this allows the PriA protein to reload the replication apparatus on the DNA and permits replication to continue.

Another potential scenario involving RecFOR is illustrated in Figure 11.14. In this scenario, the replication fork encounters a lesion in the leading strand that it cannot replicate over. The DnaB helicase might keep going on the lagging strand and allow some lagging-strand replication past the damage, but the DNA polymerase on the leading strand is stopped by its editing functions (see chapter 1). The replication apparatus might then collapse, leaving a gap in the DNA. It is also possible that DNA polymerase II (Pol II) may play a role in gap formation. The DNA Pol II might have less stringent requirements for initiating replication than does Pol III, allowing it to load on the other side of the damage and continue replicating to make a short DNA, again leaving a gap. Such gaps in DNA are the substrates for the RecFOR pathway, which then allows RecA to bind to the single-stranded DNA in the gap and invade the sister double-stranded DNA, putting a good strand opposite the damage and forming a D-loop which could migrate and be processed by other enzymes such as RecG and RuvABC. PriA could then help the replication apparatus load on the D-loop, and replication could continue as before. The details of this model are pure speculation, but there is some evidence that DNA polymerase II may be responsible for initiating DNA synthesis immediately past the damage but is replaced later by the normal Pol III (see Rangarajan et al., Suggested Reading). Also, PriA is known to play some required role in replication restarts, but the details are unclear (see Box 10.2).

Figure 11.14 Model for recombination-mediated replicative bypass of a thymine dimer in DNA. The DNA Pol III stalls at a thymine dimer in DNA but the helicase can continue on the lagging strand. Pol II may be able to initiate after the damage, producing a gap. The RecFOR proteins help RecA bind to the single-stranded DNA (ssDNA) at the gap, and the RecA nucleoprotein filament invades the sister DNA. Note that in subsequent steps, the order of the gold strands is reversed for convenience. The gap is now opposite a good strand, and the gap is filled in, leaving two Holliday junctions which are resolved by RuvABC or RecG. The PriA, PriB, PriC, and DnaT proteins then help reload the replicative polymerase Pol III on the DNA, and replication continues. Depending on how the Holliday junctions are resolved, the thymine dimer either has been transferred to the lagging strand (part 7) or remains in the leading strand template (part 7′).

388

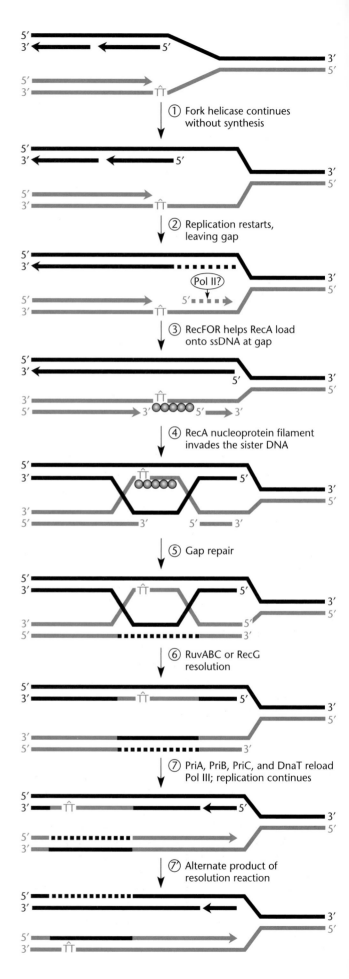

Whatever the detailed mechanism, replication restarts due to recombination apparently allow the replication apparatus to shunt back and forth between badly damaged sister DNAs, using different recombination functions as appropriate, depending upon the type of damage which is encountered. Eventually the replication fork will reach the terminus of replication, and the cell can divide. After many divisions, the damage will be diluted out of the progeny and will have been tolerated without ever having been actually repaired. These are only a few of the possible scenarios that may be followed by the cell in its use of the recombination functions to tolerate damage to the DNA. This explains why the absence of any one of these functions makes the cell more sensitive to some DNA-damaging agents. It also explains why some of the recombination functions are induced following DNA damage, to provide more of them for the needed recombination repair to follow.

REPAIR OF INTERSTRAND CROSS-LINKS IN DNA

The recombination functions may also collaborate with other repair functions to repair damage to DNA. One example might be in the repair of chemical cross-links in the DNA (see Figure 11.15). Many chemicals such as light-activated psoralens, mitomycin, cisplatin, and ethyl methanesulfonate can form **interstrand cross-links** in the DNA, in which two bases in the opposite strands of the DNA are covalently joined to each other (hence the prefix "inter," or "between"). Interstrand cross-links present special problems and cannot be repaired by either nucleotide excision repair or recombination repair alone. Cutting both strands of the DNA by the UvrABC endonuclease would cause double-strand breakage and death of the cell. Also, recombination repair by itself cannot repair DNA cross-links, because the cross-link prevents the replication fork from separating the strands.

Although DNA cross-links cannot be repaired by either excision or recombination repair alone, they can be repaired by a combination of recombination repair and nucleotide excision repair, as shown in Figure 11.15. In the first step, the UvrABC endonuclease makes nicks in one strand on either side of the interstrand cross-link, as though it were repairing any other type of damage. This will leave a gap opposite the DNA damage, as shown in the figure. In the second step, the gap is widened by the 5' exonuclease activity of DNA polymerase I. In the third step, recombination repair replaces the gap with a good strand from the other daughter DNA in the cell. The DNA damage is now confined to only one strand and is opposite a good strand; therefore, it can be repaired by the nucleotide excision repair pathway. Notice that this repair is possible only when the DNA has already replicated and there are already two copies of the DNA in the cell, since the replication apparatus cannot

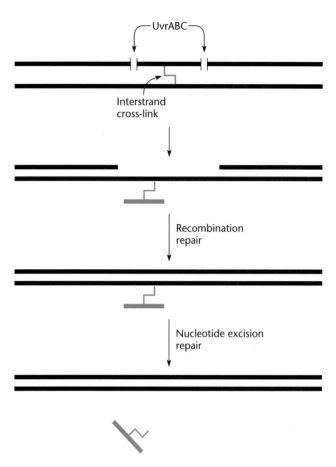

Figure 11.15 Repair of a DNA interstrand cross-link through the combined action of nucleotide excision repair and recombination repair. See the text for details.

proceed past the interstrand cross-link. However, in fast-growing cells, most of the DNA does exist in more than one copy (see chapter 1).

SOS Inducible Repair

As mentioned above, DNA damage leads to the induction of genes whose products are required for DNA repair. In this way, the cell is better able to repair the damage and survive.

The first indication that some repair systems are inducible came from the work of Jean Weigle on the reactivation of UV-irradiated λ (see Weigle, Suggested Reading). He irradiated phage λ and tested their survival by plating them on *E. coli*. He observed that more irradiated phage survived when plated on *E. coli* cells that had themselves been irradiated than when plated on unirradiated *E. coli* cells. Because the phage were restored to viability, or reactivated, by being plated on preirradiated *E. coli* cells, this phenomenon was named **Weigle reactivation** or **W-reactivation**. Apparently, a repair system was being induced in the irradiated cells that

Figure 11.16 Regulation of the SOS response regulon in *E. coli*. (A) About 30 genes around the *E. coli* chromosome are normally repressed by the binding of a LexA dimer (barbell structure) to their operators. Some SOS genes are expressed at low levels, as indicated by single arrows. (B) After DNA damage, the single-stranded DNA (ssDNA) that accumulates in the cell binds to RecA (circled A), forming a RecA nucleoprotein filament, which binds to LexA, causing LexA to cleave itself. The cleaved repressor can no longer bind to the operators of the genes, and the genes are induced as indicated by two arrows. The approximate positions of some of the genes of the SOS regulon are shown.

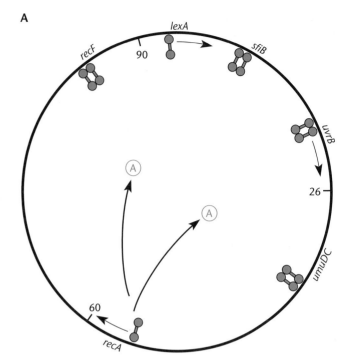

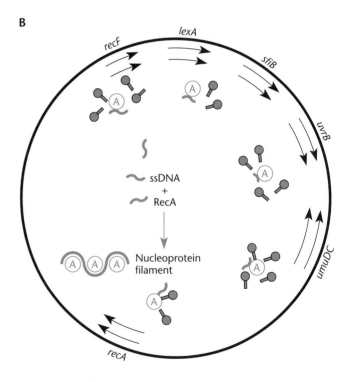

could repair the damage to the λ DNA when this DNA entered the cell.

THE SOS RESPONSE

Earlier in the chapter, we mentioned a class of genes induced after DNA damage, called the *din* genes, which includes genes that encode products that are part of the excision repair and recombination repair pathways. The products of other *din* genes help the cell survive DNA damage in other ways. For example, some *din* gene products transiently delay cell division until the damage can be repaired, and some allow the cell to replicate past the DNA damage.

Many *din* genes are regulated by the the **SOS response,** so named because this mechanism rescues cells from severe DNA damage. Genes under this type of control are called the **SOS genes.** At last count, *E. coli* contains 31 SOS genes that are regulated this way.

Figure 11.16 illustrates how genes under SOS control are induced. The SOS genes are normally repressed by a protein called the LexA repressor, which binds to sites called operators upstream of the SOS genes and prevents their transcription. However, after DNA damage, the LexA repressor cleaves itself, an action known as **autocleavage,** thereby inactivating itself and allowing transcription of the SOS genes. This is reminiscent of the induction of phage λ described in chapter 8, in which the CI repressor is also cleaved following DNA damage. The two mechanisms are in fact remarkably similar, as we discuss later.

The reason the LexA repressor no longer binds to DNA after it is cleaved is also well understood. Each LexA polypeptide has two separable domains. One, the **dimerization domain,** binds to another LexA polypeptide to form a dimer (hence its name), and the other, the **DNA binding domain,** binds to the DNA operators upstream of the SOS genes and blocks their transcription. The LexA protein will bind to DNA only if it is in the dimer state. The point of cleavage of the LexA polypep-

tide is between the two domains, and autocleavage separates the DNA binding domain from the dimerization domain. The DNA binding domain cannot dimerize and so by itself cannot bind to DNA and block transcription of the SOS genes.

Figure 11.16 also answers the next question: why is the LexA repressor cleaved only after DNA damage? After DNA damage, single-stranded DNA appears in the cell, probably owing to recombination and excision repair. This single-stranded DNA binds to RecA protein, which then binds to LexA and causes it to autocleave. This feature of the model—that LexA cleaves itself in response to RecA binding, rather than being cleaved by RecA—is supported by experimental evidence. When heated under certain conditions, even in the absence of RecA, LexA will eventually cleave. This result indicates that RecA acts as a **coprotease** to facilitate LexA autocleavage.

The RecA protein thus plays a central role in the induction of the SOS response. It senses the single-stranded DNA that accumulates in the cell as a by-product of attempts to repair damage to the DNA and then causes LexA to cleave itself, inducing the SOS genes. We have already discussed another activity of RecA, the recombination function involved in synapse formation.

We can speculate why the RecA protein might have two different roles, the one in recombination and the other in repair. It must bind to single-stranded DNA to promote synapse formation during recombination, and so it must be a sensor of single-stranded DNA in the cell. It is always present in large enough amounts to bind to all the LexA repressor and quickly promote its autocleavage. The activated RecA coprotease also promotes the autocleavage of the UmuD protein involved in SOS mutagenesis (see below).

The regulation of the SOS response through cleavage of the LexA repressor is strikingly similar to the induction of λ through cleavage of the CI repressor (see chapter 8). Like the LexA repressor, the λ repressor must be a dimer to bind to the operator sequences in DNA and repress λ transcription. Each λ repressor polypeptide consists of an N-terminal DNA binding domain and a C-terminal dimerization domain. Autocleavage of the λ repressor stimulated by the activated RecA coprotease also separates the DNA binding domain and dimerization domain, preventing the binding of the λ repressor to the operators and allowing transcription of λ lytic genes. The sequences of amino acids around the sites at which the LexA and λ repressors are cleaved are also similar. However, the cell would not have evolved the SOS system to help λ prophage induce itself and escape in the event of irreversible damage to the cell DNA. Rather, it is more likely that the cell evolved the SOS induction mechanism to help *itself* survive DNA damage by inducing the SOS *din* genes after DNA damage. By modeling its own repressor after the LexA repressor, the λ prophage allows the SOS regulatory system of the host to induce it following DNA damage, thereby allowing λ to escape a doomed host cell.

SOS INDUCIBLE MUTAGENESIS

At least one of the repair systems encoded by the SOS genes appears to be very mistake prone, and its operation often causes mutations. As we show, this system, known as **translesion synthesis** (**TLS**), allows the replication fork to proceed over damaged DNA so that the molecule can be replicated and the cell can survive. This mechanism may be a last resort that operates only when the DNA damage is so extensive that it cannot be repaired by other, less mistake-prone mechanisms.

The first indications that a mistake-prone pathway for UV damage repair is inducible in *E. coli* came from the same studies that showed that repair itself is inducible (see Weigle, Suggested Reading). In addition to measuring the survival of UV-irradiated λ when they were plated on UV-irradiated *E. coli*, Weigle counted the number of clear-plaque mutants among the surviving phage. (Recall from chapter 8 that lysogens will form in the center of wild-type phage λ plaques, making the plaques cloudy, but mutants that cannot lysogenize form clear plaques.) There were more mutants among the surviving phage if the bacterial hosts had been UV irradiated prior to infection than if they had not been irradiated. This inducible mutagenesis was named **Weigle mutagenesis** or just **W-mutagenesis**. Later studies showed that the inducible mutagenesis results from induction of the SOS genes; it will not occur without RecA and the cleavage of the LexA repressor. Thus, the inducible mutagenesis Weigle observed is called **SOS mutagenesis**.

Determining Which Repair Pathway Is Mutagenic

Although Weigle's experiments showed that one of the inducible UV damage repair systems in *E. coli* is mistake prone and causes mutations, they did not indicate which system was responsible. A genetic approach was used to answer this question (see Kato et al., Suggested Reading). To detect UV-induced mutations, the experimenters used the reversion of a *his* mutation. Their basic approach was to make double mutants with both a *his* mutation and a mutation in one or more of the genes of each of the repair pathways. The repair-deficient mutants were then irradiated with UV light and plated on media containing limiting amounts of histidine, so that only His⁺ revertants could multiply to form a colony. Under these conditions, each reversion to *his*⁺ will result in only one colony, making it possible to measure directly the number of *his*⁺ reversions that have occurred. This number, divided by the total number of surviving bacteria, will give an estimation of the susceptibility of the cells to mutagenesis by UV light.

The results of these experiments led to the conclusion that recombination repair does not seem to be mistake prone. While *recB* and *recF* mutations reduced the survival of the cells exposed to UV light, the number of *his*⁺ reversions per surviving cell was no different from that of cells lacking mutations in their *rec* genes. Also, nucleotide excision repair does not seem to be mistake prone. Addition of a *uvrB* mutation to the *recB* and *recF* mutations made the cells even more susceptible to killing by UV light but also did not reduce the number of *his*⁺ reversions per surviving cell.

However, *recA* mutations did seem to prevent UV mutagenesis. While these mutations made the cells extremely sensitive to killing by UV light, the few survivors did not contain additional mutations. We now know that two genes, *umuC* and *umuD*, must be induced for mutagenic repair and that *recA* mutations prevent their induction. Thus, the UmuC and UmuD proteins act in the opposite way from most repair systems. Rather than repairing the damage before mistakes in the form of mutations are made, the UmuC and UmuD proteins actually make the mistakes themselves. If they were not present, the cells which survive UV irradiation and some other types of DNA damage would have fewer mutations. The payoff, however, is that more cells survive. The RecA protein is also directly involved in SOS mutagenesis (see below).

Mechanism of Induction of SOS Mutagenesis

The *umuC* and *umuD* genes, whose products are required for mutagenesis after UV irradiation and some other types of DNA damage, were identified in a search for mutants of *E. coli* that are not mutagenized by UV irradiation (see below). The genes were named *umu* for *U*V-induced *mu*tagenesis. These two genes are adjacent in the *E. coli* genome and are cotranscribed in the direction *umuD*-*umuC* from a promoter immediately upstream of the *umuD* gene.

Dramatic progress has been made recently in understanding how the UmuC and UmuD proteins promote mutagenesis and allow an *E. coli* cell to tolerate damage to its DNA. The UmuC protein is a DNA polymerase which, in contrast to the normal DNA polymerase III, can replicate right over some types of damage to the DNA, including the thymine cyclobutane dimers and cytosine-thymine 6-4 dimers induced by UV irradiation as well as abasic sites in which a base has been removed (see above). Accordingly, UmuC was renamed **DNA polymerase V** and is capable of TLS. However, the damaged bases cannot pair properly, so how does the UmuC polymerase know which of the four deoxynucleotides to insert opposite each damaged base? The answer is that it incorporates the deoxynucleotide almost randomly, which is why TLS by UmuC is so mutagenic.

Figure 11.17 illustrates in more detail why SOS mutagenesis occurs only after extensive damage to the DNA. Like the other SOS genes, the *umuDC* operon is nor-

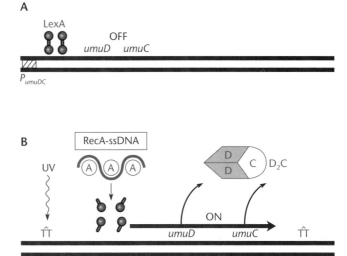

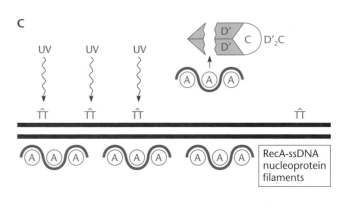

Figure 11.17 Regulation of SOS mutagenesis in *E. coli*. (A) Before DNA damage occurs, the LexA protein represses the transcription of SOS genes including the *umuDC* operon. (B) After limited DNA damage, the RecA protein binds to the single-stranded DNA (ssDNA), which accumulates, forming RecA nucleoprotein filaments. These filaments bind to LexA, promoting its autocleavage and inducing the SOS genes including the *umuDC* operon. The UmuC and UmuD proteins form a heterotrimer composed of two copies of UmuD and one copy of UmuC (UmuD′₂C). (C) More damage causes more RecA nucleoprotein filaments to accumulate, eventually promoting the autocleavage of UmuD to form UmuD′₂C. (D) The UmuC protein bound to UmuD′₂ is an active mutagenic polymerase which can replicate right over the damage, often making mistakes in the process; two G's are shown incorporated opposite a thymine dimer.

mally induced only if the DNA is damaged. The newly synthesized UmuC and UmuD proteins come together to form a heterotrimer complex with two copies of the UmuD protein and one copy of the UmuC protein (UmuD$_2$C). This complex is inactive as a DNA polymerase, although it might bind to the DNA polymerase III and temporarily arrest replication to cause a "checkpoint" (see below). However, as the RecA nucleoprotein filaments accumulate, they also cause UmuD to cleave itself (to UmuD'), in much the same way as they cause LexA and the λ repressor to cleave themselves. The autocleavage of UmuD to UmuD' requires a higher concentration of RecA nucleoprotein filaments than does the autocleavage of LexA; therefore, rather than happening immediately, it occurs only if the damage is not repaired. However, once UmuD is cleaved to UmuD', the UmuC in the UmuD'$_2$C complex is active as a translesion DNA polymerase and replicates over the damage in the DNA, making mistakes in the form of mutations. This allows replication to continue past the damage and permits the cell to survive, but it increases the frequency of mutations among the survivors.

As mentioned, the trimer complex UmuD$_2$C containing uncleaved UmuD may also play a role even before the UmuD is cleaved. Overproduction of these proteins inhibits DNA replication at low temperatures, even in the absence of DNA damage (see Sutton et al., Suggested Reading). This has been interpreted to mean that the UmuC and UmuD proteins might inhibit replication after they are first induced, creating a checkpoint and allowing more time for the other repair systems to work before the replication forks encounter the damage. Overproduction of the β-clamp protein and the ε editing function of DNA polymerase affect the inhibition, suggesting that the UmuD$_2$C complex inhibits replication by interfering with the editing functions of the DNA polymerase III.

Mechanism of Mutagenesis by the UmuD'$_2$C Complex: a Mutagenic DNA Polymerase

Figure 11.18 shows a recent model for how the UmuD'$_2$C complex performs TLS and causes mutations. This model attempts to explain a number of observations concerning SOS mutagenesis, including the observation that RecA protein is directly involved in TLS in addition to its role in cleaving LexA and UmuD (see below and Sommer et al., Suggested Reading). It also includes roles for the β clamp and clamp-loading proteins of DNA polymerase III in TLS, since some evidence indicates that these functions are also directly required for TLS. Nevertheless, this model must be considered only a "work in progress" and will undoubtedly require modification as more experiments are done and we learn more. In the first step of the model, the DNA polymerase III holoenzyme encounters damage to the DNA that has

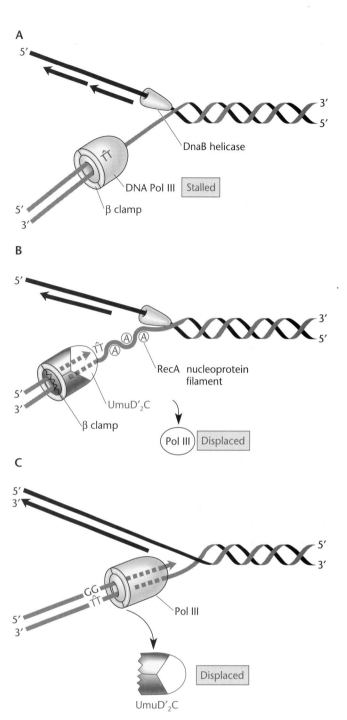

Figure 11.18 A detailed model for TLS by the UmuD'$_2$C complex. (A) In the first step, the DnaB helicase proceeds on the lagging strand past the damage, separating the strands, but the DNA polymerase III stalls, leaving a single-stranded gap. (B) The RecA protein binds to the single-stranded DNA in the gap, forming a RecA nucleoprotein filament. The bound RecA protein attracts the UmuD'$_2$C complex, which "commandeers" the β clamp from the stalled DNA polymerase III and uses it to hold itself on the DNA as it replicates over the damage, often making mistakes in the process. (C) Having done its job, the UmuD'$_2$C complex is replaced by the normal replicative DNA polymerase III and the replication continues.

not yet been repaired (in the example, a cyclobutane-thymine dimer). The editing functions on the DNA polymerase III will not let it polymerize over the damage, and so it stalls. However, the DnaB helicase which is separating the strands of the DNA ahead of the replicating polymerase keeps going for a short time on the lagging strand, opening the DNA at the damage. The RecA protein then polymerizes on the single-stranded DNA including the damaged region to form a helical RecA nucleoprotein filament. The coating of the single-stranded DNA by RecA is required for two reasons. The UmuD$'_2$C complex may only replicate DNA if it is coated in a RecA nucleoprotein filament. Also, the RecA protein in the filament may attract UmuD$'_2$C complex to the site, since it is known that RecA greatly stimulates the DNA-polymerizing activity of UmuD$'_2$C. Once the UmuD$'_2$C complex is bound to the DNA, the β clamp is transferred from the stalled DNA polymerase III to the UmuD$'_2$C complex. Some evidence suggests that the clamp is required to hold the UmuD$'_2$C complex on the DNA. The UmuD$'_2$C polymerase then replicates over the damage, inserting deoxynucleotides essentially at random opposite the damaged bases, thereby causing mutations. The UmuD$'_2$C polymerase makes as short a DNA as possible, however, before DNA polymerase III is reloaded on the DNA and normal replication continues. There is a good reason for limiting replication by UmuD$'_2$C. It not only makes mistakes opposite DNA lesions but also makes more mistakes in general, even when replicating nondamaged DNA; these are called untargeted mutations (see below).

As mentioned, this model for TLS will almost certainly have to be revised as more experimental evidence becomes available. We already know some observations about TLS that are difficult to reconcile with the model. One is an apparent continued requirement for DNA polymerase III in TLS, in addition to providing the clamp for the UmuD$'_2$C polymerase complex. There is some evidence that DNA polymerase III activity is directly required for SOS mutagenesis, and DNA polymerase III must be added to some in vitro systems for TLS by UmuD$'_2$C in the test tube. On the surface, it does not seem to make much sense that DNA polymerase III would be directly involved in TLS. Why would TLS need another DNA polymerase, when UmuC has its own DNA polymerase activity? Perhaps the very short DNA chains made by UmuD$'_2$C over the damage are too unstable unless they are immediately extended by DNA polymerase III. It is also not clear how the DNA polymerase III reloads on the DNA after TLS. PriA may play a role in this, as it does in replication restarts after recombination repair (see above), but PriA has been postulated to bind only to D-loops to recruit the DnaB

helicase and restart replication. There is no evidence that D-loops form during TLS, and in fact, as we mentioned earlier, Kato et al. had shown that the recombination functions required for D-loop formation are not required for SOS mutagenesis (Kato et al., Suggested Reading). Perhaps DNA polymerase III never leaves the DNA during TLS, so it will not need to be reloaded later.

ROLE OF MUTAGENIC REPAIR
In the discussion above, we assume that SOS mutagenic repair is a last-ditch attempt by the cell to survive extensive damage to its DNA. However, the actual function of SOS mutagenic repair remains a mystery. If this explanation for the existence of mutagenic repair were correct, then *umuC* and *umuD* mutants that lack mutagenic repair should be much more sensitive than wild-type bacteria to being killed by agents that extensively damage DNA. However, mutations that inactivate the *umuC* and/or *umuD* gene make *E. coli* only slightly more sensitive to killing by UV irradiation and other DNA-damaging agents. One way to explain this puzzling observation is to propose that UmuC and UmuD protect against types of DNA-damaging agents other than those that have been tested. Another possibility is that UmuC and UmuD offer more protection against DNA damage under conditions different from those that exist in the laboratory.

OTHER FUNCTIONS ANALOGOUS
TO UmuC AND UmuD
Whatever the real role of UmuC and UmuD, most types of bacteria seem to have similar mechanisms. Genes analogous to *umuC* and *umuD* have now been found in many bacteria including the gram-positive *Streptomyces* spp., as well as in archaea. In addition, some naturally occurring plasmids carry analogs of the *umuC* and *umuD* genes. The best studied of these genes are the *mucA* and *mucB* genes of plasmid R46, of which pKM101 is a derivative (see chapter 5). The products of these plasmid genes can substitute for UmuC and UmuD′ in mutagenic repair of *E. coli* and *Salmonella* spp. Because the pKM101 plasmid makes *Salmonella* spp. more sensitive to mutagenesis by many types of DNA-damaging agents, it has been introduced into the *Salmonella* strains used for the Ames test (see Box 11.4).

There are also other mutagenic polymerases in bacteria apart from UmuC. One the most mysterious of these is the product of the *E. coli dinB* gene, polymerase IV, which is a mutagenic polymerase related to UmuC. The *dinB* gene is also regulated by the LexA repressor, and so it is an SOS gene which is induced following DNA damage. It is not clear whether pol IV always works by itself or forms a complex with another protein like UmuC

BOX 11.4

The Ames Test

It is now well established that cancer is initiated by mutations in genes including oncogenes and tumor-suppressing genes. Therefore, chemicals that cause mutations are often carcinogenic for humans. Yet many new chemicals are being used as food additives or otherwise come into contact with humans, and each of these chemicals must be tested for its carcinogenic potential. However, such testing in animal models is expensive and time-consuming.

Because many carcinogenic chemicals damage DNA, they are mutagenic for bacteria as well as humans. Therefore, bacteria can be used in initial tests to determine if chemicals are apt to be carcinogenic. The most widely used of these tests is the Ames test, developed by Bruce Ames and his collaborators. This test uses revertants of *his* mutations of *Salmonella* spp. to detect mutations. The chemical is placed on a plate lacking histidine and on which has been spread a His$^-$ mutant of *Salmonella*. If the chemical can revert the *his* mutation, a ring of His$^+$ revertant colonies will appear around the chemi-

cal on the plate. A number of different *his* mutations must be used because different mutagens cause different mutations and they all have preferred sites of mutagenesis, or hot spots. Also, the test is made more sensitive by eliminating the non-mutagenic nucleotide excision repair system with a *uvrA* mutation and introducing the pKM101 plasmid containing the *mucA* and *mucB* genes. These genes are analogs of *umuC* and *umuD* and so increase mutagenesis (see text).

Some chemicals are not mutagenic themselves but can be converted into mutagens by enzymes in the mammalian liver. To detect these mutagens, we can add a liver extract from rats to the plates and spot the chemical on the extract. If the extract converts the chemical into a mutagen, His$^+$ revertants will appear on the plate.

Reference
McCann, J., and B. N. Ames. 1976. Detection of carcinogens as mutagens in the Salmonella/microsome test assay of 300 chemicals: discussion. *Proc. Natl. Acad. Sci. USA* **73**:950–954.

does. What is known is that it has mutagenic polymerase activity by itself, unlike UmuC, which requires UmuD′ to be active.

One thing that is mysterious about DNA polymerase IV is that it makes lots of mistakes, even when it is replicating undamaged DNA. The *dinB* gene product was first discovered to be mutagenic, even on undamaged templates, because it increases the frequency of λ mutations after λ infects cells in which the SOS functions have been induced even if the λ DNA itself has not been damaged. Such mutagenesis is called **untargeted mutagenesis** (UTM) because the mutations can occur anywhere, not just opposite to where there was damage to the DNA. The UmuD′$_2$C, polymerase can also cause UTM, but most of the mistakes due to UmuD′$_2$ occur opposite damage to the DNA.

Another mysterious thing about pol IV is that it does not increase the survival of cells after UV irradiation, even though it is induced by UV irradiation like the products of the other SOS genes. What purpose is served by inducing a polymerase that just makes mistakes and has no survival value? One possibility is that it allows replication over different types of damage from those caused by UV irradiation. Some evidence suggests that most of the mutations due to pol IV are −1 frameshifts, suggesting that the polymerase IV can replicate over bulged-out nucleotides or inserted frameshift mutagens in the DNA, causing mutations in the process.

Eukaryotes have a large number of different mutagenic polymerases related to UmuC and polymerase IV (see Box 11.5). Each type of mutagenic DNA polymerase may be required to replicate over a particular type of damage to DNA and thereby play a role in avoiding mutations due to a particular type of DNA damage. So far, only UmuC has been shown to be regulated by another protein such as UmuD, but it seems likely that other examples of such regulation will be found.

Other *din* Genes

As mentioned above, the SOS genes are only a subset of the *din* genes, many of which have been identified by making random transcriptional fusions with transposons such as Mu*dlac* (see chapter 9). Fusions to the promoters of these genes make more of the reporter gene product after DNA damage. Most *din* genes are expressed constitutively but are induced to higher levels of expression after DNA damage. The functions of many of these genes have yet to be determined.

The *groEL* and *groES* genes are examples of *din* genes that are not repressed by LexA and so are not SOS genes. The GroEL and GroES proteins are chaperones that help the folding of proteins and are also induced after heat shock (see chapters 1 and 13). In addition, these proteins play an indirect role in SOS mutagenesis, perhaps by helping to fold UmuC so that it interacts properly with UmuD′.

BOX 11.5

Translesion Synthesis and Cancer

Conversion of a normal cell to a cancer cell requires multiple mutations. In order to grow out of control without the normal constraints, a cancer cell must have mutations in genes whose products normally control the normal cell cycle checkpoints, control communication with surrounding cells, and cause the cells to kill themselves by apoptosis if they get out of hand, among others. Therefore, it is not surprising that conditions which increase the number of mutations in cells would increase the frequency of their transformation into cancer cells. We have already discussed how people with mutations in the human equivalent of the mismatch repair system have an increased frequency of some types of cancer (see Box 11.2).

As mentioned in the text, a number of genes related to *E. coli umuC* and *dinB* have been found in other organisms including humans. Once UmuC was discovered to be a mutagenic polymerase, the products of some of these genes were purified and also found to be DNA polymerases. One of the most interesting is a DNA polymerase from yeast called DNA polymerase η. It is the product of a yeast gene called *RAD30*. The *RAD* genes of yeast were isolated because they make yeast more sensitive to UV light. This DNA polymerase seems to be smarter than most of the other more mutagenic polymerases, however, in that it will incorporate A's opposite the T's in a cyclobutane thymine dimer, the most common form of UV damage, unlike UmuC, which in-

corporates two nucleotides randomly opposite such a dimer. In this way, DNA polymerase η restores the original sequence and no mutations occur. A gene encoding a homologous DNA polymerase to polymerase η has been found in humans; it is mutated in one type of heredity skin disease called xeroderma pigmentosum (XP). People with xeroderma pigmentosum are very sensitive to sunlight, and even limited exposure to UV light can cause them to develop a type of skin cancer called basal cell carcinoma. Most types of xeroderma pigmentosum are due to defects in excision repair; however, this type, called XPV (for *XP* variant because it was known to be different), is due to a mutant DNA polymerase η. Apparently, the thymine dimers which accumulate in skin cells exposed to UV light are much more mutagenic in the absence of DNA polymerase η. If DNA polymerase η is not there to accurately replicate over the thymine dimers, they must be dealt with in other ways which are much more mutagenic, leading to cancers.

References

Johnson, R. E., C. M. Kondratick, S. Prakash, and L. Prakash. 1999. hRAD30 mutations in the variant form of *Xeroderma pigmentosum*. *Science* **285**:263–265.

Masutani, C., R. Kusumoto, A. Yamada, N. Dohmae, M. Yokoi, M. Yuasa, M. Araki, S. Iwai, K. Takio, and F. Hanaoka. 1999. The XPV (*Xeroderma pigmentosum* variant) gene encodes human DNA polymerase eta. *Nature* **399**:700–704.

Summary of Repair Pathways in *E. coli*

Table 11.2 shows the repair pathways we have discussed and some of the genes whose products participate in each pathway. Some pathways, such as photoreactivation and most base excision pathways, have evolved to repair specific types of damage to DNA. Some, such as VSP repair, amend damage only in certain sequences. Others, such as mismatch repair and nucleotide excision repair, are much more general and will repair essentially any damage to DNA, provided that it causes a distortion in the DNA structure.

The separation of repair functions into different pathways is in some cases artificial. Some repair genes are inducible, and the repair enzymes themselves can play a role in their induction as well as in the induction of genes in other pathways. For example, the RecBCD nuclease is involved in recombination repair but can also help make the single-stranded DNA that activates the RecA coprotease activity after DNA damage to in-

duce SOS functions. The RecA protein is required for both recombination repair and induction of the SOS functions and is directly involved in SOS mutagenesis. Needless to say, the overlap of the functions of the repair gene products in different repair pathways has complicated the assignment of roles in these pathways.

Bacteriophage Repair Pathways

The DNA genomes of bacteriophages are also subject to DNA damage, either when the DNA is in the phage particle or when it is replicating in a host cell. Not surprisingly, many phages encode their own DNA repair enzymes. In fact, the discovery of some phage repair pathways preceded and anticipated the discovery of the corresponding bacterial repair pathways. By encoding their own repair pathways, phages avoid dependence on host pathways and repair proceeds more efficiently than it might with the host pathways alone.

TABLE 11.2	Genetic pathways for damage repair	
Repair mechanism	**Genetic loci**	**Function**
Methyl-directed mismatch repair	*dam*	DNA adenine methylase
	mutS	Mismatch recognition
	mutH	Endonuclease that cuts at hemimethylated sites
	mutL	Interacts with MutS and MutH
	uvrD (*mutU*)	Helicase
Very short patch repair	*dcm*	DNA cytosine methylase
	vsr	Endonuclease that cuts at 5′ side of T in TG mismatch
"GO" (guanine oxidizations)	*mutM*	Glycosylase that acts on GO
	mutY	Glycosylase that removes A from A:GO mismatch
	mutT	8-OxodGTP phosphatase
Alkyl	*ada*	Alkyltransferase and transcriptional activator
	alkA	Glycosylase for alkylpurines
Nucleotide excision	*uvrA*	Component of UvrABC
	urvB	Component of UvrABC
	uvrC	Component of UvrABC
	uvrD	Helicase
Base excision	*xthA*	AP endonuclease
	nfo	AP endonuclease
Photoreactivation	*phr*	Photolyase
Recombination repair	*recA*	Strand exchange
	recBCD	Helicase and nuclease at double-strand breaks
	recFOR	Recombination function
	ssb	Single-stranded DNA-binding protein
SOS system	*recA*	Coprotease
	lexA	Repressor
	umuDC	Translesion synthesis (Pol V)
	dinB	Mutagenic polymerase (Pol IV)

The repair pathways of phage T4 are perhaps the best understood. This large phage encodes at least seven different repair enzymes that help repair DNA damage due to UV irradiation, and some of these enzymes are also involved in recombination (see chapter 10). Table 11.3 lists the function of some T4 gene products and the homologous bacterial enzymes. The phage also uses some of the corresponding host enzymes to repair damage to its DNA.

TABLE 11.3	Bacteriophage T4 repair enzymes
Repair enzyme	**Host analog**
DenV	UV endonuclease of *M. luteus*
UvsX	RecA
UvsY	RecOR
UvsW	RecG
gp46/47 exonuclease	RecBCD recombination repair
gp49 resolvase	RuvABC recombination repair
gp59	PriA

One of the most important functions for repairing UV damage in phage T4 is the product of the *denV* gene (see Dodson and Lloyd, Suggested Reading), which is both an N-glycosylase and a DNA endonuclease. The N-glycosylase activity specifically breaks the bond holding one of the pyrimidines to its sugar in pyrimidine cyclobutane dimers of the *cis-syn* type (Figure 11.7). The endonuclease activity then cuts the DNA just 3′ of the pyrimidine dimer, and the dimer is removed by the exonuclease activity of the host cell DNA polymerase I. This enzyme thus works very differently from the UvrABC endonuclease of *E. coli* and, in a sense, combines both the N-glycosylase and AP endonuclease activities of some other repair pathways. The bacterium *Micrococcus luteus* has a similar enzyme. The endonuclease activity of the DenV protein will also function independently of the N-glycosylase activity and will cut apurinic sites in DNA as well as heteroduplex loops caused by short insertions or deletions. Because it cuts next to pyrimidine dimers in DNA, the purified DenV endonuclease of T4 is often used to determine the persistence of pyrimidine dimers after UV irradiation.

Genetic Analysis of Repair Pathways

Much of what is known about the repair of DNA damage and mutagenesis has come from genetic experiments with *E. coli*. In most cases, we have glossed over the genetic experiments which led to this knowledge. In this section, we discuss some of the methods used to isolate mutants defective in repair and the genetic experiments that support current models for how the different repair pathways work.

ISOLATION OF *mut* MUTANTS

As discussed earlier in the chapter, the products of the *mut* genes of *E. coli* and other bacteria help to lower spontaneous mutation rates. In *E. coli*, the *mut* genes include the genes of the mismatch repair system (*mutS*, *mutH*, and *mutL*), the genes for repair of oxidation damage due to 8-oxoG lesions (*mutM*, *mutY*, and *mutT*), and a gene encoding the DNA polymerase III holoenzyme editing function (*mutD*). Some of these same gene products also act to prevent mutations caused by some mutagens.

The *mut* genes were discovered because mutations in these genes increase spontaneous-mutation rates. The resulting phenotype is often referred to as the "mutator." Mutations in the *vsp* and *dam* genes of *E. coli* also increase the rate of at least some spontaneous mutations, although these genes were found in other ways and so were not named *mut*.

A common method for detecting mutants with abnormally high mutation rates is colony papillation. This scheme is based on the fact that all of the descendants of a bacterium with a particular mutation will be mutants of the same type. Thus, a growing colony will contain papillae, or sectors composed of mutant bacteria of various types. A *mut* mutation will increase the frequency of papillation for many types of mutants, so that the phenotype used in a colony papillation test is purely a matter of convenience.

In *E. coli*, Lac⁺ revertants of *lac* mutations are an obvious choice for papillation tests. Figure 11.19 shows a papillation test using reversion of a *lac* mutation as an indicator of mutator activity. Lac⁻ mutants growing on plates containing X-Gal will give rise to colorless colonies, but any Lac⁺ revertants appearing in the growing colony will give rise to blue papillae. If the bacteria forming a colony are *mut* mutants with a higher than normal spontaneous mutation frequency, the colony will have more blue papillae than normal. This is how many of the *mut* genes we have discussed were detected.

GENETIC ANALYSIS OF INDUCIBLE MUTAGENESIS: ISOLATION OF *umuC* AND *umuD* MUTATIONS

The early experiments by Jean Weigle, described earlier in the chapter, showed that UV mutagenesis is inducible in *E. coli*. He observed that if phage λ are UV-irradiated and then used to infect cells, the phage will suffer many

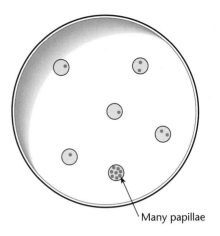

Figure 11.19 Colonies due to *mut* mutants have more papillae. A *lacZ* mutant was plated on X-Gal-containing medium. Revertants of the *lacZ* mutation will produce blue sectors or papillae (shown in gold). A *mut* mutant will give more blue papillae than normal owing to increased spontaneous mutation frequencies (arrow).

more mutations if they infect cells that had themselves been UV irradiated prior to infection than if they infect unirradiated cells. Later experiments by Kato et al. (Suggested Reading) showed that most of the mutations induced by UV light were not due to mistakes made by the recombination or excision repair pathways. If these pathways were mistake prone, mutations that inactivated a gene in the pathway should reduce the frequency of mutants among the surviving bacteria. This is not the case. Instead, evidence has indicated that these repair pathways make very few mistakes when they are repairing the damage in DNA due to UV irradiation. Repair by these pathways does, however, markedly increase the number of cells that survive the irradiation.

Once it was established that most of the mutagenesis after UV irradiation can be attributed to an inducible pathway different from recombination and nucleotide excision repair, the next step was to identify the genes of this pathway. Mutations that inactivate a *mut* gene or another repair pathway gene increase the rate of spontaneous mutations or mutagen-induced mutations because normally the gene products repair damage before it can cause mistakes in replication. However, as noted above, mutations that inactivate gene products of a mutagenic or mistake-prone repair pathway should have the opposite effect, decreasing the rate of at least some induced mutations. Because the repair pathway itself is mutagenic, mutations should be less frequent if the repair pathway does not exist than if it does exist. Cells with a mutation in a gene of the mistake-prone repair pathway may be less likely to survive DNA damage, but there should be a lower percentage of newly generated mutants among the survivors.

The first *umuC* and *umuD* mutants were isolated in two different laboratories by essentially the same method—reversion of a *his* mutation that makes cell growth require histidine to measure mutation rates—but we shall describe the one used by Kato and Shinoura (Suggested Reading). The basic strategy was to treat *his* mutants with a mutagen that induces DNA damage similar to that caused by UV irradiation and to identify mutant bacteria in which fewer *his*+ revertants occurred that could grow without histidine. These *his* mutants would have a second mutation that inactivated the mutagenic repair pathway.

Figure 11.20 illustrates the experiment in more detail. A mutant of *E. coli* with a point mutation in a *his* gene

Figure 11.20 Detection of a mutant defective in mutagenic repair. Colonies of mutagenized *his* bacteria are picked individually from a plate and patched onto a new plate containing histidine. This plate (plate 1) is then replicated onto a plate containing 4NQO (plate 2) to induce DNA damage similar to that induced by UV irradiation. The 4NQO-containing plate is then replicated onto another plate with limiting amounts of histidine (plate 3). After incubation, mottling of a patch caused by many His+ revertants indicates that the bacterium that made the colony on the original plate was capable of mutagenic repair. The colony circled in gold on the original plate may have arisen from a mutant deficient in mutagenic repair, because it gives fewer His+ revertants when replicated onto plate 3.

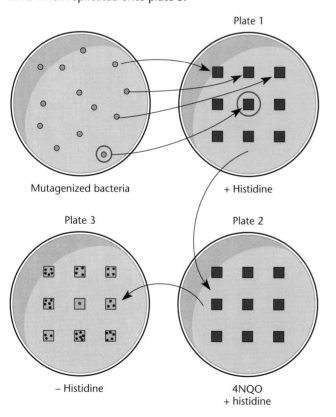

Plate 1

Mutagenized bacteria

+ Histidine

Plate 3

Plate 2

− Histidine

4NQO
+ histidine

was heavily mutagenized to try to induce mutations in the genes of the mutagenic repair pathway, individual colonies of the bacteria were patched with a loop onto plates with medium containing histidine, and the plates were incubated until patches due to bacterial growth first appeared. Each plate was then replicated onto another plate containing 4-nitroquinoline-1-oxide (4NQO), which causes DNA damage similar to that caused by UV irradiation. After the patches developed, this plate was replicated onto a third plate containing limiting amounts of histidine. Most bacteria formed patches with a few regions of heavier growth due to His+ revertants, indicating that they were being mutagenized by the 4NQO. However, a few bacteria formed patches with very few areas of heavier growth and therefore had fewer His+ revertants. The bacteria in the patches were candidates for descendants of mutants that could not be mutagenized by 4NQO or, presumably, by UV irradiation, since the types of damage caused by the two mutagens are similar.

The next step was to map the mutations in some of the mutants. Some of the mutations that prevented mutagenesis by UV irradiation mapped to either *recA* or *lexA*. These mutations could have been anticipated because mutations in these genes could prevent the induction of all the SOS genes including those for mutagenic repair. The *recA* mutations presumably inactivate the coprotease activity of the RecA protein, preventing it from causing autocleavage of the LexA repressor and thereby preventing induction of the SOS genes. The *lexA* mutations in all probability change the LexA repressor protein so that it cannot be cleaved, presumably because one of the amino acids around the site of cleavage has been altered. This is a special type of *lexA* mutation called a *lexA*(*ind*) mutation. If the LexA protein is not cleaved following UV irradiation of the cells, the SOS genes, including the genes for mutagenic repair, will not be induced.

The most interesting of the mutations that prevented UV mutagenesis mapped to a previously unknown locus at 25 min on the *E. coli* map. Complementation tests between mutations at this locus revealed two genes at this site required for UV mutagenesis, later named *umuC* and *umuD*. Later experiments also showed that these genes are transcribed into the same mRNA and so form an operon, *umuDC*, in which the *umuD* gene is transcribed first. Experiments with *lacZ* fusions revealed that the *umuDC* operon is inducible by UV light and is an SOS operon, since it is under the control of the LexA repressor (see Bagg et al., Suggested Reading).

EXPERIMENTS TO SHOW THAT ONLY *umuC* AND *umuD* MUST BE INDUCED FOR SOS MUTAGENESIS
The fact that *umuC* and *umuD* are inducible by DNA damage and are required for SOS mutagenesis does not mean that they are the *only* genes that must be induced

for this pathway. Might other genes that must be induced for SOS mutagenesis have been missed in the mutant selections? Some investigators sought an answer to this question (see Sommer et al., Suggested Reading). Their experiments used a *lexA(ind)* mutant, which, as mentioned, should permanently repress all the SOS genes. They then mutated the operator site of the *umuDC* operon so that these genes would be expressed constitutively and would no longer be under the control of the LexA repressor. Under these conditions, the only SOS gene products that should be present are those of the *umuC* and *umuD* genes, since all other such SOS genes will be permanently repressed by the LexA(*ind*) repressor. In addition, a shortened form of the *umuD* gene was used. This altered gene synthesizes only the carboxy-terminal UmuD' fragment that is the active form for SOS mutagenesis. With the altered *umuD* gene, the RecA coprotease is not required for UmuD to be autocleaved to the active form.

The experiments showed that UV irradiation induced mutations in the *lexA(ind)* mutants that expressed UmuC and UmuD' constitutively, indicating that *umuC* and *umuD* are the only genes that need to be induced by UV irradiation for SOS mutagenesis. However, this result does not entirely eliminate the possibility that other SOS gene products are involved. As discussed earlier in the chapter, the RecA protein may also be directly required for UV mutagenesis. The *recA* gene is induced to higher levels following UV irradiation but apparently is present in large enough amounts for UV mutagenesis even without induction. The GroEL and GroES proteins are also required for UV mutagenesis, but the *groEL* and *groES* genes are not under the control of the LexA repressor and so are expressed even in the *lexA(ind)* mutants.

AN ADDITIONAL ROLE FOR RecA IN UV MUTAGENESIS BESIDES ITS ROLE AS A COPROTEASE?

Similar experiments were used to show that RecA has a required role in UV mutagenesis in addition to its role as a coprotease promoting the autocleavage of LexA and UmuD. A sufficient explanation for why mutations in the *recA* gene can prevent UV mutagenesis is that they prevent the induction of all the SOS genes including *umuC* and *umuD* and that they also prevent the autocleavage of UmuD to UmuD', which is required for TLS. However, this does not eliminate the possibility that RecA plays another role in TLS besides its role as a LexA and UmuD coprotease. The mutants that express UmuC and the cleaved form of UmuD' constitutively could also be used to answer this question. If the coprotease activity of RecA alone is required for TLS, *recA* mutations should not affect UV mutagenesis in this genetic background. However, it was found that *recA* mutations still prevented UV mutagenesis even if UmuC and UmuD' were made constitutively, indicating that RecA plays another role in mutagenesis. This is what inspired the model that the UmuD'$_2$C mutagenic polymerase could replicate only over a RecA nucleoprotein filament (Figure 11.18).

SUMMARY

1. All organisms on Earth probably have mechanisms to repair damage to their DNA. Some of these repair systems are specific to certain types of damage, while others are more general and repair any damage that makes a significant distortion in the DNA helix.

2. Specific DNA glycosylases remove some types of damaged bases from DNA. Specific DNA glycosylases are known that remove uracil, hypoxanthine, some types of alkylated bases, 8-oxoG, and any A mistakenly incorporated opposite 8-oxoG in DNA. After the damaged base is removed by the specific glycosylase, the DNA is cut by an AP endonuclease and the strand is degraded and resynthesized to restore the correct base.

3. The positions of 5-methylcytosine in DNA are particularly susceptible to mutagenesis, because deamination of 5-methylcytosine produces thymine instead of uracil, and thymine will not be removed by the uracil-*N*-glycosylase. *E. coli* has a special repair system, called VSP repair, that recognizes the thymine in the thymine-guanine mismatch at the site of 5-methylcytosine and removes it, preventing mutagenesis. The products of the *mutS* and *mutL* genes of the mismatch repair system also make this pathway more efficient, perhaps by helping attract the Vsr endonuclease to the mismatch.

4. The photoreactivation system uses an enzyme called the photolyase to specifically separate the bases of one type of pyrimidine dimers created during UV irradiation. The photolyase binds to pyrimidine dimers in the dark but requires visible light to separate the fused bases.

5. Methyltransferase enzymes remove the methyl group from certain alkylated bases and phosphates and transfer it to themselves. In *E. coli*, these specific methyltransferases are inducible as part of the adaptive response. Methylation of the Ada protein converts it into a transcriptional activator for its own gene as well as for some other repair genes involved in repairing alkylation damage. Some genes of the adaptive response are also turned on when the cells reach

(continued)

SUMMARY (continued)

stationary phase, and the products of some of these genes may remove nitroasamines such as nitrosourea, which accumulate in stationary phase and can methylate DNA.

6. The methyl-directed mismatch repair system recognizes mismatches in the DNA and removes and resynthesizes one of the two strands, restoring the correct pairing. The products of the *mutL*, *mutS*, and *mutH* genes participate in this pathway in *E. coli*. The Dam methylase product of the *dam* gene helps select the strand to be degraded. The newly synthesized strand is degraded and resynthesized because it has not yet been methylated by the Dam methylase.

7. The nucleotide excision repair pathway encoded by the *uvr* genes of *E. coli* removes many types of DNA damage that cause gross distortions in the DNA helix. The UvrABC endonuclease cuts on both sides of the DNA damage, and the entire oligonucleotide including the damage is removed and resynthesized.

8. Recombination repair does not actually repair the damage but helps the cell tolerate it. Replication proceeds past the damage, leaving a gap opposite the damaged bases. Recombination with the other daughter DNA in the cell then puts a good strand opposite the damage, and the process can continue. This type of repair in *E. coli* requires the recombination functions RecBCD, RecFOR, RecA, PriA, PriB, PriC, and DnaT to restart replication forks.

9. A combination of excision and recombination repair may remove interstrand cross-links in DNA. The excision repair system cuts one strand, and the single-stranded break is enlarged by exonucleases to leave a gap opposite the dam-

age. Recombination repair can then transfer a good strand to a position opposite the damage. Excision repair can then remove the damage, since it is now confined to one strand. Interstrand cross-links can be repaired only if there are two or more daughter chromosomes in the cell.

10. The SOS regulon includes many genes that are induced following DNA damage. The genes are normally repressed by the LexA repressor, which is cleaved after DNA damage. The cleavage is triggered by RecA nucleoprotein filaments that accumulate following DNA damage.

11. SOS mutagenesis is due to the the products of genes *umuC* and *umuD*, which are induced following DNA damage. Immediately after induction, these proteins form a heterotrimer, UmuD$'_2$C, which is not active. However, the RecA single-stranded DNA nucleoprotein also promotes the autocleavage of UmuD to UmuD$'$. The UmuD$'_2$C heterotrimer is a mutagenic DNA polymerase that can replicate over abasic sites and the two forms of pyrimidine dimers formed by UV irradiation as well as some other types of DNA damage, inserting bases randomly opposite the damage and causing mutations. This may also require at least the β clamp and clamp-loading functions of the replicative DNA polymerase III to hold the UmuD$'_2$C polymerase on the DNA.

12. The RecA protein plays a direct role in SOS mutagenesis in addition to its role in promoting the autocleavage of LexA and UmuD. The UmuD$'_2$C polymerase may only be able to replicate DNA which is in the form of RecA nucleoprotein filaments.

QUESTIONS FOR THOUGHT

1. Some types of organisms, for example, yeasts, do not have methylated bases in their DNA and so would not be expected to have a methyl-directed mismatch repair system. Can you think of any other ways besides methylation that a mismatch repair system could be directed to the newly synthesized strand during replication?

2. Why do you suppose so many pathways exist to repair some types of damage in DNA?

3. Why do you think the SOS mutagenesis pathway exists if it contributes so little to survival after DNA damage?

PROBLEMS

1. Outline how you would determine if a bacterium isolated from the gut of a marine organism at the bottom of the ocean has a photoreactivation DNA repair system.

2. How would you show in detail that the mismatch repair system preferentially repairs the base in a mismatch on the strand unmethylated by Dam methylase? Hint: Make heteroduplex DNA of two mutants of your choice.

3. Outline how you would determine if the photoreactivating system is mutagenic.

4. Outline how you would determine if the nucleotide excision repair system of *E. coli* can repair damage due to aflatoxin B. Hint: Use a *uvr* mutant.

5. How would you determine if the RecA protein has a direct role in SOS mutagenesis independent of its role in cleaving LexA and UmuD?

6. Outline how you would determine if the *recN* gene is a member of the SOS regulon.

SUGGESTED READING

Bagg, A., C. J. Kenyon, and G. C. Walker. 1981. Inducibility of a gene product required for UV and chemical mutagenesis in *Escherichia coli. Proc. Natl. Acad. Sci. USA* **78:**5749–5753.

Dodson, M. L., and R. S. Lloyd. 1989. Structure-function studies of the T4 endonuclease V repair enzyme. *Mutat. Res.* **218:**49–65.

Friedberg, E. C., G. C. Walker, and W. Siede. 1995. *DNA Repair and Mutagenesis.* American Society for Microbiology, Washington, D.C.

Glickman, B. W., and M. Radman. 1980. *Escherichia coli* mutator mutants deficient in methylation instructed DNA mismatch correction. *Proc. Natl. Acad. Sci. USA* **77:**1063–1067.

Kato, T., R. H. Rothman, and A. J. Clark. 1977. Analysis of the role of recombination and repair in mutagenesis of *Escherichia coli* by UV irradiation. *Genetics* **87:**1–18.

Kato, T., and Y. Shinoura. 1977. Isolation and characterization of mutants of *Escherichia coli* deficient in induction of mutations by ultraviolet light. *Mol. Gen. Genet.* **156:**121–131.

Landini, P., and M. R. Volkert. 2000. Regulatory responses of the adaptive response to alkylation damage: a simple regulon with complex regulatory features. *J. Bacteriol.* **182:**6543–6549. (Minireview).

Lieb, M., S. Rehmat, and A. S. Bhagwat. 2001. Interaction of MutS and Vsr: some dominant-negative *mutS* mutations that disable methyladenine-directed mismatch repair are active in very-short-patch repair. *J. Bacteriol.* **183:**6487–6490.

Michaels, M. L., C. Cruz, A. P. Grollman, and J. H. Miller. 1992. Evidence that MutY and MutM combine to prevent mutations by an oxidatively damaged form of guanine in DNA. *Proc. Natl. Acad. Sci. USA* **89:**7022–7025.

Miller, J. H. 1996. Spontaneous mutations in bacteria: insights into pathways of mutagenesis and repair. *Annu. Rev. Microbiol.* **50:**1625–1643.

Nohni, T., J. R. Battista, L. A. Dodson, and G. C. Walker. 1988. RecA-mediated cleavage activates UmuD for mutagenesis: mechanistic relationship between transcriptional derepression and postranslational activation. *Proc. Natl. Acad. Sci. USA* **85:**1816–1820.

Pandya, G. A., I. Y. Yang, A. P. Grollman, and M. Moriya. 2000. *Escherichia coli* responses to a single DNA adduct. *J. Bacteriol.* **182:**6598–6604.

Pukkila, P. J., J. Petersson, G. Herman, P. Modrich, and M. Meselson. 1983. Effects of high levels of adenine methylation on methyl directed mismatch repair in *Escherichia coli. Genetics* **1044:**571–582.

Rangarajan, S., R. Woodgate, and M. F. Goodman. 1999. A phenotype for enigmatic DNA polymerase II: a pivotal role for pol II in replication restart in UV-irradiated *Escherichia coli. Proc. Natl. Acad. Sci. USA* **9:**9224–9229.

Rupp, W. D. 1996. DNA repair mechanisms, p. 2277–2294. *In* F. C. Neidhardt, R. Curtiss III, J. L. Ingraham, E. C. C. Lin, K. B. Low, B. Magasanik, W. S. Reznikoff, M. Riley, M. Schaechter, and H. E. Umbarger (ed.), Escherichia coli *and* Salmonella: *Cellular and Molecular Biology*, 2nd ed. ASM Press, Washington, D.C.

Rupp, W. D., C. E. I. Wilde, D. L. Reno, and P. Howard-Flanders. 1971. Exchanges between DNA strands in ultraviolet irradiated *E. coli. J. Mol. Biol.* **61:**25–44.

Sommer, S., K. Knezevic, A. Bailone, and R. Devoret. 1993. Induction of only one SOS operon, *umuDC*, is required for SOS mutagenesis in *E. coli. Mol. Gen. Genet.* **239:**137–144.

Sutton, M. D., M. F. Farrrow, B. M. Burton, and G. C. Walker. 2001. Genetic interactions between the *Escherichia coli umuDC* gene products and the β processivity clamp of the replicative DNA polymerase. *J. Bacteriol.* **183:**2897–2909.

Tang, M., X. Shen, E. G. Frank, M. O'Donnell, R. Woodgate, and M. F. Goodman. 1999. UmuD′(2)C is an error-prone DNA polymerase *Escherichia coli* pol V. *Proc. Natl. Acad. Sci. USA* **96:**8919–8924.

Taverna, P., and B. Sedgwick. 1996. Generation of endogenous DNA-methylating agent by nitrosation in *Escherichia coli. J. Bacteriol.* **178:**5105–5111.

Teo, I., B. Sedgwick, M. W. Kilpatrick, T. V. McCarthy, and T. Lindahl. 1986. The intracellular signal for induction of resistance to alkylating agents in *E. coli. Cell* **45:**315–324.

Weigle, J. J. 1953. Induction of mutation in a bacterial virus. *Proc. Natl. Acad. Sci. USA* **39:**628–636.

Worth, L., Jr., S. Clark, M. Radman, and P. Modrich. 1994. Mismatch repair proteins MutS and MutL inhibit RecA-catalyzed strand transfer between diverged DNAs. *Proc. Natl. Acad. Sci. USA* **91:**3238–3241.

12

Regulation of Gene Expression

THE DNA OF A CELL will contain thousands to hundreds of thousands of genes depending on whether the organism is a relatively simple single-celled bacterium or a complex multicellular eukaryote like a human. All of the features of the organism are due, either directly or indirectly, to the products of these genes. However, all the cells of an organism do not always look or act the same, even though they share essentially the same genes. Even the cells of a single-celled bacterium can look or act differently depending upon the conditions in which the cells find themselves, because the genes of cells are not always expressed at the same levels. The process by which the expression of genes is turned on and off at different times and under different conditions is called the **regulation of gene expression.**

Cells regulate the expression of their genes for many reasons. A cell may only express the genes that it needs in a particular environment so that it does not waste energy making RNAs and proteins it does not need at that time. Or the cell may turn off genes whose products might interfere with other processes going on in the cell at the time. Cells also regulate their genes as part of developmental processes, such as embryogenesis and sporulation.

As described in chapter 2, the expression of a gene moves through many stages, any one of which offers an opportunity for regulation. First, RNA is transcribed from the gene. Even if RNA is the final gene product, that molecule may require further processing to be active. If the final product of the gene is a protein, the mRNA synthesized from the gene might have to be processed before it can be translated into protein. Then the protein might need to be further processed or transported to its final location to be active. Even after the gene product is synthesized in its final form, its activity might also be modulated under certain environmental conditions.

By far the most common type of regulation occurs at the first stage, when RNA is made. Genes that are regulated at this level are said to be **transcriptionally regulated**. This form of gene regulation seems the most efficient, since synthesizing mRNA that will not be translated seems wasteful. Yet not all genes are transcriptionally regulated, at least not exclusively. Examples abound in which the expression of a gene is regulated even after RNA synthesis.

Any regulation that occurs after the gene is transcribed into mRNA is called **posttranscriptional regulation**. There are many types of posttranscriptional regulation; the most common is **translational regulation**. If a gene is translationally regulated, the mRNA may be continuously transcribed from the gene but its translation is sometimes inhibited.

Transcriptional Regulation in Bacteria

Thanks to the relative ease of doing genetics with bacteria, transcriptional regulation in bacteria is better understood than that in other organisms and has served as a framework for understanding transcriptional regulation in eukaryotic organisms. There are important differences between the mechanisms of transcriptional regulation in bacteria and higher organisms, many of which relate to the presence of a nuclear membrane. Nevertheless, many of the strategies used are similar throughout the biological world, and many general principles have been uncovered through studies of bacterial transcriptional regulation.

As discussed in chapter 2, most transcriptional regulation occurs at the level of transcription initiation at the promoter. Transcriptional regulation occurs through proteins called **transcriptional regulators**, which usually bind to DNA through helix-turn-helix motifs (see Box 12.1). Regulation of transcription initiation can be either negative or positive or both. If it is negative, it is controlled by a repressor which binds to an operator sequence in the DNA and prevents initiation of transcription by RNA polymerase. A repressor can perform this role of preventing initiation of transcription in a number of ways. The operator sequence may overlap with the promoter sequence so that binding of repressor will prevent binding of the RNA polymerase to the promoter. Alternatively, the repressor might bend the promoter so that RNA polymerase can no longer bind or the repressor might hold the RNA polymerase on the promoter so that it cannot leave as it begins to make RNA (see Rojo, Suggested Reading). If the regulation is positive, initiation of transcription is controlled by an activator which is required for initiation by RNA polymerase at the promoter. Activators may

work by increasing the binding of the RNA polymerase to the promoter, allowing it to open the strands of DNA at the promoter, or even by allowing RNA Pol to move from the promoter into the first gene and begin making RNA.

Some regulatory proteins can be either repressors or activators depending on the situation. Depending upon where it binds to the promoter region, the protein might sterically inhibit (i.e., physically get in the way of) the binding of RNA polymerase to the promoter and repress transcription. But if it binds even slightly further upstream, it might make contact with the RNA polymerase, stabilizing the binding of RNA Pol to the promoter and activating transcription. There is even one case of a phage φ29 regulatory protein that can activate transcription from some promoters and repress transcription from others even though it binds in approximately the same position in both types of promoters and makes contact with the same region of RNA polymerase bound to the promoters. In the former case, it increases the strength of binding of RNA polymerase to the promoter enough to activate transcription. In the latter case, it stabilizes the binding too much, and so the RNA polymerase cannot leave the promoter to begin making RNA! We discuss the details of repressor and activator action later in the chapter.

Genetic Evidence for Negative and Positive Regulation

Negatively and positively regulated operons behave very differently in genetic tests. One difference is in the effect of mutations that inactivate the regulatory gene for the operon. If an operon is negatively regulated, a mutation that inactivates the regulatory gene will allow transcription of the operon genes, even in the absence of inducer. If the regulation is positive, mutations that inactivate the regulatory gene will prevent transcription of the genes of the operon, even in the presence of the inducer. A mutant in which the genes of an operon are always transcribed, even in the absence of inducer, is called a **constitutive mutant**. Constitutive mutations are much more common with negatively than with positively regulated operons because any mutation that inactivates the repressor will result in the constitutive phenotype. With positively regulated operons, a constitutive phenotype can be caused only by changes that do not inactivate the activator protein but alter it so that it can activate transcription without binding to the inducer. Such changes tend to be rare.

Complementation tests reveal another difference between negatively and postively regulated operons. Constitutive mutations of a negatively regulated operon are often recessive to the wild type (see chapter 3 for genetic

BOX 12.1

The Helix-Turn-Helix Motif of DNA-Binding Proteins

Proteins that bind to DNA, including repressors and activators, often share similar structural motifs determined by the interaction between the protein and the DNA helix. One such motif is the helix-turn-helix motif. A region of approximately 7 to 9 amino acids forms an α-helical structure called helix 1. This region is separated by about 4 amino acids from another α-helical region of 7 to 9 amino acids called helix 2. The two helices are at approximately right angles to each other, hence the name helix-turn-helix. When the protein binds to DNA, helix 2 lies in the major groove of the DNA double helix while helix 1 lies crosswise to the DNA, as shown in the figure. Because they lie in the major groove of the DNA double helix, the amino acids in helix 2 can contact and form hydrogen bonds with specific bases in the DNA. Thus, a DNA-binding protein containing a helix-turn-helix motif recognizes and binds to specific regions on the DNA. Many DNA-binding proteins exist as dimers and bind to inverted repeated DNA sequences. In such cases, the two polypeptides in the dimer are arranged head to tail so that the amino acids in helix 2 of each polypeptide can make contact with the same bases in the inverted repeats.

A variant on the HTH domain is the winged helix-turn-helix (wHTH) domain, in which the "winged-turn" is 10 amino acids or more, longer than the 3–4 amino acids of the "turns" of other HTH domains.

In the absence of structural information, the existence of a helix-turn-helix motif in a protein can often be predicted from the amino acid sequence, since some sequences of amino acids cause the polypeptide to assume an α-helical form and the bent region between the two helices usually contains a glycine. The presence of a helix-turn-helix motif in a protein helps identify it as a DNA-binding protein.

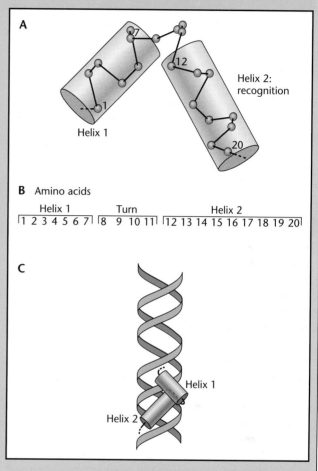

Helix-turn-helix motifs of DNA-binding proteins. (A) The structure of a helix-turn-helix domain. (B) The number of amino acids in the helix-turn-helix domain of the CAP protein. (C) The interactions of helix 1 and helix 2 with double-stranded DNA.

References

Kenney, L. J. 2002. Structure/function relationships in OmpR and other winged-helix transcription factors. *Curr. Op. Microbiol.* **5**:135–141.

Rhee, S., R. G. Martin, J. L. Rosner, and D. R. Davies. 1998. A novel DNA-binding motif in MarA: the first structure for an AraC family transcriptional activator. *Proc. Natl. Acad. Sci. USA* **95**:10413–10418.

Rosinski, J. A., and W. R. Atchley. 1999. Molecular evolution of helix-turn-helix proteins. *J. Mol. Evol.* **49**:301–309.

Steitz, T. A., D. H. Ohlendorf, D. B. McKay, W. F. Anderson, and B. W. Matthews. 1982. Structural similarity in the DNA-binding domains of catabolite gene activator and *cro* repressor proteins. *Proc. Natl. Acad. Sci. USA* **79**:3097–3100.

definitions). This is because any normal repressor protein in the cell encoded by a wild-type copy of the gene will bind to the operator and block transcription, even if the repressor encoded by the mutant copy of the gene in the same cell is inactive. In contrast, constitutive mutations in a solely positively regulated operon should be dominant to the wild type. A mutant activator protein that is active without inducer bound might activate transcription even in the presence of a wild-type activator protein. In the next sections we describe the regulation of some operons and how genetic evidence contributed to this knowledge.

Negative Regulation

The *E. coli lac* Operon

The classic example of negative regulation is regulation of the *E. coli lac* operon, which encodes the enzymes responsible for the utilization of the sugar lactose. The experiments of François Jacob and Jacques Monod and their collaborators on the regulation of the *E. coli lac* genes are excellent examples of the genetic analysis of a biological phenomenon in bacteria (see Jacob and Monod, Suggested Reading). Although these experiments were performed in the late 1950s, only shortly after the discovery of the structure of DNA and the existence of mRNA, they still stand as the framework with which all other studies of gene regulation are compared.

GENETICS OF THE *lac* OPERON

When Jacob and Monod began their classic work, it was known that the enzymes of lactose metabolism are **inducible** in that they are expressed only when the sugar lactose is present in the medium. If no lactose is present, the enzymes are not made. From the standpoint of the cell, this is a sensible strategy, since there is no point in making the enzymes for lactose utilization unless there is lactose available for use as a carbon and energy source.

To understand the regulation of the lactose genes, Jacob and Monod first isolated many mutations affecting lactose metabolism and regulation, which fell into two fundamentally different groups. Some mutants were unable to grow with lactose as the sole carbon and energy source and so were Lac⁻. Other mutants made the lactose-metabolizing enzymes whether or not lactose was present in the medium and so were constitutive mutants.

To analyze the regulation of the *lac* genes, Jacob and Monod needed to know which of the mutations affected *trans*-acting gene products—either protein or RNA—involved in the regulation and how many different genes these mutations represented. They also wished to know if any of the mutations were *cis* acting (affecting sites on the DNA involved in regulation).

To answer these questions, they needed to perform complementation tests, which require that the organisms be diploid, with two copies of the genes being tested.

Bacteria are normally haploid, with only one copy of each of their genes, but are "partial diploids" for any genes carried on an introduced prime factor. Recall that a prime factor is a plasmid into which some of the bacterial chromosomal genes have been inserted (see chapter 5). By introducing prime factors carrying various mutated *lac* genes into cells with different mutations in the chromosomal *lac* genes, Jacob and Monod performed complementation tests on each of their *lac* mutations. Their methods depended upon the type of mutation being tested.

COMPLEMENTATION TESTS WITH *lac* MUTATIONS

Whether a particular *lac* mutation is dominant or recessive was determined by introducing an F′ factor carrying the wild-type *lac* region into a strain with the *lac* mutation in the chromosome. If the partial diploid bacteria are Lac⁺ and can multiply to form colonies on minimal plates with lactose as the sole carbon and energy source, the *lac* mutation is recessive. If the partial diploid cells are Lac⁻ and cannot form colonies on lactose minimal plates, the *lac* mutation is dominant. Jacob and Monod discovered that most *lac* mutations are recessive to the wild type and so presumably inactivate genes whose products are required for lactose utilization.

The question of how many genes are represented by recessive *lac* mutations could be answered by performing pairwise complementation tests between different *lac* mutations. Prime factors carrying the *lac* region with one *lac* mutation were introduced into a mutant strain with another *lac* mutation in the chromosome (Figure 12.1). In this kind of experiment, if the partial diploid cells are Lac⁺, the two recessive mutations can complement each other and are members of *different* complementation groups or genes. If the partial diploid cells are Lac⁻, the two mutations cannot complement each other and are members of the *same* complementation group or gene. Jacob and Monod found that most of the *lac* mutations sorted into two different complementation groups, which they named *lacZ* and *lacY*. We now know of another gene, *lacA*, which was not discovered in their original selections because its product is not required for growth on lactose.

Figure 12.1 Complementation of two recessive *lac* mutations. One mutation is in the chromosome, and the other is in a prime factor. If the two mutations complement each other, the cells will be Lac⁺ and will grow with lactose as the sole carbon and energy source. The mutations will not complement if they are in the same gene or if one affects a regulatory site or is polar. See the text for more details.

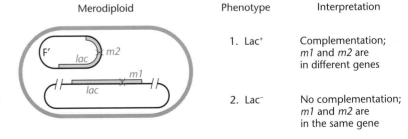

Merodiploid	Phenotype	Interpretation
	1. Lac⁺	Complementation; *m1* and *m2* are in different genes
	2. Lac⁻	No complementation; *m1* and *m2* are in the same gene

cis-Acting *lac* Mutations

Not all *lac* mutations affect diffusible gene products and can be complemented. Immediately adjacent to the *lacZ* mutations are other *lac* mutations that are much rarer and have radically different properties. These mutations cannot be complemented to allow the expression of the *lac* genes on the same DNA, even in the presence of good copies of the *lac* genes. Recessive mutations that cannot be complemented are *cis* acting and presumably affect a site on DNA rather than a diffusible gene product like RNA or protein (see chapter 3).

To show that a *lac* mutation is *cis* acting, i.e., affects only the expression of genes on the same DNA where it occurs, we could introduce an F′ factor containing the potential *cis*-acting *lac* mutation into cells containing either a *lacZ* or a *lacY* mutation in the chromosome (Figure 12.2). *trans*-acting gene products encoded by the F′ factor *lacZ* or *lacY* genes would complement the chromosomal *lacY* or *lacZ* mutations, respectively. However, if the resulting phenotypes are Lac⁻, the *lac* mutation in the F′ factor must prevent expression of both LacZ and LacY proteins from the F′ factor. The mutation in the F′ factor is therefore *cis* acting.

As we discuss below, one type of the *cis*-acting *lac* mutations affects *lacp* and is a promoter mutation that prevents transcription of the *lacZ* and *lacY* genes by changing the binding site for RNA polymerase on the DNA. Another type of *cis*-acting mutation is a polar mutation in *lacZ* that prevents the transcription of the downstream *lacY* gene.

Lac⁻ Mutants with Dominant Mutations

Some Lac⁻ mutants have mutations that affect diffusible gene products but are not recessive. Such *lac* mutations will make the cell Lac⁻ and unable to use lactose even if there is another good copy of the lactose operon in the cell, either in the chromosome or in the F′ factor. These dominant mutations are called *lacI*ˢ mutations, for superrepressor mutations. As we show below, these mutants have mutations that change the repressor so that it can no longer bind the inducer.

COMPLEMENTATION TESTS WITH CONSTITUTIVE MUTATIONS

As mentioned, some *lac* mutations do not make the cells Lac⁻ but, rather, make them constitutive, so that they express the *lacZ* and *lacY* genes even in the absence of the inducer lactose. In complementation tests between constitutive mutations, partial diploids are made in which either the chromosome or the F′ factor, or both, carry a constitutive mutation. The partial diploid cells are then tested to determine whether they express the *lac* genes constitutively or only in the presence of the inducer. If the partial diploid cells express the *lac* gene in the absence of the inducer, the constitutive mutation is dominant. However, if the partial diploid cells express the *lac* genes only in the presence of the inducer, the mutations are recessive. Using this test, Jacob and Monod found that some of the constitutive mutations, which could be recessive or dominant, affect a *trans*-acting gene product, either protein or RNA. Complementation between the recessive constitutive mutations revealed they are all in the same gene, which these investigators named *lacI* (Figure 12.3A).

cis-Acting *lacO*ᶜ Mutations

A rarer constitutive mutation is *cis* acting, allowing the constitutive expression of the *lacZ* and *lacY* genes from the DNA that has the mutation, even in the presence of a wild-type copy of the *lac* DNA. Jacob and Monod named these *cis*-acting constitutive mutations *lacO*ᶜ mutations, for *lac* operator-constitutive mutations. Figure 12.3B shows the partial diploid cells used in these complementation tests.

trans-Acting Dominant Constitutive Mutations

Some *lacI* mutations, called *lacI*⁻ᵈ mutations, are dominant, making the cell constitutive for expression of the *lac* operon even in the presence of a good copy of the *lacI* gene. These *lacI*⁻ᵈ mutations are possible because the LacI polypeptides form a homotetramer. A mixture of normal and defective subunits can be nonfunctional, causing the constitutive LacI⁻ phenotype. Hence, the

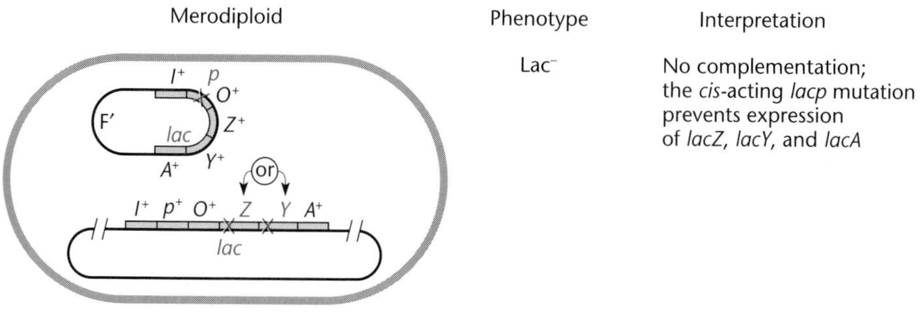

Merodiploid	Phenotype	Interpretation
	Lac⁻	No complementation; the *cis*-acting *lacp* mutation prevents expression of *lacZ*, *lacY*, and *lacA*

Figure 12.2 The *lacp* mutations cannot be complemented and are *cis* acting. A *lacp* mutation in the prime factor will prevent expression of any of the other *lac* genes on the prime factor, so that a *lac* mutation in the chromosome will not be complemented. Partial diploid cells will be Lac⁻.

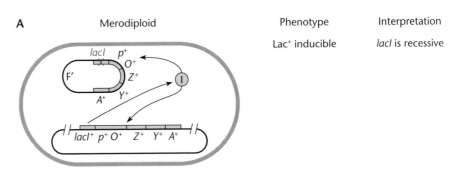

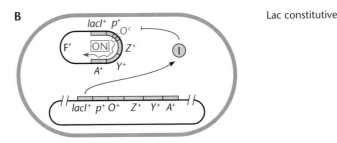

Figure 12.3 Complementation with two types of constitutive mutations. (A) The *lacI* mutation can be complemented, and so other genes on the prime factor will be inducible in the presence of a wild-type copy of the *lacI* gene. (B) In contrast, *lacO^c* mutations cannot be complemented by a wild-type *lacO* region in the chromosome and so are dominant.

lacI^-d mutations are *trans* dominant. Table 12.1 summarizes the behavior of the various *lac* mutations in complementation tests.

JACOB AND MONOD OPERON MODEL

On the basis of this genetic analysis, Jacob and Monod proposed their **operon model** for *lac* gene regulation (Figure 12.4). The *lac* operon includes the genes *lacZ* and *lacY*. These genes, known as the **structural genes** of the operon, encode the enzymes required for lactose utilization. The *lacZ* gene product is a β-galactosidase, which cleaves lactose to form glucose and galactose, which can then be used by other pathways. The *lacY* gene product is a permease that allows lactose into the cell. The operon also includes the *lacA* gene. The *lacA* gene product is a transacetylase, whose function is unknown. This enzyme was originally thought to be also encoded by the *lacY* gene.

The most important part of this model explains why the structural genes are expressed only in the presence of lactose. The product of the *lacI* gene is a repressor protein. In the absence of lactose, this repressor binds to the operator sequence (*lacO*) close to the promoter and thereby prevents the RNA polymerase from binding to the promoter and blocks the transcription of the structural genes. In contrast, when lactose is available, the inducer binds to the repressor and changes its conformation so that it can no longer bind to the operator sequence. The RNA polymerase can then bind to *lacp* and transcribe the *lacZ*, *lacY*, and *lacA* genes. The LacI repressor is very effective at blocking transcription of the

Figure 12.4 The Jacob and Monod model for negative regulation of the *lac* operon. In the absence of the inducer, lactose, the LacI repressor will bind to the operator region, preventing transcription of the other genes of the operon by RNA polymerase (RNA Pol). In the presence of lactose, the repressor can no longer bind to the operator, allowing transcription of the *lacZ*, *lacY*, and *lacA* genes. Allolactose, rather than lactose, is the true inducer molecule.

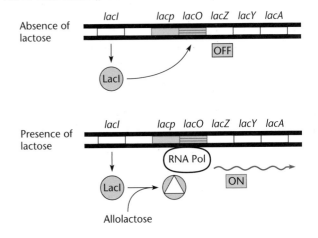

TABLE 12.1	Genetic behavior of *lac* mutations	
Mutation	**Inducibility**	**Complementation behavior**
lacZ	Noninducible	Recessive; *trans*-acting
lacI	Constitutive	Recessive; *trans*-acting
lacI^s	Noninducible	Dominant; *trans*-acting
lacI^-d	Constitutive	Dominant; *trans*-acting
lacI^q	Inducible^a	Dominant; *trans*-acting
lacO^c	Constitutive	Dominant; *cis*-acting
lacP	Noninducible	Recessive; *cis*-acting

^a Tighter on-off control.

structural genes of the operon. In the absence of repressor, transcription is about 1,000 times more active than in its presence.

It is worthwhile to emphasize how the Jacob and Monod operon model explains the behavior of mutations that affect the regulation of the *lac* enzymes. Mutants with *lacZ* and *lacY* mutations are Lac⁻ because they do not make an active β-galactosidase or permease, respectively, both of which are required for lactose utilization. These mutations are clearly *trans* acting, because they are recessive and can be complemented. An active β-galactosidase or permease made from another DNA in the same cell can provide the missing enzyme and allow lactose utilization.

The behavior of *lacp* mutations is also explained by the model. Jacob and Monod proposed that *lacp* mutations change the sequence on the DNA to which the RNA polymerase binds. The RNA polymerase normally binds to *lacp* and moves through *lacZ, lacY,* and *lacA,* making an RNA copy of these genes. This explains why *lacp* mutations are *cis* acting; if the site on DNA at which RNA polymerase initiates transcription is changed by a mutation so that it no longer binds RNA polymerase, the *lacZ, lacY,* and *lacA* genes on that DNA will not be transcribed into mRNA, even in the presence of a good copy of the *lac* region elsewhere in the cell.

Their model also explains the behavior of the two constitutive mutations: *lacI* and *lacO*^c. The *lacI* mutations affect a *trans*-acting function, because they inactivate the repressor protein that binds to the operator and prevents transcription. The LacI repressor made from a functional copy of the *lacI* gene anywhere in the cell can bind to the operator sequence and block transcription in *trans*. However, the *lacO*^c mutations change the sequence on DNA to which the LacI repressor binds to block transcription. The LacI repressor cannot bind to this altered *lacO* sequence, even in the absence of lactose. Therefore, the RNA polymerase is free to bind to the promoter and transcribe the structural genes. The *lacO*^c mutations are *cis* acting because they allow the constitutive expression of the *lacZ, lacY,* and *lacA* genes from the same DNA, even in the presence of a good copy of the *lac* operon elsewhere in the cell.

The behavior of superrepressor *lacI*^s mutations is also explained by their model. These are mutations that change the repressor molecule so that it can no longer bind the inducer. The mutated repressor will bind to the operator even in the presence of inducer, making the cells permanently repressed and phenotypically Lac⁻. The fact that this type of mutation is dominant over the wild type is also explained. The mutated repressors will repress transcription of any *lac* operon in the same cell, even in the presence of inducer, and so they will make the cell Lac⁻ even in the presence of a good *lac* operon, either in the chromosome or in an F′ factor.

The *lac* genes provide a good example of what we mean by "operon." As we stated earlier, an operon includes all the genes that are transcribed into the same mRNA plus any adjacent *cis*-acting sites that are involved in the transcription or regulation of transcription of the genes. The *lac* operon of *E. coli* consists of the three structural genes, *lacZ, lacY,* and *lacA,* which are transcribed into the same mRNA, as well as the *lac* promoter from which these genes are transcribed. It must also include the the *lac* operator, since this is a *cis*-acting regulatory sequence involved in regulating the transcription of the structural genes. However, the *lac* operon does *not* include the gene for the repressor, *lacI.* The *lacI* gene is adjacent to the *lacZ, lacY,* and *lacA* genes and regulates their transcription, but it is not transcribed onto the same mRNA as the structural genes. Moreover, its product is *trans* acting rather than *cis* acting.

UPDATE ON THE REGULATION OF THE *lac* OPERON

The operon model of Jacob and Monod has survived the passage of time. In 1965, it earned them a Nobel Prize, which they shared with Andre Lwoff. Because of its elegant simplicity, the operon model for the regulation of the *lac* genes of *E. coli* serves as the paradigm for understanding gene regulation in other organisms.

Over the years, the operon model has undergone a few refinements. As mentioned, Jacob and Monod did not know of the existence of the *lacA* gene and thought that *lacY* encoded the transacetylase. Also, most of the mutations that Jacob and Monod defined as *lacp* were not promoter mutations but, rather, strong polar mutations in *lacZ* that prevent the transcription of all three structural genes (*lacZ, lacY,* and *lacA*). Later studies also revealed that the true inducer that binds to the LacI repressor is not lactose itself but, rather, allolactose, a metabolite of lactose. In most experiments, an analog of allolactose called isopropyl-β-D-thiogalactoside (IPTG) is used as the inducer because it is not metabolized by the cells.

The most significant alteration to the Jacob and Monod model was the discovery that the LacI repressor can bind to not just one but three operators, called o_1, o_2, and o_3 (Figure 12.5). The operator closest to the promoter, o_1, seems to be the most important for repressing transcription of *lac* and acts by sterically interfering with binding of RNA polymerase to the promoter. However, deleting both o_2 and o_3 diminishes repression as much as 50-fold.

Why does the *lac* operon have more than one operator, especially since one of the operators (o_3) is so far upstream of the promoter that it seems unlikely that it could block binding of the RNA polymerase to that site? One idea is that the LacI repressor binds to two operators simultaneously (Figure 12.5), bending the DNA—and the promoter—between them. The bent promoter

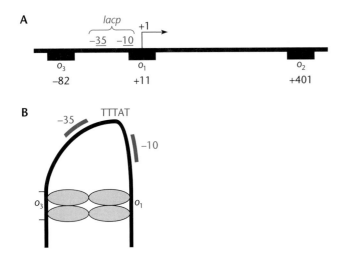

Figure 12.5 Locations of the three operators in the *lac* operon (A) and a possible model for how LacI repressor binding to these operators may help repress the operon. (B) Repressor (solid ellipses) bound to o₁ and o₃, or o₂ and o₃, could bend the DNA in the promoter region and help prevent RNA polymerase binding to the promoter. The AT-rich region may facilitate bending. Adapted from H. Choy and S. Adhya, p. 1291, *in* F. C. Neidhardt, R. Curtiss III, J. L. Ingraham, E. C. C. Lin, K. B. Low, B. Magasanik, W. S. Reznikoff, M. Riley, M. Schaechter, and H. E. Umbarger (ed.), Escherichia coli *and* Salmonella: *Cellular and Molecular Biology,* 2nd ed., ASM Press, Washington, D.C., 1996.

might not be able to bind RNA polymerase or might not undergo the changes in structure required for the initiation of transcription. As we discuss below, the *ara* and *gal* operons also contain multiple operators that may also bend the DNA in the promoter region.

CATABOLITE REPRESSION OF THE *lac* OPERON
In addition to being under the control of its own specific repressor, the *lac* operon is regulated through **catabolite repression. Catabolites** are carbon-containing molecules that are used to build other molecules. The catabolite repression system ensures that the genes for lactose utilization will not be expressed if a better carbon and energy source such as glucose is available. The name "catabolite repression" is a misnomer since the expression of operons under catabolite control requires a transcriptional activator, the catabolite activator protein (CAP), and the small molecule effector, cyclic AMP (cAMP). We defer a detailed discussion of the mechanism of catabolite repression until chapter 13, since this is a type of global regulation.

STRUCTURE OF THE *lac* CONTROL REGION
Figure 12.6 illustrates the structure of the *lac* control region in detail, showing the nucleotide sequences of the *lac* promoter and operators as well as the region to

which the CAP protein binds. The *lac* promoter is a typical σ^{70} bacterial promoter with the characteristic −10 and −35 regions (see chapter 2). One of the operators to which the LacI repressor binds (o₁) actually overlaps the mRNA start site (+1 in Figure 12.6) for transcription of *lacZ, lacY,* and *lacA.* Although not shown, the other *lacO* operator sequences lie nearby. The CAP-binding site that enhances initation by RNA polymerase in the absence of glucose is just upstream of the promoter, as shown.

EXPERIMENTAL USES OF THE *lac* OPERON
Because it is the best understood of the bacterial regulatory systems, the *lac* genes and regulatory regions have found many uses in molecular genetics. We list a few below.

The LacI Repressor
The LacI repressor protein has served as a model for the interaction of proteins with DNA. This protein is very interesting because it has more than one function: it must be able to bind to DNA at the operator sequences as well as bind the inducer, either allolactose or IPTG. The recent success in crystallizing the protein has given us a better picture of the structure of this protein. The active repressor is a homotetramer, with four identical polypeptide subunits encoded by the *lacI* gene, so that the individual polypeptides must bind to each other. Different regions or domains of LacI participate in these various functions. In chapter 14, we discuss genetic exeriments that helped identify the domains of the LacI repessor involved in its various functions. Finally, the *lac* repressor has been used in many bacteria other than *E. coli,* and has even been used in eukaryotic cells. The LacI repressor protein is very stable and can be used to block transcription by binding to the *lac* operator sequence in any cells in which it can be synthesized. Expression vectors in which the *lac* repressor is used to repress transcription of genes have even been used in mammalian cells (see chapter 2 for a description of expression vectors).

The *lac* Promoter
The *lac* promoter or its derivatives have been used to transcribe cloned genes in many expression vectors (see chapter 2). This promoter offers many advantages in these expression vectors. The *lac* promoter is fairly strong, allowing high levels of transcription of a cloned gene. The *lac* promoter is also inducible, which makes it possible to clone genes whose products are toxic to the cell. The cells can be grown in the absence of inducer, so that the cloned gene will not be transcribed. Only when the cells reach a high density is the inducer IPTG added and the cloned gene transcribed. Even if the toxic protein kills the cells, some of the protein will usually be synthesized before the cell dies.

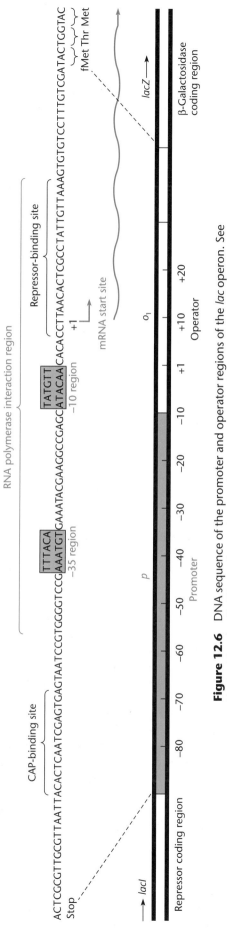

Figure 12.6 DNA sequence of the promoter and operator regions of the *lac* operon. See the text for details. The entire region is only 100 bp long. Only the o₁ operator sequence is shown.

The derivatives of the *lac* promoter used in expression vectors retain some of the desirable properties of the *lac* operon but have additional features. For example, the mutated promoter *lacUV5* is no longer sensitive to catabolite repression and so will be active even if glucose is present in the medium (see chapter 13).

A hybrid *trp-lac* promoter called the *tac* promoter has also been widely used. The *tac* promoter has the advantages that it is even stronger than the *lac* promoter but retains its inducibility. It is also insensitive to catabolite repression.

The *lacZ* Reporter Gene

The *lacZ* gene is widely used as a reporter gene in translational and transcriptional fusions. In many ways, this gene is the quintessential reporter gene and has been used in cells ranging from bacteria to fruit fly to human cells. The product of the *lacZ* gene, β-galactosidase, is easily detected by colorimetric assays using substrates such as X-Gal (5-bromo-4-chloro-3-indolyl-β-D-galactopyranoside) and ONPG (*o*-nitrophenyl-β-D-galactopyranoside) and is an unusually stable protein. The fact that this polypeptide is very large can be a disadvantage or advantage depending upon the situation.

PROSPECTUS

The *lac* operon is one of the simplest regulatory systems known, and so it is fortunate that it was one the first to be chosen to study. The relatively simple regulation of the *lac* operon encouraged attempts to understand other types of regulation. As discussed later, regulation of most other operons is more complicated and would have been even more difficult to understand if the *lac* operon had not been available as a point of reference. In the next sections, we discuss the regulation of some other representative bacterial operons.

The *E. coli gal* Operon

The operon of *E. coli* involved in the utilization of the sugar galactose, the *gal* operon, is another classic example of negative regulation. Figure 12.7 shows the organization of the genes in this operon. The products of three structural genes, *galE*, *galT*, and *galK*, are required for the utilization of galactose and convert galactose into glucose, which can then enter the glycolysis pathway.

The specific reactions catalyzed by each of the enzymes of the *gal* pathway appear in Figure 12.8. The GalK gene product is a kinase that phosphorylates galactose to make galactose-1-phosphate. The product of the *galT* gene is a transferase that transfers the galactose-1-phosphate to UDPglucose, displacing the glucose to make UDPgalactose. The released glucose can then be used as a carbon and energy source. The GalE gene

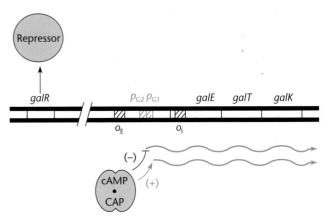

Figure 12.7 Structure of the galactose operon of *E. coli*. The *galE*, *galT*, and *galK* genes are transcribed from two promoters, p_{G1} and p_{G2}. The CAP protein with cyclic AMP (cAMP) bound turns on p_{G1} and turns off p_{G2}, as shown. There are also two operators, *galo*$_E$ and *galo*$_I$. The repressor genes are some distance away, as indicated by the broken line. Only the *galR* repressor gene is shown.

product is an epimerase that converts UDPgalactose to UDPglucose to continue the cycle. It is not clear why *E. coli* cells use this convoluted pathway to convert galactose to glucose so that the latter can be used as a carbon and energy source. However, many organisms, including both plants and animals, use this pathway.

Unlike the genes for lactose utilization, not all the genes for galactose utilization are closely linked in the *E. coli* chromosome. The *galU* gene, whose product synthesizes UDPglucose, is located in a different region of the chromosome. Also, the genes for the galactose permeases, which are responsible for transporting galactose into the cell, are not part of the *gal* operon. Furthermore, unlike the *lacI* gene, which is adjacent to the *lac* operon, the regulatory genes for the *gal* operon do not reside near this operon. The scattering of the genes for galactose metabolism may reflect the fact that galactose not only serves as a carbon and energy source but also plays

Figure 12.8 Pathway for galactose utilization in *E. coli*. See the text for details.

$$\text{Galactose + ATP} \xrightarrow{\text{GalK}} \text{galactose-1-PO}_4$$

$$\text{Galactose-1-PO}_4 + \text{UDPglucose} \xrightarrow{\text{GalT}} \text{UDPgalactose + glucose}$$

$$\text{UDPgalactose} \xrightarrow{\text{GalE}} \text{UDPglucose}$$

$$\text{UTP + glucose-1-PO}_4 \xrightarrow{\text{GalU}} \text{UDPglucose}$$

other roles. For example, the UDPgalactose synthesized by the *gal* operon donates galactose to make polysaccharides for lipopolysaccharide and capsular synthesis.

REGULATION OF THE *gal* OPERON

Everything about the *gal* operon comes in twos. There are two promoters and two operators. There are even two repressors, encoded by different genes. Either of these repressors can repress the *gal* operon, although one is more effective than the other.

The Repressors: GalR and GalS

The two repressors in control of the *gal* operon are GalR and GalS, encoded by the *galR* and *galS* genes, respectively. GalR was discovered first because mutations in the *galR* cause constitutive expression of the *gal* operon. However, the *gal* operon was also subject to other regulation. If the GalR repressor were solely responsible for regulating the *gal* operon, mutations that inactivate the *galR* gene should result in the same level of *gal* expression whether galactose is present or not. Yet some regulation of the *gal* operon could be observed, even in *galR* mutants. When galactose was added to the medium in which *galR* mutant cells were growing, more of the enzymes of the *gal* operon were made than if the cells were growing in the absence of galactose. The product of another gene, *galS*, was responsible for the residual regulation. As evidence, double mutants with mutations that inactivate both *galR* and *galS* are fully constitutive. Later studies showed that the product of the *galS* gene is also a repressor that negatively regulates the *gal* operon.

The GalS and GalR repressor proteins are closely related, and they both bind the inducer galactose. Even so, they may play somewhat different roles. The GalR repressor is responsible for most of the repression of the *gal* operon in the absence of galactose. The GalS repressor plays only a minor role in regulating the *gal* operon but solely controls the genes of the galactose transport system, which transports galactose into the cell. The reason for this two-tier regulation is unclear but also may be related to the diverse roles of galactose in the cell (see above).

The Two *gal* Operators

There are also two operators in the *gal* operon. One is upstream of the promoters, and the other is internal to the first gene, *galE* (Figure 12.7). The two operators are named o_E and o_I for operator *external* to the *galE* gene and *internal* to the *galE* gene, respectively. The discovery of the o_I operator involves some interesting genetics, so we discuss it in some detail.

1. Isolating gal *operator mutants.* The first mutant with an o_I mutation was isolated as part of a collection of constitutive mutants of the *gal* operon (see Irani et al., Suggested Reading). These mutants are easier to isolate in strains with superrepressor *galR*^s mutations than in wild-type *E. coli*. The *galR*^s mutations are analogous to *lacI*^s mutations. The superrepressor mutation will make a strain Gal⁻ and uninducible because galactose cannot bind to the mutated repressor. Therefore, *E. coli* with a *galR*^s mutation cannot multiply to form colonies on plates containing only galactose as the carbon and energy source. However, a constitutive mutation that inactivates the GalR^s repressor or changes the operator sequence will prevent the mutant repressor from binding to the operator and allow the cells to use the galactose and multiply to form a colony. Thus, if bacteria with a *galR*^s mutation are plated on medium with galactose as the sole carbon and energy source, only constitutive mutants will multiply to form a colony. However, most of the constitutive mutants isolated this way will have mutations in the *galR* gene that inactivate the GalR^s repressor rather than operator mutations, since the operator is by far the smaller target. Many *galR* mutants would have to be screened before a single operator mutant was found. Therefore, to make this method practicable for isolating constitutive mutants with operator mutations, the frequency of *galR* mutants must be decreased until it is not too much higher than that of constitutive mutants with mutations in the operator sequences.

One way to reduce the frequency of *galR* mutants is to use a strain that is a partial diploid for (has two copies of) the *galR*^s gene. Then, even if one *galR*^s gene is inactivated by a mutation, the other *galR*^s gene will continue to make the GalR^s protein, making the cell phenotypically Gal⁻. Only two independent mutations, one in each *galR*^s gene, can make the cell constitutive. Therefore, since two independent *galR* mutations should be no more frequent than single operator mutations, cells with operator mutations should now be a significant fraction of the constitutive mutants. Moreover, constitutive mutants with operator mutations can be distinguished from the double mutants with mutations in both the *galR*^s genes. The operator mutations will map in the *gal* operon, unlike mutations in either copy of the *galR*^s gene.

Accordingly, a partial diploid was constructed that had one copy of the *galR*^s gene in the normal position and another copy in a specialized transducing λ phage integrated at the λ attachment site (see chapter 8). When this strain was plated on medium containing galactose as the sole carbon and energy source, a few Gal⁺ colonies arose from constitutive mutants. The mutations in two of these constitutive mutants mapped in the region of the *gal* operon and so were presumed to be operator mutations. When the DNA of the two mutants was sequenced, it was discovered that one mutation had changed a base pair in the known operator region upstream of the promoters (o_E), as expected. However, the other operator mutation had changed a base pair in the *galE* gene, suggesting that a sequence in that gene—o_I—also functions as an operator. Furthermore, this mutation occurred in a sequence homologous to the 15 bp making up the known operator. In fact, 12 bp of this sequence is identical in o_I and o_E. Moreover, the mutation in the *galE* gene was *cis* acting for constitutive expression of the *gal* operon, one of the criteria for an operator mutation.

2. Escape synthesis of the Gal enzymes. The mutations in the *galE* gene cause *cis*-acting constitutive expression of the *gal* genes, suggesting that the o_I sequence functions as an operator. However, the constitutive mutants could be explained in other ways.

If the sequence in the *galE* gene is truly an operator, it would bind the GalR repressor protein. To test this, the experiment illustrated in Figure 12.9 was performed. First, DNA containing the *galE* gene is cloned into a multicopy plasmid. When this multicopy plasmid is transformed into a cell, that cell will contain many copies of the *galE* gene. Then, if the sequence in the *galE* gene does bind the repressor, these extra copies should bind most of the GalR protein in the cell, leaving too little to completely repress expression of the *gal* operon. Thus, the cells would appear to be constitutive mutants.

This general method is called **titration**, and the enzymes of the *gal* operon are synthesized through **escape synthesis**, so named since the operon is "escaping" the effects of the repressor. In the actual experiment, cells containing many copies of the *galE* gene region did exhibit a partially constitutive phenotype. In contrast, multiple copies of mutant DNA with the putative operator mutation (i.e., o_I^c) do not cause escape synthesis of the Gal enzymes. This result was interpreted to confirm the presence in *galE* of a second binding site for repressor, which is inactivated by the mutation.

3. Why does the gal *operon have two operators?* There are two general hypotheses for why the *gal* operon has two operators. According to one, the two operators function independently to block transcription of the *gal* operon. The other proposes that the operators cooperate to block transcription.

Genetic evidence supports the second general hypothesis—the two operators cooperate to block transcription. If the two operators functioned independently, the effect of mutations in both operators would be additive; in other words, the level of expression of the genes in the operon when both operators were mutated would be the sum of the levels of operon expression when each of the operators was mutated separately. However, genetic experiments showed that the level of expression in the double mutant is higher than the sum of the expression levels

Figure 12.9 Escape synthesis of the enzymes of the galactose operon caused by additional copies of the operator regions. Clones of the operator regions in a multicopy plasmid dilute out the repressor, inducing the operon even in the absence of galactose. (A) The cell contains the normal number of operators and the operon is not induced. (B) The multicopy plasmid contains only o_E, and the operon is only partially induced. (C) The multicopy copy plasmid contains both o_E and the *galE* gene containing o_I, and the operon is fully induced.

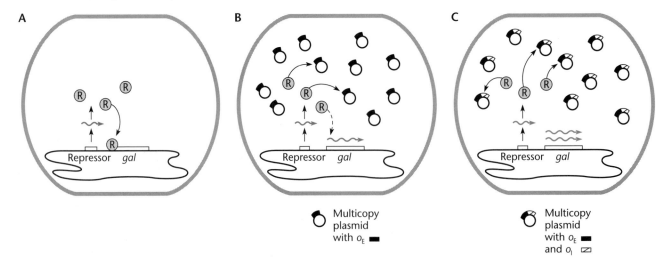

in each single mutant. This observation indicated cooperation between the two operators.

Figure 12.10 shows a model for this cooperation. Repressor molecules bound to the two operators interact with each other to bend the DNA of the promoter that is between the two operators, helping to prevent initiation of RNA synthesis at the promoter. A similar model had been proposed for the *ara* operon, and another was proposed later for the extra *lac* operators (Figure 12.5).

This model leads to a specific prediction: the spacing in the DNA between the two operators should be very important for the repression. As illustrated in Figure 12.10, the significance of the spacing is related to the structure of DNA. A region of DNA looks different depending on the face of the molecule. For example, on one face you might see the major groove plus one of the two strands, while on another you might see the minor groove, and so on. The two repressor molecules presumably recognize the same face of both operator sequences, since the two operator sequences are almost identical. Furthermore, if the two repressor molecules, actually dimers of the polypeptide product of the *galR*

gene (see below), are to interact with the two operator sequences, the spacing on the DNA between the operator sequences should matter since double stranded DNA is stiff over short distances and it is difficult to twist it. Figure 12.10 shows a simple picture to illustrate this bending. The two operators are normally a multiple of 10 bp apart, which is something of a surprise because without any twisting, the bending of the DNA would put the two operator sequences in parallel with each other so that the two repressors would be bound to opposite sides of the operators. However, as we see below, another protein, called HU, binds between the two operators and introduces a 180° turn in the DNA, which puts the two operators in the antiparellel configuration so that the repressor can be bound to the same side of both of them. In any case, if the repression is due to the repressor binding to both operators and bending the DNA between them, the spacing between the two operators should matter and increasing or decreasing the spacing should relieve the repression.

In an experiment to determine the effect of the spacing on repression, the two operators were moved farther apart by inserting extra DNA sequences between them. The results were dramatic and strongly supported the model. The two operators still functioned normally if they were moved farther apart, but only if the spacing was increased in multiples of 10 bp. If the spacing was changed to a multiple of 15 bp, the cells were partially constitutive.

Further work has confirmed this model and has contributed some details to the looping structure (see Figure 12.11). To loop the DNA between the two operators, a dimer of GalR repressor binds to each operator and these dimers bind to each other to form a tetramer. A histone-like protein called HU binds between the two repressor dimers and negatively supercoils the DNA 180° to put the two operator sequences in an antiparallel configuration so that the repressor dimers can bind to the same face of both of them. This large structure with the loop of promoter DNA held by the repressor has been called a repressosome (see Geanacopoulos et al., Suggested Reading).

Two Gal Promoters and Catabolite Repression of the *gal* Operon

As mentioned, the *gal* operon also has two promoters called p_{G1} and p_{G2} (Figure 12.7). The *gal* operon may have two promoters because the enzymes are needed even when a better carbon source is available, since they are involved in making polysaccharides as well as in utilizing galactose. Like *lac*, the *gal* operon is regulated by catabolite repression so that the transcription of the *gal* genes is repressed if a better carbon source such as glucose is

Figure 12.10 Repressor molecules (circled Rs) bound to the two operators of the *gal* operon can interact more easily if they are bound on the same side of the helix. (A) Multiples of 10 bp separating o_E and o_I allow R binding to the same side of the helix. (B) Arrows denote twisting of the molecule that is necessary when the operators are a multiple of 15 bp apart. See the text for details.

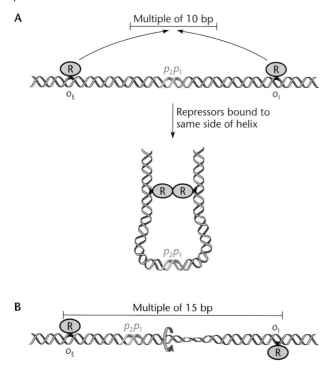

A

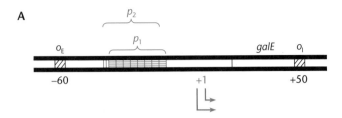

B

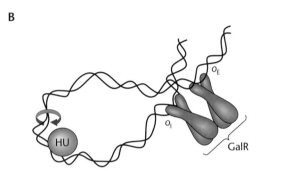

Figure 12.11 Formation of the *gal* operon repressosome. Two dimers of the *galR* repressor gene product bind to each other and to the operators o_E and o_I (A) to bend the DNA between the operators. (B) A histone-like DNA-binding protein, HU, introduces a twist in the DNA (shown by the arrow), so that the two operators are in an antiparallel configuration. Bending of the DNA in the promoter regions inactivates the promoters. Details are given in the text.

available, but only one of the two *gal* operon promoters is repressed. The other promoter is active even in the presence of glucose and continues making the Gal enzymes, which are needed to make galactose-containing cell ways even if glucose is present. We discuss the differential regulation of the two *gal* promoters in more detail in chapter 13 under catabolite repression.

Regulation of Biosynthetic Operons: Aporepressors and Corepressors

The enzymes encoded by the *lac*, *gal*, and *ara* operons are involved in degrading compounds to obtain catabolites in order to build other molecules. Consequently, these operons are called **catabolic operons** or **degradative operons**. Not all operons are involved in degrading compounds, however. The enzymes encoded by some operons synthesize compounds needed by the cell, such as nucleotides, amino acids, and vitamins. These operons are called **biosynthetic operons**.

The regulation of a biosynthetic operon is essentially opposite to that of a degradative operon. The enzymes of a biosynthetic pathway should *not* be synthesized in the presence of the end product of the pathway, since the product is already available and energy should not be wasted in synthesizing more. However, the mechanisms by which degradative and biosynthetic operons

are regulated operate in essentially the same way. Biosynthetic operons can be regulated negatively by repressors. If the genes of a biosynthetic operon are expressed in the absence of the regulatory gene product, even if the compound is present in the medium, the operon is negatively regulated. Biosynthetic operons that are negatively regulated are constitutively expressed if the genes are expressed even if the compound is *present* in the medium.

The terminology we use to describe the negative regulation of biosynthetic operons differs somewhat from that used for catabolic operons, despite shared principles. The effector that binds to the repressor and allows it to bind to the operators is called the corepressor. A repressor that negatively regulates a biosynthetic operon is not active in the absence of the corepressor and in this state is called the **aporepressor**. However, once the corepressor is bound, the protein is able to bind to the operator and so is now called the repressor.

THE *trp* OPERON OF *E. COLI*

The tryptophan (*trp*) operon of *E. coli* is the classic example of a biosynthetic operon that is negatively regulated by a repressor. The enzymes encoded by the *trp* operon (Figure 12.12) are responsible for synthesizing the amino acid L-tryptophan, which is a constituent of most proteins and so must be synthesized if none is available in the medium. The products of five structural genes in the operon are required to make tryptophan from chorismic acid. These genes are transcribed from a single promoter, p_{trp}, shown in Figure 12.12. The *trp* operon is negatively regulated by the TrpR repressor protein, whose gene, like the *gal* repressor gene, is unlinked to the rest of the operon. Also, like the *galR* gene, this may reflect the fact that TrpR regulates more than one operon. In addition to the *trp* operon, it regulates the *aroH* operon to make chorismate and it regulates its own gene, *trpR* (see below).

Figure 12.13 shows the model for the regulation of the *trp* operon by the TrpR repressor. By binding to the operator, the TrpR repressor can prevent transcription from the p_{trp} promoter. However, the TrpR repressor can bind to the operator only if the corepressor tryptophan is *present* in the medium. The tryptophan binds to the TrpR aporepressor protein and changes its conformation so that it *can bind* to the operator.

The TrpR repressor has been crystallized and its structure has been determined in both the aporepressor form, when it cannot bind to DNA, and the repressor form with tryptophan bound, when it can bind to the operator. These structures have led to a satisfying explanation of how tryptophan corepressor binding changes the TrpR protein so that it can bind to the operator (Figure 12.14). The TrpR repressor is a dimer, and each copy of

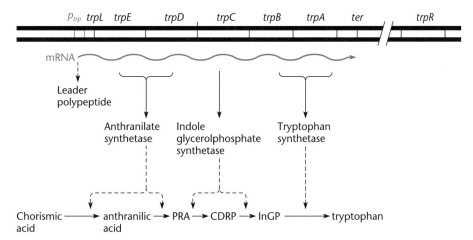

Figure 12.12 Structure of the tryptophan biosynthetic (*trp*) operon of *E. coli*. The structurural genes *trpEDCBA* are transcribed from the promoter p_{trp}. Upstream of the structural genes is a short coding sequence for the leader peptide called *trpL*. The *trpR* repressor gene is unlinked, as shown by the broken line. PRA, phosphoribosyl anthranilate; CDRP, 1-(*o*-carboxyphenylamino)-1-deoxyribulose-5-phosphate.

the *trpR* polypeptide has α-helical structures (shown as cylinders). Helices D and E form the helix-turn-helix (HTH) DNA binding domain (see Box 12.1). Helix D corresponds to helix 1, the nonspecific DNA binding helix and helix E corresponds to helix 2, the DNA sequence specific recognition helix. In the aporepressor state, the conformation of the two HTH domains in the dimer do not allow proper interactions with successive major grooves in the DNA helix. Binding of the typtophan corepressor alters the HTH conformations, allowing repressor to bind to the operators.

Autoregulation of the *trpR* Gene

As mentioned, the TrpR repressor negatively regulates not only the transcription of the *trp* operon but also the transcription of its own gene, *trpR*. In the absence of the TrpR protein, the transcription of the *trpR* gene is about five times higher than in its presence. When the product of a gene regulates the expression of its own gene, it is called **autoregulation**.

It is perhaps surprising that the TrpR repressor would negatively regulate the transcription of its own gene, since there would be less repressor in the cell when tryptophan

A   In the absence of tryptophan

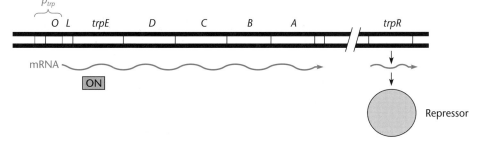

Figure 12.13 Negative regulation of the *trp* operon by the TrpR repressor. See the text for details.

B   In the presence of tryptophan

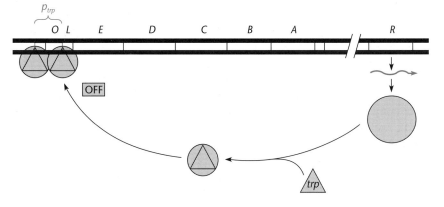

A Aporepressor dimer

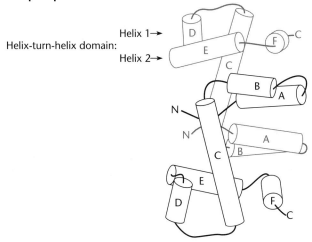

Helix-turn-helix domain:
Helix 1→
Helix 2→

B Aporepressor - repressor conformational change

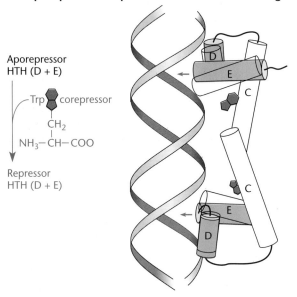

Aporepressor
HTH (D + E)

Trp corepressor

Repressor
HTH (D + E)

Figure 12.14 The structure of the TrpR repressor and an illustration of how tryptophan binding allows it to convert from the aporepressor to the repressor that binds to the operator. (A) The helices (shown as cylinders) of the aporepressor dimer in the inactive state with no tryptophan bound. (B) The gold cylinders represent the HTH domains of the active repressor with tryptophan bound.

is present than when tryptophan is absent. It would seem advantageous to have more repressor present when tryptophan is present to better repress the *trp* operon. One possible answer to this riddle is that through negative autoregulation of transcription of the *trpR* gene, the cell ensures that repression can be established more quickly if tryptophan suddenly appears in the medium.

Isolation of *trpR* Mutants

Like other negatively regulated operons, constitutive mutations of the *trp* operon are quite common and most map in *trpR*, inactivating the product of the gene. Mutants with constitutive mutations of the *trp* operon can be obtained by selecting for mutants resistant to the tryptophan analog 5-methyltryptophan in the absence of tryptophan. This tryptophan analog also binds to the TrpR repressor and acts as a corepressor. However, 5-methyltryptophan cannot be used in place of tryptophan to make active proteins. Therefore, in the presence of the analog, the *trp* operon will not be induced even in the absence of tryptophan and the cells will starve for this amino acid. Only constitutive mutants that continue to express the genes of the operon in the presence of 5-methyltryptophan can multiply to form colonies on plates with this analog but without tryptophan.

Other Types of Regulation of the *trp* Operon

The *trp* operon is also subject to a completely different type of regulation called **attenuation**. This type of regulation is discussed later in the chapter. Also, as in many biosynthetic pathways, the first enzyme of the *trp* pathway is subject to **feedback inhibition** by the end product of the pathway, tryptophan. We also return to feedback inhibition later in the chapter.

Positive Regulation

In the first part of the chapter, we covered the classic examples of negative regulation by repressors, but many operons are regulated positively by activators. An operon under the control of an activator protein will be transcribed only in the presence of that protein with the inducer bound. We devote the next portion of the chapter to positive regulation in bacteria.

The *E. coli* L-*ara* Operon

The L-*ara* operon was the first example of positive regulation in bacteria to be discovered. The L-*ara* operon, usually called the *ara* operon, is responsible for the utilization of the five-carbon sugar L-arabinose. The genes of this operon are responsible for converting L-arabinose into D-xylulose-5-phosphate, which can be used by other pathways. The *E. coli* bacterium can also utilize D-arabinose, an isomer of L-arabinose, but the enzymes for D-arabinose utilization are encoded by a different operon, which lies elsewhere in the chromosome.

Figure 12.15A illustrates the structure of the *ara* operon. Three structural genes in the operon, *araB*, *araA*, and *araD*, are transcribed from a single promoter, p_{BAD}. Upstream of the promoter is the activator region, *araI*, where the activator protein AraC binds to activate transcription in the presence of L-arabinose, and the CAP site, at which the catabolic activator protein binds. There are also two operators, *araO₁* and *araO₂*, at which the AraC protein binds to repress transcription. The *araC* gene that encodes the regulatory protein is also shown. This gene is transcribed from the promoter p_C in the opposite direction from *araBAD*, as shown by the arrows in the figure. As described below, the AraC protein is a positive activator of transcription. As such, it is a member of a large family of activator proteins (see Box 12.2).

GENETIC EVIDENCE FOR POSITIVE REGULATION OF THE *ara* OPERON

Early genetic evidence indicated that the *lac* and *ara* operons are regulated by very different mechanisms (see

Englesberg et al., Suggested Reading). One observation was that loss of the regulator proteins results in different phenotypes. For example, deletions and nonsense mutations in the the *araC* gene—mutations that presumably inactivate the protein product of the gene—lead to a "superrepressed" phenotype in which the genes of the operon are not expressed, even in the presence of the inducer arabinose. Recall that deletion or nonsense mutations in the regulatory gene of a negatively regulated operon such as *lac* result in a constitutive phenotype.

Another difference between *ara* and negatively regulated operons is in the frequency of constitutive mutants. Mutants that constitutively express a negatively regulated operon are relatively common because any mutation that inactivates the repressor gene will cause

---

BOX 12.2

Families of Activators

As the sequences of the genes for more and more bacterial repressors and activators become known, it is becoming increasingly obvious that some of them are related to each other, even though they regulate operons of very different functions and respond to different effectors. There are at least 15 different families of activators. For example, the *E. coli* activators LysR and CysB, as well as the *S. enterica* serovar Typhimurium activator MetR and the ClcR and TfdR/S activators of chlorocatechol-degrading operons on plasmids of *Pseudomonas* spp. and *Alcaligenes* spp., respectively, are all closely related to each other and are members of the LysR family of activators. The AraC family of activators includes many other activator proteins including the SoxS activator, involved in turning on genes following oxidative stress, and the XylS activator protein of the Tol plasmid of *P. putida*, described in this chapter. They have a homologous stretch of 100 amino acids involved in DNA binding, which is usually at their C terminus. This family has more than 100 known members. Members of another family of activators are related to the NtrC activator that activates transcription of the *glnA* operon involved in nitrogen regulation (discussed in chapter 13). The NtrC family of activators has been studied most extensively and includes the XylR activator of the *tol* operon and the DmpR activator that activates an operon involved in phenol degradation in *Pseudomonas* sp. strain CF600. These activators function only with σ⁵⁴-type promoters and seem to be organized in distinguishable domains (see Box 13.4). The domain of the activator protein that either binds the inducer or is phosphorylated is at the N terminus and the DNA-binding domain is at the C terminus. The middle region of the polypeptide contains a region that interacts with RNA polymerase and has an ATPase activity re-

quired for activation. All of the activators in this family may function in a similar way, with the only difference being that the ATPase domain can be unmasked either by binding an effector to the N-terminal domain or by phosphorylating it.

Hybrid activators, made by fusing the C-terminal DNA-binding domain of one type of activator protein to the N-terminal inducer-binding domain of another activator protein, provide a graphic demonstration of the commonality of activator action. Sometimes, such hybrid activators can still activate the transcription of an operon, but which operon is activated by the hybrid activator depends on the source of the carboxy-terminal DNA-binding domain, while which inducer induces the operon depends on the source of the N-terminal inducer-binding domain. This leads to a situation where an operon is induced by the inducer of a different operon. It is intriguing to think that all activator proteins may have evolved from a few different precursor proteins through simple changes in their effector binding and, to some extent, their DNA-binding regions yet continue to activate the RNA polymerase by the same basic mechanism.

References

Gallegos, M. T., R. Schleif, A. Bairoch, K. Hoffman, and J. L. Ramos. 1997. AraC/XylS family of transcriptionl regulators. *Microbiol. Mol. Biol. Rev.* **61**:393–410.

Henikoff, S., G. W. Haughn, J. M. Calvo, and J. C. Wallace. 1988. A large family of bacterial activators. *Proc. Natl. Acad. Sci. USA* **85**:6602–6606.

Parek, M. R., S. M. McFall, D. L. Shinabarger, and A. M. Chakrabarty. 1994. Interaction of two LysR-type regulatory proteins CatR and ClcR with heterologous promoters: functional and evolutionary implications. *Proc. Natl. Acad. Sci. USA* **91**:12393–12397.

Schell, M. A. 1993. Molecular biology of the LysR family of transcriptional regulators. *Annu. Rev. Microbiol.* **47**:597–626.

constitutive expression. However, mutants that constitutively express *ara* are very rare, which suggests that mutations that result in the constitutive phenotype do not inactivate *araC*.

Isolating Constitutive Mutations of the *ara* Operon

Because constitutive mutations of the *ara* operon are so rare, special tricks are required to isolate them. One method for isolating rare constitutive mutations in *araC* uses the anti-inducer D-fucose. This anti-inducer binds to the AraC protein and prevents it from binding L-arabinose, thereby preventing induction of the operon. As a consequence, wild-type *E. coli* cannot multiply to form colonies on agar plates containing D-fucose with L-arabinose as the sole carbon and energy source. Only mutants that constitutively express the genes of the *ara* operon will form colonies under these conditions.

A more clever trick for isolating constitutive mutations in the *ara* operon depends on the presence of the other *E. coli* operon responsible for the utilization of D-arabinose, the isomer of L-arabinose. The enzymes produced by the L- and D-*ara* operons cannot use each other's isomer, with one exception. The product of the *araB* gene—the ribulose kinase enzyme of the L-*ara* operon pathway, which phosphorylates L-ribulose as the second step of the path-

way—will also phosphorylate D-ribulose, so that the L-*ara* kinase can substitute for that of the D-*ara* operon. Nevertheless, *E. coli* mutants that lack the D-*ara* kinase cannot multiply to form a colony on plates containing only D-arabinose as a carbon and energy source, because D-arabinose is not an inducer of the L-*ara* operon. Therefore, constitutive mutants with mutations of the L-*ara* operon can be isolated merely by plating D-*ara* kinase-deficient mutants on agar plates containing D-arabinose as the sole carbon and energy source.

A MODEL FOR THE POSITIVE REGULATION OF THE *ara* OPERON

The contrast in phenotypes between mutations that inactivate the *lacI* and *araC* genes led to an early model for the regulation of the *ara* operon. According to this early model, the AraC protein can exist in two states, called P1 and P2. In the absence of the inducer, L-arabinose, the AraC protein is in the P1 state and inactive. If L-arabinose is present, it binds to AraC and changes the protein conformation to the P2 state. In this state, AraC binds to the DNA at the site called *araI* (Figure 12.15A) in the promoter region and activates transcription of the *araB*, *araA*, and *araD* genes.

This early model explained some, but not all, of the behavior of the *araC* mutations. It explained why muta-

Figure 12.15 (A) Structure and function of of the L-arabinose operon of *E. coli.* (B) Binding of the inducer L-arabinose converts the AraC protein from an antiactivator P1 form to an activator P2 form. See the text for details.

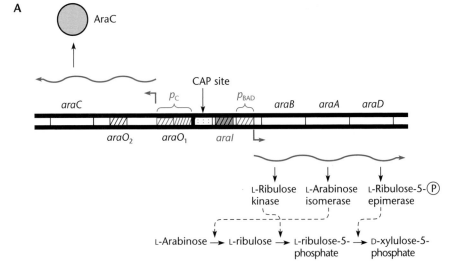

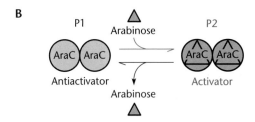

tions in *araC* that cause the constitutive phenotype are rare but do occur at a very low frequency. According to this model, these mutations, called *araC*^c mutations, change AraC so that it is permanently in the P2 state, even in the absence of L-arabinose, and thus the operon will always be transcribed. Such mutations would be expected to be very rare because only a few amino acid changes in the AraC protein could specifically change the conformation of the AraC protein to the P2 state.

AraC IS NOT JUST AN ACTIVATOR
One prediction of this model for regulation of the *ara* operon is that *araC*^c mutations should be dominant over the wild-type allele in complementation tests. If AraC acts solely as an activator, partial diploid cells that have both an *araC*^c allele and the wild-type allele would be expected to constitutively express the *araB*, *araA*, and *araD* genes. In other words, *araC*^c mutations should be dominant over the wild type, since the mutant AraC in the P2 state should activate transcription of *araBAD*, even in the presence of wild-type AraC protein in the P1 state.

The prediction of the model was tested with complementation. An F′ factor carrying the wild-type *ara* operon was introduced into cells with an *araC*^c mutation in the chromosome. Figure 12.16 illustrates that the partial diploid cells were inducible, not constitutive, indicating that *araC*^c mutations were recessive rather than dominant. This observation was contrary to the prediction of the model. Therefore, the model had to be changed.

Figure 12.17 illustrates a detailed model to explain the recessiveness of *araC*^c mutations (see Johnson and Schleif, Suggested Reading). In this model, the P1 form of the AraC protein that exists in the absence of arabinose is not simply inactive but takes on a new identity as an antiactivator (Figure 12.15B). The P1 state is called an antiactivator rather than a repressor because it does not repress transcription like a classical repressor but, rather, acts to prevent activation by the P2 state of the protein. In the P1 state, the AraC protein preferentially binds to the operator *araO*₂ and another site, *araI*₁, bending the DNA between the two sites like the GalR repressor. Because AraC in the P1 form preferentially binds to *araO*₂, it cannot bind to *araI*₂ and activate transcription from the *p*_{BAD} promoter. In the presence of L-arabinose, however, the AraC protein changes to the P2 form and now preferentially binds to *araI*₁ and *araI*₂.

This model explains why the *araC*^c mutations are recessive to the wild type in complementation tests (Figure

Figure 12.16 Recessiveness of *araC*^c mutations. See text for details.

A Absence of arabinose

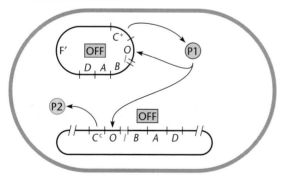

P1 form binds to operators of both operons, preventing *araBAD* expression

B Presence of arabinose

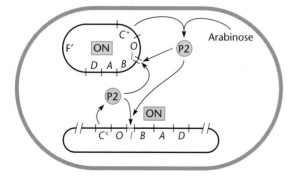

P2 form and/or the AraC^c (=P2) form binds to *I* (induction) sites of both operons, activating *araBAD* transcription

A   Absence of L-arabinose

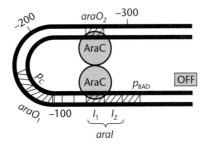

B   Presence of L-arabinose

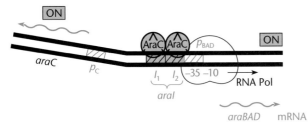

C

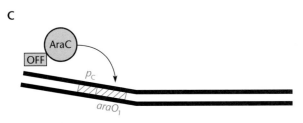

Figure 12.17   A model to explain how AraC can be a positive activator of the *ara* operon in the presence of arabinose and an antiactivator in the absence of this compound, as well as how AraC can negatively autoregulate transcription of its own gene. (A) In the absence of arabinose, AraC molecules in the P1 state preferentially bind to *araI*₁ and *araO*₂, preventing any P2 form from binding to *araI*₂ and activating transcription from *p*ᵦₐ𝒹, thereby shutting off the operon. Bending of the DNA between the two sites may also inhibit transcription of the *araC* gene by inhibiting transcription from the *p*𝒸 promoter. (B) In the presence of arabinose, AraC shifts to the P2 state and preferentially binds to *araI*₁ and *araI*₂. AraC bound to *araI*₂ activates transcription from *p*ᵦₐ𝒹. The bend in the DNA occurs owing to binding by the CAP protein (see chapter 13). (C) If the AraC concentration becomes too high, it will also bind to *araO*₁, thereby repressing transcription from its own promoter *p*𝒸.

12.16), because AraCᶜ in the P2 form can no longer bind to *araI*₁ and *araI*₂ to activate transcription as long as wild-type AraC in the P1 state is already bound to *araO*₂ and *araI*₁. It also explains the behavior of certain deletion mutations, known as the Englesberg deletions, which we have not mentioned yet. These deletions remove the *araO*₂ region but leave the *araI*₁ and *araI*₂ regions intact. In this case, *araCᶜ* mutations are no longer

recessive to the wild-type allele of *araC*. Without *araO*₂ to bind to, the AraC protein in the P1 form seems unable to antiactivate transcription of the operon.

AUTOREGULATION OF AraC

The AraC protein not only regulates the transcription of the *ara* operon but also negatively autoregulates its own transcription. Like TrpR, the AraC protein seems to repress its own synthesis, so that less AraC protein is synthesized in the absence of arabinose than in its presence. However, if the concentration of AraC becomes too high, its synthesis will again be repressed.

Figure 12.17 includes a model for the autoregulation of AraC synthesis. In the absence of arabinose, the interaction of two AraC monomers bound at *araO*₂ and *araI*₁ bend the DNA in the region of the *araC* promoter *p*𝒸, thereby inhibiting transcription from this promoter. In the presence of arabinose, the AraC protein will no longer be bound to *araO*₂, and so the promoter will no longer be bent and transcription from *p*𝒸 will occur. However, if the AraC concentration becomes too high, the excess AraC protein binds to the operator *araO*₁, preventing further transcription of *araC* from the *p*𝒸 promoter.

CATABOLITE REGULATION OF THE L-*ara* OPERON

The *ara* operon is also regulated through catabolite repression, so the genes for arabinose utilization will not be expressed if the medium contains a better carbon source. The CAP protein that regulates the transcription of genes subject to catabolite repression is also a positive activator, like the AraC protein. By binding to the CAP-binding site shown in Figure 12.15A, the CAP protein may help open the loop of DNA created when AraC binds to *araO*₂ and *araI*₁. Opening the loop may prevent AraC from binding to *araO*₂ and *araI*₁, facilitating the binding of AraC to *araI*₁ and *araI*₂ and the activation of transcription from *p*ᵦₐ𝒹. Thus, the absence of glucose or another carbon source better than arabinose may enhance the transcription of the *ara* operon.

The *E. coli* Maltose Operons

Other well-studied positively regulated operons in bacteria include those for the utilization of the sugar maltose in *E. coli*. Figure 12.18 shows that the genes for maltose transport and metabolism are organized in four clusters at 36, 75, 80, and 91 min on the *E. coli* genetic map. The operon at 75 min has two genes, *malQ* and *malP*, whose products are involved in converting maltose and other polymers into glucose and glucose-1-phosphate once maltose and the polymers are in the cytoplasm. This cluster also includes the regulatory gene *malT*. The *malS* gene at 80 min encodes an enzyme that breaks down

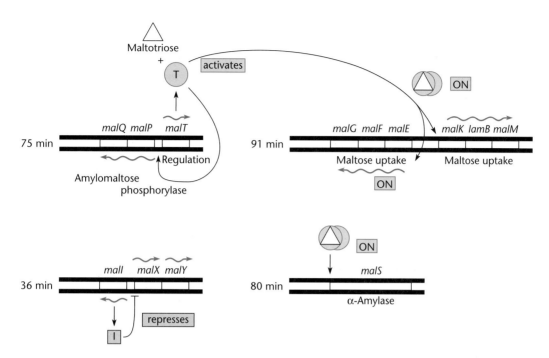

Figure 12.18   The maltose operons in *E. coli*. The MalT activator protein regulates both operons at 75 min and the operons at 80 and 91 min on the *E. coli* map. Another operon at 36 min is also induced by maltose. MalY and MalK bind to MalT, inactivating it (see text).

polymers of maltose. The other cluster, at 91 min, has two operons whose gene products can transport maltose into the cell. An operon at 36 min encodes enzymes that degrade polymers of maltose.

Although they allow the cell to use maltose as a carbon source, the more significant function of the products of these operons is probably to enable the cell to transport and degrade polymers of maltose called **maltodextrins**. These compounds are products of the breakdown of starch molecules, which are very long polysaccharides stored by cells to conserve energy. The sugar maltose is itself a disaccharide composed of two glucose residues with a 1–4 linkage, and the enzymes of the *malP-malQ* operon can break the maltodextrins down into maltose and then into glucose-1-phosphate, which can enter other pathways. Some bacteria, including species of *Klebsiella*, excrete extracellular enzymes that degrade long starch molecules and allow the bacteria to grow on starch as the sole carbon and energy source. *E. coli* lacks some of the genes needed to degrade starch to maltodextrins; therefore, in nature it probably depends on neighboring microorganisms to break the starch down to the maltodextrins so that it can use them as a carbon and energy source.

THE MALTOSE TRANSPORT SYSTEM

Most of the protein products of the *mal* operons are involved in transporting maltodextrins and maltose

through the outer and inner membranes (Figure 12.19). The product of the *lamB* gene resides in the outer membrane, where it can bind maltodextrins in the medium. This protein forms a large channel in the outer membrane through which the maltodextrins can pass. LamB is not required for growth on maltose, probably because maltose is small enough to pass through the outer membrane without its help. The LamB protein in the outer membrane also serves as the cell surface receptor for phage λ, and so the gene name is derived from the phage name (*lamB* from lambda). Mutants of *E. coli* resistant to λ have mutations in the *lamB* gene.

Once maltodextrins are through the outer membrane, the MalS protein in the periplasm may degrade them into smaller polymers before they can be transported through the inner membrane. The smaller polymer of maltose binds to MalE in the periplasm between the outer and inner membranes. The MalF, MalG, and MalK genes then transport the maltodextrin through the inner membrane.

EXPERIMENTAL USES OF THE *mal* GENES

Because they have been studied so extensively, the *mal* operons have been put to much use in molecular genetics. The ability of the MalE protein, sometimes called the maltose-binding protein (MBP), to tightly bind maltose has made it useful in some biotechnological applications.

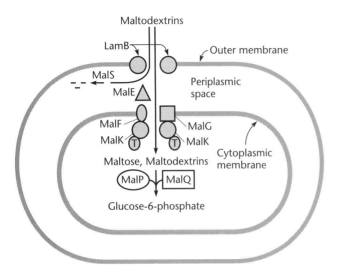

Figure 12.19 Function of the genes of the maltose regulon in the transport and processing of maltodextrins in *E. coli.* The LamB protein binds maltodextrins and transports them across the outer membrane. The MalE protein in the periplasmic space then passes them through a pore in the cytoplasmic or inner membrane formed by the MalF, MalB, and MalK proteins. Once in the cytoplasm, the maltodextrins and maltose are degraded by MalP and MalQ to glucose-1-phosphate and glucose, respectively. These compounds can then be converted into glucose-6-phosphate for use as energy and carbon sources. Adapted from M. Schwartz, p. 1484, *in* F. C. Neidhardt, R. Curtiss III, J. L. Ingraham, E. C. C. Lin, K. B. Low, B. Magasanik, W. S. Reznikoff, M. Riley, M. Schaechter, and H. E. Umbarger (ed.), Escherichia coli *and* Salmonella: *Cellular and Molecular Biology,* 2nd ed., ASM Press, Washington, D.C., 1987.

In some cloning vectors, the *malE* gene is fused to other proteins, making it possible to purify the fusion protein on maltose affinity columns. Because the products of the *E. coli mal* operons are involved in transporting long molecules into the cell, the *mal* operons have been of particular interest in studies of large-molecule transport systems. Also, many of the proteins encoded by the operons are localized in the inner and outer membranes or the periplasmic space. These proteins must be transported into or through the inner membrane and so serve as models for the study of protein transport through cellular membranes. We shall discuss such experiments in more detail in chapter 14.

REGULATION OF THE *mal* OPERONS

The regulation of the *mal* operons is also illustrated in Figure 12.18. The inducer of the *mal* operons is **maltotriose**, which is composed of three molecules of glucose held together by the maltose linkage. Maltotriose can be synthesized from maltose brought into the cell by some of the enzymes encoded by the operons. Also, the

cell normally contains polymers of maltose that were synthesized in the cell from glucose (and so do not need to be transported in) and can be broken down into the inducer, maltotriose. The enzymes that degrade maltose polymers therefore play an indirect role in regulating the operons.

The genes in all three clusters are regulated by an activator encoded by the *malT* gene in the first cluster. The MalT activator protein specifically binds the inducer, maltotriose, and activates transcription of the operons. The activation of the genes of the *mal* regulon by MalT involves the DNA wrapping around many copies of the protein, thereby changing the DNA conformation. The MalT activator is also a member of a large family of activators that includes SoxS, which regulates genes that relieve oxygen toxicity (see Box 12.2).

The genetic analysis of the regulation of the *mal* operons has been complicated because maltose polymers are natural components of the cell and play many roles including protecting the cell against high osmolarity. In such situations, caution must be exercised in concluding that constitutive mutations are located in regulatory genes. For example, preliminary genetic evidence suggested that the MalK protein is a repressor of the *mal* operons, since *malK* mutants appear to be constitutive for the expression of the other genes of the operons. It now appears that MalK, which is normally part of the transport system (see Figure 12.19) may bind to MalT and hold it in an inactive state if maltose is not present. This may be a way of coupling the regulation of genes to transport (see Boos and Bohm, Suggested Reading).

The *tol* Operons

Many of the bacterial operons of soil bacteria involved in the degradation of cyclic hydrocarbons are also subject to positive regulation. Cyclic hydrocarbons are based on the conjugated ring structure of benzene. Many do not exist naturally and are pesticides and other industrially important manufactured chemicals. Moreover, many of the manufactured chemicals are also chlorinated. Chlorinated compounds were very rare until the advent of modern chemistry, yet despite this short time interval, some types of bacteria have evolved enzymatic pathways to degrade some of these compounds. We have taken advantage of this quick adaptation by using bacteria and other microorganisms to remove contaminating chemicals in a process called **bioremediation**. Understanding the regulation of operons involved in degrading cyclic compounds may lead to yet more rational approaches to bioremediation of toxic waste.

The *tol* operons in plasmid pWWO, originally isolated from the soil bacterium *Pseudomonas putida*, encode enzymes that degrade toluene and the closely related compound xylene. Toluene itself is not chlorinated and has presumably always existed in nature, but operons related to *tol* have been discovered that degrade similar compounds that are chlorinated, including chlorinated catechols.

Figure 12.20 diagrams the *tol* operons. The plasmid contains two, separated by a few thousand base pairs of DNA. The first operon encodes the enzymes of the "upper pathway," which converts toluene into benzoate; the other operon encodes enzymes of the "lower pathway," which breaks the ring of benzoate and degrades it to intermediates of the tricarboxylic acid (TCA) cycle to be used for energy and to make carbon-containing compounds.

REGULATION OF THE *tol* OPERONS

The regulation of the *tol* operons is also illustrated in Figure 12.20. The regulation of the upper and lower *tol* operons is both coordinated and independent. Both operons must be coordinately turned on if toluene is present in the medium, since both operons are required to degrade toluene to TCA cycle intermediates. However, only the lower operon should be turned on if only benzoate is present in the medium, since there is no need to induce the upper pathway under these conditions.

The coordinate regulation of the upper and lower *tol* operons in the presence of toluene is achieved through two activators, one of which activates transcription of the other's gene. The activator for the upper operon is

Figure 12.20 The structure of the *tol* operons of the Tol plasmid pWWO of *P. putida*. The upper pathway converts toluene (or xylene) to benzoate. The lower encodes a meta cleavage pathway that converts the benzoate to intermediates of the TCA cycle. The two operons are separated by a few thousand base pairs of DNA, as indicated by the broken line. The promoters activated by each of the activator proteins XylS and XylR are indicated by the arrows and the direction of mRNA transcription is indicated by the blue arrows. p_U is the promoter for the upper operon; p_L is the promoter for the lower operon.

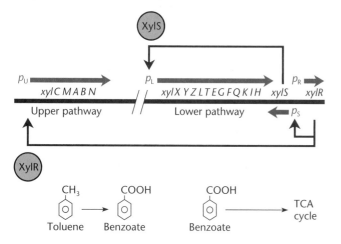

XylR, a member of the NtrC family of activators, and the activator for the lower pathway is XylS, a member of the LysR family of activators (see Box 12.2). In the presence of toluene (or xylene), the XylR activator activates transcription of the upper operon from the promoter called p_U. Toluene is thus degraded into benzoate, but not enough benzoate is produced to bind to XylS and activate transcription of the lower operon. However, the XylR activator also activates transcription of the *xylS* gene from the promoter p_S. At higher concentrations, XylS can activate transcription of the lower operon even without benzoate being bound. If, however, high levels of benzoate are present in the medium, the benzoate will bind to XylS and induce the lower pathway; much less XylS is required for this type of activation. This regulatory interaction ensures that both the upper and lower pathways will be induced if toluene is present but only the lower pathway will be induced in the presence of benzoate alone.

The regulation of the *tol* operons and their regulatory genes is similar to the regulation of the genes of other catabolic pathways in bacteria. For instance, the XylR protein also regulates its own transcription; that is, it is transcriptionally autoregulated, like AraC. Therefore, in the presence of toluene, more XylR will be present to activate transcription of the other genes. Another similarity is that the promoters p_U and p_L are recognized by an RNA polymerase containing the alternate sigma factor σ^{54}, the same sigma factor used to transcribe the nitrogen-regulated genes discussed in chapter 13. Why these operons should use an alternate sigma factor is not known, but the expression of operons in bacteria as distantly related as the nitrogen-regulated genes in *E. coli* and the toluene-degrading operons in *P. putida* share this feature.

GENETICS OF THE *tol* OPERONS

The above picture of the organization of the genes of the *tol* operons came from molecular genetic experiments. These studies of the organization and regulation of the *tol* operons were greatly aided by the fact that plasmid pWWO carrying the *tol* operons is a broad-host-range, self-transmissible plasmid that can transfer itself from the original *P. putida* strain into *E. coli*, where more sophisticated genetic tests have been developed. Once in *E. coli*, the plasmid could be easily mutagenized with transposon Tn5 and then transferred back into *P. putida* to determine whether a particular transposon insertion inactivates a gene required for growth on toluene. Insertions that inactivate *tol* genes could then be located by restriction endonuclease mapping as discussed in chapter 9. In this way, the maps of the *tol* operons shown in Figure 12.20 were obtained.

Molecular genetic tests also yielded the picture for the regulation of the *tol* operons outlined above. The XylS protein was first identified as a positive activator of the lower pathway because clones that did not include the *xylS* gene failed to express the genes of the lower pathway, even in the presence of benzoate. Similarly, clones of the upper pathway that excluded the *xylR* gene did not express the upper-pathway operon, and *xylS* was transcribed at a lower rate if *xylR* was missing. These observations implicated XylR as a positive activator of the upper operon and the *xylS* gene. Clones that overexpress XylS turn on the lower pathway, even in the absence of benzoate, leading to the model that XylS can activate transcription of the lower pathway in the absence of benzoate, provided that XylS is present at a high concentration.

USING SELECTIONAL GENETICS TO BROADEN THE RANGE OF INDUCERS OF THE *tol* LOWER OPERON

A promising avenue of research is to use genetic selections to alter known pathways so that they can use alternate substrates. For example, while the *tol* lower pathway can degrade some substituted forms of benzoate, including 3-methylbenzoate and 4-methylbenzoate, and allow the cell to use them as carbon and energy sources, it cannot use other derivatives of benzoate such as 4-ethylbenzoate. This particular substrate cannot be used because the second enzyme of the pathway cannot use it as a substrate and because it does not function as an inducer of the operon. Even if the enzymes encoded by an operon can degrade a derivative of the normal substrates for the operon, the enzymes will not be present if the derivative does not function as an inducer of the pathway.

Gene fusion techniques were used to select *xylS* mutants that can use 4-ethylbenzoate and other derivatives of benzoate as an inducer of the lower *tol* operon (see Ramos et al., Suggested Reading). These studies could be performed in *E. coli* because the XylS protein can activate transcription of the lower operon even in *E. coli*. In these experiments, the promoter p_L for the lower operon was fused to a tetracycline resistance gene on a plasmid so that the tetracycline resistance gene would not be transcribed unless the p_L promoter was activated. This plasmid was then used to transform *E. coli* containing a second compatible plasmid expressing the XylS protein. When a particular derivative of benzoate caused XylS to activate transcription from the p_L promoter, the cells became tetracycline resistant (Tet^r) and grew on plates containing tetracycline. However, when the derivative did not induce the operon, the cells remained tetracycline sensitive (Tet^s). As expected, benzoate made the cells Tet^r. Moreover, some derivatives of benzoate, including 2-chlorobenzoate, functioned well as inducers, making the cells Tet^r. However, other derivatives, such as 4-ethylbenzoate and 2,4-dichlorobenzoate, were not inducers and the cells remained Tet^s.

Selectional genetics was used to try to isolate mutants with altered XylS proteins in which 4-ethylbenzoate or similar noninducing derivatives could function as inducers. The bacteria described above containing the two plasmids were mutagenized, and large numbers were spread on plates containing tetracycline and a potential inducer, for example, 4-ethylbenzoate. Most of the bacteria did not multiply to form a colony, but a few colonies of Tet^r mutants appeared. Some of these were constitutive mutants that were Tet^r even in the absence of inducer, and so they were discarded. However, some had mutations changing the XylS protein so that it could use the new inducer. Mutants with *xylS* mutations that allowed induction by 4-ethylbenzoate were separated from other, unwanted types by isolating the *xylS*-containing plasmid and transforming it into new bacteria containing the *tet* fusion. Only if the mutation was in the plasmid containing the *xylS* gene, and therefore presumably in the *xylS* gene itself, were transformants Tet^r in the presence of ethylbenzoate. The mutation was shown to be in the *xylS* gene and not somewhere else in the plasmid by recloning the *xylS* gene into a new plasmid and showing that this plasmid also confers the Tet^r phenotype in the presence of 4-ethylbenzoate. Finally, the mutated *xylS* gene could be sequenced to determine what amino acid changes in the XylS protein can allow it to use 4-ethylbenzoate as an inducer. The success of these experiments with 4-ethylbenzoate and other benzoate derivatives revealed that the inducer specificity of activator proteins can sometimes be changed by simple mutations. This may be the origin of families of activators (see Box 12.2).

Regulation by Attenuation of Transcription

In the above examples, the transcription of an operon is regulated through the initiation of RNA synthesis at the promoter of the operon. However, this is not the only known means of regulating operon transcription. Another mechanism is the **attenuation of transcription**. Unlike repressors and activators, which turn on or off transcription from the promoter, the attenuation mechanism works by terminating transcription—which begins normally—before the RNA polymerase reaches the first structural gene of the operon. The classic examples of regulation by attenuation are the *his* and *trp* operons of *E. coli*. Closely related mechanisms regulate such *E. coli* biosynthetic operons as leucine (*leu*), phenylalanine (*phe*), threonine (*thr*), and isoleucine-valine (*ilv*) and the *Bacillus subtilis* tRNA synthetase genes and *trp* operon (see Box 12.3). In this section, we discuss the regulation of the *E. coli trp* operon by attenuation.

BOX 12.3

Regulation by Attenuation: the Aminoacyl-tRNA Synthetase Genes and the *trp* Operon of *Bacillus subtilis*

Several bacterial operons are now known to be regulated by some form of attenuation and antitermination of transcription. Many of these use mechanisms different from those described for the *trp* and *his* operons. Some types of attenuation control use *trans*-acting proteins to regulate transcription through termination signals. Examples include the *bgl* operon in *E. coli* and the *trp* operon in *Bacillus subtilis*. In these operons, regulatory proteins bind to the leader RNA and prevent the formation of secondary structures, thereby enhancing or reducing transcription termination at downstream termination signals.

Regulation of Aminoacyl-tRNA Synthetase Genes

One of the most dramatic examples of regulation through attenuation is the regulation of the transcription of the genes for the aminoacyl-tRNA synthetase genes in *B. subtilis*. Bacteria synthesize higher levels of aminoacyl-tRNA synthetases in response to amino acid deprivation. The higher levels of the synthetases presumably allow more efficient attachment of the amino acids to their cognate tRNAs.

In *B. subtilis*, the synthetase genes are regulated by attenuation of transcription. In high concentrations of the amino acid, transcription of the synthetase gene often terminates in the leader region, so that less synthetase is synthesized. If the amino acid for that synthetase is limiting, however, transcription terminates less often in the leader sequence, and more synthetase is made.

Whether or not transcription terminates in the leader sequence of each synthetase gene is determined by the relative levels of the unaminoacylated cognate tRNA for that synthetase. At high levels of the amino acid, most of the tRNA will have its amino acid attached (i.e., be aminoacylated). However, at low concentrations of the amino acid, more of the tRNA will lack its amino acid (i.e., be unaminoacylated). The anticodon of the unaminoacylated tRNA can bind to a strategically placed codon for that amino acid upstream of the transcription terminator in the leader region. This converts the region of the codon into an antiterminator, allowing increased transcription of the synthetase gene. In this way, the synthetase genes can all be regulated by the same mechanism, but each synthetase gene will respond only to levels of its own cognate amino acid.

Regulation of the *trp* Operon

It is always interesting to compare the regulation of the same operon in different types of bacteria. Another operon whose regulation has been studied extensively is the *trp* operon of *B. subtilis* (see Babitzke and Gollnick below). There is no repressor regulating the *trp* operon in this species, and all of the transcriptional regulation occurs through antitermination. The mechanism of antitermination is also easier to visualize. Rather than depending upon pausing by the ribosome at trptophan codons to sense limiting tryptophan and alter the secondary structure of the mRNA, the *B. subtilis trp* operon uses a protein called TRAP (for "*trp* RNA-binding attenuation protein"). In the presence of tryptophan, this protein binds to an antiterminator hairpin in the mRNA and prevents its formation. This antiterminator hairpin is just upstream of a terminator hairpin, and the inverted repeated sequences encoding the two hairpins overlap, so that they cannot both form at the same time. Disrupting the antiterminator hairpin by TRAP binding allows the terminator hairpin to form and transcription to terminate, preventing transcription of the structural genes of the *trp* operon.

The regulation of the *B. subtilis trp* operon also uses another protein which allows it also to respond to the amount of uncharged trytophan tRNA (tRNA with no amino acid attached) in the cell. In the presence of uncharged trptophan tRNA, this protein, called anti-TRAP, binds to TRAP and inactivates it, preventing its binding to the antitermination hairpin and causing termination. In this way, the *trp* operon of *B. subtilis* behaves much like the *trp* operon of *E. coli* in that it is turned on either if tryptophan is limiting or if most of the tryptophan tRNA in the cell is not charged with tryptophan, even though the mechanisms which are used are very different.

In addition to regulating transcription of the *trp* operon, the TRAP protein regulates *trp* translation. In the presence of tryptophan, TRAP binding to the leader region of the *trp* mRNA also prevents its translation (see figure).

References

Babitzke, P., and P. Gollnick. 2001. Posttranscription initiation control of tryptophan metabolism in *Bacillus subtilis* by the *trp* RNA-binding attenuation protein (TRAP) anti-TRAP and RNA structure. *J. Bacteriol.* **183:**5795–5802. (Minireview.)

Grundy, F. C., J. C. Collins, S. M. Rollins, and T. M. Henkin. 2000. tRNA determinants for antitermination of the *Bacillus subtilis tyrS* gene. *RNA* **6:**1131–1141.

(continued)

BOX 12.3 (continued)

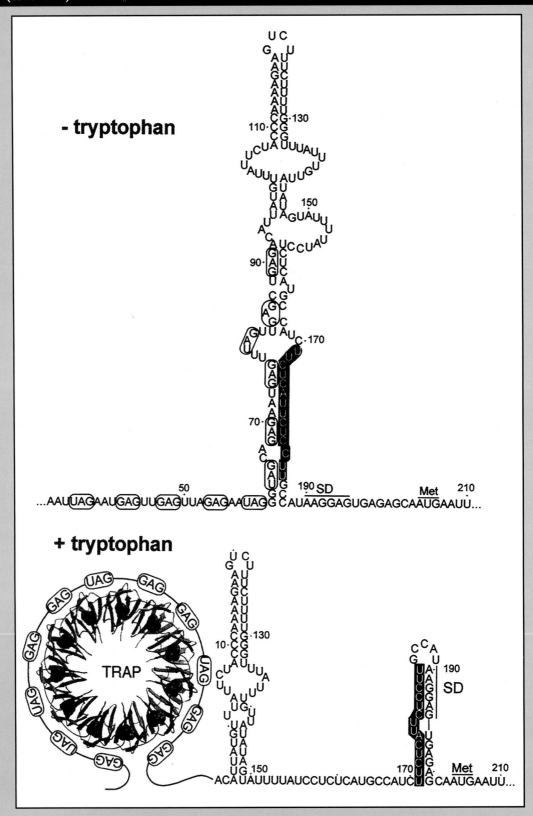

trpE translational control model. Under tryptophan-limiting conditions, TRAP is not activated and is unable to bind to the *trp* leader transcript. In this case the *trp* leader RNA adopts a structure such that the *trpE* SD sequence is single stranded and available for translation. Under excess tryptophan conditions, TRAP is activated and binds to the (G/U)AG repeats. As a consequence, the *trpE* SD blocking hairpin forms, which prevents ribosome binding and translation. The overlap between the two alternative structures is shown. Numbering is from the start of transcription. Reprinted with permission from Babitzke and Gollnick (see above).

Genetic Evidence for Attenuation

As discussed earlier in the chapter, the *trp* operon is negatively regulated by the TrpR repressor protein. However, early genetic evidence suggested that this is not the only type of regulation for *trp*. If the *trp* operon were regulated solely by the TrpR repressor, the levels of the *trp* operon enzymes in a *trpR* mutant would be the same in the absence and the presence of tryptophan. However, even in a *trpR* null mutant, the expression of these enzymes is higher in the absence of tryptophan than in its presence, indicating that the *trp* operon is subject to another regulatory system.

Early evidence suggested that tRNA^{Trp} plays a role in the regulation of the *trp* operon in the absence of TrpR (see Morse and Morse, Suggested Reading). Mutations in the trytophanyl-tRNA synthetase (the enzyme responsible for transferring tryptophan to tRNA^{Trp}) and mutations in the structural gene for the tRNA^{Trp}, as well as mutations in genes whose products are responsible for modifying the tRNA^{Trp} increase the expression of the operon. All these mutations presumably lower the amount of aminoacylated-tRNA^{Trp} in the cell, suggesting that this other regulatory mechanism is not sensing the amount of free tryptophan in the cell but, rather, the amount bound to the tRNA^{Trp}.

Other evidence suggested that the region targeted by this other type of regulation is not the promoter but a region downstream of the promoter called the **leader region**, or *trpL* (Figures 12.12 and 12.21). Deletions in this region, which lies between the promoter and *trpE*, the first gene of the operon, eliminate the regulation. Double mutants with both a deletion mutation of the leader region and a *trpR* mutation are completely constitutive for expression of the *trp* operon. Deletions of the leader region are also *cis* acting and affect only the expression of the *trp* operon on the same DNA. Later evidence indicated that transcription terminated in this leader region in the presence of tryptophan because of an excess of aminoacylated-tRNA^{Trp}. Because the regulation seemed to be able to stop, or attenuate, transcription that had already initiated at the promoter, it was called attenuation of transcription, in agreement with an analogous type of regulation already discovered for the *his* operon.

MODEL FOR REGULATION OF THE *trp* OPERON BY ATTENUATION

Figures 12.21 and 12.22 illustrate a current model for regulation of the *trp* operon by attenuation (see Yanofsky and Crawford, Suggested Reading). According to this model, the percentage of the tRNA^{Trp} that is aminoacylated (i.e.,

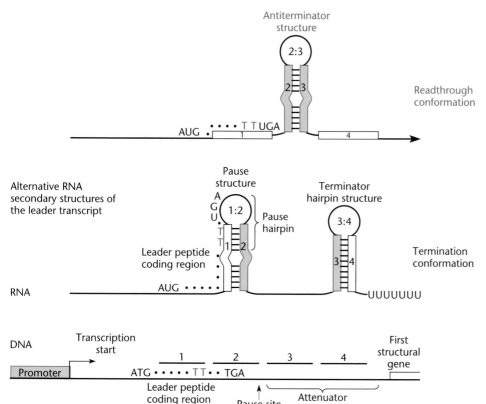

Figure 12.21 Structure and relevant features of the leader region of the *trp* operon involved in regulation by attentuation. TT (in blue) indicates the two *trp* codons in the leader region. See the text for details. Adapted from R. Landick, C. Turnbough, Jr., and C. Yanofsky, *in* F. C. Neidhardt, R. Curtiss III, J. L. Ingraham, E. C. C. Lin, K. B. Low, B. Magasanik, W. S. Reznikoff, M. Riley, M. Schaechter, and H. E. Umbarger (ed.), Escherichia coli *and* Salmonella: *Cellular and Molecular Biology,* 2nd ed., ASM Press, Washington, D.C., 1996, and reprinted with permission from Landick, R., and C. L. Turnbough, Jr. 1992. Transcriptional attenuation, p. 407–446, *in* S. L. McKnight and K. R. Yamamoto (ed.), *Transcriptional Regulation.* Cold Spring Harbor Laboratory, Cold Spring Harbor, N.Y. © 1992, Cold Spring Harbor Laboratory.

A   RNA Pol pauses at 1:2 pause site

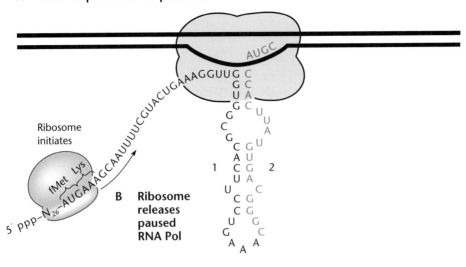

Ribosome initiates

B   Ribosome releases paused RNA Pol

C   Attenuation in presence of Trp

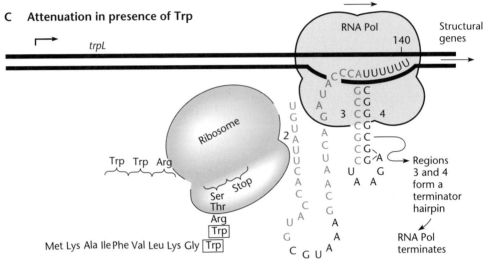

trpL

RNA Pol

140

Structural genes

Ribosome

Trp Trp Arg

Stop

Ser
Thr
Arg
Trp

Met Lys Ala Ile Phe Val Leu Lys Gly Trp

Regions 3 and 4 form a terminator hairpin

RNA Pol terminates

D   Transcription elongation in absence of Trp

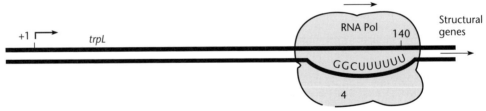

+1   trpL

RNA Pol

140

Structural genes

GGCUUUUUU

4

Lack of *trp* tRNA causes ribosome to stall

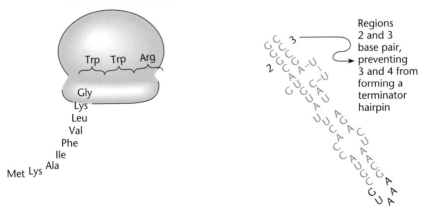

Trp Trp Arg

Gly
Lys
Leu
Val
Phe
Ile
Ala
Met Lys

Regions 2 and 3 base pair, preventing 3 and 4 from forming a terminator hairpin

has tryptophan attached) determines which of several alternative secondary-structure hairpins will form in the leader RNA. Figure 12.21 shows that four RNA regions can form three different hairpins. Recall from chapter 2 that that the secondary structure of an RNA, or a hairpin, results from complementary pairing between the bases in RNA transcribed from inverted repeated sequences.

Whether or not transcription termination occurs depends on whether the attenuation mechanism senses relatively low or high levels of tryptophan. The *trpL* region, which contains two adjacent *trp* codons, provides the signal. The *trp* codons are there to allow the ribosome to test the water before the RNA polymerase is allowed to plunge into the structural genes of the operon. If levels of tryptophan are low, the levels of tryptophanyl-tRNA^{Trp} (tRNA^{Trp} with tryptophan attached) will also be low. When a ribosome encounters one of the *trp* codons, it will temporarily stall, unable to insert the amino acid. This stalled ribosome in the *trpL* region therefore communicates that the tryptophan concentration is low and that transcription should continue (Figure 12.22).

Figures 12.21 and 12.22 also show how the hairpins operate in attenuation. Four different regions in the *trpL* leader RNA—regions 1, 2, 3, and 4—can form three different hairpins, 1:2, 2:3, and 3:4, as shown in Figure 12.21. The formation of hairpin 3:4 causes RNA polymerase to terminate transcription because this hairpin is part of a factor-independent transcription termination signal (see chapter 2).

Whether hairpin 3:4 forms is determined by the dynamic relationship between ribosomal translation of the *trp* codons in the *trpL* region and the progress of RNA polymerase through the *trpL* region (also illustrated in Figure 12.22). After RNA polymerase initiates transcription at the promoter, it moves through the *trpL* region to a site located just after region 2, where it pauses. The hairpin formed by mRNA regions 1 and 2 is an important part of the signal to pause. The pause is short, probably less than 1 s, but it ensures that a ribosome has time to load on the mRNA before the RNA polymerase proceeds to region 3. The moving ribosome may help release the paused RNA polymerase by colliding with it.

The process of ribosome translation through the *trp* codons of *trpL* then determines whether hairpin 3:4 will form, causing termination, or 2 will pair instead with 3, preventing formation of the 3:4 hairpin. Region 3 will pair with region 2 if the ribosome stalls at the *trp* codons because of low tryptophan concentrations (Figure 12.22D). If the ribosome does not stall at the *trp* codons, it will continue until it reaches the UGA stop codon at the end of *trpL*. By remaining at the stop codon while region 4 is synthesized, the ribosome will prevent hairpin 2:3 from forming. Therefore hairpin 3:4 can form and terminate transcription (Figure 12.22C).

GENETIC EVIDENCE FOR THE MODEL
No model is satisfactory unless it is supported by experimental evidence. The existence and in vivo functioning of hairpin 2:3 were supported by the phenotypes produced by mutation *trpL75*. This mutation, which changes one of the nucleotides and prevents pairing of two of the bases holding the hairpin together, should destabilize the hairpin. In the *trpL75* mutant, transcription terminates in the *trpL* region, even in the absence of tryptophan, consistent with the model that formation of hairpin 2:3 normally prevents formation of hairpin 3:4.

That translation of the leader peptide from the *trpL* region is essential to the regulation is supported by the phenotypes of mutation *trpL29*, which changes the AUG start codon of the leader peptide to AUA, preventing initiation of translation. In *trpL29* mutants, termination also occurs even in the absence of tryptophan. The model also explains this observation as long as we can assume that the RNA polymerase paused at hairpin coding sequence 1:2 will eventually move on, even without a translating ribosome to nudge it, and will eventually transcribe the 3:4 region. Without a ribosome stalled at the *trp* codons, however, hairpin 1:2 will persist, preventing the formation of hairpin 2:3. If hairpin 2:3 does not form, hairpin 3:4 will form and transcription will terminate.

One final prediction of the model is that stopping translation at other codons in *trpL* should also relieve attenuation. The codon immediately downstream of the second tryptophan codon in the *trpL* region is for arginine.

Figure 12.22 Details of regulation by attenuation in the *trp* operon. (A) RNA polymerase pauses after transcribing regions 1 and 2. (B) A ribosome has time to load on the mRNA and begin translating, eventually reaching the RNA polymerase and bumping it off the pause site. (C) In the presence of tryptophan, the ribosome translates through the *trp* codons and prevents the formation of hairpin 2:3, thereby allowing the formation of hairpin 3:4, which is part of a transcription terminator. Transcription terminates. (D) In the absence of tryptophan, the ribosome stalls at the *trp* codons, and hairpin 2:3 forms, preventing the formation of hairpin 3:4 and allowing transcription to continue through the terminator. See the text for more details.

Starving the cells for arginine also prevents attenuation of the *trp* operon, fulfilling this prediction of the model. Once antitermination was discovered in the *trp* operon, it was found to be used in a number of other regulatory situations as well (see Box 12.3 and Yanofsky, Suggested Reading).

Feedback Inhibition

Biosynthetic pathways are not regulated solely through transcriptional regulation of their operons; they are also often regulated by feedback inhibition of the enzymes once they are made. In feedback inhibition, the end product of a pathway binds to the first enzyme of the pathway, inhibiting its activity. Feedback inhibition is common to many types of biosynthetic pathways and is a more sensitive and rapid mechanism for modulating the amount of the end product than is transcriptional regulation, which responds only slowly to changes in the concentration of the end product of the pathway.

Tryptophan Operon

The Trp biosynthetic pathway is subject to feedback inhibition. Tryptophan binds to the first enzyme of the Trp synthesis pathway, anthranilate synthetase, and inhibits it, thereby blocking the synthesis of more tryptophan. We can isolate mutants defective in feedback inhibition of tryptophan biosynthesis by using the tryptophan analog 5-methyltryptophan to study this process. At high concentrations, 5-methyltryptophan will bind to anthranilate synthetase in lieu of tryptophan and inhibit the enzyme's activity, starving the cells for tryptophan. Only mutants defective in feedback inhibition because of a missense mutation in the *trpE* gene that prevents the binding of tryptophan (and 5-methyltryptophan) to the anthranilate synthetase enzyme will multiply to form a colony.

A similar method was described earlier for isolating constitutive mutants with mutations of the *trp* operon, but selection of constitutive mutants requires lower concentrations of 5-methyltryptophan. If the concentration of this analog is high enough, even constitutive mutants will be starved for tryptophan.

Isoleucine-Valine Operon

Feedback inhibition is also responsible for valine sensitivity of *E. coli*. If *E. coli* cells are presented with high concentrations of valine, they will die, as long as isoleucine is not provided in the medium. The reason is that valine and isoleucine are synthesized by the same pathway, encoded by the *ilv* (isoleucine-valine) operon. The first enzyme of the pathway, acetohydroxy acid synthase, is feedback inhibited by valine, so that if the concentration of valine is high, the cells can make neither valine nor isoleucine. The cells will then starve for isoleucine unless this amino acid is provided in the medium. Such a situation seldom occurs in nature, since degraded proteins are the usual source of amino acids and since isoleucine and valine are two of the most common amino acids and so are present in most proteins.

While most *E. coli* strains are valine sensitive, mutants that are valine resistant are easily isolated by plating *E. coli* in the presence of high concentrations of valine with no isoleucine. Any colonies that arise are due to the multiplication of valine-resistant mutants. These mutants are about as frequent as mutants resistant to the tryptophan analog 5-methyltryptophan, and a priori one might assume they had the same molecular basis, in this case, an altered acetohyroxy acid synthase that is still active but is no longer feedback inhibited by valine. However, mutations to valine resistance are often revertants of a mutation that normally inactivates a gene for another acetohydroxy acid synthase that is not feedback inhibited by valine and so performs the first step in the synthesis of isoleucine, regardless of the valine concentration.

SUMMARY

1. Regulation of gene expression can occur at any stage in the expression of a gene. If the amount of mRNA synthesized from the gene differs under different conditions, the gene is transcriptionally regulated. If the regulation occurs after the mRNA is made, the gene is posttranscriptionally regulated. A gene is translationally regulated if the mRNA is made but not always translated at the same rate.

2. In bacteria, more than one gene is sometimes transcribed into the same mRNA. Such a cluster of genes, along with their adjacent controlling sites, is called an operon.

3. The regulation of operon transcription can be negative, positive, or a combination of the two. If a protein blocks the transcription of the operon, the operon is negatively regulated and the regulatory protein is a repressor. If a protein is required for transcription of an operon, the operon is positively regulated and the regulatory protein is an activator.

4. If an operon is negatively regulated, mutations that inactivate the regulatory gene product will result in constitutive mutants in which the operon genes are always expressed. If the operon is positively regulated, mutations that inactivate the regulatory protein will cause permanent repression of the expression of the operon. Therefore in general, constitu-

(continued)

SUMMARY (continued)

tive mutations are much more common with negatively regulated operons than with positively regulated operons.

5. Sometimes the same protein can be both a repressor and an activator in different situations, which complicates the analysis of the regulation.

6. The regulation of transcription of bacterial operons is often achieved through small molecules called effectors, which bind to the repressor or activator protein, changing its conformation. If the presence of the effector causes the operon to be transcribed, it is often called an inducer; if its presence blocks trancription of the operon, it is called a corepressor. The substrates of catabolic operons are usually inducers, whereas the end products of biosynthetic pathways are usually corepressors.

7. The regions on DNA to which repressors bind are called operators. Some repressors act by physically interfering with the binding of the RNA polymerase to the promoter (closed-complex formation). Others allow repressor binding but prevent opening of the DNA at the promoter (open complexes). Yet others prevent the RNA polymerase from leaving the promoter to begin RNA synthesis (promoter clearance). Some repressors act by binding to two operators simultaneously, bending the DNA between them and inactivating the promoter.

8. The regions to which activator proteins bind are called activator sequences. The activator proteins seem to activate transcription by binding directly to the RNA polymerase at the promoter. Some activators act by making additional contacts with the RNA polymerase at the promoter allowing stronger binding at the promoter (closed complex). Others enhance strand separation at the promoter (open complex).

9. Some regulatory proteins seem to bind to different regions on DNA depending upon whether the effector is present, in this way acting as both repressors and activators.

10. Some operons are transcriptionally regulated by a mechanism called attenuation. In operons regulated by attenuation, transcription will begin on the operon but then terminate after a short leader sequence has been transcribed if the enzymes encoded by the operon are not needed. Attenuation of biosynthetic and degradative operons is sometimes determined by whether certain codons in the leader sequence are translated. Pausing of the ribosome at these codons can cause secondary structure changes in the leader mRNA, leading to termination of transcription by RNA polymerase before it reaches the first gene of the operon.

11. The activity of biosynthetic operons is often not regulated solely through the transcription of the genes of the operon. It may also be regulated by controlling the translation of the mRNA or through reversible regulation of the activity of the enzymes of the pathway. This reversible regulation, called feedback inhibition, usually results from binding of the end product of the biosynthetic pathway to the first enzyme of the pathway.

QUESTIONS FOR THOUGHT

1. Why do you suppose both negative and positive mechanisms of transcriptional regulation are used to regulate bacterial operons?

2. Why are regulatory protein genes sometimes autoregulated?

3. Why do you suppose the *galU* gene, whose product synthesizes UDPglucose, which is required for galactose utilization (Figure 12.8), is not part of the *gal* operon? Note: UDPglucose is the donor of glucose in biosynthetic reactions including cell wall synthesis.

4. What advantages or disadvantages are there to regulation by attenuation? Would you expect operons other than amino acid biosynthetic/degradative operons to be regulated by attenuation? Why or why not?

PROBLEMS

1. Outline how you would isolate a *lacI*s mutant of *E. coli*.

2. Is the AraC protein in the P1 or P2 state with D-fucose bound?

3. What would the phenotype of the following merodiploid *E. coli* cells be with one form of the operon region in the F′ factor and the other in the chromosome?

a. F′ *lac*+/*lacI*s

b. F′ *lac*+/*lacO*c *lacZ*(Am) [The *lacZ*(Am) mutation is an N-terminal polar nonsense mutation]

c. F′ *ara*+/*araC* (the *araC* mutation is inactivating)

d. F′ *ara*+/*araI*

e. F′ *ara*+/*araP*$_{BAD}$

4. The *phoA* gene of *E. coli* is turned on only if phosphate is limiting in the medium. What kind of genetic experiments would you do to determine whether the *phoA* gene is positively or negatively regulated?

5. Outline how you would use 5-methyltryptophan to isolate constitutive mutants of the *trp* operon. Then use these to isolate feedback inhibition mutants.

SUGGESTED READING

Bhende, P. M., and S. M. Egan. 2000. Genetic evidence that transcription activation by RhaS involves specific amino acid contacts with sigma 70. *J. Bacteriol.* **182**:4959–4969.

Boos, W., and A. Bohm. 2000. Learning new tricks from an old dog: MalT of the *Escherichia coli* maltose system is part of a complex regulatory network. *Trends Genet.* **16**:404–409.

Busby, S., and R. H. Ebright. 1999. Transcription activation by catabolite activator protein (CAP). *J. Mol. Biol.* **293**:199–213.

Englesberg, E., C. Squires, and F. Meronk. 1969. The arabinose operon in *Escherichia coli* B/r: a genetic demonstration of two functional states of the product of a regulator gene. *Proc. Natl. Acad. Sci. USA* **62**:1100–1107.

Geanacopoulos, M., G. Vasmatzis, V. B. Zhurkin, and S. Adhya. 2001. Gal repressosome contains an antiparallel DNA loop. *Nat. Struct. Biol.* **8**:432–436.

Hochschild, A., and S. L. Dove. 1998. Protein-protein contacts that activate and repress prokaryotic transcription. *Cell* **92**:597–600. (Minireview.)

Irani, M. H., L. Orosz, and S. Adhya. 1983. A control element within a structural gene: the *gal* operon of *Escherichia coli*. *Cell* **32**:783–788.

Jacob, F., and J. Monod. 1961. Genetic regulatory mechanisms in the synthesis of proteins. *J. Mol. Biol.* **3**:318–356.

Johnson, C. M., and R. F. Schleif. 1995. *In vivo* induction kinetics of the arabinose promoters in *Escherichia coli*. *J. Bacteriol.* **177**:3438–3442.

Morse, D. E., and A. N. C. Morse. 1976. Dual control of the tryptophan operon is mediated by both tryptophanyl-tRNA synthetase and the repressor. *J. Mol. Biol.* **103**:209–226.

Oxender, D. L., G. Zurawski, and C. Yanofsky. 1979. Attenuation in the *Escherichia coli* tryptophan operon. Role of RNA secondary structure invoving the tryptophan codon region. *Proc. Natl. Acad. Sci. USA* **76**:5524–5528.

Ramos, J. L., C. Michan, F. Rojo, D. Dwyer, and K. Timmis. 1990. Signal-regulator interactions: genetic analysis of the effector binding site of *xylS*, the benzoate-activated positive regulator of *Pseudomonas* Tol plasmid *meta*-cleavage pathway operon. *J. Mol. Biol.* **211**:373–382.

Rojo, F. 1999. Repression of transcription initiation in bacteria. *J. Bacteriol.* **181**:2987–2991. (Minireview.)

Roy, S., S. Garges, and S. Adhya. 1998. Activation and repression of transcription by differential contact: two sides of a coin. *J. Biol. Chem.* **273**:14059–14062.

Schlegal, A., O. Danot, E. Richet, T. Ferenci, and W. Boos. 2002. The N-terminus of the *Escherichia coli* transcription activator MalT is the domain of interaction with MalY. *J. Bacteriol.* **184**:3069–3077.

Summers, A. O. 1992. Untwist and shout: a heavy metal-responsive transcriptional regulator. *J. Bacteriol.* **174**:3097–3101.

Yanofsky, C. 2000. Transcription attenuation: once viewed as a novel regulatory strategy. *J. Bacteriol.* **182**:1–8. (Guest commentary.)

Yanofsky, C., and I. P. Crawford. 1987. The tryptophan operon, p. 1453–1472. *In* F. C. Neidhardt, R. Curtiss III, J. L. Ingraham, E. C. C. Lin, K. B. Low, B. Magasanik, W. S. Reznikoff, M. Riley, M. Schaechter, and H. E. Umbarger (ed.), *Escherichia coli and Salmonella: Cellular and Molecular Biology*, 2nd ed. ASM Press, Washington, D.C.

Global Regulatory Mechanisms

BACTERIA MUST BE ABLE TO ADAPT to a wide range of environmental conditions to survive. Nutrients are usually limiting, so bacteria must be able to protect themselves against starvation until an adequate food source becomes available. Different environments also vary greatly in the amount of water or in the concentration of solutes, so bacteria must also be able to adjust to desiccation and differences in osmolarity. Temperature fluctuations are also a problem for bacteria. Unlike humans and other warm-blooded animals, bacteria cannot maintain their own cell temperature and so must be able to function over wide ranges of temperature.

Mere survival is not enough for a species to prevail, however. The species must compete effectively with other organisms in the environment. Competing effectively might mean being able to use scarce nutrients efficiently or taking advantage of plentiful ones to achieve higher growth rates and thereby become a higher percentage of the total population of organisms in the environment. Moreover, different compounds may be available for use as carbon and energy sources. The bacterium may need to choose the carbon and energy source it can use most efficiently and ignore the rest, so that it does not waste energy making extra enzymes.

Conditions not only vary in the environment, but their changes can be abrupt. The bacterium may have to adjust the rate of synthesis of its cellular constituents quickly in response to the change in growth conditions. For example, different carbon and energy sources allow different rates of bacterial growth. Different growth rates require different rates of synthesis of cellular macromolecules such as DNA, RNA, and proteins, which in turn require different concentrations of the components of the cellular macromolecular synthesis machinery such as ribosomes, tRNA, and RNA polymerase. Moreover, the relative rates of synthesis of the different cellular components

435

must be coordinated so that the cell does not accumulate more of some components than it needs.

Adjusting to major changes in the environment requires regulatory systems that simultaneously regulate numerous operons. These systems are called **global regulatory mechanisms**. Often in global regulation, a single regulatory protein controls a large number of operons, which are then said to be members of the same **regulon**.

Table 13.1 lists some global regulatory mechanisms known to exist in *Escherichia coli*. If the genes are under the control of a single regulatory gene (and so are members of the same regulon), the regulatory gene is also listed. We have discussed some examples of regulons in previous chapters. For example, all the genes under the control of the TrpR repressor, including the *trpR* gene itself, are part of the TrpR regulon. The Ada regulon comprises the adaptive response genes, including those encoding the methyltransferases that repair alkylation damage to DNA (see chapter 11); all of these genes are under the control of the Ada protein. Similarly, the SOS genes induced after UV irradiation and some other types of DNA-damaging treatments are all under the control of the same protein, the LexA repressor, and so are part of the LexA regulon. In other cases, the molecular basis of the global regulation is less well understood and may involve a complex interaction between several cellular signals.

In this chapter, we discuss what is known about how some global regulatory mechanisms operate on the molecular level and describe some of the genetic experiments that have contributed to this knowledge. The ongoing studies on the molecular basis of global regulatory mechanisms represent one of the most active areas of research involving bacterial molecular genetics; therefore, new developments will probably have occurred in many of these areas by the time you read this book.

Catabolite-Sensitive Operons

One of the largest global regulatory systems in bacteria coordinates the expression of genes involved in carbon and energy source utilization. All cells must have access to high-energy, carbon-containing compounds, which they degrade to generate ATP for energy and smaller molecules needed for cellular constituents. Smaller molecules resulting from the metabolic breakdown of larger molecules are called **catabolites**.

In times of plenty, bacterial cells may be growing in the presence of several different carbon and energy sources, some of which can be used more efficiently than others. Energy must be expended to synthesize the enzymes needed to metabolize the different carbon sources, and the utilization of some carbon compounds requires more enzymes than does the utilization of others. By making only the enzymes for the carbon and energy source that yields the highest return, the cell will get the most catabolites and energy, in the form of ATP, for the energy it expends. The mechanism for ensuring that the cell will preferentially use the best carbon and energy source available is called **catabolite repression**, and operons subject to this type of regulation are **catabolite sensitive**. The name "catabolite repression" originates from the fact that cells growing in better carbon sources have more catabolites, which seem to repress the transcription of operons for the utilization of poorer carbon sources. However, as we shall see, the name "catabolite repression" is a misnomer because, at least in some of the regulatory systems in *E. coli*, the genes under catabolite control are *activated* when poorer carbon sources are the only ones available. Catabolite repression is sometimes called the **glucose effect** because glucose, which yields the highest return of ATP per unit of expended energy, usually strongly represses operons for other carbon sources. To use glucose, the cell need only convert it to glucose-6-phosphate, which can enter the glycolytic pathway. Thus, glucose is the preferred carbon and energy source for most types of bacteria.

Figure 13.1 illustrates what happens when *E. coli* cells are growing in a mixture of glucose and galactose. The cells first use the glucose, and only after it is depleted do they begin using the galactose. When the glucose is gone, the cells stop growing briefly while they synthesize the enzymes for galactose. Growth then resumes but at a slightly lower rate.

cAMP and the cAMP-Binding Protein

Most bacteria and lower eukaryotes are known to have systems for catabolite repression. The best understood is the **cyclic AMP (cAMP)**-dependent system of *E. coli* and other enteric bacteria. cAMP is like AMP (see chapter 2), with a single phosphate group on the ribose sugar, but the phosphate is attached to both the 5′-hydroxyl and the 3′-hydroxyl groups of the sugar, thereby making a circle out of the phosphate and sugar. Only *E. coli* and other closely related enterics seem to use this cAMP-dependent system. Some bacteria have an entirely different system, which does not involve cAMP. Even *E. coli* has a second, cAMP-independent system for catabolite repression, which we discuss in Box 13.1.

REGULATION OF cAMP SYNTHESIS

Catabolite regulation in *E. coli* is achieved through fluctuation in the levels of cAMP, which vary inversely with the levels of cellular catabolites. In other words, cellular concentrations of cAMP are higher when catabolite levels are lower, the situation that prevails when the bacteria are growing in a relatively poor carbon source such as lactose or maltose.

TABLE 13.1 Some *E. coli* global regulatory systems

System	Response	Regulatory gene(s) (protein[s])	Category of mechanism	Some genes and operons regulated
A. Nutrient limitation				
Carbon	Catabolite repression	*crp* (CAP, also called CRP)	DNA-binding activator or repressor	*lac, ara, gal, mal,* and numerous other C source operons
	Control of fermentative vs oxidative metabolism	*cra* (*fruR*) (CRA)	DNA-binding activator or repressor	Enzymes of glycolysis, Krebs cycle
Nitrogen	Response to ammonia limitation	*rpoN* (NtrA)	Sigma factor (σ^{54})	*glnA* (GS) and operons for amino acid degradation
		ntrBC (NtrBC)	Two-component system	
Phosphorus	Starvation for inorganic orthophosphate (P_i)	*phoBR* (PhoBR)	Two-component system	>38 genes, including *phoA* (bacterial alkaline phosphatase) and *pst* operon (P_i uptake)
B. Growth limitation				
Stringent response	Response to lack of sufficient aminoacylated-tRNAs for protein synthesis	*relA* (RelA), *spoT* (SpoT)	(p)ppGpp metabolism	rRNA, tRNA
Stationary phase	Switch to maintenance metabolism and stress protection	*rpoS* (RpoS)	Sigma factor (σ^s)	>40 genes, including *otsBA* (trehalose synthesis) and *dps* (DNA-binding protein)
Oxygen	Response to anaerobic environment	*fnr* (Fnr)	CAP family of DNA-binding protein	>31 transcripts, including *narGHJI* (nitrate reductase)
		arcAB (ArcAB)	Two-component system	>20 genes, including *cob* (cobalamin synthesis)
C. Stress				
Osmoregulation	Response to abrupt osmotic upshift	*kdpDE* (KdpD, KdpE)	Two-component system	*kdpFABC* (K⁺ uptake system)
	Adjustment to osmotic environment	*rpoS* (RpoS)	Sigma factor (σ^s)	>16 genes, including *osmB* (an outer membrane lipoprotein)
		envZ/ompR (EnvZ/OmpR)	Two-component system	OmpC and OmpF outer membrane porins
		micF	Antisense RNA	*ompF* (porin)
Oxygen stress	Protection against reactive oxygen species	*soxS* (SoxS)	AraC family of DNA-binding proteins	>10 genes, including *sodA* (superoxide dismutase) and *micF* (antisense RNA regulator of *ompF*)
		oxyR (OxyR)	LysR family of DNA-binding proteins	>10 genes, including *katG* (catalase)
Heat shock	Tolerance to abrupt temperature increase	*rpoH* (RpoH)	Sigma factor (σ^{32})	>20 genes encoding Hsps (heat shock proteins), including *dnaK, dnaJ,* and *grpE* (chaperones), and *lon, clpP, clpX,* and *hflB* (proteases)
		rpoE (RpoE)	Sigma factor (σ^{24})	>10 genes, including *rpoH* (σ^{32}) and *degP* (a periplasmic protease)
pH shock	Tolerance of acidic environment	*cadC* (CadC)	ToxR-related DNA-binding protein	Many genes, including *cadBA* (amino acid decarboxylase)

The synthesis of cAMP is controlled through the regulation of the activity of adenylate cyclase. This enzyme, which makes cAMP from ATP, is more active when cellular concentrations of catabolites are low and less active when catabolite concentrations are high. The adenylate cyclase enzyme is associated with the inner membrane and is the product of the *cya* gene.

Figure 13.2 outlines the current picture of the regulation of adenylate cyclase activity. An important factor in the regulation is the phosphoenolpyruvate (PEP)-dependent sugar phosphotransferase system (PTS), which, as the name implies, is responsible for transporting certain sugars, including glucose, into the cell. One of the protein components of the PTS, named IIA^Glc, can exist in either an

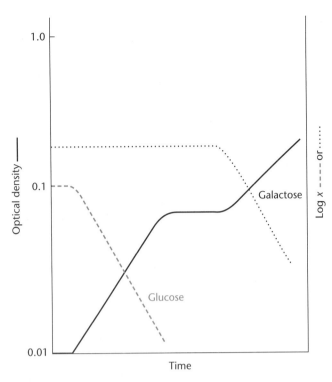

Figure 13.1 Growth of *E. coli* in a mixture of glucose and galactose. The concentration of the sugars in the medium is shown as the dashed lines. The optical density, a measure of cell density, is shown in black. Only after the cells deplete all the glucose will they begin to use the galactose, and then only after a short lag while they induce the *gal* operon (plateau in optical density). They then grow more slowly on the galactose. See the text for details.

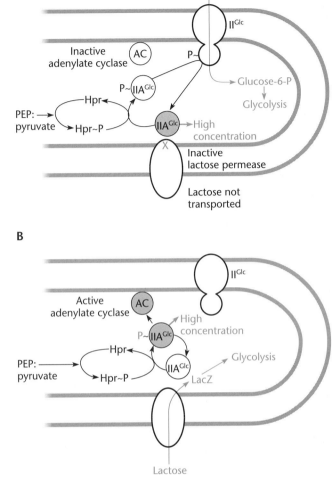

Figure 13.2 Exogenous glucose inhibits both cAMP synthesis and the uptake of other sugars, such as lactose. (A) In the presence of glucose, the ratio of IIA^{Gl} to $IIA^{Glc}{\sim}P$ is high, as glucose is phosphoylated upon transport by the glucose transporter, II^{Glc}. IIA^{Glc} inhibits the lactose permease, resulting in "inducer exclusion." (B) In the absence of glucose, the $IIA^{Glc}{\sim}P$ concentration is high, and it activates adenylate cyclase. Also, lactose transport is permitted.

unphosphorylated (IIA^{Glc}) or a phosphorylated ($IIA^{Glc}{\sim}P$) form. The $IIA^{Glc}{\sim}P$ form activates adenylate cyclase to make cAMP. However, $IIA^{Glc}{\sim}P$ forms in low concentration when one of the sugars it is involved in transport of, such as glucose, is present in the medium. Thus, little of the $IIA^{Glc}{\sim}P$ form will then exist to activate the adenylate cyclase, and cAMP levels will drop.

The ratio of $IIA^{Glc}{\sim}P$ to IIA^{Glc} is largely determined by the ratio of PEP to pyruvate in the cell. When a rapidly metabolizable substrate such as glucose is available, the PEP:pyruvate ratio will be low. This information is signaled to IIA^{Glc} by a phosphorylation cascade that includes a phosphotransferase named Hpr (for *histidine protein* [see Box 13.1]).

The unphosphorylated form of IIA^{Glc} also inhibits other sugar-specific permeases that transport sugars such as lactose. Therefore, less of these other sugars will enter the cell if glucose or another better carbon source is available, and so less inducer will be present to initiate synthesis from their respective operons (see chapter 12). This effect is called **inducer exclusion** (Figure 13.2). In fact, it is

difficult to distinguish this efect on sugar transport, called inducer exclusion from effects of cAMP on the promoter (see Inada et al., Suggested Reading and below).

THE CATABOLITE ACTIVATOR PROTEIN CAP

The mechanism by which cAMP turns on catabolite-sensitive operons in *E. coli* is quite well understood. The cAMP binds to the protein product of the *crp* gene, which is an activator of transcription of catabolite-sensitive operons. This activator protein goes by two names, CAP (for catabolite gene activator protein) and CRP (for cAMP receptor protein). We often call it CAP. CAP with cAMP bound (CAP-cAMP) functions like other activator proteins discussed in chapter 12 in that it interacts with the RNA polymerase to activate transcription from promoters for operons under its control, including *lac*, *gal*, *ara*, and *mal*. These operons are all members of the **CAP regulon**, or the catabolite-sensitive regulon (Table 13.1). However, the mechanism of CAP-cAMP regulation varies. As discussed below, CAP can function not only as an activator but also as a repressor, depending on where it binds relative to the promoter (see the section on regulation of other catabolite-sensitive operons by CAP-cAMP, below).

REGULATION OF *lac* BY CAP-cAMP

A detailed model for CAP activation of transcription from promoters is shown in Figure 13.3. This model incorporates much of what has been learned from in vitro experiments and genetic studies about the requirements for CAP activation. Upstream of the promoter is a short sequence called the **CAP-binding site**, which is similar in all catabolite-sensitive operons and so can be easily identified. CAP can bind to this site only when it is bound to cAMP, so that this site will be occupied only when cAMP

Figure 13.3 Model for CAP activation at class I and class II CAP-dependent promoters. (A) Sequence of the CAP-binding site upstream of the class I *lac* promoter. (B) The binding and location interactions of CAP with the C-terminal and N-terminal domains of the α subunit with class I and class II promoters, respectively. Adapted with permission from N. J. Savery, G. S. Lloyd, S. J. W. Busby, M. S. Thomas, R. H. Ebright, and R. L. Gourse, *J. Bacteriol.* **184**:2273–2280, 2002.

A

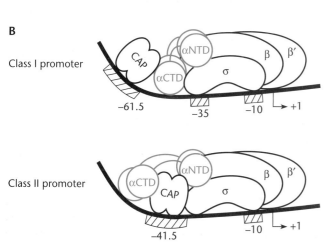

CAP-binding site

B

levels are high. RNA polymerase binds very weakly to the *lac* promoter in the absence of CAP and cAMP at the CAP-binding site. However, the cAMP-CAP bound on this site will make contact with an incoming RNA polymerase, facilitating its binding to the promoter and thereby making the *lac* promoter much stronger.

Recent observations shed light on the details of CAP activation (see Busby and Ebright, Suggested Reading). The CAP protein functions like other activators to make contact with the RNA polymerase at the promoter and stimulate one or more of the steps in the initiation of transcription. We discussed transcriptional activators in chapter 12. The CAP protein is interesting in that it can contact diferent regions of the RNA polymerase and stimulate different steps in initiation, depending upon where it is bound relative to the promoter. This is illustrated in Figure 13.3. At so-called class I CAP-dependent promoters, for example, the *lac* promoter, a dimer of CAP with cAMP, binds upstream of the promoter and contacts the C-terminal end of the α subunit of RNA polymerase. This contact strengthens the binding of RNA polymerase to the promoter (closed complex). In other promoters the so-called class II CAP-dependent promoters such as the *gal* promoter *pG1*, the CAP dimmer-binding site slightly overlaps that of RNA polymerase and it contacts a region in the N terminus of the α subunit. In this position, it stimulates the opening of the DNA at the promoter (the open complex). There are even promoters in which more than one CAP dimer binds and that stimulate both steps of initiation. CAP also bends the DNA when it binds to the CAP-binding site, although the significance of this bending is unknown.

REGULATION OF OTHER CATABOLITE-SENSITIVE OPERONS BY CAP-cAMP

The positioning of the CAP-binding sequence relative to the promoter is different in operons other than *lac*. In the *ara* operon, the CAP-binding site is far upstream, with the AraC-binding site between it and the promoter (see Figure 13.4). Nevertheless, it can still make contact with the C terminus of the α subunit of RNA polymerase, which can reach far up the DNA, as shown. In cases such as these, the CAP protein can also stimulate transcription by interacting with the activator or can stimulate transcription by preventing the binding of a repressor.

Other operons in the CAP regulon, such as *gal*, are less sensitive to catabolite repression than others. As discussed in chapter 12, the *gal* operon is not totally repressed by glucose in the medium because it has two promoters, p_{G1} and p_{G2} (see Figure 12.7). p_{G2} does not require CAP for its activation. In fact, CAP-cAMP represses this promoter, probably because it binds to the −35 sequence of the promoter. The p_{G2} promoter permits some expression of the *gal* operon even in the presence of

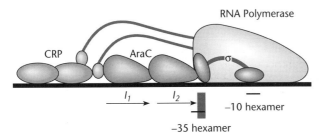

Figure 13.4 Polymerase-promoter and activator-promoter interactions at p_{BAD}. The σ^{70} subunit of RNA polymerase contacts the –35 and –10 hexamers. Occupancy of the I_1 and I_2 half-sites by AraC activates transcription with the aid of the CRP protein, most likely utilizing the α-subunit-activator interactions as shown. The binding sites of σ^{70} and AraC overlap by 4 bp at p_{BAD}. The nucleotides in the –35 hexamer that lie outside the region of overlap are shaded. Reprinted with permission from A. Dhiman and R. Schlief, *J. Bacteriol.* **182**:5076–5081, 2000.

glucose. This low level of expression is necessary to allow the synthesis of cell wall components that include galactose, since the UDP-galactose synthesized by the operon serves as the donor of galactose in biosynthetic reactions. However, the level of expression of the *gal* operon from p_{G2} is not high enough for the cells to grow well on galactose as a carbon and energy source.

RELATIONSHIP OF CATABOLITE REPRESSION TO INDUCTION

An important point about CAP regulation of catabolite-sensitive operons is that it occurs in addition to any other regulation to which the operon is subject (Figure 13.5). Two conditions must be met before catabolite-sensitive operons can be transcribed: better carbon sources such as glucose must be absent, and the inducer of the operon must be present. Take the example of the *lac* operon. If a carbon source better than lactose is available, cAMP levels will be low and so CAP-cAMP will not bind upstream of the *lac* promoter to activate transcription. Also, the *lac* transport system will be inhibited, excluding the inducer from the cell. However, even at high cAMP levels, the *lac* operon will not be transcribed unless the inducer allolactose is also present. In the absence of inducer, the LacI repressor will be bound to the operator and prevent the RNA polymerase from binding to the promoter and transcribing the operon (see chapter 12).

Genetic Analysis of Catabolite Regulation in *E. coli*

The above model for the regulation of catabolite-sensitive operons is supported by both genetic and biochemical analysis of catabolite repression in *E. coli*. This analysis

has involved the isolation of mutants defective in the global regulation of all catabolite-sensitive operons, as well as mutants defective in the catabolite regulation of specific operons.

ISOLATION OF *crp* AND *cya* MUTATIONS

According to the model presented above, mutations that inactivate the *cya* and *crp* genes for adenylate cyclase and CAP, respectively, should prevent transcription of all the catabolite-sensitive operons. In these mutants, there is no CAP with cAMP attached to bind to the promoters. In other words, *cya* and *crp* mutants should be Lac⁻, Gal⁻, Ara⁻, Mal⁻, and so on. In genetic terms, *cya* and *crp* mutations are **pleiotropic**, because they cause many phenotypes, i.e., the inability to use many different sugars as carbon and energy sources.

The fact that *cya* and *crp* mutations should prevent cells from using several sugars was used in the first isolations of *crp* and *cya* mutants (see Schwartz and Beckwith, Suggested Reading). The selection was based on the fact that colonies of bacteria will turn tetrazolium salts red while they multiply, provided that the pH remains high. However, bacteria that are fermenting a carbon source will give off organic acids such as lactic acid that will lower the pH, preventing the conversion to red. As a consequence, wild-type *E. coli* cells growing on a fermentable carbon source form white colonies on tetrazolium-containing plates, whereas mutant bacteria that cannot use the fermentable carbon source will use a different carbon source and so will form red colonies. Some of these red-colony-forming mutants might have *crp* or *cya* mutations, although most would have mutations that inactivate a gene within the operon for the utilization of the fermentable carbon source. Thus, without a way to increase the frequency of *crp* and *cya* mutants among the red-colony-forming mutants, many red-colony-forming mutants would have to be tested to find any with mutations in either *cya* or *crp*.

For these experiments, the investigators reasoned that they could increase the frequency of *crp* and *cya* mutants by plating heavily mutagenized bacteria on tetrazolium agar containing *two different* fermentable sugars, for example, lactose and galactose or arabinose and maltose. Then, to prevent the utilization of both sugars, either two mutations, one in each operon, or a single mutation, in *cya* or *crp*, would have to occur. Since mutants with single mutations should be much more frequent than mutants with two independent mutations, the *crp* and *cya* mutants should be a much higher fraction of the total red-colony-forming mutants growing on two carbon and energy sources. Indeed, when the red-colony-forming mutants that could not use either of the two sugars provided were tested, most of them were found to

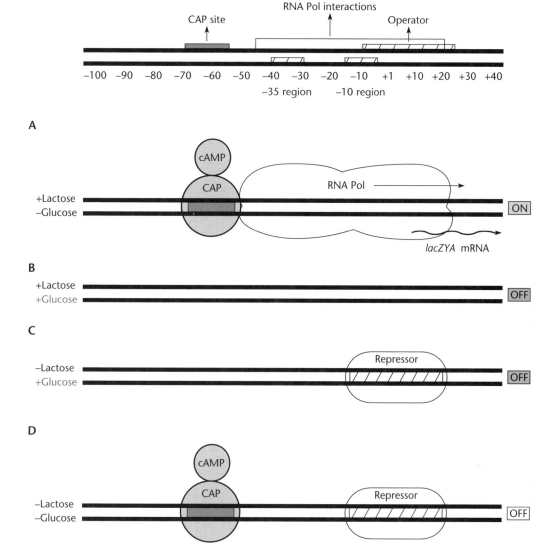

Figure 13.5 Regulation of the lactose operon by both glucose and the inducer lactose. (A) The operon will be ON only in the absence of glucose and the presence of lactose, which is converted into the inducer, allolactose. (B and C) The operon will be OFF in the presence of glucose whether or not lactose is present, because the CAP-cAMP complex will not be bound to the CAP site. (D) The operon will be OFF if lactose is not present, even if glucose is also not present, because the repressor will be bound to the operator. The relative positions of the CAP-binding site, operator, and promoter are shown. The entire regulatory region covers about 100 bp of DNA.

be deficient either in adenylate cyclase activity or lack a protein, now named CAP, later shown to be required for the activation of the *lac* promoter in vitro.

PROMOTER MUTATIONS THAT AFFECT CAP ACTIVATION

Three classes of mutations have been isolated in the *lac* promoter. Those belonging to class I change the CAP-binding site so that CAP can no longer bind to it. The *lac* promoter mutation L8 is an example (Figure 13.6). By preventing the binding of CAP-cAMP upstream of the promoter, this mutation weakens the *lac* promoter. As a result, the *lac* operon is expressed poorly, as measured by β-galactosidase activity, even when cells are growing in lactose and cAMP levels are high. However, the low level of expression of the *lac* operon is less affected by the carbon source and is not reduced much more if glucose is added and cAMP levels drop.

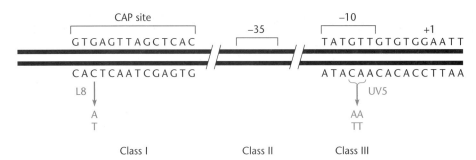

Figure 13.6 Mutations in the *lac* regulatory region that affect activation by CAP. The class I mutation L8 changes the CAP-binding site so that CAP can no longer bind and the promoter cannot be turned on even in the absence of glucose (high cAMP). The class III mutation UV5 changes 2 bp in the –10 sequence of the promoter so that the promoter no longer requires activation by CAP and the operon can be induced even in the presence of glucose (low cAMP). No class II mutations are shown. The changes in the sequence in each mutation appear in gold.

Class II mutations change the –35 region of the RNA polymerase-binding site so that the promoter is less active even when cAMP levels are high. However, with this type of mutation, the residual expression of the *lac* operon will still be sensitive to catabolite repression. Consequently, the amount of β-galactosidase synthesized when cells are growing in the presence of lactose plus glucose while cAMP levels are low will be smaller than the amount synthesized when the cells are growing in the presence of a poorer carbon source plus lactose and cAMP levels are high.

A third, very useful mutated *lac* promoter, class III, was found by isolating apparent Lac⁺ revertants of class I mutations such as p_{L8} or of *cya* or *crp* mutations. One such mutation is called p_{lacUV5}. This mutant promoter is almost as strong as the wild-type *lac* promoter but no longer requires cAMP-CAP for activation. As shown in Figure 13.6, the *lacUV5* mutation changes a 2-bp stretch of the –10 region of the *lac* promoter, so that the sequence reads TATAAT instead of TATGTT. This mutant –10 sequence more closely resembles the sequence of a consensus σ^{70} promoter (see chapter 2). Some expression vectors use the *lacUV5* promoter rather than the wild-type *lac* promoter, so that the promoter can be induced even if the bacteria are growing in glucose-containing media.

INTERACTION OF RNA POLYMERASE WITH CAP
Biochemical experiments first identified the carboxy terminus of the α subunit as a region of the RNA polymerase that contacts CAP for some promoters (see Igarashi and Ishihama, Suggested Reading). In these experiments, the α subunits of RNA polymerase were synthesized in vitro by using a clone of the *rpoA* gene, the gene for the α subunit. If the region of the *rpoA* gene encoding the carboxy terminus of the α subunit is deleted from the clone, an RNA polymerase containing the trun-

cated α subunits can be assembled and is still active for transcription from most promoters. However, the defective RNA polymerase cannot initiate transcription from catabolite-sensitive promoters, even in the presence of CAP-cAMP. These experiments therefore suggested that CAP interacts with the carboxy-terminal portion of the α subunit of RNA polymerase to activate transcription from class I catabolite-sensitive promoters. Genetic experiments confirmed this role of the carboxy-terminal region of the α subunit and identified some of the amino acids important for the interaction (see Savery et al., Suggested Reading).

Uses of cAMP in Other Organisms

The use of cAMP to regulate catabolite-sensitive operons seems unique to enteric bacteria such as *E. coli*. Other bacteria use different mechanisms to regulate catabolite-sensitive operons (see Box 13.1). Many bacteria do not appear to make cAMP at all, whereas in others the levels of cAMP do not vary depending on the carbon sources available. Furthermore, the CAP-cAMP-mediated catabolite regulatory system is not the only system used by *E. coli* and other enteric bacteria (see Box 13.1).

In eukaryotes, cAMP does not regulate carbon source utilization but does have many other uses, including the regulation of G proteins and roles in cell-to-cell communication. In view of the general importance of cAMP in various biological systems, it is somewhat surprising that, among the bacteria, only the enteric bacteria seem to use this nucleotide to regulate carbon source utilization.

Regulation of Nitrogen Assimilation

Nitrogen is a component of many biological molecules, including nucleotides, vitamins, and amino acids. Thus, all organisms must have a source of nitrogen atoms for

BOX 13.1

cAMP-Independent Catabolite Repression

Not all catabolite repression in bacteria is mediated by cAMP. In fact, most gram-positive bacteria, including *Bacillus subtilis,* do not even have cAMP. Even in *E. coli* and other enteric bacteria, there is a mechanism of catabolite repression that does not depend upon cAMP. This mechanism involves the Cra protein, named for its function as a *catabolite repressor/activator* (sometimes called FruR, for fructose repressor). The Cra protein is encoded by the *cra* gene and is a DNA-binding protein similar to LacI and GalR. Cra was discovered during the isolation of mutants with mutations that suppress *ptsH* mutations of *E. coli* and *Salmonella enterica* serovar Typhimurium. The *ptsH* gene encodes the HPr protein that phosphorylates many so-called PTS sugars, including glucose, during transport (Figure 13.2). Therefore, *ptsH* mutants cannot use these sugars, because phosphorylation is the first step in the glycolytic pathway. The *cra* mutations suppress *ptsH* mutations and allow growth on PTS sugars by allowing the constitutive expression of the fructose catabolic operon, one of whose genes encodes a protein that can substitute for HPr. The *cra* mutants were found to be pleiotropic in that they are unable to synthesize glucose from many substrates, including acetate, pyruvate, alanine, and citrate. However, they demonstrate elevated expression of genes involved in glycolytic pathways.

The pleiotropic phenotype of *cra* mutants suggested that Cra functioned as a global regulatory protein, activating the transcription of some genes and repressing the synthesis of others. As for other regulation proteins, whether Cra activates or represses transcription depends on where it binds to the operon. If its binding site is upstream of the promoter, it activates transcription of the operon; if its binding site overlaps or is downstream of the promoter, it represses transcription. In either case, Cra will come off the DNA if it binds the sugar catabolites fructose-1-phosphate and fructose-1,6-bisphosphate that are present in high concentrations during growth in the presence of sugars such as glucose. What effect this will have on the transcription of a particular operon depends on whether Cra functions as a repressor or activator of that operon. If it functions as a repressor, the transcription of the operon will increase; if it functions as an activator, the transcription of the operon will decrease—a process called antiactivation. In general, the Cra protein functions as a repressor of operons whose products are involved in alternate pathways for sugar catabolism, such as the Embden-Meyerhof and Entner-Doudoroff pathways, so that the transcription of these genes will increase when glucose and other good carbon sources are available. In contrast, it usually activates operons whose products are involved in synthesizing glucose from acetate and other metabolites (gluconeogenesis), and so it will not activate the transcription of these operons if glucose is present. Why make glucose if some is already available in the medium?

It is in *B. subtilis* where cAMP-independent catabolite repression has been most extensively studied. As mentioned, this bacterium does not have camp and so depends exclusively on cAMP-independent pathways to regulate its carbon source utilization pathways. In fact, diauxic growth (Figure 13.1) was first discovered with this bacterium. In some ways, the pathway for cabolite repression in *B. subtilis* is similar to the cAMP-independent pathway in *E. coli,* in that there is a repressor protein called CcpA, for *catabolite control protein A,* which is a member of the LacI and GalR family of regulators. The CcpA repressor binds to operator sites called *cre* sites for (catabolite repressor sites) in the promoters of many catabolite-sensitive genes, repressing their transcription. Almost 100 genes in *B. subtilis* are under the control of this repressor. However, whether the CcpA repressor will bind to the *cre* operators depends upon the state of phosphorylation of another protein called Hpr, for *histidine protein.* If cells are growing in a good carbon source such as glucose, high levels of intermediates in the glycolytic pathway accumulate, including fructose-1,6-bisphosphate (FBP). High levels of FBP stimulate the phosporylation of Hpr on one of its serines, which converts it into a corepressor which binds to the CcpA and allows it to bind to the *cre* operators, repressing transcription. Interestingly, the same Hpr protein which serves as a corepressor with CcpA when it is phosporhylated on the serine also serves as the phosphate donor in the PTS for sugar transport when it is phosphorylated on a histidine. Apparently, phosphorylation of Hpr at the serine can inhibit phosphorylation at the histidine and inhibit the transport of sugars which use this system. This allows the close coordination of sugar transport and the regulation of catabolite-sensitive operons. Another corepressor of CcpA called Crh, which is also phosphorylated on one of its serines in the presence of high FBP levels, also exists in *B. subtilis* and acts on some of the catabolite-sensitive operons. However, this second corepressor may exist only in strains of *Bacillus,* and its significance is unclear.

References

Deutscher, J., A. Galinier, and I. Martin-Verstraete. 2002. Carbohydrate uptake and metabolism, p. 129–150. *In* A. L. Sonenshein, J. A. Hoch, and R. Losick (ed.), Bacillus subtilis *and its Closest Relatives.* ASM Press, Washington, D.C.

Saier, M. H., Jr., and T. M. Ramseier. 1996. The catabolite repressor/activator (Cra) protein of enteric bacteria. *J. Bacteriol.* **178:**3411–3417.

growth to occur. For bacteria, possible sources include ammonia (NH_3) and nitrate (NO_3^-), as well as nitrogen-containing organic molecules such as nucleotides and amino acids. Some bacteria can even use atmospheric nitrogen (N_2) as a nitrogen source, a capability that makes them apparently unique on Earth (see Box 13.2).

NH_3 is the preferred source of nitrogen for most types of bacteria. All biosynthetic reactions involving the addition of nitrogen either use NH_3 directly or transfer nitrogen in the form of an NH_2 group from glutamate and glutamine, which in turn are synthesized by directly adding NH_3 to α-ketoglutarate and glutamate, respectively.

Thus, because NH_3 is directly or indirectly the source of nitrogen in biosynthetic reactions, most other forms of nitrogen must be reduced to NH_3 before they can be used in biosynthetic reactions. This process is called **assimilatory reduction** of the nitrogen-containing compounds, because the nitrogen-containing compound converted into NH_3 will be introduced, or assimilated, into biological molecules. In another type of reduction, **dissimilatory reduction**, oxidized nitrogen-containing molecules such as NO_3^- are reduced when they serve as electron acceptors in anaerobic respiration (in the absence of oxygen). However, the compounds are generally not reduced all the way to NH_3 in this process, and the nitrogen is not assimilated into biological molecules. Here, we discuss only the assimilatory uses of nitrogen-containing compounds. The genes whose products are required for anaerobic respiration are members of a different regulon (Table 13.1).

Pathways for Nitrogen Assimilation

Enteric bacteria use different pathways to assimilate nitrogen depending on whether NH_3 concentrations are low or high (Figure 13.7). When NH_3 concentrations are low, for example when the nitrogen sources are amino acids, which must be degraded for their NH_3, an enzyme named **glutamine synthetase (GS)**, the product of the *glnA* gene, adds the NH_3 directly to glutamate to make glutamine. Most of this glutamine is then converted to glutamate by another enzyme, **glutamate synthase**, which removes an NH_2 group from glutamine and adds it to α-ketoglutarate to make two glutamates. These glu-

---

BOX 13.2

Nitrogen Fixation

Some bacteria can use atmospheric nitrogen (N_2) as a nitrogen source by converting it to NH_3 in a process called nitrogen fixation, which appears to be unique to bacteria. However, N_2 is a very inconvenient source of nitrogen. The bond holding the two nitrogen atoms together must be broken, which is very difficult. Bacteria that can fix nitrogen include the cyanobacteria and members of the genera *Klebsiella*, *Azotobacter*, *Rhizobium*, and *Azorhizobium*. These organisms play an important role in nitrogen cycles on Earth.

Some types of nitrogen-fixing bacteria, including members of the genera *Rhizobium* and *Azorhizobium*, are symbionts that fix N_2 in nodules on the roots or stems of plants and allow the plants to live in nitrogen-deficient soil. In return, the plant furnishes nutrients and an oxygen-free atmosphere in which the bacterium can fix N_2. This symbiosis therefore benefits both the bacterium and the plant. An active area of biotechnology is the use of N_2-fixing bacteria as a source of natural fertilizers.

The fixing of N_2 requires the products of many genes, called the *nif* genes. In free-living nitrogen-fixing bacteria such as *Klebsiella* spp., the fixing of nitrogen requires about 20 *nif* genes arranged in about eight adjacent operons. Some of the *nif* genes encode the nitrogenase enzymes directly responsible for fixing N_2. Other *nif* genes encode proteins involved in assembling the nitrogenase enzyme and in

regulating the genes. Plant symbiotic bacteria also require many other genes whose products produce the nodules on the plant (*nod* genes) and whose products allow the bacterium to live and fix nitrogen in the nodules (*fix* genes).

Because atmospheric nitrogen is such an inconvenient source of nitrogen, the genes involved in N_2 fixation are part of the Ntr regulon and are under the control of the NtrC activator protein. In *Klebsiella pneumoniae*, in which the regulation of the *nif* genes has been studied most extensively, the phosphorylated form of NtrC does not activate all eight operons involved in N_2 fixation directly. Instead, the phosphorylated form of NtrC activates the transcription of another activator gene, *nifA*, whose product is directly required for the activation of the eight *nif* operons. In addition, in the presence of oxygen, the *nif* operons are negatively regulated by the product of the *nifL* gene, which senses oxygen. Because the nitrogenase enzymes are very sensitive to oxygen, the bacteria can fix nitrogen only in an oxygen-free (anaerobic) environment, such as exists in the nodules on the roots of plants.

References

Fischer, H.-M. 1994. Genetic regulation of nitrogen fixation in rhizobia. *Microbiol. Rev.* **58**:352–386.

Margolin, W. 2000. Differentiation of free-living Rhizobia into endosymbiotic bacteroids, p. 441–466. *In* Y. V. Brun and L. J. Shimkets (ed.), *Prokaryotic Development.* ASM Press, Washington, D.C.

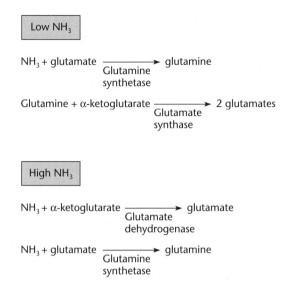

Figure 13.7 Pathways for nitrogen assimilation in *E. coli* and other enteric bacteria. With low NH_3 concentrations, the glutamine synthetase (GS) enzyme adds NH_3 directly to glutamate to make glutamine. Glutamate synthase can then convert the glutamine plus α-ketoglutarate into two glutamates, which can reenter the cycle. In the presence of high NH_3 concentrations, the NH_3 is added directly to α-ketoglutarate by the glutamate dehydrogenase to make glutamate, which can be subsequently converted to glutamine by GS.

tamates can, in turn, be converted into glutamine by glutamine synthetase. Because the NH_3 must all be routed by glutamine synthetase to glutamine when NH_3 concentrations are low, the cell needs a lot of the glutamine synthetase enzyme under these conditions. We shall return to the significance of this later.

If NH_3 concentrations are high because the medium contains NH_3 in the dissolved form of NH_4OH, the nitrogen is assimilated through a very different pathway. In this case, the enzyme **glutamate dehydrogenase** adds the NH_3 directly to α-ketoglutarate to make glutamate. Some of the glutamate is subsequently converted into glutamine by glutamine synthetase. Much less glutamine is required for protein synthesis and biosynthetic reactions than for assimilation of limiting nitrogen from the medium. Therefore, cells need much less glutamine synthetase when growing in high concentrations of NH_3 than when growing in low concentrations.

The reaction catalyzed by glutamate dehydrogenase when NH_3 concentrations are high is very efficient, which is why bacteria that have a glutamate dehydrogenase enzyme prefer NH_3 as a nitrogen source. The rapid assimilation of nitrogen under these conditions allows them to multiply rapidly. However, a few bacteria, including *Bacillus subtilis*, lack glutamate dehydrogenase. These bacteria prefer glutamine to NH_3 as a nitrogen source and cannot assimilate NH_3 as quickly.

REGULATION OF NITROGEN ASSIMILATION PATHWAYS BY THE Ntr SYSTEM

The operons for nitrogen utilization are part of a nitrogen-regulated regulon (the **Ntr system**). Ntr regulation ensures that the cell will not waste energy making enzymes for the use of nitrogen sources such as amino acids or nitrate when NH_3 is available. In this section, we discuss what is known about how the Ntr global regulatory system works. As usual, geneticists led the way, by identifying the genes whose products are involved in the regulation, so that a role could eventually be assigned to each one. The Ntr regulatory system is remarkably similar in all genera of gram-negative bacteria, including *Escherichia, Salmonella, Klebsiella,* and *Rhizobium,* but with important exceptions, some of which we point out. Early indications are that this system may be different in gram-positive bacteria, although work on Ntr in these bacteria is in its early stages.

Regulation of the *glnA-ntrB-ntrC* Operon by a Signal Transduction Pathway

Since cells need more glutamine synthetase while growing at low NH_3 concentrations than when growing at high NH_3 concentrations, the expression of the *glnA* gene, which encodes glutamine synthetase, must be regulated according to the nitrogen source that is available. This gene is part of an operon called *glnA-ntrB-ntrC*. The products of the two other genes in the operon, NtrB and NtrC (*n*itrogen *r*egulator B and C), are involved in regulating the operon. (These proteins are also called NR_{II} and NR_I, respectively, but we use the Ntr names in this chapter.) Because the *ntrB* and *ntrC* genes are part of the same operon as *glnA*, their products are also synthesized at higher levels when NH_3 concentrations are low.

Figure 13.8 illustrates the regulation of the *glnA-ntrB-ntrC* operon and other Ntr genes in detail. In addition to NtrB and NtrC, the proteins GlnD and P_{II} participate in regulation of the operon. These four proteins form a **signal transduction pathway**, in which news of the available nitrogen source is passed, or transduced, from one protein to another until it gets to its final destination, the regulator protein.

The availability of nitrogen is sensed through the level of glutamine in the cell. If the cell is growing in a nitrogen rich environment, the levels of glutamine will be high, if the cell is growing in limiting nitrogen the levels of glutamine will be low. How the levels of glutamine affect the regulation of the Ntr genes involved in using alternative nitrogen sources is probably best explained by working backwards from the last protein in the signal transduction pathway, NtrC. The NtrC protein can be phosphorylated to form NtrC~P, the form in which it is a transcriptional activator that will activate transcription of the Ntr genes, which are turned on in limiting nitrogen (Figure 13.8).

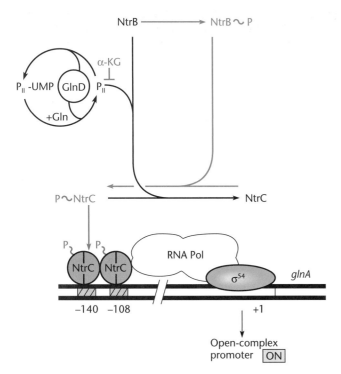

Figure 13.8 Regulation of Ntr operons by a signal transduction pathway in response to NH₃ levels. At low NH₃ concentrations, the reactions shown in gold will predominate. At high NH₃ concentrations, the reactions shown in black will predominate. α-KG, α-ketoglutarate. The effect of α-KG on P_{II} is indirect. See the text for details.

The penultimate protein in the pathway is NtrB. Together, NtrB and NtrC form a two-component phospho-relay system (see Box 13.3 and below) in which the NtrB protein is the sensor-kinase and NtrC is the response regulator. NtrB can phosphorylate itself to form NtrB~P and transfer this phosphate to NtrC to form NtrC~P and activate transcription of the Ntr genes but only if nitrogen is limiting. Whether or not NtrB can phosphorylate itself depends upon the state of modification of another protein called P_{II}. The P_{II} protein can be modified not by phosphorylation but by having UMP attached to it (P_{II}~UMP). If nitrogen is limiting, making the glutamine level low, most of this protein will exist as P_{II}~UMP. However, if nitrogen is in excess and the glutamine level is high, the glutamine will stimulate an enzyme called GlnD to remove the UMP from P_{II}. The unmodified P_{II} protein will bind to NtrB and inhibit its autokinase activity so that it cannot phosphorylate itself to form NtrB~P. If it cannot phosphorylate itself, it cannot transfer a phosphate to NtrC to activate transcription of the Ntr genes involved in using alternate nitrogen sources.

The level of α-ketoglutarate (α-KG) in the cell may also play a role in this process. The level of a α-KG will tend to be higher if nitrogen is limiting, depending upon the carbon sources available and α-KG may indirectly counteract the activity of unmodified P_{II} protein (see Reitzer and Schneider, Suggested Reading). The purpose of superimposing the regulation by α-KG on the regulation by nitrogen may be to coordinate the utilization of alternate carbon

BOX 13.3

Sensor-Response Regulator Two-Component Systems

Many regulation mechanisms require that the cell sense changes in the external environment and change the expression of its genes or the activity of its proteins accordingly. Examples of such regulation include motility in response to chemical attractants and induction of pathogenesis operons upon entry into a suitable host. The mechanisms regulating many of these systems are remarkably similar, and large bacteria can have hundreds of these systems. They usually involve two proteins, one of which is a sensor histidine kinase which phosphorylates itself in response to a particular environmental stimulus. In only a few cases is this stimulus known. The phosphorylation occurs on a histidine which is conserved among the various histidine kinases. The second protein is a response regulator, which can be a transcriptional regulator such as a repressor or an activator. Others activate another cellular function; for example, CheY modulates flagellar movement. The response

regulator transfers the phosphate from the sensor kinase to a conserved aspartate within itself, thereby altering its activity. For example, if it is a transcriptional regulator, it will turn the operon(s) under its control on or off.

Not only do many sensor and response regulator pairs of proteins work in remarkably similar ways, but also they share considerable amino acid sequence homology, as though they were evolutionarily derived from each other.

The way these two-component systems operate in general is illustrated in the Figure. Panel A, part I, shows that sensor kinases are often integral membrane proteins responsive to external signals. Panel A, part II, shows the functions of the protein domains. The C-terminal domain of a sensor kinase has the conserved histidine that is phosphorylated (step 1). The response regulators are similar in their N-terminal region, which includes the asparatate which is phosphorylated

(continued)

BOX 13.3 (continued)

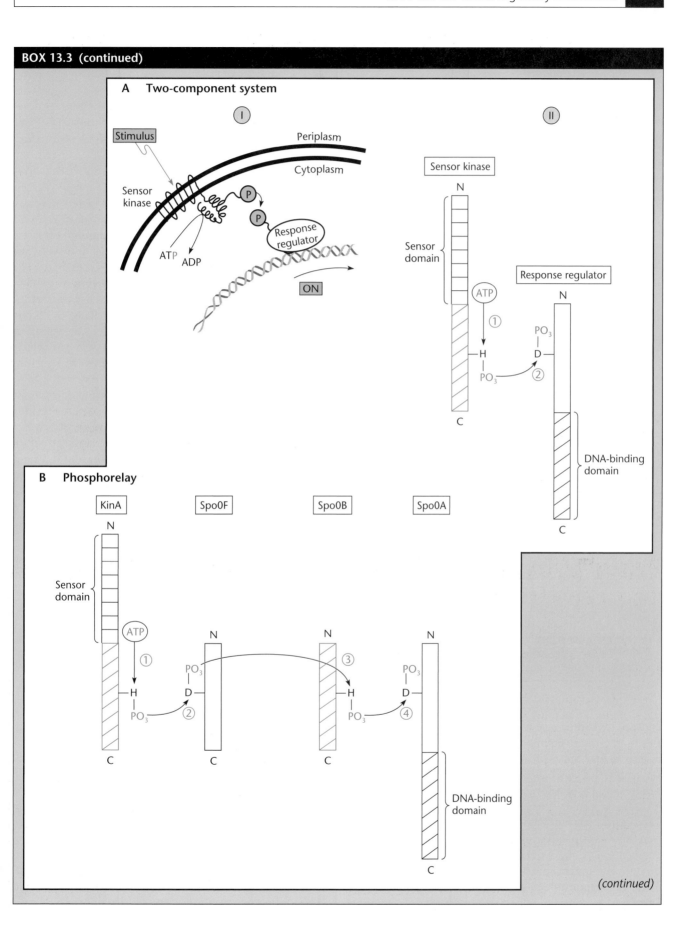

A   Two-component system

B   Phosphorelay

(continued)

BOX 13.3 (continued)

Sensor-Response Regulator Two-Component Systems

C   SpoOB-SpoOF cocrystal structure

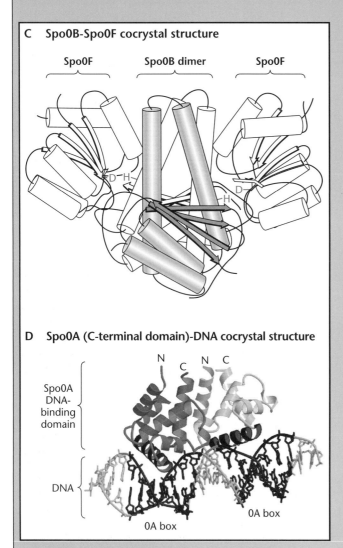

SpoOF   SpoOB dimer   SpoOF

D   SpoOA (C-terminal domain)-DNA cocrystal structure

SpoOA
DNA-
binding
domain

DNA

0A box

0A box

(step 2). The remainder of the protein differs depending on its function, although different subfamilies of response regulators show regions of high homology in other parts of the protein, including the helix-turn-helix motif of transcriptional regulators.

Some bacterial pathways respond to multiple signal inputs. An example is the sporulation phosphorelay of *B. subtilis* (see chapter 14). As illustrated in panel B of the Figure, the histidine kinase and the phosphorylated aspartate domains can be found on separate polypeptides. The phosphoryl group is transferred from one protein to the next as shown in steps 1 through 4. Panel C illustrates a cocrystal structure of a SpoOB dimer containing the conserved histidine residues with two SpoOF polypeptides which contain the aspartate residues. The close proximity of the histidine- and aspartate-containing active sites allows phosphoryl group transfer. Adapted with permission from Hoch and Varughese below. Panel D shows a cocrystal structure of the transcription regulator SpoOA with its DNA-binding site. Only the C-terminal domain of SpoOA was used in the crystallization experiment. Adapted with permission from Bourret et al., below, courtesy of K. I. Varughese. In this response regulator, the unphosphorylated N terminal domain inhibits C-terminal domain binding to DNA. When the phosphorelay is operating, phosphorylation of the N-terminal domain relieves the N-terminal domain-mediated inhibition.

References

Bourret, R. B., N. W. Charon, A. M. Stock, and A. H. West. Bright lights, abundant operons—fluorescence and genomic technologies advance studies of bacterial locomotion and signal transduction: review of the BLAST meeting, Cuernavaca, Mexico, 14 to 19 January 2001. *J. Bacteriol.* **184**:1–17, 2002. (Minireview.)

Hoch, J. A., and K. I. Varughese. Keeping signals straight in phosphorelay signal transduction. *J. Bacteriol.* **183**:4941–4949, 2001.

sources with their use as nitrogen sources, since sometimes the same compounds in the medium, such as amino acids, must provide both types of atoms, and other times a nitrogen containing molecule might be needed as a carbon source even if ammonia is present.

NtrB and NtrC: a Two-Component Sensor-Response Regulator System

As mentioned, the NtrB and NtrC proteins form a two-component system in which NtrB is a **sensor-kinase** and NtrC is a **response-regulator**. Such protein pairs are now known to be common in bacteria, and the corresponding members of such pairs are remarkably similar to each other (see Box 13.3). Typically, one protein of the pair "senses" an environmental parameter and phosphorylates itself at a histidine. This phosphoryl group is then passed on to an aspartate in the second protein. The activity of this second protein, the response regulator, depends on whether it is phosphorylated. Many response regulators are transcriptional regulators. In chapter 6, we discussed another example of a sensor-response regulator pair, ComP and ComA, involved in the development of transformation competence in *B. subtilis*, and later in this chapter we describe other pairs.

The Other Ntr Operons

The other operons besides *glnA-ntrB-ntrC* that are activated by NtrC~P depend on the type of bacterium and the other nitrogen sources they can use. In general, operons under the control of NtrC~P are those involved in using poorer nitrogen sources. For example, genes for the uptake of the amino acids glutamine in *E. coli* and histidine and arginine in *Salmonella enterica* serovar Typhimurium are under the control of NtrC~P. An operon for the utilization of nitrate as a nitrogen source in *Klebsiella pneumoniae* is activated by NtrC~P, but neither *E. coli* nor *S. enterica* serovar Typhimurium has such an operon.

In some bacteria, the Ntr genes are not regulated directly by NtrC~P but are under the control of another gene product whose transcription is activated by NtrC~P. For example, operons for amino acid degradative pathways in *Klebsiella aerogenes* use σ^{70} promoters that do not require activation by NtrC~P. However, they are indirectly under the control of NtrC~P because transcription of the gene for their transcriptional activator, *nac*, is activated by NtrC~P. The nitrogen fixation genes of *K. pneumoniae* are similarly under the indirect control of NtrC~P because this protein activates transcription of the gene for their activator protein, *nifA* (see Box 13.2).

Another Ntr Regulatory Pathway?

The need for so many steps in the regulation of the Ntr system is not clear, but it may reflect the central role that nitrogen utilization plays in the physiology of the cell. Other pathways may intersect with the various steps of the signal transduction pathway for nitrogen utilization, so that expression of the genes can be coordinated with many other cellular functions. In support of this idea, some Ntr regulation occurs even in *ntrB* mutants (see the section on genetics of Ntr regulation below). At least one other pathway besides the one involving NtrB must lead to the phosphorylation and dephosphorylation of NtrC in response to changes in the nitrogen source.

TRANSCRIPTION OF THE *glnA-ntrB-ntrC* OPERON

The *glnA-ntrB-ntrC* operon is transcribed from three promoters, only one of which is NtrC~P dependent. The positions of the three promoters and the RNAs made from each are shown in Figure 13.9. Of the three promoters, only the p_2 promoter is activated by NtrC~P and is responsible for the high levels of glutamine synthetase and NtrB and NtrC synthesis in low NH_3. We defer discussion of the other two promoters, p_1 and p_3, until later.

The p_2 promoter is immediately upstream of the *glnA* gene, and RNA synthesis initiated at this promoter continues through all three genes, as shown in Figure 13.9. However, some transcription terminates at a transcriptional terminator located between the *glnA* and *ntrB*

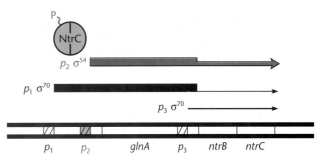

Figure 13.9 The *glnA-ntrB-ntrC* operon of *E. coli*. Three promoters service the genes of the operon. The arrows show the mRNAs that are made from each promoter. The gold arrow indicates the mRNA expressed from the nitrogen-regulated promoter, p_2. The thickness of the lines indicates how much RNA is made from each promoter in each region. See the text for details.

genes, so that much less NtrB and NtrC is made than glutamine synthetase.

The "Nitrogen Sigma" σ^{54}

The p_2 promoter and other Ntr-type promoters activated by NtrC~P are unusual in terms of the RNA polymerase holoenzyme that recognizes them. Most promoters are recognized by the RNA polymerase holoenzyme with σ^{70} attached, but the Ntr-type promoters are recognized by holoenzyme with σ^{54} attached (Box 13.4). As shown in Figure 13.10, promoters that are recognized by the σ^{54}

Figure 13.10 Sequence comparison of the promoters recognized by RNA polymerase holoenzyme carrying the normal sigma factor (σ^{70}), the nitrogen sigma factor (σ^{54}), and the heat shock sigma factor (σ^{32}). Instead of consensus sequences at −10 and −35 bp, with respect to the RNA start site, the σ^{54} promoter has consensus sequences at −12 and −24 bp. The σ^{32} promoter has consensus sequences at approximately −10 and −35 bp, but they are different from the consensus sequences of the σ^{70} promoters. X indicates that any base pair can be present at this position. +1 is the start site of transcription.

σ^{70} promoter

```
        -35                    -10        +1
                                           A →
    ═══ T T G A C A ══//══ T A T A A T ═══ T ═══
    ═══ A A C T G T ══//══ A T A T T A ═══
```

σ^{54} promoter

```
        -24                    -12        +1
                                           A →
    ═══ C T G G X A ══//══ T T G C A ═══ T ═══
    ═══ G A C C X T ══//══ A A C G T ═══
```

σ^{32} promoter

```
        -35                          -10        +1
                                                 A →
    ═══ T C T C X C C C T T G A A ══//══ C C C C A T X T A ═══ T ═══
    ═══ A G A G X G G G A A C T T ══//══ G G G G T A X A T ═══
```

BOX 13.4

The Alternate Sigma Factor σ⁵⁴: the Nitrogen Sigma

Sigma (σ) factors seem to be unique to bacteria and their phages and are not found in eukaryotes. We have discussed sigma factors in previous chapters. These proteins cycle on and off RNA polymerase and help direct it to specific promoters. They also help RNA polymerase melt the promoter to initiate transcription, and they help in promoter clearance after initiation. They also often contain the contact points of activator proteins that help the activator protein stabilize the RNA polymerase on the promoter and activate transcription (see chapter 12). Promoters are often identified by the σ they use; for example, a "σ⁷⁰ promoter" is one which uses the RNA polymerase holoenzyme with σ⁷⁰ attached while a σˢ promoter uses σˢ RNA polymerase.

There are two major classes of sigma factors: the σ⁷⁰ class and σ⁵⁴, which seems to form a class by itself. There is no sequence similarity between members of these two classes, and they seem to differ fundamentally in their mechanism of action (see below). Members of the σ⁷⁰ class are found in all bacteria and play many diverse roles, some of which we have discussed in previous chapters. The σ⁵⁴-type promoters are widely distributed among both gram-positive and gram-negative bacteria, but they are not universal. This sigma was originally named the "nitrogen sigma," σᴺ, because it was first found in the genes for Ntr regulation that were turned on during nitrogen-limited growth in *E. coli* (see the text). However, there is no common theme for σ⁵⁴-expressed genes.

The major distinction between these two classes of sigmas is in the way their promoters are activated. For example, many promoters that use a sigma of the σ⁷⁰ family can initiate transcription without the help of an activator protein. If a σ⁷⁰ promoter requires an activator, it generally binds adjacent to the promoter and helps recruit RNA polymerase to the promoter (see chapter 12). However, all σ⁵⁴ promoters studied thus far absolutely require a specialized activator protein with ATPase activity, which binds to an enhancer sequence which can be hundreds of base pairs upstream of the promoter (see Box 12.2 and below). In some ways this makes σ⁵⁴ promoters more like the RNA polymerase II promoters of eukaryotes. They also differ from σ⁷⁰ promoters in their mechanism of activation. The activation of a σ⁵⁴ promoter is illustrated in the Figure. Panel A shows the functionally important domains of σ⁵⁴ that allow bind-

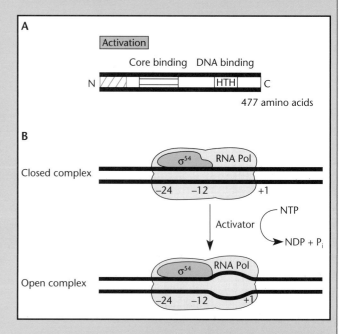

ing to core RNA polymerase and DNA. The N terminal domain allows σ⁵⁴ to respond to activators. Panel B shows that the σ⁵⁴-RNA polymerase forms a stable but closed complex with the promoter, even in the absence of the activator bound to the upstream enhancer. The activator has a latent ATPase activity, which becomes activated by phosphorylation that is often passed down from a regulatory cascade. Many of the activators of σ⁵⁴ promoters are at the end of a phosphorylation cascade, with NtrC being the prototype (see the text). Phosphorylation of the N terminus of the activator alters its affinity for enhancer sites. Multimerization and formation of a DNA-bound complex activates the ATPase activity in its central domain. Once activated, the ATPase can cause the σ⁵⁴ to undergo a conformational change that overcomes its inhibition of open-complex formation by the σ⁵⁴-RNA polymerase and allows initiation at the promoter (see also Box 12.2).

Reference

Buck, M., M.-T. Gallegos, D. J. Studholme, Y. Guo, and J. D. Gralla. 2000. The bacterial enhancer dependent σ⁵⁴ (σᴺ)transcription factor. *J. Bacteriol.* **182:**4129–4136. (Minireview.)

holoenzyme look very different from promoters recognized by the σ⁷⁰ holoenzyme. Unlike the typical σ⁷⁰ promoter, which has RNA polymerase-binding sequences at –35 and –10 bp upstream of the RNA start site, the σ⁵⁴ promoters have very different binding sequences at –24

and –12 bp. Because promoters for the genes involved in Ntr regulation are recognized by RNA polymerase with σ⁵⁴, this sigma factor was first named the "nitrogen sigma" and the gene for σ⁵⁴ was named *rpoN* (Table 13.2). However, σ⁵⁴-type promoters have been found in

TABLE 13.2	Genes for nitrogen regulation		
Gene	**Alternate name**	**Product**	**Function**
glnA		Glutamine synthetase	Synthesize glutamine
glnB		P_{II}, P_{II}-UMP	Inhibit phosphatase of NtrB, activate adenylyltransferase
glnD		Uridylyltransferase (UTase)/Uridylyl-removing enzyme (UR)	Transfer UMP to and from P_{II}
glnE		Adenylyltransferase (ATase)	Transfer AMP to glutamine synthetase
glnF	*rpoN*	σ^{54}	RNA polymerase recognition of promoters of Ntr operons
ntrC	*glnG*	NtrC, NtrC-PO_4	Activator of promoters of Ntr operons
ntrB	*glnL*	NtrB, NtrB-PO_4	Autokinase, phosphatase; phosphate transferred to NtrC

operons unrelated to nitrogen utilization, including in the flagellar genes of *Caulobacter* spp. and some promoters of the toluene-biodegradative operons of the *Pseudomonas putida* Tol plasmid (see chapter 12). Interestingly, all of the known σ^{54}-type promoters require activation by an activator protein, which usually belongs to the NtrC family (see Box 13.4).

NtrC

The polypeptide chains of NtrC-type activators have the basic structure shown in Figure 13.11. A DNA-binding domain that recognizes the σ^{54}-type promoter lies at the carboxy-terminal end of the polypeptide. A regulatory domain that either binds inducer or is phosphorylated is present at the amino-terminal end. The region of the

Figure 13.11 Model for the activation of the p_2 promoter by phosphorylated NtrC protein. (A) Functions of the various domains of NtrC. P denotes a phosphate. (B) Two dimers of phosphorylated NtrC bind to the inverted repeats in the upstream activator sequence (UAS). The DNA is bent between the UAS and the promoter, allowing contact between NtrC and the RNA polymerase bound at the promoter more than 100 bp downstream.

A

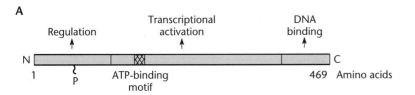

B

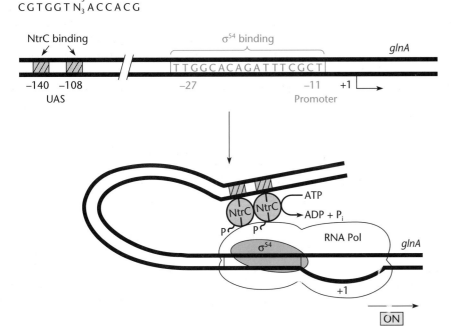

polypeptide responsible for transcriptional activation is in the middle. This region has an ATP-binding domain and an ATPase activity that cleaves ATP to ADP. As mentioned in Boxes 12.2 and 13.4), the N-terminal domain somehow masks the middle domain for activation unless the N-terminal domain has bound inducer or has been phosphorylated.

Detailed Model for Activation of Ntr-Type Promoters

The NtrC-activated promoters, including p_2 of the *glnA-ntrB-ntrC* operon, are unusual in that NtrC binds to an **upstream activator sequence (UAS)**, which lies more than 100 bp upstream of the promoter. For most prokaryotic promoters, the activator protein binding sequences are adjacent to the promoter. Activation at a distance, such as occurs with the NtrC-activated promoters, is much more common in eukaryotes, where many examples are known.

Figure 13.11 also shows one detailed model for how NtrC~P activates transcription from p_2 and other Ntr-activated promoters. NtrC oligomers bind strongly to the UAS only when the protein is in the NtrC~P form. Because the UAS has an inverted repeat and one NtrC~P polypeptide can bind to each copy of the repeated sequence, multimers of NtrC~P bind to the UAS. The DNA bends between the UAS and the promoter so that the molecules of NtrC~P bound at the UAS make contact with both the RNA polymerase and the promoter. Like other activators, NtrC~P must touch the RNA polymerase before the RNA polymerase can initate transcription. However, σ^{54}-type promoters may require that the activator protein also help separate the strands of the DNA at the promoter, which the RNA polymerase cannot do by itself. This may be the function of the ATPase activity of the NtrC protein. Cleaving ATP to ADP may generate the energy required to separate the strands of the DNA in the promoter as shown.

Function of the Other Promoters of the *glnA-ntrB-ntrC* Operon

As mentioned, the p_2 promoter is only one of three promoters that service the *glnA-ntrB-ntrC* operon (Figure 13.9). The other promoters are p_1 and p_3. The p_1 promoter is further upstream of *glnA* than is p_2, and most of the transcription that initiates at the p_1 promoter terminates at the transcription termination signal just downstream of *glnA*, as shown in Figure 13.9. The p_3 promoter is between the *glnA* and *ntrB* genes and services the *ntrB* and *ntrC* genes.

The p_1 and p_3 promoters use σ^{70} and so do not require NtrC~P for their activation. In fact, they are repressed by NtrC~P, so they will not be used if NH$_3$ concentra-

tions are low. The function of these promoters is presumably to ensure that the cell will have some glutamine synthetase and NtrB and NtrC when NH$_3$ concentrations are high. Unless glutamine is provided in the medium, the cell must have glutamine synthetase to make glutamine for use as the NH$_2$ group donor in some biosynthetic reactions and in protein synthesis. The cell must also have some NtrB and NtrC in case conditions change suddenly from a high-NH$_3$ to a low-NH$_3$ environment, in which the products of the Ntr genes are suddenly needed.

ADENYLYLATION OF GLUTAMINE SYNTHETASE

Regulating the transcription of the *glnA* gene is not the only way the activity of glutamine synthetase is regulated in the cell. This activity is also modulated by the adenylylation of (transfer of AMP to) a specific tyrosine in the enzyme by an adenylyltransferase (ATase) enzyme when NH$_3$ concentrations are high. The adenylylated form of the glutamine synthetase enzyme is less active and is also much more susceptible to feedback inhibition by glutamine than is the unadenylylated form (see chapter 12 for an explanation of feedback inhibition). This makes sense, considering how glutamine synthetase plays different roles when NH$_3$ concentrations are high and when they are low. When NH$_3$ concentrations are high, the primary role of the glutamine synthetase is to make glutamine for protein synthesis, which requires less enzyme; the little enzyme activity that remains should be feedback inhibited to ensure that the cells will not accumulate too much glutamine. When NH$_3$ concentrations are low, more enzyme is required. In this situation, the enzyme should not be feedback inhibited, because its major role is to assimilate nitrogen.

The state of adenylylation of the glutamine synthetase enzyme is also regulated by GlnD and the state of the P$_{II}$ protein. When the cells are in low NH$_3$ so that the P$_{II}$ protein has UMP attached (Figure 13.8), the adenylyltransferase will remove AMP from glutamine synthetase. When the cells are in high NH$_3$, so that the P$_{II}$ protein does not have UMP attached, the P$_{II}$ protein will bind to and stimulate the adenylyltransferase to add more AMP to the glutamine synthetase.

The P$_{II}$ protein may not be the only way to stimulate the deadenylylation of glutamine synthetase in *E. coli*, however. Some work has suggested that another protein, GlnK, may form GlnK-UMP in response to ammonia deprivation. The GlnK-UMP protein can also stimulate the adenylyltransferase to adenylylate glutamine synthetase when NH$_3$ concentrations are high (see van Heeswijk et al., Suggested Reading). Clearly, many pathways interact to regulate nitrogen source utilization by bacteria.

Coordination of Catabolite Repression, the Ntr System, and the Regulation of Amino Acid Degradative Operons

Not only must bacteria sometimes use one or more of the 20 amino acids as a nitrogen source, but also they must sometimes use amino acids as carbon and energy sources. The type of amino acids that different bacteria can use varies. For example, *E. coli* can use only alanine, proline, arginine, and tryptophan as carbon and nitrogen sources, whereas some *Salmonella* serovors can also use proline, arginine, and histidine. Like the sugar-utilizing and biosynthetic operons, the amino acid-utilizing operons not only have their own specific regulatory genes, so that they are transcribed only in the presence of their own inducer, but often are also part of larger regulons. As discussed in this section, they are often under Ntr regulation and so will not be induced in the presence of their inducer while a better nitrogen source such as NH_3 is in the medium. In some bacteria, these operons are also under the control of the catabolite regulon and will not be expressed in the presence of better carbon sources, such as glucose.

Sometimes multiple levels of operon control can be a disadvantage to the bacterium. For example, *S. enterica* serovar Typhimurium cells will starve for nitrogen if they are growing in the presence of histidine plus glucose but no other nitrogen source. The glucose apparently prevents CAP from activating the promoter for the *hut* operon, so that the cells cannot use the histidine as a nitrogen source. Glucose itself contains no nitrogen, and so, although the cells can use it as a carbon source, they will starve for nitrogen.

In addition to their potential as nitrogen, carbon, and energy sources, amino acids are necessary for other purposes, including protein synthesis and, in the case of proline, osmoregulation. Therefore, the use of amino acids as carbon and nitrogen sources can present strategic problems for the cell. The way in which all these potentially conflicting regulatory needs are resolved is bound to be complicated.

Genetic Analysis of Nitrogen Regulation in Enteric Bacteria

The present picture of nitrogen regulation in bacteria began with genetic studies. Most of this work was first done in *K. aerogenes,* although some genes were first found in *S. enterica* serovar Typhimurium or *E. coli.*

The first indications of the central role of glutamine and glutamine synthetase in Ntr regulation came from the extraordinary number of genes that, when mutated, could affect the regulation of the glutamine synthetase or give rise to an auxotrophic growth requirement for glut-

amine (see Magasanik, Suggested Reading). The *gln* genes were originally named *glnA, glnB, glnD, glnE, glnF, glnG,* and *glnL* (Table 13.2). These genes are not lettered consecutively because, as often happens in genetics, genes presumed to exist because of a certain phenotype were later found to not exist or to be the same as another gene, and so their letters were retired. Sorting out the various contributions of the *gln* gene products to arrive at the model for nitrogen regulation outlined above was a remarkable achievement. It took many years and required the involvement of many people. In this section, we describe how the various *gln* genes were first discovered and how the phenotypes caused by mutations in the genes led to the model.

THE *glnB* GENE

The *glnB* gene encodes the P_{II} protein. The first *glnB* mutations were found among a collection of *K. aerogenes* mutants with the Gln⁻ phenotype, the inability to multiply without glutamine in the medium. Mutations in *glnB* apparently can prevent the cell from making enough glutamine for growth. However, early genetic evidence indicated that these *glnB* mutations do not exert their Gln⁻ phenotype by inactivating the P_{II} protein. As evidence, transposon insertions and other mutations that should totally inactivate the *glnB* gene (null mutations) do not result in the Gln⁻ phenotype. In fact, null mutations in *glnB* are intragenic suppressors of the Gln⁻ phenotype of the original *glnB* mutations (see chapter 3 for a discussion of the different types of suppressors).

We now know that the original *glnB* mutations do not inactivate P_{II} but, rather, change the binding site for UMP so that UMP cannot be attached to it by GlnD. This should have two effects. P_{II} without UMP will bind to NtrB~P and prevent phosphorylation of NtrC. Therefore, even in low NH_3 concentrations, little glutamine will be synthesized. By itself, however, this effect does not explain the Gln⁻ phenotype, since the *glnA* gene can also be transcribed from the p_1 promoter, which does not require NtrC~P for activation (see above). It is the second function of P_{II}—the stimulation of the adenylyltransferase—that causes the Gln⁻ phenotype. In the mutants, enough P_{II} without UMP attached accumulates to stimulate the adenylyltransferase to the extent that glutamine synthetase will be too heavily adenylated to synthesize enough glutamine for growth. This also explains why null mutations in *glnB* do not cause the Gln⁻ phenotype. By inactivating P_{II} completely, null mutations prevent the P_{II} protein from stimulating the adenylyltransferase so that less glutamine synthetase is adenylylated and enough glutamine is synthesized for growth.

THE *glnD* GENE

The *glnD* gene was also discovered in a collection of mutants with mutations that cause the Gln⁻ phenotype. However, in this case, experiments showed that null mutations in *glnD* cause the Gln⁻ phenotype. Furthermore, null mutations in *glnB* suppress the Gln⁻ phenotype of null mutations in *glnD*. These observations are consistent with the above model. Since the GlnD protein is the enzyme that transfers UMP to P_{II}, null mutations in *glnD* should behave like the original *glnB* mutations and prevent UMP attachment to P_{II} but leave the P_{II} protein intact. This will make the cells Gln⁻ for the reasons given above. Null mutations in *glnD* are suppressed by null mutations in *glnB* because the absence of GlnD protein does not matter if the cell contains no P_{II} protein to bind to the adenylyltransferase and stimulate the attachment of AMP to glutamine synthetase. The glutamine synthetase without AMP attached will remain active and synthesize enough glutamine for growth.

THE *glnL* GENE

The *glnL* gene was later renamed *ntrB* to better reflect its function. Mutations in this gene were first discovered not because they cause the Gln⁻ phenotype but because they are extragenic suppressors of *glnD* and *glnB* mutations. These mutations do not inactivate NtrB completely; instead, they leave NtrB intact and prevent the binding of P_{II} so that NtrB will transfer its phosphate to NtrC regardless of the presence of P_{II}-UMP. Mutants with null mutations in *ntrB* also retain some regulation of the *glnA-ntrB-ntrC* operon. As previously mentioned, this evidence supports the hypothesis of an alternative NtrB-independent pathway of NtrC phosphorylation (see the section on another Ntr regulatory pathway, above).

THE *glnG* GENE

The *ntrC* gene, originally called *glnG*, was discovered because mutations in it can suppress the Gln⁻ phenotype of *glnF* mutations. The *glnF* gene encodes the nitrogen sigma σ^{54} (see below). Further work showed that the suppressors of *glnF* are null mutations in *ntrC* because transposon insertions and other mutations that inactivate the *ntrC* gene also suppress *glnF* mutations. Null mutations in *ntrC* do not cause the Gln⁻ phenotype, because the *glnA* gene can also be transcribed from the p_1 promoter, which does not require NtrC for activation. They do, however, prevent the expression of other Ntr operons that require NtrC~P for their activation, many of which do not have alternative promoters.

Some mutations in *ntrC* do cause the Gln⁻ phenotype, however. These presumably are mutations that change NtrC so that it can no longer activate transcription from p_2, but it retains the ability to repress transcription from p_1 because it can be phosphorylated.

THE *glnF* GENE

As mentioned, the glnF gene, now renamed *rpoN*, encodes the nitrogen sigma, σ^{54}. The gene was also discovered in a collection of Gln⁻ mutants. Without σ^{54}, the p_2 promoter cannot be used to transcribe the *glnA-ntrB-ntrC* operon. By itself, inactivation of p_2 would not be enough to cause the Gln⁻ phenotype, since the *glnA* gene can also be transcribed from the p_1 promoter, which does not require σ^{54}. However, the NtrC~P form of NtrC will repress the p_1 promoter if the cells are growing in low NH_3 concentrations (see above). In high NH_3 concentrations, the p_1 promoter will not be repressed, but the small amount of glutamine synthetase synthesized will be heavily adenylylated, preventing the synthesis of sufficient glutamine for growth. This interpretation of the Gln⁻ phenotype of *rpoN* mutations is consistent with the fact that null mutations in *ntrC* suppress the Gln⁻ phenotype of *rpoN* mutations as discussed above. Without NtrC~P present to repress the p_1 promoter, sufficient glutamine synthetase will be synthesized from the p_1 promoter, even in low NH_3 concentrations.

Regulation of Porin Synthesis

Most bacteria can survive outside a eukaryotic host. Because they are free living, they often find themselves in media of changing solute concentrations and consequently variable osmolarity. The osmotic pressure is normally higher inside the inner membrane than outside it. This pressure would cause the bacterium to swell, but the rigid cell wall keeps the cell from expanding. Even the cell wall is not invincible, however, and bacteria must keep the difference in osmotic pressure inside and outside the cell from becoming too great. The ability to monitor osmolarity can be important to bacteria for a second reason: bacteria also sometimes sense changes in their external environment by the changes in osmolarity. In fact, one way in which pathogenic bacteria sense that they are inside a host, and induce their virulence genes, is by the much higher osmolarity inside the host (see the section on global regulation in pathogenic bacteria, below). The systems by which bacterial cells sense these changes in osmolarity and adapt are global regulatory mechanisms, and many genes are involved.

Much is known about how bacteria respond to media of different osmolarities. One way they regulate the differences in osmotic pressure across the membrane is by excreting or accumulating K^+ ions and other solutes such as proline and glycine betaine. Gram-negative bacteria such as *E. coli* have the additional problem of maintaining an equal osmotic pressure across the outer membrane. They achieve this in part by synthesizing oligosaccharides in the periplasmic space to balance solutes in the external environment.

One of the major mechanisms by which gram-negative bacteria balance osmotic pressure across the outer membrane is by synthesizing pores to let solutes into and out of the periplasmic space. These pores are composed of outer membrane proteins called **porins**. To form pores, these proteins trimerize in the outer membrane.

The two major porin proteins in *E. coli* are OmpC and OmpF. Pores composed of OmpC are slightly smaller than those composed of OmpF, and the size of the pores can affect the rate at which solutes pass through the pores and thus confer protection under some conditions. For example, the smaller pores, composed of OmpC, may prevent the passage of some toxins such as the bile salts in the intestine. The larger pores, composed of OmpF, may allow more rapid passage of solutes and so confer an advantage in dilute aqueous environments. Accordingly, *E. coli* cells growing in a medium of high osmolarity, such as the human intestine, have more OmpC than OmpF, whereas *E. coli* cells growing in a medium of low osmolarity, such as dilute aqueous solutions, have less OmpC than OmpF.

Other environmental factors besides osmolarity can alter the ratio of OmpC to OmpF. This ratio increases at higher temperatures or when the cell is under oxidative stress due to the accumulation of reactive forms of oxygen (see chapter 11). The ratio also increases when the bacterium is growing in the presence of organic solvents such as ethanol or some antibiotics. Presumably, the smaller size of OmpC pores limits the passage of many toxic chemicals into the cell. The reason for the temperature effect is more obscure, however. There is no obvious reason why the size of the pores should help defend against higher temperatures. One possibility is that the bacterium normally uses a temperature increase as an indication that it has passed from the external environment into the intestine of a warm-blooded vertebrate host, its normal habitat. It then must synthesize mostly OmpC-containing pores to keep out toxic materials such as bile salts, as mentioned above. While not yet completely understood, the regulation of OmpC and OmpR porin synthesis in *E. coli* has served as a model for systems that allow the cell to sense the external environment and adjust their gene expression accordingly; therefore, we shall discuss this subject in some detail.

Genetic Analysis of Porin Regulation

As in the genetic analysis of any regulatory system, the first step in studying the osmotic regulation of porin synthesis in *E. coli* was to identify the genes whose products are involved in the regulation. The isolation of mutants defective in the regulation of porin synthesis was greatly aided by the fact that the some of the porin proteins also serve as receptors for phages and bacteriocins, so that mutants that lack a particular porin will be resistant to a given phage or bacteriocin. This fact offers an easy selection for mutants defective in porin synthesis. Only mutants that lack a certain porin in the outer membrane will be able to form colonies in the presence of the corresponding phage or bacteriocin.

Using such selections, investigators isolated mutants that had reduced amounts of the porin protein, OmpF, in their outer membrane. These mutants were found to have mutations in two different loci, which were named *ompF* and *ompB*. Mutations in the *ompF* locus can completely block OmpF synthesis, whereas mutations in *ompB* only partially prevent its synthesis. The quantitative difference in the effect of mutations in the two loci suggested that *ompF* is the structural gene for the OmpF protein and that an *ompB*-encoded protein(s) is required for the expression of the *ompF* gene. Complementation experiments showed that the *ompB* locus actually consists of two genes, *envZ* and *ompR*. Using *lacZ* fusions to *ompF* to monitor the transcription of the *ompF* gene (see chapter 2), investigators confirmed that EnvZ and OmpR are required for optimal transcription of the *ompF* gene (see Hall and Silvy, Suggested Reading).

EnvZ AND OmpR: A SENSOR-RESPONSE REGULATOR PAIR OF PROTEINS

The *envZ* and *ompR* genes were cloned and sequenced by methods such as those discussed in chapter 1. Similarities in amino acid sequence between EnvZ and OmpR and other sensor-response regulators, including NtrB and NtrC, suggested that these proteins are also a sensor-response regulator pair of proteins. Like many sensor proteins, the EnvZ protein is an inner membrane protein, with its N-terminal domain in the periplasm and its C-terminal domain in the cytoplasm (see the figure in Box 13.3). Current evidence indicates the N-terminal domain of EnvZ apparently senses an unknown signal in the periplasm that indicates that the osmolarity is low and transfers this information to the cytoplasmic domain. The information is then transferred to the OmpR protein, a transcriptional regulator that regulates transcription of the porin genes. Like the NtrB protein, the EnvZ protein is known to be autophosphorylated, and its phosphate is also known to be transferred to OmpR.

An Early Model for *ompC* and *ompF* Regulation

Because of the similarity of NtrB and NtrC to EnvZ and OmpR, respectively, it was tempting to speculate that EnvZ and OmpR might regulate the transcription of *ompC* and *ompF* by the following simple model. According to this model, the EnvZ protein would be an autokinase that phosphorylates itself when the osmolarity is high. This phosphate is then transferred to OmpR. The phosphorylated form of OmpR then activates transcription of *ompC*, whereas the unphosphorylated form of OmpR activates transcription of *ompF*. Therefore, the relative rates of transcription of *ompC* and *ompF* would depend on the state of phosphorylation of OmpR.

Genetic Evidence Fails To Support a Simple Model

Several lines of genetic evidence indicate that the phosphorylated form (OmpR~P) does not simply activate transcription of *ompC* while the unphosphorylated form of OmpR activates the *ompF* gene. The regulation is more complicated; therefore, at this point, we will outline some of the genetic evidence.

1. Mutant phenotypes. Table 13.3 lists several relevant phenotypes. According to the simple model, mutations that totally inactivate the EnvZ protein (*envZ* null mutations) should completely prevent the phosphorylation of OmpR under any conditions, freezing it in the form that activates transcription of *ompF*. Therefore, *envZ* null mutations would be predicted to completely prevent the transcription of *ompC* but allow high-level transcription of *ompF*. However, the evidence shows that while *envZ* null mutations do prevent the transcription of *ompC*, they allow only very limited transcription of *ompF*

TABLE 13.3	Phenotypes of *envZ* and *ompR* mutations
Genotype	**Phenotype**
envZ⁺ ompR⁺	OmpC⁺ OmpF⁺
envZ⁺ ompR1	OmpC⁻ OmpF⁻
envZ(null) *ompR⁺*	OmpC⁻ OmpF⁺⁻ᵃ
envZ⁺ ompR2(Con)	OmpC⁻ OmpF⁺ (low osmolarity)
	OmpC⁻ OmpF⁺ (high osmolarity)
envZ(null) *ompR2*(Con)	OmpC⁻ OmpF⁺ (low osmolarity)
	OmpC⁻ OmpF⁺ (high osmolarity)
envZ⁺ ompR3(Con)	OmpC⁺ OmpF⁻ (low osmolarity)
	OmpC⁺ OmpF⁻ (high osmolarity)
envZ⁺ ompR3(Con)/*envZ⁺ ompR⁺*	OmpC⁺ OmpF⁻ (low osmolarity)
	OmpC⁺ OmpF⁻ (high osmolarity)

ᵃ+ – indicates that OmpF levels are reduced but not eliminated.

(shown as OmpF⁺⁻ in Table 13.3; see Slauch et al., Suggested Reading). The EnvZ protein is apparently required to activate the transcription of *both* porin genes, perhaps because some OmpR~P is required to activate the transcription of both genes.

2. Constitutive mutations in ompR. One type of constitutive mutation, called *ompR2*(Con) in Table 13.3, prevents the expression of *ompC* but causes the constitutive expression of *ompF*, even when coupled with a null mutation in *envZ*. A second type of constitutive mutation, called *ompR3*(Con), causes the constitutive expression of *ompC* but prevents the expression of *ompF*. The existence of these constitutive mutations could have been predicted from the simple model. In analogy to the AraC activator (see chapter 12), the OmpR activator may exist in two forms (see chapter 12), one when it is active and another when it is not. The *ompR3*(Con) mutations could change OmpR into the form in which it normally exists when phosphorylated, even without phosphorylation. It could then activate the transcription of *ompC* but not *ompF*.

3. Diploid analysis. The behavior of the *ompR*(Con) mutations in complementation tests is hard to reconcile with the simple model. According to the simple model, in partial diploids carrying both a constitutive allele and a wild-type allele (*envZ⁺, ompR3*(Con)/*envZ⁺, ompR⁺* in Table 13.3), both *ompC* and *ompF* would be predicted to be expressed at high levels, even at low osmolarity, because the mutant OmpR protein should activate the transcription of *ompC* whereas the wild-type OmpR protein should activate transcription of *ompF*. In genetic terms, the constitutive mutations should be dominant over the wild-type allele for the expression of *ompC* but recessive for the expression of *ompF* under conditions of low osmolarity. Instead, as shown in Table 13.3, *ompF* is not expressed in the partial diploids, even in media of low osmolarity, although *ompC* is constitutively expressed.

The Affinity Model

A more current model for how EnvZ and OmpR regulate the transcription of the porin genes incorporated these and other observations about the regulation. The EnvZ protein is known to have both phosphotransferase and phosphatase activities, allowing it to both donate a phosphoryl group to and remove one from OmpR. Whether the phosphorylated or unphosphorylated form of OmpR predominates depends on whether the phosphotransferase or phosphatase activity of EnvZ is most active. Under conditions of high osmolarity, the phosphotransferase activity predominates and the levels of phosphorylated OmpR (OmpR~P) are high. Under conditions of low osmolarity, the phosphatase activity predominates, and most of the OmpR is unphosphorylated.

To explain how *envZ* null mutations can prevent the optimal transcription of both genes, we must propose that the phosphorylated form of OmpR is required to activate transcription of both the *ompC* and *ompF* genes. But how can the relative levels of phosphorylated OmpR regulate the transcription of *ompC* and *ompF* so that higher levels of OmpR~P allow the transcription of one gene to predominate whereas lower levels allow transcription of the other to predominate? One solution to this question was to propose the existence of multiple binding sites for OmpR~P upstream of the promoters for the *ompF* and *ompC* genes (see Pratt et al., Suggested Reading). Some of these sites bind OmpR~P more tightly than others; in other words, some are high-affinity sites whereas others are low-affinity sites. By binding to these sites, the OmpR~P protein can either activate or repress transcription from the promoters, depending on the position of the binding site relative to the promoter (see chapter 2). However, more recent evidence suggests that the binding affinities of phosphorylate OmpR to *ompF* and *ompC* are too similar to account for the regulation. The molecular mechanism of regulation by OmpR is the subject of current investigation (see Mattison et al., Suggested Reading).

The EnvZ protein is a transmembrane protein that can communicate information about what is happening outside the cell to the internal regulatory pathways. Therefore, studies with this system should further our understanding of how cells sense the external environment. How the EnvZ protein achieves this feat should tell us much about how information is transferred across cellular membranes in general.

MicF
As mentioned, the OmpC/OmpF ratio increases not only when the osmolarity increases but also when the temperature increases or when toxic chemicals including organic solvents such as ethanol are in the medium. A different mechanism is responsible for this regulation. It involves a regulatory RNA called MicF. The *micF* gene is immediately upstream of the *ompC* gene and is transcribed in the opposite direction from *ompC*.

There are many examples of regulatory RNAs in bacteria (see Box 13.5). We have discussed RNA I, which regulates the replication and incompatibility of ColE1 plasmids (see chapter 4). Other regulatory RNAs inhibit the translation of mRNAs such as that for the transposase of the transposon Tn*10*. In general, regulatory RNAs have sequences of nucleotides that are complementary to the sequences of a target RNA, so that they

can inhibit the function of the target RNA by base pairing with it. The double-stranded region thus formed may either mask important sequences in the target RNA or create a substrate for a cellular RNase and thereby cause the target RNA to be degraded. Regulatory RNAs that can base pair with an mRNA and thereby inhibit its translation are called **antisense RNAs** by analogy to the antisense strand of DNA, which has the complementary sequence of the sense, or coding, strand of DNA (see chapter 1).

The MicF RNA is an antisense RNA to the mRNA for OmpF. It is not complementary to the entire OmpF mRNA but only to a short sequence including the translational initiation region (TIR), which contains the Shine-Dalgarno sequence and the AUG initiation codon. By forming double-stranded RNA with the TIR, MicF can prevent access by ribosomes to the TIR and hence translation of OmpF. As a consequence, *E. coli* cells containing higher concentrations of MicF RNA, the situation that prevails when they are growing at higher temperatures or in media containing organic solvents or certain antibiotics, will make less OmpF protein and therefore have a higher OmpC to OmpF ratio.

The cellular levels of MicF RNA increase under certain conditions because the promoter for the *micF* gene contains binding sites for many transcriptional activators. Each seems to work independently and activates transcription from the *micF* promoter under its own particular set of conditions. For example, the transcriptional activator SoxS activates transcription from the *micF* promoter when the cell is under oxidative stress. Another, MarA, activates transcription of *micF* when weak acids or some antibiotics are present. Even OmpR may activate the transcription of *micF*, allowing another level of regulation of OmpF synthesis by osmolarity. Other proteins of unknown function may also bind close to the *micF* promoter and activate or repress transcription of the *micF* gene. The regulation of porin synthesis by osmolarity and other environmental factors is obviously central to cell survival, which is why it is so complicated and involves so many interacting systems.

Regulation of Virulence Genes in Pathogenic Bacteria

The virulence genes of pathogenic bacteria represent another type of global regulon. Virulence genes allow pathogenic bacteria to adapt to their eukaryotic hosts and cause disease. Most pathogenic bacteria express their virulence genes only in the eukaryotic host; somehow, the conditions inside the host turn on the expression of these genes. Virulence genes can be identified because mutations that inactivate them render the bacterium

BOX 13.5

Regulatory RNAs

It is becoming increasingly evident that much of the work of regulation in cells is done by small regulatory RNAs (sRNAs) (see Altuvia and Wagner, below). This mode of regulation is sometimes called "riboregulation" and can occur at many different levels. Some regulatory RNAs are involved in transcriptional regulation. For example, an sRNA plays a role in the pheromone-responsive plasmid transfer in *Enterococcus faecalis,* where the 200-nucleotide mD RNA enhances transcriptional termination in *trans* (see Tomita and Clewell, below). Other sRNAs might bind to proteins and regulate them, for example a small RNA that binds to RNA polymerase (see Wasserman and Storz, below). However, most sRNAs regulate gene expression posttranscriptionally as antisense RNAs, and their activities depend upon base-pairing interactions with their targets. We have already discussed some antisense RNAs in connection with the regulation of plasmid replication. Others are known to exist in phages and transposons. In these cases, the antisense RNA-encoding regions overlap with the DNA regions encoding the target of their regulation but are transcribed from the other strand and therefore are called counter-transcribed RNAs (ctRNAs). For ctRNAs, the complementarity between the antisense and target RNAs is exact and may extend as far as 100 nucleotides. Because of their extensive complementarity, usually only one target is regulated by a particular ctRNA.

Antisense ctRNAs can also exist in the chromosome of bacteria. An example of a chromosomal ctRNA that is countertranscribed from its regulatory target is the Crp Tic RNA, which overlaps the *crp* gene of *E. coli,* encoding the activator CAP protein for catabolite-sensitive genes. Many chromosomal antisense RNA genes do not overlap their target genes, however, and may even be distantly located on the chromosome. For these RNAs, the region of complementarity is short, less than 12 bp. Because of such short interactions, a single antisense RNA may be able to regulate more

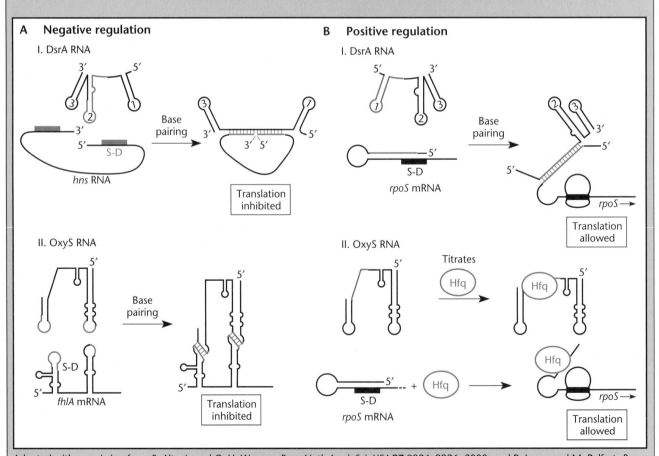

Adapted with permission from S. Altuvia and G. H. Wagner, *Proc. Natl. Acad. Sci. USA* **97:**9824–9826, 2000, and R. Lease and M. Belfort, *Proc. Natl. Acad. Sci. USA* **97:**9919–9924, 2000. Copyright 2000 National Academy of Sciences, U.S.A.

(continued)

BOX 13.5 (continued)

Regulatory RNAs

than one target gene, with different regions of the sRNA base pairing with the various target sequences.

The DsrA and OxyS sRNAs are examples of antisense RNAs that regulate multiple targets. The DsrA and OxyS RNAs act in *trans* and can either increase or decrease the expression of the target gene depending on the specific situation (see the figure which is adapted with permission from Altuvia and Wagner [2000] and Lease and Belfort [2000] on page 458). The DsrA RNA has at least two targets: *hns*, whose product is a histone-like silencer of genes; and *rpoS*, which encodes the stationary-phase sigma factor σ^S (see Table 13.1). Short base-pairing interactions of DsrA with its target mRNAs in their Shine-Dalgarno regions either inhibit or allow translation of the mRNA (see Lease and Belfort, below). Interaction of DsrA with its target mRNA also affects the stability of the target mRNA, either decreasing it, as in the case of *hns* mRNA stability, or increasing it, as in the case of *rpoS* mRNA. The Figure illustrates the binding of DsrA and its targets via different regions of DsrA.

The OxyS sRNA, which is induced in response to oxidative stress, also has multiple targets of regulation. Like DsrA, it regulates translation of the *rpoS* mRNA. A second target is *fhlA*, a transcriptional activator of formate metabolism. The mechanism of regulation of *fhlA* expression is similar to that of other sRNAs in that OxyS binds to the *fhlA* translation initiation region, inhibiting ribosome binding. However, its mechanism of regulation of *rpoS* is more indirect. OxyS does not itself bind to the *rpoS* mRNA but, rather, sequesters a positive regulator of *rpoS* translation, the Hfq protein. Hfq is needed for *rpoS* translation because it binds the *rpoS* leader sequence, exposing the Shine-Dalgarno sequence. Hfq also binds to and stabilizes other RNAs including the DsrA sRNA described above. However, Hfq binding to some mRNAs can also have the opposite effect, decreasing their translation and destablizing them.

Previously, sRNAs were found the hard way, by genetic analysis of the regulation of a gene. Now, however, new sRNAs can be found by comparing the genomes of closely related bacteria and microarray analysis (see the text). For example, sRNAs have been found for *E. coli* by comparing the *E. coli* genome to the genomes of other closely related bacteria, *Klebsiella pneumoniae* and *Salmonella* spp., using a combined computational and microarray approach (see Wassarman et al., below). The reasoning is that the sequences of intergenic regions, which normally differ, will be conserved if they encode important regulatory sRNAs. The criteria used for identifying candidate sRNA-encoding genes included a recognizable promoter and a ρ-independent terminator. Candidates could then be tested directly by Northern blotting. Short ORFs which potentially encode peptides were also found this way, but many of these could be excluded from the potential sRNA pool on the basis of ORF-predicting programs that use criteria such as codon usage and higher variation in the third base of putative codons. Seventeen potential new sRNAs were found this way.

References

Altuvia, S., and G. H. Wagner. 2000. Switching on and off with RNA. *Proc. Natl. Acad. Sci. USA.* **97:**9824–9826. (Commentary.)

Lease, R., and M. Belfort. 2000. A *trans*-acting RNA as a control switch in *Escherichia coli*: DsrA modulates function by forming alternative structures. *Proc. Natl. Acad. Sci. USA* **97:**9919–9924.

Tomita, H., and D. B. Clewell. 2000. A pAD1-encoded small RNA molecule, mD, negatively regulates *Enterococcus faecalis* pheromone response by enhancing transcription termination. *J. Bacteriol.* **182:**1062–1073.

Wassarman, K. M., and G. Storz. 2000. 6S RNA regulates *E. coli* RNA polymerase activity. *Cell* **101:**613–623.

Wassarman, K. M., F. Repoila, C. Rosenow, G. Storz, and S. Gottesman. Identification of novel small RNAs using comparative genomics and microarrays. *Genes Dev.* **15:**1637–1651.

nonpathogenic but do not affect its growth outside the host. We discuss some examples of the regulation of virulence regulons in this section.

Diphtheria

Diphtheria is caused by the bacterium *Corynebacterium diphtheriae*, a gram-positive bacterium that colonizes the human throat. It is spread from human to human through aerosols created by coughing or sneezing. The colonization of the throat by itself results in few symptoms. However, strains of *C. diphtheriae* that harbor a prophage named β (see chapter 8) produce diphtheria toxin, which is responsible for most of the symptoms.

Excreted from the bacteria in the throat, the toxin enters the bloodstream, where it does its damage.

DIPHTHERIA TOXIN

The diphtheria toxin is a member of a large group of A-B toxins, so named because they have two subunits, A and B. In most A-B toxins, the A subunit is an enzyme that damages host cells and the B subunit helps the A subunit enter the host cell by binding to specific cell receptors. The two parts of the diphtheria toxin are first synthesized from the *tox* gene as a single polypeptide chain, which is cleaved into the two subunits A and B as it is excreted from the bacterium. These two subunits are held together by a disulfide bond until they are translocated into the host cell, where the disulfide bond is reduced and broken, releasing the individual A subunit into the cell.

The action of the diphtheria toxin A subunit on eukaryotic cells is quite well understood. The A subunit enzyme specifically ADP-ribosylates (adds ADPribose to) a modified histidine amino acid of the translation elongation factor EF-2 (called EF-G in bacteria; see chapter 2). The ADP-ribosylation of the translation factor blocks translation and kills the cell. Interestingly, the opportunistic pathogen *Pseudomonas aeruginosa* makes a toxin that is identical in action to the diptheria toxin, although it has a somewhat different sequence.

Regulation of the *tox* Gene

Like many virulence genes, the *tox* gene is turned on only under the conditions present in the infected organism. One factor that characterizes the eukaryotic environment is the lack of available iron. All of the iron in the body is tied up in other molecules, such as transferrins and hemoglobin. The bacterium senses this lack of free iron, concludes that it is in the eukaryotic host, and turns on the *tox* gene and other genes required for pathogenesis.

Interestingly, even though the *tox* gene encoding the toxin of *C. diphtheriae* is carried on the lysogenic phage β, it is regulated by the product of a chromosomal gene called *dtxR,* as shown in Figure 13.12 (see Schmidt and

Figure 13.12 Regulation of the *tox* gene of the *Corynebacterium diphtheriae* prophage β. The DtxR repressor protein binds to the operator for the *tox* gene only in the presence of ferrous ions (Fe^{2+}).

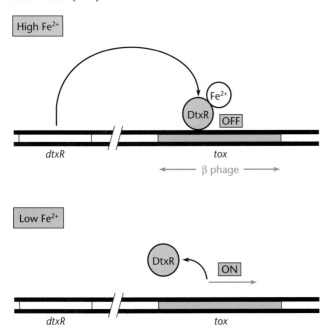

Holmes, Suggested Reading). The DtxR protein is a transcriptional repressor that binds to the operator region for the *tox* gene only in the presence of ferrous ions (Fe^{2+}) and some other divalent cations. Therefore, the *tox* gene is expressed only when ferrous ions are limiting, the condition that prevails in the eukaryotic host. This is an interesting example of close cooperation between normal chromosomal genes of a bacterium and genes encoded by a foreign DNA element.

IRON REGULATION

Iron is not only a signal of the internal eukaryotic environment but also an essential nutrient for bacteria. Thus, iron limitation presents a problem for all bacteria but for pathogenic bacteria in particular. To multiply in a eukaryotic host, pathogenic bacteria must extract the iron from the transferrins and other proteins to which it is bound. For this purpose, *C. diphtheriae* and many other bacteria use siderophores, which are small molecules excreted from the cells. Siderophores bind Fe^{2+} more tightly than do other molecules and so can wrest Fe^{2+} from them. They then are transported back into the bacterial cell with their "catch" of Fe^{2+}.

The genes for making the siderophores and a high-efficiency transport system for iron are also regulated by the DtxR protein in *C. diphtheriae*. Therefore, these components are synthesized only when iron is limiting.

THE Fur REGULON OF *E. COLI*

The DtxR protein of *C. diphtheriae* is highly homologous to a protein called Fur, which is responsible for iron regulation in *E. coli*. Over 30 *E. coli* genes, including toxin genes in pathogenic strains of *E. coli*, are regulated by Fur. Like DtxR, Fur is an iron-dependent repressor that binds to operators only in the presence of Fe^{2+}. Genes highly homologous to the *E. coli fur* repressor gene and the *dtxR* gene of *C. diphtheriae* have been found in many other bacteria, including *P. aeruginosa, Yersinia pestis,* and *Vibrio cholerae,* and presumably also encode iron-dependent repressors.

Cholera

The disease cholera is another well-studied example of the global regulation of virulence genes. *V. cholerae,* the causative agent, is a gram-negative bacterium that is spread through water contaminated with human feces. The disease continues to be a major health problem worldwide, with periodic outbursts, especially in countries with poor sanitation. When ingested by a human, *V. cholerae* colonizes the small intestine, where it synthesizes cholera toxin, which acts on the mucosal cells to cause a severe form of diarrhea. Like the diphtheria toxin, the cholera toxin is encoded by a lysogenic phage, in this case a relative of the single-stranded DNA phage

M13 (see chapter 8). Other virulence determinants are the flagellum that allows the bacterium to move in the mucosal layer of the small intestine and pili that allow it to stick to the mucosal surface.

CHOLERA TOXIN

The mechanism of action of the cholera toxin has been the subject of intense investigation, in part because of what it reveals about the normal action of eukaryotic cells. The cholera toxin is composed of two subunits, CtxA and CtxB. Like diphtheria toxin, the CtxA subunit of cholera toxin is an ADP-ribosylating enzyme. However, rather than ADP-ribosylating an elongation factor for translation, the cholera toxin ADP-ribosylates a mucosal cell membrane protein called Gs, which is part of a signal transduction pathway that regulates the activity of the adenylcyclase enzyme that makes cAMP. The ADP-ribosylation of Gs causes cAMP levels to rise and alters the activity of transport systems for sodium and chloride ions. The loss of sodium and chloride ions causes loss of water from the cells, resulting in severe diarrhea.

Regulation of Cholera Toxin Synthesis

The *ctxA* and *ctxB* genes encoding the cholera toxin are part of a large regulon including as many as 20 genes. In addition to the *ctx* genes, the genes of this regulon include those encoding pili, colonization factors, and outer membrane proteins related to osmoregulation. Although some of these genes, including the *ctx* genes, are carried on the prophage, others, including the pilin genes, are carried on the bacterial chromosome. The transcription of the genes of this regulon are activated only under conditions of high osmolarity and in the presence of certain amino acids, conditions that may mimic those in the small intestine. Here we describe the cascade of genes involved in *ctx* regulation.

1. ToxR-ToxS. The cholera virulence regulon was first found to be under the control, either directly or indirectly, of the activator protein ToxR, the product of the *toxR* gene (see Yu and DiRita, Suggested Reading). The ToxR protein combines elements that are somewhat related to both proteins of the two-component sensor and response regulator type of system in the same polypeptide. The ToxR polypeptide traverses the inner membrane so that the carboxy-terminal part of the protein is in the periplasm, where it may sense the external environment. The amino-terminal part is in the cytoplasm, where it contains an OmpR-like DNA-binding domain that can activate the transcription of genes under its control.

However, the ToxR protein is not activated by phosphorylation, unlike orthodox response regulator proteins. A clue to its activation lies with another protein, ToxS. The *toxS* gene is immediately downstream of *toxR* in the same operon, and the gene was discovered because mutations that inactivate it also prevent expression of the genes of the ToxR regulon. The ToxS protein is also anchored in the inner membrane, but a large domain sticks out into the periplasm.

One model for how ToxS might activate ToxR came from the experiments designed to investigate the membrane topology of ToxR. Fusion proteins composed of the reporter gene PhoA fused to both the N-terminal and C-terminal portions of ToxR were created. The PhoA protein is an alkaline phosphatase enzyme that must dimerize in the periplasm to be active, and fusions to PhoA can be used to determine whether a particular domain in a protein is in the cytoplasm or in the periplasm (see chapter 14). The results revealed that the N-terminal part of the ToxR protein is in the cytoplasm while the C-terminal part in in the periplasm. A surprising result was that some ToxR-PhoA fusions exhibited ToxR activation of transcription independent of ToxS. It was hypothesized that the PhoA part of these fusions was dimerizing in the periplasm, driving the dimerization of the ToxR portion and activating transcription. This suggested that the normal function of ToxS is to dimerize ToxR, which is not needed in the fusions.

Other evidence suggests that dimerization, by itself, may not be enough to activate the ToxR protein. Some feature related to the membrane anchoring of ToxR may also be required. If dimerization were sufficient, attaching other dimerization domains, such as the ones from the CI repressor protein of λ phage, to the cytoplasmic domain of the ToxR protein should cause ToxR to dimerize in the cytoplasm and activate transcription. However, fusion proteins composed of ToxR and other dimerization domains are still inactive unless the ToxR protein retains its transmembrane domain. Perhaps the ToxR protein must be at least partly in the membrane to be active and ToxS plays some other role in stabilizing ToxR.

2. ToxT. As mentioned, *V. cholerae* has many virulence genes besides the toxin genes that are also considered part of the ToxR regulon, since *toxR* mutations prevent their expression. However, most of the ToxR regulon genes are not directly controlled by the ToxR protein but, rather, are controlled by an intermediary, the ToxT protein, which is a member of the AraC family of activators (see Box 12.2). Transcription of the *toxT* gene is activated by ToxR, and the transcription of the other genes is then activated by ToxT. Therefore, this is a regulatory cascade.

The *toxT* gene is located on *V. cholerae* pathogenicity island (VPI). The genes for additional regulators of *toxT* transcription, *tcpP* and *tcpH,* are also located on VPI.

Also, inner membrane proteins, TcpP and TcpH function as the transcriptional regulator and effector, respectively. As with the ToxR-ToxS pair, the mechanism by which TcpH activates TcpP, which in turn acts as the transcriptional activator of *toxT*, is not known. Transcription of the *tcpP-tcpH* operon responds to environmental cues, but we do not understand how these two genes, together with the *toxR-toxS* genes, transduce environmental signals that the bacterium is in the intestine of its host into activation of ToxT. A general outline of the various regulatory pathways to turn on the genes required for *V. cholerae* pathogenicity is given in Figure 13.13. Obviously, much more needs to be done before we can begin to understand this important model system of bacterial pathogenicity.

CLONING THE *toxR* GENE

The current level of understanding of the virulence regulon in *V. cholerae* would not have been possible without clones of the *toxR* gene (see Miller and Mekalanos, Suggested Reading). When this work was begun, little was known except that mutations in a gene of *V. cholerae* called *toxR* caused the cells to produce little toxin. This suggested that the *toxR* gene encoded a positive regulatory protein that was required to activate transcription of the toxin genes.

For convenience, the cloning was attempted in *E. coli* rather than in *V. cholerae*. As discussed in chapters 1 and 4, many cloning vectors have been developed for *E. coli*,

and cloning techniques are much more advanced for this species than for other bacteria. The investigators had reason to hope that the *toxR* gene would be expressed and function in *E. coli*, since *V. cholerae* and *E. coli* are fairly closely related and since most *V. cholerae* genes that had been tested are expressed well in *E. coli*.

Their strategy was to screen a library (see chapter 1) for clones that expressed a protein that could activate transcription from the promoter for the toxin genes, the *ctxA* promoter, in *E. coli*. They first made a partial *Sau*3A library of wild-type *V. cholerae* DNA in an *E. coli* plasmid cloning vector. This library was then transformed into a strain of *E. coli* harboring a lysogenic phage carrying a transcriptional fusion of *ctxA* to a reporter gene, *lacZ*. The *ctxA* and *lacZ* genes were arranged so that the *lacZ* gene would be transcribed only from the *ctxA* promoter. The investigators then plated the transformants on agar containing X-Gal and observed whether the cells synthesized β-galactosidase, the product of the *lacZ* gene. As mentioned in other chapters, X-Gal turns blue if it is cleaved by β-galactosidase. Most transformants made only faintly blue colonies. However, 2 of about 5,000 transformants made deep blue colonies. The clones in these transformants were presumed to contained the *toxR* gene, which was activating transcription from the *ctxA* promoter.

Further confirmation that the clones contained the *toxR* gene came from complementation tests in which plasmids containing the clones were mobilized into *toxR* mutants of *V. cholerae* with a ColE1-derived plasmid

Figure 13.13 Regulatory cascade for *Vibrio cholerae* virulence factors. The ToxR-ToxS activator-effector pair directly regulate *omp* virulence factors and the *toxT* regulatory gene located on a *V. cholerae* pathogenicity island (VPI) (indicated in gold). The VPI-encoded TcpP-TcpH activator-effector pair also regulates *toxT* transcription. ToxT activates the Ctx prophage-encoded *ctxAB* toxin genes and the toxin-coregulated pilus (TCP) pilus genes. ToxT also positively regulates its own expression from the promoter for *tcpA–F* transcription.

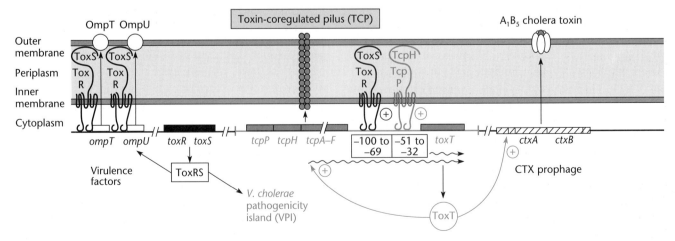

cloning vector, so that it could be mobilized by the F-plasmid Tra functions (see chapter 5). The narrow-host-range ColE1-derived plasmids can nevertheless replicate to some extent in *V. cholerae*, since, as mentioned, *V. cholerae* is closely related to *E. coli*. Because the plasmid clones complemented *toxR* mutations in *V. cholerae* to allow normal synthesis of the toxin, they presumably contained the *toxR* gene. With clones of the *toxR* gene, it was also possible to construct PhoA fusions to investigate the membrane topology of the ToxR protein (see above).

Whooping Cough

Another well-studied disease involving global regulation of virulence genes is whooping cough, caused by the gram-negative bacterium *Bordetella pertussis*. Whooping cough is mainly a childhood disease and is characterized by uncontrolled coughing, hence the name. The bacteria colonize the human throat and are spread through aerosols resulting from the coughing. Effective vaccines have been developed, but the disease continues to kill hundreds of children worldwide, mainly in areas where the vaccines are not available.

In spite of their very different symptoms, the diseases caused by *V. cholerae* and *B. pertussis* have a similar molecular basis. *B. pertussis* also makes a complex A-B toxin (pertussis toxin) that is in some ways remarkably similar to the cholera toxin. The pertussis toxin has six subunits, although only two of them are identical. One of the subunits (S1) is the enzyme, while the others (S2 to S5) are involved in adhesion to the mucosal surface of the throat. The pertussis toxin is also similar to the cholera toxin in that it ADP-ribosylates a G protein in a signal transduction pathway involved in deactivating the adenylate cyclase, leading to elevated levels of cAMP. However, rather than causing a loss of water from the cells, the elevated cAMP levels seem to cause an increase in mucus production, presumably because the pertussis toxin attacks cells of the throat rather than of the small intestine.

In addition to pertussis toxin, *B. pertussis* also synthesizes a number of other toxins. One is an adenylate cyclase enzyme that enters host cells and presumably directly increases intracellular cAMP levels by synthesizing cAMP. This observation supports the importance of increased cAMP levels to the pathogenesis of the bacterium, although the contribution of this adenylate cyclase to the symptoms is unknown. Other known toxins include one that causes necrotic lesions on the skin of mice and a cytotoxin that is not a protein but a peptidoglycan fragment and that kills ciliated cells of the throat. Other virulence factors are involved in the adhesion of the bacterium to the mucosal layer. The pertussis toxin itself may play a role in such adhesion.

REGULATION OF PERTUSSIS VIRULENCE GENES

Like the virulence genes of *C. diphtheriae* and *V. cholerae*, many of the virulence genes of *B. pertussis* are presumably expressed only when the bacterium enters the eukaryotic host and others are repressed. The regulation of the virulence genes of *B. pertussis* is achieved by a sensor-response regulator pair of proteins encoded by linked genes, *bvgA* and *bvgS* (for *b*ordetella *v*irulence *g*enes); *bvgS* encodes the sensor, and *bvgA* encodes the transcriptional activator.

The BvgS-BvgA system is similar to many other sensor-response regulators in that the BvgS protein is a transmembrane protein, with its N terminus in the periplasm and its C terminus in the cytoplasm, allowing it to communicate information from the external environment across the membrane to the inside of the cell. Also, like many other sensor proteins that work in two-component systems, the BvgS protein is autophosphorylated in response to some signal from the external environment and donates this phosphate to the BvgA protein, which then activates transcription of the virulence genes. One difference between BvgS-BvgA and other sensor-response regulator pairs is that in the BvgS sensor protein, other amino acids are phosphorylated in addition to the histidine that is autophosphorylated in the other sensors. The significance of these other sites of phosphorylation for the activation of the pertussis virulence genes is not clear.

In vitro, the signal transduction pathway to transcribe the pertussis toxin gene can be activated by growing the bacteria in media with low nicotinamide and magnesium concentrations and possibly low iron concentrations, conditions that presumably mimic those in the throat. Also, expression of the pertussis toxin is highest at 37°C, the temperature of the human body. At first, the BvgA protein seemed to be both a transcriptional activator and a repressor, since the signal transduction pathway seems to activate some virulence genes and repress others. It is now known that it activates the transcription of a repressor gene, bvgR, immediately downstream of itself and that this repressor represses other genes (see Boucher et al., Suggested Reading).

CLONING THE *bvgA-bvgS* OPERON

As with the ToxR regulon, the results described above would not have been possible without clones of the *bvg* gene region. The *bvg* locus, where the two genes lie, was first found by Tn5 transposon insertion mutations that prevent the synthesis of many of the virulence factors of *B. pertussis* (see Stibitz et al. [1988], Suggested Reading). By cloning the kanamycin resistance gene on Tn5 and the flanking chromosomal sequences, the investigators constructed a clone of the *bvg* region. This clone was then used as a probe in colony hybridizations to screen a library made from wild-type *B. pertussis* DNA to identify bacteria containing the wild-type *bvg* locus.

When the *bvg* locus was sequenced, it was found to contain an operon with two open reading frames (ORFs), one of which encoded a protein with sequence homology to sensor autokinases and so was presumably a gene encoding a sensor autokinase. The other ORF encoded a protein with a DNA-binding motif and other features of a transcriptional activator and so presumably encoded one of these proteins. The genes were later named *bvgS* and *bvgA*, respectively.

Clones of the *bvg* genes facilitated the genetic analysis of the activation mechanism. Some mutations in *bvgS* cause the constitutive expression of the virulence genes, even in high concentrations of nicotinamide and magnesium. Also, deletions of the N-terminal periplasmic domain of the protein forestall the requirement for conditions of low nicotinamide and magnesium concentrations to activate the regulon. Apparently, these mutations change BgvS so that it no longer requires a signal from the environment to be autophosphorylated.

PHASE VARIATION OF *BORDETELLA* VIRULENCE

B. pertussis cells spontaneously lose the ability to express many of their virulence factors at a frequency of about 1 in 10^6, which is fairly high for spontaneous mutations. The loss of virulence is reversible, and some bacteria also regain the ability to make the virulence factors at a fairly high frequency.

The switch from virulence to nonvirulence is often called **phase variation**, by analogy to the phenomenon in *S. enterica* serovar Typhimurium (see chapter 9). Like *S. enterica* serovar Typhimurium cells, which periodically express different flagellin proteins, *B. pertussis* cells may undergo phase variation to avoid the host immune system. Many of the virulence factors, such as the proteins for flagella and pili, are strong antigens, and the reversible loss of the ability to make these virulence factors may help the bacteria establish themselves in the host.

Figure 13.14 illustrates the mechanism of phase variation in *B. pertussis*. Rather than result from inversion of a DNA segement as in *S. enterica* serovar Typhimurium, phase variation in *B. pertussis* is caused by a mutation that inactivates the *bvgS* gene, thereby blocking synthesis of all the virulence factors (see Stibitz et al. [1989], Suggested Reading). These mutations add a GC base pair in a string of six GC base pairs in the *bvgS* gene, causing a frameshift and inactivating the product of the *bvgS* gene. The *bvgS* mutations are reversible because the activity of the gene can be restored by a second mutation that removes a GC base pair somewhere in the

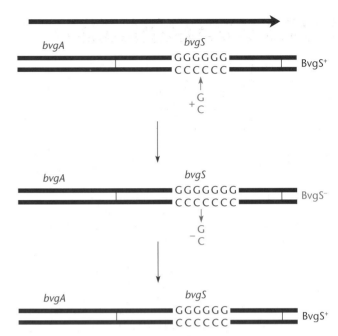

Figure 13.14 Operon structure of the *bvg* locus of *Bordetella pertussis* and the molecular basis for phase variation. The toxin genes of *B. pertussis* are regulated by a two-component system consisting of the sensor (BvgS) and the response regulator (BvgA). The phase variation that causes the reversible loss of virulence determinants results from a frameshift mutation in a string of six GC base pairs in the *bvgS* gene. Adapted from J. F. Miller, S. A. Johnson, W. J. Black, D. T. Beattie, J. J. Mekalanos, and S. Falkow, *J. Bacteriol.* **174**:970–979, 1992.

same string of GC base pairs, restoring the original sequence of the gene. The sequence around the string of GCs in *bvgS* may make the region a hot spot for frameshift mutations, thereby increasing the frequency of the phase shift.

The Heat Shock Regulon

The heat shock regulon is one of the most extensively studied regulons in bacteria and other organisms. One of the major challenges facing cells is to survive abrupt changes in temperature. To adjust to abrupt increases, cells induce at least 30 different genes encoding proteins called the **heat shock proteins** (**Hsps**). The concentrations of these proteins quickly increase in the cell after a temperature upshift and then slowly decline, a phenomenon known as the **heat shock response**. Besides being induced by abrupt increases in temperature, the heat shock genes are induced by other types of stress, such as the presence of ethanol and other organic solvents in the medium or DNA damage. Therefore, these proteins are more of a general stress response rather than a specific response to an abrupt increase in temperature. Nevertheless, the name heat shock has stuck, and we will use it here.

Unlike most shared cellular processes, the heat shock response was observed in cells of higher organisms long before it was seen in bacteria. Some of the Hsps are remarkably similar in all organisms on Earth and presumably play similar roles in protecting cells against heat shock. The mechanism of regulation of the heat shock response may also be similar in organisms ranging from bacteria to higher eukaryotes (see Craig and Gross, and Mager and de Kruijff, Suggested Reading).

Heat Shock Response in *E. coli*

The molecular basis for the heat shock response is better understood in *E. coli* than in any other organism. In *E. coli*, the genes for the Hsps form two regulons, the **heat shock regulons**. The major one, the σ^{32} regulon, includes about 30 genes encoding 30 different Hsps. The minor regulon, the σ^E regulon, encodes only three proteins, and we do not discuss it in detail here.

The functions of many Hsps are known (see Table 13.1 and chapter 2). Most Hsps play roles during the normal growth of the cell and so are always present at low concentrations, but after a heat shock, their rate of synthesis increases markedly and then slowly declines to normal levels.

Some Hsps, including GroEL, DnaK, DnaJ, and GrpE, are chaperones that direct the folding of newly translated proteins (see chapter 2). The names of these proteins do not reflect their function but, rather, how they were orginally discovered. For example, DnaK and DnaJ were found because they affect the assembly of a protein complex required for λ phage DNA replication, but they themselves have little to do with replication. Chaperones may help the cell survive a heat shock by binding to proteins denatured by the abrupt rise in temperature and either helping them to refold properly or targeting them for destruction. As mentioned, the chaperones are among the most highly conserved proteins in cells, being largely unchanged from bacteria to humans.

Other Hsps, including Lon and Clp, are proteases, which may degrade proteins that are so badly denatured by the heat shock that they are irreparable and so are best degraded before they poison the cell. Some other Hsps are proteins normally involved in protein synthesis, including special aminoacyltransferases that are induced after a heat shock. The function of this type of Hsp in protecting the cell after a temperature rise is not clear.

Knowing that many Hsps are involved in helping proteins fold properly or in destroying denatured proteins helps explain the transient nature of the heat shock response. Immediately after the temperature increases, the concentrations of salts and other cellular components are not appropriate for protein stability at the higher temperature, leading to massive protein unfolding. Later, after the temperature has been elevated for some time, the internal conditions will have had time to adjust, and so proteins will not continue to be denatured and the increased numbers of chaperones and other Hsps will no longer be necessary. Hence, the synthesis of the Hsps declines.

Genetic Analysis of the *E. coli* Heat Shock Regulon

As with other regulatory systems, the analysis of the heat shock response was greatly aided by the discovery of mutants with defective regulatory genes. The first of such mutants was found in a collection of temperature-sensitive mutants. It was later shown to be unable to induce the Hsps after a shift to high temperature (see Zhou et al., Suggested Reading). This mutant, which failed to make the regulatory gene product, made it possible to clone the regulatory gene by complementation. A library of wild-type *E. coli* DNA was introduced into the mutant strain, and clones that permitted the cells to survive at high temperatures were isolated. When the sequence of cloned gene for the regulatory Hsp was compared with that of other proteins, it was found to encode a new type of sigma factor, σ^{32}, and was named the heat shock sigma. The RNA polymerase holoenzyme with σ^{32} attached recognizes promoters for the heat shock genes that are different from the promoters recognized by the normal σ^{70} and the nitrogen sigma, σ^{54} (Figure 13.10). The gene for the heat shock sigma was named *rpoH* for RNA *p*olymerase subunit *h*eat shock.

Regulation of σ^{32} Synthesis

Normally, very few copies of σ^{32} exist in the cell. However, immediately after an increase in temperature from 30 to 42°C, the amount of σ^{32} in the cell increases 15-fold. This increase in concentration leads to a significant rise in the rate of transcription of the heat shock genes, since they are transcribed from σ^{32}-type promoters. Understanding how the heat shock genes are turned on after a heat shock requires an understanding of how this increase in the amount of σ^{32} occurs.

An abrupt increase in temperature might increase the amount of σ^{32} through several mechanisms. One possibility is that the *rpoH* gene for σ^{32} is transcriptionally autoregulated. According to this hypothesis, the cell would normally contain a small amount of σ^{32}, which is somehow activated after heat shock, and this activated σ^{32} would direct more of the RNA polymerase to the *rpoH* gene, leading to the synthesis of more σ^{32}, and so forth. For this hypothesis to be correct, the *rpoH* gene would have to be strongly transcriptionally regulated. However, although the rate of transcription of the *rpoH* gene increases slightly after a heat shock, it does not increase enough to explain the large rise in σ^{32} levels. Moreover, if the *rpoH* gene is transcriptionally autoregulated, at least one of the promoters servicing the *rpoH* gene should be of the type that uses the heat shock sigma. However, of the four different promoters from which *rpoH* is transcribed, none are recognized by σ^{32}. Three promoters are used by RNA polymerase with the normal σ^{70}, and another is probably used by RNA polymerase with another type of sigma factor, σ^E (also called σ^{24}), which is more active at higher temperatures, perhaps explaining the slight increase in *rpoH* transcription after a heat shock.

Because the transcription of the *rpoH* gene does not increase significantly after heat shock, the amount of σ^{32} in the cell must be posttranscriptionally regulated. In fact, immediately after temperature upshift, σ^{32} stability increases markedly and the translation rate of the *rpoH* mRNA increases 10-fold.

DnaK: THE CELLULAR THERMOMETER?

Recent evidence indicates that at least some of the posttranscriptional regulation of σ^{32} levels after a heat shock is due to the protein chaperones DnaK and DnaJ. These chaperones normally bind to nascent proteins in the process of being synthesized and help them to fold properly (see chapter 2). Under heat shock conditions, they can also bind to denatured proteins and help them refold.

The protein-binding ability of these Hsps may allow them to indirectly regulate their own synthesis and that of the other Hsps. According to this hypothesis, at least one of these Hsps is a "cellular thermometer" that senses the change in temperature and induces the transcription of the heat shock genes. The Hsp DnaK is the prime suspect for the cellular thermometer, and so we will assume that it is responsible, although DnaJ and GrpE could also be involved (see Craig and Gross, Suggested Reading).

Figure 13.15 shows a model for how the protein-binding ability of the DnaK chaperone may indirectly regulate the synthesis of the heat shock proteins. One of the proteins to which DnaK binds is σ^{32} (see Liberek et al., Suggested Reading). By binding to σ^{32}, the DnaK protein regulates the transcription of the heat shock genes in two ways. First, it affects the stability of σ^{32}. The σ^{32} protein with DnaK bound is more susceptible to a cellular protease called FtsH than is free σ^{32} (see Blaszczak et al., Suggested Reading). Therefore, DnaK can cause σ^{32} to be degraded almost as soon as it is made, keeping the amount of σ^{32} at low levels and therefore reducing the transcription of the heat shock genes. Second, DnaK inhibits the activity of σ^{32}. With DnaK protein bound, σ^{32} may be less active in transcription, either because it is less able to bind to RNA

polymerase or because the complex of RNA polymerase, σ^{32}, and DnaK is less able to bind to the heat shock promoters. By inhibiting the activity of σ^{32}, the DnaK protein will lower the transcription of the heat shock genes.

According to the model, the release of DnaK inhibition of σ^{32} activity is responsible for the increase in transcription of the heat shock genes after an abrupt increase in temperature. However, this seems to be backward. The concentration of DnaK increases after a heat shock, but its effects on σ^{32} activity are all negative. How could the increase in a protein that inhibits the activity of σ^{32} lead to an increase in the transcription of the heat shock genes?

The answer to this dilemma lies in the chaperone role of DnaK; that is, in addition to binding to the σ^{32} protein, DnaK binds to denatured proteins to help them refold properly. As mentioned, salt concentrations and other conditions in the cell influence the stability of proteins, and the optimal conditions for protein stability may change with the temperature. If the temperature increases too rapidly, the cellular conditions may not have time to adjust, and many proteins may be denatured. In this case, most of the DnaK protein in the cell will be commandeered to help renature the unfolded proteins, leaving less DnaK available to bind to σ^{32}. The σ^{32} protein will then be more stable and will accumulate in the cell. It will also be more active, increasing the transcription of the heat shock genes, including the *dnaK* gene itself.

This model also explains the transient nature of the heat shock response, in which the concentration of the Hsps slowly declines after a temperature increase. When enough DnaK has accumulated to bind to all the unfolded proteins and internal conditions have adjusted so that no more proteins are denatured at the higher temperature, extra DnaK will once again be available to bind to and inhibit σ^{32}, leading to the observed drop in the rate of synthesis of the Hsps.

The σ^{32} protein may also be **translationally autoregulated**, in other words, able to repress its own translation. Generally, such proteins bind to their own TIR in the mRNA, thereby blocking access by ribosomes. After an upshift in temperature, the translational repression of σ^{32} synthesis is less complete but then returns, suggesting that DnaK (and DnaJ) may be involved. The translational repression of σ^{32} synthesis also involves secondary structures that are formed in the σ^{32} mRNA at lower temperatures and removed at higher temperatures.

In addition to sensing a shift in temperature indirectly, through the presence of denatured proteins in the cell, the DnaK protein might be able to sense the temperature more directly, since its ATPase is much more active at higher temperatures. It is interesting how the protein binding and other activities of some of the Hsps may be used to regulate their own synthesis as well as that of other Hsps.

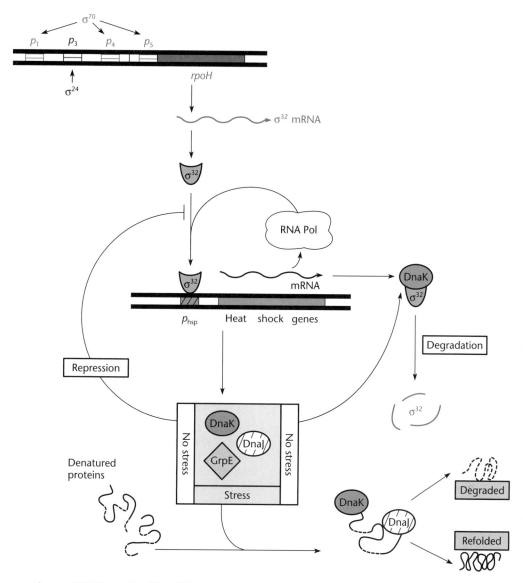

Figure 13.15 Role of DnaK in the induction of the heat shock genes after cells are exposed to an abrupt increase in temperature or other type of stress. The σ^{32} with DnaK bound is susceptible to a protease and is quickly degraded. In addition, σ^{32} with DnaK bound is less active for the initiation of transcription from the heat shock promoters. After an abrupt increase in temperature, many other proteins are denatured, and DnaK binds to these to help them refold properly. This action frees σ^{32}, stabilizing it and making it more active for transcription initiation. When the cell adjusts to the higher temperature and DnaK accumulates to the point where some is again available to bind to σ^{32}, the activity of σ^{32} in the cell again drops, and the transcription of the heat shock genes returns to basal levels. Adapted from W. H. Mager and A. J. J. de Kruijff, *Microbiol. Rev.* **59**:506–531, 1995.

Regulation of Ribosome and tRNA Synthesis

To compete effectively in the environment, cells must make the most efficient use possible of the available energy. One of the major ways in which cells conserve energy is by regulating the synthesis of their ribosomes and tRNAs, so that they make only enough to meet their needs. More than half of the RNA made at any one time comprises rRNA and tRNA. Moreover, each ribosome is composed of about 50 different proteins, and there are about as many different tRNAs.

The number of ribosomes and tRNA molecules needed by the cell varies greatly, depending on the growth rate. Fast-growing cells require many ribosomes and tRNA molecules to maintain the high rates of protein synthesis required for fast cellular growth. Cells growing more slowly, either because they are using a relatively poor carbon and energy source or because some nutrient is limiting, need fewer ribosomes and tRNAs. For example, a rapidly growing *E. coli* cell contains as many as 70,000 ribosomes, but slowly growing cells have fewer than 20,000. As with most of the global regulatory systems we have discussed, the regulation of ribosome and tRNA synthesis is much better understood for *E. coli* than any other organism on earth. Even in *E. coli*, however, some major questions remain unanswered. In this section, we confine our discussion to *E. coli*, with occasional references to other bacteria when information is available.

Ribosomal Proteins

Ribosomes are composed of both proteins and RNA (see chapter 2). The ribosomal proteins are designated by the letter L or S, to indicate whether they are from the *large* (50S) or *small* (30S) subunit of the ribosome, followed by a number for the particular protein. Thus, protein L11 is protein number 11 from the 50S subunit of the ribosome, whereas protein S9 is protein number 9 from the 30S subunit. The gene names begin with *rp*, for ribosomal protein, followed by a lower case *l* or *s* to indicate whether the protein product resides in the large or small subunit. Another capital letter designates the specific gene. For example, the gene *rplK* is ribosomal protein gene K encoding the L11 protein; note that *K* is the 11th letter of the alphabet. Similarly, *rpsL* encodes the S12 protein; *L* is the 12th letter of the alphabet.

MAPPING RIBOSOMAL PROTEIN GENES

A total of 54 different genes encode the 54 polypeptides that compose the *E. coli* ribosome, and mapping these genes was a major undertaking. Some ribosomal protein genes were mapped simply by mapping mutations that caused resistance to antibiotics such as streptomycin, which binds to the ribosomal protein S12, blocking the translation of other genes (see chapter 2). More complex techniques involving specialized transducing phages and DNA cloning (see chapters 1 and 8) were needed to map the other ribosomal protein genes. Often, clones containing these genes were identified because they synthesize a particular ribosomal protein in coupled transcription-translation systems (see chapter 2).

The final map revealed some intriguing aspects to the organization of the ribosomal protein genes in the chromosome of *E. coli*. Rather than being randomly scattered around the chromosome, the 54 genes are organized into large clusters of operons, with the largest at 73 and 90 min in the *E. coli* genome. Furthermore, these operons also often contain genes for other compo-

nents of macromolecular synthesis, including subunits of RNA polymerase, tRNAs, and genes for proteins of the DNA replication apparatus.

In the cluster shown in Figure 13.16, four genes for tRNAs, *thrU*, *tryD*, *glyT*, and *thrT*, are followed by *tufB*, a gene for the translation elongation factor EF-Tu. These five genes constitute one operon; they are all cotranscribed into one long precursor RNA, from which the individual tRNAs are cut out later. The next operon in the cluster contains two genes for ribosomal proteins, *rplK* and *rplA*, and a third operon has four genes, *rplJ* and *rplL* (encoding two more ribosomal proteins) and *rpoB* and *rpoC* (encoding the β and β' subunits of the RNA polymerase, respectively).

Several hypotheses have been proposed to explain why genes involved in macromolecular synthesis would be clustered in *E. coli* DNA. First, the products of these genes must all be synthesized in large amounts to meet the cellular requirements. Some clusters are near the origin of replication, *oriC*, and cells growing at high growth rates have more than one copy of the genes near this site (see chapter 1), which allows higher rates of synthesis of the gene products. Other possible reasons have to do with the structure of the bacterial nucleoid (see chapter 1). Genes clustered together will probably be on the same loop of the nucleoid. A loop for the macromolecular synthesis genes might be relatively large and extend out from the core of the nucleoid to allow RNA polymerase and ribosomes to gain easier access to the genes. A third possible explanation is that the genes for macromolecular synthesis must be coordinately regulated with the growth rate and that their assembly in clusters of operons facilitates their coordinate regulation, although why this should be the case is not clear.

REGULATION OF THE SYNTHESIS OF RIBOSOMAL PROTEINS

The regulation of ribosomal protein synthesis is best understood in *E. coli*. However, there is every reason to believe that the regulation is similar in all bacteria. The

Figure 13.16 Arrangement of a gene cluster in *E. coli* encoding ribosomal proteins and other gene products involved in macromolecular synthesis. The cluster contains three operons, transcribed in the direction shown by the arrows. The genes *thrU*, *tryU*, *glyT*, and *thrT* all encode tRNAs. The *tufB* gene encodes translation elongation factor EF-Tu. The genes *rplK*, *rplA*, *rplJ*, and *rplL* encode proteins of the large subunit of the ribosome. The genes *rpoB* and *rpoC* encode subunits of the RNA polymerase.

ribosomal proteins and rRNAs are synthesized independently and then assembled into mature ribosomes. Nevertheless, there is never an excess of either free ribosomal proteins or free rRNA in the cell, suggesting that their synthesis is somehow coordinated. Either the rate of ribosomal protein synthesis is adjusted to match the rate of synthesis of the rRNA, or vice versa. As it turns out, the rate of protein synthesis is adjusted to match the rate of rRNA synthesis.

Like the synthesis of the heat shock sigma (σ^{32}), the synthesis of the ribosomal proteins is translationally autoregulated. The ribosomal proteins bind to TIRs in the mRNA and repress their own translation. Rather than each ribosomal gene of the operon translationally regulating itself independently, the protein product of only one of the genes of the operon is designated as responsible for regulating the translation of all the ribosomal proteins encoded by the operon. This designated protein can also bind to free rRNA in the cell, which, as we discuss later, is what coordinates the synthesis of the ribosomal proteins with the synthesis of rRNA.

Figure 13.17 illustrates this regulation with the relatively simple operon *rplK-rplA*. Of the two proteins encoded by the operon, L1 is the one believed to function in regulation of the operon. The L1 protein can bind to both free rRNA and the TIR on the *rplK* mRNA. By binding to the *rplK* TIR and thereby inhibiting the trans-

lation of this gene, the L1 protein simultaneously inhibits the translation of its own gene, since the two genes are translationally coupled (see chapter 2). However, if the cell contains free rRNA, the L1 protein will preferentially bind to it, allowing translation of L11 and L1 to resume. The same thing happens to the other ribosomal proteins, and so their synthesis also resumes. When there is no longer any free rRNA in the cell—because it is all taken up by the ribosomes—free L1 protein will begin to accumulate, bind to the TIR for the *rplK* gene, and once again repress the translation of itself and the L11 protein.

The protein in each operon designated to be the regulator may have been chosen because it normally binds to rRNA during the assembly of the ribosome and so already has an rRNA-binding ability. In at least some cases, the sequence to which the designated protein binds in the rRNA may be mimicked in the TIR for the first gene in the mRNA.

Experimental Support for the Translational Autoregulation of Ribosomal Proteins

Evidence that the synthesis of ribosomal proteins is translationally autoregulated came from a series of experiments called **gene dosage experiments**, in which the number of copies, or dosage, of a gene in the cell is increased and the effect of this increase on the rate of synthesis of the protein product of the gene is determined (see Yates and Nomura [1980 and 1981], Suggested Reading). Figure 13.18 illustrates the principle behind a

Figure 13.17 Translational autoregulation of the ribosomal proteins, as illustrated by the *rplK-rplA* operon shown in Figure 13.16. See the text for details.

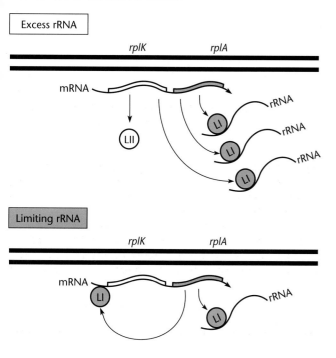

Figure 13.18 A gene dosage experiment to determine if a gene is autoregulated. The number of copies of the gene for the A protein is doubled. If the gene is not autoregulated, twice as much of protein A will be made. If gene *A* is autoregulated, the same amount of protein A will be made.

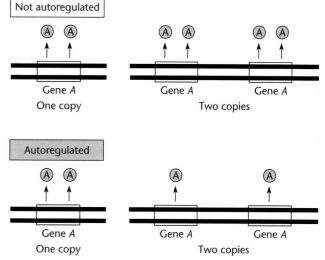

gene dosage experiment. If the gene is not autoregulated, the rate of synthesis of the gene product should be approximately proportional to the number of copies of the gene. However, if the gene is autoregulated, the product of the gene should repress its own synthesis, and the rate of synthesis of the gene product should not increase, regardless of the number of gene copies.

In the actual experiment, specialized transducing phage carrying operons encoding one or more ribosomal proteins were integrated at the normal phage attachment site (see chapter 8). As a result, the cell contained two copies of these ribosomal protein genes, one at the normal position and another copy at the site of integration of the transducing phage DNA. The same amount of the ribosomal proteins was synthesized with two copies of the ribosomal protein genes as with one, indicating that the synthesis is autoregulated.

The next step was to determine if the autoregulation occurs at the level of transcription or translation. If the synthesis of the ribosomal proteins were transcriptionally autoregulated, the rate of transcription of the ribosomal genes would not increase when the number of copies of the genes increased. However, if the autoregulation occurred only at the level of translation, the rate of gene transcription would increase. The investigators found that the rate of transcription approximately doubled when the gene dosage was doubled, indicating that the autoregulation did not occur transcriptionally. If genes are not transcriptionally autoregulated, they must be translationally autoregulated. Therefore, the ribosomal proteins are capable of repressing their own translation.

The next step was to determine whether all the proteins in the operon independently repress their own translation or whether some of the proteins in each operon are responsible for regulating translation of themselves as well as the others. To answer this question, the investigators systematically deleted genes for some of the proteins encoded by each operon in the transducing phage and evaluated the effect of their absence on the synthesis of the other proteins. Deleting most of the genes in each operon had no effect on the synthesis of the proteins encoded by the other genes. However, when one particular gene in each operon was deleted, the rate of synthesis of the other proteins doubled. Therefore, this one protein was known to repress the translation of itself and the other proteins encoded by the same operon.

rRNA and tRNA Regulation

As discussed in chapter 2, the 16S, 23S, and 5S rRNAs are synthesized together as a long precursor RNA, often with intermingled tRNA sequences. After synthe-

sis, the long precursor RNA is processed into the individual rRNAs and tRNAs. Every ribosome contains one copy of each of the three types of rRNAs, and the synthesis of the three rRNAs as part of the same precursor RNA ensures that all three will be made in equal amounts.

Each cell has tens of thousands of ribosomes, requiring the synthesis of large amounts of rRNA. To meet this need, bacteria have evolved many ways to increase the output of their rRNA genes. For example, many bacteria have more than one copy of the gene for rRNA. *E. coli* and *S. enterica* serovar Typhimurium have seven copies of the genes for the rRNAs. Many bacteria also have very strong promoters for their rRNA genes, so that as many as 50 RNA polymerase molecules can be transcribing each rRNA operon simultaneously, almost as many as a DNA of this length will hold. The RNA polymerase molecules initiate trancription and start down the operon, one immediately after another. Some of the rRNA promoters are so strong because they have a sequence called the UP element upstream of the promoter. This sequence enhances initiation of transcription from the promoter by interacting with the carboxyl terminus of the α subunit of RNA polymerase, much like CAP and some other activator proteins, which interact with the polymerase to activate transcription (see Hirvonen et al., Suggested Reading).

The rate of rRNA gene transcription is also increased because rRNA operons have antitermination sequences lying just downstream of the promoter and in the spacer region between the 16S and 23S coding sequences. These antitermination sequences reduce pausing by RNA polymerase and prevent termination at ρ-dependent transcription termination sites (chapter 2), allowing the synthesis of a complete rRNA to be completed in a shorter time (see Condon et al., Suggested Reading).

The rRNA and tRNA genes in *E. coli* seem subject to at least two types of regulation. In the first type, **stringent control**, rRNA and tRNA synthesis ceases when cells are starved for an amino acid. In the second, **growth rate regulation**, the rate of rRNA and tRNA synthesis decreases when cells are growing slowly on a poor carbon source. As we discuss next, these two types of regulation may have some features in common.

STRINGENT CONTROL

Protein synthesis requires all 20 amino acids. A cell is said to be **starved** for an amino acid when the amino acid is missing from the medium and the cell cannot make it. The ribosomes of the cell will then stall whenever they encounter a codon for the missing amino acid because it will not be available for insertion into the growing polypeptide chain.

In principle, rRNA synthesis can continue in cells starved for an amino acid, since RNA does not contain amino acids. However, in *E. coli*, and probably other types of cells, the synthesis of rRNA and tRNA ceases when an amino acid is lacking. This coupling of the synthesis of rRNA and tRNA to the synthesis of proteins is called stringent control. Stringent control saves energy; there is no point in making ribosomes and tRNA if one or more of the amino acids are not available for protein synthesis.

Synthesis of ppGpp during Stringent Control

The stringent control of rRNA synthesis results from the accumulation of an unusual nucleotide, **guanosine tetraphosphate (ppGpp)**. A similar nucleotide, **guanosine pentaphosphate (pppGpp)**, also accumulates under these conditions, and some is subsequently converted to ppGpp. These nucleotides are made by transferring two phosphates from ATP to the 3′ hydroxyl of GDP or GTP, respectively. These nucleotides were originally called magic spots I and II (MSI and MSII) because they show up as a distinct spots during some types of chromatography (see Cashel and Gallant, Suggested Reading). All evidence indicates that the two nucleotides have similar effects, so we refer to them collectively as ppGpp.

1. RelA. Figure 13.19 shows a model for how amino acid starvation stimulates the synthesis of ppGpp. The nucleotides are made by an enzyme called RelA (for *re*laxed control gene A), which is bound to the ribosome. When *E. coli* cells are starved for an amino acid (lysine in the example), the tRNAs for that amino acid, tRNA^Lys, are uncharged. The uncharged tRNA will not bind EF-Tu (see chapter 2) and so will not enter the ribosome. Consequently, when a ribosome moving along an mRNA encounters a codon for that amino acid (the codon AAA in the example), the ribosome will stall. If the ribosome stalls long enough, an uncharged tRNA may eventually enter the A site of the ribosome even though it is not bound to EF-Tu. Uncharged tRNA entering the A site stimulates the RelA protein on the ribosome to synthesize ppGpp.

The intracellular levels of ppGpp during amino acid starvation are also regulated by the SpoT (for magic spot) protein, which is the product of the *spoT* gene. The SpoT protein normally degrades ppGpp, but its degradation activity is inhibited after amino acid starvation, leading to more accumulation of ppGpp. Therefore, after amino acid starvation, the cellular concentration of ppGpp is determined both by the activation of the

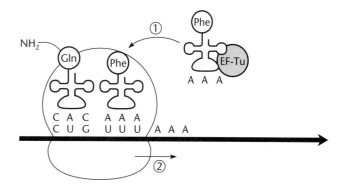

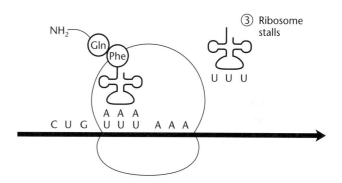

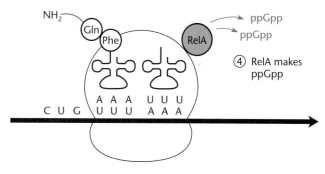

Figure 13.19 A model for synthesis of ppGpp after amino acid starvation. Cells are starved for the amino acid lysine. The tRNA^Lys will then have no lysine attached, and EF-Tu cannot bind to a tRNA that is not aminoacylated. A ribosome moving along the mRNA will then stop when it arrives at a codon for lysine (AAA), because it has no aminoacylated-tRNA to translate the codon. However, if the ribosome remains stalled at the codon long enough, a tRNA^Lys (anticodon UUU in the example) will bind to the A site of the ribosome even though EF-Tu is not bound. This binding will cause RelA to synthesize ppGpp.

ppGpp synthesis activity of RelA and the inhibition of the ppGpp-degrading activity of SpoT. The *spoT* gene product not only can degrade ppGpp but also may synthesize ppGpp during growth rate regulation (see below).

2. *Isolating* relA *mutants.* Evidence that the accumulation of ppGpp is responsible for stringent control came from the behavior of *relA* mutants of *E. coli* that lack the RelA enzyme activity. These mutants do not accumulate ppGpp after amino acid starvation and also do not shut off rRNA and tRNA synthesis. Because rRNA synthesis and tRNA synthesis are not stringently coupled to protein synthesis in *relA* mutants, strains with *relA* mutations are called **relaxed strains**, which is the orgin of the gene name.

As is often the case, the first *relA* mutant was isolated by chance. A mutant strain of *E. coli* was observed to have a difficult time recovering after amino acid starvation, whereas the wild-type parent could start growing almost immediately after the amino acid was restored. A later study showed that rRNA and tRNA synthesis continued after this mutant was starved for an amino acid, and so this mutant was called a relaxed mutant.

The relative inability of the original *relA* mutant to recover from amino acid starvation suggested a way of enriching for *relA* mutants (see Fiil and Frieson, Suggested Reading). This procedure is based on the fact that growing cells are killed by ampicillin but cells that are not growing will survive (see chapter 3 for methods). *E. coli* cells that were auxotrophic for an amino acid were mutagenized, washed, and resuspended in a medium without the amino acid, so that they stopped growing. After a long period of incubation, the amino acid was restored and ampicillin was added to the medium. The ampicillin was removed shortly thereafter, and the process was repeated. After a few cycles of this treatment, the bacteria were plated. A high percentage of the surviving bacteria that multiplied to form colonies were *relA* mutants, as evidenced by the fact that most of the surviving strains continued to synthesize rRNA and tRNA after amino acid starvation.

How Does ppGpp Block rRNA and tRNA Synthesis?

The mechanism by which ppGpp blocks rRNA and tRNA synthesis is still not clearly established. Most evidence suggests that ppGpp blocks the initiation of transcription of the genes for these RNAs by acting directly on their promoters. Mutations in the promoters for rRNA and tRNA genes can render them insensitive to ppGpp. The presence of ppGpp can also inhibit the binding of RNA polymerase to rRNA promoters in vitro and can inhibit in vitro rRNA synthesis.

However, other evidence suggests that ppGpp does not block initiation but, rather, the elongation of transcription. In support of this hypothesis is the fact that transcription of the rRNA genes from promoters that are normally not sensitive to stringent control, including the T7 or λ phage promoters, is still sensitive to ppGpp. Thus, the target of ppGpp may not be exclusively the promoter of the rRNA genes, and ppGpp may also block rRNA transcription after it is under way. Why different experimental methods produce apparently conflicting results on the mechanism by which ppGpp blocks rRNA and tRNA synthesis is not clear.

Although ppGpp seems to block rRNA and tRNA synthesis, it also stimulates the transcription of some other genes, including the *his* operon and some other biosynthetic operons. Failure to adequately express these biosynthetic operons in the absence of ppGpp makes *relA spoT* double mutants auxotrophic for several growth substances. High levels of ppGpp may also decrease the rate of transcription and translation and lower the translation error rates of many other genes. More experiments are needed to determine how ppGpp affects transcription and whether all the effects are related.

GROWTH RATE REGULATION OF RIBOSOMAL AND tRNA SYNTHESIS

As mentioned above, cells growing fast in rich medium have many more ribosomes and a higher concentration of tRNA than do cells growing slowly in poor medium. This regulation of rRNA and tRNA synthesis is called growth rate regulation.

An interesting model has been proposed to explain the growth rate dependence of rRNA synthesis (see Barker and Gourse, Suggested Reading). According to this model, the rate of initiation at rRNA promoters is dependent on the concentration of the initiating nucleotide. Like most transcription, rRNA synthesis begins with ATP or GTP. The RNA polymerase may form weak open complexes at rRNA promoters unless the concentration of the initiating nucleotide is high enough to stabilize the complex. The concentration of ATP and GTP increases when growth rates are high; this would lead to more frequent initiations at the rRNA promoters and therefore more synthesis of rRNAs and ribosomes. In support of this model, the *rrn* promoters that are least sensitive to initiating nucleoside triphosphate concentration in vitro are also least sensitive to growth rate regulation in vivo.

Guanosine tetraphosphate (ppGpp) levels are also higher when cells are growing more slowly in poorer medium, suggesting that ppGpp may also play a role in growth rate regulation. Surprisingly SpoT which plays a role in degrading ppGpp may also play a role in synthesizing it during growth rate regulation (see Hernandez and Breiner, Suggested Reading). It is possible that this

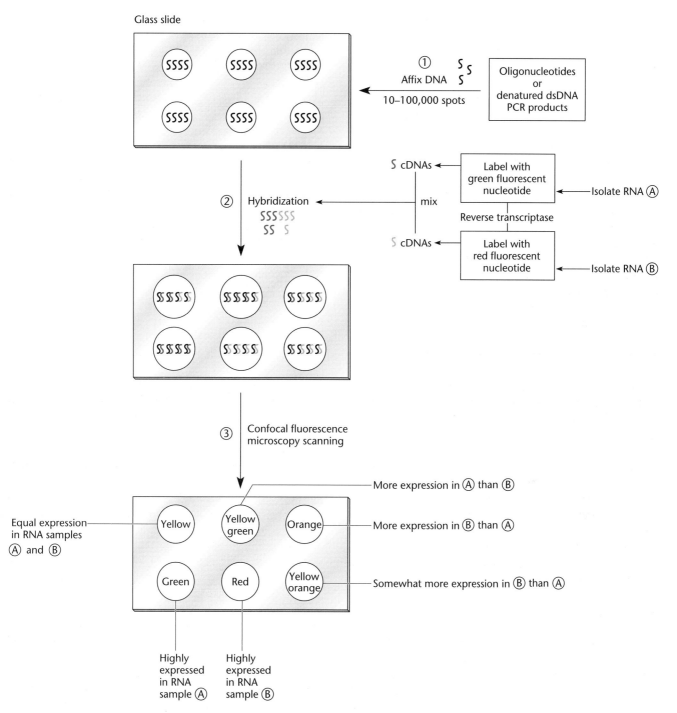

Figure 13.20 Two-color microarray transcription profiling. (Step 1) DNA probes representing the entire genome sequence are affixed to a glass slide. (Step 2) Fluorescently labeled cDNAs, produced by reverse transcriptase primed by random hexamers from template RNA isolated from different cultures (A and B), are hybridized to the glass slide. (Step 3) Scanning of the slide produces dots that vary in color from red to green to yellow to intermediate shades. The colors are interpreted as shown. dsDNA, double-stranded DNA.

role is indirect and that ppGpp could affect the concentrations of ATP and GTP or that it could compete with binding of the initiating nucleoside triphosphate to the RNA polymerase, preventing the latter from stabilizing the open complex.

Microarray and Proteomic Analysis of Regulatory Networks

Each new genome sequence tempts us with a renewed challenge: to pursue the quest for complete understanding of cellular regulation. This analysis is called genomics. The (almost) complete gene content of an organisms can be displayed on a computer screen, usually in multicolor. The ever-growing sequence databases allow the assignment of presumptive functions, enzymatic or structural, to most genes of an organism. Two-dimensional protein electrophoresis can separate all of the proteins of the cell into individual spots on a membrane, and the spots can be identified by mass spectrometry and correlated with the ORFs in the genome sequence (see Graves and Haystead, Suggested Reading). However, as in any unsolved "connect the dots" game, the meaningful patterns do not "jump out." Nevertheless, our curiosity is piqued. If only we could fill in the connecting lines and regulatory arrows in all the right places, what intriguing new pictures of interconnected regulatory networks would emerge?

The *E. coli* and *B. subtilis* genomes contain more than 4,290 and 4,107 ORFs, respectively. Genome-wide protein expression profiling of proteins can be accomplished by proteomic techniques (see Graves and Haystead, Suggested Reading). RNA profiling can be efficiently accomplished with high-density DNA microarrays (Figure 13.20). These microarrays can be produced by robotically spotting ORF-specific PCR products representing more than 95% of the ORFs on nylon membranes or glass slides (see Richmond et al., Suggested Reading). Only three slides are needed for the entire *E. coli* genome. Glass chips containing hundreds of thousands of oligonucleotide probes, called microarrays representing the genomic sequence, can be manufactured.

Some examples of experiments using microarray technology have compared bacterial cultures incubated in different growth conditions. In one study (see Zheng et al. [2001a], Suggested Reading), the response of *E. coli* to hydrogen peroxide stress revealed the induction of at least 140 mRNAs. Because previous genetic analysis had identified the OxyR protein as a peroxide response regulator, comparison of isogenic wild-type and *oxyR* deletion strains not only confirmed the known OxyR-regulated genes but also identified new candidates for the OxyR regulon. The addition of a "bioinformatics" computational search for OxyR-binding sites revealed several more OxyR-regulated genes (see Zheng et al. [2001b], Suggested Reading). In another study, microarray analysis of *E. coli* genes induced by nitrogen limitation identified at least 7 new NtrC-regulated genes (see Zimmer et al., Suggested Reading).

From Genes to Regulons to Networks to Genetic Analysis

Relatively sophisticated genomic analyses have also been carried out with *B. subtilis*. For example, the phenomenon of transformation competence (see chapter 6) involves regulation by the ComK transcription factor. Microarray analysis revealed at least 180 new members of the ComK regulon beyond those previously known (see Berka et al., Suggested Reading). How many of these genes are essential for transformation? Gene knockout experiments, performed so as to avoid polar effects, will be necessary to determine their loss-of-function phenotypes. However, in genetics, unlike in boxing, a knockout is not the end of the bout. As we see in many of the examples of genetic analysis discussed in this textbook, an understanding of gene function often requires study of a variety of allelic forms that result from base pair changes. With genome sequences, we can better direct base pair changes to the regions of proteins with presumptive functions by using recently developed methods of site-specific mutagenesis which involve the λ phage Red system (see Box 10.3). The job of the geneticist in understanding regulation has not reached a plateau but is just beginning.

SUMMARY

1. The coordinate regulation of a large number of genes is called global regulation. Operons that are regulated by the same regulatory protein are part of the same regulon.

2. In catabolite repression, the operons for the use of alternate carbon sources cannot be induced when a better carbon and energy source, such as glucose, is present. In *E. coli* and other enteric bacteria, this catabolite regulation is achieved, in part, by cAMP, which is made by adenylate cyclase, the product of the *cya* gene. When the bacteria are growing in a poor carbon source such as lactose or galactose, the adenylate cyclase is activated and cAMP levels are high. When the bacteria are growing in a good carbon source such as glucose, cAMP levels are low. The cAMP acts through a protein called CAP, the product of the *crp* gene. CAP is a transcriptional activator, which, with cAMP bound, will activate transcription of catabolite-sensitive operons such as *lac* and *gal*.

3. Bacterial cells induce different genes depending on the nitrogen sources available. Genes that are regulated through the nitrogen source are called Ntr genes. Most bacteria, including *E. coli*, prefer NH_3 as a nitrogen source and will not transcribe genes for using other nitrogen sources when growing in NH_3. Glutamine concentrations are low when NH_3 concentrations are low. A signal transduction pathway is then activated, culminating in the phosphorylation of NtrC. This signal transduction pathway begins with the GlnD protein, a uridylyl transferase, which is the sensor of the glutamine concentration in the cell. At low concentrations of glutamine, GlnD transfers UMP to the P_{II} protein, inactivating it. However, at high concentrations of glutamine, the GlnD protein removes UMP from P_{II}. The P_{II} protein without UMP attached can bind to NtrB, somehow preventing the transfer of phosphate to NtrC and causing the removal of phosphates from NtrC. The phosphorylated NtrC protein activates the transcription of the *glnA* gene, the gene for glutamine synthetase, as well as the *ntrB* gene and its own gene, *ntrC*, since they are part of the same operon as *glnA*. It also activates the transcription of operons for using other nitrogen sources.

4. The NtrB and NtrC proteins form a sensor-response regulator pair and are highly homologous to other sensor-kinase/response-regulator pairs in bacteria.

5. NtrC-regulated promoters of *E. coli* and other enterics require a special sigma factor called σ^{54}, which is also used to transcribe the flagellar genes and some biodegradative operons in other types of bacteria.

6. The cell also regulates the activity of glutamine synthetase by adenylylating the glutamine synthetase enzyme. The enzyme is highly adenylylated at high glutamine to α-ketoglutarate ratios, which makes it less active and subject to feedback inhibition.

7. One of the ways that bacteria adjust to changes in the osmolarity of the medium is by changing the ratio of their porin proteins, which form pores in the outer membrane through which solutes can pass to equalize the osmotic pressure on both sides of the membrane. The major porins of *E. coli* are OmpC and OmpF, which make pores of different sizes, thereby allowing the passage of different-sized solutes. The relative amounts of the OmpC and OmpF change in response to changes in the osmolarity of the medium. The *ompC* and *ompF* genes in *E. coli* are regulated by a sensor-response regulator pair of proteins, EnvZ and OmpR, which are similar to NtrB and NtrC. The EnvZ protein is an inner membrane protein with both kinase and phosphatase activities that, in response to a change in osmolarity, can transfer a phosphoryl group to or remove one from OmpR, a transcriptional acivator. The state of phosphorylation of OmpR affects the relative rates of transcription of the *ompC* and *ompF* genes.

8. The ratio of OmpF to OmpC porin proteins is also affected by an antisense RNA named MicF. A region of the MicF RNA can base pair with the translation initiation region of the OmpF mRNA and block access by ribosomes, thereby inhibiting OmpF translation. The *micF* gene is regulated by a number of transcriptional regulatory proteins, including SoxS, which induces the oxidative stress regulon, and MarA, which induces genes involved in excluding toxic chemicals and antibiotics from the cell.

9. The virulence genes of pathogenic bacteria are also global regulons and are normally trancribed only when the bacterium is in its vertebrate host.

10. The diphtheria toxin gene, *dtxR*, encoded by a prophage of *Corynebacterium diphtheriae*, is turned on only when iron is limiting, a condition mimicking that in the host. The *dtxR* gene is regulated by a chromosomally encoded repressor protein, DtxR, which is similar to the Fur protein involved in regulating the genes of iron availability pathways in *E. coli* and other enteric pathogens.

11. The toxin genes of *Vibrio cholerae* are also carried on a prophage and are regulated by a regulatory cascade that begins with a transcriptional activator, ToxR. The ToxR protein traverses the inner membrane and is activated by a second protein, ToxS. ToxR/ToxS act in concert with another gene pair, TcpP/TcpH, to activate the transcription of *toxT*, whose gene product in turn activates the transcription of virulence genes.

12. The virulence genes of *Bordetella pertussis* are regulated by a sensor-response regulator pair of proteins, BvgS and BvgA. Phase variation in *B. pertussis* results from frameshift mutations in *bvgS*.

13. Bacteria induce a set of proteins called the heat shock proteins in response to an abrupt increase in temperature. Some of the heat shock proteins are chaperones, which as-

(continued)

SUMMARY (continued)

sist in the refolding of denatured proteins. Others are proteases, which degrade denatured proteins. The heat shock response is common to all organisms, and the heat shock proteins have been highly conserved throughout evolution.

14. The promoters of the heat shock genes are recognized by RNA polymerase holoenzyme with an alternative sigma factor called the heat shock sigma, or σ^{32}. The amount of this sigma factor markedly increases following a heat shock, leading to increased transcription of the heat shock genes. The increase in σ^{32} following a heat shock involves DnaK, a chaperone that is one of the heat shock proteins.

15. The genes encoding the ribosomal proteins, rRNAs, and tRNAs are part of the largest regulon in bacteria, with hundreds of genes that are coordinately regulated. A large proportion of the cellular energy goes into making the rRNAs, tRNAs, and ribosomal proteins; therefore, regulating these genes saves the cell considerable energy.

16. The synthesis of ribosomal proteins is coordinated by coupling the translation of the ribosomal protein genes to the amount of free rRNA that is not yet in a ribosome. The ribosomal protein genes are organized into operons, and one ribosomal protein of each operon plays the role of translational repressor. The same protein also binds to free rRNA, so that when there is excess rRNA in the cell, all of the repressor protein binds to the free rRNA and none is available to repress translation.

17. The synthesis of rRNA and tRNA following amino acid starvation is inhibited by guanosine tetraphosphate (ppGpp), synthesized by an enzyme associated with the ribosome called RelA. All types of bacteria contain ppGpp, and so the regulation may be universal. However, it is not yet clear how higher levels of ppGpp inhibit transcription of the genes for rRNA and tRNA and stimulates others.

18. Cells contain fewer ribosomes when they are growing more slowly in poorer media. This is called growth rate control and may be due to the lower concentration of the initiating ribonucleosides, ATP and GTP, in slower-growing cells. RNA polymerase forms weak open complexes on the promoters for the rRNA genes, and these may have to be stabilized by high concentrations of ATP and GTP. Guanosine tetraphosphate may also play a role in growth rate control, perhaps by reducing the concentration of ATP and GTP or by competing with ATP and GTP for the initiating complex.

19. It is becoming increasingly apparent that small RNAs play an important role in gene regulation in bacteria. These small RNAs can pair with complementary sequences in mRNA and block translation or target the mRNA for degradation by RNAses.

20. Techniques such as microarrays have made it possible to monitor the expression of many genes simultaneously. This has led to the discovery of many more genes belonging to the same regulon.

QUESTIONS FOR THOUGHT

1. Why do you suppose that proteins involved in gene expression (i.e., transcription and translation) are among the heat shock proteins?

2. Why do you think genes for the utilization of amino acids as a nitrogen source are not under Ntr regulation in *Salmonella* spp. but they are in *Klebsiella* spp.?

3. Why are the corresponding sensor and response regulator genes of the various two-component systems so similar to each other?

4. Why is the enzyme responsible for ppGpp synthesis in response to amino acid starvation different from the one responsible for ppGpp synthesis during growth rate control? Why might SpoT be used to degrade ppGpp made by RelA after amino acid starvation but be used to synthesize it during growth rate control?

PROBLEMS

1. You have isolated a mutant of *E. coli* that cannot use either maltose or arabinose as a carbon and energy source. How would you determine if your mutant has a *cya* or *crp* mutation or whether it is a double mutant with mutations in both the *ara* operon and a *mal* operon?

2. What would you expect the phenotypes of the following mutations to be?

a. a *glnA* (glutamine synthetase) mutation.

b. an *ntrB* mutation.

c. an *ntrC* mutation.

d. a *glnD* null mutation that inactivates the UTase so that P_{II} has no UMP attached.

e. a constitutive *ntrC* mutation that changes the NtrC protein so that it no longer needs to be phosphorylated to be active.

f. a *dnaK* mutation.

g. a *dtrR* mutation of *Corynebacterium diphtheriae*.

h. a *relA spoT* double mutant.

3. How would you show that the toxin gene of a pathogenic bacterium is not a normal chromosomal gene but is carried on a prophage not common to all the bacteria of the species?

4. Explain how you would use gene dosage experiments to prove that the heat shock sigma (σ^{32}) gene is not transcriptionally autoregulated.

5. Explain how you would show which of the ribosomal proteins in the *rplJ-rplL* operon is the translational repressor.

SUGGESTED READING

Barker, M. M., and R. L. Gourse. 2001. Regulation of rRNA transcription correlstes with nucleoside triphosphate sensing. *J. Bacteriol.* **183**:6315–6323.

Berka, R. M., J. Hahn, M. Albano, I. Draskovic, M. Persuh, X. Cui, A. Sloma, W. Widmer, and D. Dubnau. 2002. Microarray analysis of the *Bacillus subtilis* K-state: genome-wide expression changes dependent on ComK. *Mol. Microbiol.* **43**:1331–1345.

Blaszczak, A., C. Georgopoulos, and K. Liberek. 1999. On the mechanism of FtsH-dependent degradation of the sigma 32 transcriptional regulator of *Escherichia coli* and the role of the DnaK chaperone machine. *Mol. Microbiol.* **31**:157–166.

Boucher, P. E., M.-S. Yang, D. A. Schmidt, and S. Stibitz. 2001. Genetic and biochemical analyses of BvgA interaction with the secondary binding region of the fha promoter of *Bordetella pertussis*. *J. Bacteriol.* **183**:536–544.

Busby, S., and R. H. Ebright. 1999. Transcription activation by catabolite activator protein (CAP). *J. Mol. Biol.* **293**:199–213.

Cashel, M., and J. Gallant. 1969. Two compounds implicated in the function of the RC gene in *Escherichia coli*. *Nature* (London) **221**:838–841.

Condon, C., C. Squires, and C. L. Squires. 1995. Control of rRNA transcription in *Escherichia coli*. *Microbiol. Rev.* **59**:623–645.

Craig, E., and C. A. Gross. 1991. Is Hsp70 the cellular thermometer? *Trends Biochem. Sci.* **16**:135–140.

Eisen, M. B., and P. D. Brown. 1999. DNA arrays for analysis of gene expression. *Methods Enzymol.* **303**:179–205.

Fiil, N., and J. D. Frieson. 1968. Isolation of relaxed mutants of *Escherichia coli*. *J. Bacteriol.* **95**:729–731.

Fisher, S. H. 1999. Regulation of nitrogen metabolism in *Bacillus subtilis*: vive la difference! *Mol. Microbiol.* **32**:223–232.

Gall, T., M. S. Bartlett, W. Ross, C. L. Turnbough, Jr., and R. L. Gourse. 1997. Transcription initiation by initiating NTP concentration: rRNA synthesis in bacteria. *Science* **278**:2092–2097.

Graves, P. R., and T. A. J. Haystead. 2002. Molecular biologist's guide to proteomics. *Microbiol. Mol. Biol. Rev.* **66**:39–63.

Hall, M. N., and T. J. Silhavy. 1981. Genetic analysis of the *ompB* locus in *Escherichia coli* K-12. *J. Mol. Biol.* **151**:1–15.

Henngge-Aronis, R. 1996. Back to log phase: σ^s as a global regulator of osmotic control of gene expression in *Escherichia coli*. *Mol. Microbiol.* **21**:887–893.

Hernandez, V. J., and H. Bremer. 1991. *Escherichia coli* ppGpp synthetase II activity requires *spoT*. *J. Biol. Chem.* **266**:5991–5999.

Hirvonen, C. A., W. Ross, C. E. Wozniak, E. Marasco, J. R. Anthony, S. E. Aiyer, V. H. Newburn, and R. L. Gourse. 2001. Contributions of UP elements and the transcription factor FIS to expression from the seven *rrn* P1 promoters in *Escherichia coli*. *J. Bacteriol.* **183**:6305–6314.

Igarashi, K., and A. Ishihama. 1991. Bipartite functional map of *E. coli* RNA polymerase α subunit: involvement of the C-terminal region in transcription activation by cAMP-CRP. *Cell* **65**:1015–1022.

Inada, T., K. Kimata, and H. Aiba. 1996. Mechanisms responsible for glucose-lactose diauxie in *Escherichia coli*: challenge to the cAMP model. *Genes Cells* **1**:293–301.

Krukonis, E. S., R. R. Yu, and V. J. DiRita. 2000. The *Vibrio cholerae* ToxR/Tcp/ToxT virulence cascade: distinct roles for the two membrane-localized transcriptional activators on single promoter. *Mol. Microbiol.* **38**:67–84.

Liberek, K., T. P. Galitski, M. Zyliez, and C. Georgopoulos. 1992. The DnaK chaperon modulates the heat shock response of *E. coli* by binding to the σ^{32} transcription factor. *Proc. Natl. Acad. Sci. USA* **89**:3516–3520.

Magasanik, B. 1982. Genetic control of nitrogen assimilation in bacteria. *Annu. Rev. Genet.* **16**:135–168.

Mattison, K., R. Oropeza, N. Byers, and L. J. Kenney. 2002. A phosphorylation site mutant of OmpR reveals different binding conformations at *ompF* and *ompC*. *J. Mol. Biol.* **315**:497–511.

Merkel, T. J., C. Barros, and S. Stibitz. 1998. Characterization of the *bvgR* locus of *Bordetella pertussis*. *J. Bacteriol.* **180**:1682–1690.

Miller, V. L., and J. J. Mekalanos. 1984. Synthesis of cholera toxin is positively regulated at the transcriptional level by *toxR*. *Proc. Natl. Acad. Sci. USA* **81**:3471–3475.

Nagai, H., H. Yuzawa, and T. Yura. 1991. Interplay of two *cis*-acting mRNA regions in translational control of σ^{32} synthesis during the heat shock response of *Escherichia coli*. *Proc. Natl. Acad. Sci. USA* **88**:10515–10519.

Nomura, M., J. L. Yates, D. Dean, and L. E. Post. 1980. Feedback regulation of ribosomal protein gene expression in *Escherichia coli*: structural homology of ribosomal RNA and ribosomal protein mRNA. *Proc. Natl. Acad. Sci. USA* **77**:7084–7088.

Pratt, L. A., W. Hsing, K. E. Gibson, and T. J. Silhavy. 1996. From acids to *osmZ*: multiple factors influence synthesis of the OmpF and OmpC porins in *Escherichia coli*. *Mol. Microbiol.* **20**:911–917.

Reitzer, L., and B. L. Schneider. 2001. Metabolic context and possible physiological themes of σ^{54}-dependent genes in *Escherichia coli*. *Microbiol. Mol. Biol. Rev.* **65**:422–444

Richmond, C. S., J. D. Glasner, R. Mau, H. Jin, and F. R. Blattner. 1999. Genome wide expression profiling in *Escherichia coli* K–12. *Nucleic. Acids Res.* **27**:3821–3835.

Savery, N. J., G. S. Lloyd, S. J. W. Busby, M. S. Thomas, R. H. Ebright, and R. L. Gourse. 2002. Determinants of the C-terminal domain of the *Escherichia coli* RNA polymerase α subunit important for transcription at class I cyclic AMP receptor protein-dependent promoters. *J. Bacteriol.* **184**:2273–2280.

Schmidt, M., and R. K. Holmes. 1993. Analysis of diphtheria toxin repressor-operator interactions and the characterization of mutant repressor with decreasing binding activity for divalent metals. *Mol. Microbiol.* **9:**173–181.

Schwartz, D., and J. R. Beckwith. 1970. Mutants missing a factor necessary for the expression of catabolite-sensitive operons in *E. coli,* p. 417–422. *In* J. R. Beckwith and D. Zipser (ed.), *The Lactose Operon.* Cold Spring Harbor Laboratory Press, Cold Spring Harbor, N.Y.

Skorupski, K., and R. K. Taylor. 1997. Control of the ToxR virulence regulon in *Vibrio cholerae* by environmental stimuli. *Mol. Microbiol.* **25:**1003–1009. (Microreview.)

Slauch, J. M., S. Garrett, D. E. Jackson, and T. J. Silhavy. 1988. EnvZ functions through OmpR to control porin gene expression in *Escherichia coli* K-12. *J. Bacteriol.* **170:**439–441.

Stibitz, S., W. Aaronson, D. Monack, and S. Falkow. 1989. Phase variation in *Bordetella pertussis* by frameshift mutation in a gene for a novel two-component system. *Nature* (London) **338:**266–269.

Stibitz, S., A. A. Weiss, and S. Falkow. 1988. Genetic analysis of a region of *Bordetella pertussis* chromosome encoding filamentous hemagglutinin and the pleiotropic regulatory locus *vir. J. Bacteriol.* **170:**2904–2913.

Storz, G. T., and R. Hengge-Aronis (ed.). 2000. Bacterial stress responses. ASM Press, Washington, D.C.

van Heeswijk, W. C., S. Hoving, D. Molenaar, B. Stegeman, D. Kahn, and H. V. Westerhoff. 1996. An alternative P_{II} protein in the regulation of glutamine synthetase in *Escherichia coli. Mol. Microbiol.* **21:**133–146.

Yates, J. L., and M. Nomura. 1980. *E. coli* ribosomal protein L4 is a feedback regulatory protein. *Cell* **21:**517–522.

Yates, J. L., and M. Nomura. 1981. Localization of the mRNA binding sites for ribosomal proteins. *Cell* **24:**243–249.

Yu, R. R., and V. J. DiRita. 2002. Regulation of gene expression in *Vibrio cholerae* by ToxT involves both antirepression and RNA polymerase stimulation. *Mol. Microbiol.* **43:**119–134.

Zheng, M., X. Wang, L. J. Templeton, D. R. Smulski, R. A. LaRossa, and G. Storz. 2001a. DNA microarray-mediated transcriptional profiling of the *Escherichia coli* response to hydrogen peroxide. *J. Bacteriol.* **183:**4562–4570.

Zheng, M., X. Wang, B. Doan, K. A. Lewis, T. D. Schneider, and G. Storz. 2001b. Computation-directed identification of OxyR DNA binding sites in *Escherichia coli. J. Bacteriol.* **183:**4571–4579.

Zhou, Y., and S. Gottesman. 1998. Regulation of proteolysis of the stationary phase sigma factor RpoS. *J. Bacteriol.* **180:**1154–1158.

Zhou, Y. N., N. Kusukawa, J. W. Erickson, C. A. Gross, and T. Yura. 1988. Isolation and characterization of *Escherichia coli* mutants that lack the heat shock sigma factor σ^{32}. *J. Bacteriol.* **170:**3640–3649.

Zimmer, D. P., E. Soupene, H. L. Lee, V. F. Wendisch, A. B. Khodursky, B. J. Peter, R. A. Bender, and S. Kustu. 2000. Nitrogen regulatory protein C-controlled genes of *Escherichia coli:* scavenging as a defense against nitrogen limitation. *Proc. Natl. Acad. Sci. USA* **97:**14674–14679.

PART IV

Genes in Practice

14   Molecular Genetic Analysis in Bacteria

I N PREVIOUS PARTS, we reviewed many of the facts about bacterial cells that have been revealed through molecular genetic studies. However, we have not gone into detail about how molecular genetic data are assembled and analyzed. In this last part, we discuss the analysis of genetic data in more detail and give some particularly elegant and historically important examples of molecular genetic analysis in bacteria. It is our hope that with the background assimilated in the previous parts, you will be able to imagine yourself actually participating in these experiments.

14

Molecular Genetic Analysis in Bacteria

AS WE'VE DISCUSSED in previous chapters, some types of bacteria are among the most amenable organisms on Earth for genetic analysis. To reiterate, they multiply rapidly and have no obligate sexual cycle but can multiply from a single cell by parthenogenesis, giving rise to clonal growth. Large numbers of these bacteria can grow on a single petri plate, and they grow in discrete colonies, making possible the selection of even very rare mutants and recombinants. They are haploid, so that the phenotypes of even recessive mutations are immediately expressed. They are easy to store in large quantities. A few types of bacteria which share some of the most desirable features have been chosen as model organisms to study basic cellular functions. A number of investigators have concentrated on these few model organisms and have developed a large number of genetic tools to use with them. Because of all these advantages, studies with model system bacteria have laid the groundwork for the development of modern molecular genetics and the discovery of many features common to all living cells, making this one of the major accomplishments in the history of science.

In previous chapters, we presented many of the advances in molecular genetics which have come from the use of model system bacteria. While presenting these advances, we attempted, when appropriate, to describe the types of experiments which led to them. We conclude this book with a few examples of basic bacterial molecular genetic analyses which combine many of the concepts and methods that we discussed previously. We chose these particular examples because they have historical importance in bacterial genetics and/or, in our opinion, they illustrate particularly well the application of molecular genetic analysis to the study of gene structure, function, and development in bacteria.

The *lacI* Gene of *Escherichia coli*

Some of the most elegant early experiments in bacterial genetics involved analysis of the structure of the *lacI* gene of *Escherichia coli* (see Miller and Schmeissner and Schmeissner et al., Suggested Reading). These experiments demonstrated the domain structure of proteins, with different regions of the protein being dedicated to different functions of the protein.

As we discussed in chapter 12, the *lac* operon encodes the enzymes responsible for the utilization of lactose. The *lac* operon is regulated by a repressor, encoded by the *lacI* gene. The repressor is a tetramer, formed from four polypeptides encoded by the *lacI* gene, and binds to operator regions to prevent transcription of the structural genes of the operon. However, when the inducer allolactose or one of its analogs binds to the LacI repressor, LacI can no longer bind to the operators, and transcription ensues. Because of its many functions, missense mutations in *lacI* can cause various phenotypes (Table 14.1). For example, a *lacI* mutation may inactivate the repressor by preventing binding to the operators or preventing tetramer formation or both. Alternatively, it might prevent binding of the inducer. The *lacI* mutations can also be subdivided into two groups: recessive and dominant. Any *lacI* mutation that completely inactivates the repressor so that it no longer binds to the operators is recessive to the wild type and is called simply a *lacI* mutation. However, some *lacI* mutations inactivate the operator-binding domain of the repressor but do not affect the tetramerization domain and so can lead to the formation of tetramers. These *lacI* mutations can be dominant to the wild type if a mixed tetramer, made up of both wild-type and mutant subunits (like most of them will be), cannot bind to the operators. They are therefore called *lacI*$^{-d}$ mutations, where the "d" stands for "dominant." Other dominant lacI mutations, called *lacI*^{s}, result in permanent repression, even in the presence of inducer. These mutations prevent the binding of the inducer to the repressor. With some mutations, called *lacI*^{rc}, the situation is reversed. These mutations change the repressor so that it binds to the operators only in the presence of inducer.

In the papers we are discussing, investigators reasoned that the position in the *lacI* gene of missense mutations which cause a particular phenotype may reveal information about which regions of the LacI protein are involved in its various functions. Accordingly, they isolated numerous different types of lacI missense mutations and mapped them to determine their position in the lacI gene.

Isolating Deletions into *lacI* Gene

The most efficient way to map a large number of mutations in a particular gene is by deletion mapping. We have already discussed deletion mapping of *r*II mutations in phage T4 (see chapter 7), and *thyA* mutations by marker rescue with M13 phage clones (see chapter 3), and the principles are the same here. No wild-type recombinants will appear in crosses between a mutant with a point mutation and a mutant with a deletion mutation if the point mutation lies within the deleted region.

The first step in deletion mapping is to obtain a large number of deletion mutations extending into different regions of the gene. One property of deletion mutations is that they can inactivate more than one gene simultaneously (chapter 3). Therefore, one way to select *lacI* deletions is to select mutations in a nearby gene and then screen them to identify those which have deletions which extend into *lacI*. The investigators decided to use the *tonB* gene to select such deletions because they had a positive selection for mutations in the *tonB* gene. We have already discussed the *tonB* gene in chapter 3 in connection with the classic experiments of Luria and Delbrück and of Newcombe on inheritance in bacteria; these investigators used E. coli resistance to phage T1 as the mutant phenotype in their selections. The *tonB* gene product is part of the receptor for phage T1 and some types of bacteriocins, and so mutations that inactivate the *tonB* gene will make the cells resistant to these agents. Therefore, any mutant bacteria which multiply to form a colony in the presence of phage T1 have a *tonB* mutation. A subset of these *tonB* mutations will be deletions, and some of these deletions will extend into the nearby *lacI* gene. Actually, the *lacI* gene is not normally near the *tonB* gene, and so it had to be moved there on a prophage. The genomic structure of this lysogenic strain, E. coli X7800, with the *lacI* gene close to *tonB*, is diagrammed in Figure 14.1. These bacteria with *tonB* mutations were then selected by plating with phage T1, and those with deletions extending into *lacI* were detected by plating the *tonB* mutants on X-Gal plates without inducer. Only constitutive *lac* mutants will make blue colonies on X-Gal plates in the absence of the in-

TABLE 14.1	Types of *lacI* mutations	
Mutation	**Function affected**	**Phenotype**
lacI	Operator binding or tetramer formation	Constitutive; recessive
lacI^d	Operator binding	Constitutive; dominant
lacI^s	Inducer binding	Permanently repressed; dominant
lacI^{rc}	Conformational change after inducer binding	Repressed only with inducer bound; dominant

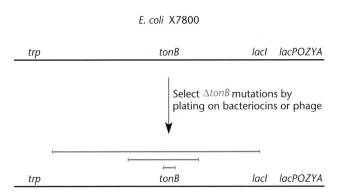

Figure 14.1 Selecting *tonB* deletions that extend into the *lacI* gene in *E. coli* X7800. Mutant bacteria that have *tonB* deletion mutations extending into *lacI* will form blue colonies on X-Gal plates in the absence of inducer.

ducer. As illustrated in Figure 14.1, any *tonB* mutants which were also constitutive for *lac* expression probably had deletion mutations extending through *tonB* and into *lacI*, inactivating both genes simultaneously, since deletions will be much more frequent than double point mutations that inactivate both genes. The deletions probably end in the *lacI* gene, however, and do not extend into *lacZ*, because the product of the *lacZ* gene, β-galactosidase, is required to cleave X-Gal and turn the colonies blue. The end points of the deletions in the *lacI* gene were then mapped by crossing them with a few *lacI* point mutations that had been mapped previously by three-factor crosses.

Isolation of *lacI* Missense Mutations

The next step in their analysis was to isolate a large number of the different types of *lacI* missense mutations. To facilitate mapping, these *lacI* mutations were isolated in an F′ factor rather than in the chromosome. F′ *lac-pro*, the F′ factor used, contains the *lac* genes as well as the wild-type *proB* gene as a selectable marker. This prime factor was maintained in a strain that had the chromosomal *lac* genes deleted, so that any *lacI* mutations would occur in the F′ factor.

Isolation of different types of *lacI* missense mutations required different selection procedures. To isolate mutants with constitutive *lacI* mutations, the bacteria containing the F′ factor were mutagenized and mutants that formed blue colonies on X-Gal plates in the absence of inducer were selected. As discussed above, under these conditions only constitutive *lacI* mutants will express the *lacZ* gene and form blue colonies.

Isolation of strains with *lacI*s mutations was somewhat more difficult. These mutants form colorless colonies on X-Gal plates, even with inducer, but *lacZ* mutants also form colorless colonies under these conditions and are much more common. However, unlike

most other mutations that make the colonies colorless, *lacI*s mutations are dominant over the wild type. In other words, they will make a strain colorless even if there is a wild-type *lac* operon in the chromosome. Therefore, the mutagenized F′ factor was mated into a strain with a functional *lac* operon in the chromosome and the transconjugants were plated on X-Gal plates with the inducer. Under these conditions, many of the cells that form colorless colonies will have a *lacI*s mutation in the F′ factor.

Mapping of the *lacI* Missense Mutations

Figure 14.2 shows how, once a collection of *lacI* mutations with different properties was assembled, they could be mapped. The F′ factor containing the particular point mutation to be mapped, as well as the wild-type *proB* gene, was crossed into derivatives of *E. coli* X7800 that each contained one of the chromosomal deletions extending into the *lacI* gene. After selection for the ability to grow without proline (Pro⁺), the partial diploid strains containing the F′ factor were grown to allow possible

Figure 14.2 Deletion mapping of mutations in the *lacI* gene of *E. coli*. Step 1: cells lacking a chromosomal *lac* region but with an F′ factor containing this region are mutagenized. Step 2: the F′ factor is mated into a strain with one of the *lacI* deletions isolated, as in Figure 14.1, and Pro⁺ bacteria are selected. A few *lacI*⁺ recombinants will appear if the deleted region in the chromosome does not extend into the region of the point mutation in the prime factor.

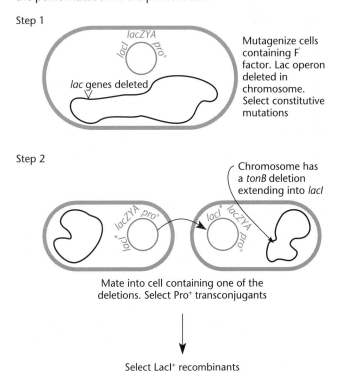

recombination between the mutant *lacI* gene on the F′ factor and the partially deleted *lacI* gene in the chromosome. The presence of any *lacI*+ recombinants in the population was evidence that the *lacI* mutation in the F′ factor lay outside the deleted region in the chromosome. However, any *lacI*+ recombinants, even if they exist in a particular cross, will be rare and must be selected. The selection method used also depends on the type of mutation being mapped.

MAPPING OF *lacI* CONSTITUTIVE MUTATIONS

To detect *lacI*+ recombinants in crosses with mutants that had constitutive *lacI* point mutations, the investigators used the fact that *galE* mutants are killed by galactose (see chapter 3). The *lac* deletion mutant strains were made mutant for *galE* by methods such as those described above for strain construction. The investigators could then use phenyl-β-D-galactoside (P-Gal) to select *lacI*+ recombinants. The selection was based on the fact that the β-galactosidase product of the *lacZ* gene can cleave the galactose off P-Gal and kill the *galE* mutant strains but P-Gal itself is not an inducer of the *lac* operon. In the absence of inducer, only strains that are still constitutive for the expression of the *lacZ* gene will cleave P-Gal and kill themselves while any *lacI*+ recombinants will not make β-galactosidase and will survive. As a consequence, if P-Gal is added to the partial-diploid *galE* mutant strain, all the bacteria that are constitutive for *lac* gene expression will be killed and only *lacI*+ recombinants will multiply to form a colony. Therefore, when the transconjugants in step 2 of Figure 14.2 were plated on agar containing P-Gal, the appearance of colonies on the plates was evidence that the *lacI* point mutation lies outside the deleted region.

MAPPING OF *lacI*ˢ MUTATIONS

The selection method outlined above could not be used to detect *lacI*+ recombinants with *lacI*ˢ mutant on their F′ plasmid. The *lacI*ˢ mutations are dominant, and so the *lacZ* gene will not be expressed to make β-galactosidase to cleave P-Gal and kill the *galE* mutant cells, even in partial diploids with a *lacI* deletion in the chromosome. Therefore, the transconjugants will grow on the P-Gal plates regardless of whether recombination with the chromosome occurs, rendering the detection of rare *lacI*+ recombinants impossible.

One way to detect possible *lacI*+ recombinants in crosses with *lacI*ˢ mutations depends on the fact that the partial diploids with a *lacI*ˢ mutation in the F′ factor and a *lacI* deletion in the chromosome cannot use lactose because the *lacZ* and *lacY* genes cannot be induced. Therefore, if the transconjugants are plated on minimal plates with lactose as the sole carbon and energy source, any *lacI*+ cells in which the *lacI*ˢ gene in the F′ factor has re-

combined with the *lacI* deletion in the chromosome will become inducible and will multiply to form a colony. However, another type of recombinant also will exhibit the Lac+ phenotype. In this recombinant, both the F′ factor and the chromosome have the *lacI* deletion, because the *lacI* deletion has been transferred to the F′ factor by recombination. This latter type of recombinant can occur even if the *lacI*ˢ mutation lies in the deleted region, but it can be distinguished from the other type because it is constitutive. Therefore, to determine whether the *lacI*ˢ mutation lies outside the deleted region, only Lac+ recombinants that are *lacI*+ and not constitutive should be counted. In the experiment, the two types of Lac+ recombinants were distinguished by replicating the colonies onto X-Gal plates in the presence and absence of the inducer isopropyl-β-D-thiogalactopyranoside (IPTG). Only the *lacI*+ recombinants should form blue colonies in the presence of IPTG and colorless colonies in the absence of IPTG. The *lacI* constitutive recombinants should form blue colonies on both types of plates.

IMPLICATIONS OF THE MAP POSITIONS OF DIFFERENT TYPES OF *lacI* MUTATIONS FOR THE STRUCTURE OF THE LacI REPRESSOR

The locations of the various types of mutations in the *lacI* gene suggested that the LacI repressor protein is divided into domains much like the λ repressor. The fact that *lacI*ᵈ mutations mostly map in the region encoding the N terminus of the polypeptide suggested that the N-terminal domain probably binds to DNA. Other types of *lacI* mutations are scattered throughout the gene, suggesting that the domain of the protein involved in tetramer formation is less well defined. The *lacI*ˢ mutations are also somewhat scattered throughout the gene but tend to be clustered in a few places. Some of these sites presumably form the binding site on the repressor for the inducer, although some may also cause more general conformational changes that interfere with inducer binding. This analysis of the structure of the LacI repressor is still under way and serves as a prototype for mutational analyses of proteins. The three-dimensional structure of the LacI repressor protein is now known, and structural studies have now largely confirmed the conclusions of this elegant genetic analysis.

Isolation of Tandem Duplications of the *his* Operon in *Salmonella enterica* Serovar Typhimurium

The process of selecting tandem duplication mutations is a good example of the genetic analysis of chromosome structure in bacteria. As discussed in a general way in chapter 3 and in connection with *r*II duplications of phage T4 in chapter 7, tandem-duplication mutations can occur by recombination between directly repeated sequences, causing the duplication of the DNA between the repeated sequences. However, once they form, tandem-duplication mutations are usually very unstable because recombination anywhere within the duplicated segments can destroy the duplication.

Most long tandem-duplication mutations do not cause easily detectable phenotypes because they do not inactivate any genes. Even the genes in which the mistaken recombination occurred to create the duplication exist in a functional copy at the other end of the duplication. Therefore, special methods must be used to select bacteria with duplication mutations.

Transduction has been used to select tandem duplications of the *his* region of *Salmonella enterica* serovar Typhimurium (see Anderson et al., Suggested Reading). The products of the genes of the *his* operon are required to make the amino acid histidine. Their selection depends on the properties of two deletion mutations in the *his* region, Δ*his2236* and Δ*his2527* (Figure 14.3). These deletion mutations complement each other because one

ends in *hisC* and the other ends in *hisB* so they are not overlapping. However, because the endpoints of the two deletion mutations are very close to each other, crossovers between them will occur very infrequently.

The process of using these deletions to select duplications of the *his* region is illustrated in Figure 14.4. P22 transducing phage were propagated on a strain with one of the deletion mutations and used to transduce a strain with the other deletion mutation. The His⁺ tranductants were then selected by plating on minimal plates without histidine. A few His⁺ transductants arose, even though there should be very little recombination between the deletions. Moreover, many of the His⁺ transductants that arose were unusual. Most were very unstable, spontaneously giving off His⁻ segregants at a high frequency when they multiplied in the presence of histidine. Also, some of the His⁺ transductants were slimy (mucoid) in appearance but the His⁻ segregants had lost this mucoidy. On the basis of these observations, these investigators

Figure 14.3 Mechanism for generation and destruction of *his* duplications. One of the two copies of the *his* region of bacteria with a preexisting duplication (in the example, Δ*his2527*) recombines with incoming transducing DNA carrying the other deletion (in the example, Δ*his2236*) to replace one copy of the duplicated region with the corresponding region of the donor. The two deletions complement each other to make the bacteria His⁺. The duplication can be destroyed by looping out one of the two duplicated segments, giving rise to His⁻ segregants with one or the other of the two deletions.

Figure 14.4 Using transduction to select bacteria with duplication mutations of the *his* region of *Salmonella enterica* serovar Typhimurium. The P22 transducing phage are grown on bacteria with one deletion and used to transduce bacteria with the other deletion. The transductants are plated on medium without histidine to select for the His⁺ phenotype. See the text for details and conclusions.

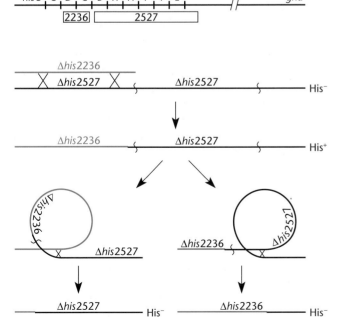

Step 1   Grow phage P22 in *S. enterica* serovar Typhimurium with one *his* deletion

Step 2   Use phage to transduce bacteria with the other *his* deletion

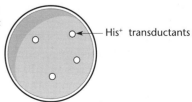

Step 3   Plate on medium without histidine; incubate — His⁺ transductants

concluded that unstable His⁺ transductants had tandem duplications of the *his* region (Figure 14.3), in which one copy has the Δ*his2236* deletion whereas the other has the Δ*his2527* deletion. The two deletion mutations can complement each other, making the transductants His⁺.

Figure 14.3 also shows how these duplications might have arisen. Recombination might have occurred between directly repeated sequences in some of the recipient bacteria while they were multiplying as shown. When a bacterium that contains a duplicated *his* operon with one deletion mutation is transduced with the *his* region having the other deletion, the *his* region of the transducing DNA can replace one of the *his* regions of the recipient cell, giving rise to the duplication in which one copy has one deletion and the other copy has the other deletion. This type of His⁺ transductant, although rare, might be more frequent than His⁺ transductants produced from recipients that do not have a duplicated *his* operon, since the latter would require recombination in the small region between the two deletions.

The unusual properties of the His⁺ transductants observed in the experiment are explained by the duplication structure. As with other tandem duplication mutations, growing the duplications in the absence of selective pressure (i.e., in the presence of histidine) will give rise to His⁻ segregants at a high frequency because recombination anywhere in the duplicated regions will leave only one copy of the *his* region, as also illustrated in Figure 14.3. These cells are called haploid segregants because they have only one copy of the duplicated region and occur spontaneously without the need for genetic crosses (see chapter 3 for definitions). The mucoidy (see above) of some of the His⁺ transductants can also be explained. The duplicated region in some of the His⁺ transductants may contain another gene that makes colonies appear mucoid when twice as much of the gene product is synthesized.

A *recA* Mutation Helps Stabilize the Duplications

If the instability of the putative duplications results from recombination between the repeated segments, then *recA* mutations, which prevent homologous recombination, should stabilize the duplications. To test this hypothesis, the investigators introduced a *recA* mutation by Hfr crosses into cells containing some of the putative duplications. They introduced the *recA* mutation by selecting for a *serA* marker closely linked to *recA* by Hfr crosses as discussed in chapter 5. Ser⁺ transconjugants that had the *recA* mutation were identified by their sensitivity to UV light (see chapter 11). The tandem duplication in these *recA* recombinants was now stable and did not

give His⁻ haploid segregants even when grown under nonselective conditions in media with histidine.

Determination of the Length of Tandem Duplications

The investigators also used genetic experiments to determine the length of the segment duplicated in some of the strains. In particular, they wanted to know if the duplicated regions in some of the strains ever extended as far as the *metG* gene, about 2 min (~100,000 bp) away from the *his* region in the *E. coli* chromosome. For this experiment, the investigators selected *his* duplications in a strain that also had a *metG* mutation. Strains containing these duplications were propagated without histidine in the medium to maintain the duplication and then transduced a second time with phage propagated on a *metG*⁺ donor. The Met⁺ transductants were selected and allowed to segregate the haploids by growing them with histidine and methionine in the medium. The His⁻ segregants were then tested to determine if any were also Met⁻. The reasoning was that if the *metG* gene is included in the duplicated region in a particular duplication, there should be two copies of the *metG* gene and only one of these will have been transduced to *metG*⁺ (Figure 14.5). Some of the His⁻ haploid segregants should also be Met⁻, depending on whether they lose the copy which has been transduced to *metG*⁺ or the copy which retains the *metG* mutation. The results of this work revealed that many duplications of the *his* operon did include the *metG* gene. In fact, some duplications extended much farther, even as far as the *aroD* gene, which is 10 min (or 10% of the entire genome) away from *his*.

Figure 14.5 Determining if the duplicated segment in a *his* duplication extends as far as the *metG* gene. A duplication is made as before but in a strain with a *metG* mutation. A strain with the duplication to be tested is transduced with phage grown on *metG*⁺ bacteria. If the *metG* gene is in fact duplicated, one of the two copies of the *metG* gene will be transduced to *metG*⁺ and the strain will spontaneously give rise to Met⁻ segregants.

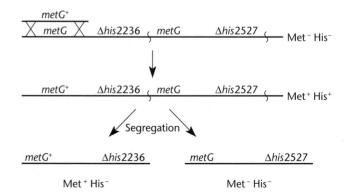

Frequency of Spontaneous Duplications

The investigators also attempted to estimate the frequency of spontaneous tandem duplication mutations in a growing population of bacteria. They did this by comparing tranduction frequencies when the donor and recipient bacteria had different deletion mutations, with the normal transduction frequencies, when only the recipient had one of the deletions. In the first case, most of the recipient cells that were transduced must have had a tandem duplication of the *his* region, whereas in the second case, most of the recipients will have had only one copy of the *his* region. The investigators estimated that duplications occur as frequently as once every 10^4 times a cell divides, which is hundreds of times greater than normal mutation frequencies. Apparently, spontaneous duplications in bacteria occur quite often during cell multiplication and can occur over very large regions of the chromosome. However, because they cause few phenotypic changes, most of these large duplications have little effect on the organism.

Analysis of Protein Transport in *E. coli*

Another elegant and historically important example of genetic analysis in bacteria was the isolation of mutants defective in protein transport. These studies allowed the identification of genes whose products are required for protein transport, and the processes turn out to be remarkably similar in bacteria and higher organisms. As discussed in chapter 2, about one-fifth of the proteins made in a bacterium do not remain in the cytoplasm but are transported into or through the surrounding membranes. Proteins that remain in either the inner or outer membrane are membrane proteins, while those that remain in the periplasmic space are periplasmic proteins. Proteins that are passed out of the cell into the surrounding environment are secreted proteins.

The pathway that is used to transport a protein depends upon its final destination (see chapter 2). Proteins which are transported into the periplasmic space and beyond generally use the SecB-SecA pathway and the SecYEG channel in the membrane. These proteins usually contain a short hydrophobic signal sequence on their N terminus which is cleaved off by proteases as they pass through the SecYEG channel. However, proteins destined for the inner membrane of bacteria are targeted by the SRP system, which includes the Ffh protein, a small 4.5S RNA, and a membrane receptor or "docking protein," FtsY. At least some inner membrane proteins may also use SecA and the SecYEG channel components of the *sec* system as well as another protein, YidC. Inner membrane proteins generally lack a signal sequence but have long hydrophobic transmembrane domains which traverse the membrane one or more times. Interspersed with these transmembrane domains are more hydrophilic periplasmic and cytoplasmic domains which face the periplasm and cytoplasm, respectively. Because they are very hydrophobic, these proteins are cotranslated as they are inserted into the membrane, to prevent their premature folding in the more aqueous polar environment of the cytoplasm.

We have mentioned a number of exported proteins in previous chapters. For example, the β-lactamase enzyme that makes the cell resistant to penicillin resides in the periplasm and so must be secreted through the inner membrane. The protein disulfide isomerases also reside in the periplasm and form disulfide linkages in some periplasmic or extracellular enzymes when they are in the periplasm. The maltose-binding protein, MalE, also resides in the periplasm, where it can help transport maltodextrins into the cell, while MalF is in the inner membrane. The *tonB* gene product of *E. coli* must also pass through the inner membrane to its final destination in the outer membrane, where it participates in transport processes and serves as a receptor for some phages and colicins. Sensor kinases like EnvZ reside in part in the inner membrane, where they can sense the external environment in the periplasm and communicate this information to response regulators in the cytoplasm.

Use of the *mal* Genes To Study Protein Transport: Signal Sequences; *sec*; and SRP

The products of many of the genes of the *mal* operons of *E. coli* (discussed in chapter 12) are membrane or periplasmic proteins and so must be transported through the inner membrane. In this section, we discuss how the *mal* genes have been used to identify the important features of proteins that allow them to be secreted as well as to identify other functions required for the secretion of proteins. These studies continue to be very important in informing our view of how protein transport occurs in all organisms.

ISOLATION OF MUTATIONS THAT AFFECT THE SIGNAL SEQUENCE OF THE MalE AND LamB PROTEINS

The products of the *lamB* and *malE* genes reside in the outer membrane and periplasm of *E. coli*, respectively (see Figure 12.19). To reach their final destination, these proteins must be secreted through the inner membrane. Like most secreted proteins, LamB and MalE are first made as precursor polypeptides with approximately 25 extra amino acids at their N-terminal ends. These amino acids, known as the signal sequence, are cleaved off as the polypeptides traverse the inner membrane (see chapter 2).

Gene fusion techniques and selectional genetics were used to determine which amino acids in the signal sequence of MalE and LamB were important for the secretion of these proteins (see Bassford and Beckwith, Suggested Reading). Mutations that cause amino acid changes in a short sequence like a signal sequence are rare and require a positive selection. The method capitalized on translational fusions that joined the N terminus of the MalE or LamB protein, including the signal sequence, to the LacZ protein, which is normally a cytoplasmic protein (see chapter 2 for an explanation of translational fusions). The N-terminal signal sequence will direct the fusion protein to the membrane, and the transport machinery will attempt to export it. However, for unknown reasons, the large fusion protein will become trapped in the membrane, killing the cells, perhaps by causing them to lyse.

The sensitivity of cells to transport of the fusion proteins offers a means of positively selecting transport-defective mutants (Figure 14.6). Since the fusion protein is made in large amounts only in the presence of the inducer maltose, which turns on transcription from the *mal* promoter (see Figure 12.18), bacteria containing these fusions are maltose sensitive (Mal^s) and are killed by the addition of maltose to the growth medium. However, bacteria with mutations that prevent transport of the fusion proteins will be resistant to maltose (Mal^r) and will survive. Therefore, any colonies that form when bacteria containing the gene fusions are plated on maltose-containing plates may be due to mutant bacteria that no longer can transport the fusion proteins, provided that they still make the fusion protein in large amounts.

Some of the Mal^r mutants may have mutations that change the coding region for the signal sequence in the MalE or LamB portion of the fusion protein so that the signal sequence no longer functions in transport. Then the fusion protein will no longer enter the membrane and kill the cell, and the cell will be resistant to maltose. Determining which amino acids in the signal sequence have been changed in these Mal^r mutants should reveal which amino acids in the signal sequence are required for signal sequence function. These mutations can be distinguished, by their map positions, from other mutations that cause defects in secretion. Signal sequence mutations should map in the *malE-lacZ* fusion gene, while mutations that affect other proteins required for transport should map elsewhere in the *E. coli* genome. Furthermore, mutations that change the signal or some other sequence in the MalE or LamB part of the fusion protein should specifically affect the transport of the fusion protein and should not affect the transport of other transported proteins.

On the basis of this selection, a number of signal sequence mutations were isolated and later sequenced to

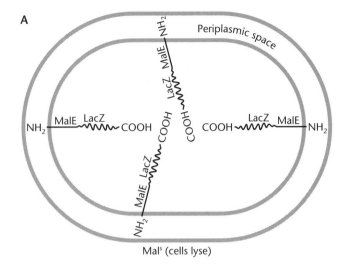

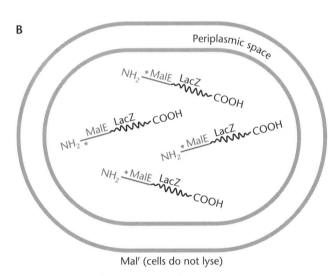

Figure 14.6 Model for the maltose sensitivity (Mal^s) of cells containing a *malE-lacZ* gene fusion. (A) In Mal^s cells, the presence of maltose induces the synthesis of the fusion protein, which cannot be transported completely through the membrane and so lodges in the membrane, causing the cell to lyse. (B) In Mal^r cells, a mutation in the region encoding the signal sequence (asterisk) prevents transport of the fusion protein into the membrane.

determine which amino acid changes could prevent the function of the signal sequence. Many of these were changes from hydrophobic to charged amino acids that might interfere with the insertion of the signal sequence into the hydrophobic membrane.

ISOLATION OF MUTATIONS IN *sec* GENES
The *mal* genes were also used to select mutants with mutations in genes whose products are part of the protein transport machinery (see Oliver and Beckwith, Sug-

gested Reading). The products of these genes, named the *sec* genes (for protein *sec*retion), are required for the transport of some or maybe even all of the membrane, periplasmic, and exported proteins. Therefore, unlike signal sequence mutations, which affect the transport of only one protein, *sec* mutations should cause defects in the transport of many proteins into or through the membranes and map in many places on the genome.

A different, somewhat more sensitive selection was used to isolate *sec* mutants from the selection that was used to isolate signal sequence mutations. This selection was based on the observation that cells containing a *malE-lacZ* fusion do make some MalE-LacZ fusion protein in the absence of maltose in the medium but do not make enough to kill the cells. Nevertheless, even though they do make some fusion protein and the fusion protein has β-galactosidase activity, the cells are not Lac⁺ and able to multiply with lactose as the sole carbon and energy source. The investigators reasoned that the cells may not exhibit the β-galactosidase activity because the fusion protein is being transported to the periplasmic space, where the normally cytoplasmic β-galactosidase may be inactivated by the formation of disulfide bonds between its cysteines by disulfide isomerases in the periplasm.

The fact that transport of the fusion protein inactivates its β-galactosidase activity, thereby preventing growth on lactose, offers a positive selection for *sec* mutations that prevent its transport. Any mutation that prevents the transport of the fusion protein through the membrane should make the cells Lac⁺ and able to form colonies on lactose minimal plates, because the MalE-LacZ fusion protein should remain in the cytoplasm and retain its β-galactosidase activity. Therefore, to isolate *sec* mutants, these investigators could merely plate cells containing the *malE-lacZ* fusion on minimal plates containing lactose but no maltose. They identified six different *sec* genes this way and named them *secA*, *secB*, *secD*, *secE*, *secF*, and *secY*. As we discussed above and in chapter 2, some *sec* gene products are required to form a channel in the membrane through which other proteins pass. Others are chaperones that are required to target signal sequence-containing proteins to the membrane channel. Figure 14.7 illustrates the role of SecB and shows why a *secB* mutant could be selected by this procedure.

ISOLATION OF MUTATIONS IN THE SRP PATHWAY FOR INNER MEMBRANE PROTEINS

The *mal* genes have also been used to isolate mutations in genes of the signal recognition particle (SRP) pathway which targets proteins to the inner membrane (see Tian and Beckwith, Suggested Reading). The method used to isolate mutations in genes that target proteins to the

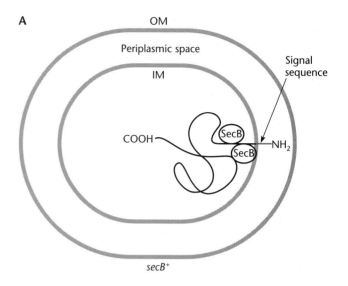

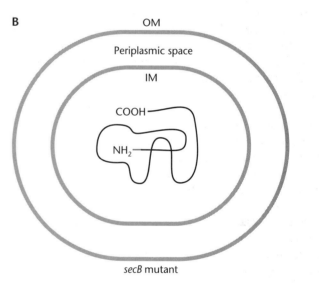

Figure 14.7 Function of SecB in protein transport. (A) The SecB protein (black) prevents premature folding of a protein, keeping the signal sequence (gold) exposed so that it can enter the membrane after protein synthesis is complete. (B) In a *secB* mutant, the signal sequence folds into the interior of the protein and so the protein cannot be transported. OM, outer membrane; IM, inner membrane.

inner membrane is similar to that used above to isolate mutations in genes required for transport and signal sequence mutations, except that rather than using the MalE protein, which resides in the periplasm, they used the MalF protein, which resides in the inner membrane (see Figure 12.19). They fused the N-terminal coding region for MalF, including the first transmembrane domain and first periplasmic domain, to *lacZ* and introduced this gene fusion into a phage λ vector, which they then integrated into the chromosome. Now, however, rather than selecting for sensitivity to maltose or ability to grow on lactose as a sole carbon source, they looked for blue colonies on X-Gal plates. Cells containing this fusion normally make white colonies on X-Gal plates because the MalF portion of the fusion protein is transported through the inner membrane, dragging the β-galactosidase portion along with it into the periplasm. Once in the periplasm, the β-galactosidase is inactive because its cysteines bcome cross-linked, as discussed above. The inactive β-galactosidase cannot cleave the X-Gal on the plates and turn the colonies blue, and so they remain white.

These investigators reasoned that they might be able to use this system to isolate mutants of *E. coli* with mutations in genes required to transport proteins into the inner membrane. If a mutation in such a gene prevents the transport of the MalF portion of the fusion protein into the inner membrane so that it remains in the cytoplasm, the β-galactosidase to which it is fused will also remain in the cytoplasm. If the β-galactosidase remains in the cytoplasm, it will not be cross-linked by the Dsb proteins and will remain active, making the mutant colonies blue on X-Gal plates. The mutants they isolated this way had mutations in the *ffs* gene, encoding the protein of the SRP, in the *ffh* gene, encoding the 4.5S RNA part of the SRP, and in the *ftsY* gene, encoding the docking protein. Surprisingly, the only mutations they found in a *sec* gene were in *secM*, whose product plays a role in regulating SecA. Why they did not find mutations in other *sec* genes, if the products of these genes also play a role in transporting proteins into the inner membrane, is unclear.

These are just a few examples of how genetic analysis has been used to study the requirements for protein transport in bacteria. As mentioned, many aspects of protein transport are now known to be similar in bacteria and eukaryotes (see chapter 2). The *sec* channels are similar in the inner membrane of bacteria and the endoplasmic reticulum of eukaryotic cells, which play related roles. Both secreted proteins of bacteria and eukaryotes have signal sequences (see below). An analog to the RNA-containing SRP, which targets proteins to the endoplasmic reticulum of eukaryotes, also exists in *E. coli*, where its role is to target proteins to the inner membrane. The highly specialized structures involved in protein secretion in eukaryotes such as the endoplasmic reticulum and the Golgi apparatus may be necessitated by the larger size of eukaryotic cells (see the introductory chapter).

Genetic Analysis of Transmembrane Domains of Inner Membrane Proteins

Gene fusions have also been used to determine the topography of proteins in the inner membrane of *E. coli*. Rather than being completely buried in the membrane, most inner membrane proteins have regions that are exposed to the cytoplasm, the periplasm, and/or the external environment. These exposed regions allow the membrane protein to pass information from the external environment to the interior of the cell or from one cellular component to another. Proteins that are exposed at both surfaces of a membrane are called transmembrane proteins, and the regions of the polypeptide that traverse the membrane from one surface to the other are called the transmembrane domains. Some transmembrane proteins traverse the membrane many times. The transmembrane domains alternate with those exposed to the cytoplasm (cytoplasmic domains) and the periplasm (periplasmic domains). The term "membrane topology" refers to the way the different sections of the protein are distributed in the membrane.

The transmembrane domains of a polypeptide can often be distinguished from the periplasmic or cytoplasmic domains merely by the primary sequence of the gene. Because the transmembrane domains are embedded in the hydrophobic membrane, they are composed mostly of hydrophobic amino acids such as phenylalanine and tyrosine (see the inside cover of this book), which makes them more soluble in hydrophobic environments. The periplasmic and cytoplasmic domains have a larger number of charged amino acids, such as arginine, lysine, or glutamate, which make them more soluble in the ionic environments outside the membrane. However, whether a domain is in the periplasm or the cytoplasm cannot be determined from the amino acid sequence alone.

Gene fusions to the alkaline phosphatase gene (*phoA*) of *E. coli* have been used to identify *E. coli* genes that encode transmembrane proteins and to study the membrane topology of inner membrane proteins of *E. coli* (see Hoffman and Wright, and San Milan et al., Suggested Reading). The alkaline phosphatase product of the *phoA* gene is a scavenger enzyme, which can obtain phosphates from larger molecules for cellular reactions. To fulfill this role, the alkaline phosphatase resides in the periplasmic space, where it can obtain phosphates from molecules that are too large or too ionic to be easily transported. Alkaline phosphatase removes the phosphate, which can then be transported into the cell by itself, leaving the remainder of the molecule outside.

The property of *E. coli* alkaline phosphatase which makes it so useful for studying the membrane topology of proteins is that it is active only in the periplasm and not in the cytoplasm. To be active, the enzyme must form a homodimer of two identical polypeptide products of the *phoA* gene which are held together by disulfide bonds between their cysteines. Disulfide bonds form only in the oxidizing environment of the periplasm, where the protein disulfide isomerase enzymes that form disulfide bonds reside, and not in the reducing environment of the cytoplasm (see chapter 2). The PhoA enzyme is also easy to assay, and bacteria which synthesize active PhoA make blue colonies on plates containing the chromogenic compound 5-bromo-4-chloro-3-indolylphosphate (XP), which turns blue when the phosphate is cleaved off by alkaline phosphatase.

The way *phoA* translational fusions can be used to determine the domains of a transmembrane protein that are in the periplasm is illustrated in Figure 14.8. Briefly, the carboxy-terminal coding region for PhoA, without its signal sequence, is fused to various lengths of the coding sequence for the N terminus of the transmembrane protein. If the region of the protein to which PhoA is fused is in the periplasm, i.e., is a periplasmic domain, cells containing the fusion will make blue colonies on XP plates. However, if the region of the protein is a cytoplasmic domain, the colonies will be colorless.

Identification of Genes for Transported Proteins by Random *phoA* Fusions

Some transposons have been engineered to generate random gene fusions by transposon mutagenesis (see, for example, Mud*lac* in chapter 9). These transposons contain a reporter gene that will be expressed only if the transposon hops into an expressed gene in the correct orientation. One such transposon, Tn*phoA*, was developed to identify genes whose protein products are transported into or through the inner membrane (see Gutierrez et al., Suggested Reading). Tn*phoA* has the *phoA* reporter gene inserted so that it does not have its own promoter or translation initiation region and lacks its own signal sequence coding region. A fusion protein with PhoA fused to another protein will be synthesized whenever the transposon hops into an expressed open reading frame (ORF) in the right reading frame. In addition, the fusion protein will have alkaline phosphatase activity and will turn colonies blue if the Tn*phoA* has integrated into a gene whose protein product is translocated and the *phoA* gene happens to be fused to a periplasmic domain of the protein.

Figure 14.8 Using *phoA* fusions to determine the membrane topology of a transmembrane protein. (A) A transmembrane protein. (B and C) A fusion that joins the transmembrane protein to alkaline phosphatase (AP) at x or z leaves the alkaline phosphatase in the periplasm where it will be active. The bacteria form blue colonies on XP plates. (D) The transmembrane protein is fused at y to alkaline phosphatase, leaving the alkaline phosphatase in the cytoplasm, where it is inactive. The bacteria form colorless colonies on XP plates.

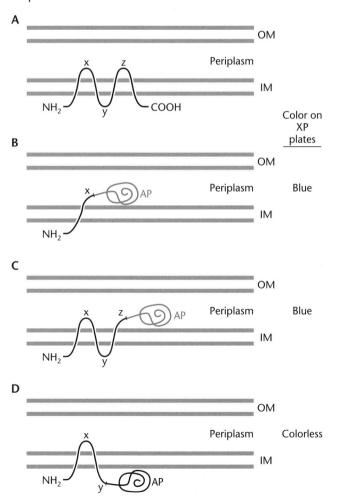

Replication Control of ColE1-Derived Plasmids

Many molecular genetic analyses have addressed questions of plasmid function. One such analysis addressed the copy number control and incompatiblity (Inc) grouping of ColE1 related plasmids (see Lacatena and Cesareni, Suggested Reading). Chapter 4 contains an in-depth discussion of the subject of ColE1 regulation, but to reiterate briefly, the ColE1-type plasmids have a high copy number and use two plasmid-encoded RNAs, RNA I and RNA II, to regulate their replication. RNA II is the primer for initiation of replication; it must be processed before it can function. RNA I inhibits replication by binding to RNA II and preventing its processing. The more copies of the plasmid the cell contains, the more RNA I is made and the more plasmid replication is inhibited. Thus, the plasmid replicates until its concentration reaches its copy number. Then the high concentration of RNA I inhibits further replication of the plasmid.

The RNA I of a ColE1 plasmid is also responsible for determining the incompatibility (Inc) group of the plasmid. Two plasmid members of the same Inc group do not coexist for long in the same cell because they synthesize the same RNA I and inhibit each other's replication.

Genetic studies of plasmid replication control are complicated because mutations that inactivate the plasmid origin of replication or allow its regulation to run uncontrolled cannot be isolated since they will make production of the plasmid impossible or kill the host cell, respectively. Therefore, a different approach was necessary to study the copy number control and incompatibility determinants of ColE1-type plasmids. To isolate mutations in the ColE1 plasmid origin of replication, the investigators constructed a **phasmid** (part phage and part plasmid) with two *ori* regions, one from ColE1 and the other from λ, by inserting the ColE1 plasmid into a λ

cloning-vector phage (Figure 14.9). The phasmid also contained *attB* and *attP* sites on either side of the plasmid so that the plasmid could be excised from the phasmid merely by introducing the phasmid into cells containing excess λ integrase enzyme. In such cells, the λ integrase will promote recombination between the *attB* and *attP* sites excising the plasmid. Once out of the phasmid, the mutant plasmids could be studied to determine whether they could regulate their own replication and that of other plasmids and phasmids.

With two origins of replication, the phasmid can replicate and form plaques under conditions that inactivate either one of its origins of replication, as long as the other origin remains functional (Figure 14.9). The plas-

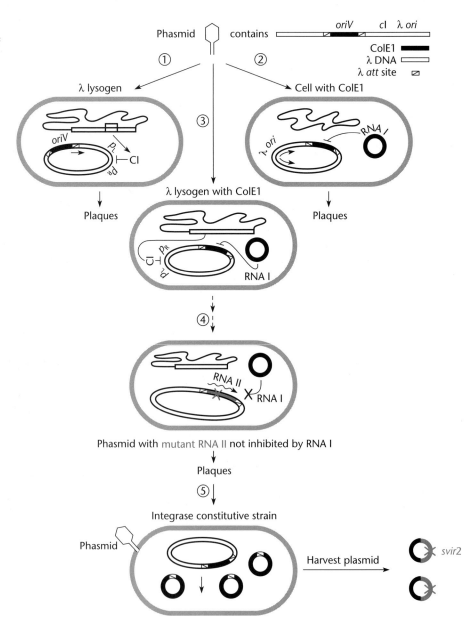

Figure 14.9 A phasmid method for isolating copy number mutants of the ColE1 plasmid. The phasmid has two origins of replication and so can replicate in either a λ lysogen or cells harboring a ColE1-derived plasmid.

mid origin of replication will be inactive if the cells contain another ColE1-derived plasmid of the same Inc group, and the λ phage origin of replication will be inactive if the phasmid is plated on a λ lysogen. In the former case, the RNA I from the resident ColE1 plasmid will inhibit replication from the ColE1 *ori* site of the phasmid. In the latter case, the repressor made by the λ prophage prevents the synthesis of the λ O and P proteins as well as the transcription of the λ *ori* region, which are all required for replication from the λ origin (see chapter 8). The phasmid can replicate and form plaques on cells that are λ lysogens or that contain a ColE1-derived plasmid of the same Inc group, but it is replicating from different origins under these conditions (Figure 14.9, steps 1 and 2). Only if the cells harbor both a λ prophage and a ColE1-derived plasmid will plaques not form (step 3).

The investigators reasoned that it might be possible for the phasmid to replicate on a λ lysogen carrying a ColE1-type plasmid if the ColE1 replication origin were mutated so that its replication was no longer inhibited by the resident plasmid. Accordingly, they plated millions of mutagenized phasmids on a λ lysogen carrying a ColE1 plasmid. A few plaques formed, presumably from phasmids with mutations in either the plasmid or the λ origin of replication (step 4). Phasmids picked from these plaques were then used to infect cells with excess λ integrase (step 5) in order to excise the plasmid, as described above, and determine whether its control of replication had been altered. Two types of mutations in the ColE1 *ori* region were observed and are discussed below.

Mutations That Prevent Interaction of RNA I and RNA II

No transformants appear when cells are transformed with plasmids with this type of mutation. Presumably, these mutations cause the transformants to be killed, apparently by uncontrolled, runaway replication of the plasmid. Presumably, the RNA I is no longer able to interact with RNA II to inhibit primer formation, and the plasmid continues to replicate until the cells are killed. Note that such a mutation will not prevent plaque formation by the phasmid, since the phasmid replication does not need to be controlled for a plaque to form.

Mutations That Change the Inc Group

The other type of mutation, illustrated in Figure 14.10A, is more intriguing. The plasmids in these mutant phasmids can be excised and used to transform bacteria, where they are able to maintain their copy number.

Figure 14.10 (A) Specificity tests for incompatibility groupings. Some RNA I mutants of ColE1-derived plasmids form their own Inc group. The mutant phasmid will not form plaques when plated on cells containing its own excised mutant plasmid but will form plaques on cells containing the original ColE1 plasmid or any other mutant plasmid. (B) Location of *svir* copy number control mutations in RNA I of the ColE1-type plasmid. The mutations lie in what would be the anticodon region of a tRNA-like structure. Adapted with permission from R. M. Lacatena and G. Cesareni, *Nature* (London) **294:**623–626, 1981. Copyright 1981, Macmillan Publishers, Ltd.

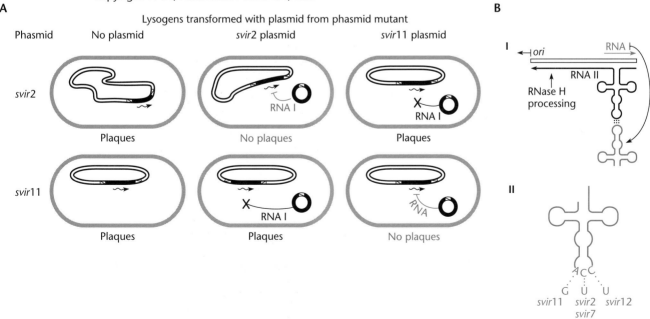

However, the mutant phasmids from which these plasmids are excised will not form plaques on cells containing their own excised plasmid, even though they will form plaques on any cell containing a different excised plasmid. For example, a phasmid containing the *svir2* mutation in its ColE1 *ori* region will not be able to form plaques on a λ lysogen containing the plasmid with the *svir2* mutation but will be able to form plaques on cells containing the plasmid with the *svir11* mutation and vice versa, as illustrated in the figure.

We can deduce what is happening by examining the model for copy number control of ColE1 plasmids shown in Figure 4.6. According to the model, RNA I must bind to RNA II through complementary base pairing to inhibit plasmid replication by interfering with the processing of the primer RNA II. Because the two RNAs are made from opposite strands of the DNA in the same region, a mutation which changes RNA I will make the complementary change in RNA II, and so the two will still be able to pair with each other and inhibit plasmid replication. Therefore, the mutant phasmid which depends upon its ColE1 replication origin to replicate and form a plaque in a λ lysogen will not be able to do so in the presence of the mutant plasmid excised from it, which has the same mutation and therefore the same mutant RNA I. However, it will be able to replicate and form a plaque in any cell containing a different mutant plasmid which makes a different mutant RNA I. Apparently, the origin of replication of the mutant plasmid now belongs to a different Inc group from other ColE1-derived plasmids. Indeed, it now defines its own Inc group! In any case, the fact that changing RNA I and RNA II in complementary ways can create new Inc groups offers a dramatic confirmation of the model for ColE1 regulation presented in chapter 4.

The positions of some of the changes that alter the incompatibility group are also intriguing. Figure 14.10B shows some of these changes. RNA I can be drawn as a cloverleaf structure reminiscent of a tRNA, and the changes occur in what would correspond to the anticodon loop. However, it is not clear whether this cloverleaf structure ever forms and what relationship, if any, this structure (if it does form) has to the regulation. In any case, this loop in RNA I must be important for the pairing of RNA I with RNA II and forms the first "kissing complex" with RNA II (see chapter 4).

Identification and Mapping of the *tra* Genes on a Plasmid

Another plasmid function which has been analyzed genetically is conjugation. Genetic analysis allowed the detection and mapping of the *tra* genes on plasmids and preliminary studies on their functions in conjugation. As discussed in chapter 5, the products of the *tra* genes are required for plasmid transfer during conjugation. Every self-transmissible plasmid must encode its own Tra functions, and a large percentage of these plasmids are devoted to encoding such functions. Some *tra* genes encode proteins required for mating-pair formation (Mpf); these include proteins required for pilus assembly and for formation of the mating channel and the coupling protein. Others are required for DNA transfer (Dtr) and include the relaxase and primases.

To illustrate the genetic analysis of conjugation, we shall describe experiments for the identification and mapping of *tra* genes on plasmid pKM101 (see Winans and Walker, Suggested Reading, and Figure 5.1).

Isolation of *tra* Mutant Plasmids

The first step in identifying the *tra* genes of plasmid pKM101 was to mutagenize the plasmid with the Tn5 transposon to obtain a large number of transferless (Tra⁻) mutants. One of the two methods used was illustrated in Figure 9.17 in the discussion of random transposon mutagenesis of a plasmid. Cells containing plasmid pKM101 were infected with a λ suicide vector containing transposon Tn5, and plasmids with transposon insertion mutations were isolated by preparing the plasmids and using them for transformation. Kanamycin-resistant transformants, each of which contained the plasmid with a transposon insertion in a different region of the plasmid, were isolated. The second method was the same initially, in that the λ suicide vector carrying the transposon was used to infect cells containing the plasmid. However, after being infected, the cells were incubated only long enough to give the transposon time to hop. The cells were then mixed with recipient bacteria, and kanamycin-resistant transconjugants were isolated. In general, only cells in which the transposon had hopped into the plasmid would be able to transfer kanamycin resistance to the recipient, since usually only the plasmid would be transferred. Even if the transposon had happened to hop into a *tra* gene in the plasmid, the plasmid would still transfer, because after such a short incubation, the cells would still contain some of the Tra function owing to phenotypic lag (see chapter 3). Using conjugation rather than transformation to isolate plasmids with transposon insertions has the advantage that conjugation is often more efficient than transformation, making it easier to isolate many mutant plasmids.

Once the investigators had isolated several transposon insertions in the plasmid, they needed to determine which of the transposon insertions had inactivated a *tra* gene. If the transposon had hopped into, and therefore inactivated, a *tra* gene, the plasmid would no longer be able to transfer itself. Therefore, each of the mutagenized plasmids was tested individually for its ability to transfer.

To streamline the process of screening the mutants, the investigators devised the replica-plating test for transfer illustrated in Figure 14.11. In the example, a number of plasmid-containing bacteria were patched onto a plate so that each patch would have bacteria containing a different mutagenized plasmid. After incubation to allow multiplication of the bacteria in the patches, this plate was replicated onto another plate containing rifampin to counterselect the donor (see the discussion of Hfr mapping in chapter 5) and rifampin-resistant recipient bacteria. After incubation to allow any potential transconjugants to form and multiply, the second plate was replicated onto a third plate containing medium with both kanamycin and rifampin. If the Tn5 insertion in the plasmid did not inactivate a *tra* gene, the plasmid would have transferred into some of the recipient bacteria on the second plate, yielding

transconjugants that are both kanamycin and rifampin resistant. However, if the transposon had inserted into a *tra* gene, no kanamycin- and rifampin-resistant transconjugants should appear on the third plate. In this way, several mutant plasmids were identified that had a transposon insertion in one of their *tra* genes. The site of the transposon insertions in these *tra* plasmids could then be physically mapped by methods such as those described in chapter 9 in the discussion of transposon mutagenesis of plasmids.

Complementation Tests To Determine the Number of *tra* Genes

Mapping the site of insertion of the transposon in a *tra* mutant plasmid reveals where a *tra* gene is located on the plasmid but does not, by itself, reveal the number of *tra* genes. Only complementation tests can determine how many genes are represented in the collection of *tra* mutants. To do complementation tests, plasmids containing two different tra mutations must be introduced into the same cell. If the mutations inactivate different genes, each mutant plasmid will furnish the Tra function that the other one lacks, and one or both plasmids will be able to transfer. If, however, the two mutations inactivate the same *tra* gene, neither plasmid will be able to make the product of that gene and neither plasmid will be able to transfer. To save time, these complementation tests need be done only between *tra* mutants with insertion mutations close to each other on the plasmid. Any two transposon insertion mutations which are not closely linked or which have transposon insertions between them that do not inactivate a *tra* gene are almost certainly not in the same *tra* gene, and so they do not need to be tested by complementation.

Complementation tests between different *tra* mutations in the same plasmid are complicated by the fact that they are in the same Inc group so will not coexist for long. Figure 14.12 shows one way to overcome this difficulty. The region of plasmid pKM101 containing one of the *tra* mutations (tra_1 in the figure) was inserted into a plasmid cloning vector from a different Inc group. The plasmid containing the clone was then used to transform cells carrying the pKM101 plasmid with the other *tra* mutation (tra_2). Since the plasmids carrying the two *tra* mutations are members of different Inc groups, they will stably coexist, and if the two *tra* mutations are in different genes, they will complement each other and the mutant PKM101 plasmid will transfer into recipient cells. However, the cloned tra_1 region must include the entire gene, and the gene must be expressed on the cloning vector plasmid.

The other way to avoid the problem of incompatability is by "transient heterozygosis." Even if plasmids are in the same Inc group, they can coexist transiently, allowing time for complementation. Therefore, if one

Figure 14.11 Replica plating to find *tra* plasmid mutants resulting from insertion of a transposon. Cells containing the plasmid with the transposon inserted in different places are patched onto a plate without antibiotics, and the plates are incubated to allow the formation of colonies. The following steps are as indicated in the illustration. Failure of a colony to develop on the third plate after incubation indicates that the transposon is in a *tra* gene of the plasmid, so that no transconjugants had formed on the second plate.

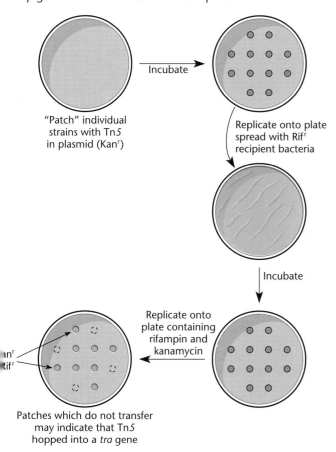

"Patch" individual strains with Tn5 in plasmid (Kan^r)

Incubate

Replicate onto plate spread with Rif^r recipient bacteria

Incubate

Replicate onto plate containing rifampin and kanamycin

an^r
Rif^r

Patches which do not transfer may indicate that Tn5 hopped into a *tra* gene

① Clone one mutant *tra* region from a *tra* mutant plasmid

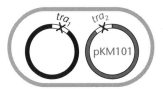

② Transform into a cell with a different *tra* mutant plasmid

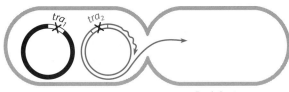

Figure 14.12 Testing for complementation between two *tra* mutations. The region containing one of the *tra* mutations is cloned into a cloning vector from a different Inc group. If the two mutations complement each other, the plasmid can transfer, as determined by methods such as that outlined in Figure 14.11. The cloning vector is shown in black.

③ Test for transfer to a recipient. If transfer occurs, the mutations complement

Recipient

mutant plasmid is used to transform cells containing the other mutant plasmid and the cells are immediately mixed with a recipient, the presence of any kanamycin-resistant transconjugants will be evidence of complementation between the insertion mutations in the two plasmids.

By using these procedures, the investigators estimated the number of *tra* genes on pKM101. However, this number may be an underestimation of the total number of *tra* genes because it is possible that not all the *tra* genes have transposon insertions. Moreover, transposon insertion mutations are sometimes polar (see chapter 2 for an explanation of polarity), so that two mutations in different genes can behave as though they were in a single gene in complementation tests. In spite of these difficulties, the investigators identified most of the *tra* genes of plasmid pKM101 and, by mapping the site of insertions that inactivated each of the *tra* genes, obtained the map of pKM101 shown in Figure 5.1. Subsequent experiments were done to find the *oriT* region of the plasmid and to begin to determine the functions of the products of each of the *tra* genes in conjugation. Such experiments are discussed in a general way in chapter 5.

Genetic Analysis of Sporulation in *Bacillus subtilis*

As mentioned in the introductory chapter, many bacteria undergo complex developmental cycles. In their development, some bacteria perform many functions reminiscent of higher organisms. They undergo regulatory cascades. Their cells communicate with each other and differentiate and form complex multicellular structures. The cells in these multicellular structures often perform different distinct functions which requires compartmentalization and cell-cell communication. The cells use phosphorelays to respond to changes in communication with other cells and with the external environment. Because of the relative ease of molecular genetic analysis with some bacteria, some of these developmental processes have been extensively investigated as potential model systems for even more complex developmental processes in higher organisms (see Braun and Shimkets, Suggested Reading).

The best understood bacterial developmental system is sporulation in *Bacillus subtilis*. When starved, *B. subtilis* cells undergo genetically programmed developmental changes. The cells first attempt to obtain nutrients from neighboring organisms by producing antibiotics and extracellular degradative enzymes. If starvation conditions persist, the cells sporulate, producing endospores that are metabolically dormant and highly resistant to environmental stresses.

The process of sporulation starts with an asymmetric division that produces two cell types with different morphological fates. The larger cell, which is called the mother cell, engulfs the smaller forespore and then nurtures it. Eventually the mother cell lyses, releasing the endospore.

Many of the changes that occur in the sporulating cell can be visualized by electron microscopy. These are schematized in Figure 14.13. The figure also shows the

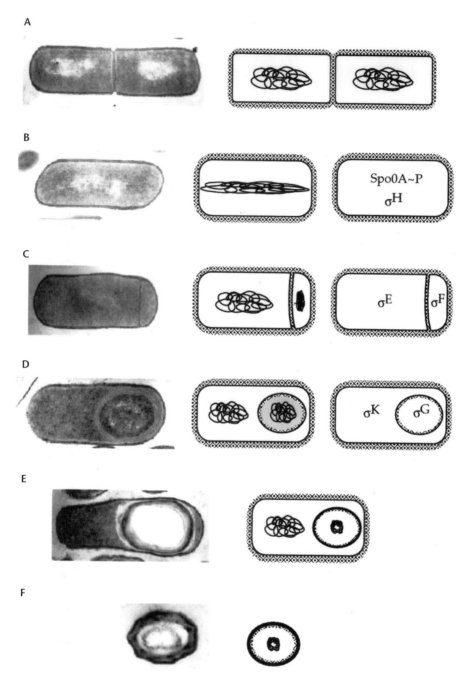

Figure 14.13 Stages of sporulation. The left of each panel shows an electron micrograph (EM) of the stage of sporulation, and the right shows, in cartoon form, the disposition of the chromosomes and the time and site of action of the principal regulatory proteins that govern sporulation gene expression. (A) Vegetative cells. (B through E) Sporangia at entry into sporulation (stage 0) (B), at polar division (also called polar septation) (stage II) (C), at engulfment (stage III) (D), and at cortex and coat formation (stages V to VI) (E). (F) A free spore. Reprinted with permission from P. A. Levin and R. Losick, Asymmetric division and cell fate during sporulation in *B. subtilis,* p. 167–190, *in* Y. V. Brun and L. J. Shimkets (ed.), *Prokaryotic Development,* ASM Press, Washington, D.C., 2000, and reprinted with permission from A. Driks, Spatial and temporal control of gene expression in prokaryotes, p. 23, Fig. 2 (A–F), *in* V. E. A. Russo, D. J. Cove, L. G. Edgar, R. Jaenisch, and F. Salamini (ed.), *Development: Genetics, Epigenetics and Environmental Regulation,* Springer-Verlag, Berlin, Germany, 1999. EMs are provided courtesy of A. Driks.

proteins that have been identified as key regulators of specific stages of development. We describe the experiments that identified these key regulators below.

Identification of Genes That Regulate Sporulation

Isolation of mutants was crucial to the process of identifying the important regulators of sporulation. Many mutants were isolated on the basis of a phenotype referred to as Spo⁻ (for *sporulation-minus*). Such mutants could be identified as nonsporulating colonies, because plate-grown cultures of the wild type develop a dark brown spore-associated pigment whereas the nonsporulators remain unpigmented.

Spo⁻ mutants were phenotypically characterized by electron microscopy and then were grouped according to the stage at which development was arrested (Figure 14.14). Some of the key regulatory genes defined by analysis of the mutants are listed in Table 14.2. The names of *B. subtilis* sporulation genes reflect three aspects of the genetic analysis of these genes. The roman numerals refer to the results of phenotypic categorization of the mutant stains, with the numbers 0 through V indicating the stages of sporulation at which mutants were found to be blocked. The gene names also contain one or two letters. The first letter designates the separate loci that mutated to cause similar phenotypes. Each such locus was defined by the set of mutations that caused the same morphological block and that were genetically closely linked. The second letter in the names indicate the individual ORFs that were found when DNA sequencing revealed that a locus contained several ORFs.

Regulation of Initiation of Sporulation

Much of what we understand about the mechanism of sporulation initiation is based on studies of the class of

TABLE 14.2	*Bacillus subtilis* sporulation regulators	
Stage of mutant arrest	**Gene**	**Function**
0	spo0A	Transcription regulator
	spo0B	Phosphorelay component
	spo0F	Phosphorelay component
	spo0E	Phosphatase
	spo0L	Phosphatase
	spo0H	Sigma H
II	spoIIAA	Anti-anti-sigma
	spoIIAC	Sigma F
	spoIIE	Phosphatase
	spoIIGA	Protease
	spoIIGB	Pro-sigma E
III	spoIIIG	Sigma G
IV	spoIVCB-spoIIIC^a	Pro-sigma K
	spoIVF	Regulator of sigma K

^{a}In *B. subtilis* and some other bacilli, two gene fragments undergo recombination to produce the complete sigma K coding region.

sporulation-minus mutants, which were designated *spo0* because of their failure to begin the sporulation process. Many of these mutants have pleiotropic phenotypes, meaning that they are altered in several characteristics. Besides being unable to sporulate, they fail to produce the antibiotics or degradative enzymes that are characteristically produced by starving cultures and they do not develop competence for transformation (see chapter 6).

Two of the *spo0* genes, *spo0A* and *spo0H*, encode transcriptional regulators. *spo0A* encodes a "two-component system" response regulator (see Box 13.3) that is responsible for regulating the cellular response to starvation. The product of *spo0H* is a sigma factor (σ^H). Many of the genes that are targets for Spo0A regulation are

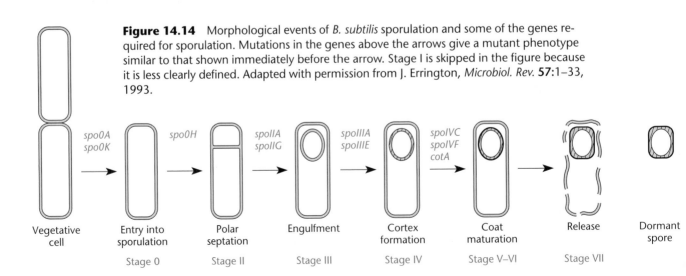

Figure 14.14 Morphological events of *B. subtilis* sporulation and some of the genes required for sporulation. Mutations in the genes above the arrows give a mutant phenotype similar to that shown immediately before the arrow. Stage I is skipped in the figure because it is less clearly defined. Adapted with permission from J. Errington, *Microbiol. Rev.* **57**:1–33, 1993.

| spo0A spo0K | spo0H | spoIIA spoIIG | spoIIIA spoIIIE | spoIVC spoIVF cotA | | |

| Vegetative cell | Entry into sporulation | Polar septation | Engulfment | Cortex formation | Coat maturation | Release | Dormant spore |

| | Stage 0 | Stage II | Stage III | Stage IV | Stage V–VI | Stage VII | |

transcribed by the σ^H-containing RNA polymerase holoenzyme.

Like most response regulators, Spo0A must be phosphorylated in order to carry out its transcriptional regulatory functions. Phosphorylation of Spo0A involves a "phosphorelay" system (Figure 14.15) that includes another two of the spo0 gene products, Spo0F and Spo0B. The phosphorelay also involves several protein kinases: these each phosphorylate Spo0F under certain growth conditions. Spo0B is a phosphotransferase enzyme that transfers phosphoryl groups from Spo0F~P to Spo0A. Spo0A~P then regulates its target genes by binding to their promoter regions, activating some and repressing others (see Box 13.3).

The regulatory effect of Spo0A on a given target gene depends on the amount of Spo0A~P in the cell. At low levels, Spo0A~P positively regulates genes involved in synthesis of antibiotics and degradative enzymes. This positive regulation results from what is actually a "double-negative" series of events, in which the direct effect of Spo0A~P action is repression of a gene called *abrB*, which itself encodes a repressor that acts on the antibiotic and degradative enzyme genes. At higher levels, Spo0A~P directly activates several sporulation operons, including *spoIIA*, *spoIIE*, and *spoIIG*. Activation of these genes irreversibly commits a cell to the sporulation process.

REGULATION OF THE Spo0A PHOSPHORELAY SYSTEM

Numerous genes participate in regulating the amount of Spo0A~P that is produced in a cell. Several of these encode the kinases mentioned above, which phosphorylate Spo0F and therefore increase Spo0A~P levels. Others encode phosphatases that can dephosphorylate Spo0F~P, thereby draining phosphate out of the phosphorelay and diminishing Spo0A~P levels, or dephosphorylate Spo0A~P directly (see below). These kinases and phosphatases respond to physiological and environmental signals, only a few of which are known.

Negative Regulation of the Phosphorelay by Phosphatases

Genetic analysis of two of the *spo0* loci revealed that their gene products functioned as negative regulators of sporulation. The combined results from sequencing mutant alleles, construction of gene knockouts, obtaining enhanced expression from many copies of the genes, and isolation of suppressor mutations were all important to deciphering the gene functions (See Perego et al., Suggested Reading). We list these regulators below.

1. For the *spo0E* locus, sequence analysis determined that two mutant alleles in Spo⁻ strains contained nonsense mutations. This result would usually be interpreted as an indication that the *spo0E* gene product played a positive role in initiating sporulation, since nonsense mutations usually inactivate the gene product and a requirement of the regulatory gene product for expression is the genetic definition of positive regulation (see chapter 2). However, deletion analysis of *spo0E* was contradictory, since Δ*spo0E* strains were capable of sporulation—in fact, they hypersporulated. Furthermore, multiple cloned copies of the *spo0E* gene inhibited sporulation. These latter two observations exemplify behavior typical of negative regulators. A resolution of this paradox came from following the trail of a clue provided by one aspect of the Δ*spo0E* mutant phenotype: Δ*spo0E* mutant strains had a tendency to segregate Spo⁻ papillae that were visible as translucent patches on the surface of sporulating colonies. Genetic analysis of these Spo⁻ papillae showed that they contained suppressor mutations, several of which mapped to the *spo0A* gene. This result suggested the hypothesis that cells lacking *spo0E* experienced an especially strong pressure to sporulate because increased expression or activity of the phosphorelay components produced an exceptionally high level of Spo0A~P. The *spo0E* mutants were found to have no alteration in phosphorelay

Figure 14.15 The phosphorelay activation (phosphorylation) of the transcription factor Spo0A. The phosphorelay (see Box 13.3) is initiated by at least three histidine kinases, which autophosphorylate on a histidine residue in response to unknown signals. The phosphate is transferred to Spo0F, to Spo0B, and finally to Spo0A. Spo0A~P regulates transcription as described in the text.

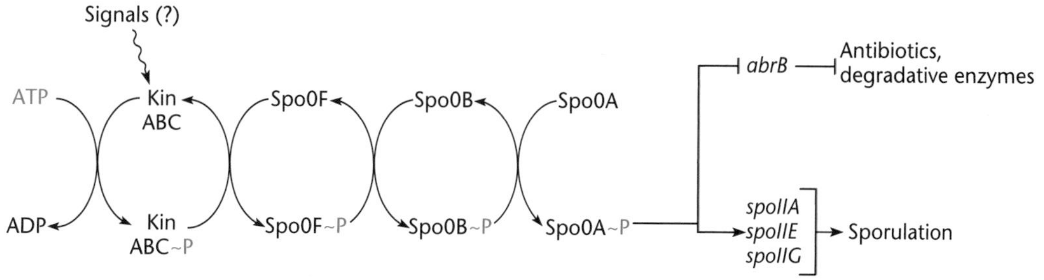

gene transcription, but a biochemical study of the Spo0E protein showed that it functioned as a specific phosphatase of Spo0A~P. Such an activity would indeed provide a negative regulatory function.

But what would explain the finding that nonsense *spo0E* mutants had a Spo⁻ phenotype? Both of these mutations were found to affect the C terminus of Spo0E, leaving most of the protein intact. One hypothesis is that the Spo0E C terminus is a regulatory domain, perhaps one that binds a signal molecule which regulates the phosphatase activity. If so, the mutations may prevent signal binding and hence lock Spo0E into the phosphatase mode, thereby preventing Spo0A~P accumulation and sporulation.

2. The *spo0L* locus shared some genetic properties with *spo0E*: multiple copies of *spo0L* caused a sporulation deficiency, and Δ*spo0L* mutations caused hypersporulation, as well as accumulation of Spo⁻ segregants. Like Spo0E, Spo0L behaved like a negative regulator of the phosphorelay. Because the *spo0L* mutants isolated on the basis of their Spo⁻ phenotype contained missense mutations of *spo0L*, it was reasoned that isolation of suppressors of these mutations might identify the target of Spo0L activity. Accordingly, a plan was made to mutagenize a Spo⁻ *spo0L* mutant strain and then look for Spo⁺ colonies. One problem that arose was that the most frequent class of sporulating mutants contained null mutations of *spo0L*. To overcome this problem, a strain containing two copies of the *spo0L* missense allele was constructed; this strain was mutagenized for isolation of extragenic suppressor mutations. The result of this experiment was isolation of a suppressor mutation in the *spo0F* gene, one of the phosphorelay components. When tested for phophatase activity, Spo0L proved to be a phosphatase of Spo0F~P.

An additional phosphatase of Spo0F~P was found in the *B. subtilis* genome sequence as a Spo0L homolog. Named Spo0P, this homologous protein inhibited sporulation when hyperexpressed from multicopies of the gene and encoded a phosphatase with 60% identical amino acid residues to Spo0L. The recent renaming of Spo0L and Spo0P, to RapA and RapB, respectively, reflects their roles as *r*esponse regulator *a*spartyl-phosphate *p*hosphatases.

Figure 14.16 illustrates inhibition of the phosphorelay by the RapA, RapB, and Spo0E phosphatases. Since these phosphatases function to reduce accumulation of Spo0A~P, their activities must be inhibited under conditions that promote antibiotic synthesis and sporulation. A regulator of Spo0E has been hypothesized, as discussed above. For RapA and RapB, the known regulatory signals are peptide molecules named PhrA and competence-

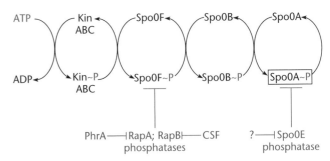

Figure 14.16 Regulation of the phosphorelay by phosphatases. RapA and RapB are phosphatases for SpoF~P and Spo0E is a phosphatase for Spo0A~P (see Stephenson and Perego, Suggested Reading). RapA and RapB are inhibited by the PhrA and CSF (PhrC) pentapeptides, respectively. It is not known what controls the activity of the phosphatase, Spo0E. Adapted with permission from B. Lazazzera, T. Palmer, J. Quisnel, and A. D. Grossman, Cell density control of gene expression and development in *Bacillus subtilis*, p. 27–46, *in* G. M. Dunny and S. C. Winans (ed.), *Cell-Cell Signaling in Bacteria,* ASM Press, Washington, D.C., 1999.

stimulating factor (CSF). Produced by the bacilli themselves, they may function as indicators of population density or "quorum sensors." The quorum sensors of gram-negative bacteria are typically homoserine lactones, while those of gram-positives are more typically peptides.

Regulation of the phosphorelay involves additional signals, besides those mentioned above. Starvation, cell density, metabolic states, cell cycle events, and DNA damage are all known to influence Spo0A~P levels. Many of the signals involved and the mechanisms by which they affect Spo0A~P are poorly understood.

Compartmentalized Regulation of Sporulation Genes

The mother cell and the forespore are genetically identical, but certain proteins must be made specifically in the developing spore and others (such as those that form the sturdy spore coat) must be made in the surrounding mother cell cytoplasm. Thus, the set of genes that is transcribed from the mother cell DNA must differ from the set transcribed from the forespore DNA. What mechanisms account for transcription of different sporulation genes in the two compartments?

REGULATION OF SPORULATION GENES BY SEQUENTIAL AND COMPARTMENT-SPECIFIC ACTIVATION OF RNA POLYMERASE SIGMA FACTORS

The entire collection of sporulation genes can be sorted into a handful of classes on the basis of transcription by a specific sigma factor. The sporulation sigma factors

TABLE 14.3	Sporulation sigma factors and their genes		
Sigma factor	***spo* gene name**	***sig* gene name**	**Compartment of activity**
σ^H	*spo0H*	*sigH*	Predivisional sporangium
σ^F	*spoIIAC*	*sigF*	Forespore
σ^E	*spoIIGB*	*sigE*	Mother cell
σ^G	*spoIIIG*	*sigG*	Forespore
σ^K	*spoIVCB-spoIIIC*	*sigK*	Mother cell

replace the principal vegetative cell sigma factor A (σ^A) in RNA polymerase holoenzyme, possibly by outcompeting σ^A for RNA polymerase. The σ^A of *B. subtilis* plays a role similar to the σ^{70} of *E. coli* (see chapter 2). As shown in Table 14.3, there are five distinct sigma factors, sigma H (σ^H), sigma E (σ^E), sigma F (σ^F), sigma G (σ^G), and sigma K (σ^K); these associate with RNA polymerase to transcribe the sporulation genes. Each of the sigma factors is active at a specific time during sporulation. Four of the sigmas are regulated such that they are active in only one of the two developing cell compartments: σ^E and σ^K are sequentially active in the mother cell, and σ^F and σ^G are sequentially active in the forespore.

Analysis of the Role of Sigma Factors in Sporulation Regulation

Four kinds of information have been important to understanding gene regulation in *Bacillus subtilis*.

TEMPORAL PATTERNS OF REGULATION
Measurements of the times of expression of the sporulation genes indicated that many of the genes underwent dramatic increases in expression at specific times after the sporulation process started. Use of gene fusions allowed large-scale comparisons of the complete set of sporulation genes. The most commonly used reporter gene was *lacZ*, from *E. coli* (Figure 14.17). The product of the *lacZ* gene, β-galactosidase, could be assayed by adding an "artificial" substrate (o-nitrophenyl-β-D-galactopyronoside [ONPG] or methylumbelliferyl-β-glucuronide [MUG]) to samples of the test culture at various times after sporulation was induced by a nutritional downshift. The appearance of β-galactosidase activity indicated the onset of gene expression. Direct measurements of mRNA of various sporulation genes correlated well with the results of *lacZ* fusion experiments. Therefore, the use of *lacZ* fusions became widespread, because of the relative ease and convenience of fusion assays.

A significant outcome of comprehensive fusion experiments was the extensive assessment and comparison of

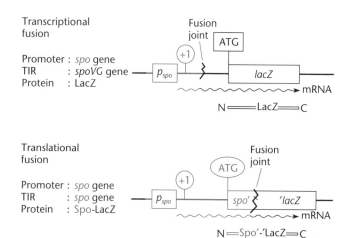

Figure 14.17 Reporter gene *lacZ* fusions to sporulation genes. Translational and transcriptional gene fusions are both transcribed from a *B. subtilis spo* promoter. In translational fusions, a fusion protein is expressed from the TIR of the gene being studied. In a transcriptional fusion, *lacZ* is translated from the translation initiation region (TIR) of a *B. subtilis spo* gene, *spoVG*.

the times of expression of many sporulation genes. Moreover, the timing of *lacZ* expression could be correlated with the timing of morphological changes visible as sporulation progressed.

DEPENDENCE PATTERNS OF EXPRESSION
Fusions with *lacZ* were also used to determine whether the expression of one gene depended on the activity of a second gene. If the expression of one gene depends on a second gene, the second gene may encode a direct or indirect regulator of the first. The use of *spo* mutations in combination with *spo-lacZ* fusions (Figure 14.18) allowed the testing of many regulatory dependencies. An example of a set of experimental data is shown in Figure 14.18B. In this example, expression of a *spoIIA::lacZ* fusion was dependent on all of the *spo0* loci but not on any of the "later" loci. Tests of many pairwise combinations of *spo* mutations and gene fusions are also summarized in

A

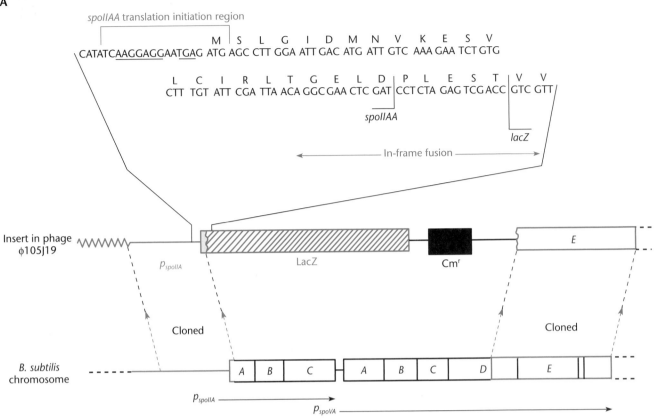

spoIIAA translation initiation region

```
                        M   S   L   G   I   D   M   N   V   K   E   S   V
CATATCAAGGAGGAATGAG ATG AGC CTT GGA ATT GAC ATG ATT GTC AAA GAA TCT GTG
```

```
        L   C   I   R   L   T   G   E   L   D   P   L   E   S   T   V   V
       CTT TGT ATT CGA TTA ACA GGC GAA CTC GAT CCT CTA GAG TCG ACC GTC GTT
```

spoIIAA

lacZ

In-frame fusion

Insert in phage φ105J19

p_{spoIIA} LacZ   Cm^r E

Cloned

Cloned

B. subtilis chromosome

A B C   A B C D   E

p_{spoIIA} ———→

p_{spoVA} —————————————→

B

Mutation	β-Galactosidase activity (units ml^{-1})		No. of determinations	Mutation	β-Galactosidase activity (units ml^{-1})		No. of determinations
	$t_{1.5}$	t_4			$t_{1.5}$	t_4	
spo0				**spoIII**			
spo0A43	0.016	0.037	2	spoIIIA65	0.18	0.22	4
spo0B136	0.023	0.054	2	spoIIIB2	0.65	0.13	2
				spoIIIC94	0.24	0.23	4
spo0E11	0.018	0.015	2				
spo0F221	0.021	0.050	2	**spoIV**			
				spoIVA67	0.29	0.080	3
spo0H17	<0.01	0.018	2	spoIVB165	0.12	0.19	3
spo0J93	0.048	0.25	2	spoIVC23	0.29	0.15	3
spo0K141	0.013	0.050	2				
				spoV			
spoII				spoVA89	0.44	0.34	2
spoIIAA562	0.41	0.38	2	spoVB91	0.49	0.21	2
spoIIAC1	0.44	0.51	2	spoVC134	0.61	0.46	2
spoIIB131	0.36	0.17	3				
spoIID298	0.43	0.11	2				
spoIIE48	0.67	0.77	3				
spoIIG55	0.69	0.58	3				

Figure 14.18 Testing the regulatory dependencies of *spo* genes. (A) Use of a *B. subtilis* transducing phage to create translational *lacZ* fusions to a chromosomal *spo* gene. Shown is the structure of phage φ 105J19 carrying the *spoIIAA-lacZ* gene fusion. The lower part of the figure shows the region of the *B. subtilis* chromosome containing the *spoIIA* operon (three genes) and the adjacent *spoVA* operon (five genes). The phage contains a cloned fragment of chromosomal DNA covering these operons, but the central portion of the insert has been replaced by the *E. coli lacZ* gene (colored hatches) and a chloramphenicol resistance gene (Cm^r)(in black). The insertion is arranged so that the region encoding the *lacZ* gene is fused in frame to the N terminus of the *spoIIAA* gene. (B) Effect of *spo* mutations on the production of β-galactosidase by the *spoIIAA-lacZ* gene fusion during sporulation. Phage φ105J19 (see panel A) was transduced into a series of isogenic strains carrying *spo* mutations. Sporulation was induced, and samples were taken for assay of β-galactosidase. Results shown are mean activities in samples at 1.5 and 4 hours after induction of sporulation. The wild-type Spo$^+$ strain containing the phage produced 0.58 units of β-galactosidase per ml at $t = 1.5$ h and 0.17 units at $t = 4$ h. With no *lacZ* fusion, the Spo$^+$ strain produced 0.013 units per ml at $t = 1.5$ h and 0.032 units per ml at $t = 4$ h. Adapted with permission from J. Errington and J. Mandelstam, *J. Gen. Microbiol.* **132**:2967–2976, 1986.

TABLE 14.4	Timing and dependence patterns of gene expression						
Gene or operon fusion[a]	Time of expression (min)	Expression in mutants of genes or operons:					
		spo0	spoIIA (σ^F)	spoIIE	spoIIG (σ^E)	spoIIIG (σ^G)	spoIVC (σ^K)
spoIIA	40	−	+	+	+	+	+
spoIIE	30–60	−	+	+	+	+	+
spoIIG	0–60	−	+	+	+	+	+
gpr	80–120	−	−	−	+	+	+
spoIIIG	120	−	−	−	−	+	+
ssp	>120	−	−	−	−	−	+
spoIVC	150	−	−	−	−	+	+
cotA	240	−	−	−	−	−	−

[a]The functions of these are described in the following sections.

Table 14.4. Besides the data for the genes shown, data for many dozens of additional genes have contributed to our understanding of gene regulation.

TRANSCRIPTION FACTOR DEPENDENCE

Once cloning and sequencing of *spo* genes had been accomplished, it was possible to determine the functions of some of the proteins because of their amino acid sequence similarities to known families of regulatory proteins such as sigma factors, which share characteristic amino acid motifs. The *spo0H* gene could be seen to encode a sigma factor, as could ORFs of the *spoIIA*, *spoIIG*, *spoIIIG*, and *spoIVC* loci (Table 14.3). In vitro experiments confirmed the functions of these proteins as transcription factors.

For many of the other sporulation genes, it was possible to infer their sigma factor dependence on the basis of sequence comparisons around the transcription start sites. In some cases, allele-specific suppressors of promoter mutations could be isolated in a particular sigma factor gene. Remember from chapter 3 that an allele-specific suppressor will suppress only a particular type of mutation in a gene. Another important type of experiment involved in vitro transcription studies with RNA polymerase containing specific sigma factors.

CELLULAR LOCALIZATION

Several methods have been used to determine the cellular location of expression of sporulation genes. For example, expression of β-galactosidase in the forespore can be distinguished from that in the mother cell on the basis that the forespore is more resistant to lysozyme. Immunoelectron microscopy has been useful for visualizing the expression of β-galactosidase, and, more recently, the use of green fluorescent protein fusions has allowed the determination of the cellular locations of numerous sporulation proteins (see front of book for the distribution of σ^E).

From studies like these, it could be seen that all of the genes turned on after septation were expressed in only one compartment. The genes transcribed by RNA polymerase with σ^F and σ^G were expressed only in the forespore compartment, and the genes transcribed by RNA polymerase with σ^E and σ^K were expressed only in the mother cell.

Intercompartmental Regulation during Development

When the observations on timing, dependence relationships, and localization of sporulation gene expression are combined, a complex pattern of regulation that includes a cascade of sigma factors and signaling between the developing compartments is revealed.

Figure 14.19 shows that after septation, gene expression in the forespore depends at first on σ^F and later on σ^G. An early σ^F-dependent transcript, *gpr*, encodes a protease which is important during spore germination. Besides its dependence on σ^F, *gpr* requires functional *spo0* genes, because σ^F expression and activity depends on them. Another σ^F-transcribed operon is *spoIIIG*, which encodes the late forespore sigma, σ^G. Transcription of *spoIIIG* differs from *gpr* in that it occurs later and, although confined to the forespore, requires functioning of the *spoIIG* locus (Table 14.4), which encodes the mother cell-specific sigma, σ^E.

Once σ^G is produced in the forespore, it transcribes a set of *ssp* genes, which encode spore-specific proteins that condense the nucleoid. As shown in Table 14.4, *ssp* transcription is blocked in *spoIIIG* mutants as well as in mutants with mutations in all of the genes discussed above, such as the *spoIIG* gene, because they are involved in σ^G production.

Gene expression in the mother cell also reveals intercompartmental regulation. Figure 14.19 shows that a gene transcribed relatively early in the mother cell by σ^E

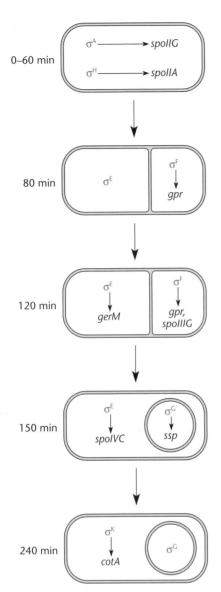

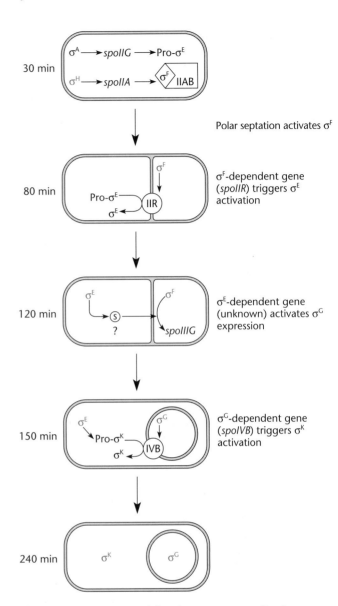

Figure 14.19 Compartmentalization of sigma factors and temporal regulation of transcription within compartments. The genes for σ^E and σ^F are transcribed before polar septation. σ^F is active in the forespore compartment and is required for transcription of the gene for σ^G, which succeeds it. σ^E is active in the mother cell and is required for transcription of the gene for its successor, σ^K. Within their compartments, σ^F and σ^E are required for transcription of their target genes at various times.

Figure 14.20 Sequential and compartmentalized activation of the *B. subtilis* sporulation sigma factors (gold). A series of signals allows communication between the two developing compartments, as described in the text, at approximately the times shown after induction of sporulation.

RNA polymerase is the *gerM* gene, which encodes a germination protein. Later, σ^E RNA polymerase transcribes the gene for σ^K. σ^K RNA polymerase then transcribes *cotA*, one of a set of *cot* genes that encode proteins incorporated into the spore coat. Table 14.4 shows that *cotA* transcription also requires activity of the late forespore sigma, σ^G.

As we describe in the next sections, activation of the sigmas alternates between the two developing compartments. As shown in Figure 14.20, each successive activation step requires intercompartmental communication. The critical information is whether morphogenesis and/or gene expression in the other compartment has progressed beyond a "checkpoint." The way in which the two compartments communicate their status to each other is a fascinating area of current research.

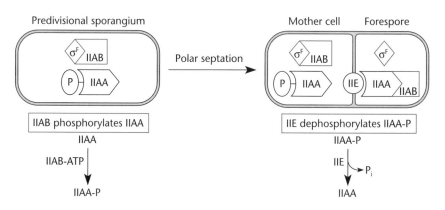

Figure 14.21 Model for regulation of σ^F activity. SpoIIAB holds σ^F in an inactive state in both the predivisional sporangium and the mother cell. SpoIIAB phosphorylates SpoIIAA. SpoIIAA~P cannot bind SpoIIAB. The SpoIIE phosphatase controls the generation of SpoIIAA, which complexes with SpoIIAB. These reactions allow the release of σ^F.

TEMPORAL REGULATION AND COMPARTMENTALIZATION OF σ^E AND σ^F

Both the *sigE* and *sigF* genes, encoding σ^E and σ^F, respectively (Table 14.3), are transcribed in the developing cell before the sporulation septum divides off the forespore compartment (Figure 14.19). However, neither sigma factor starts to transcribe its target genes until after the septum forms, and then, as mentioned above, each sigma becomes active in only one compartment: σ^E in the mother cell and σ^F in the forespore. Before septation, the sigma factors are held in inactive states, with a different inhibitory mechanism acting on each sigma. In the case of σ^E, the active form of the protein must be proteolytically released from an inactive precursor, Pro-σ^E. In the case of σ^F, the active protein must be released from a complex that contains an inhibitory "anti-sigma" factor SpoIIAB (Figure 14.21). Once the sporulation septum forms, σ^F becomes active first, in the forespore. Subsequently, σ^E becomes active in the mother cell.

Activation of σ^F in the forespore requires the interplay of a set of proteins. Two of these are binding partners and are named SpoIIAA and SpoIIAB. It is SpoIIAB that is the above-mentioned "anti-sigma" factor that binds to and inactivates σ^F. SpoIIAA is an anti-anti-sigma factor which nullifies the anti-sigma activity of SpoIIAB; it does this because of its own ability to bind SpoIIAB.

A cycle of phosphorylation and dephosphorylation of SpoIIAA modulates the binding of SpoIIAA to SpoIIAB. Only when it is unphosphorylated can SpoIIAA bind to SpoIIAB. Before septation, SpoIIAA is in a *phosphorylated* state and so does not bind to SpoIIAB in the preseptational sporangium. Unbound by SpoIIAA, SpoIIAB is free to bind to and inactivate σ^F. After septation, SpoIIAA is in the unphosphorylated state in the forespore; hence, it binds SpoIIAB, thereby releasing σ^F.

The enzymes that phosphorylate and dephosphorylate SpoIIAA are SpoIIAB and SpoIIE, respectively. Their opposing activities, before and after septation, determine the balance between the two forms of SpoIIAA. Before septation, SpoIIAB kinasing of SpoIIAA predominates, whereas once the spore septum has formed, SpoIIE phosphatase activity predominates in the forespore. Regulation of the SpoIIE phosphatase by septum formation is a current area of investigation.

Regulation of σ^F

Important progress in understanding the regulation of σ^F activity came from studying the two genes, *spoIIAA* and *spoIIAB*, that are cotranscribed with the *sigF* gene in the same operon (Figure 14.22A). Two key findings were that the *spoIIAB* gene product was a protein that inhibited σ^F activity and that SpoIIAB was, in turn, inhibited

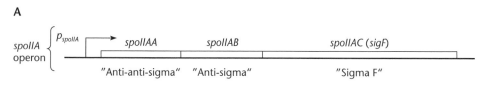

Figure 14.22 The *spoIIA* operon and its gene products. (A) The three genes that are cotranscribed. (B) The inhibitory effects of SpoIIAA and SpoIIAB, as inferred from genetic experiments described in the text.

by SpoIIAA. These conclusions were drawn from experiments on the effects of *spoIIAA* and *spoIIAB* mutations on σ^F activity (see Schmidt et al., Suggested Reading).

The test for σ^F activity in these experiments was to measure the expression of genes with σ^F-dependent promoters. Two such genes are *spoIIIG* and *gpr*. (Table 14.4). The use of *lacZ* fusions to these genes allowed gene expression to be monitored by β-galactosidase assays. Control experiments showed that *sigF* transcription and translation were normal, ensuring that differences in β-galactosidase activity from the σ^F-dependent *lacZ* fusions reflected activity of σ^F and not its expression.

The comparisons of β-galactosidase activity levels in *spoIIAA* and *spoIIAB* mutant cultures and wild-type cultures showed that *spoIIIG::lacZ* and *gpr::lacZ* expression was substantially higher in the *spoIIAB* mutant strain than in the wild type. Conversely, the tester fusions were not expressed at all in the *spoIIAA* mutant cultures. Outcomes like these could occur if SpoIIAA function was required for σ^F activity and SpoIIAB function inhibited σ^F activity (Figure 14.22B). An additional genetic experiment showed that SpoIIAA was required because it was needed to counteract SpoIIAB. This experiment was performed to assay the tester fusions in double *spoIIAA spoIIAB* mutant cultures; these cultures overexpressed the tester fusions, like the *spoIIAB* mutant had. Thus, SpoIIAA was required only if SpoIIAB was active.

Further study showed that *spoIIAB* inhibition of σ^F was an essential event in the normal course of sporulation. One observation was that the *spoIIAB* mutant strains could not survive sporulation and could be maintained only in media that suppressed sporulation. The proposed explanation for this phenotype was that unregulated transcription by σ^F was lethal to the cell. It is important to note here that the *spoIIAB* mutant strain used in the experiments above was not isolated in a mutant hunt for Spo$^-$ strains but, rather, was a constructed deletion mutant. Lethality caused by deregulated σ^F activity could also be observed if *sigF* was artificially induced in vegetative cells from the P$_{spac}$ promoter (Figure 14.23).

Another important advance in our understanding of σ^F regulation came from biochemical studies on SpoIIAA and SpoIIAB. The SpoIIAB amino acid sequence suggested that it might have protein kinase activity. It was indeed able to phosphorylate a protein: its substrate turned out to be SpoIIAA! Additional studies examined the binding interactions of the three proteins SpoIIAA, SpoIIAB, and σ^F, and found that (i) SpoIIAA could bind to SpoIIAB, but only if SpoIIAA was not phosphorylated; and (ii) SpoIIAB could bind to σ^F. Genetic evidence included site-directed mutagenesis of the SpoIIAA phosphorylation target, a serine residue, changing it to aspartate or alanine, and so mimicking phosphorylated and non-phosphorylated states, respectively (see Diederich et al., Suggested Reading). Together, these observations formed the basis for the following model for σ^F regulation in a sporulating cell (Figure 14.21). As soon as the three *spoIIA* operon genes are expressed in the preseptation cell, SpoIIAB binds to and therefore inactivates σ^F. SpoIIAB also phosphorylates SpoIIAA and so prevents SpoIIAA-SpoIIAB binding. Thus, the SpoIIAB-σ^F complex is stable and σ^F cannot direct transcription. However, σ^F could become active in the forespore if SpoIIAB released it after septation. Thus, the model proposed that SpoIIAA is dephosphorylated *in the forespore*, so that it can bind SpoIIAB and cause the release of active σ^F.

The necessity that SpoIIAA be dephosphorylated in order to release σ^F activity predicted that the sporulating cell must express a SpoIIAA~P phosphatase. The collection of *spo* mutants was evaluated for the possibility that one of the known genes might encode the hypothesized phosphatase. A candidate for such a phosphatase was the *spoIIE* gene product, because *spoIIE* mutants had a phenotype consistent with a defect in σ^F activation: they expressed the *spoIIA* (*sigF*) operon but failed to express σ^F-dependent genes (see Table 14.4). This prediction was borne out by in vitro studies in which SpoIIE dephosphorylated SpoIIAA~P.

The SpoIIE protein associates with the polar septum. This and other poorly understood factors contribute to the limitation of σ^F activity to the forespore (see Levin and Losick, Suggested Reading).

Regulation of σ^E

Mother cell transcription depends on σ^E, which is encoded by the *spoIIGB* gene of the *spoIIG* operon. The primary product of *spoIIGB* is an inactive precursor, named Pro-σ^E, which is processed to form the active sigma factor. The protease that cleaves Pro-σ^E is the product of the first gene in the *spoIIG* operon, *spoIIGA*. Although the *spoIIG* operon is expressed in the predivisional sporangium, the *spoIIGA* product does not process SpoIIGB immediately but waits about an hour, until after septation has occurred. Then, notification

Figure 14.23 Induction of *spoIIAC* (*sigF*) from the P$_{spac}$ promoter, which consists of RNA polymerase recognition sequences of *B. subtilis* phage SPo1 and *lac* operator sequences. Insertion of *lacI* accompanies insertion of P$_{spac}$, and so *spoIIAC* transcription is inducible with IPTG.

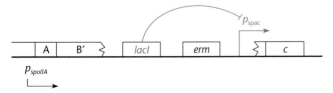

from the forespore that development is proceeding and that σ^F has become active comes via the messenger SpoIIR, which is the product of a σ^F-transcribed gene. SpoIIR is secreted into the spaces between the cell membranes, where it signals to SpoIIGA to process Pro-σ^E.

An important clue to the explanation for the time lag in Pro-σ^E processing was the observation that mutants which lacked σ^F activity failed to process Pro-σ^E to σ^E. Could a σ^F-transcribed gene be required for the processing mechanism? A genetic search for such a gene was undertaken (see Karow et al., Suggested Reading) with the rationale that a strain with a mutant in the hypothetical gene might have a SigF$^+$ SigE$^-$ phenotype, i.e., it would express σ^F-dependent fusions but would not express σ^E-dependent fusions. Accordingly, mutants that expressed the σ^F-dependent fusion *gpr-gus* but did not express two σ^E-dependent fusions, *spoIID-lacZ* and *spoVID-lacZ*, were sought. The use of two *lacZ* fusions was intended to reduce the likelihood of isolating Lac$^-$ strains that were mutant in *lacZ* itself rather than in the desired regulatory gene. The Gus$^+$ Lac$^-$ mutants isolated from this screen were of two types: mutants with mutations of the *spoIIG* locus, as would be expected, and mutants with mutations of a new locus, which was named *spoIIR*. Evaluation of the *spoIIR* mutants for Pro-σ^E synthesis and processing showed that Pro-σ^E was indeed synthesized but was not processed. Thus, the *spoIIR* gene seemed to have the predicted properties. Subsequently, the SpoIIR protein was found to be secreted into the spaces between the membranes, with the outcome that SpoIIGA is activated to process Pro-σ^E. There is an additional factor, forespore-specific degradation of σ^E, that operates to restrict σ^E accumulation to the mother cell (Figure 14.24).

σ^G, A SECOND FORESPORE-SPECIFIC SIGMA FACTOR

The σ^G-encoding gene, *spoIIIG*, is transcribed in the forespore by σ^F RNA polymerase. Its transcription lags behind that of other σ^F-transcribed genes, evidently because it requires a signal from the mother cell sigma, σ^E. The evidence for the existence of such a signal is indirect at present—the *spoIIIG* gene is not transcribed in a *spoIIG* (σ^E) mutant—and the signal has not yet been identified. Another aspect of σ^G regulation involves a mechanism related to σ^F regulation. Indeed, the two sigmas are closely related in their amino acid sequences, and, like σ^F, σ^G is inactivated by the anti-sigma SpoIIAB. However, the mechanism that releases σ^G from SpoIIAB differs, since the phosphorylation cycle described above for σ^F does not appear to affect σ^G regulation.

σ^K, A MOTHER CELL SIGMA

The last sigma to be made, σ^K, is expressed only in the mother cell. σ^E RNA polymerase transcribes the *sigK*

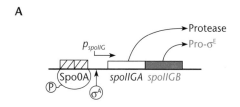

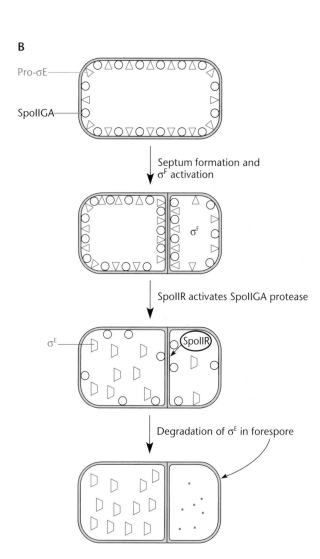

Figure 14.24 A model for activation of σ^E in the mother cell compartment. (A) The SpoIIG operon is transcribed by σ^A RNA polymerase and requires activation by Spo0A~P (hatched boxes indicate Spo0A~P-binding sites [see Box 13.3]). (B) Pro-σ^E and SpoIIGA are associated with the cytoplasmic membrane in the sporangium. After septum formation, both proteins are associated with all cell membranes. Then SpoIIR is expressed in the forespore under the control of σ^F and SpoIIR activates SpoIIGA protease, which cleaves Pro-σ^E to form active σ^E, which is distributed in the cytoplasm. Finally, σ^E is degraded in the forespore.

gene. Like σ^E, σ^K is cleaved from a precursor protein, Pro-σ^K. Also, like σ^E processing, σ^K processing depends on a signal from the forespore. In this case, the signal is expression of the *spoIVB* gene under the control of the forespore sigma, σ^G (Figure 14.20).

The SpoIVB protein is thought to be secreted across the innermost membrane surrounding the forespore and to communicate with the Pro-σ^K processing factors across the membrane, thus activating the protease that cleaves Pro-σ^K (Figure 14.25). SpoIVB activation of the σ^K-specific protease does not occur directly but, rather, by deactivation of proteins that inhibit the protease. The complexity of this mechanism was revealed by isolation of "bypass suppressor" mutations. These were mutations that bypassed the requirement for σ^G involvement in σ^K activation.

σ^K Activation

The motivation for isolating suppressor mutations was the observation that late mother cell gene expression depended not only on the mother cell sigma, σ^K, but also on σ^G, the forespore sigma, and other forespore proteins. Moreover, the σ^G requirement was manifested at the step of Pro-σ^K processing, since Pro-σ^K protein accumulated in *spoIIIG* mutant cells. It was hypothesized that mutations that bypassed the σ^G requirement might be informative as to the mechanism of σ^G involvement.

Figure 14.25 Model for regulation of Pro-σ^K processing based on the genetic data discussed in the text.

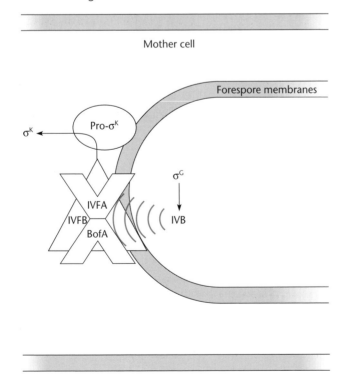

The isolation of suppressor mutations involved a screen for mutations that allowed expression of a σ^K-dependent fusion, *cotA::lacZ*, in a $\Delta sigG$ strain (see Cutting et al., Suggested Reading). The *sigG* mutant strain was mutagenized with nitrosoguanidine (NTG) to produce a broad spectrum of mutations. The *cotA::lacZ* fusion was then introduced by specialized transduction, and lysogens were screened for a blue colony phenotype on X-Gal plates. Two classes of mutations were isolated and localized, defining loci that were named *bofA* and *bofB* (for bypass of forespore). Characterization of the *bof* mutants showed that *cotA* expression still required σ^K and, importantly, that Pro-σ^K processing was restored in *bof* mutants. The *bofA* mutations defined a new gene. However, the *bofB* mutations were missense or nonsense mutations in a previously discovered gene, *spoIVFA*. The second gene of this operon, *spoIVFB*, is thought to be the protease that processes Pro-σ^K. Recent work has also shown that SpoIVFA and the BofA protein work together to inhibit SpoIVFB.

The identity of SpoIVB as the σ^G-dependent signal was inferred from the observation that *bof* mutations restored *cotA::lacZ* expression to a *spoIVB* mutant, just as for the *spoIIIG* mutant. The SpoIVB protein is not yet fully characterized, but it is hypothesized that this protein, produced in the forespore as discussed above, can cross the innermost membrane surrounding the forespore to trigger the Pro-σ^K proteolysis mechanism.

Finding Sporulation Genes: Mutant Hunts, Suppressor Analysis, and Functional Genomics

The discussions of the preceding sections show that a variety of approaches have been useful for identifying *B. subtilis* sporulation genes. A very large percentage of the regulatory genes were defined by mutations that caused a Spo⁻ phenotype. However, an important class of regulatory genes was not well represented in the Spo⁻ mutant collection. These genes are the negative regulators. Loss-of-function mutations—which are the most frequent kind of mutation—in negative regulators would not cause a Spo⁻ phenotype, as discussed above for genes such as *spoIIAB*. "Special" alleles, e.g., gain-of-function mutations, might cause a Spo⁻ phenoype, as for *spo0E* and *spo0L*, the negative regulators of the phosphorelay, but these would generally be found at a low frequency. However, suppression analysis is a powerful tool in such cases and has revealed the existence of important negative regulators. As one example, the *abrB* gene (Figure 14.15) was defined by mutations that restored degradative enzyme and antibiotic synthesis, but not sporulation, to *spo0A* mutants.

Another type of gene underrepresented in the sporulation mutant collections encodes a function that is redundant, or overlapping, with that encoded by another gene.

An example is the set of *kin* genes, which encode the kinases that initiate the phosphorelay. Inactivating mutations in any one of these cause only a weak Spo⁻ phenotype. Only one of these, *kinA*, was found as a very leaky *spo* mutant in a collection of Tn*917*-induced mutants.

Additional genes involved in sporulation have been defined by gene knockout analysis of ORFs annotated in the *Bacillus* genome sequence (http://bacillus.genome. ad.jp) and by microarray analysis. These experiments include radioactively labeled cDNAs synthesized by random priming from RNAs isolated from the wild type, a *spo0A* mutant and a Δ*sigF* mutant. Hybridization experiments with PCR products representing more than 90% of the *B. subtilis* ORFs identified many dozens of genes dependent on both Spo0A and σᶠ. Gene knockouts of at least two ORFs gave late sporulation defects, thus identifying two new genes required for sporulation (see Fawcett et al., Suggested Reading).

PROBLEMS

1. You have isolated a constitutive mutant of the *trp* operon of *E. coli*. When you introduce the wild-type *trpR* repressor gene on an F′ factor into cells with the mutant *trpR* gene, it is still constitutive. Describe how you would determine if it has a *cis*-acting operator mutation or if it has a dominant repressor *trpR*ᵈ mutation.

2. The product of the *envZ* gene is a transmembrane protein in the inner membrane of *E. coli*. (see chapter 13). Describe how you would determine which of the regions of the EnvZ protein are in the periplasm and which are in the cytoplasm.

3. Outline how you would isolate suppressors of mutations in the signal sequence coding region of the *malE* gene.

4. How were *B. subtilis* regulatory genes identified?

5. Contrast the similarities and differences of the *B. subtilis* phosphorelay and typical two-component systems (see Box 13.4).

6. What types of *spo0E* and *spo0L* mutations suggested that the genes were positive regulators? Negative regulators? How did suppression analysis help in understanding that Spo0E and Spo0L are actually negative regulators?

7. What is the difference between *spo-lacZ* transcriptional and translational fusions?

8. The *tlp* and *cotD* genes encode proteins that are structural components of the endospore. Their dependence patterns of expression are similar to the *ssp* genes and *cotA*, respectively. What sigma factor is *tlp* transcription dependent on? *cotD*? Which compartments would you predict these genes are expressed in?

SUGGESTED READING

Anderson, R. P., C. G. Miller, and J. R. Roth. 1976. Tandem duplications of the histidine operon observed following generalized transduction in *Salmonella typhimurium*. *J. Mol. Biol.* 105:201–218.

Bassford, P., and J. Beckwith. 1979. *Escherichia coli* mutants accumulating the precursor of a secreted protein in the cytoplasm. *Nature* (London) 277:538–541.

Bourret, R. B., N. W. Charon, A. M. Stock, and A. H. West. 2002. Bright lights, abundant operons—fluorescence and genomic technologies advance studies of bacterial locomotion and signal transduction: review of the BLAST meeting, Cuernavaca, Mexico, 14 to 19 January 2001. *J. Bacteriol.* 184:1–17. (Minireview.)

Brun, Y. V., and L. J. Shimkets (ed.). 2000. *Prokaryotic Development*. ASM Press, Washington, D.C.

Cutting, S., V. Oke, A. Driks, R. Losick, S. Liu, and L. Kroos. 1990. A forespore checkpoint for mother cell gene expression during development in *B. subtilis*. *Cell* 62:239–250.

Diederich, B., J. F. Wilkinson, T. Magnin, S. M. A. Najafi, J. Errington, and M. D. Yudkin. 1994. Role of interactions between SpoIIAA and SpoIIBB in regulating cell-specific transcription factor σᶠ of *Bacillus subtilis*. *Genes Dev.* 8:2653–2663.

Errington, J. 1993. *Bacillus subtilis* sporulation: regulation of gene expression and control of morphogenesis. *Microbiol. Rev.* 57:1–33.

Errington, J., and J. Mandelstam. 1986. Use of a *lacZ* gene fusion to determine the dependence pattern of sporulation operon *spoIIA* in *spo* mutants of *Bacillus subtilis*. *J. Gen. Microbiol.* 132:2967–2976.

Fawcett, P., P. Eichenberger, R. Losick, and P. Youngman. 2000. The transcriptional profile of early to middle sporulation in *Bacillus subtilis*. *Proc. Natl. Acad. Sci. USA* 97:8063–8068.

Feucht, A., T. Magnin, M.D. Yudkin, and J. Errington. 1996. Bifunctional protein required for asymmetric cell division and cell-specific transcription in *Bacillus subtilis*. *Genes Dev.* 10:794–803.

Gutierrez, C., J. Barondess, C. Manoil, and J. Beckwith. 1987. The use of transposon Tn*phoA* to detect genes for cell envelope proteins subject to a common regulatory stimulus. *J. Mol. Biol.* 195:289–297.

Hill, C. W., J. Foulds, L. Soll, and P. Berg. 1969. Instability of a missense suppressor resulting from a duplication of genetic material. *J. Mol. Biol.* 39:563–581.

Hoch, J. A., and K. I. Varughese. 2001. Keeping signals straight in phosphorelay signal transduction. *J. Bacteriol.* 183:4941–4949.

Hoffman, C. S., and A. Wright. 1985. Fusions of secreted proteins to alkaline phosphatase: an approach for studying protein secretion. *Proc. Natl. Acad. Sci. USA* 82:5107–5111.

Karow, M., P. Glaser, and P. J. Piggot. 1995. Identification of a gene, *spoIIR*, that links the activation of σᴱ to the transcriptional activity of σᶠ during sporulation in *Bacillus subtilis*. *Proc. Natl. Acad. Sci. USA* 92:2012–2016.

Lacatena, R. M., and G. Cesareni. 1981. Base pairing of RNA 1 with its complementary sequence in the primer precursor inhibits ColE1 replication. *Nature* (London) **294**:623–626.

Lazazzera, B., T. Palmer, J. Quisel, and A. D. Grossman. 1999. Cell density control of gene expression and development in *Bacillus subtilis*, p. 27–46. *In* G. M. Dunny and S. C. Winans (ed.), *Cell-Cell Signaling in Bacteria*. ASM Press, Washington, D.C.

Levin, P. A., and R. Losick. 2000. Asymmetric division and cell fate during sporulation in *B. subtilis*, p. 167–190. *In* Y. V. Brun and L. J. Shimkets (ed.), *Prokaryotic Development*. ASM Press, Washington, D.C.

Miller, J. H. 1992. *A Short Course in Bacterial Genetics*. Cold Spring Harbor Laboratory, Cold Spring Harbor, N.Y.

Miller, J. H., and U. Schmeissner. 1979. Genetic studies of the *lac* repressor. X. Analysis of missense mutations in the *lacI* gene. *J. Mol. Biol.* **131**:223–248.

Oliver, D. B., and J. Beckwith. 1981. *E. coli* mutant pleiotropically defective in the export of secreted proteins. *Cell* **25**:2765–2772.

Perego, M., C. Hanstein, K. M. Welsh, T. Djavakhishvili, P. Glaser, and J. A. Hoch. 1994. Multiple protein-aspartate phosphatases provide a mechanism for the integration of diverse signals in the control of development in *B. subtilis*. *Cell* **79**:1047–1055.

San Milan, J. L., D. Boyd, R. Dalbey, W. Wickner, and J. Beckwith. 1989. Use of *phoA* fusions to study the topology of the *Escherichia coli* inner membrane protein leader peptidase. *J. Bacteriol.* **171**:5536–5541.

Schatz, P. J., and J. Beckwith. 1990. Genetic analysis of protein export in *Escherichia coli*. *Annu. Rev. Genet.* **24**:215–248.

Schmeissner, U., D. Ganem, and J. H. Miller. 1977. Genetic studies of the *lac* repressor. II. Fine structure deletion map of the *lacI* gene, and its correlation with the physical map. *J. Mol. Biol.* **109**:303–326.

Schmidt, R., P. Margolis, L. Duncan, R. Coppolecchia, C. P. Moran, Jr., and R. Losick. 1990. Control of developmental transcription factor σ^F by sporulation regulatory proteins SpoIIAA and SpoIIAB in *Bacillus subtilis*. *Proc. Natl. Acad. Sci. USA* **87**:9221–9225.

Sonenshein, A. L., J. A. Hoch, and R. Losick (ed.) 2001. Bacillus subtilis *and its Closest Relatives: from Genes to Cells*. ASM Press, Washington, D.C.

Stephenson, S. J., and M. Perego. 2002. Interaction surface of the Spo0A response regulator with the SpoOE phosphatase. *Mol. Microbiol.* **44**:1455–1467.

Tian, H., and J. Beckwith. 2002. Genetic screen yields mutations in genes encoding all known components of the *Escherichia coli* signal recognition particle pathway. *J. Bacteriol.* **184**:111–118.

Winans, S. C., and G. C. Walker. 1985. Conjugal transfer system of the Inc N plasmid, pKM101. *J. Bacteriol.* **161**:402–410.

Answers to Questions for Thought and Problems

Chapter 1

Questions for Thought

1. The two strands of bacterial DNA probably aren't replicated in the 3′-to-5′ direction simultaneously because replicating a DNA as long as the chromosome in this manner would leave single-stranded regions that were so long that they would be too unstable or susceptible to nucleases.

2. DNA molecules may be very long because if cells contained many short pieces of DNA, each one would have to be segregated individually into the daughter cells.

3. DNA may be the hereditary material instead of RNA because double-stranded DNA has a slightly different structure than double-stranded RNA. The B-form structure of DNA may have advantages for replication, etc. The use of DNA instead of RNA may allow the primer, which is made of RNA, to be more easily identified and removed from a DNA molecule by the editing functions. The editing functions do not operate when the 5′ end is synthesized. By removing and resynthesizing any RNA regions, using upstream DNA as primer, mistakes can be minimized.

4. A temperature shift should cause the rate of DNA synthesis to drop but not too abruptly. Each cell would complete the rounds of replication that were under way at the time of the shift but would not begin another round. The synthesis rate would drop exponentially. If cells are growing rapidly, the drop would be less steep because a number of rounds of DNA replication would be under way in each molecule and each would have to complete the cycle.

5. The gyrase of *Streptomyces sphaeroides* might be naturally resistant to novobiocin. You could purify the gyrase and test its ability to introduce supercoils into DNA in vitro in the presence of novobiocin.

6. How chromosome replication and cell division are coordinated in bacteria like *Escherichia coli* is not well understood. One hypothesis is that a protein required for cell division might be encoded by a gene located close to the termination region and this gene is transcribed into RNA only when it replicates. Cell division could then begin only when the termination region has replicated. You could move the terminus of replication somewhere else in the chromosome and see if this affects the timing of cell division.

7. No known answer, but maybe this allows for more genome rearrangements such as inversions and duplications, which may play an important role in evolution. If chromosome replication obligatorily stopped at a certain *ter* sequence, a major chromosome rearrangement may make it impossible to replicate the entire DNA because then the *ter* sequence could be encountered before the entire DNA had replicated.

Problems

1. 5′GGATTA3′

2. 5′GGAddT3′

   5′GGATddT3′

   5′GGATTACGGddT3′

   5′GGATTACGGTAAGGddT3′

3. $I = 25$ min, $C = 40$ min, $D = 20$ min

4. $I = 90$ min, $C = 40$ min, $D = 20$ min

5. The *topA* mutant lacks topoisomerase I, which removes negative supercoils, so that there should be more negative supercoils in the DNA of the mutant.

6. You could use PCR to amplify a clone containing the region to be mutated. One primer could be complementary to the region to be mutated but have the desired change. The primers could also introduce restriction sites to make the fragment easier to clone afterward. This method would depend upon the presence of a convenient restriction site close enough to the site to be mutated that two sequences can be carried on the same PCR primer. Other methods that are more universally applicable and do not depend upon restriction sites close to the region to be mutated use PCR primers complementary to the two strands in the region to be mutated and having the desired change. The DNA to be mutated is cloned in a circular plasmid. After amplification, the linear amplified DNA is blunt-end ligated and transformed into cells. It is also useful to have some means of eliminating the unmutated template DNA, for example by cutting the DNA after amplification with a restriction endonuclease which cuts only methylated DNA, since the amplified DNA will not be methylated. The major problem with PCR-based mutagenesis is that the DNA polymerases used for PCR amplification are very error prone and so random mutations may be introduced, making it necessary to carefully sequence the mutated DNA to ensure that it has only the desired change.

Chapter 2

Questions for Thought

1. Some people think RNA came first because the peptidyl-transferase that links amino acids together to make protein is the 23S rRNA, and some other enzymes are also RNA. However, the question remains open to speculation.

2. The genetic code may be universal because once the code was established, too many components—aminoacyl-tRNA synthetases, etc.—were involved in translating the code to change all of them.

3. Why eukaryotes do not have polycistronic mRNAs is open to speculation.

4. The genetic code of mitochondrial genes differs from the chromosomal code of eukaryotes because mitochondria were once bacteria with their own simple translation apparatus, ribosomes, etc. This translation apparatus has remained independent of that for the chromosomal genes.

5. Selenocysteine may be a relic of what was once a useful process in an earlier organism from which all other organisms evolved.

6. The translation apparatus is very highly conserved evolutionarily, so that an antibiotic that inhibits the translation apparatus of one type of bacteria is apt to inhibit the translation apparatus of all bacteria.

7. It has been proposed that the reason has to do with the relatively larger size of eukaryotic cytoplasmic proteins. Eukaryotic proteins are too large to fit into the chamber of a GroEL-type chaperonin, making this type of folding impractical in the eukaryotic cytoplasm.

8. The *sec* system transports proteins only through the inner membrane of gram-negative bacteria, and so secreted proteins must have their own system to get through the outer membrane. It might be best to have the transport system through the outer membrane directly coupled to the transport system through the inner membrane. Also, it might be more efficient to have their own system because they do not have to compete for transport with other transported proteins. Another possibility is that the presence of their own specialized transport system makes secretion of a protein encoded by a prophage less dependent upon the bacterium the prophage finds itself in. The protein can secrete itself from the cell even if the cell has an incompatible *sec* system.

9. One reason is to save energy by not making a particular protein unless it is needed. At least two GTPs and one ATP are required to translate each amino acid, not to mention the hundreds of nucleoside triphosphates required to transcribe a gene. Another reason is to prevent interference between intermediates of pathways. For example, the intermediates in some degradative pathways might inhibit other degradative pathways. A third reason is that it might help in replication of the chromosome. Transcribing RNA polymerase can interfere with the replication fork, and so the transcription of only a few genes at any one time may help speed up chromosome replication.

Problems

1. 5′CUAACUGAUGUGAUGUCAACGUCCUACU-CUAGCGUAGUCUAA3′

2. Translation should begin in the second triplet GUG because this is followed by a long ORF, although this hypothesis would have to be tested.

3. The answer is (a). Both sequences have a string of A's, but only in (a) is this preceded by an inverted repeat that could form a hairpin loop in the mRNA.

4. A good primer would be CC GGA TCC ATG TTG CGA TTT. A good way to think of it is that the *Bam*HI restriction site in the cloning vector is read GGA TCC, so that the *Bam*HI site in the fragment being cloned will also be read this way after it is cloned. Reading the *Bam*HI site in this way in the primer will put the downstream coding region in the correct frame. You also have to add at least two random deoxynucleotides to the 5′ end before the amplified fragment can be cut with *Bam*HI. We added two C's. Some restriction endonucleases require more than two deoxynucleotides at the 5′ end before they will cut at a site.

5. a. Expression of the operon will be constitutive and the genes will be expressed even if inducer is not added.

b. Expression of the operon is turned off, and the operon cannot be induced.

Chapter 3

Questions for Thought

1. Why the genetic maps of *Salmonella* and *Escherichia* spp. are similar is unknown. Perhaps there is an optimal way to arrange genes on a chromosome, with genes that are expressed at high levels closest to the origin of replication and transcribed in the same direction in which they are replicated. Inversions would then be selected against.

2. Duplication mutations would allow the number of genes of an organism to increase. The duplicated genes could then evolve so that their products could perform novel functions. Sometimes organisms with a duplication of a particular region may have a selective advantage in a particular environment, and so the duplication would be preserved.

3. The cells of higher organisms may be more finely tuned because runthrough proteins resulting from translation through the ends of genes may create more problems for eukaryotic cells, which, being more complicated, may be less tolerant of aberrant proteins.

4. It is not known how directed, or adaptive, mutations might occur. In their purest form, adaptive mutations would require that the cell somehow sense that a mutation would be desirable and change the DNA sequence accordingly.

Problems

1. The cultures with the largest number of mutants probably had the earliest mutation.

2. Arginine auxotrophy would have a higher mutation rate because many genes encode enzymes to make arginine and any mutation that inactivates the product of one of these genes would make the cell Arg⁻. Rifampin resistance, however, can be caused by only a few mutations in the gene for the β subunit of RNA polymerase because very few amino acids can be changed and have the RNA polymerase no longer bind rifampin but still be active for transcription.

3. Approximately 8×10^{-10}

4. 5.5×10^{-8}

5. a. *arg-1* is probably a leaky missense or another type of base pair change mutation since it seems to retain some activity of the gene product and it reverts.

b. *arg-2* is probably a deletion mutation since it is not leaky and does not revert.

c. *arg-1* could be a frameshift mutation close to the carboxy end of the gene, so that the gene product is not totally inactivated. The mutant with *arg-2* could be a double mutant with two missense mutations in *arg* genes, so both of them would seldom revert.

6. You could make a double mutant with a *dam* mutation and an *arg* mutation like *arg-1*. Then you could compare the reversion frequency to Arg⁺ of this double mutant with that of the single *arg* mutant without the *dam* mutation.

7. If the mutants are isolated from the same culture they could be siblings with the same original mutation so they will be the same and not representative of all the mutations that can cause the phenotype.

8. a. Plate large numbers of the bacteria on plates containing the antibiotic coumermycin and all the other needed growth supplements.

b. Plate large numbers of the Trp⁻ bacteria on minimal plates lacking tryptophan but containing all the other needed growth supplements.

c. Plate large numbers of the *dnaA*(Ts) mutant on plates at the high, nonpermissive temperature to isolate mutants that can multiply to form a colony at the high temperature. Test by crossing with the wild type to show that there are temperature-sensitive recombinant types but that the two mutations are closely linked.

d. Plate large numbers of the *araD* mutant on plates containing L-arabinose plus another carbon source. Test mutants that can multiply by crossing with the wild type to show that there are some arabinose-sensitive recombinants.

e. Make a partial diploid that has a polar *hisC* mutation in one copy of the operon and a *hisB* mutation in the other copy. Then plate large numbers of the partial diploid on minimal plates lacking histidine but having all the other necessary growth supplements. Mutants that can grow to form a colony could have a polarity-suppressing mutation in *rho* that allows HisB expression from the first copy of the operon, complementing the *hisB* mutation in the other copy.

9. Grow large numbers of the cells at the low, permissive temperature and then shift to the high, nonpermissive temperature before adding ampicillin. After incubating the cells for a period, wash out the ampicillin and plate the cells at the lower temperature. Test colonies which arise to see which ones are due to mutant bacteria which cannot form colonies at the high temperature.

10. Nonsense suppressors are dominant. The mutant tRNA will still insert an amino acid at the nonsense codon even in the presence of the normal tRNA.

11. The mutation is probably a nonsense suppressor in a tRNA gene. You could test it to see if the strain will propagate a phage with a nonsense mutation in an essential gene or if it will suppress nonsense mutations in other genes.

12. Introduce the gene with the Cm^r cassette cloned in a plasmid cloning vector into the cell in a way such that the cloning vector cannot replicate (i.e., is a suicide vector), and plate on plates containing chloramphenicol. Test any Cm^r cells for the presence of the cloning vector. For example, if the cloning vector carries ampicillin resistance (Ap^r), you could screen for calls which are Cm^r but not Ap^r. These are presumably strains which have had two crossovers, replacing the normal *arg* gene with the *arg* gene containing the Cm^r cassette. You would expect the strain to be Arg$^-$.

13. Many more than 4,000.

14. The clone probably does not contain all of the *hemA* gene. Recombination between the clone and the *hemA* gene in the chromosome seems to be required to make a functional *hemA* gene, explaining why not all the bacteria containing the clone are HemA$^+$. If the clone contained the entire *hemA* gene, it would complement the mutation in the chromosome and all the bacteria would be HemA$^+$.

Chapter 4

Questions for Thought

1. If essential genes were carried on a plasmid, cells that were cured of the plasmid would die. By having nonessential genes on plasmids, the chromosome can be smaller so that the cells can multiply faster yet the species can adapt to a wide range of environments because of selection for those cells with a particular plasmid. You would not expect genes encoding enzymes of the tricarboxylic acid cycle such as isocitrate dehydrogenase or genes for proteins involved in macromolecular synthesis such as RNA polymerase to be carried on plasmids. Genes involved in using unusual carbon sources such as the herbicide 2,4-D or genes required for resistance to antibiotics such as ampicillin might be expected to be on plasmids.

2. Why some but not all plasmids have a broad host range is unknown. A broad-host-range plasmid can parasitize more species of bacteria. A narrow-host-range plasmid can develop a better commensal relationship with its unique host.

3. Perhaps the plasmid binds to a unique site on the membrane of the cell. When the cell is ready to divide, there are two copies of this site, each of which can bind a copy of the plasmid. To ensure that the number of copies of the plasmid never exceeds the number of such membrane sites, perhaps the plasmid can replicate only once and then only when it is bound to one of these sites.

4. If the genes required for replication of the plasmid are not all closely linked to the *ori* site, you could find them by isolating temperature-sensitive mutants of the plasmid that cannot replicate at high temperature and then look for pieces of plasmid DNA that can help the mutant plasmid replicate at the high temperature when introduced into the cell in a cloning vector.

5. You could determine which of the replication genes of the host *E. coli* (*dnaA*, *dnaC*, etc.) are required for replication of a given plasmid by introducing the plasmid into cells with temperature-sensitive mutations in each of the genes and then determining whether the plasmid can replicate at the high nonpermissive temperature for the mutant.

6. One advantage to having the leader sequence degraded, rather than the mRNA for the RepA protein itself, is that it may help prevent the synthesis of defective RepA protein that could interfere with replication. If the mRNA for the RepA protein is itself degraded, defective translation products of RepA due to running off the end of the degraded mRNA might compete with normal RepA for replication.

7. It actually seems more consistent with the pairing model. The daughter plasmids may remain paired until just before cell division, when they separate to opposite poles.

Problems

1. You introduce your plasmid into cells with the other plasmid and grow the cells in the absence of kanamycin. After a few generations, you plate the cells and test the colonies for kanamycin resistance. If more of the cells are kanamycin sensitive than would be the case if your plasmid were replicating in the same type of cell without the other plasmid, your plasmid is probably an IncQ plasmid.

2. 1/2,048

3. You take the colonies due to the transformants on ampicillin plates and test them on tetracycline plates to determine whether they are tetracycline resistant. The plasmids in the cells that are ampicillin resistant but tetracycline sensitive probably have an insert in the *Bam*HI site, since the *Bam*HI site of pBR322 is in the Tet^r gene.

4. Make plates containing higher and higher concentrations of ampicillin until bacteria containing the RK2 plasmid can no longer multiply to form a colony. Plate large numbers of bacteria containing the plasmid on this concentration of ampicillin. Any bacteria which form colonies may contain high-copy-number mutants of RK2. The copy number of the plasmid in these bacteria could be determined, and the plasmid could be introduced into new bacteria to show that the new bacteria are also made resistant to higher concentrations of ampicillin by the plasmid, thereby showing that the mutation is in the plasmid itself. The *repA* gene in the mutant plasmid could then be sequenced to ensure that this is the region of the responsible mutation.

5. The plasmid should have an easily selectable gene such as for resistance to an antibiotic. Cells containing the plasmid are grown through a number of generations in the absence of the antibiotic and plated without the antibiotic. The number of cells cured of the plasmid is then determined by replicating these plates onto a plate containing the antibiotic. Any colonies that do not transfer onto the new plate are due to bacteria that have been cured of the plasmid. If this number is much smaller than predicted from the normal distribution based on the copy number of the plasmid, the plasmid may have a partitioning system.

Chapter 5

Questions for Thought

1. If plasmids of the same Inc group all have the same Tra functions, the exclusion functions will prevent a plasmid from transferring into a cell that already contains a plasmid of the same Inc group, where one or the other would be lost.

2. If the *tra* genes and the *oriT* site on which they act are close to each other, recombination will seldom occur between the *oriT* site and the genes for the Tra functions. Separating the Tra functions from the *oriT* site on which they act will render them nonfunctional.

3. Perhaps plasmids with certain *mob* sites can be transferred only by certain corresponding Tra functions because only these functions will allow the mobilizable plasmid to recognize the signal that the cell has made contact with a potential recipient cell and be cut.

4. Self-transmissible plasmids encode their own primases, so they can transfer themselves into a host cell with an incompatible primase and still synthesize their complementary strand.

5. The absence of an outer membrane in gram-positive bacteria may make a pilus unnecessary. The question is open to speculation.

6. The answer is not known. Perhaps the role of the pilus in helping transmit DNA into the cell can be easily subverted by the phage to transfer its own DNA into the cell.

7. Plasmids are generally either self-transmissible or mobilizable because if they were neither, they would not be able to move to cells that did not already contain them. By being promiscuous, plasmids can expand their host range.

8. They can be mobilized by plasmids of a number of different Inc groups. Apparently, they are designed so that their Dtr system can communicate with a number of different coupling factors. A self-transmissible plasmid lacking its Mpf functions can be mobilized only by members of its own Inc group.

Problems

1. The recipient strain is the one that becomes recombinant and retains most of the characteristics of the original strain. If the transfer is due to a prime factor, the apparent recombinants will become donors of the same genes.

2. Determining which of the *tra* genes of a self-transmissible plasmid encodes the pilin protein is not easy. For example, phage-resistant mutants will not necessarily have a mutation in the pilin gene. They could also have a mutation in a gene whose product is required to assemble the pilus on the cell surface. You could purify the pili and make antibodies to them. Then *tra* mutants in the pilin gene won't make an antigen that will react with the antibody. Similarly, to determine which *tra* mutants don't make the DNase that nicks the DNA at the *ori* region or the helicase that separates the DNA strands, you may have to develop assays for these enzymes in crude extracts and determine which *tra* mutants don't make the enzyme in your assay.

3. You can show that only one strand of donor DNA enters a recipient cell by using a recipient that has a temperature-sensitive mutation in its primase gene. The plasmid DNA should remain single stranded after transfer into such a strain, provided, of course, that the plasmid can't make its own primase. Single-stranded DNA is more sensitive to some types of DNases and behaves differently from double-stranded DNA during gel electrophoresis.

4. If the tetracycline resistance gene is in a plasmid, the tetracycline-resistant recipient cells should have acquired a plasmid. If it is in a conjugative transposon, no transferred plasmid will be in evidence.

5. Male-specific phage cannot infect cells containing only a mobilizable plasmid because mobilizable plasmids do not encode a pilus, which serves as the adsorbtion site for the phage.

6. The *trpA* mutation is the selected marker, and the *argH* and *hisG* mutations are the unselected markers.

7. Very close to the *thyA* gene.

8. The selected marker is the *metB1* mutation, and the unselected markers are the *leuA5* mutation and the Tn5 insertion mutation. A little before 44 min.

9. Yes, I expect them to be sensitive to M13 because, with such a short mating, most of the Leu⁺ recombinants will be due to an F′ factor and so will become male.

10. Close to *argG*.

Chapter 6

Questions for Thought

1. The chapter lists some possible reasons for the development of competence in bacteria. At this time we do not know the correct answer.

2. To discover whether the competence genes of *Bacillus subtilis* are turned on by UV irradiation and other types of DNA damage, you could make a gene fusion with a reporter gene to one of the competence factor-encoding genes and see if the reporter gene is induced following UV irradiation.

3. To determine whether antigenic variation in *Neisseria gonorrhoeae* results from transformation between bacteria or recombination within the same bacterium, you could introduce a

selectable gene for antibiotic resistance into one of the antigen genes and see if it is transferred naturally under conditions where antigenic variation occurs.

Problems

1. To determine whether a given bacterium is naturally competent, you would isolate an auxotrophic mutant, such as a Met⁻ mutant, and mix it with DNA extracted from the wild-type bacterium. The mixture would then be plated on medium without methionine. The appearance of colonies due to Met⁺ recombinants would be evidence of transformation.

2. To isolate mutants defective in transformation, you would take your Met⁻ mutant, mutagenize it, and repeat the test above on individual isolates. Any mutants that do not give Met⁺ recombinants when mixed with the wild-type DNA might be mutants with a second mutation in a competence gene.

3. To discover whether a naturally transformable bacterium can take up DNA of only its own species or any DNA, you could make radioactive DNA and mix it with your competent bacteria. Any DNA taken up by the cells would become resistant to added DNase, and the radioactivity would be retained with the cells on filters. Try this experiment with radioactive DNA from the same species as well as from different species.

4. If the bacterium can take up DNA of only the same species, it must depend upon uptake sequences from that species. The experiment should be done as in problem 3 but with only a piece of DNA instead of the entire molecule.

5. If the DNA of a phage successfully transfects competent *E. coli*, plaques will appear when the tranfected cells are plated with bacteria sensitive to the phage.

Chapter 7

Questions for Thought

1. If phages made the proteins of the phage particle at the same time as they made DNA, the DNA might be prematurely packaged into phage heads, leaving no DNA to replicate.

2. Phages that make their own RNA polymerase can shut off host transcription by inactivating their host RNA polymerase without inactivating their own RNA polymerase. However, phages that use the host RNA polymerase can take advantage of the ability of the host molecule to interact with other host proteins, allowing more complex regulation.

3. In this way, they can infect a wider range of hosts and still replicate their DNA.

4. No known answer. Perhaps again it has something to do with the range of hosts they can infect. In one range of hosts, the Pri proteins may be more compatible, while in another range of hosts, the RNA polymerases may be more compatible.

5. No known answer, but if the phage injects a protein with the DNA that is intended to be used early, this protein could present problems for the phage which is in later states of development. Also, the DNA of the second infecting phage would be in a different stage of replication, which might interfere with the replication of the DNA of the first infecting phage.

Problems

1. You could mix a known amount of the virus with the cells and then measure the fraction of survivors. From the Poisson distribution, you could then measure the effective MOI. e^{-MOI} = fraction of surviving bacteria. The ratio of the effective MOI to the actual MOI is the fraction of the viruses that actually infected the cells.

2. The regulatory gene is probably gene *M*, because mutations in this gene can prevent the synthesis of many different gene products. The other genes probably encode products required for the assembly of tails and heads.

3. Amber mutations introduce a nonsense UAG codon into the coding sequence of an mRNA, stopping translation and leading to synthesis of a shortened gene product. The *ori* sequence does not encode a protein, and so an amber mutation could not be isolated in it.

4. The order is *A-Q-M*, because with this order most of the Am⁺ recombinants with a crossover between *amA* and *amQ* would have the Ts mutation in gene *M*. A second crossover between *Q* and *M* would be required to give the wild-type recombinant.

5. T1 has a linear genetic map which is expanded for the genes at the end of the linear DNA.

6. A phage with a mutation in its lysozyme gene will not lyse the cells and release phage unless egg white lysozyme is added. Infect the cells and allow the infection to proceed long enough for the cells to lyse if they had been infected by the wild-type phage. Then divide the culture in half and add lysozyme to one of the divided cultures. Plate both cultures with indicator bacteria. If the culture with the added lysozyme yields many more plaques than the culture without lysozyme, the phage that had infected the cells had a mutation in its lysozyme gene. Addition of the lysozyme caused the cells to lyse, releasing their phage and making many more plaques rather than just one plaque where the original infected cell was located.

7. The cotransduction frequency of the *argH* and *rif* markers is about 34%. The cotransduction frequency of the *argH* and *metA* markers is about 14%. They are almost the same.

8. The cotransduction frequency of the *argH* and *metB* markers is 37%, and that of the *argH* and *rif* markers is 20%. The order is probably *metB1–argH5–rif-8,* which is also consistent with the three-factor cross data. The order *argH5–metB1–rif-8* is consistent with some of the data, but the *argH* marker is probably in the middle since most of the Rif⁺ transductants are not also Met⁺, as they would be if they were on the same side of the *argH* marker.

9. First, transduce the Tn*10* transposon insertion mutation into your *hemA* mutant, selecting the tetracycline resistance gene on the transposon by plating on plates containing tetracycline. Identify a tetracycline-resistant transductant that still has the *hemA* mutation because it will not grow on plates lacking δ-aminolevulinic acid. Use this strain as a donor to transduce other strains, selecting for tetracycline resistance. Many of these transductants will have your *hemA* mutation, as determined by their inability to grow without δ-aminolevulinic acid.

Chapter 8
Questions for Thought

1. Perhaps the λ prophage uses different promoters to transcribe the *cI* repressor gene immediately after infection and in the lysogenic state because this may allow the repressor gene to be transcribed from a strong, unregulated promoter immediately after infection but then be transcribed from a weaker, regulated promoter in the lysogenic state.

2. By making two proteins, the cell can use the Int protein to promote recombination for both integration and excision. Then the smaller Xis protein need only recognize the hybrid *att* sites at the ends of the prophage.

3. Toxins and other proteins that confer pathogenicity may often be encoded by phages for the same reason that only conditionally essential genes are carried on plasmids (see chapter 4). Many of the same bacteria can also live outside the host, and by having the virulence genes carried on prophages, not all the bacteria of the species need carry them and so the bacteria can have a smaller genome and replicate more rapidly.

4. Why some types of prophage can be induced only if another phage of the same type infects the lysogenic cell containing them is unknown. Perhaps there is some way of inducing them that hasn't been tried.

5. There is not much to distinguish them. P4 encodes very few of its own gene products to make a phage and depends upon P2 for its head and tail proteins and most of its other functions.

Problems

1. If the clear mutant has a *vir* mutation, it will form plaques on a λ lysogen.

2. A specialized transducing phage carrying the *bio* operon of *E. coli* would be isolated in the same way as the λd*gal* phage in the text, except that a Bio⁻ mutant would be infected and plated on medium without biotin. The Bio⁺ bacteria would be isolated, and the phage would be induced. It might be necessary to add a wild-type helper phage before induction since *bio* substitutions extend into the *int* and *xis* genes. The λp*bio* should form plaques because no replication genes should be substituted.

3. A λ phage with *vir* mutations in the o_1 sites of o_L and o_R should form plaques on a λ lysogen because the repressor must bind first to the o_1 sites of the operators.

4. Both Int and Xis are required to integrate phage λ transducing particles next to an existing prophage because the recombination occurs between two hybrid *attP-attB* sites.

5. Amber mutations lead to termination of translation, so that you cannot make the reminder of the protein carrying the other domain for intragenic complementation.

6. Infect a P4 lysogen with your phage, and see if the lysate contains any P4 phage. They can be detected because their head is smaller and they can form plaques on a P2 lysogen but not on a nonlysogen.

7. To determine whether the *Staphylococcus aureus* toxin is encoded by a prophage, you could take the strain of *S. aureus* and try to induce a phage from it by UV irradiation, etc. The cells could be filtered out, and the cell-free medium could be plated on a closely related *S. aureus* strain to see if plaques form. You could then use the phage in the plaques to try to isolate lysogens of the related *S. aureus* strain to determine whether the lysogens produced the toxin, as evidenced by the acquired ability to kill human cells in culture.

Chapter 9
Questions for Thought

1. Perhaps replicative transposons do not occur in multiple copies around a genome because their resolution functions cause deletions between repeated copies of the transposon, resulting in the death of cells with more than one copy. Also, a poorly understood phenomenon called target immunity inhibits transposition of a transposon into a DNA that already contains the same transposon.

2. The transposon Tn3 and its relatives may have spread throughout the bacterial kingdom on promiscuous plasmids.

3. Transposons sometimes, but not always, carry antibiotic resistance or other traits of benefit to the host. They may also help the host move genes around, as in the construction of plasmids carrying multiple drug resistance.

4. The origin of the genes within integrons is not known. Perhaps they come from plasmids or other transposons.

5. It is possible that invertible sequences rarely invert because very little of the invertase enzyme is made or because the enzyme works very inefficiently on the sites at the ends of the invertible DNA sequence.

Problems

1. You would integrate a λd*gal* at the normal λ attachment site close to the *gal* operon with one of the *gal* mutations by selecting for Gal⁺ transductants. Sometimes the *gal* genes in the chromosome would recombine with the *gal* genes on the integrated λ and the cell would become Gal⁻ by gene conversion (homogenoting). When the λd*gal* are induced, their DNA should be longer and the phage denser owing to the inserted DNA.

2. The colonies should not be sectored because only one strand, either the *lac* or *lac*⁺ strand, of the original heteroduplex transposon will have been inserted.

3. The advantages are that transposon insertions almost always inactivate the gene and are not leaky, so that the phenotypes of a null mutant are known. They also mark the site of the mutation both genetically and physically, so that the site of the mutation is easier to determine by either genetic or physical

mapping techniques. The disadvantage is that they almost always inactivate the gene, so that the effects of other types of mutations in the gene cannot be studied. You also cannot get transposon insertion mutations in an essential gene in a haploid organisms because they will be lethal.

4. Follow the procedure for the mutagenesis and mapping of pAT153 outlined in the text. Only the sizes of the junction fragments will be different.

5. Perform Southern blot analyses using a sequence within the transposon as a probe. Isolate a number of strains with independent transposon insertions. Digest DNA with a restriction endonuclease that has no sites in the transposon. If different-size fragments always light up in the different strains, the transposon integrates randomly.

6. The Kan^r gene should be on the side closest to the *Sal*I site on the plasmid. The transposon is inserted 0.372 kb from the *Sal*I site on the plasmid, so that, with this orientation, the smallest *Sal*I site should be 2.6 + 0.372 = 2.972 kb and the largest should be 9.2 − 2.972 = 6.228 kb.

7. a. Mobilize a plasmid suicide vector containing Tn5 into the strain of *P. putida*. Select the Kan^r transconjugants on rich medium containing kanamycin, and then replicate them onto minimal medium with 2,4-D as the sole carbon source, looking for transposon insertion mutants that cannot use 2,4-D as a sole carbon and energy source. Isolate the DNA from such mutants, and randomly ligate pieces of the DNA into an *E. coli* plasmid cloning vector. Use the ligation mix to transform *E. coli*, selecting Kan^r transformants. These should contain a plasmid clone with at least part of a gene whose product is required to use 2,4-D. The corresponding wild-type gene could be found by using the clone with the transposon as a probe to screen a wild-type library of the *P. putida* strains by plate hybridizations.

b. Make a library of the DNA of the *P. putida* strain in a broad-host-range plasmid containing the *oriT* of a promiscuous plasmid such as RP4. Mobilize the plasmid library into mutants of the *P. putida* strain that cannot use 2,4-D, and select transconjugants that can form colonies on minimal plates containing 2,4-D as the sole carbon source. The plasmid-cloning vector in the bacteria in these colonies should contain the gene that was mutated to prevent growth on 2,4-D.

8. Perhaps MuB or some other protein remains attached to the end of the Mu DNA when it enters the cell, allowing only one round of a cut-and-paste transposition event.

9. You could introduce an amber mutation into the *int* gene of the conjugative transposon and use an amber suppressor strain for the donor cell and a nonsuppressor for the recipient cell. If the transposon can still integrate into the chromosome of the recipient, it must have transferred the Int protein made in the donor cell since it cannot make Int protein in the recipient.

10. Start a culture from a single colony, isolate the DNA from the culture, and perform a Southern hybridization after cutting with a restriction nuclease that cuts off center in the G segment and using a probe complementary to the G segment. If you get two bands, the segment has inverted. Another way might be to map transposon insertion mutations to antibiotic resistance in the prophage G segment with respect to markers in the neighboring chromosomal DNA by three-factor crosses. If you get a consistent order, the G segment is not inverting in the prophage.

11. You can use mutants of the phages that lack a functional DNA invertase of their own and propagate them in isogenic cells that are lysogenic for e14 and cured of e14. Pick plaques, and test the host range of the phage in the plaques to determine if their invertible sequence has inverted and changed their host range. To test for *Salmonella* phase shift, again use a *Salmonella* mutant which lacks the invertase and introduce the e14 prophage. Test whether it can now shift from one cell surface antigen to the other.

12. By methods such as those outlined in chapter 1, you could clone the pigment gene and use it as a probe in Southern blot analyses to see whether the physical map of the DNA around the gene changes when it is in the pigmented as opposed to the nonpigmented form.

13. You could make a plasmid that has two copies of the Mu phage by cloning a piece of DNA containing Mu into a plasmid cloning vector also containing Mu. You could then see if the two repeated Mu elements could resolve themselves in the absence of the host recombination functions.

Chapter 10

Questions for Thought

1. Recombination might help speed up evolution by allowing new combinations of genes to be tried or may help in repair of DNA damage.

2. The reason that RecBCD recombination is so complicated is unknown. Its real role might be to repair double-strand breaks caused when replication encounters damage in the DNA.

3. The different pathways of recombination may function under different conditions for recombination between short and long DNAs or at breaks and gaps, etc.

4. The RecF pathway may be preferred under conditions different from the ones normally used in the laboratory to measure recombination. The SbcB and SbcC functions may interfere only under the conditions normally used in laboratory crosses.

5. By encoding their own recombination functions, phages can increase their rate of recombination. Also, some phages use the recombination functions for replication and, by encoding their own functions, can inhibit the host recombination functions to prevent them from interfering with phage replication, as in the case of the RecBCD function and λ rolling-circle replication.

6. There could be another as yet undetected X-phile in the cell that cuts the Holliday junctions migrated by the RecG helicase. Alternatively, the RecG helicase may allow replication restarts to replicate out the Holliday junctions rather than resolving them by using an X-phile.

Problems

1. To determine if recombinants in an Hfr cross have a *recA* mutation, you could take the individual recombinants and make a streak with them on a plate. Then half of each streak could be covered with glass (glass is opaque to UV) and the plate irradiated before incubation. If the recombinant is RecA⁻, it will grow only in the part of the streak that was covered by the plate.

2. To determine what, if any, other genes participate in the *recG* pathway, start with a *ruvABC* mutant and isolate mutants with additional mutations that make them deficient in recombination. See if any have mutations in genes other than the *recG* gene.

3. The recombination promoted by homing double-stranded nucleases to insert an intron occurs in the same manner as in Figure 9.4, except that the double-stranded break that initiates the recombination occurs at the site in the target DNA into which the intron will home. The invading DNA will then pair with the homologous flanking DNA on one side of the transposon in the donor DNA and replicate over the transposon until it meets the other 5′ end.

4. In a RecB⁺C⁺D⁻ host, compare recombination between the same two markers in λ DNAs, one with a *chi* mutation and another without. If there is no difference, *chi* sites only stimulate recombination because of the RecD function.

Chapter 11

Questions for Thought

1. The answer is unknown. Perhaps the alternative to methylation of the bases is some other type of covalent modification, such as methylation of the phosphates, or a protein or RNA may remain bound to the newly synthesized DNA strand for a period after it has been synthesized.

2. Different repair pathways may work better depending upon where the damage occurs or the extent of the damage. For example, if damage is so extensive that lesions occur almost opposite each other in the two strands of the DNA, it might be easier to repair the lesions with excision repair than with recombination repair.

3. The SOS mutagenesis pathway might exist to allow the cells to survive damage other than that due to UV irradiation, or it might be more effective under culture conditions different from those used in the laboratory.

Problems

1. Assuming that you can grow the organism in the laboratory (you might have to grow it under high pressure), you could irradiate it in the dark and then divide the culture in half and expose half of the cells to visible light before diluting and measuring the surviving bacteria. If more of the cells survive after they have been exposed to visible light, the bacterium has a photoreactivation system.

2. The procedure is explained in the text. Briefly, to show that the mismatch repair system preferentially repairs the unmethy-

lated base, you could make heteroduplex DNA of λ phage. One strand should be unmethylated and heavier than the other because the λ from which this strand was derived were propagated on Dam⁻ *E. coli* cells grown in heavy isotopes. The two λ used to make the heteroduplex DNA should also have mutations in different genes, so that there will be mismatches at these positions. After the heteroduplex λ DNA is transfected into cells, test the progeny phage to determine which genotype prevails, the genotype of the phage from which the unmethylated DNA was prepared, or the genotype of the phage with methylated DNA.

3. To determine whether the photoreactivating system is mutagenic, perform an experiment similar to that in problem 1 but with *E. coli*. Measure the frequency of mutations (such as reversion of a *his* mutation) among the survivors of UV irradiation in the dark as opposed to those that have been exposed to visible light after UV irradiation. More cells should survive if they are exposed to visible light, but a higher *frequency* of these survivors should be His⁺ revertants if photoreactivation is mutagenic. If photoreactivation is not mutagenic, a lower frequency should be His⁺ revertants because the photoreactivation system will have removed some of the potentially mutagenic lesions. It may be better to do this experiment with a *umuCD* mutant to lower the background mutations due to SOS mutagenesis.

4. To find out whether the nucleotide excision repair system can repair damage due to aflatoxin B, treat wild-type *E. coli* cells and a *uvrA*, *uvrB*, or *uvrC* mutant with aflatoxin B. Dilute and plate. Compare the survivor frequencies of the *uvr* mutant and the wild type.

5. Express *umuC* and *umuD* from a clone that has a constitutive operator mutation so that the cloned genes are not repressed by LexA. Also, be sure that part of the *umuD* gene has been deleted so that UmuD′, rather than the complete UmuD, will be synthesized. This clone can be put into isogenic RecA⁻ and RecA⁺ strains of *E. coli* that have a *his* mutation. After UV irradiation, the frequency of His⁺ revertants among the surviving bacteria can be compared for each strain. If the RecA⁺ strain shows a higher frequency of His⁺ revertants, the RecA protein may have a role in UV mutagenesis other than cleaving LexA and UmuD.

6. You could make a transcriptional fusion of the *recN* gene to a reporter gene such as *lacZ* and then determine whether more of the reporter gene product is synthesized after UV irradiation, as it should be if *recN* is an SOS gene. A strain that also has a *lexA*(Ind⁻) mutation should not show this induction if *recN* is an SOS gene.

Chapter 12

Questions for Thought

1. Why operons are regulated both positively and negatively is not clear. However, there may be different advantages to the two types of regulation. For example, negative regulation might require more regulatory protein but might allow more complete repression, while it might be easier to achieve intermediate levels of expression with positive regulation. There may be less interaction between negative regulatory systems than between

positive regulatory systems, in which a regulatory protein might inadvertently turn on another operon. Also, constitutive mutants will be rarer with positive regulation.

2. The genes for regulatory proteins may be autoregulated to save energy. If they are autoregulated, only the amount of regulatory protein needed will be synthesized. Also, in the case of positive regulators, more can be made after induction to further increase the expression of the operons under their control.

3. The *galU* gene product is required to make the donor for glucose, UDPglucose, needed in many necessary biosynthetic reactions, including cell wall synthesis, and not just in galactose metabolism. If it were in the *gal* operon, *galU* would not be expressed if there were no galactose in the medium, preventing these other necessary biosynthetic reactions.

4. Regulation by attenuation of transcription of amino acid biosynthetic operons offers the advantage that the ability of the cell to translate codons for that amino acid can be exploited to regulate the operon. The regulatory system can be designed so that transcription will continue into the structural genes of the operon if ribosomes stall in the leader region at codons for the amino acid because not enough of the amino acid is available. A major disadvantage of this type of regulation is that it is wasteful. A short RNA is always made from the operon, even if it is not needed.

Problems

1. To isolate a *lacI*^{s} mutant, take advantage of the fact that *lacI*^{s} mutations, while rare, are dominant Lac$^-$ mutations. Mutagenize a *galE* mutant that contains an F′ factor with the *lac* operon, and plate it on P-Gal medium containing another carbon source such as maltose. The survivors that form colonies are good candidates for *lacI*^{s} mutants because inactivation of the *lacZ* genes in both the chromosome and the F′ factor requires two independent mutations, which should be even rarer. The mutants can be further tested by mating the F′ factor from them into other strains whose chromosome contains a wild-type *lac* operon. If the F′ factor makes the other strain Lac$^-$, the F′ factor must contain a *lacI*^{s} mutation.

2. AraC must be in the P1 state, since it represses the *ara* operon.

3. a. Lac$^-$ permanently repressed. The LacI^s repressor will bind to the operators of both operons, even in the presence of the inducer, and will prevent transcription of the *lacZYA* structural genes.

b. Inducible Lac$^+$. In other words it will be wild type for the *lac* operon. The *lacO*^{c} mutation in the chromosome will make the *lac* operon on the chromosome constitutive, but LacZ and LacY will not be made anyway because of the polar mutation in *lacZ*.

c. Inducible Ara$^+$. The inactivating *araC* mutation will be recessive to the wild-type *araC* allele, and the cell will be wild type for the *ara* operon and inducible by L-arabinose.

d. Inducible Ara$^+$. The *araI* mutation will prevent transcription of the operon in the chromosome, but the mutation is *cis*

acting and so will not prevent transcription of the operon on the F′ factor.

e. Inducible Ara$^+$. Same reason as (d). The *cis*-acting *pBAD* promoter mutation will prevent transcription of the chromosomal operon but not the operon on the F′ factor.

4. To determine whether *phoA* is negatively or positively regulated, you could first isolate constitutive mutants to determine how frequent they are. Since PhoA turns XP blue, you could mutagenize cells and isolate mutants that form blue colonies on XP-containing medium even in the presence of excess phosphate in the medium. If *phoA* is negatively regulated, constitutive mutants should be much more frequent than if it is positively regulated. Also, at least some of these constitutive mutants should have null mutations, deletions, etc., that inactivate the regulatory gene.

5. Plate wild-type *E. coli* cells in the presence of low concentrations of 5-methyltryptophan in the absence of tryptophan. Only constitutive mutants can multiply to form colonies under these conditions, because 5-methyltryptophan is a corepressor of the *trp* operon but cannot be used for protein synthesis, so that the nonmutant wild-type *E. coli* cells will starve for tryptophan. To isolate mutants defective in feedback inhibition of tryptophan synthesis, plate a constitutive mutant in the presence of higher concentrations of 5-methyltryptophan in the absence of tryptophan. Even constitutive mutants cannot multiply to form colonies under these conditions, because the first enzyme of tryptophan synthesis will be feedback inhibited by the 5-methyltryptophan. Only mutants that are defective in feedback inhibition will form colonies.

Chapter 13

Questions for Thought

1. Perhaps it is important to synthesize more of the proteins involved in synthesizing new proteins so that the rate of protein synthesis will increase after heat shock, allowing more rapid replacement of the proteins irreversibly denatured as a result of the shock.

2. Perhaps *Salmonella* species, which are normal inhabitants of the vertebrate intestine, are usually in an environment where amino acids are in plentiful supply but NH$_3$ is limiting. *Klebsiella* species may usually be free-living, where NH$_3$ is present but amino acids are not.

3. Perhaps the genes for corresponding sensor and response regulator genes are similar to allow cross talk between regulatory pathways. If the genes are similar, a signal from one pathway can be passed to the other pathway, allowing coordinate regulation in response to the same external stimulus. However, there is no good evidence for the importance of cross talk. Another possible explanation is that the genes had a common ancestor in evolution and still retain many of the same properties.

4. The enzymes responsible for ppGpp synthesis during amino acid starvation and during growth rate control may be different because the enzymes involved in stringent control and growth rate regulation must be in communication with different cellular constituents. The RelA protein works in association with the ribosome, where it can sense amino acid starvation, while the enzyme involved in synthesizing ppGpp during growth rate regulation might have to sense the level of catabolites. SpoT might be involved both in synthesizing and degrading ppGpp if the equilibrium of the reaction is somehow shifted. All enzymes function by lowering the activation energy and so in a sense catalyze both the forward and backward reaction. However, because the equilibrium favors the forward reaction, this is the reaction that predominates. If the equilibrium were shifted, perhaps by sequestering the ppGpp as it is made, SpoT could synthesize ppGpp rather than degrade it.

Problems

1. Determine if the mutant can use other carbon sources such as lactose and galactose. If it has a *cya* or *crp* mutation, it should not be able to induce other catabolite-sensitive operons and so will not be able to grow on these other carbon sources.

2. a. Gln⁻ (glutamine requiring); Ntr constitutive (express Ntr operons even in presence of NH_3)

b. Ntr⁻ (can't express Ntr operons even at low NH_3 concentrations). Make intermediate levels of glutamine synthetase independent of the presence or absence of NH_3.

c. Gln⁻, Ntr⁻

d. Gln⁻, Ntr⁻

e. Ntr constitutive

f. Temperature sensitive (won't grow at high temperature)

g. Constitutive expression of diphtheria toxin and other virulence determinants, even in Fe^{2+}

h. No ppGpp; grows very slowly and is auxotrophic for some amino acids

3. As described in chapter 1, you could clone the gene for the toxin and then perform Southern blot analysis to show that the gene is carried on a large region of DNA that is not common to all the members of the species. Using the methods in chapter 8, you could also try to induce a phage from the cells and show that production of the toxin requires lysogeny by the phage.

4. If the *rpoH* gene for σ^{32} is transcriptionally autoregulated, the same amount of RNA should be made from the gene when it exists in two or more copies as is made when it exists in only one copy. Introduce a clone of *rpoH* in a multicopy plasmid into cells, and measure the amount of RNA made on the gene by DNA-RNA hybridization. If more RNA is made from *rpoH* under these conditions, the gene is not transcriptionally autoregulated.

5. To show which ribosomal protein is the translational repressor, introduce an in-frame deletion into the *rplJ* gene and determine if the synthesis of L12 increases. Similarly, introduce a mutation (any inactivating mutation will do) into *rplL* and determine if L10 synthesis increases.

Chapter 14

Problems

1. If it has a *cis*-acting operator mutation, it will be constitutive only for the operon in the chromosome. Introduce a mutation into one of the structural genes in the chromosome, and it will no longer express the product of that gene constitutively, even from the copy on the F′ factor. If the mutation is a dominant constitutive mutation in the *trpR* gene, it will be *trans* acting and the cells will still be constitutive for expression of the gene from the F′ factor. Incidentally, *trpR*ᵈ mutations are possible because the TrpR repressor functions as a dimer and repressors made with one or more copies of the mutant *trpR* polypeptide are inactive and will not bind to the operator, making the cell constitutive for *trp* operon transcription.

2. Fuse various regions of the *envZ* gene to the *phoA* gene and introduce the fusion genes into a *phoA* E. coli mutant. Plate the cells on plates containing XP. If the colonies are blue, the region in the EnvZ protein to which the PhoA protein is fused is in the periplasm; if they are colorless, it is in the cytoplasm.

3. A signal sequence mutation in *malE* will make the cells Mal⁻ and unable to transport maltose for use as a carbon and energy source. A suppressor of the signal sequence mutation will make them Mal⁺ and able to grow on minimal plates with maltose as the sole carbon and energy source. Spread millions of bacteria with the *malE* signal sequence mutations on minimal plates with maltose. Any colonies which arise are Mal⁺ and may be revertants or may have suppressors of the signal sequence mutation. To distinguish those which have suppressors from the true revertants, you could use some of them as donors to transduce the *mal* operon into another strain, selecting for a nearby marker. If any of the transductants are Mal⁻, the *mal* operon in that Mal⁺ apparent revertant had a mutation somewhere else in the chromosome which was suppressing the signal sequence mutation.

4. By isolating mutants blocked in sporulation. The regulatory genes could then be identified because mutations in these genes blocked the expression of many other genes as determined by *lacZ* fusions. Also, some were similar in sequence to other known regulators, including sigma factors and transcriptional activators.

5. Both involve histidine kinases, which phosphorylate on histidine residues. Transfer of phosphoryl groups occurs between histidine and aspartate residues and transcriptional activators are regulated by phosphorylation. The *B. subtilis* phosphorelay contains a series of proteins that carry out discrete steps. Each of these is regulated by other phosphatases and kinases.

6. The original mutations in *spo0E* and *spo0L* both conferred the Spo⁻ phenotype, suggesting that the products of the genes were positive regulators. However, these mutations did not inactivate the gene products, and deletion mutations of the genes caused hypersporulation, suggesting that they were in fact negative regulators. Also, multiple copies of the genes inhibited sporulation, as expected of negative regulators. Spo⁻ suppressors of *spo0E* deletion mutations were in Spo0A a phosphorylated protein. Also, Spo⁺ suppressors of *spo0L* missense mutations were in spo0F, another phosphorylated protein. This suggested that Spo0E and Spo0L reduced the phosphorylation of these proteins by acting as phosphatases, and this was subsequently confirmed.

7. In a *spo-lacZ* transcriptional fusion, the *lacZ* gene has its own TIR and so is translated independently of the upstream *spo* sequences. Trancriptional fusions can be used to determine when the *spo* gene is transcribed. In the *spo-lacZ* translational fusions, the coding region of the *spo* gene is fused to the *lacZ* gene to encode a fusion protein with β-galactosidase activity. These fusions can be used to determine when the *spo* gene product is made.

8. The *tlp* gene is dependent upon σ^G, and *cotD* is dependent upon σ^K. The *tlp* gene is expressed in the forespore, and *cotD* is expressed in the mother cell.

Glossary

A site. The site on the ribosome to which the incoming aminoacylated tRNA binds.

Activator. A protein that positively regulates transcription of an operon by interacting with RNA polymerase at the promoter and allowing RNA polymerase to begin transcribing.

Activator site. Sequence in DNA upstream of the promoter to which the activator protein binds.

Adaptive response. Activation of transcription of the genes of the Ada regulon, which is involved in the repair of some types of alkylation damage to DNA.

Adenine (A). One of the two purine (two-ringed) bases in DNA and RNA.

Affinity tag. A polypeptide that binds tightly to some other molecule. If the polypeptide coding sequence is translationally fused to the coding sequence of another protein, it will allow the protein to be purified more easily.

Alkylating agent. A chemical that reacts with DNA and thereby forms a carbon bond to one of the atoms in DNA.

Allele. One of the forms of a gene, e.g., the gene with a particular mutation. Can refer to the wild-type or mutant form.

Allele-specific suppressor. A second-site mutation that restores the wild-type phenotype but only in strains with a particular type of mutation in a gene.

Allelism test. A complementation test to determine if two mutations are in the same gene, e.g., are in different alleles of the same gene.

Amber codon. The codon UAG.

Amber mutation. A mutation that causes the codon UAG to appear in frame in the coding region of an mRNA.

Amino group. The NH_2 chemical group.

Amino terminus. *See* N terminus.

Antibiotic. A chemical that kills cells or inhibits their growth.

Antibiotic resistance gene cassette. A fragment of DNA, usually bracketed by restriction sites, that contains a gene whose product confers resistance to an antibiotic.

Anticodon. The three-nucleotide sequence in a tRNA that pairs with the codon in mRNA by complementary base pairing.

Antiparallel. Referring to the fact that moving in one direction along a double-stranded DNA or RNA, the phosphates in one strand are attached 3′ to 5′ to the sugars while the phosphates in the other strand are attached 5′ to 3′.

Antisense RNA. An RNA that contains a sequence complementary to a sequence in an mRNA.

Antitermination. A process in which changes in the RNA polymerase allow it to transcribe through transcription termination signals in DNA.

AP endonuclease. A DNA-cutting enzyme that cuts next to a deoxynucleotide that has lost its base, i.e., an apurinic or apyrimidinic site.

Aporepressor. A protein that can be converted into a repressor if a small molecule called the corepressor is bound to it.

Archaea. A separate kingdom of prokaryotic single-celled organisms that share some of the features of both eukaryotes and prokaryotes and usually inhabit extreme environments.

Assimilatory reduction. Addition of electrons to nitrogen-containing compounds to convert them into NH_3 for incorporation into cellular constituents.

Attenuation. Regulation of an operon by premature termination of transcription.

Autocleavage. Process by which a protein cuts itself.

Autokinase. A protein able to transfer a PO_4 group from ATP to itself.

Autophosphorylation. Process by which a protein transfers a PO_4 group to itself, independent of the source of the PO_4 group.

Autoregulation. Process through which a gene product controls the level of its own synthesis.

Auxotrophic mutant. A mutant that cannot make or use a growth substance that the normal or wild-type organism can make or use.

Backbone. The chain of phosphates linked to deoxyribose sugars that holds the DNA chain together.

Bacterial lawn. The layer of bacteria on an agar plate that forms when many bacteria are plated and the bacterial colonies grow together.

Bacteriophage. A virus that infects bacteria.

Base. Carbon-, nitrogen-, and hydrogen-containing chemical compounds with structures composed of one or two rings that are constituents of the DNA or RNA molecule.

Base analog. A chemical that resembles one of the bases and so is mistakenly incorporated into DNA or RNA during synthesis.

Base pair. Each set of opposing bases in the two strands of double-stranded DNA or RNA that are held together by hydrogen bonds and thereby help hold the two strands together. Also used as a unit of length.

Base pair change. A mutation in which one type of base pair in DNA (e.g., an AT pair) is changed into a different base pair (e.g., a GC pair).

Binding. Process by which molecules are physically joined to each other by noncovalent bonds.

Bioremediation. The removal of toxic chemicals from the environment by microorganisms.

Biosynthesis. Synthesis of chemical compounds by living organisms.

Biosynthetic operon. An operon composed of genes whose products are involved in synthesizing compounds rather than degrading them.

Blot. The process of transferring DNA, RNA, or protein from a gel or agar plate to a filter. Also refers to the filter that contains the DNA, RNA, or protein.

Blunt end. A double-stranded DNA end in which the 3′ and 5′ termini are flush with each other, that is, with no overhanging single strands.

Branch migration. The process by which the site at which two double-stranded DNAs held together by crossed-over strands (such as in a Holliday junction) moves, changing the regions of the two DNAs that are paired in heteroduplexes.

Broad host range. The ability of a phage, plasmid, or other DNA element to enter and/or replicate in a wide variety of bacterial species.

Bypass suppression. A suppressor mutation that bypasses the need for a gene product.

CAP-binding site. The sequence on DNA to which the catabolite activator or CAP protein binds.

CAP protein. Also called the Crp protein. The protein that activates catabolite-sensitive genes.

CAP regulon. All of the operons that are regulated, either positively or negatively, by the CAP protein.

Capsid. The protein and/or membrane coat that surrounds the genomic nucleic acid (DNA or RNA) of a virus.

Carboxyl group. The chemical group COOH of amino acids.

Carboxyl terminus. *See* C terminus.

Catabolic operon. An operon composed of genes whose products degrade organic compounds.

Catabolism. The degradation of an organic compound, such as a sugar, to make smaller molecules with the concomitant production of energy.

Catabolite. A small molecule produced by the degradation of larger organic compounds such as sugars.

Catabolite activator protein (CAP). The DNA and cAMP binding protein that regulates catabolite-sensitive operons in enteric bacteria by binding to their promoter regions. Also called Crp protein.

Catabolite repression. The reduced expression of some operons in the presence of high cellular levels of catabolites.

Catabolite-sensitive operons. Operons whose expression is regulated by the cellular levels of catabolites.

Catenenes. Formed when two or more circular DNA molecules are joined like links in a chain.

CCC. *See* Circular and covalently closed.

Cell division. The splitting of a cell into two daughter cells.

Cell division cycle. The events occurring between the time a cell is created by division of its mother cell and the time it divides.

Cell generation. The making of a new cell by the growth and division of its mother cell.

Central Dogma. The tenet that protein is translated from RNA that was transcribed from DNA.

Chaperone. A protein that binds to other proteins and helps them fold correctly or prevents them from folding prematurely.

Chaperonin. A protein which forms chambers to take up denatured proteins and refold them. Represented by the Hsp60 protein GroEL in *E. coli*.

***chi* (χ) mutation.** A mutation that causes the sequence of a *chi* site to apppear in the DNA.

***chi* (χ) site.** The sequence 5′GCTGGTGG3′ in *E. coli* DNA. Stimulates recombination by the RecBCD nuclease in *E. coli* by inhibiting the 3′-to-5′ nuclease activity of RecBCD.

Chromosome. In a bacterial cell, the DNA molecule that contains most of the genes required for cellular growth and maintenance.

CI repressor. The phage λ-encoded protein that binds to the phage operator sequences close to the p_R and p_L promoters and prevents transcription of most of the genes of the phage.

Circular and covalently closed (CCC). Referring to a circular double-stranded DNA with no breaks or discontinuities in either of its strands.

***cis* acting.** Referring to a mutation that affects only the DNA molecule in which it occurs and not other DNA molecules in the same cell.

***cis*-acting site.** A functional region on a DNA molecule that does not encode a gene product and so affects only the DNA molecule in which it resides (e.g., an origin of replication).

Classical genetics. The study of genetic phenomena by using only intact living organisms.

Clonal. Referring to a situation in which all the descendants of an organism or replicating DNA molecule remain together, as in colonies on an agar plate.

Clone. A collection of DNA molecules or organisms that are all identical to each other because they result from replication or multiplication of the same original DNA or organism.

Cloning vector. An autonomously replicating DNA, usually a phage or plasmid, into which can be introduced other DNA molecules that are not capable of replicating themselves.

Closed complex. The complex which forms when RNA polymerase first binds to a promoter and before the strands of DNA at the promoter separate.

Cochaperone.   A smaller protein which helps chaperones fold proteins or cycle their adenine nucleotide. Represented by DnaJ and GrpE in *E. coli*.

Cochaperonin.   A smaller protein which helps chaperonins fold proteins by forming the cap on the chamber once the protein has been taken up. Represented by the Hsp10 GroES protein in *E. coli*.

Coding strand.   The strand of DNA in a gene that has the same sequence as the mRNA transcribed from the gene.

Codon.   A 3-base sequence in mRNA that stipulates one of the amino acids.

Cognate aminoacyl-tRNA synthetase.   The enzyme that attaches a particular amino acid onto a tRNA.

Cointegrate.   A DNA molecule with another DNA molecule inserted into it.

Cold-sensitive mutant.   A mutant that cannot live and/or multiply in the lower temperature ranges at which the normal or wild-type organism can live and/or multiply.

Colony.   A small lump or pile made up of millions of multiplying cells on an agar plate.

Colony papillation.   Process leading to sectors or sections in a colony that appear different from the remainder of the colony.

Colony purification.   Isolation of individual bacteria on an agar plate so that all the cells in a colony that forms after incubation will be descendants of the same bacterium.

Compatible restriction endonucleases.   Restriction endonucleases that leave the same overhangs after cutting a DNA molecule. The resulting ends can pair, allowing the molecules to be ligated to each other.

Competence pheromones.   Small peptides given off by bacterial cells. Required to induce competence in neighboring cells.

Competent.   The state of cells capable of taking up DNA.

Complementary base pair.   A pair of nucleotides that can be held together by hydrogen bonds between their bases, e.g., dGMP and dCMP or dAMP and dTMP.

Complementation.   The restoration of the wild-type phenotype when two DNAs containing different mutations that cause the same mutant phenotype are in the cell together. Usually means the two mutations affect different genes.

Complementation group.   A set of mutations of which none will complement any of the others.

Composite transposon.   A transposon made up of two insertion (IS) elements plus any DNA between them.

Concatemer.   Two or more almost identical DNA molecules linked tail to head.

Condensation.   A way of making the chromosome occupy a smaller space, for example by supercoiling.

Condensins.   Proteins that bind chromosomal DNA in two different places, folding it into large loops and thereby making it more condensed. Represented by the Smc protein in *B. subtilis* and by the MukB protein in *E. coli*.

Conditional lethal mutation.   A mutation that inactivates an essential cellular component, but only under a certain set of circumstances, for example, a temperature-sensitive mutation that inactivates RNA polymerase only at relatively high temperatures.

Conjugation.   The transfer of DNA from one bacterial cell to another by the transfer functions of a self-transmissible DNA element such as a plasmid.

Conjugative transposon.   A transposon that encodes functions that allow it to transfer itself into other bacteria.

Consensus sequence.   A nucleotide sequence in DNA or RNA, or an amino acid sequence in protein, in which each position in the sequence has the nucleotide or amino acid that has been found most often at that position in molecules with the same function.

Conservative.   Referring to a reaction involving double-stranded DNA in which the molecule retains both of its original strands.

Constitutive mutant.   A mutant in which the genes of an operon are transcribed whether or not the inducer of the operon is present.

Context.   The sequence of nucleotides in DNA or RNA surrounding a particular sequence.

Cooperative binding.   Process in which the binding of one protein molecule to a site greatly enhances the binding of another protein molecule of the same type to an adjacent site. For example on DNA.

Coprotease.   A protein that binds to another protein and thereby activates the second protein's protease activity.

Copy.   A molecule of a particular type identical to another in the same cell.

Copy number. The number of copies of a plasmid per cell immediately after cell division. Also the ratio of the number of plasmids of a particular type in the cell to the number of copies of the chromosome.

Core polymerase. The part of the DNA or RNA polymerase that actually performs the polymerization reaction and functions independently of accessory and regulatory proteins that cycle on and off the protein.

Corepressor. A small molecule that binds to an aporepressor and converts it into a repressor.

***cos* site.** The sequence of deoxynucleotides at the ends of λ DNA in the phage head. A staggered cut in this sequence at the time the phage DNA is packaged gives rise to complementary or cohesive ends that can base pair with each other.

Cosmid. A plasmid that carries the sequence of a *cos* site so that it can be packaged into λ phage heads.

Cotranscribed. Referring to two or more contiguous genes transcribed by a single RNA polymerase molecule from a single promoter.

Cotransducible. Referring to two genetic markers that are close enough together on the DNA that they can be carried in the same phage head during transduction.

Cotransduction. Occurs when tranductants that were selected for being recombinant for one marker in DNA are also recombinant for a second marker.

Cotransduction frequency. The percentage of transductants selected for one genetic marker from the donor that have also received another genetic marker from the donor. A measure of how far apart the markers are on DNA.

Cotransformable. As in cotransducible, but the regions of two genetic markers are close enough together to be carried on the same piece of DNA during transformation.

Cotransformation frequency. As in cotransduction frequency, except that the percentage is of transformants. A measure of how far apart the markers are on DNA.

Cotranslational translocation. Occurs when a protein is translated as it is inserted into the membrane by the SRP system.

Counterselection of donor. Selection of transconjugants under conditions in which the donor bacterium cannot multiply to form colonies.

Coupling model. A model for the regulation of replication of iteron plasmids in which two or more plasmids are joined by binding to the same Rep protein through their iteron sequences. *See* Handcuffing model.

Coupling protein. A protein that is part of the Mpf system of self-transmissible plasmids which communicates to the Dtr system that contact has been made with a recipient cell.

Covalent bond. Bonds that hold two atoms together by a sharing of their electron orbits.

Covalently closed circular. *See* Circular and covalently closed.

Cross. Any means of exchange of DNA between two organisms.

Crossing. Allowing the DNAs of two strains of an organism to enter the same cell so they can recombine with each other.

Crossover. The site of the breaking and rejoining of two DNA molecules during homologous recombination.

C-terminal amino acid. The amino acid on one end of a polypeptide chain that has a free carboxyl group unattached to the amino group of another amino acid.

C terminus. The end of a polypeptide chain with the free carboxyl (COOH) group.

Cured. Referring to a cell that has lost a DNA element such as a plasmid.

Cut and paste. A mechanism of transposition in which the entire transposon is excised from one place in the DNA and inserted into another.

Cyclic AMP (cAMP). Adenosine monophosphate with the phosphate attached to both the 3′ and 5′ carbons of the ribose sugar.

Cyclobutane ring. A ring structure of four carbons held together by single bonds. Present in some types of pyrimidine dimers in DNA.

Cytoplasmic domain. The regions of a polypeptide of a transmembrane protein that are in the interior or cytoplasm of the cell.

Cytosine (C). One of the pyrimidine (one-ringed) bases in DNA and RNA.

D-loop. The three-stranded structure that forms when a single strand of DNA invades a double-stranded DNA, displacing one of the strands.

Damage tolerance mechanism. A way of dealing with damage to DNA which does not involve repairing the damage, for example replication restart or translesion synthesis.

Daughter cell.   One of the cells arising from division of another cell.

Daughter DNA.   One of the two DNAs arising from replication of another DNA.

Deaminating agent.   A chemical that reacts with DNA, causing the removal of amino (NH_2) groups from the bases in DNA.

Deamination.   The process of removing amino (NH_2) groups from the bases in DNA.

Decatenation.   The process performed by type II topoisomerases of passing DNA strands through each other to resolve catenenes.

Defective prophage.   A DNA element in the bacterial chromosome that presumably was once capable of being induced to form phages but has lost genes essential for lytic development.

Degenerate probe.   A chemically synthesized oligonucleotide that is made to be complementary to a certain protein-coding sequence in DNA or RNA but in which the third base in some codons has been randomized to include all the codons that could encode each amino acid.

Degradative operon.   Like a catabolic operon, an operon whose genes encode enzymes required for the breakdown of molecules into smaller molcules with the concomitant release of energy.

Deletion mapping.   Procedure in which mutants that have mutations to be mapped are crossed with mutants that have deletion mutations with known endpoints. Wild-type recombinants will appear only if the unknown mutation lies outside the deleted region.

Deletion mutation.   A mutation in which a number of base pairs have been removed from the DNA.

Deoxyadenosine.   An adenine base attached to a deoxyribose sugar.

Deoxyadenosine methylase (Dam methylase).   An enzyme that attaches a CH_3 (methyl) group to the adenine base in DNA. The enzyme from *E. coli* methylates the A in the sequence GATC.

Deoxycytidine.   A cytosine base attached to a deoxyribose sugar.

Deoxyguanosine.   A guanine base attached to a deoxyribose sugar.

Deoxynucleoside.   A base (A, G, T, or C) attached to a deoxyribose sugar.

Deoxyribose.   A sugar similar to the five-carbon sugar ribose but with a hydrogen atom (H) rather than a hydroxyl (OH) group attached to the 2′ carbon.

Deoxythymidine.   The thymine base attached to deoxyribose sugar.

Dimer.   A protein made up of two polypeptides.

Dimerization domain.   The region of a polypeptide that binds to another polypeptide of the same type to form a dimer.

Dimerize.   To bind two identical polypeptides to each other.

Diploid.   The state of a cell containing two copies of each of its genes, which are not derived from replication of the same DNA. *See* Haploid.

Direct repeat.   A short sequence of deoxynucleotides in DNA closely followed by an almost identical sequence on the same strand.

Directed-change hypothesis.   The hypothesis that mutations in DNA occur preferentially when they benefit the organism.

Directional cloning.   Cloning a piece of DNA into a cloning vector in such a way that it can be inserted in only one orientation, for example by using two different restriction endonucleases.

Dissimilatory reduction.   The reduction of nitrogen-containing compounds that occurs when they are used as terminal electron acceptors in anaerobic respiration. The nitrogen is not necessarily incorporated into the cellular molecules.

Disulfide bonds.   Covalent bonds between two sulfur atoms, such as those that occur between the sulfur atoms in two cysteine amino acids.

Disulfide oxidoreducatase.   An enzyme which can form or break disulfide bonds between cysteines.

Division septum.   The crosswall that forms between two daughter cells just before they separate.

Division time.   The time that elapses between successive cell divisions in an exponentially growing culture.

DNA binding domain.   The region of a polypeptide in a DNA binding protein that binds to DNA.

DNA box.   Sequence on DNA to which a protein binds.

DNA clone.   A fragment of DNA inserted in a cloning vector such that many identical copies of the fragment will be made when the cloning vector replicates.

DNA glycosylase.   An enzyme that removes bases from DNA by cleaving the bond between the base and the deoxyribose sugar.

DNA helicase.   An enzyme that uses the energy of ATP to separate the strands of double-stranded DNA.

DnaK. *See* Hsp70.

DNA library. A collection of clones of the DNA of an organism that together represent all the DNA sequences of that organism.

DNA ligase. An enzyme that can join the phosphate-terminated 5′ end of one DNA strand to the 3′ hydroxyl end of another.

DNA polymerase accessory proteins. Proteins that travel with the DNA polymerase during replication.

DNA polymerase III holoenzyme. The replicative DNA polymerase in *E. coli* including all the accessory proteins, sliding clamp, editing functions, etc.

DNA polymerase V. The product of the UmuC gene of *E. coli*; DNA polymerase capable of translesion synthesis.

DNA replication complex. The entire complex of proteins, including the DNA polymerase, that moves along the DNA at the replication fork.

DNA transfer functions (Dtr component). The *tra* gene functions of a plasmid responsible for preparing the DNA for transfer.

DnaA box. The sequence 5′TTATCCACA3′ in DNA to which the DnaA protein binds. The DnaA protein is required for the initiation of chromosome replication in *E. coli*.

Domain. A region of a polypeptide with a particular function or localization.

Dominant. When a mutation or other genetic marker exerts its phenotype even in an organism that is diploid for the region because it also contains the corresponding region from the wild-type organism.

Dominant mutation. A mutation that affects the phenotype, even in a diploid organism containing a wild-type allele of the gene.

Donor. The bacterial strain that is the source of the transferred DNA in a bacterial cross. For example, in a transductional cross, it is the strain in which the phage was previously propagated. In conjugation, it is the strain harboring the self-transmissible plasmid.

Donor allele. The form of the gene that exists in the donor strain if the donor and recipient in a cross have different forms of a gene.

Donor DNA. DNA that is extracted from the donor bacterium and used to transform a recipient bacterium.

Double mutant. A mutant with two mutations.

Downstream. From a given point, sequences that lie in the 3′ direction on RNA or in the 3′ direction on the coding strand of a DNA region from which an RNA is made.

Dtr component. DNA transfer component of a plasmid transmission system. The *tra* or *mob* genes of the plasmid involved in preparing the plasmid DNA for transfer.

Duplication junction. The point at which a crossover occurred, resulting in a tandem duplication mutation in DNA.

E site. The site on the ribosome at which the tRNA binds after it has contributed its amino acid to the growing polypeptide and just before it exits the ribosome.

Early gene. A gene expressed early during a developmental process, for example, during bacterial sporulation or phage infection.

Ectopic recombination. Homologous recombination occurring between two sequences in different regions of the two DNAs participating in the recombination. "Out-of-place" recombination.

Editing. Process of removing and replacing a wrongly inserted deoxynucleotide during replication, for example, a C inserted opposite a template A, to reduce the frequency of mutations.

Editing functions. The 3′ exonuclease activities that remove nucleotides erroneously incorporated during replication. Such activities can be part of the DNA polymerase polypeptide itself or can be associated with proteins that travel with the DNA polymerase during replication.

Effector. A small molecule that binds to a protein and changes its properties.

EF-G. *See* Translation elongation factor G.

EF-Tu. *See* Translation elongation factor Tu.

Eight-hitter. A type II restriction endonuclease that recognizes and cuts in an 8-bp sequence in DNA.

8-OxoG. A commonly damaged base in DNA caused by reactive forms of oxygen, in which an oxygen atom has been added to the 8-position of the small ring of the base guanine.

Electroporation. The introduction of nucleic acids or proteins into cells through exposure of the cells to an electric field.

Elongation factor G. *See* Translation elongation factor G.

Elongation factor Tu. *See* Translation elongation factor Tu.

Endonuclease. An enzyme that can cut internal phosphodiester bonds in a polynucleotide.

Enrichment. The process of increasing the frequency of a particular type of mutant in a population, often by using an antibiotic that kills growing but not stationary-phase cells.

Epistasis. A type of interaction in which a mutation at one locus can affect the phenotype of a different locus.

Escape synthesis. Induction of transcription of an operon as a result of titration of its repressor owing to an increase in the number of operators to which the repressor binds.

Essential genes. Genes whose products are required for maintenance and/or growth of the cell under all conditions.

Eubacteria. "True" bacteria. A member of the kingdom of organisms characterized by a relatively simple cell structure free of many cellular organelles, the presence of 16S and 23S rRNAs, and usually a four-component core RNA polymerase, among other features.

Eukaryotes. Members of the kingdom of organisms whose cells contain a nucleus surrounded by nuclear membrane and many other cellular organelles, including a Golgi apparatus and an endoplasmic reticulum. Have 18S and 28S rRNAs.

Exonuclease. An enzyme which removes nucleotides one at a time from the end of a polynucleotide.

Exported proteins. Proteins which leave the cytoplasm after they are made and end up in a membrane, in the periplasmic space, or outside the cell.

Expression vector. A cloning vector in which a cloned gene can be transcribed and sometimes also translated from a vector promoter and translation initiation region, respectively.

Extracellular protein. A protein that is excreted from cells after it is made.

Extragenic. Involving a different gene.

Extragenic suppressor. A suppressor mutation that is in a gene different from the gene encoding the mutation that it suppresses.

Factor-dependent transcription termination site. A sequence on DNA that causes transcription termination only in the presence of a particular protein such as the Rho protein of *E. coli*.

Factor-independent transcription termination site. A sequence in DNA that causes transcription termination by RNA polymerase alone, in the absence of other proteins. In bacteria, characterized by a GC-rich region with an inverted repeat followed by a string of A's on the template strand.

Feedback inhibition. Inhibition of the synthesis of the product of a pathway that results from binding of the end product of the pathway to the first enzyme of the pathway, thereby inhibiting activity of the enzyme.

Filamentous phage. A phage with a long, floppy appearance. The nucleic acid genome of these phages is merely coated with protein, making the phage as long as the genome and giving the floppy appearance. In contrast, the nucleic acids of most phages are encapsulated in a rigid, almost spherical, icosahedral head.

Filter mating. Procedure in which two bacteria are trapped on a filter to hold them in juxtaposition so that conjugation can occur.

5′ end. The end of a nucleic acid strand (DNA or RNA) in which the 5′ carbon of the ribose sugar is not attached through a phosphate to another nucleotide.

5′ exonuclease. A deoxyribonuclease (DNase) that degrades DNA starting with a free 5′ end.

5′ overhang. A short, single-stranded 5′ end on an otherwise double-stranded DNA molecule.

5′ phosphate end. In a polynucleotide, a 5′ end that has a phosphate attached to the 5′ carbon of the ribose sugar of the last nucleotide.

5′ untranslated region. The sequence of nucleotides that extends from the 5′ end of an mRNA to the 5′ end of the first protein-coding region.

Flanking sequences. The sequences that lie on either side of a gene or other DNA element.

Formylmethionyl-tRNA$_f^{Met}$. The special tRNA in prokaryotes that is activated by formylmethionine and is used to initiate translation at prokaryotic translation initiation regions (TIRs). It binds to translation initiation factor IF2 and responds to the initiator codons AUG and GUG and, more rarely, to other codons in a TIR.

Four-hitter. A type II DNA restriction endonuclease that recognizes and cuts at a 4-bp sequence in DNA.

4.5S RNA. The RNA component of the signal recognition particle (SRP) of bacteria.

Frameshift mutation. Any mutation that adds or removes one or more (but not a multiple of 3) base pairs from DNA, whether or not it occurs in the coding region for a protein.

Functional domain. The region of a polypeptide chain that performs a particular function in the protein.

Functional genomics. The use of techniques such as reverse genetics, microarrays, and proteomics to study the functions of a sequenced genome.

Fusion protein. A protein created when coding regions from different genes are fused to each other so that one part of the protein is encoded by sequences from one gene and another part is encoded by sequences from a different gene.

Gel electrophoresis. Procedure for separating proteins, DNA, or other macromolecules that involves the application first of the macromolecules to a gel made of agarose, acrylamide, or some other gelatinous material and then of an electric field, forcing the electrically charged macromolecules to move toward one or the other electrode. How fast the macromolecules move will depend upon their size and their charge.

Gene. A region on DNA encoding a particular polypeptide chain or functional RNA such as an rRNA or a tRNA.

Gene conversion. Nonreciprocal apparent recombination associated with mismatch repair on heteroduplexes that are formed between two DNA molecules during recombination. The name comes from genetic experiments with fungi in which the alleles of the two parents were not always present in equal numbers in an ascus, as though an allele of one parent had been "converted" into the allele of the other parent.

Gene disruption. Any alteration of the structure or activity of a gene. Usually refers to changes that inactivate the gene.

Gene dosage experiment. An experiment in which the number of copies of a gene in a cell is increased to determine what effect this has on the amount of gene product that is synthesized.

Gene replacement. A molecular genetic technique in which a cloned gene is altered in the test tube and then reintroduced into the organism, selecting for organisms in which the altered gene has replaced the corresponding normal gene in the organism.

Generalized recombination. *See* Homologous recombination.

Generalized transduction. The transfer, via phage transduction, of essentially any region of the DNA from one bacterium to another.

Generation time. The time it takes for the number of cells in an exponentially growing culture to double.

Genetic code. The assignment of each mRNA nucleotide triplet to an amino acid.

Genetic linkage map. An ordering of the genes of an organism solely on the basis of recombination frequencies between mutations in the genes in genetic crosses.

Genetic marker. A difference in sequence of the DNAs of two strains of an organism that causes the two strains to exhibit different phenotypes that can be used for genetic mapping.

Genetic recombination. The joining of genetic markers or DNA sequences in new combinations.

Genetics. The science of studying organisms on the basis of their genetic material.

Genome. The nucleic acid (DNA or RNA) of an organism or virus that includes all the information necessary to make a new organism or virus.

Genomics. The process of using the sequence of the entire DNA of an organism to study its physiology and relationship to other organisms.

Genotype. The sequence of nucleotides in the DNA of an organism, usually discussed in terms of the alleles of its genes.

Global regulatory mechanism. A regulatory mechanism that affects many genes scattered around the genome.

Glucose effect. The regulation of genes involved in carbon source utilization based on whether glucose is present in the medium.

Glutamate dehydrogenase. An enzyme that adds ammonia directly to α-ketoglutarate to make glutamate. Responsible for assimilation of nitrogen in high ammonia concentrations.

Glutamate synthase. An enzyme that transfers amino groups from glutamine to α-ketoglutarate to make glutamate.

Glutamine synthetase (GS). An enzyme that adds ammonia to glutamate to make glutamine. Responsible for the assimilation of nitrogen in low ammonia concentrations.

Gradient of transfer. In a conjugational cross, the decrease in the transfer of chromosomal markers the farther they are from the origin of transfer of an integrated plasmid.

Gram-negative bacteria. Bacteria characterized by an outer membrane and a thin peptidoglycan cell wall that stains poorly with a stain invented by the Danish physician Hans Christian Gram in the 19th century.

Gram-positive bacteria. Bacteria characterized by having no outer membrane and a thick peptidoglycan layer that stains well with the Gram stain.

GroEL. *See* Hsp60.

GroES. *See* Hsp10.

Growth rate regulation of ribosomal synthesis. The regulation of ribosomal synthesis that ensures that cells growing more slowly have fewer ribosomes.

Guanine (G). One of the two purine (two-ringed) bases in DNA and RNA.

Guanosine. The base guanine with a ribose sugar attached. A nucleoside.

Guanosine pentaphosphate (pppGpp). The nucleoside guanosine with two phosphates attached to the 3′ carbon and three phosphates attached to the 5′ carbon of the ribose sugar. Responsible for stringent control.

Guanosine tetraphosphate (ppGpp). The nucleoside guanosine with two phosphates attached to each of the 3′ and 5′ carbons of the ribose sugar. Responsible for stringent control.

Gyrase. A type II topoisomerase capable of introducing negative supercoils two at a time into DNA with the concomitant cleavage of ATP. Apparently unique to bacteria.

Hairpin. A secondary structure formed in RNA or single-stranded DNA when one region of the polynucleotide chain folds back on itself and pairs by complementary base pairing with another region a few nucleotides away.

Handcuffing model. A model for the regulation of replication of iteron plasmids in which two plasmid molecules are held together by binding to the same Rep protein through their iteron sequences. *See* Coupling model.

Haploid. The state of a cell that has only one copy of each of its chromosomal genes. *See* Diploid.

Haploid segregant. A haploid cell or organism derived from multiplication of a partially or fully diploid or polyploid cell.

Headful packaging. A mechanism of encapsulation of DNA in a virus head in which the concatemeric DNA is cut after uptake of a length of DNA sufficient to fill the head.

Heat shock protein (Hsp). One of a group of proteins whose rate of synthesis markedly increases after an abrupt increase in temperature or certain other stresses on the cell.

Heat shock regulon. The group of *E. coli* genes under the control of σ^{32}, the heat shock sigma.

Heat shock response. The cellular changes that occur in the cell after an abrupt rise in temperature.

Helix-destabilizing protein. A protein that preferentially binds to single-stranded DNA and so can help keep the two complementary strands of DNA separated during replication, etc.

Helper phage. A wild-type phage that furnishes gene products that a deleted form of the phage cannot make, thereby allowing the deleted form to multiply and form phage.

Hemimethylated. When only one strand of DNA is methylated at a sequence with twofold symmetry, such as when only one of the two A's in the sequence GATC/CTAG is methylated.

Heterodimer. A protein made of two polypeptide chains that are encoded by different genes. *See* Homodimer.

Heteroduplex. A double-stranded DNA region in which the two strands come from different DNA molecules.

Heteroimmune. Referring to related lysogenic phages that carry different immunity regions and therefore cannot repress each other's transcription.

Heterologous probe. A DNA or RNA hybridization probe taken from the same gene or region of a different organism. It is usually not completely complementary to the sequence being probed. *See* Hybridization probe.

Heteromultimer. A protein made of several polypeptide chains that are encoded by different genes. *See* Heterodimer.

Hfr strain. A bacterial strain that contains a self-transmissible plasmid integrated into its chromosome and which thus can transfer its chromosome by conjugation.

HFT lysate. The lysate of lysogenic phage containing a significant percentage of transducing phage with bacterial DNA substituted for phage DNA.

High multiplicity of infection (MOI). Referring to a virus infection in which the number of viruses greatly exceeds the number of cells being infected.

High negative interference. A phenomenon in which a crossover in one region of the DNA greatly increases the probability of an apparent second crossover close by.

Holliday junction. An intermediate in homologous recombination in which one strand from each of two DNAs crosses over and is joined to the corresponding strand on the opposite DNA.

Holliday model. A model for homologous recombination developed by Robin Holliday, stating that one strand of each DNA is cut at exactly the same place and crosses over to be joined to the corresponding strand on the other DNA. The resulting Holliday junction can then isomerize and be cut in the crossed strands to recombine the flanking DNA sequences.

Homeologous recombination. Homologous recombination in which the deoxynucleotide sequences of two participating regions are somewhat different from each other, usually because they are in different regions of the DNA or because the DNAs come from different species. *See* Ectopic recombination.

Homing. The process of double-strand break repair and gene conversion and by which an intron or intein in a gene enters the same site in the same gene in a new DNA which lacks it.

Homing endonuclease. The sequence-specific DNA endonuclease that makes a double-strand break in the target DNA to initiate the homing of an intron or intein.

Homodimer. A protein made up of two polypeptide chains that are identical to each other, usually because they are encoded by the same gene. *See* Homomultimer.

Homologous proteins. Proteins encoded by genes derived from a common ancestral gene.

Homologous recombination. A type of recombination that depends on the two DNAs having identical or at least very similar sequences in the regions being recombined because complementary base pairing between strands of the two DNAs must occur as an intermediate state in the recombination process.

Homomultimer. A protein made up of more than one identical polypeptide, usually encoded by the same gene. *See* Homodimer.

Host range. All of the types of host cells in which a DNA element, plasmid, phage, etc., can multiply.

Hot spot. A position in DNA that is particularly prone to mutagenesis by a particular mutagen.

Hsp10. Heat shock protein of 10 kDa. The cochaperonin to Hsp60 that forms a cap on the cylinder in which denatured proteins are folded. Represented by GroES in bacteria.

Hsp60 chaperonin. A heat shock-induced chaperonin of 60 kDa, represented by GroEL in bacteria.

Hsp70. A heat shock-induced chaperone of 70 kDa, represented by DnaK in bacteria.

Hybridization. The process by which two complementary strands of DNA or RNA, or a strand of DNA and a strand of RNA, are allowed to base pair with each other and form a double helix.

Hybridization probe. A DNA or RNA that can be used to detect other DNAs and RNAs because it shares a complementary sequence with the DNA or RNA being sought and so will hybridize to it by base pairing.

Hypoxanthine. A purine base derived from the deamination of adenine.

IF2. *See* Initiation factor 2.

Incompatibility. The interference of plasmids with one another's replication and/or partitioning.

Incompatibility (Inc) group. A set of plasmids that interfere with each other's replication and/or partitioning and so cannot be stably maintained together in the same culture.

Induced mutations. Mutations that are caused by irradiating cells or treating cells or DNA deliberately with a mutagen such as a chemical.

Inducer. A small molecule that can increase the transcription of an operon.

Inducer exclusion. The process by which the inducer of an operon such as a sugar is kept out of the cell by inhibiting its transport through the membrane.

Inducible. Referring to an operon capable of having its transcription increased by an inducer.

Induction. In gene regulation, the turning on of the expression of the genes of an operon. In phage, the initiation of lytic development of a prophage.

In-frame deletion. A deletion mutation in an ORF that removes a multiple of 3 bp and so does not cause a frameshift. Particularly useful because they cannot be polar and can remove a specific domain without removing the rest of the protein.

Initiation codon. The 3-base sequence in an mRNA that specifies the first amino acid to be inserted in the synthesis of a polypepetide chain; in prokaryotes, the 3-base sequence (usually AUG or GUG) within a translation initiation region for which formylmethionine is inserted to begin translation. In eukaryotes, the AUG closest to the 5' end of the mRNA is usually the initiation codon and methionine is inserted to begin translation.

Initiation factor 2 (IF2). The specialized EF-Tu-like protein that binds formylmethionyl-tRNA ($tRNA_f^{Met}$) and brings it into the ribosome in response to an initiation codon that is part of a translation initiation region.

Initiation mass. The size of a bacterial cell at which initiation of a new round of chromosome replication occurs.

Inner membrane protein. A protein that resides, at least in part, in the inner membrane of a gram-negative bacterium.

Insertion element. *See* Insertion sequence (IS) element.

Insertion mutation. A change in a DNA sequence due to the incorporation of another DNA sequence such as a transposon or antibiotic resistance cassette.

Insertion sequence element (IS element). A small transposon in bacteria that carries only genes for the enzymes needed to promote its own transposition.

Insertional inactivation. Inactivation of the product of a gene by an insertion mutation, for example a transposon or the cloning of an antibiotic resistance cassette into the gene.

Integrase. A type of site-specific recombinase that promotes recombination between two defined sequences in DNA, causing the integration of one DNA into another DNA, such as the integration of a phage DNA into the chromosome.

Intergenic. In different genes.

Intergenic suppressor. A suppressor mutation that is in a gene different from that containing the mutation it suppresses. Also called extragenic suppressor.

Internal fragments. Fragments, created by cutting DNA containing a transposon or other DNA element with a restriction endonuclease, that come from entirely within the DNA element.

Interstrand cross-links. Covalent chemical bonds between the two complementary strands of DNA in a double-stranded DNA.

Intragenic. In the same gene.

Intragenic complementation. Complementation between two mutations in the same gene. Rare and allele specific. Usually occurs only if the protein product of the gene is a homomultimer.

Intragenic suppressor. A suppressor mutation that occurs in the same gene as the mutation it is suppressing.

Inversion junctions. The points where the recombination event occurred that inverted a sequence.

Inversion mutation. A change in DNA sequence as a result of flipping a region within a longer DNA so that it lies in reverse orientation. Usually due to homologous recombination between inverted repeats in the same DNA molecule.

Inverted repeat. Two nearby sequences in DNA that are the same or almost the same when read 5′ to 3′ on the opposite strands.

Invertible sequence. A sequence in DNA that inverts often owing to the action of a site-specific recombinase protein that promotes recombination between inverted repeats at the ends of the sequences.

In vitro mutagen. A mutagen that reacts only with purified DNA or with viruses. Cannot be used to mutagenize intact cells.

In vitro packaging. The incorporation of DNA or RNA into virus heads in the test tube.

In vivo mutagen. A mutagen that will enter and mutagenize the DNA of intact cells.

IS element. *See* Insertion sequence element.

Isogenic. Strains of an organism that are almost identical genetically except for one small region or gene.

Isolation of mutants. The process of obtaining a pure culture of a particular type of mutant from among a myriad of other types of mutants and the wild type.

Isomerization. The changing of the spatial conformation of a molecule without breaking any bonds.

Iteron sequences. Short DNA sequences, often repeated many times in the origin region of some types of plasmids, that play a role in the regulation of replication of the plasmid.

Junction fragments. Fragments, created by cutting a DNA containing a transposon or other DNA element, that contain sequences from one of the ends of the DNA element as well as flanking sequences from the DNA into which the element has inserted.

Kinase. An enzyme that transfers a phosphate group from ATP to another molecule.

Knockout mutation. A mutation that eliminates the function of a gene.

Lagging strand. During DNA replication, the newly synthesized strand that must be synthesized in the direction opposite the overall movement of the replication fork.

Late gene. A gene that is expressed only relatively late in the course of a developmental process, e.g., a late gene of a phage.

Lawn. *See* Bacterial lawn.

Leader region. The region 5′ of the coding region of the first structural gene on the mRNA for an operon.

Leader sequence. An RNA sequence within a leader region.

Leading strand. During DNA replication, the newly synthesized strand that is made in the same direction as the overall direction of movement of the replication fork.

Leaky mutation. A mutation in a gene that does not completely inactivate the product of the gene, hence leaving some residual activity.

Lep protease. One of the enzymes that cleaves the signal sequence off of secreted proteins as they pass through the SecYEG channel.

Lesion. Any change in a DNA molecule as a result of chemical alteration of a base, sugar, or phosphate.

Linkage. Occurs when two genetic markers are sufficiently close together on the DNA that recombination between them is less than optimum.

Linked. A genetic term referring to the fact that two markers are close enough together that they are usually not separated by recombination.

Locus. A region in the genome of an organism.

Low multiplicity of infection (MOI). Referring to a virus infection in which the number of cells almost equals or exceeds the number of viruses, so that most cells remain uninfected or are infected by at most one or very few viruses.

Lyse. To break open cells and release their cytoplasm into the medium.

Lysogen. A strain of bacterium that harbors a prophage.

Lysogenic conversion. A property of a bacterial cell caused by the presence of a particular prophage.

Lysogenic cycle. The series of events that follow infection by a bacteriophage and culminate in the formation of a stable prophage.

Lysogenic phage. A phage which is capable of entering a prophage state.

Lytic cycle. The series of events that follow infection by a bacteriophage or induction of a prophage and culminate in lysis of the bacterium and the release of new phage into the medium.

Macromolecule. A large molecule such as DNA, RNA, or protein.

Major groove. In the DNA double helix, the larger of the two grooves between the two strands of DNA wrapped around each other.

Male bacterium. A bacterium harboring a self-transmissible plasmid.

Male-specific phage. A phage that infects only cells carrying a particular self-transmissible plasmid. The plasmid produces the sex pilus used by the phage as its adsorbtion site.

Maltodextrins. Short chains of glucose molecules held together by α1-4 linkages. Breakdown products of starch.

Maltotriose. A chain of three glucose molecules held together by α1-4 linkages.

Map distance. The distance between two markers in the DNA as measured by recombination frequencies.

Map expansion. A phenomenon that occurs in genetic linkage experiments in which two markers appear to be farther apart than they are because of hyperactive apparent recombination.

Map unit. A distance between genetic markers corresponding to a recombination frequency of 1% between the markers.

Marker effect. A difference in the apparent genetic linkage between the site of a mutation and other markers depending on the type of mutation at the site.

Marker rescue. The acquisition of a genetic marker by the genome of an organism or virus through recombination with a cloned DNA fragment containing the marker.

Maxam-Gilbert sequencing. A method for DNA sequencing that depends on the ability of certain chemicals to react with and cleave DNA at particular bases.

Membrane protein. A protein that at least partially resides in, or is tightly bound to, one of the cellular membranes.

Membrane topology. The distribution of the various regions of a membrane protein between the membrane and the two surfaces of the membrane.

Merodiploid. Referring to a bacterial cell that is mostly haploid but is diploid for some regions of the genome. *See* Partial diploid.

Messenger RNA (mRNA). An RNA transcript that includes the coding sequences for at least one polypeptide.

Methionine aminopeptidase. An enzyme that removes the N-terminal methionine from newly synthesized polypeptides.

Methyl-directed mismatch repair system. A repair system in enteric bacteria that recognizes mismatches in newly replicated DNA and specifically removes and resynthesizes the new strand, which is distinguishable from the old strand because it is not methylated at nearby GATC sequences.

Methyltransferase. In DNA repair, an enzyme that removes a CH_3 (methyl) or CH_3CH_2 (ethyl) group from a base in DNA by attaching the group to itself.

Microarray. Hybridization of the RNA or DNA of an organism synthesized under a given set of conditions with a major part or all of the genome of the organism displayed in discrete pieces on a glass plate or nylon membrane.

Migration. *See* Branch migration.

Mini-Mu. A shortened version of phage Mu DNA in which most of the phage DNA has been deleted except the inverted repeat ends and the transposase genes, leaving it unable to replicate or be packaged into a phage head without the assistance of a helper wild-type phage mu. Other DNAs, such as genes for antibiotic resistance and a plasmid origin of replication, can be inserted between copies of the mini-Mu.

Minor groove. In double-stranded DNA, the smaller of the two gaps between the two strands of DNA wrapped around each other in a helix.

Minus (–) strand. In a virus with a single-stranded nucleic acid genome (DNA or RNA), the strand that is complementary to the strand in the virus head.

–10 sequence. In a bacterial σ^{70}-type promoter, a short sequence that lies about 10 bp upstream of the transcription start site. The canonical or consensus sequence is TATAAT/ATATTA.

–35 sequence. In a bacterial σ^{70}-type promoter, a short sequence that lies about 35 bp upstream of the transcription start site. The canonical or consensus sequence is TTGACA/AACTGT.

Mismatch. When the normal bases in DNA are not paired properly, e.g., an A opposite a C.

Mismatch repair system. Any pathway for removing mismatches in DNA and replacing them with the correctly paired bases.

Missense mutation. A base pair change mutation in a region of DNA encoding a polypeptide that changes an amino acid in the polypeptide.

***mob* region.** A region in DNA carrying an origin of transfer (*oriT* sequence) and genes whose products allow the plasmid to be mobilized by self-transmissible plasmids.

Mobilizable. A plasmid or other DNA element that cannot transfer itself into other bacteria but can be transferred by other self-transmissible elements.

Molecular genetic analysis. Any study of cellular or organismal functions that involves manipulations of DNA in the test tube.

Molecular genetic techniques. Methods for manipulating DNA in the test tube and reintroducing the DNA into cells.

Mother cell. A cell that divides or differentiates to give rise to a new cell or spore.

Mpf component. Mating-pair formation. This component is made up of *tra* gene products of a self-transmissible plasmid involved in making the surface structures (pilus, etc.) that contact another cell and transfer the DNA during conjugation.

mRNA. *See* Messenger RNA.

Multimeric protein. A protein that consists of more than one polypeptide chain.

Multiple cloning site. A region of a cloning vector that contains the sequences cut by many different type II restriction endonucleases. Also called polyclonal site.

Multiplicity of infection (MOI). The ratio of viruses to cells in an infection.

Mutagen. A chemical or type of irradiation that causes mutations by damaging DNA.

Mutagenic chemicals. Chemicals that cause mutations by damaging DNA.

Mutagenic repair. A pathway for repairing damage to DNA that sometimes changes the sequence of deoxynucleotides as a consequence.

Mutant. An organism that differs from the normal or wild type as a result of a change in the sequence (mutation) of its DNA.

Mutant allele. The mutated gene of a mutant organism that makes it different from the wild type.

Mutant enrichment. A procedure for increasing the frequency of a particular type of mutant in a culture.

Mutant phenotype. A characteristic that makes a mutant organism different from the wild type.

Mutation. Any heritable change in the sequence of deoxynucleotides in DNA.

Mutation rate. The probability of occurrence of a mutation causing a particular phenotype each time a newborn cell grows and divides.

N-terminal amino acid. The amino acid on the end of a polypeptide chain whose amino (NH_2) group is not attached to another amino acid in the chain through a peptide bond.

N terminus. The end of a polypeptide chain with the free amino (NH_2) group.

Narrow host range. If the range of hosts in which a DNA element can replicate includes only a few closely related types of cells.

Natural competence.   The ability of some types of bacteria to take up DNA at a certain stage in their growth cycle without chemical or other treatments.

Naturally transformable bacteria.   Bacteria that have a growth stage during which they are naturally competent.

Negative regulation.   A type of regulation in which a protein or RNA molecule, in its active form, inhibits a process such as the transcription of an operon or translation of an mRNA.

Negative selection.   The process of detecting a mutant on the basis of the inability of the mutant to multiply under a certain set of conditions in which the normal or wild-type organism can multiply.

Negatively supercoiled.   Referring to two strands of a DNA molecule that are wrapped around each other less than about once every 10.5 bp.

Nicked DNA.   Double-stranded DNA in which one strand contains a broken phosphate-deoxyribose bond in the phosphodiester backbone.

Noncomposite transposon.   A transposon in which the transposase genes and the inverted-repeat ends are included in the minimum transposable element and are not part of autonomous IS elements.

Noncovalent change.   Any change in a molecule that does not involve the making or breaking of a covalent bond in the molecule.

Nonhomologous recombination.   The breaking and rejoining of two DNAs into new combinations which does not necessarily depend on the two DNAs having similar sequences in the region of recombination.

Nonpermissive conditions.   Those conditions under which a mutant organism or virus cannot multiply but the wild type can multiply.

Nonpermissive host.   A host organism in which a mutant virus cannot multiply but the wild type can multiply.

Nonpermissive temperature.   A temperature at which the wild-type organism or virus but not the mutant organism or virus can multiply.

Nonselective.   Referring to conditions or media in which both the mutant and wild-type strains of an organism or virus can multiply.

Nonsense codons.   Usually the codons UAG, UGA, and UAA. These codons do not stipulate an amino acid in most types of organisms but, rather, trigger the termination of translation.

Nonsense mutation.   In a region of DNA encoding a protein, a base pair change mutation that causes one of the nonsense codons to be encountered in frame when the mRNA is translated.

Nonsense suppressor.   A suppressor mutation that allows an amino acid to be inserted at some frequency for one or more of the nonsense codons during the translation of mRNAs.

Nonsense suppressor tRNA.   A mutation in the gene for a tRNA that allows the tRNA to pair with one or more of the nonsense codons in mRNA during translation and therefore causes an amino acid to be inserted for the nonsense codon. This type of mutation usually changes the anticodon on the tRNA.

Northern blotting.   Transfer of RNA from a gel to a filter for hybridization to a sequence-specific probe.

Ntr (nitrogen regulation) system.   A global regulatory system that regulates a number of operons in response to the nitrogen sources available.

Nuclease.   An enzyme that cuts the phosphodiester bonds in DNA or RNA polymers.

Nucleoid.   A compact, highly folded structure formed by the chromosomal DNA in the bacterial cell and in which the DNA appears as a number of independent supercoiled loops held together by a core.

Nucleotide excision repair.   A system for the repair of DNA damage in which the damaged nucleotide is removed. A cut is made on either side of the damage on the same strand, and the damaged strand is removed and resynthesized.

Ochre codon.   The nonsense codon UAA.

Ochre mutation.   In a region of DNA encoding a polypeptide, a base pair change mutation that causes the codon UAA to appear in frame in the mRNA for the polypeptide.

Ochre suppressor.   A mutation that causes an amino acid to be inserted at some frequency wherever the codon UAA is encountered during mRNA translation.

Okazaki fragments.   The short pieces of DNA that are initially synthesized during replication of the lagging strand at the replication fork.

Oligopeptide.   A short polypeptide only a few amino acids long.

Opal codon.   The codon UGA. Also called umber codon.

Open complex.   The complex of RNA polymerase and DNA at a promoter in which the strands of the DNA have been separated.

Open reading frame (ORF). A sequence on DNA, read three nucleotides at a time, that is unbroken by any nonsense codons.

Operator. Usually a sequence on DNA to which a repressor protein binds. More generally, any sequence in DNA or RNA to which a negative regulator binds.

Operon. A region on DNA encompassing genes that are transcribed into the same mRNA, as well as any adjacent *cis*-acting regulatory sequences.

Operon model. The model proposed by Jacob and Monod for the regulation of the *lac* operon in which transcription of the structural genes of the operon is prevented by the LacI repressor binding to the operator region and thereby preventing access of the RNA polymerase to the promoter. In the presence of lactose, the inducer binds to LacI and changes its conformation so that it can no longer bind to the operator, and as a result, the structural genes are transcribed.

oriC. The site in the bacterial chromosome at which initiation of a round of replication occurs.

Origin of replication. The site on a DNA at which replication initiates, including all of the surrounding *cis*-acting sequences required for initiation.

Ortholog. A gene or protein that has a common ancestor with another gene or protein. Usually refers to proteins in different species that have similar functions.

P site. The site on the ribosome to which the peptidyl-tRNA is bound.

pac **site.** The sequence in phage DNA at which packaging of the phage DNA into heads begins.

Packaging sites. *See pac* site.

Papilla. A section of a bacterial colony with an appearance different from that of most of the colony.

par **function.** A site or a gene that encodes a product required for the partitioning of a plasmid.

Paralog. A gene or protein that has common ancestor with another gene or protein as a result of gene duplication within a species. Usually refers to proteins that have somewhat different functions.

Parent. An organism participating in a genetic cross.

Parental types. Progeny of a genetic cross that are genetically identical to one or the other of the parents.

Partial digest. A restriction endonclease digestion of DNA in which not all the available sites are cut, either because the amount of endonuclease enzyme is limiting or because the time of incubation is too short.

Partial diploid. A bacterium that has two copies of part of its genome, usually because a plasmid or prophage in the bacterium contains some bacterial DNA. Also called merodiploid.

Partitioning. An active process by which at least one copy of a replicon (plasmid, chromosome, etc.) is distributed into each daughter cell at the time of cell division.

Pathogenicity island (PAI). A DNA element integrated into the chromosome of a pathogenic bacterium which carries genes whose products are required for pathogenicity and which shows evidence of carrying, at least originally, genes for its own integration.

PCR. *See* Polymerase chain reaction.

Peptide bond. A covalent bond between the amino (NH_2) group of one amino acid and the carboxyl (COOH) group of another.

Peptide deformylase. An enzyme that removes the formyl group from the amino-terminal formylmethionine of newly synthesized polypeptides.

Peptidyltransferase. The enzyme activity of the 23S rRNA (28S in eukaryotes) which forms a bond between the carboxyl group of the growing polypeptide and the amino group of the incoming amino acid.

Peptidyl tRNA hydrolase. An enzyme that removes polypeptides from tRNA. It is not associated with the ribosome, and so it may be a scavenger enzyme that is involved in the recycling of polypeptides that are prematurely released during translation.

Periplasm. The space between the inner and outer membranes in gram-negative bacteria.

Periplasmic domain. A region of a polypeptide that is located in the periplasm of the cell.

Periplasmic protein. A protein that is located in the periplasm.

Permissive conditions. Conditions under which a mutant organism or virus can multiply.

Permissive host. A strain of an organism which can support the multiplication of a particular mutant virus.

Permissive temperature. A temperature at which both a temperature-sensitive mutant (or cold-sensitive mutant) and the wild type can multiply.

Phage. *See* Bacteriophage.

Phage genome. The nucleic acid (DNA or RNA) that is packaged into the phage head.

Phase variation. The reversible change of one or more of the cell surface antigens of a bacterium at a high frequency.

Phasmid. A hybrid DNA element containing both plasmid and phage sequences.

Phenotype. Any identifiable characteristic of a cell or organism that can be altered by mutation.

Phenotypic lag. The delay between the time a mutation occurs in the DNA and the time the resulting change in the phenotype of the organism becomes apparent.

Phosphate. The chemical group PO_4.

Photolyase. An enzyme that uses the energy of visible light to split pyrimidine butane dimers in DNA, restoring the original pyrimidines.

Photoreactivation. The process by which cells exposed to visible light after DNA damage achieve greater survival rates than cells kept in the dark.

Physical map. A map of DNA showing the actual distance in deoxynucleotides between sites.

Pilin. A protein in pili. *See* Pilus.

Pilus. A protrusion or filament attached to the surface of a bacterial cell. *See* Sex pilus.

Plaque. The clear spots in a bacterial lawn caused by phage killing and lysing the bacteria as the bacterial lawn is forming.

Plaque purification. The isolation of a pure strain of a phage by plating to obtain individual plaques.

Plasmid. Any DNA molecule in cells that replicates independently of the chromosome and regulates its own replication so that the number of copies of the DNA molecule remains relatively constant.

Plasmid incompatibility. *See* Incompatibility.

Pleiotropic mutation. A mutation that causes many phenotypic changes in the cell.

Plus (+) strand. In a virus with a single-stranded genome (DNA or RNA), the strand that is packaged in the phage head.

Polarity. A condition in which a mutation in one gene reduces the transcription of a downstream gene that is cotranscribed into the same mRNA.

Polycistronic mRNA. An mRNA that contains more than one translational initiation region so that more than one polypeptide can be translated from the mRNA.

Polyclonal site. *See* Multiple cloning site.

Polymerase chain reaction (PCR). A technique that uses the DNA polymerase from a thermophilic bacterium and two primers to make many copies of a given region of DNA occurring between sequences complementary to the primers.

Polymerization. A reaction in which small molecules are joined together in a chain to make a longer molecule.

Polymorphism. A difference in DNA sequence between otherwise closely related strains.

Polypeptide. A long chain of amino acids held together by peptide bonds. The product of a single gene.

Porin. A protein that forms channels in the outer membrane of gram-negative bacteria.

Positive regulation. A type of regulation in which the gene is expressed only if the active form of a regulatory protein (or RNA) is present.

Positive selection. Conditions under which only the mutant being sought and not the normal or wild type can multiply.

Positively supercoiled. Referring to a DNA in which the two strands of the double helix are wrapped around each other more than once every 10.5 bp.

Postreplication repair. *See* Recombination repair.

Posttranscriptional regulation. Any regulation in the expression of a gene that occurs after the mRNA has been synthesized from the gene.

Precise excision. The removal of a transposon or other foreign DNA element from a DNA in such a way that the original DNA sequence is restored.

Precursor. The smaller molecules that are polymerized to form a polymer.

Presecretory protein. A secreted protein with its signal sequence still attached.

Primary structure. The sequence of nucleotides in an RNA or of amino acids in a polypeptide.

Primase. An enzyme that synthesizes short RNAs to prime the synthesis of DNA chains.

Prime factor. A self-transmissible plasmid carrying a region of the bacterial chromosome.

Primer. A single-stranded DNA or RNA that can hybridize to a single-stranded template DNA and provide a free 3′ hydroxyl end to which DNA polymerase can add deoxynucleotides to synthesize a chain of DNA complementary to the template DNA.

Primosome. A complex of proteins involved in making primers for the initiation of synthesis of DNA strands.

Probe. A short oligonucleotide (DNA or RNA) that is complementary to a sequence being sought and so will hybridize to the sequence and allow it to be identified from among many other sequences.

Prokaryotes. Organisms whose cells do not contain a nuclear membrane and visible nucleus or many of the other organelles characteristic of the cells of higher organisms. They include the eubacteria and archaea.

Promiscuous plasmid. A self-transmissible plasmid that can transfer itself into many types of bacteria, some of which are only distantly related to each other.

Promoter. A region on DNA to which RNA polymerase binds to initiate transcription.

Prophage. The state of phage DNA in a lysogen in which the phage DNA is integrated into the chromosome of the bacterium or replicates as a plasmid.

Protein disulfide isomerase. An enzyme that catalyzes the oxidation of the sulfhydryl groups of cysteines in polypeptides, cross-linking the cysteines to each other.

Protein export. The transport of proteins into or through the cellular membranes.

Protein secretion. The transport of proteins through the cellular membranes.

Proteomics. The large-scale characterization of all the proteins of an organism.

Pseudoknot. An RNA tertiary structure with interlocking loops held together by regions of hyrogen bonding between the bases.

Purine. A base in DNA and RNA with two ring structures.

Pyrimidine. A base in DNA and RNA with only one ring.

Pyrimidine dimer. A type of DNA damage in which two adjacent pyrimidines are covalently joined by chemical bonds.

Quaternary structure. The complete three-dimensional structure of a protein including all the polypeptide chains making up the protein and how they are wrapped around each other.

R-loop. A three stranded structure formed by the invasion of a double-stranded DNA by an RNA, displacing one of the strands of the double-stranded DNA.

Random gene fusions. A technique in which transposon mutagenesis is used to fuse reporter genes to different regions in the chromosome. A transposon containing a reporter gene hops randomly into the chromosome, resulting in various transposon insertion mutants that have the reporter gene on the transposon fused either transcriptionally or translationally to different genes or to different regions within each gene.

Random-mutation hypothesis. A hypothesis explaining the adaptation of organisms to their environment. It states that mutations occur randomly, free of influence from their consequences, but that mutant organisms preferentially survive and reproduce themselves if the mutations inadvertently confer advantages under the circumstances.

RC plasmid. *See* Rolling-circle plasmid.

Reading frame. Any sequence of nucleotides in RNA or DNA read three at a time in succession, as during translation of an mRNA.

Rec$^-$ (recombination-deficient) mutant. A mutant in which DNA shows a reduced capacity for recombination.

Recessive. A mutation or other genetic marker that does not exert its phenotype in an organism that is diploid for the region because it also contains the corresponding region from the wild-type organism.

Recessive mutation. Referring to complementation tests, a mutation that does not exhibit its phenotype in the presence of a wild-type copy of the same region of DNA.

Recipient. In a genetic cross between two bacteria, the bacterium that receives DNA from another bacterium.

Recipient allele. The sequence of a gene or allele as it occurs in the recipient bacterium.

Recipient strain. Bacteria with the genotype of those that were used as recipients in a genetic cross.

Reciprocal cross. A genetic cross in which the alleles of the donor and recipient strain are reversed relative to an earlier cross. An example would be a transduction in which the phage were grown on the strain that had the alleles of what was previously the recipient strain and used to transduce a strain with the alleles of what was previously the donor strain.

Recombinant DNA. A DNA molecule derived from the sequences of two different DNAs joined to each other in a test tube.

Recombinant type. In a genetic cross, progeny that are genetically unlike either parent in the cross because they have DNA sequences that are the result of recombination between the parental DNAs.

Recombinase. An enzyme that specifically recognizes two sequences in DNA and breaks and rejoins the strands to cause a crossover within the sequences.

Recombination. The breakage and rejoining of DNA into new combinations.

Recombination frequency. In a genetic cross, the number of progeny that are recombinant types for the two parental markers divided by the total number of progeny of the cross.

Recombination repair. Another name for postreplication repair. A mechanism by which the cell tolerates DNA damage by using recombination to transfer an undamaged strand to the other daughter DNA containing a damaged strand, so that the damaged strand will be opposite a good complementary strand.

Redundancy. *See* Terminal redundancy.

Regulation of gene expression. Control of the rate of synthesis of the active product of a gene, so that the gene product can be synthesized at different rates, depending, for example, on the developmental stage of the organism or the state in which the organism finds itself.

Regulatory cascade. A strategy for regulating the expression of genes during developmental processes in which the products of genes expressed during one stage of development regulate the expression of genes for the next stage of development.

Regulatory gene. A gene whose product regulates the expression of other genes.

Regulon. The set of operons that are all regulated by the product of the same regulatory gene.

Relaxase. The protein of a self-transmissible or mobilizable plasmid that makes a cut at the *oriT nic* site, remains attached to the 5′ end at the cut, is secreted into the recipient cell, and rejoins the cut ends in the recipient cell.

Relaxed control. Occurs when the synthesis of rRNA and other stable RNAs continues even after protein synthesis is blocked by starvation for an amino acid.

Relaxed plasmid. A plasmid that has a high copy number, so that its replication need not be tightly controlled.

Relaxed strain. A strain of bacteria that exhibits relaxed control of rRNA and other stable RNA synthesis. Does not synthesize ppGpp in response to amino acid starvation.

Relaxosome. The complex of proteins, including the relaxase, which is bound to the *oriT* sequence of a self-transmissible or mobilizable plasmid in the donor cell.

Release factors. Proteins that are required for the termination of polypeptide synthesis when the ribosome encounters a specific in-frame nonsense codon in the mRNA. Also required for the release of the newly synthesized polypepetide from the ribosome.

Replica plating. A technique in which bacteria grown on one plate are transferred to a fuzzy cloth and then are transferred from the fuzzy cloth onto another plate so that the bacteria on the first plate will be transferred to the corresponding position on the second plate.

Replication fork. The region in a replicating double-stranded DNA molecule where the two strands are separating to allow synthesis of the complementary strands.

Replication restart. The process of reloading the replication apparatus on the DNA after it has been dissociated, for example at a nick or damage to the DNA.

Replicative bypass. Occurs when the replication fork moves past damage to the DNA that interferes with proper base pairing, either by skipping over the damage and leaving a gap in the newly synthesized strand or by inserting deoxynucleotides at random opposite the damaged bases.

Replicative form (RF). The double-stranded DNA or RNA that forms by synthesis of the complementary minus strand after infection by a virus that has a single-stranded genome.

Replicative transposition. A type of transposition in which single-stranded nicks are made at each end of the transposon and a staggered double-strand break is made in the target DNA. The free 3′ ends at the ends of the transposon are ligated to the free 5′ ends of the target DNA and the free 3′ ends of the target DNA are used as primers to synthesize over the transposon, giving rise to a cointegrate.

Replicon. A DNA molecule capable of autonomous replication because it contains an origin of replication that functions in the cell.

Reporter gene. A gene whose product is stable and easy to assay and so is convenient for detecting and quantifying the expression of genes to which it is fused.

Repressor. A protein that, in its active state, binds to operator sequences close to the promoter for an operon and thereby prevents transcription of the operon. More generally, a protein or RNA that negatively regulates transcription or translation.

Resolution of a cointegrate. Separation of the two DNAs joined in a cointegrate by recombination between repeated sequences in the two copies of the transposon, leaving each DNA with one copy of the transposon.

Resolution of Holliday junctions. Cutting the two crossed strands of DNA in a Holliday junction, for example by an X-phile, so that the DNA molecules, held together by the Holliday junction, are separated.

Resolvase. A type of site-specific recombinase that breaks and rejoins DNA in *res* sequences in the two copies of the transposon in a cointegrate, thereby resolving the cointegrate into separate DNAs, each with one copy of the transposon.

Response regulator. A protein that is part of a two component system and that upon receiving a signal from another protein, the sensor protein, performs a regulatory function, e.g., activates transcription of operons.

Restriction fragment. A piece of DNA obtained by cutting a longer DNA with a restriction endonuclease.

Restriction fragment length polymorphism (RFLP). A difference in the size of restriction fragments obtained by cutting DNA from two different strains with the same restriction endonuclease.

Restriction modification system. A complex of proteins that, among them, recognize specific sequences in DNA, methylate a base in the sequence, and cut in or near the sequence if it is not methylated.

Retroregulation. Occurs when the amount of an RNA in the cell is determined by an event that occurs at the 3′ end of an RNA rather than at the 5′ end.

Retrotransposon. A type of transposon that hops to a new location in DNA by first making an RNA copy of itself, then making a DNA copy of this RNA with a reverse transcriptase, and then inserting its DNA copy into the target DNA.

Reverse genetics. The process by which the function of the product of a gene is determined by first altering its sequence in vitro and then reintroducing it into a cell to see what effect the alteration has on the organism.

Reversion. The restoration of a mutated sequence in DNA to the wild-type sequence.

Reversion rate. The probability that a mutated sequence in DNA will change back to the wild-type sequence each time the organism multiplies.

Revert. *See* Reversion.

Revertant. An organism in which the mutated sequence in its DNA has been restored to the wild-type sequence.

RF. *See* Replicative form.

RFLP. *See* Restriction fragment length polymorphism.

Ribonucleoside triphosphate. A base (usually A, U, G, or C) attached to a ribose sugar with three phosphate groups attached in tandem to the 5′ carbon of the sugar.

Ribonucleotide reductase. An enzyme that catalyzes the reduction of nucleoside diphosphates to deoxynucleoside diphosphates by removing the hydroxy group at the 2′ carbon of the ribonucleoside diphosphate and replacing it with a hydrogen.

Riboprobe. A hybridization probe made of RNA rather than DNA.

Ribosomal proteins. The proteins that, in addition to the rRNAs, make up the structure of the ribosome.

Ribosomal RNA (rRNA). Any one of the three RNAs (16S, 23S, and 5S in bacteria) that make up the structure of the ribosome.

Ribosome. The cellular organelle, made up of about 50 different proteins and 3 different RNAs, that is the site of protein synthesis.

Ribosome-binding site. *See* Translational initiation region.

Ribosome cycle. The association and dissociation of the 30S and 50S ribosomes during initiation and termination of translation.

Ribosome release factor. *See* Release factors.

RNA modification. Any covalent change to RNA that does not involve the breaking and joining of phosphate-phosphate or phosphate-ribose bonds in the backbone of the RNA.

RNA polymerase. An enzyme that polymerizes ribonucleoside triphosphates to make RNA chains by using a DNA or RNA template.

RNA polymerase holoenzyme. The $\alpha_2\beta\beta'$ RNA polymerase with a σ factor attached.

RNA processing. Covalent changes to RNA that involve the breaking and joining of phosphate-phosphate or phosphate-ribose bonds in the backbone of the RNA.

Rolling-circle plasmid. RC plasmid. A plasmid which replicates by a rolling-circle mechanism.

Rolling-circle replication. A type of replication of circular DNAs in which a single-stranded nick is made in one strand of the DNA and the 3′ hydroxyl end is used as a primer to replicate around the circle, displacing the old strand.

Round of replication. Cycle of replication of a circular DNA in which one copy of the DNA is made.

Sanger dideoxy sequencing. A method for DNA sequencing in which the chain-terminating property of the dideoxynucleotides is used.

Satellite virus. A naturally occurring virus that depends on another virus for its multiplication.

Screening. The process of testing a large number of organisms for a particular mutant type.

sec **gene.** One of the genes whose products are required for transport of proteins through the inner membrane.

Sec system. The general system encoded by the *sec* genes for secreting proteins across the inner membrane of gram-negative bacteria, consisting of the targeting factors SecA and SecB and including the components of the SecYEG channel in the inner membrane.

Secondary structure. A structure of a polynucleotide or polypeptide chain that results from noncovalent pairing between nucleotides or amino acids in the chain.

Secreted protein. A protein which leaves the cell and moves into the outside environment.

Segregation. The process by which newly replicated DNAs are separated into daughter cells.

Selected marker. A difference in DNA sequence between two strains in a bacterial or phage cross that is used to select recombinants. The cross is plated so that only recombinants that have received the donor sequence can multiply.

Selection. Procedure in which bacteria or viruses are grown under conditions in which only the wild type or the desired mutant or recombinant can multiply, allowing the isolation of even very rare mutants and recombinants.

Selectional genetics. Genetic analysis in which selection of mutants or recombinants is used.

Selective conditions. Conditions under which only the wild type or the desired mutant can multiply.

Selective media. Media that have been designed to allow multiplication of only the desired mutant or wild type. Such media often lack one or more nutrients or contain a substance that is toxic.

Selective plate. An agar plate made with selective media.

Self-transmissible. Referring to a plasmid or other DNA element that encodes all the gene products needed to transfer itself to other bacteria through conjugation.

Semiconservative replication. A type of DNA replication in which the daughter DNAs are composed of one old strand and one newly synthesized strand.

Sensitive cell. A type of cell that can serve as the host for a particular type of virus.

Sensor. The protein in a two-component system that detects changes in the environment and communicates this information to the response regulator, usually in the form of a phosphoryl group.

Sequestration. The entry of the origin of replication (*oriC*) of the bacterial chromosome into an incompletely understood dormant state after a round of chromosome replication has initiated.

Serial dilution. A procedure in which an aliquot of a solution is first diluted into one vessel and then an aliquot of the solution in this vessel is diluted into a second vessel, and so forth. The total dilution is the product of each of the individual dilutions.

7,8-Dihydro-8-oxoguanine. *See* 8-OxoG.

Sex pilus. A rod-like structure that forms on the surface of bacterial cells containing a self-transmissible plasmid and that facilitates transfer of the plasmid or other DNAs into another bacterium.

Shine-Dalgarno sequence. A short sequence, usually about 10 nucleotides upstream of the initiation codon in a bacterial translation initiation region, that is complementary to a sequence in the 3′ end of the 16S rRNA. It helps position the ribosome for initiation of translation. Named after the persons who first noticed it.

Shuttle vector. A plasmid cloning vector that contains two origins of replication that function in different types of cells so that the plasmid can replicate in both types of cells.

Siblings. Two cells or viruses that arose from the multiplication of the same cell or virus.

Signal recognition particle (SRP). A particle consisting of a small RNA and one or more proteins that bind to the N terminus of a newly synthesized protein to be exported and target the protein to the membrane.

Signal sequence. A short sequence, composed of mostly hydrophobic amino acids, that is located at the N terminus of some secreted proteins and that is removed as the protein passes into or through the membrane.

Signal transduction pathway. A set of proteins that pass a signal, usually a phosphoryl group, from one to the other in reponse to a change in the cellular conditions.

Silent mutation. A change in a DNA sequence of a gene encoding a protein that does not change the amino acid sequence of the protein, usually because it changes the last base in a codon, thereby changing it to another codon but one that encodes the same amino acid.

Single mutant. A mutant organism that has only one of the two or more mutations being studied.

Site-specific (site-directed) mutagenesis. One of many methods for mutagenizing DNA in such a way that the change is localized to a predetermined base pair or small region in the DNA.

Site-specific recombinase. An enzyme that catalyzes the breaking and rejoining of DNA molecules but only at certain sequences in the DNA.

Site-specific recombination. Recombination that occurs only between defined sequences in DNA. Usually performed by recombinases.

6-4 lesion. A type of damage to DNA in which the carbon at the 6 position of a pyrimidine is covalently bound to the carbon at the 4 position of an adjacent pyrimidine.

Six-hitter. A type II restriction endonuclease that recognizes and cuts at a specific 6-bp sequence in DNA.

SOS gene. A gene that is a member of the LexA regulon, so that its transcription is normally repressed by LexA repressor.

SOS mutagenesis. Mutagenesis that occurs at an increased rate in cells in which the SOS genes are induced.

SOS response. Induction of transcription of the SOS genes in response to DNA damage.

Southern blot hybridization. A procedure for transferring DNA from an agarose gel to a filter for hybridization. Named after the person who developed the procedure.

Specialized transduction. A type of transduction caused by lysogenic phages in which only DNA sequences close to the attachment site of the prophage in the chromosome are transduced. It results from mistakes made during the excision of the prophage that substitute some chromosomal DNA sequences for some of the phage DNA sequences that are normally packaged in the phage head.

Spontaneous mutations. Mutations that occur in organisms without deliberate atempts to induce them by irradiation or chemical treatment.

Sporulation. A developmental process that leads to the development of spores, which are dormant cells containing the DNA of the organism and are often resistant to desiccation and other harsh environmental conditions.

Start point of transcription. The deoxynucleotide in DNA with which the first ribonucleotide pairs to begin making an RNA.

Starve. To deprive an organism of an essential nutrient that it cannot make for itself.

Sticky end. The short single-stranded DNA that sticks out from the end of the DNA molecule after it has been cut with a type II restriction endonuclease that makes a staggered break in the DNA.

Strain. A subdivision of species. A group of organisms that are identical to each other but differ genetically from other organisms of the same species.

Strand exchange. The process by which a strand of a double-stranded DNA changes partners so that it pairs with a different complementary strand.

Strand passage. A reaction performed by topoisomerases in which one or two strands of a DNA are cut and the ends of the cut DNA are held by the enzyme to prevent rotation while other strands of the same or different DNAs are passed through the cuts.

Stringent control. Occurs when the synthesis of rRNA and other stable RNAs in the cell ceases when the cells are starved for an amino acid.

Stringent plasmid. A plasmid that exists in only one or very few copies per cell, so that its replication must be very tightly controlled.

Structural gene. One of the genes in an operon for a pathway that encodes one of the enzymes of the pathway.

Subclone. A smaller clone obtained by cutting a clone with restriction endonucleases and cloning one of the pieces.

Sugar. A simple carbohydrate with the general formula $(CH_2O)_n$; as found in nature, n is 3 to 9.

Suicide vector. A DNA element, usually plasmid or phage DNA, that cannot replicate in the cells into which it is being introduced.

Supercoiled. Referring to DNA in which the two strands of the double helix are wrapped around each other either more or less often than predicted from the Watson-Crick structure of DNA.

Superinfection. The infection of cells by a virus when the same cells are already infected by the same type of virus.

Suppression. The alleviation of the effects of a mutation by a second mutation elsewhere in the DNA.

Suppressive effect. Occurs if a clone, when introduced into a cell in high copy number, alleviates the effect of a mutation indirectly, rather than complementing the mutation.

Suppressor mutation. A mutation elsewhere in the DNA that alleviates the effects of another mutation.

Symmetric sequence. A sequence of deoxynucleotides in double-stranded DNA that reads the same in the 5′-to-3′ direction on both strands.

Synapse. In recombination, a structure in which two DNAs are held together by pairing between their strands.

Synchronize. To treat a culture of cells so that they will all be at approximately the same stage in their cell cycle at the same time.

Tag. A sequence of amino acids added to a protein so that the protein can be purified more easily.

Tag vector. A cloning vector designed so that an amino acid sequence that is easy to purify will be added to a protein if the gene for the protein is cloned into the vector in frame with the translation initiation region on the vector and with no intervening nonsense codons.

Tandem duplication. A type of mutation that causes a sequence in DNA to be followed immediately by the same sequence.

Target DNA. The DNA into which a transposon hops.

Tautomer. A form of a molecule in which the electrons are distributed differently among the atoms.

Temperature-sensitive (Ts) mutant. A mutant that cannot grow in the higher-temperature ranges in which the wild type can multiply.

Template strand. The strand of DNA in a region from which RNA is synthesized that has the complementary sequence of the RNA and so serves as the template for RNA synthesis.

Terminal redundancy. A term used to describe a DNA molecule that has direct repeats at both ends; that is, the sequences at both ends are the same in the direct orientation.

Termination of replication. The process by which the replication apparatus leaves the DNA when DNA replication is completed and the daughter DNAs are separated.

Termination of transcription. The process by which the RNA polymerase leaves the DNA and the RNA chain is released at a transcription termination site in the DNA.

Termination of translation. The process by which the ribosome leaves the mRNA and the polypeptide is released when a nonsense codon in the mRNA is encountered in frame.

Tertiary structure. The three-dimensional structure of a polypeptide or polynucleotide.

Tetramer. A protein made up of four polypeptides.

Theta replication. A type of replication of circular DNA in which the replication apparatus initiates at an origin of replication and proceeds in one or both directions around the circle with leading and lagging strands of replication. The molecule in an intermediate state of replication resembles the Greek letter theta (θ).

Three-factor cross. A type of genetic cross used to order three closely linked mutations and in which one parent has two of the mutations and the other parent has the third mutation.

3′ end. The terminus of a polynucleotide chain (DNA or RNA) ending in the nucleotide that is not joined at the 3′ carbon of its ribose to the phosphate of another nucleotide.

3′ exonuclease. An enzyme that degrades a polynucleotide from its 3′ end by removing nucleotides one at a time.

3′ hydroxyl end. In a polynucleotide, a 3′ end that has a hydroxyl group on the 3′ carbon of the ribose sugar of the last nucleotide without a phosphate group attached.

3′ overhang. An unpaired single strand extending from the strand with a free 3′ end in a double-stranded DNA.

3′ untranslated region. In an mRNA, the sequences downstream or toward the 3′ end of the last sequence that encodes a protein.

Thymine (T). One of the pyrimidine (one-ringed) bases.

TIRs. *See* Translational initiation region.

Titration. A process of increasing the concentration of one of two types of molecules that bind to each other until all of the other type of molecule is bound.

Topo IV. A type II topoisomerase of *E. coli* that is responsible for decatenation of daughter chromosomes after replication and for relieving positive supercoils ahead of the replication fork.

Topoisomerase. An enzyme that can alter the topology of a DNA molecule by cutting one or both strands of DNA, passing other DNA strands through the cuts while holding the cut ends so that they are not free to rotate, and then resealing the cuts.

Topology of DNA. The relationship of the strands of DNA to each other in space.

Tra functions. Gene products encoded by the *tra* genes of self-transmissible DNA elements that allow the plasmids to transfer themselves into other bacteria.

***trans*-acting function.** A gene product that can act on DNAs in the cell other than the one from which it was made.

***trans*-acting mutation.** A mutation that affects a gene product that leaves the DNA from which it is made and so can be complemented.

Transconjugant. A recipient cell that has received DNA from another cell by conjugation.

Transcribe. To make an RNA that is a complementary copy of a strand of DNA.

Transcribed strand. In a region of a double-stranded DNA that is transcribed into RNA, the strand of DNA that is used as a template and so is complementary to the RNA.

Transcript. An RNA made from a region of DNA.

Transcription antitermination. *See* Antitermination.

Transcription bubble. A region that follows the RNA polymerase during transcription of DNA in which the two strands of the DNA are separated and the newly synthesized RNA forms a short RNA-DNA duplex with the transcribed strand of DNA.

Transcription termination site. A sequence on DNA at which the RNA polymerase falls off the template, stopping transcription. It can be either factor independent or dependent upon a transcription termination factor such as ρ.

Transcription vector. A cloning vector which contains a promoter from which a cloned DNA can be transcribed.

Transcriptional activator. A protein that is required for transcription of an operon. The protein makes contact with the RNA polymerase and allows the RNA polymerase to initiate transcription from the promoter of the operon.

Transcriptional autoregulation. The process by which a protein regulates the transcription of its own gene.

Transcriptional fusion. The introduction of a gene downstream of the promoter for another gene or genes so that it will be transcribed from the promoter for the other gene(s) but will be translated from its own translational initiation region.

Transcriptional regulation. When the amount of product of a gene that is synthesized under certain conditions is determined by how much mRNA is made from the gene.

Transducing phage. A type of phage that sometimes packages bacterial DNA during infection and introduces it into other bacteria upon infection of the other bacteria.

Transductant. A bacterium that has received DNA from another bacterium by transduction.

Transduction. A process in which DNA other than phage DNA is introduced into a bacterium by infection by a phage containing the DNA.

Transfection. Initiation of a virus infection by introducing virus DNA into a cell by transformation rather than by infection.

Transfer RNA (tRNA). The small stable RNAs in cells to which specific amino acids are attached by aminoacyl tRNA synthetases. The tRNA with the amino acid attached enters the ribosome and base pairs through its anticodon sequence with a 3-nucleotide codon sequence in the mRNA to insert the correct amino acid in the growing polypeptide chain.

Transformant. A cell that has received DNA by transformation.

Transformasomes. Globular structures that appear on the surfaces of some types of bacteria into which DNA first enter during natural transformation of the bacteria.

Transformation. The introduction of DNA into cells by mixing the DNA and the cells.

Transformylase. The enzyme that transfers a formyl (CHO) group to the amino group of methionine to make formylmethionine.

Transgenic organism. An organism that has inherited foreign DNA sequences that have been experimentally introduced into its ancestors. The introduced DNA sequences are passed down from generation to generation because they are inserted into a stably inherited DNA, such as a chromosome.

Transgenics. The process of introducing foreign DNA into the chromosome of an organism to make a transgenic organism.

Transition mutation. A type of base pair change mutation in which the purine base has been changed into the other purine base and the pyrimidine base has been changed into the other pyrimidine base (e.g., AT to GC or GC to AT).

Translated region. A region of an mRNA that encodes a protein.

Translation elongation factor G (EF-G). The protein required to move the peptidyl tRNA from the A site to the P site on the ribosome, with the concomitant cleavage of GTP to GDP after the peptide bond has formed.

Translation elongation factor Tu (EF-Tu). The protein that binds to aminoacyl tRNA and accompanies it onto the ribosome. It then cycles off the ribosome with the concomitant cleavage of GTP to GDP.

Translation termination site. Any one of the nonsense codons for the organism in the frame being translated.

Translation vector. A cloning vector which contains a TIR from which a cloned DNA sequence can be translated.

Translational coupling. A gene arrangement in which the translation of one protein coding sequence on a polycistronic mRNA is required for the translation of the second, downstream coding sequence.

Translational fusion. The fusion of parts of the coding regions of two genes so that translation initiated at the translational initiation region for one polypeptide on the mRNA will continue into the coding region for the second polypeptide in the correct reading frame for the second polypeptide. A polypeptide containing amino acid sequences from the two genes that were joined to each other will be synthesized.

Translational initiation region (TIR). The initiation codon, the Shine-Dalgarno sequence, and any other surrounding sequences in mRNA that are recognized by the ribosome as a place to begin translation. Also called a ribosome-binding site (RBS).

Translational regulation. The variation, under different conditions, in the amount of synthesis of a polypeptide due to variation in the rate at which the polypeptide is translated from the mRNA.

Translationally autoregulated. Referring to a protein that can affect the rate of translation of its own coding sequence on its mRNA. Usually in such cases the protein binds to its own translational initiation region or that of an upstream gene to which it is translationally coupled; hence, the protein represses its own translation.

Translesion synthesis (TLS). Synthesis of DNA over a template region containing a damaged base or bases that are incapable of proper base pairing.

Translocase. The SecYEG membrane channel.

Translocation. During translation, the movement of the tRNA with the polypeptide attached from the A site to the P site on the ribosome after the peptide bond has formed.

Transmembrane domain. The region in a polypeptide between a region that is exposed to one surface of a membrane and a region that is exposed to the other surface. This region must traverse and be embedded in the membrane. Usually, transmembrane domains have a stretch of at least 20, mostly hydrophobic, amino acids that is long enough to extend from one face of a bilipid membrane to the other.

Transmembrane protein. A membrane protein that has surfaces exposed at both sides of the membrane.

Transposase. An enzyme encoded by a transposon that cuts the target DNA and the DNA at both ends of the transposon and joins the cut ends to each other during transposition.

Transposition. The act of a transposon moving from one place in DNA to another.

Transposon. A DNA sequence that can move from one place in DNA to a different place with the help of transposase enzymes. Should be distinguished from homing DNA elements that usually move only into the same sequence in another DNA and depend upon homologous recombination.

Transposon mutagenesis. Technique in which a transposon is used to make random insertion mutations in DNA. The transposon is usually introduced into the cell in a suicide vector, and so it must transpose into another DNA in the cell to become established.

Transversion. A type of base pair change mutation in which the purine in the base pair is changed into the pyrimidine and vice versa, e.g., GC to TA or GC to CG.

Trigger factor. A chaperone in *E. coli* that is closely associated with the ribosome and that helps proteins fold as they emerge from the ribosome. Can substitute for DnaK.

Triparental mating. A conjugational mating for introducing mobilizable plasmids into cells, in which three strains of bacteria are mixed. One strain contains a self-transmissible plasmid, and the second strain contains the mobilizable plasmid, which is then mobilized into the third strain.

tRNA$_f^{Met}$. The tRNA to which formylmethionine is added and that pairs with the initiator codon in a translational initiation region to initiate translation of a polypeptide in bacteria.

Two-component regulatory system. A pair of proteins, one of which, the sensor, undergoes a change in response to a change in the environment and communicates this change, usually in the form of a phosphate, to another protein, the response regulator, which then causes the appropriate cellular response. Different two-component systems are often highly homologous to each other.

Umber codon. The codon UGA. Also known as the opal codon.

Unselected marker. A difference between the DNA sequences of two bacteria or phages involved in a genetic cross that can be used for genetic mapping. Is a difference other than the difference used to isolate recombinants. Mapping information can be obtained by testing recombinants that have been selected for being recombinant for one marker, the selected marker, to determine if they have the sequence of the donor or the recipient for another marker, an unselected marker.

Untargeted mutations (UTM). Mutations that occur elsewhere in the DNA than at the sites of DNA damage.

Upstream. From a given point, sequences that lie in the 5′ direction on RNA or in the 5′ direction on the coding strand of a DNA region from which an RNA is made.

Upstream activator sequence or site (UAS). A sequence in DNA upstream of a promoter that increases transcription from the promoter. Removal of a UAS from the DNA reduces transcription from the promoter.

Uptake sequence. A short sequence in DNA that allows DNA containing the sequence to be taken up by some types of bacteria during natural transformation.

Uracil (U). One of the pyrimidine (one-ringed) bases.

Uracil-N-glycosylase. An enzyme that removes the uracil base from DNA by cleaving the bond between the base and the deoxyribose sugar.

UvrABC endonuclease (UvrABC exinuclease). A complex of three proteins that cuts on both sides of any DNA lesion causing a significant distortion of the helix, as a first step in excision repair of the damage.

Very short patch (VSP) repair. A type of repair in enteric bacteria that removes the mismatched T in the sequence CT(A/T)GG/GG(T/A)CC and replaces it with a C. This prevents mutagenesis at this site due to deamination of the C, which is methylated in enteric bacteria, to a T, which will not be recognized by the uracil-N-glycosylase. A very short stretch of the DNA strand around the mismatched T is removed and resynthesized during the repair, hence the name.

VSP repair. *See* Very short patch repair.

Watson-Crick structure of DNA. The double helical structure of DNA first proposed by James Watson and Francis Crick. The two strands of the DNA are antiparallel and held together by hydrogen bonding between the bases.

Weigle mutagenesis (W-mutagenesis). Another name for SOS mutagenesis. Named after Jean Weigle, who first observed it.

Weigle reactivation (W-reactivation). Applies to the increased ability of phages to survive UV irradiation damage to their DNA if the cells they infect have been previously exposed to UV irradiation. Named after Jean Weigle, who first observed it.

Western blot. A membrane onto which proteins have been transferred from a gel.

Wild type. The organism as it was first isolated from nature. In a genetic experiment, the strain from which mutants were derived. The normal type.

Wild-type allele. The form of a gene as it exists in the wild-type organism.

Wild-type phenotype. The particular outward trait characteristic of the wild type that is different in the mutant.

W-mutagenesis. *See* Weigle mutagenesis.

Wobble. The property of the genetic code that codons for the same amino acid often differ only in the last (third) nucleotide. Reflects the fact that the base of the first nucleotide in the anticodon of a tRNA can often pair with more than one base in the third nucleotide of a codon in the mRNA.

X-phile. One of a group of enzymes that can cut the crossed DNA strands at a Holliday junction.

Xanthine. A purine base that results from deamination of guanine.

Zero frame. In the coding region of a gene, the sequence of nucleotides, taken three at a time, in which the polypeptide encoded by the gene is translated.

Index